BIOLOGY

A GUIDE TO THE NATURAL WORLD

ACADEMIC ADVISORS

BIOLOGY

A GUIDE TO THE NATURAL WORLD

David Krogh
Berkeley, California

Prentice Hall
Upper Saddle River, New Jersey 07458

Library of Congress Cataloging-in-Publication Data

Krogh, David.
  Biology : a guide to the natural world / David Krogh.
    p. cm.
  ISBN 0-02-366891-1
  1. Biology. I. Title.
QH308.2.K76 2000
570--dc21   99-40292
                                   CIP

Executive Editor: Sheri L. Snavely
Project Development Editor and Editor in Chief of Development: Ray Mullaney
Production Editor: Donna F. Young
Formatting Supervisor: Jim Sullivan
Electronic Production Specialist/Electronic Page Makeup: Joanne Del Ben
Art Director: Ann France
Art Manager: Gus Vibal
Editor in Chief: Paul F. Corey
Executive Marketing Manager: Jennifer Welchans
Assistant Vice President of Production and Manufacturing: David W. Riccardi
Executive Managing Editor: Kathleen Schiaparelli
Creative Director: Paul Belfanti
Associate Creative Director: Amy Rosen
Manufacturing Manager: Trudy Pisciotti
Media Editor: Andrew T. Stull
Media Production Editor: Charlene M. Barr
Project Manager: Karen Horton
Editorial Assistants: Nancy Bauer, Lisa Tarabokjia
Copy Editor: Chris Thillen
Proofreader: Jennefer Vecchione
Cover Designer: Luke Daigle
Interior Designer: Lynne Stiles
Illustrators: Imagineering, Medical and Scientific Illustrations, Inc., Laurie O'Keefe
Art Editor: Karen Branson
Art Project Support: Xiaohong Zhu
Photo Researcher: Yvonne Gerin
Cover Photos—*Front cover:* Gorilla, Parc des Volcanoes, Africa © Michael Nichols/Aurora & Quanta Productions; Giant Lobelia, Ruwenzori Mountains © Robert Caputo/Aurora & Quanta Productions; Forest, Sabattus Mountain area © Jose Azel/Aurora & Quanta Productions. *Spine:* Great Gray Owl/Spanande lappuggla © Tero Niemi. *Back cover:* A pair of Kuda, of the antelope family, Etosha National Park, Namibia © Robert Caputo/Aurora & Quanta Productions; Red leaf, Baxter State Park, Maine © Jose Azel/Aurora & Quanta Productions; Cactus, Sonoran Desert, Arizona © Joanna Pinneo/Aurora & Quanta Productions.

Printed in the United States of America

10 9 8 7 6 5 4 3 2 1

ISBN 0-02-366891-1

Prentice-Hall International (UK) Limited, *London*
Prentice-Hall of Australia Pty. Limited, *Sydney*
Prentice-Hall Canada Inc., *Toronto*
Prentice-Hall Hispanoamericana, S.A., *Mexico*
Prentice-Hall of India Private Limited, *New Delhi*
Prentice-Hall of Japan, Inc., *Tokyo*
Prentice-Hall (Singapore) Pte. Ltd., *Singapore*
Editora Prentice-Hall do Brasil, Ltda., *Rio de Janeiro*

For the lovely Anne
And for my darling Tessa

MediaLabs

There are twelve MediaLabs throughout the book. The topics were carefully chosen not only for student interest but also because they highlight issues that students may come across in their daily lives. Each MediaLab takes the reader on a journey of discovery through CD-ROM activities and web investigations.

Essays

Thirty-two essays in the book have an applied slant, dealing with such topics as acid rain, DNA fingerprinting, and even how to choose the right fertilizer for houseplants.

In the How Did We Learn? essays, students can come to understand the inventiveness and the plain hard work that generally are prerequisites to scientific discovery.

Brief Contents

Contents

A portrait of the most famous molecule in all of biology. **59**

Inside the cell. **71**

Contents

A cellular sandwich. **94**

Photosynthesis in action. **157**

Unit 3   How Life Goes On: Genetics   171

Protein synthesis in action. **175**

Contents

Sister chromatids line up. **195**

cross-pollination

Cross-pollination. **211**

Essays
HOW DID WE LEARN?
Thomas Hunt Morgan: Using Fruit
Flies to Look More Deeply into
Genetics **234**
Testing for Genetic Trouble **242**

The wild and the mutant flies. **236**

Essay
HOW DID WE LEARN?
Getting Clear about What Genes
Do: Beadle and Tatum **256**

The double helix. **254**

Contents

Mass production. **275**

How it was done. **290**

Contents

Unit 4 Life's Organizing Principle: Evolution and the Diversity of Life 307

Chapter 16 An Introduction to Evolution: Charles Darwin, Evolutionary Thought, and the Evidence for Evolution **308**

Essay
Can Darwinian Theory Make Us Healthier? **318**

Charles Darwin. **311**

Contents

This different allele frequency is reflected in today's population.

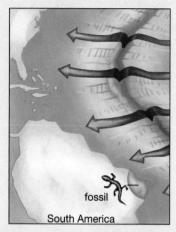
Essays
The Unsolvable Taxonomy
Problem 395
Modes of Nutrition: How Organisms Get What They Need to Survive 400

An important type of cell. **412**

Contents

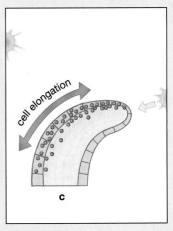

The power of light. **438**

Contents

Food carriers. **462**

Unit 6 What Makes the Organism Tick?
Animal Anatomy and Physiology 481

Chapter 23 Introduction to Animal Anatomy and Physiology: The Integumentary, Skeletal, and Muscular Systems 482

Essay
Doing Something about
Osteoporosis While You Are
Young **496**

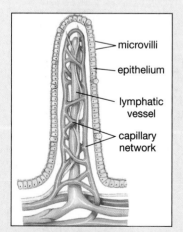

microvilli

epithelium

lymphatic
vessel

capillary
network

Contents

What are somites? **575**

b

Many sperm come
in contact with the
accessory cells and
are capacitated.

Fertilization begins. **596**

RESOURCE PARTITIONING

Cape May
warbler

That's my branch. **629**

Contents

Essays
A Cut for the Middleman: Livestock and Food **658**
Good News about the Environment **670**

The taiga. **663**

Preface

David Krogh has been writing about science for 20 years in newspapers, magazines, books, and for educational institutions. He is the author of *Smoking: The Artificial Passion,* an account of the pharmacological and cultural motivations behind the use of tobacco, which was nominated for the Los Angeles *Times* Book Prize in Science and Technology. David has written on physics and on technology issues, but his primary interest has been in biology. He has written on the possible effect methane may be having on global warming; on early research into the role that growth factors may play in neural regeneration following injury; on the synthesis of naturally occurring neurotoxins and their possible use in heart disease; on the use of imported drugs to treat cancer; and on the relationship between alcohol and mood states in women. He has a particular interest in the history of biology and in the relationship between biological research and modern American culture. He holds bachelor's degrees in both journalism and history from the University of Missouri. In another facet of his writing career, he is the director of communications for the Academic Senate of the University of California.

From the Author

Book titles may be the first thing any reader sees in a book but they often are the last thing an author ponders. Not so with *Biology: A Guide to the Natural World.* The title arrived fairly early on, courtesy of the muse, and then stuck because it so aptly expresses what I think is special about this book.

Flip through these pages and you will see all the elements that students and teachers look for in any modern introductory textbook: rich, full-color art, an extensive study apparatus, a full complement of digital learning tools. Leaf slowly through the book and start to read a little of it, however, and I think that something a little more subtle will start coming through. This second quality essentially has to do with a sense of connection with students. The sensibility that I hope is apparent in *A Guide to the Natural World* is that there is a wonderful living world to be explored; that we who produced this book would like nothing better than to show this world to students; and that we want this to be an instructive walk for them, rather than a difficult march.

All the members of the team who produced *A Guide to the Natural World* worked with this idea in mind. We felt that we were taking students on a journey through the living world and that, rather like tour guides, we needed to be mindful where students are at any given point. Would they remember this term from earlier in the chapter? Had we created enough of a bridge between one subject and the next? The idea was never to leave students with the feeling that they were wandering alone through terrain that lacked signposts. Rather, we aimed to give them the sense that they had a companion—this

book—that would guide them through the subject of biology. This book, then, really is intended as a kind of guide to the natural world, with its audience being students who are taking biology but not majoring in it.

Biology is complex, however, and if students are to understand it at anything beyond the most superficial level, details are necessary. It won't do to make what one faculty member called "magical leaps" over the difficult parts of complex processes. Our goal was to make the difficult comprehensible, not to make it disappear altogether. Thus the reader will find in this book fairly detailed accounts of such subjects as cellular respiration, photosynthesis, immune system function, and population dynamics. It is in the coverage of such topics that our concern for student comprehension was put to its greatest test. We like the way we handled these subjects and other key topics, however, and we hope readers will feel the same way.

Science as a Way of Learning

In one respect, *A Guide to the Natural World* is fairly conventional in that, in its coverage of biology topics, it goes from large to small—from atoms to the biosphere. It deviates from the norm, however, in that the very first chapter is devoted not only to the usual introduction to science and biology, but to the subject of science and society as well. Nonmajors science courses are intended to bring scientific literacy to those students who will not be majoring in science, the idea being that, to be informed citizens, students need to have such a grounding. In general, however, science textbooks make little effort to lay out for students the nature of the relationship between science and society. We thought a good number of faculty would find coverage of such a topic helpful and thus have devoted much of the first chapter to it and then revisit this connection often in the text.

Introductory textbooks such as this have a special obligation to inform students of the process of scientific inquiry. The means by which scientific knowledge is gained is important—once again for an informed citizenry—and this subject tends to be a bigger mystery to students than do scientific findings themselves. *A Guide to the Natural World* has numerous boxed sections whose general theme is "How Did We Learn?" In these sections, students can come to understand the inventiveness and the plain hard work that generally are the prerequisites to scientific discovery. Many of the other boxes in the book have an applied slant to them, dealing with such topics as acid rain, DNA fingerprinting, and even how to choose the right fertilizer for houseplants.

As for general coverage of topics in the main text, no textbook can provide ideal coverage for all courses—indeed, it may be impossible for a textbook to provide ideal coverage for any given course. In deciding what to have more or less of, however, we let biology faculty from across the country be our guides. In shaping *A Guide to the Natural World*, we queried teachers, looked at their syllabi, and asked them what they thought of our ideas on coverage. The result is our best estimation of what most faculty would actually use in the courses they teach.

Notable Features in *A Guide to the Natural World*

Design and Illustrations
Each chapter of *A Guide to the Natural World* is divided up into numbered sections (1.1, 1.2, and so on), so that instructors can easily assign selected parts of a given chapter. The chapter sections are listed at the start of each chapter, and end-of-chapter summaries are indexed by section.

Flip quickly through the pages of *A Guide to the Natural World* and you'll note one useful design element right away: Text almost always occupies the top of a page, with illustrations at the bottom. As a result, text continued from one page to the next is almost never broken up by a photo or illustration. Students reading text will not have their concentration broken by graphics when they turn to new pages. Each chapter begins with a visual "filmstrip" that provides an intriguing advance look at what is to come in the chapter.

As for the illustrations themselves, I think *A Guide to the Natural World* is first-rate for reasons of both process and personnel. With respect to process, the illustrations were constructed chapter by chapter in very satisfying collaboration between myself and artist Kim Quillin. This would not be notable except that the usual case in textbook production is for an author to provide sketches to a sometimes-distant art house after completion of the chapters. In this case, revision of the chapters followed initial design of the illustrations, such that text and art are integrated to a very high degree in the book. Moreover, this method of operation meant that Kim could fully share in the vision of the book as a user-friendly guide for students. To see what I mean, look at Figure 2.1 on pages 20–21, Figure 21.26 on page 436, Figure 22.36 on page 470 , or for something on the more dramatic side, Figure 19.3 on page 369.

With respect to art and personnel, I've introduced Kim as an artist, but what I haven't said is that she is a biologist as well. She was finishing up her Ph.D. at Berkeley in biomechanics while we were working together, and her knowledge of biology informed all that she did for the book.

Media: CD-ROM, Website, and the MediaLabs
The explosion in digital media that has occurred over the last few years is having a great impact in the world of textbooks. Whereas once CD-ROMs and websites were minor adjuncts to textbooks, they have now become real partners in a broad-based learning package.

A Guide to the Natural World gives students and faculty a first-rate set of digital learning tools. Each book comes with its own

Preface

student CD-ROM, and *A Guide to the Natural World* has its own website as well (http://www.prenhall.com/krogh). One of the things these digital tools provide is a set of animations that cover material on a wide range of biology topics. Any writer has to be humbled by the difficulties that come with trying to make static words on a page adequately describe the dynamic realities of the living world. Animations, therefore, are a great help. *A Guide to the Natural World* allows students to see in action such dynamic processes as meiosis, genetic translation, and water molecules working as a solvent.

At the heart of the digital learning tools in the book, however, are the twelve MediaLabs it contains. Look, for example, on page 206 to see what's included in the MediaLab on cell division and human health. Better yet, start out on some of the investigations to get a sense of the interactive nature of these learning modules. Contributor and editor Peggy Brickman of the University of Georgia worked closely with Paul Ramp (Pellissippi State Technical Community College), Michael Golden (Grossmont College), and Michael Farrabee (Estrella Mountain Community College) to share ideas and make the MediaLabs an active learning experience. Students not only receive information—in all the forms that digital media provide—but then they are called upon to synthesize what they've received and produce reports or participate in group activities related to this material. Estimated completion times are given for each of the exercises, lest students end up wandering the byways of the web or the CD-ROM for too long. With the MediaLabs, students should be able to make an easy transition from information on the printed page, to digital information on a CD-ROM, to the universe of digital information out there on the World Wide Web.

The *Guide to the Natural World* Team

There is no such thing as a solo effort on Broadway, in a moon shot, or in the development and production of a textbook. Any book is only as good as the team that put it together, and it has been my good fortune to have been given a great team for *A Guide to the Natural World*. So large is an effort such as this that there are many people I've never met who have worked on this book, so let me just name those with whom I've had the most contact. Kim Quillin has already been mentioned as a careful and creative life-sciences artist, and I can't thank her enough for her contributions. Ray Mullaney is a development editor supreme who leaves no stone unturned in making sure that everything is right and in its place. Donna Young is a great production editor who somehow manages to keep her eye on the details and the big picture at the same time. The book's inventive design is the work of Lynne Stiles, and, as noted, Peggy Brickman was responsible for the MediaLabs. Finally, I must note Prentice Hall's estimable biology editor, Sheri Snavely, who has been the heart and soul of *A Guide to the Natural World* from its inception and a tireless champion of quality in this book.

Apart from these team members, over one hundred biology faculty provided careful critiques of every word and figure in *A Guide to the Natural World*. Their names can be found beginning on page xxx. These reviewers were then joined by a team of academic advisors who provided expert guidance on each unit, making sure not only that what we included was right, but that we included the right things. The advisors are Gary Anderson, University of California, Davis (reproduction); Tania Baker and Leslie Roldan, Massachusetts Institute of Technology (genetics); David Berrigan, University of Washington (evolution); Leon W. Browder, University of Calgary (development); Anthony Ives, University of Wisconsin (ecology); Kevin M. Kelley, California State University, Long Beach (animal anatomy and physiology); Michelle Murphy, University of Notre Dame (genetics); Anu Singh-Cundy, Western Washington University (plants); Erica Suchman, Colorado State University (cell biology); Sara Via, University of Maryland (evolution); Nicholas Wade, *The New York Times* (science and society); John Whitmarsh, University of Illinois (energy and photosynthesis). In shaping the book, we were greatly aided by focus groups composed of college faculty listed on page xxx. Lastly, we received valuable input from college students themselves who read final drafts of selected chapters and told us what they thought. Happily, these students liked what they read. I hope you will too.

David Krogh
Berkeley, California

Kim Quillin received her B.A. in biology at Oberlin College and her Ph.D. in integrative biology from the University of California, Berkeley. Her teaching experience ranges from elementary school science to undergraduate biology at both Oberlin College and UC, Berkeley. She has studied birds in the Smithsonian Museum of Natural History, howler monkey social behavior in Costa Rica, and restoration ecology of aquatic plants in Ohio. She is currently studying the effects of water quality on the mechanics of seagrasses as a visiting research scientist at the Horn Point Laboratory in Maryland. Kim has studied art for over two decades. Her formal art training ranges from the Maryland Summer Center for the Arts to college courses and professional workshops. This book is evidence of her dedicated efforts toward the effective visual communication of biological principles.

Ray Mullaney has spent his entire career in the college textbook division of Prentice Hall. Beginning in 1966 as a development editor he is now vice president, editor in chief for development in the Engineering, Science, and Mathematics division of Prentice Hall's Higher Education Group. Ray divides his time between hands-on editing of books and management responsibilities. He has devoted much of the last three years to *Biology: A Guide to the Natural World,* working closely with the Krogh team and playing a key role in development of the text, photo and art program, and supplements. In 1996, Prentice Hall recognized Ray's many contributions to its college publishing program by awarding him the prestigious Frank Enenbach Award for distinguished service, an honor that has been bestowed on only a handful of employees since the company's founding in 1913.

Marguerite (Peggy) Brickman received her B.A. from Columbia (College) University and her Ph.D. in genetics from the University of California, Berkeley. In teaching non-science majors, she focuses on making the material both relevant and entertaining. As a lecturer at the University of Georgia, she has been ranked number one by student evaluations in the Division of Biology for the past three semesters. One of the reasons her classes are so popular is her ability to integrate media into her lectures. Peggy teaches her students to view electronic information critically, so that after they finish the class they can approach these same media with intelligence and savvy. As MediaLab editor for the book, Peggy has designed the labs to encourage students to think through the material logically and critically.

Lynne Stiles has specialized in information design for print and digital media for over 10 years. Her work has included a variety of educational, informational, and marketing projects. In designing this book, Lynn focused on the way in which a student would use it. Her goal was to help the reader navigate intuitively through the various features—text, essays, multiple visual images—that together introduce the nonmajor to the wonders of the living world. Formerly a partner in YO, a graphic design firm, her clients have included Apple, Netscape, and a variety of publishers. Her work has been featured in *Print, Communication Arts*, and *Graphic Design USA* 17 (American Institute of Graphic Arts). She is the co-author of *Elements of Web Design,* 2nd ed. (Peachpit Press).

Preface

Reviewers

We express sincere gratitude to the expert reviewers who worked closely with the author in reviewing final pages to ensure the scientific accuracy of the text and art.

Anthony Ives,
University of Wisconsin, Madison

Leslie Roldan
Massachusetts Institute of Technology

Anu Singh-Cundy
Western Washington University

Ellen Smith
Arizona State University West

Erica Suchman
Colorado State University

Christine Tachibana
University of Washington

Sara Via
University of Maryland

John Whitmarsh
University of Illinois

The end-of-chapter questions were carefully crafted by a team of dedicated instructors, and we wish to acknowledge their contribution.

Ed Bartholomew,
Maui Community College

David Berrigan
University of Washington

Gail E. Gasparich
Towson University

Carol A. Hurney
Virginia Commonwealth University

Kate Lajtha
Univerity of Oregon

Michelle Murphy
University of Notre Dame

Rhoda E. Perozzi
Virginia Commonwealth University

Heidi Rottschafer
University of Notre Dame

Anu Singh-Cundy
WesternWashington University

Sara Via
University of Maryland

Dawn Adams, *Baylor University*

David L. Alles, *Western Washington University*

Gary Anderson, *University of California, Davis*

Michael F. Antolin, *Colorado State University*

Kerri Armstrong, *Community College of Philadelphia*

Mary Ashley, *University of Illinois, Chicago*

Kemuel Badger, *Ball State University*

Michael C. Bell, *Richland College*

William J. Bell, *University of Kansas*

David Berrigan, *University of Washington*

Lois A. Bichler, *Stephens College*

A. W. Blackler, *Cornell University*

Robert S. Boyd, *Auburn University*

Bonnie L. Brenner, *Wilbur Wright College (City College of Chicago)*

Mimi Bres, *Prince George's Community College*

Peggy Brickman, *University of Georgia*

Leon Browder, *University of Calgary*

Arthur L. Buikema, *Virginia Polytechnic Institute and State University (Allegheny College)*

Steven K. Burian, *Southern Connecticut State University*

Janis K. Bush, *University of Texas, San Antonio*

Linda Butler, *University of Texas, Austin*

W. Barkley Butler, *Indiana University of Pennsylvania*

William S. Cohen, *University of Kentucky*

Tricia Cooley, *Laredo Community College*

Karen A. Conzelman, *Glendale Community College*

Patricia B. Cox, *University of Tennessee*

John Crane, *Washington State University*

Brent DeMars, *Lakeland Community College*

Jean DeSaix, *University of North Carolina, Chapel Hill*

Matthew M. Douglas, *Grand Rapids Community College (University of Kansas)*

Lee C. Drickamer, *Southern Illinois University*

Charles Duggins, Jr., *University of South Carolina*

Susan A. Dunford, *University of Cincinnati*

Ron Edwards, *University of Florida*

Douglas J. Eernisse, *California State University, Fullerton*

Jamin Eisenbach, *Eastern Michigan University*

George Ellmore, *Tufts University*

Patrick E. Elvander, *University of California, Santa Cruz*

Michael Emsley, *George Mason University*

David W. Essar, *Winona State University*

Michael Farabee, *Estrella Mountain Community College*

Rita Farrar, *Louisiana State University*

John Philip Fawley, *Westminster College*

Eugene J. Fenster, *Longview Community College*

John L. Frola, *University of Akron*

Larry Fulton, *American River College*

Gail E. Gasparich, *Towson University*

Claudette Giscombe, *University of Southern Indiana*

Jack M. Goldberg, *University of California, Davis*

Glenn A. Gorelick, *Citrus College*

Melvin H. Green, *University of California, San Diego*

Gail Hall, *Trinity College*

Linnea S. Hall, *California State University, Sacramento*

Madeline Hall, *Cleveland State University*

Steven C. Harris, *Clarion University*

Walter Hewitson, *Bridgewater State College*

Jane Aloi Horlings, *Saddleback College*

Eva Horne, *Kansas State University*

Terry L. Hufford, *The George Washington University*

Carol A. Hurney, *James Madison University*

Andrea Huvard, *California Lutheran University*

Martin Ikkanda, *Los Angeles Pierce College*

Rose M. Isgrigg, *Ohio University*

Anthony Ives, *University of Wisconsin, Madison*

Tom Jurik, *Iowa State University*

Anne Keddy-Hector, *Austin Community College*

Kathleen Keeler, *University of Nebraska*

Nancy Keene, *Pellissippi State Technical Community College*

Kevin M. Kelley, *California State University, Long Beach*

Tom Knoedler, *Ohio State University, Lima Campus*

Preface

Don E. Krane, *Wright State University*

Erica Ann Lawson, *Columbia College*

Mike Lawson, *Missouri Southern State College*

Ann Lumsden, *Florida State University*

Michael M. Martin, *University of Michigan, Ann Arbor*

Michel Masson, *Santa Barbara City College*

Mary Colleen McNamara, *Albuquerque T-VI A Community College*

Scott M. Moody, *Ohio University*

Joseph Moore, *California State University, Northridge*

Deborah A. Morris, *Portland State University*

Allison Morrison-Shetlar, *Georgia Southern University*

Richard Mortensen, *Albion College*

Michelle Murphy, *University of Notre Dame*

Royden Nakamura, *California Polytechnic State University*

Jane Noble-Harvey, *University of Delaware*

Marcy P. Osgood, *University of Michigan*

Andrea Ostrofsky, *University of Maine*

Patricia A. Peroni, *Davidson College*

Rhoda E. Perozzi, *Virginia Commonwealth University*

John S. Peters, *College of Charleston*

Kim M. Peterson, *University of Alaska, Anchorage*

Raleigh K. Pettegrew, *Denison University*

Gary W. Pettibone, *State University of New York, College at Buffalo*

Barbara Pleasants, *Iowa State University*

John M. Pleasants, *Iowa State University*

Don Pribor, *University of Toledo*

Louis Primavera, *Hawaii Pacific University*

Paul Ramp, *Pellissippi State and Technical Community College*

Sonia J. Ringstrom, *Loyola University*

Leslie Ann Roldan, *Massachusetts Institute of Technology*

Heidi Rottschafer, *University of Notre Dame*

John Rueter, *Portland State University*

Nancy Sanders, *Northeast Missouri State University*

Gary Sarinsky, *City University of New York, Kingsborough Community College*

Edna Seaman, *University of Massachusetts, Boston*

Ralph W. Seelke, *University of Wisconsin, Superior*

Prem P. Sehgal, *East Carolina University*

C. Thomas Settlemire, *Bowdoin College*

Robert Shetlar, *Georgia Southern University*

Mark A. Shotwell, *University of Slippery Rock*

Linda Simpson, *University of North Carolina, Charlotte*

Anu Singh-Cundy, *Western Washington University*

Ellen Smith, *Arizona State University West*

Philip J. Snider, *University of Houston*

Frederick W. Spiegel, *University of Arkansas*

Kathleen M. Steinert, *Bellevue Community College*

Donald P. Streubel, *Idaho State University*

Erica Suchman, *Colorado State University*

Gerald Summers, *University of Missouri, Columbia*

Joanne Tornow, *University of Southern Mississippi*

Robin W. Tyser, *University of Wisconsin, LaCrosse*

Sara Via, *University of Maryland*

Dennis Vrba, *North Iowa Area Community College*

Nicholas Wade, *The New York Times*

John H. Wahlert, *Baruch College, The City University of New York*

Timothy S. Wakefield, *Auburn University*

Sarah Ward, *Colorado State University*

R. Barry Welch, *San Antonio College*

John Whitmarsh, *University of Illinois at Urbana-Champaign*

Sandra Winicur, *Indiana University, South Bend*

William Wischusen, *Louisiana State University*

Deborah Wisti-Peterson, *University of Washington*

Rachel Witcher, *University of Central Florida*

Wade B. Worthen, *Furman University*

Robert Yost, *Indiana University Purdue University Indianapolis*

Student Reviewers

The following students provided helpful feedback during the development of the book. Their comments offered us valuable insight into the student experience. Thanks to Linda Smith-Staton of Pellissippi State Technical Community College.

Chris Adams
Connie Bettencourt
Christy Boling
Holly Bryson
Lisa Bryson
April Cross
Laramy Gregory
Sarah Grooms
Becky Johnson
Kevin Keith
Sarah Livesey
Jessika Livingston
Shannon Morris
Kelly Mize
Keeley Nash
Trey Petro
Amanda Reagan
Whitney Spann
Becca Treadway

Print and Media Resources for Students and Instuctors

FOR INSTRUCTORS

Instructor's Toolbox
This easy-to-use lecture organizer will help you integrate the transparencies and instructional material for Krogh by chapter.

Instructor's Guide to the Natural World: Print and Media Resources
by Peggy Brickman, University of Georgia and Anu Singh-Cundy at Western Washington University. (0-02-366892-X)

Instructor's Image Bank/ Presentation Manager 3.0
(0-13-014237-9)

Instructor's Website
A groundbreaking resource, available January 2000, providing chapter outlines, additional internet activities, instructional material for the MediaLabs as well as all of the art and captions from the text in Powerpoint. **www.prenhall.com/krogh**

Transparency Pack
Includes all of the line art and tables with clear, easy-to-read labels. (0-02-366896-2)

Slides
All of the full color acetates are available in the slide set. (0-02-366897)

WebCT
A tool that facilitates the creation of World Wide Web-based courses.

Test Items File and Prentice Hall Test Manager
by Charles Good, Ohio State University, Lima; Tom Knoedler, Ohio State University, Lima; Kate Lajtha, Oregon State University; Kathleen M. Steinert, Bellevue Community College; Erica Suchman, Colorado State University; Christine Tachibana, University of Washington. (0-02-366893). Computerized testing available in both MAC and Windows versions. (MAC 0-02366895-4) (Windows 0-02-366894-6)

More than 1700 questions all identified by orientation (conceptual, applied and recall).

ABCNEWS
ABC News/Prentice Hall Video Library for Biology
(0-13-514191-5)

The Prentice Hall Laser Disc for Biology
(0-13-378100-3)

FOR STUDENTS

Student CD-ROM
Beyond the twelve animated tutorials (launched from in-text MediaLabs), the CD-ROM also contains over 60 concept animations, one self-grading quiz per chapter, a glossary with audio pronunciations, and an integrated link to the student's Companion Website. (0-13-014238-7)

Companion Website
Includes chapter-specific resources for the student. These include 29 self-grading chapter quizzes, links to relevant hyperlinked resources, a student bulletin board, and a search tool for mining the Web for biological information.

Study Guide
by Lydia Daniels, Michele Shuster, and Laurel Roberts, all of the University of Pittsburgh. (0-13-014239-5)

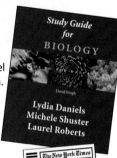

New York Times Themes of the Times
Coordinated by Jane Horlings, Saddleback College. (0-13-022461-8)

Science on the Internet: A Student's Guide
By Andrew Stull. (0-13-021308-X)

FOR THE LABORATORY

Biological Explorations: A Human Approach
by Stanley E. Gunstream. (0-13-575770-3)

Explorations in Basic Biology
by Stanley E. Gunstream. (0-13-079990-4)

Thinking About Biology: An Introductory Biology Lab Manual
by Mimi Bres and Arnold Weisshaar (0-13-633033-9)

In-Text MediaLabs—Launching Point for Web and CD Activities

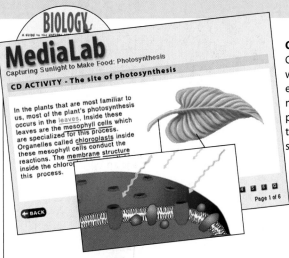

CD Activity
Guided CD-ROM activities with animated tutorials engage students. The animations allow key processes described in the text to come alive on screen.

The MediaLabs are introduced to the student in a two page spread in the textbook. The topics were carefully chosen not only for student interest but also because they highlight information that students may very well come across in their daily lives. All of the necessary background information, including an estimated time for completion, is included in the text. The static becomes dynamic once the student accesses the CD for activities, including detailed animations, and is launched to the website for further investigation of the topic. As a summation, students are asked to communicate the results of their investigations.

168

MediaLab

Capturing Sunlight to Make Food: Photosynthesis

You've probably heard songs in which the singers attribute their life, the act of eating and breathing, to their romantic interest. Love is a powerful emotion, but you'd think at least one songwriter would give homage to the event that really does make our lives possible—the process of photosynthesis. Perhaps it wouldn't

sell. The *CD Activity* reviews photosynthesis, so that in the following *Web Investigations* you are prepared to gather information on topics closely associated with it. In the *Communicate Your Results* section, you can interpret for yourself how important photosynthesis is in our lives.

This *MediaLab* can be found in Chapter 8 on your CD-ROM and Companion Website (http://www.prenhall.com/krogh).

CD ACTIVITY

The overall process of photosynthesis is often summarized in the following reaction:

water + carbon dioxide + light energy -> glucose + oxygen

As with most biological processes, photosynthesis does not occur in one step, but in a series of reactions. In this *CD Activity*, you'll review the three phases of photosynthesis, focusing on reactants and products.

Activity

1. First, you will learn about how light energy is absorbed by pigment molecules in plants.
2. Next, you will discover how light energy is converted to a stored chemical form for transport and later storage.
3. Finally, you will investigate the storage of energy in glucose.

After this review, you'll be prepared to apply, extend, and experience some of the consequences of the photosynthetic process in the following *Web Investigations*.

WEB INVESTIGATION

Investigation 1
Estimated time for completion = 10 minutes

If photosynthesis is to occur at all, the plant must be able to intercept light. The better a plant is in intercepting the light, the greater opportunities the plant has to produce food. Select the Keyword **LEAVES** on your CD or Website for this *MediaLab* to read an article describing how plants maximize the amount

of light they can intercept. At the bottom of that page is a forward link to a discussion of leaf position in a canopy. Review this page. Why have so many plants developed broad, flattened leaves? What is heliotropism, and how does it make plants more efficient? What are sleep movements? Describe the model for leaf orientation in the upper, middle, and lower levels of a leaf canopy.

Investigation 2
Estimated time for completion = 10 minutes

Plants absorb light energy using special molecules called *photosynthetic pigments* (predominantly chlorophylls). These pigments absorb most of the visible light except for wavelengths in the green part of the spectrum. Consequently, plants appear green. But what happens in the fall when leaves change colors? Why are there so many different colors? And why are the colors more intense some years? Select the Keyword **COLORS** on your CD or Website for this *MediaLab* to access a site that will help you answer these questions.

Investigation 3
Estimated time for completion = 15 minutes

You've probably heard of global warming and the greenhouse effect, in which certain gases cause heat to be trapped in the Earth's atmosphere in the same way heat is trapped in a greenhouse. How hot will it be in your community in 100 years? Select the Keyword **HOTSPOT** on your CD or Website for this *MediaLab* to review predicted temperatures for the year 2100. One of the major greenhouse

gases is carbon dioxide, which has been added to the atmosphere by the burning of fossil fuels. Of course, plants in photosynthesis trap CO_2 out of the atmosphere. Could plants help with global warming? Select the Keyword CO_2 to view changing CO_2 levels over 25 years in Barrow, Alaska; Mauna Loa, Hawaii; Samoa, and the South Pole. What might account for the yearly fluctuations in CO_2 levels? Where fluctuations are the greatest, is there a seasonal difference in plant activity?

COMMUNICATE YOUR RESULTS

Exercise 1
Estimated time for completion = 15 minutes

Using the knowledge from the articles in *Web Investigation 1*, design a plant that can effectively intercept light while keeping the amount of plant tissues to a minimum. Here are some characteristics of this hypothetical plant to keep in mind: The plant grows in a seasonal environment, water conservation is important but not critical, and the plant typically has ten leaves located at five levels on the stem. Draw the leaves for this plant. Also draw a view of the plant from the side and a view of the plant from the top. Describe the decisions you made while creating this plant.

Exercise 2
Estimated time for completion = 15 minutes

If it's fall and the leaves are changing, make a collection of leaves from different species in your area. Otherwise, find pictures of fall leaves and clip or copy them to create a collection.

Place these leaves on a poster or in a bound collection. Identify the species of plant as well as possible using available references. For each species, using the leaf's color as a guide, identify the dominant pigment present in the leaf. Is there a pattern of leaf color and plant type? Describe any patterns you observe.

Exercise 3
Estimated time for completion = 15 minutes

In *Web Investigation 3*, you saw an annual fluctuation in CO_2 levels. One possible explanation for the yearly cycle is that increased plant growth decreases atmospheric CO_2 levels during the growing season. How would you test this hypothesis? Prepare a short presentation or poster explaining the initial observation and hypothesis and your test, observations, and conclusions; and include your conclusions on how photosynthesis could be used to fight global warming, and how deforestation would contribute to the problem.

169

Web Investigations
Step-by-step investigations help guide students through thought provoking web-based exercises. An estimated time for completion helps students plan their time.

Communicate Your Results
Perfect for individuals or groups, communication assignments can help assess true understanding of the material.

1

Science as a Way of Learning
A Guide to the Natural World

You don't know what this insect is, but it scares you. Why?
(page 13)

From atom to rain forest, the hierarchy of life.
(page 10)

About 20 years ago, a friend of mine showed me a magazine article that was published which asserted that science and technology were becoming such important parts of our society that people would either have to jump on an approaching technological train or get run over by it. At the time, it just didn't look that way to me. Why, I wondered, would members of my generation have to know any more about science than members of my parents' generation did?

Looking back on this, it's easy to see that the magazine article had it about right. Indeed, the beginning of the 1980s can be seen as about the time when science and technology began knocking on the average person's door with great frequency. The fundamental breakthrough that brought about modern electronics, including the computer, was the invention of the transistor at Bell Labs in 1947. Average American workers then *heard* about computers for 30 years, but by the mid 1980s, they were *using* computers right on their desktops and having other electronic innovations, such as VCRs and CD players, thrust at them from every angle (**Figure 1.1**). The fundamental breakthrough that brought about the biotechnology industry was the

a

b

Figure 1.1
Then and Now

a A technician enters data into the world's first programmable computer, run initially in 1948. Called "Baby," the computer was more than 2 meters (6 feet) tall and almost 5 meters (15.5 feet) wide, but had a total memory of only 128 bytes—less than one-ten-thousandth the amount of data that can be put on a common floppy disk today.

b A businesswoman using a laptop computer participates in a video conference while viewing colleagues in a window on her computer screen. A small video camera sits on top of her computer.

Nature's finery—it's evolutionary.
(page 12)

How scientists think about the world—
the scientific method.
(page 5)

A famous experiment by a famous scientist.
(page 6)

1

description of the DNA molecule in 1953 by James Watson and Francis Crick. People heard more about genes in subsequent years, but only in 1978 did we get the first genetically engineered medicine (human insulin).

What once was a trickle in both these fields became a stream in the 1980s and a torrent in the 1990s. Indeed, in the 1990s it would have been quite a challenge to *avoid* the technological changes going on all around. (You might decline a computer at home, but try evading one at work.) Strictly biological innovations didn't have this kind of presence for the average person, but it was not hard to see signs of a coming wave. Consider in vitro fertilization ("test-tube babies"), DNA testing for paternity, the wide popularity of new antidepressant drugs such as Prozac, the use of a genetically engineered human growth hormone, the controversy over a genetically engineered *bovine* (cow) growth hormone for milk production, and genetic testing for such diseases as cystic fibrosis. Then, there was the cloning of Dolly the sheep from a single cell of an adult sheep (**see Figure 1.2** and page 289), an innovation that has so far been removed from the personal realm, but who knows what's to come?

If we ask, then, whether science and technology are becoming ever more important in our society, the answer is plain. To fully participate in the workforce, to make everyday decisions, and to make informed choices at the ballot box, the average person must now be more technologically and scientifically literate than at any time in the past, and this trend only stands to accelerate in coming years.

Given this, it seems worthwhile to take a look at the relationship between science and society. What practical effect does science have on us? What do average citizens think of science and scientists? These questions will be examined at the beginning of this chapter, with the United States serving as the society in question. This will be followed by a review of the nature of science and of biology.

1.1 How Does Science Impact the Everyday World?

A Look at the News

How is science in general—and biology in particular—likely to be relevant to a person's life in contemporary society? To get an idea, let's look at some of the biology-related news that came to Americans through one magazine (*Time*) during one randomly selected period (six months in 1998). More than two dozen stories connected to biology appeared during this period (**Figure 1.3**). Of the five noted here, four touch on either personal or societal issues, while the fifth is of more intellectual concern.

Boy? Girl? Up to You said a headline from *Time* in its September 21 issue. The subject was a breakthrough a Virginia firm had made in allowing parents to choose the sex of their children, by sorting a father's sperm by type. This obviously is a real-world issue for couples who are contemplating having children, but it's also an issue for society in general, because it adds to the catalogue of reproductive techniques that some people view as being "contrary to nature."

In describing the technique, *Time* writer Frederic Golden noted that "sperm have a gender, in a manner of speaking: some carry an X chromosome and some carry a Y. If the former fertilizes an egg, the baby will be a girl; if the latter, a boy." As it happens, Golden noted, X chromosomes have about 2.8 percent more DNA than Ys, and this difference allows technicians to separate Xs from Ys by staining both kinds with a fluorescent dye and then bouncing a laser light off them. But what's an X chromosome or a Y chromosome? For that matter, what's a chromosome? *Time* did not have space to say, and a reader

Figure 1.2
Something New
Dolly was the first mammal ever cloned from a cell of an adult mammal.

without some background on this subject would simply have to leap over the central word "chromosome" in puzzlement. Readers of this *textbook* who find themselves in a similar position can find out what a chromosome is beginning on page 179.

Sugar Busters! was a two-word headline that appeared in *Time*'s July 6 issue regarding 1998's version of a "miracle diet." This one held out the prospect of weight loss not through exercise or calorie restriction, but through avoidance of foods that cause surges of insulin levels. *Time* writer Christine Gorman did a fine job of explaining what the insulin hormone is and how authors of the best-selling 1998 diet books allege it is linked to weight control. As you might guess, dietary experts were not persuaded by these claims. Their advice was the usual: Exercise, watch calories, cut down on fats, and eat plenty of vegetables and whole-grain carbohydrates. Readers who want to know more about fad diets can see page 148. An introduction to fats begins on page 49.

Global Warming: It's Here appeared in *Time*'s August 24 issue. It noted a just-released report by an international scientific panel that projected a possible rise in Earth's surface temperature of about 1° to 3.5° Celsius (1.8° to 6.3° Fahrenheit) by 2100, a scenario that would threaten low-lying coastal areas, such as parts of Louisiana, with flooding. Background information on global warming can be found on pages 168 and 657.

The Personality Genes appeared in the April 27 issue of *Time* and was a profile of the work of biologist Dean Hamer of the National Institutes of Health. Hamer has done a good deal of research in "behavioral genetics," meaning the linking of genes to human traits and behaviors. His work supports the idea that sexual orientation, the personality trait of "novelty-seeking," and the ability to quit smoking are influenced by genes, though not completely controlled by them.

This work has some obvious down-to-earth relevance in that it might, for example, lead to better ways to help people stop smoking. It also could have great bearing on a broad and very contentious issue in society, which is the question of the origins of sexual orientation. Specifically, is an inclination to-ward homosexuality something that people are born with—is it partly under genetic control, in other words—or is it a "lifestyle" that some people choose to adopt? Gay rights activists believe it is inborn, while other others in society maintain it is a question of choice. The distinction matters greatly because legal civil rights protections in America are based in part on whether a characteristic is inborn or chosen. Central to an understanding of behavioral genetics is an understanding of what genes do. You can find an introduction to this topic beginning on page 174.

Dinosaurs of a Feather appeared in *Time*'s July 6 issue, recounting the recent discovery in China of two new species of dinosaurs, each of which had feathers. These findings provided evidence for a theory that has won general support among evolutionary biologists over the last 25 years: That birds are the direct descendants of dinosaurs or, depending on how you look at it, the last living dinosaurs. Some scientists remain unconvinced, however, with their reservations centering on whether dinosaur and bird features merely look alike (they are "analogous") or whether they are common structures that result from having shared ancestors (they are "homologous"). The difference between analogy and homology in evolution is reviewed on page 360, while the evolutionary line of dinosaurs and birds can be seen on page 362.

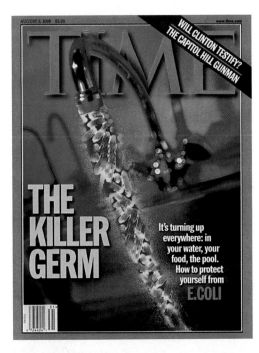

Figure 1.3
The Popular Press Is Filled with Stories on Science in General and Biology in Particular
The text contains a summary of five important stories in *Time* magazine in 1998. Here the cover story of August 3, 1998, was on a rash of incidents in which the bacterium *E. coli*, found in food and water, sickened many people around the country.

1.2 What Does the Public Think, and Know, about Science?

Public Attitudes toward Science

Science issues may be thrust upon the public today, but what do Americans think about science, and how "scientifically literate" are they? A wealth of information on these topics was contained in a report titled *Science & Engineering Indicators—1998*. This document was one in a series of reports on the state of science and engineering in America produced by the National Science Foundation (NSF) over the last 28 years. In a section of the 1998 report that dealt with the American public and its relation to science, two things stood out. One is that Americans are positive about science in many ways; the other is that their *knowledge* of science is uneven—surprisingly high in some areas and surprisingly low in others.

On American attitudes toward science, surveys used in the NSF report revealed that Americans were more interested in science in the late 1990s than at any time in the previous 20 years. They have more confidence in American scientific leadership than they do in the leadership of Congress, corporations, the press, organized religion, even the U.S. Supreme Court. (Only leaders in medicine ranked higher in the public's esteem.) And they believe science has brought much to them already, and that it stands to bring a lot more. More than 85 percent of all Americans believe that the world is a better place because of science.

Public Knowledge of Science

Given Americans' attitudes toward science, it's somewhat surprising to learn what they

know about science. If you look at **Figure 1.4**, you can see the proportion of a representative sample of adult Americans who could correctly answer some basic scientific questions. Better than 75 percent understand that light travels faster than sound, but only about 20 percent understand what DNA is. Almost 80 percent of Americans know that the continents are moving about the face of the Earth, but less than half know that it takes a year for the Earth to go around the sun. (The alternatives were that it made the journey in a day or a month.) Meanwhile, another question revealed that more than a quarter of all Americans think the sun goes around the Earth. With respect to how science works, 27 percent of those questioned were classified as having at least a minimal understanding of scientific inquiry, which is to say they understood that science involves formulating hypotheses and testing them with experiments, a topic that you'll learn more about later in this chapter.

Science Education Makes for Informed Citizens

Connected to the issue of scientific literacy is the question of science education. According to the NSF study, there is a high correlation between how much American adults know about general topics in science and how many science courses they have taken in school. In short, science education is retained in a significant way in that it makes for better-informed citizens.

1.3 What Is Science?

Science as a Body of Knowledge

Having looked a little at the ways science affects our everyday world, let's now review something about science in general and about biology as one of its disciplines. The point here is to give you some sense of the underpinnings of science and biology—to review something about the how and why of them before getting to what they have revealed. This discussion starts with the big picture and progressively gets more specific.

Science is in one sense a process—a *way* of learning. In this respect, it is an activity carried out under certain loosely agreed-to rules, which you'll get to shortly. **Science** is also a body of knowledge, however. It is a collection of unified insights about nature,

Figure 1.4
What Do Americans Know about Science?
Some results published by the National Science Foundation in 1998.

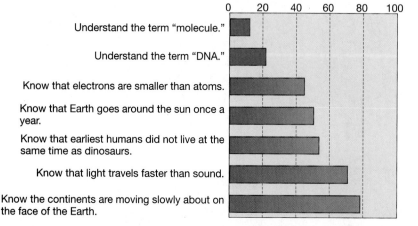

Percent of Americans who...

- Understand the term "molecule."
- Understand the term "DNA."
- Know that electrons are smaller than atoms.
- Know that Earth goes around the sun once a year.
- Know that earliest humans did not live at the same time as dinosaurs.
- Know that light travels faster than sound.
- Know the continents are moving slowly about on the face of the Earth.

the evidence for which is an array of facts. The unified insights of science are commonly referred to as *theories*.

It's unfortunate but true that *theory* means one thing in everyday speech and something almost completely different in scientific communication. In everyday speech, a theory can be little more than a hunch. It is an unproven idea that may or may not have any evidence to support it. In science, meanwhile, a **theory** is a general set of principles, supported by evidence, that explains some aspect of nature. There is, for example, a Big Bang theory of the universe. It is a general set of principles that explains how our universe came to be and how it developed. Among its principles are that a cataclysmic explosion occurred 10–15 billion years ago; and that, after it, matter first developed in the form of gases that then coalesced into the stars we can see all around us. There are numerous facts supporting these principles, such as the current size of the universe and its average temperature.

As you might imagine, with any theory this grand there are *pieces* of it that are in dispute; some facts don't fit with the theory, and scientists disagree about how to *interpret* this piece of information or that. On the whole, though, these general insights have withstood the questioning of critics, and together they stand as a scientific theory.

The Importance of Theories

Far from being a hunch, a scientific theory actually is a much more valued entity than is a scientific fact, for the theory has an *explanatory* power, while a fact is generally an isolated piece of information. That the universe is at least 10 billion years old is a wonderfully interesting fact, but it explains very little in comparison with the Big Bang theory. Facts are important; theories could not be supported or refuted without them. But science is first and foremost in the theory-building business, not the fact-finding business.

Science as a Process: Arriving at Scientific Insights

So how does a body of facts and theories come about? What is the process of scientific investigation, in other words? Looking at **science** as a process, it could be defined as a means of coming to understand the natural world that involves the testing of hypotheses through observation and experiment. This process generally is referred to as the **scientific method**. The starting state for scientific inquiry is always *observation*: A piece of the natural world is observed to work in a certain way. Then follows the *question*, which broadly speaking is one of three types: a "what" question, a "why" question, or a "how" question. Biologists have asked, for example, What are genes made of? Why does the number of species decrease as we move from the equator to the poles? How does the brain make sense of visual images?

Formulating Hypotheses, Performing Experiments

Following the formulation of the question, various hypotheses are proposed that might answer it. A **hypothesis** is a tentative, testable explanation for an observed phenomenon. In almost any scientific question, several hypotheses are proposed to account for the same observation. Which one is correct? Most frequently in science, the answer is provided by a series of *experiments*, meaning controlled tests of the question at hand (**see Figure 1.5**). It may go without saying that

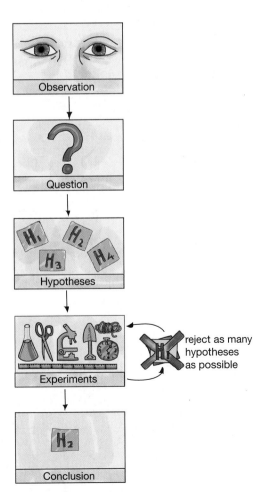

reject as many hypotheses as possible

Figure 1.5
Scientific Method
The scientific method enables us to answer questions by testing hypotheses.

scientists don't regard all hypotheses as being equally worthy of undergoing experimental test. By the time scientists arrive at the experimental stage, they usually have an idea of which is the most promising hypothesis among the contenders, and they then proceed to put it to the test. Let's see how this worked in an example from history.

The Test of Experiment: Pasteur and Spontaneous Generation

Does life regularly arise from anything *but* life, or can it be created "spontaneously," through the coming together of basic chemicals? The latter idea had a wide acceptance from the time of the ancient Romans forward, and as late as the nineteenth century it was championed by a number of the leading *scientists* of the day. So how could the issue be decided? The famous French chemist and medical researcher Louis Pasteur formulated a hypothesis to address this question (**see Figure 1.6**). He believed that many purported examples of life arising spontaneously were simply instances of airborne microscopic organisms landing on a suitable substance and then multiplying in such profusion that they could be seen. Life came from life, in other words, not from spontaneous generation. But how could this be

Figure 1.6
Pasteur's Experiments
Pasteur's spontaneous generation experiments and the scientific method. Nineteenth-century observation made clear that life would appear in a medium, such as broth, that had been sterilized. But what was the source of this life? One hypothesis was that it arose through "spontaneous generation," meaning it formed from the simple chemicals in the broth. Conversely, Pasteur hypothesized that it originated from airborne microorganisms. He was able to design an experiment that offered evidence for this hypothesis. The device he used was an S-shaped flask, which enabled air to enter the flask freely while trapping all particles (including invisible microorganisms) in a bend in the neck.

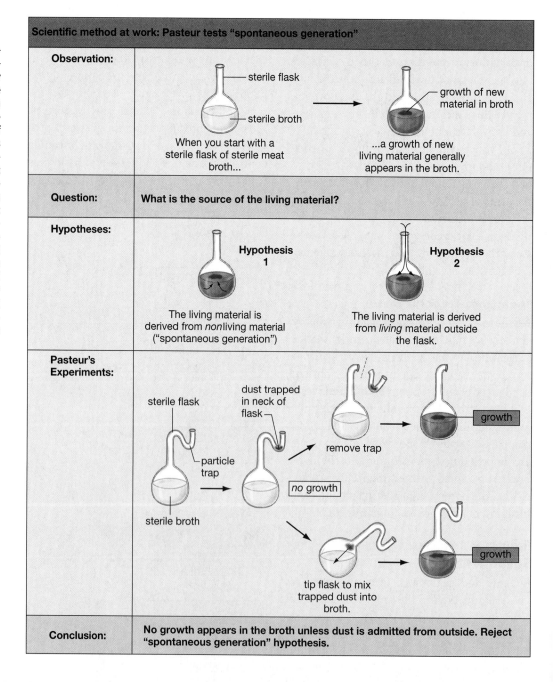

demonstrated? In 1860, Pasteur sterilized a meat broth in glass flasks by heating it, while at the same time heating the glass *necks* of the flasks, after which he bent the necks into a "swan" or S-shape. The ends of the flasks remained open to the air, but inside the flasks there was not a sign of life. Why? The broth remained sterile because microbe-bearing dust particles got trapped in the bend of the flask's neck. If Pasteur broke the neck off before the bend, however, the flask soon had a riot of bacterial life growing within it. In another test, Pasteur tilted the flask so that the broth *touched* the bend in the neck, a change that likewise got the microbes growing.

Elements in Pasteur's Experiments

Now, note what was at work here. Pasteur had a preconceived notion of what the truth was, and designed experiments to test his hypothesis. Critically, he performed the same set of steps several times in the experiments, keeping all the elements the same each time—except for one. The nutrient broth was the same in each test; it was heated the same amount of time and in the same kind of flask. What *changed* each time was one critical **variable**, meaning an adjustable condition in an experiment. In this case, the variable was the access airborne microbes had to the broth. Given the fact that all other elements of the experiments were kept the same, the experiments had rigorous **controls**: All conditions were held constant over several trials, except for a single variable.

Note also that the idea of spontaneous generation was not banished with this one set of experiments—nor should it have been. Pasteur's experiments provided one of the *facts* mentioned earlier, in this case the fact that flasks of liquid will remain sterile under certain conditions. The idea that life arises only from life is, however, one of the scientific *theories* noted earlier, meaning that it requires the accumulation of many facts pointing in the same direction.

Other Kinds of Support for Hypotheses

Some scientific questions are difficult or impossible to test purely through experiment. For example, recall the controversy noted earlier about whether birds are the direct descendants of dinosaurs. What kind of experiment could be run to test this hypothesis? Certain modern-day evidence is available to us—the DNA of living birds, for example—but examining DNA does not amount to an experiment. Instead it is observation, which is another valid way to test a hypothesis. Evidence from the past can also be observed, of course, which in this case means the observation of dinosaur and bird fossils. Indeed, fossils have been the key evidence in convincing most experts that birds are the descendants of dinosaurs. Joining experiment and observation, statistics is a tool used frequently in science as you can see in "A Chance Event or Not?" on page 8.

From Hypothesis to Theory

When does an idea move from hypothesis to theory? One of the ironies of the orderly undertaking called science is that there's nothing orderly about the change from hypothesis to theory. No scientific supreme court exists to make a decision. Scientists aren't polled for their views on such questions; and even if they were, at what point would we say something had been "proven"? When more than 50 percent of the experts in the field assent to it? When there are no dissenters left?

Provisional Assent to Findings: Legitimate Evidence and Hypotheses

One of the tenets of science is that nothing is ever finally proven. Instead, every finding is given only *provisional* assent, meaning it is believed to be true for now, pending the addition of new evidence. In practice, some theories are so well established that no one expects them to be overturned . . . but you never know! At the moment, one of the most established beliefs in biology—that nearly all of Earth's living things derive their energy ultimately from the sun—is being called into question by some respected scientists. (It may be that the Earth houses a vast underground world of microbes who derive their energy from chemicals found in soil and rocks.)

So in the final analysis, scientists must always be open to the possibility that what we "know" is wrong. Indeed, this is one of the fundamental principles of science. Here it is, stated another way, along with some other

A Chance Event or Not? Anecdote, Common Sense, and Statistics in Science

Valuable as they are, experimental and observational tests often are not enough to provide answers to scientific questions. In countless instances, scientists employ an additional tool in coming to comprehend reality—a mathematical tool—as you'll see in the following example.

The evidence that cigarette smoking causes lung cancer (and heart disease and emphysema and on and on) has been around for so long that most people have no idea why smoking was looked into as a health hazard in the first place. You might think that scientists were suspicious of tobacco decades ago and thus began experimenting with it in the laboratory, but this wasn't the case. Instead, the trail that led to tobacco as a health hazard started with a mystery about disease.

When the lung-cancer pioneer Alton Ochsner was in medical school in 1919, his surgery professor brought both the junior and senior classes in to see an autopsy of a man who had died of lung cancer. The disease was then so rare that the professor thought the young medical students might never see another case during their professional lifetimes. Prior to the 1920s, lung cancer was among the rarest forms of cancer, because cigarette smoking itself was rare before the twentieth century. It did not become the dominant form of tobacco use in the United States until the 1920s. This made a difference in lung-cancer rates because cigarette smoke is inhaled, while pipe and cigar smoke generally are not.

If you look at **Figure 1**, you can see the rise in lung-cancer mortality in U.S. males from 1930 forward. A graph for women would show a later rise in lung-cancer deaths, because women started smoking *en masse* later.

Nevertheless, this trend resulted in lung cancer becoming the most deadly form of cancer among women—a distinction that it retains to this day. (More women contract breast cancer, but more women die of lung cancer.)

Given the lung-cancer trends that were apparent by the 1930s, the task before scientists was to explain the alarming increase in this disease. What could the cause of this scourge be, medical detectives wondered. The effects of men being gassed in World War I? Increased road tar? Pollution from power plants? Through the 1940s, cigarette smoking was only one suspect among many.

Laboratory experiment eventually would play a part in fingering tobacco as the lung-cancer culprit, but the original indictment of smoking was written in numbers—in statistical tables showing that smokers were contracting lung cancer at much higher rates than nonsmokers.

The importance of probability and statistics to science can hardly be overstated: These tools are used frequently in nearly every scientific discipline.

It has sometimes been said that "science is measurement," and the phrase is a marvel of compact truth. For centuries, people had an idea that smoking might be causing serious harm, but this information fell into the realm of guessing or of *anecdote*, meaning personal stories. The important scientific principles regarding hypotheses and evidence:

- Every assertion regarding the natural world is subject to challenge and revision.
- Results obtained in experiments must be *reproducible*. Different investigators must be able to obtain the same results from the same sets of procedures and materials.
- Any scientific hypothesis or claim must be *falsifiable*, meaning open to negation through means of scientific inquiry. The assertion that "UFOs are visiting the Earth" does not rise to the level of a scientific claim, because there is no way to prove that this is *not* so.

1.4 Biology
What Is Life?

Let us shift now from an overview of science to a more narrow focus on **biology**, which can be defined as the study of life. But what is life? It may surprise you to learn that there is no standard short answer to this question. Indeed, the only agreement among scholars seems to be that there is not, and perhaps cannot be, a short answer to this question.

problem with anecdote is that there is no measurement in it; there is no way of judging the validity of one story as opposed to the next. Related to anecdote is the notion of "common sense," which is valuable in many instances, but which also had us believing for centuries that the sun moved around the Earth. In the case of smoking, it took the extremely careful measurement provided by science—through a discipline called *epidemiology*—to separate truth from fiction.

Probability in Science

Note that "measurement" in this instance was a matter of calculating *probability*, which is often the case in science. Epidemiologists found a linkage between smoking and lung cancer—in the sense that those who smoked were more likely to get the disease. But, having seen this, scientists then had to ask: Could this result be a matter of pure chance? A person tossing a coin might get heads seven times in a row, and it might be written off to chance. But would it be the same if the person came up with heads *seventy* times in a row? No; at that point there would be justification for assuming that some force other than chance was in operation (such as a rigged coin). When the epidemiologists looked at their statistical tables and saw so many more smokers than nonsmokers getting lung cancer, they had to ask whether this result fell into the realm of seven heads in a row, or seventy. Even in the earliest studies they concluded that more than chance was at work in the results; after many studies, they concluded that smoking was *causing* lung cancer. But how did they judge what was probable and what was not in an issue as complicated as this one? The researchers relied on tech-

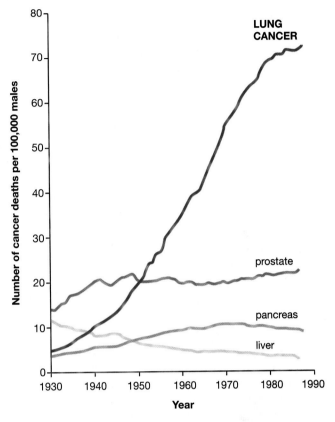

Figure 1
Rise in Lung-Cancer Mortality in U.S. Males from 1930 Forward

niques developed in the branch of mathematics called *statistics*.

The importance of probability and statistics to science can hardly be overstated: These tools are used frequently in nearly every scientific discipline.

The main impediment to such a definition is that any one quality common to all living things is likely to exist in some nonliving things as well. Some living things may "move under their own power," but so does the wind. Living things may grow, but crystals and fire do the same thing. Therefore, biologists generally define life in terms of a group of characteristics possessed by living things. Looked at together, these characteristics are sufficient to separate the living world from the nonliving. We can say that living things:

- Can assimilate and use energy
- Can respond to their environment
- Can maintain a relatively constant internal environment
- Possess an inherited information base, encoded in DNA, that allows them to function
- Can reproduce, through use of the information encoded in DNA
- Are composed of one or more cells
- Evolved from other living things
- Are highly organized compared to inanimate objects

Every one of these qualities exists in all the varieties of Earth's living things. The

atom
(hydrogen)

molecule
(water)

organelle
(mitochondrion)

cell
(muscle)

tissue
(muscle)

organ
(heart)

organism
(human)

simplest bacterium needs an energy source no less than any human being. Our energy source is the food that's familiar to us; the bacterium's might be the remains of vegetation in the soil. The bacterium responds to its environment, just as we do. You would take action if you smelled gas in your house; the bacterium would move away if it encountered something it regarded as noxious. Humans maintain **homeostasis** or a relatively stable internal environment by, for example, sweating when they get hot; when the bacterium's external environment gets too hot, it has certain genes that will switch on to keep it functioning. Both humans and bacteria use the molecule DNA as a repository of the information necessary to allow them to live. Bacteria and human beings both reproduce—bacteria by simple cell division, human beings through the use of two kinds of reproductive cells (egg and sperm). A bacterium is a single-celled life form, while humans are a 10-trillion-celled life form. Bacteria and humans both evolved from complex living things and ultimately share a single common ancestor.

There are some exceptions to these "universals." For example, the overwhelming majority of bees and ants are sterile females; they can't reproduce, but no one would doubt that they're alive. In the main, however, if something is living, it has all these qualities.

Life Is Highly Organized, in a Hierarchical Manner

One item on the list of qualities requires a little more explanation. It is that living things are highly organized compared to inanimate matter. More specifically, they are organized in a "hierarchical" manner, meaning one in which lower levels of organization are progressively integrated to make up higher levels. The main levels in this hierarchy could be compared to the organization of a business. In a corporation, there may be individuals making up an office, several offices making up a department, several departments making up a division, and so forth. In life, there is one set of organized "building blocks" making up another (**see Figure 1.7**).

Actually life is not just "highly" organized. Nothing else comes *close* to it in terms of organizational complexity. The sun is a large thing, but it is an uncomplicated thing compared to the most simple organism. Consider that you have about 10 trillion cells in your body and that, with some exceptions, each of these cells has in it a complement of DNA that is made up of chemical building blocks. How many building blocks? Three billion of them. Now, you probably know that most cells divide regularly, one cell becoming two, the two becoming four, and so on. Each time this happens, each of the 3 billion DNA building blocks

Figure 1.7
Levels of Organization in Living Things

population
(Yanomamö tribe)

community
(Amazon rain forest)

biosphere

must be faithfully *copied*, so that both cells that result from cell division will have their own complete copy of DNA. And this is just the copying of the molecule, before anything is actually done with it. Complex indeed. Let's see what life's levels of organization are.

Levels of Organization in Living Things

The building blocks of matter, called *atoms*, lie at the base of life's organizational structure. (See Chapter 2 for an account of them.) Atoms come together to form *molecules*, meaning entities consisting of a defined number of atoms in a defined spatial relationship to one another. A molecule of water is one atom of oxygen bonded to two atoms of hydrogen, with these atoms *arranged* in a very precise way. Molecules in turn form what are called *organelles*, meaning "tiny organs" in a cell. Each of your cells has, for example, hundreds of organelles in it, called *mitochondria*, that transform the energy from food into an energy form your body can use. Such an organelle is not just a collection of molecules that exist close to one another. It is a highly organized structure, as you can tell just from looking at the rendering of it in Figure 1.7.

At the next step up the organizational chain are entities that are actually *living*, as opposed to entities that are components of life. *Cells* are units that can do all of the things listed earlier: assimilate energy, reproduce, react to their environment, etc. Indeed, most experts would agree that cells are the *only* place that life exists. You may say: But isn't there a lot of material in between my cells? The answer is yes; it's mostly water with a good number of other molecules in the mix. But if all the cells were removed from this watery milieu, there would be nothing resembling life left in it.

The next step up is to a *tissue*, meaning a collection of cells that serve a common function. Your body contains collections of muscle cells that serve the same function (contraction). Each concentration of these cells constitutes muscle tissue. Several *kinds* of tissues can come together to form a functioning unit known as an *organ*. Your heart, for example, is a collection of nerve, muscle, and circulatory tissues. An assemblage of cells, tissues, and organs can then form a multicelled *organism*. (Of course, back down at the cell level, a one-celled bacterium is also an organism; it's just not one with organs and so forth.)

From here on out, life's levels of organization all involve *many* organisms. Members of a single type of living thing (a species) living together in a defined area, make up what is known as a *population*. When you look at *all* the kinds of living things in a given area, you are looking at a *community*. Finally, all the communities of the Earth—and the physical environment with which they interact—make up the *biosphere*.

1.5 Special Qualities of Biology

Almost all scientific disciplines can trace their origins to the ancient Greeks, and biology is no exception. In the work of such Greeks as Hippocrates and Galen, we can find the origins of modern medical science. In the work of Aristotle and others, we can find the origins of "natural history," which led to what we think of today as mainstream biology and the larger category of the **life sciences**, which includes not only biology, but medicine, forestry, and the like.

Despite these ancient origins, biology is, in a sense, a much younger science than, say, physics, which is one of the **physical sciences**, meaning the natural sciences not concerned with life. Western Europe's revolution in the physical sciences probably can be dated from the sixteenth century, when Nicholas Copernicus published his work *On the Revolution of Heavenly Spheres*, which demonstrated that Earth moves around the sun.

Prior to the nineteenth century, biology was almost purely *descriptive*, meaning that the "naturalists" that we would today call biologists largely confined themselves to describing living things—what kinds there were, where they lived, what features they had, and so forth. Beginning in about the 1820s, however, biologists began to formulate biological *theories* as that term was defined earlier. They began to postulate that all life exists within cells, that life comes only from life, that life is passed on through small packets of information that we now call genes, and so forth. To put this another way, biologists in the nineteenth century began describing the *rules* of the living world, whereas before they were largely describing *forms* in the living world.

This change moved biology closer to the same scientific footing as physics, but biology was then, and remains now, a very different kind of science from any of the physical sciences, with physics being a clear case in point. One reason for this difference is that the constituent parts of physics are very uniform and far fewer in number than is the case in biology. Physics deals with only 92 stable elements, such as hydrogen and gold, and to a first approximation, if you've seen one electron, you've seen them all.

Meanwhile, in biology, if you've seen one species you've seen just that—one species. Each species is at least marginally different from another, and many are greatly dissimilar. Moreover, there are thought to be at least 10 million species on Earth. And each of these species has all the organizational levels of elements in physics *and more*. (They have organelles, cells, tissues, etc.) Biology is concerned with the rules that govern all species, and you've seen that there are some biological "universals." But when cancer researchers are looking for the principles that underlie cell division, they are likely to be looking at only one of two main kinds of cells; when ecologists are looking at what causes dry grassland to turn into desert, their findings are likely to have little relevance to the rainforest. Put simply, the living world is tremendously diverse compared to the nonliving world, and such diversity means that universal rules in biology are likely to be few and far between. Biology is concerned with the *particular* to a far greater degree than is the case in the physical sci-

ences. Note also that "universals" in biology may not apply beyond Earth; we don't know if life even exists anywhere else, much less what its rules are. Meanwhile, the rules of physics truly are universal in that they are equally applicable on Earth or in the farthest reaches of the cosmos.

Biology's Chief Unifying Principle

Almost all biologists would agree that the most important thread that runs through biology is **evolution**, meaning the gradual modification of populations of living things over time, with this modification sometimes resulting in the development of new species. Evolution is central to biology because every living thing has been *shaped* by evolution. (There are no exceptions to this universal.) Given this, the explanatory power of evolution is immense. Why do peacocks have their finery, or frogs their coloration, or trees their height (**Figure 1.8**)? All these things stand as wonders of nature's diversity, but with knowledge of evolution they are wonders of diversity that *make sense*. For example, why do so many unrelated stinging insects (such as bees) look alike? Evolutionary principles suggest they *evolved* to look alike because of the general protection this provides from predators. Think of yourself for a moment as a bee predator. Having once gotten stung, would you annoy *any* roundish insect that had a black-and-yellow-striped coloration? You probably learned your

Figure 1.8
Evolution Has Shaped the Living World

a A peacock displaying his plumage

b A poison dart frog in Colombia

c New Caledonia pine trees towering above palm trees in the New Caledonia islands, east of Australia

a

b

c

lesson about this in connection with one species, but *many* species of insects are now protected from you simply by virtue of the coloration they share with the others (**Figure 1.9**). Thus, there were reproductive benefits to individuals who, through genetic chance, happened to get a slightly more striped coloration: They left more offspring, because they were bothered less by predators. Over time, entire populations moved in this direction. They evolved, in other words.

The means by which one kind of living thing can evolve to look like another is a topic this book takes up beginning in Chapter 16. Suffice it to say for now that a consideration of evolution is never far from most biological observations. So strong is evolution's explanatory power that, in uncovering something new about, say, a sequence of DNA or the life cycle of a given organism, one of the first things a biologist will ask is: Why would evolution shape things in this way?

The Organization of This Book

This book has something in common with the levels of organization you looked at ear-

lier, in that it too goes from constituent parts to the larger whole. It begins with atoms, moves on to the biological molecules that atoms make up, and then goes to cells. The end of the book covers the highest levels of biological organization, which is to say natural communities and Earth's biosphere. In between, however, are tours of such facets of life as energy, DNA-encoded information, and reproduction. Even here, however, you'll be moving in a general way from the small to the large, because much of the first part of the book is given over to **molecular biology**—meaning the study of individual molecules and their interactions as they affect living things. Then you'll move into evolution, which touches on **organismal biology**, meaning the study of whole organisms. Next is the **physiology** or physical functioning of plants and animals, which largely concerns tissues and organs. Finally there is **ecology**, which is the study of the interactions of organisms with each other and with their physical environment. And so, let's begin to look at biology—as a body of knowledge and a way of learning.

a

b

Figure 1.9
Similar Enough to Yield a Benefit

a The golden northern bumble-bee

b The Sandhills hornet

These are two of the many stinging insects that have the black-and-yellow striped coloration that offers protection from predators.

Chapter Review

Key Terms

biology 8
control 7
ecology 13
evolution 12
homeostasis 10
hypothesis 5
life science 11
molecular biology 13

Summary

1.1 How Does Science Impact the Everyday World?

- Science is playing an increasingly important role in the everyday lives of Americans, as evidenced by weekly news regarding such issues as reproductive techniques, environmental dilemmas, and the impact of genes on behavior.

1.2 What Does the Public Think, and Know, about Science?

- Americans are interested in science, have a great deal of confidence in American scientific leadership, and overwhelmingly believe that the world is a better place because of science.

- Americans have an uneven knowledge about science. Almost 80 percent of adult Americans know that the continents are moving about the face of the Earth, for example, but more than a quarter think the sun goes around the Earth.

1.3   What Is Science?

- Science is a body of knowledge, a collection of unified insights about nature, the evidence for which is an array of facts.

- The unified insights of science are known as theories. A theory is a general set of principles, supported by evidence, that explains some aspect of nature.

- Science can also be defined as a process that involves the testing of hypotheses through observation and experiment.

- Science works through the scientific method, in which an observation leads to the formulation of a question about the natural world. Then comes a hypothesis—an explanation that has not been proven to be true. The hypothesis may be tested through a series of experiments.

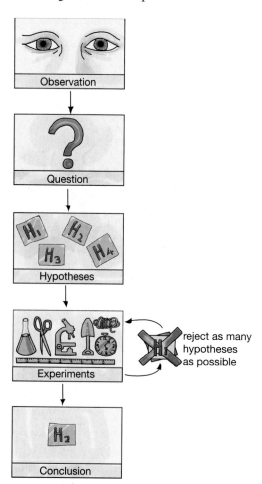

Observation

Question

Hypotheses

Experiments — reject as many hypotheses as possible

Conclusion

- Every assertion regarding the natural world is subject to challenge and revision. Results obtained in experiments must be reproducible. Any scientific hypothesis or claim must be falsifiable, meaning open to negation through means of scientific inquiry.

1.4   Biology

- Biology is the study of life. Life is defined by a group of characteristics possessed by living things. Living things can assimilate energy, respond to their environment, maintain a relatively constant internal environment, and possess an inherited information base, encoded in DNA, that allows them to function. Living things can also reproduce, are composed of one or more cells, are evolved from other living things, and are highly organized compared to inanimate objects.

- Life is organized in a hierarchical manner, running in increasing complexity, from atoms to molecules and then in sequence to organelles, cells, tissues, organs, organisms, populations, communities, and the biosphere.

1.5   Special Qualities of Biology

- Until the early nineteenth century, biology was largely a descriptive science, meaning it largely catalogued and described the Earth's living things. Beginning about the 1820s, however, life-science researchers began to formulate biological theories, such as that life comes only from life and exists only within cells.

- Biology's subject matter—the living world—is notable for its diversity.

- Biology's chief unifying principle is evolution, which can be defined as the gradual modification of populations of living things over time, with this modification sometimes resulting in the development of new species.

Understanding the Basics

Multiple-Choice Questions

1. Which of the following statements best describes the nature of a scientific hypothesis?
   a. A hypothesis is an idea that is widely accepted as a description of objective reality by a majority of scientists.
   b. A hypothesis must stand alone, and not be based on prior knowledge.
   c. A scientific hypothesis must be testable through experimentation, observation, or mathematical demonstration.
   d. Experiments can be designed that will prove the validity of a hypothesis.
   e. A hypothesis when accepted becomes a scientific law.

2. Those who wish to berate a scientific theory sometimes say "that's only a theory." The use of the word "theory" for a biological concept means that
   a. there is absolute certainty about the validity of the concept
   b. most scientists would agree that there is a preponderance of evidence in support of the concept
   c. the concept is in doubt among most scientists
   d. the concept is no more than a hypothesis
   e. the concept has no basis in fact

3. It may be argued that an automobile constitutes living matter because the burning of gasoline represents metabolism. Also, because a car picks up speed when the accelerator is depressed, one might claim it is responding to stimuli. Which of these reasons would you give for definitively concluding that a car is a nonliving entity?
   a. It does not store or use energy.
   b. It does not reproduce by transmitting genetic information through DNA.
   c. Carbon is not a major component of its chemical make-up.
   d. It exhibits cellular organization.
   e. It breaks down sometimes.

4. Evolution is a central, unifying theme in biology because
   a. evolutionary principles can explain ecological processes, although they are not relevant in epidemiology
   b. humans have evolved from ancestors we share with present-day monkeys
   c. evolution has occurred in the past, even though it no longer operates today
   d. the enormously diverse forms of life on Earth have all been shaped by evolution
   e. almost all biologists believe in it

5. Biologists generally define life in terms of a group of characteristics possessed by living things. Which of the following is not a characteristic of living things?
   a. All living things possess an inherited information base, encoded in DNA, that allows them to function.
   b. All living things can respond to their environment.
   c. All living things can maintain a relatively constant internal environment.
   d. All living things evolved from other living things.
   e. All living things are composed of two or more cells.

Brief Review

1. What is science? In what ways is science similar to, and different from, belief systems such as religious faith?

2. What is a controlled experiment? Why is it important to keep all variables but one constant in a scientific experiment?

3. How did Louis Pasteur disprove the idea of spontaneous generation through experimentation?

4. Describe the defining features of life as we know it on the Earth.

5. Living systems can be described at various hierarchical levels. List as many levels of biological organization as you can think of, from the microscopic to the largest levels imaginable.

Applying Your Knowledge

1. Would you agree that it is valuable for a nation to have a citizenry that is reasonably well versed in science? Give reasons for your answer. Would you say this need has become especially urgent in the last two decades? If so, why?

2. Is it harder to prove a hypothesis than to disprove it, or do you always achieve one when you achieve the other? Imagine you wanted to establish that cheetahs are the fastest land animals; assume you have the ability to clock any animal moving at its top speed. Now, what would it take to disprove the idea that cheetahs are the fastest land animals? What would it take to prove that cheetahs are the fastest land mammals, meaning no other land mammal could run faster than they?

3. If you were sent on an interplanetary mission to investigate the presence of life on Mars, what would you look for? Would you explore the land and the atmosphere? Imagine you discover an entity you suspect is a living being. Realizing that life elsewhere in the universe may not be organized by the same rules as on Earth, which of the features of life on Earth, if any, would you insist that the entity display before you would declare it living?

Group Activity

Over a week or two, collect as many stories related to biology as you find in your local newspaper, a national daily, a weekly news magazine, and CNN's homepage. Each member of the group may want to take up one of these news sources as his or her "beat." Arrange the news items in categories of your own making—medicine and health, agriculture, environmental issues, child development, for example. Decide which of the stories have a societal dimension; that is, which of them is likely to have a direct impact on the daily lives of ordinary people. Debate among members of your group whether decision making in that arena should be left up to technical experts alone or whether the voting public should have a say in the issues the news story is concerned with.

Essential Parts: Atoms, Molecules, and Cells

An abundance of life in a drop of water from a pond.

2

The Fundamental Building Blocks
Chemistry and Life

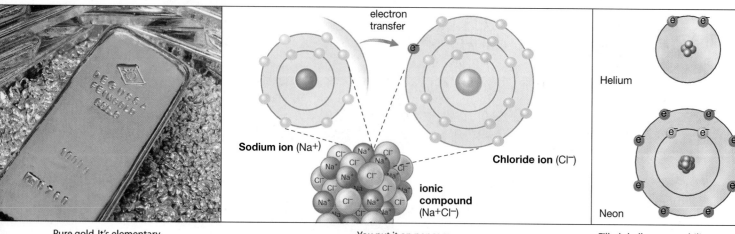

Pure gold. It's elementary.
(page 22)

electron transfer

Sodium ion (Na+)

Chloride ion (Cl⁻)

ionic compound
(Na+Cl⁻)

You put it on popcorn.
(page 29)

Helium

Neon

Filled shells mean stability.
(page 24)

Cities are made of

buildings and buildings are made of bricks, bricks are made of earth, earth is made of . . .? To answer this question, in this chapter we will look at what the material world is made of. Biology is our subject, but to fully understand it, you need to learn a little about what underlies biology. You need to learn a little about what *biology* is made of, in a sense. And to do this, you need to understand some of the basics in two other fields: chemistry and physics.

How are these disciplines relevant to biology? Well, consider chemistry. The average person probably is aware that living things are made up of individual units called cells. But beyond this bit of knowledge, reality fades and a kind of fantasy takes over. In it, the cells that populate people or plants or birds carry on their activities under the direction of their own low-level consciousness. A cell *decides* to move, it *decides* to divide, and so on. Not so. By the time our story is finished, many chapters from now, it will be clear to you that the cells that make up living things have no such consciousness, but rather do what they do as the result of a

chain of chemical reactions. Repulsion and bonding, latching on and reforming, depositing and breaking down—all these things take place without a scrap of consciousness underlying them. What makes people and plants and birds function at this level is *chemistry*.

What is chemistry concerned with? Look around you. Do you see a table, light from a lamp, a patch of night or daytime sky? Everything that exists can be viewed as falling into one of two categories: matter or energy. You will learn something about energy in this chapter, but we are most concerned here with *matter and its transformations*, which is the subject of chemistry. Matter can be defined as anything that takes up space and has **mass**. This latter term is a measure of the *quantity* of matter in any given object. How much space does an object occupy—how much "volume," to put it another way—and how *dense* is the matter within that space? These are the things that define mass. For our purposes, we may think of mass as being equivalent to weight, though in physics they make a distinction between these two things.

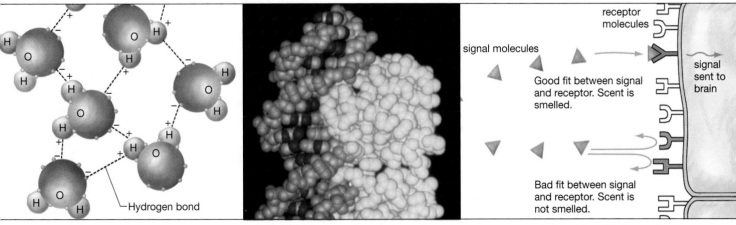

The ties that bond water molecules together.
(page 30)

Many molecules are large and complex.
(page 31)

Do I smell bread in the oven?
(page 31)

2.1 The Nature of Matter: The Atom

Beholding matter all around us, it is natural to ask, what is its nature? A child sees a grain of sand, pounds it with a rock, sees the smaller bits that result and wonders: What is this stuff like at the end of these divisions? Not surprisingly, adults too have wondered about this question—for centuries. About 2,400 years ago, the Greek philosopher Plato accepted the notion that all matter is made up of four primary substances: earth, air, fire, and water. A near-contemporary of his, Democritus, believed that these substances were in turn made up of smaller units that were both invisible and in*div*isible—they could not be broken down further. He called these units atoms (**see Figure 2.1**).

Well, at least one cheer for Democritus, because he had it partly right. Centuries of painstaking work lying between his time and ours has confirmed that matter is indeed composed of tiny pieces of matter, which we still call atoms. But these atoms are not indivisible, as Democritus thought. Rather, they are themselves composed of constituent parts. A superficial account of *all* the parts scientists have discovered to date would go on for pages and still be incomplete. Physicists are continually slamming together parts of atoms with ever-greater force in an effort to determine what *else* there may be at the heart of matter. (This is what the machines called "atom smashers" do. The physicists who run them could be compared to people who, in trying to find out what parts a watch has, throw it on the ground and record the way its various mechanisms fly out upon impact (**see Figure 2.2**).

Interesting stuff, but it is purely the business of physics, with little relation to biology. We are not concerned here with what's at the very end of these divisions. We do care a good deal, however, about what's *nearly* at the end of them.

Protons, Neutrons, and Electrons

For our purposes there are three important constituent parts of an atom: **protons**, **neutrons**, and **electrons**. These three parts exist in a spatial arrangement that is uniform in all matter. Protons and neutrons are packed tightly together in a core (the atom's **nucleus**), and electrons move around this core some distance away (**see Figure 2.3**). The one variation on this theme is the substance hydrogen, the lightest of all the kinds of matter we will run into. Hydrogen has no neutrons, but rather only one proton in its nucleus and one electron in motion around it.

These three "subatomic" particles have mind-bending sizes and proportions. As P. W. Atkins has pointed out, an atom is so small that 100 million carbon atoms would lie end to end in a line of carbon about this long ——————— (3 centimeters). Things are just as disorienting when we consider the size of the atom as a whole, relative to the nucleus. The whole atom, with electrons at its edge, is 100,000 times bigger than

Figure 2.1
The Building Blocks of Life
Viewing this idealized feather at different levels of magnification, we eventually arrive at the building block of all matter, the atom. The atom selected here, from among a multitude that make up the feather, is a single hydrogen atom, composed of one proton and one electron.

the nucleus. If you were to draw a model of an atom *to scale* and began by sketching a nucleus of, say, half an inch, you'd have to draw some of its electrons more than three-quarters of a mile away.

Although the nucleus accounts for very little of the *space* an atom takes up, it accounts for almost all of the *mass* an atom has. So negligible are electrons in this regard, in fact, that all of the mass (or weight) of an atom is considered to reside with the nucleus' protons and neutrons.

Atoms have another quality that interests us: **electrical charge**. Protons are positively charged and electrons are negatively charged. Meanwhile, neutrons—as their name implies—have no charge, but are electrically neutral. Because all these particles do not exist separately, but *combine* to form an atom, as a whole the atom may be electrically neutral as well. The negative charge of the electrons balances out the positive charge of the protons. Why? Because in this state the *number* of protons an atom has is exactly equal to the number of electrons it has (though we'll see a different, "ionic" state later in this chapter). In contrast, the number of *neutrons* an atom has can vary in relation to the other two particles.

With this picture of atoms in mind, we can begin to answer the question that has been handed down to us through history: What is matter? We certainly have a common-sense answer to this question. Matter is any substance that exists in our everyday experience. For example, the iron that goes into cars is matter. But what is it that differentiates this iron from, say, gold? The answer is that an iron atom has 26 protons in its nucleus, while a gold atom has 79.

Figure 2.2
Getting to the Heart of Matter
Understanding what the tiny objects called atoms are made of requires the use of extremely large particle accelerators or "atom smashers," such as this one at Fermilab in Batavia, Illinois. Subatomic particles are accelerated around Fermilab's magnetic ring, some 6.3 kilometers or 3.8 miles in circumference, and then slammed into each other or into a fixed target. How big is this operation? The detectors that record the collisions weigh 5,000 tons apiece and are three stories high.

Figure 2.3
Representations of Atoms
One conceptualization of two separate atoms, hydrogen and helium. The model is not drawn to scale; if it were, the electrons would be perhaps a third of a mile away from the nuclei. The model also is simplified, giving the appearance that electrons exist in track-like orbits around an atom's nucleus. In fact, electrons spend time in volumes of space that have several different shapes.

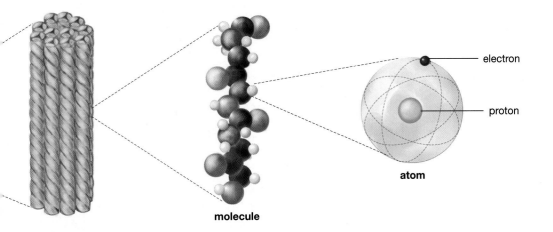

Fundamental Forms of Matter: The Element

Gold is an **element**—a substance that is "pure" because it cannot be reduced to any simpler set of component substances through chemical processes. And what defines each element is the number of protons it has in its nucleus. A solid-gold bar, then, represents a huge collection of identical atoms, each of which has 79 protons in its nucleus (**Figure 2.4**). In making gold jewelry, an artist may combine gold with another metal such as silver or copper to form an alloy that is stronger than pure gold, but the gold atoms are still present, all of them retaining their 79-proton nuclei. Given what you've just read about protons, neutrons, and electrons, you may wonder why gold—or any other element—cannot be reduced to any "simpler set of component substances." Aren't protons and neutrons components of atoms? Yes, but they are not component *substances*, because they cannot exist by themselves as matter. Rather, protons and neutrons must *combine* with each other to make up atoms.

Assigning Numbers to the Elements

In the same way that buildings can be defined by a location, and thus have a street number assigned to them, these elements, which are defined by protons in their nuclei, have an **atomic number** assigned to them. We have observed that hydrogen has but one proton in its nucleus, and it turns out that scientists have constructed the atomic numbering system so that it goes from smallest number of protons to largest. Thus, hydrogen has the atomic number 1. The next element, helium, has two protons, so it is assigned the atomic number 2. Continuing on this scale all the way up through the elements found in nature, we would end with uranium, which has an atomic number of 92.

Figure 2.4
Pure Gold
Gold is an element because it cannot be reduced to any simpler set of substances through chemical means. Each gold bar and nugget is made up of a vast collection of identical atoms—those with 79 protons in their nuclei.

Figure 2.5
Constituent Elements
The major chemical elements found in Earth's crust (including the oceans and the atmosphere) and in the human body.

Earth's crust

other 8%

oxygen 50%

silicon 26%

aluminum 8%
calcium 3%
iron 5%

Human body

other 7%

hydrogen 10%

oxygen 65%

carbon 18%

Given this view of the nature of matter, we are now in a position to answer the question posed at the beginning of the chapter: What is a handful of earth—or anything else—made of? The answer is one or more elements. If you look at **Figure 2.5**, you can see the most important elements that go into making up both the earth's crust and human beings.

Isotopes

All this seems like a nice, tidy way to identify elements—one element, one atomic number, based on number of protons—except that we're leaving out something. Recall that atoms also have *neutrons* in their nuclei, that these neutrons add weight to the atom, and that the number of neutrons can vary independently of the number of protons. What this means is that in thinking about an element in terms of its *weight*, we have to take neutrons into account. Furthermore, since the number of neutrons in an element's nucleus may vary, we can have various *forms* of elements, called **isotopes**. Most people have heard of one example of an isotope, whether or not they recognize it as such. The element carbon has six protons, giving it an atomic number of 6. In its most common form, it also has six neutrons. However, a relatively small amount of carbon exists in a form that has *eight* neutrons. Well, the element is still carbon, and in this form the number of its protons and neutrons equals 14, so the *isotope* is carbon-14, which is used in determining the ages of fossils and geologic samples.

Most elements have several isotopes. Hydrogen, for example, which usually has one proton and one electron, also exists in two other forms: deuterium, which has one proton and one neutron and tritium, which has one proton and two neutrons (**see Figure 2.6**). **Figure 2.7** shows you how isotopes are used in medicine.

The Importance of Electrons

In our account so far of the subatomic trio, we have had much to say about protons and neutrons, but little to say about electrons. This was necessary because we needed to go over the nature of matter, but in a sense you can regard what has been set forth to this point as so much stage-setting, because what's most important in biology is the way elements *combine* with other elements. And in this combining, it is the outermost electrons that play a critical role. Just as you

come into contact with the world through what lies at your surface—your eyes, your ears, your hands—so atoms link up with one another through what lies at their outer edges. The interior life of an atom is very quiet in a sense, while the atom's outer electrons live in a world that can be one of continual forming and breaking of alliances.

2.2 Matter Is Transformed through Chemical Bonding

The process of chemical combination and rearrangement is called **chemical bonding**, and for us it represents the heart of the story in chemistry. When the outermost electrons of two atoms come into contact, it becomes possible for these electrons to reshuffle themselves in a way that allows the atoms to become attached to one another. This can take place in two ways: One atom can *give up* one or more electrons to another; or one

Hydrogen (H)
1 proton
0 neutrons

Deuterium
1 proton
1 neutron

Tritium
1 proton
2 neutrons

Figure 2.7
Diagnosis with Isotopes
One of the most common uses of isotopes is in medicine. Radioactive isotopes, which occur naturally or are created in medical laboratories, are used as an aid in locating disease in the human body. How? Their radioactivity acts like a signal that says "isotope here now, in this quantity." The figure shows the skeleton of a healthy person who has been injected with a radioactive substance that concentrates in bone. Because cancer cells take up more of the isotope, any cancers in the bone would show up as brighter "hot spots" on the image.

Figure 2.6
Same Element, Different Forms
Pictured are three isotopes of hydrogen which, like all isotopes, differ in the number of neutrons they have.

0005 POST   (ORIGINAL)   2.99

atom can *share* one or more electrons with another atom. Giving up electrons is called ionic bonding; sharing electrons is called covalent bonding. A third type of bond, which we'll get to shortly, also is important for our purposes: the hydrogen bond.

Energy Always Seeks Its Lowest State

Atoms that undertake bonding with one another do so because they are in a more *stable* state after the bonding than before it. A frequently used phrase is helpful in understanding this kind of stability: Energy always seeks its lowest state. Imagine a boulder perched precariously on a hill. A mere shove might send it rolling toward its lower energy state—at the bottom of the hill. It would not then roll *up* the hill, either spontaneously or with a light shove, because it is now existing in a lower energy state than it did before— one that is clearly more stable than its former precarious perch. When we turn to electrons, the energy is not gravitational, but electrical. Atoms bond with one another to the extent that doing so moves them to a lower, more stable energy state.

But what determines the *likelihood* of any given atom taking part in this bonding? An analogy might be helpful here. At one time, social dances that regarded themselves as, shall we say, high-class employed what were known as *dance cards* to make sure the young men and women present actually danced. For the first dance, you'd agree in advance with so-and-so to dance together (and write that down on your card); for the second dance, you'd agree with someone else, and so on. Now the important point was, *everyone wanted to have a full dance card*. Otherwise you risked standing around by yourself on the sidelines feeling foolish. But, having filled up your card, you were no longer free to pair up with anyone else.

Seeking a Full Outer Shell: Covalent Bonding

What an atom "seeks" to fill is not a dance card, but its outer shell. It does so because, with a filled outer shell, it is in a lower energy state. Having achieved this, however, it is far less likely to "pair up" with any other atom.

What are these "outer shells"? As it happens, electrons reside in certain well-defined "energy levels" outside the nuclei of atoms. The number of these energy levels varies depending on the element in question. Here we need only note the practical effect of these levels on bonding: *Two* electrons are required to fill the first energy level (or shell) of any given atom, and *eight* are required to fill all the levels thereafter. If you look at the electron configurations pictured in **Figure 2.8**, you can see that two elements—hydrogen and helium—have so few electrons in orbit around them that they have nothing *but* a first energy level, while the other elements pictured have two or three energy levels. This means that hydrogen and helium require only two electrons in orbit around their nuclei to have filled outer shells, but that all other elements require eight electrons to have this kind of complete outer electron complement—to have full dance cards, we might say.

How Chemical Bonding Works In One Instance: Water

To see how chemical bonding works in connection with this concept of filled outer shells, take a look at the bonding that occurs with the

Figure 2.8
Electron Configurations in Some Representative Elements
The concentric rings represent energy levels or "shells" of the elements, and the dots on the rings represent electrons. Hydrogen has but a single shell and a single electron within it, while carbon has two shells with a total of six electrons in them. Helium, neon, and argon have filled outer shells and are thus unreactive. Hydrogen, carbon, and sodium do not have filled outer shells and are thus reactive—they readily combine with other elements.

OUTERMOST ELECTRON SHELL UNFILLED	OUTERMOST ELECTRON SHELL FILLED
Hydrogen	Helium
Carbon	Neon
Sodium	Argon
unstable, very reactive	stable, unreactive

constituent parts of one of the most simple (and important) substances on Earth: water. In so doing, you'll see one of the kinds of bonding we talked about—covalent bonding.

The familiar chemical symbol for water is H_2O. This means that two atoms of hydrogen (H_2) have combined with one atom of oxygen (O) to form water. (See "Notating Chemistry" on page 26 for an explanation of symbols in chemistry.) Recall that hydrogen has but one electron running around in its single energy level. Also recall, however, that this first level is not completed until it has *two* electrons in it. Hydrogen could fill this shell in any number of ways. It might, for example, come into contact with another hydrogen atom. These two atoms can then form a **covalent bond** by sharing a *pair* of electrons with each other—one electron from each atom. These electrons can now be found orbiting the nuclei of both atoms. As a result, both have two electrons in their outer energy levels, which makes them filled.

Our hydrogen atom might also, however, come into contact with an oxygen atom, which has eight electrons. Looking at **Figure 2.9**, you can see what this means: Two electrons fill oxygen's first energy level, which leaves six left over for its second. But remember that the second shell of any atom is not completed until it holds *eight* electrons. Thus oxygen, like hydrogen, would welcome a partner. Only it needs *two* electrons to fill its outer shell, which means that two atoms of hydrogen would do. Once again, pairs of electrons are shared. The oxygen atom and first hydrogen atom donate one electron each for the first pair; and the oxygen and second hydrogen atom each donate one electron for the second pair. The result? H_2O: Two atoms of hydrogen and one atom of oxygen, covalently bonded together and all of them "satisfied" to be in that condition. (Occasionally in nature, covalent bonding will take place in a way that leaves one atom with an unpaired electron, a potentially harmful phenomenon you can read about in "Free Radicals" on page 28.)

Matter Is Not Gained or Lost in Chemical Reactions

Note that when this pairing up of electrons happens, no matter has been gained or lost. We started with two atoms of hydrogen and one atom of oxygen, and we finish that way; the difference is that these atoms are now bonded. This points up an important principle, known as the **law of conservation of mass**, which states that matter is neither created nor destroyed in a chemical reaction.

What Is a Molecule?

When two or more atoms combine in this kind of covalent reaction, the result is a **molecule**: a compound of a defined number of atoms in a defined spatial relationship. Here, one atom of oxygen has combined with two atoms of hydrogen to create *one water molecule*. (What we commonly think of as water, then, is an enormous collection of these individual water molecules.) A molecule need not be made of two different elements, however. Two hydrogen atoms can covalently bond to form one *hydrogen* molecule. On the other side of the coin, a molecule could contain many different elements bonded together. Consider sucrose, or regular table sugar, which is $C_{12}H_{22}O_{11}$—12 carbon atoms bonded to 22 hydrogen atoms and 11 oxygen atoms.

a   hydrogen molecule

b   water molecule

Figure 2.9
Covalent Bonding
A covalent bond is formed when two atoms share one or more pairs of electrons.

a Two atoms of hydrogen have come together, and each shares its lone electron with the other; this gives both atoms a filled outer shell—and stability.

b Two hydrogen atoms have linked with one oxygen atom; in this case, two pairs of electrons are shared, one pair between each of the hydrogen atoms and the oxygen atom.

Notating Chemistry

One of the "languages of science" is the system of symbols that chemistry uses. Somewhat forbidding at first viewing, this system soon comes to serve its intended purpose of conveying a lot of information very quickly.

Our starting place is that each chemical element has its own symbol, so that hydrogen becomes H, carbon C, and platinum Pt.

When we begin to combine these elements into molecules, it is necessary to specify how *many* atoms of each element are part of the molecule. If we have two atoms of oxygen together—which is the way oxygen is packaged in our atmosphere—we have the molecule O_2. Three molecules of O_2 is written as $3O_2$. This kind of notation is known as a **molecular formula**. It is very helpful in stipulating the makeup of molecules from the simple, such as oxygen, to the complex, such as chlorophyll, which is notated $C_{55}H_{72}MgN_4O_5$.

To see a molecular formula is to learn a lot about what atoms are in a molecule, but nothing about the way the atoms are *arranged* in relation to one another. (Look at chlorophyll's formula. Is there a line of 55 carbon atoms followed by 72 hydrogen atoms? From the molecular formula, how could you tell?) To convey this ordering information, chemists and biologists use what are known as **structural formulas**—two-dimensional representations of a given molecule. Methane (CH_4) is a very simple molecule composed, as the molecular formula shows, of one atom of carbon and four of hydrogen. In a structural formula, these constituent parts are conceptualized like this:

$$H - C - H$$

methane

Note that there is a single line between each hydrogen atom and the central carbon atom. This has not been done just because a single line is easy to draw. The bond between any of the hydrogen atoms and carbon is a **single bond**: Each line represents *one* pair of electrons being shared. Thus:

$$H : C : H$$

— electron pairs

There can also be **double bonds** and **triple bonds**. When carbon dioxide forms, there are two oxygen atoms. Each shares two pairs of electrons with a lone carbon atom. Here's how we would notate the double bond in a carbon dioxide (CO_2) molecule:

$$O = C = O$$

carbon dioxide

Reactive and Unreactive Elements

The elements considered so far all welcome bonding partners, because all of them have incomplete outer shells. This is not true of all elements, however. There is, for example, the helium atom, which has two electrons. It thus *comes equipped*, we might say, with a filled outer shell. As such, it is extremely stable—it is unreactive with other elements. It is so unreactive that it is part of a family of elements that at one time were known as the inert gases, because it was thought that these elements never combined with anything. At the opposite end of the spectrum are elements that are extremely *reactive*. Look again at the representation of the sodium molecule in Figure 2.8. It has 11 electrons, two in the first shell and eight in the second, which leaves but one electron in the third shell—a very unstable state. Between the extremes of sodium and helium are elements with a range of outer (or *valence*) electrons. Thus, there is a spectrum of stability in the chemical elements, based on the number of outer-shell electrons each element has—from 1 to 8.

Polar and Nonpolar Bonding

Not all covalent bonds are created alike. When four hydrogen atoms come together with one carbon atom, the result is a molecule called methane (CH_4), which is an important gas on Earth. Now, in the methane molecule as a whole, the electrons shared by the carbon and hydrogen atoms are shared *equally*. That is, the shared electrons are equally attracted to both carbon and hydrogen. This is not the case, however, with the water molecule. Look at the representation

Though structural (or "skeletal") formulas can tell us a good deal about the ordering of atoms in a molecule, they tell us very little about the *three-dimensional* arrangements of atoms. For this, we rely on two other kinds of representations, the **ball-and-stick model** and the **space-filling model**. An ammonia molecule (NH_3) is pictured next in both forms, with the molecular and structural formulas added to show the progression:

NH₃   H—N̈—H with H below

molecular formula   structural formula   ball-and-stick model   space-filling model

Note that the ball-and-stick model gives us a better idea of molecular angles of the atoms, but that the space-filling model gives us a better idea of the relative size of these atoms and how one actually hugs the other.

Finally, it is useful to have a way to notate the "before" and "after" stages of a chemical reaction. This is done by employing a simple arrow, as when carbon reacts with hydrogen to form methane: $C + 4H \rightarrow CH_4$. The carbon and hydrogen atoms on the left are **reactants**; the arrow means "yields"; and the methane molecule on the right is

the **product** of the reaction. Here's a graphic representation of what is happening:

C + 4H → CH₄

reactants   product

An important thing to keep in mind about notation goes back to the law of conservation of mass (page 25): Matter is neither created nor destroyed in a chemical reaction. It follows that reactions such as the one above must be *balanced*: We must have the same number of atoms when the reaction is finished (on the right) as when it started (on the left). Thus, we could not have $C_2 + 4H \rightarrow CH_4$. This is an unbalanced reaction. We started out with two atoms of carbon on the left, but somehow ended up with one atom of carbon on the right. Nature doesn't play that way.

of it in **Figure 2.10**. As it turns out, the oxygen atom has a greater power to attract electrons to itself than do the hydrogen atoms. The term for measuring this kind of pull is **electronegativity**. Because the oxygen atom has more electronegativity than do the hydrogen atoms, it tends to pull the shared electrons away from the hydrogen and toward itself. When this happens, the molecule takes on a **polarity** or a difference in electrical charge at one end as opposed to the other. Because electrons are negatively charged, and because they can be found closer to the oxygen nucleus, the oxygen end of the molecule becomes slightly negatively charged, while the hydrogen regions become slightly positively charged. We still have a covalent bond, but it is a specific type: a **polar covalent bond**. Conversely, with the methane

molecule—where electrons are being shared equally—we have a **nonpolar covalent bond**.

To grasp the importance of this, consider the water molecule, with its positive and negative regions. What's going to happen when it comes into contact with *other* polar molecules? Right: The oppositely charged parts of

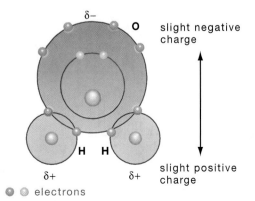

δ− O slight negative charge

H H

δ+   δ+   slight positive charge

electrons

Figure 2.10
Polar Covalent Bonding
In the water molecule, the oxygen atom exerts a greater attraction on the shared electrons than do the hydrogen atoms. Thus the electrons are shifted toward the oxygen atom, giving the oxygen atom a partial negative charge (because electrons are negatively charged) and the hydrogen atoms a partial positive charge. ("Partial" here is indicated by the Greek delta symbol δ.) The molecule as a whole has taken on a polarity or a difference in charge at one end as opposed to the other.

Free Radicals

Their name makes them sound like a group of sixties activists set loose after years in jail, but **free radicals** aren't people at all; they are atoms or molecules. They *have* been set loose, however—to damage human bodies in illnesses that may range from cancer to coronary heart disease.

The way the public normally hears about these culprits is through the recommended means of limiting their harm: a diet rich in vitamins C, E, and the substance beta carotene. Looking at free radicals from another angle, however, we can see how they represent a damaging exception to the rules of chemical bonding.

You know that atoms "seek" to have a full outer energy shell, which in most cases means eight electrons. In covalent bonding, atoms achieve this state by sharing *pairs* of electrons with one another, one electron of each pair coming from each of the atoms involved. Occasionally in nature, however, atoms come together to create a molecule in which one of the component atoms has an *unpaired electron* in its outer shell. Nitrogen can come together with oxygen, for example, to form nitric oxide (NO). Recall that oxygen has six outer electrons (and thus needs two for stability), while nitrogen has five outer electrons (and thus needs three). When oxygen and nitrogen hook up, they can share only two electrons with one another before oxygen's outer shell is filled. This, however, leaves nitrogen with an unpaired electron.

Unstable molecules like this usually exist only briefly, as intermediate molecules in chemical reactions. And in people, that's just the problem. Human beings are among the many species that use a terrific amount of oxygen to extract energy from food. In this process, oxygen is constantly picking up electrons. Through the many steps in metabolism, oxygen may come together with other substances to create a type of free radical called reactive oxygen. Though any one reactive oxygen molecule is short-lived, the damage comes by way of a destructive chain reaction: Seeking partners, one free radical begets more, which beget more.

Where's the harm? Well, for one thing, free radicals may irritate or scar artery walls, which invites artery-clogging fatty deposits around the damage. They also may

> *Free radicals may irritate or scar artery walls, which invites artery-clogging fatty deposits around the damage, and they may have a mutation-causing effect on human DNA.*

have a mutation-causing or "mutagenic" effect on human DNA, which can be a factor leading to cancer. Some primary sites of free-radical generation and damage are the "powerhouses" in cells—structures called mitochondria—which are primary sites at which energy is transferred from food. Indeed, a growing body of evidence supports a long-standing theory of human aging, which holds that many of the things we associate with getting older—memory loss, hearing impairment—can be traced to the cumulative effects of free radicals damaging DNA in the mitochondria, thus diminishing the body's energy supply.

Free radicals are the natural product of metabolism in human beings; they are the price we pay for being alive. However, they can be created in us in *greater* numbers in accordance with our behavior. Some of the usual suspects seem to be involved here—cigarette smoking, alcohol consumption, and sunlight exposure.

Against this production of free radicals, however, nature has also provided its own set of free-radical scavengers, among them the aforementioned beta carotene and vitamins C and E. We can control these, in the sense of making sure we have plenty of them in our diets. We can buy them in pill form, of course, but the jury is still out on the effect of very large doses of these substances. The best bet is to eat the right kinds of foods—meaning a lot of citrus fruits, whole grains, and vegetables of the green leafy, orange, and yellow variety.

the molecules will attract and the similarly charged parts will repel. It's like having a bar magnet and trying to bring its positive end into contact with the positive end of another magnet: Left on its own, the second magnet just flips around, so that positive is now linked to negative. In the same way, molecules flip around in connection with their polarity.

Ionic Bonding: When Electrons Are Lost or Gained

So we've gone from nonpolar covalent bonding, where electrons are shared equally, to polar covalent bonding, where they're pulled to one side of the resulting molecule. What if we carried this just one step further and had instances in which the electronegativity differences between two atoms were so extreme that electrons were pulled *off* of one atom altogether, only to latch on to the atom that was attracting them? This is what happens in our second type of bonding, **ionic bonding**. The classic illustration of this type of bonding involves the sodium we looked at earlier and the element chlorine. Recall that sodium has 11 electrons, meaning that there's a lone electron flying around in its third electron shell. Chlorine, meanwhile, has 17 electrons, meaning it has 7 electrons in the third shell. Remember that 8 is the magic number for outer-shell stability. Sodium could get to this number by *losing* one electron, while chlorine could get to it by *gaining* one electron. That's just how this encounter goes: Sodium does in fact lose its one electron, chlorine gains it, and both parties become stable in the process (**see Figure 2.11**).

What Is an Ion?

But this story has a postscript. Having lost an electron (with its negative charge), sodium (Na) then takes on an overall *positive* charge. Having gained an electron, chlorine (Cl) takes on a negative charge. Each is then said to be an **ion**: a charged atom. Or, to put it another way, an atom whose number of electrons differs from its number of protons. We denote the ionized forms of these atoms like this: Na^+, Cl^-. Were an atom to gain or lose more electrons than this, we would put a number in front of the charge sign. For example, to show that the magnesium atom has lost two electrons and thus become a positively charged magnesium ion, we would write Mg^{2+}.

Note that we now have two ions, Na^+ and Cl^-, with differing charges in proximity to one another. They are thus attracted to one another through an *electrostatic attraction*, and have an ionic bond between them. This hardly ever happens with just *two* atoms, of course. Many billions of atoms are bonded together in this way, up, down, and sideways from each other. This whole collection is, likewise, called an ion; or, if two or more elements are mixed together this way, an **ionic compound**. The particular ionic compound just described actually is very familiar. Sodium and chlorine combine to create sodium chloride, which is better known as table salt. The notation should properly be written Na^+Cl^-, but it is usually denoted as just plain NaCl.

Is it apparent how an ionic compound differs from a molecule? In an ionic compound, there is no fixed number of atoms linked up in a defined spatial relationship, as in H_2O. Rather, an undefined number of charged atoms are bonded together, as in NaCl.

**Figure 2.11
Ionic Bonding**

a Sodium has but a single electron in its outer shell, while chlorine has seven, meaning it *lacks* only a single electron to have a completed outer shell.

b When these two atoms come together, sodium loses its third-shell electron to chlorine, becoming, in the process, a sodium ion with a net positive charge (because it now has more protons than electrons). Having gained an electron, the chlorine atom becomes a chloride ion, with a net negative charge (because it has more electrons than protons).

c The sodium and chloride ions are now attracted to each other because they are oppositely charged.

d The result of this "electrostatic" attraction, involving many sodium and chloride ions, is a sodium chloride crystal (NaCl), better known as table salt.

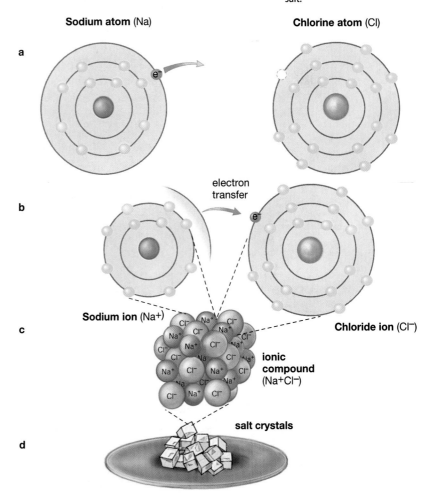

Sodium atom (Na) Chlorine atom (Cl)

a

electron transfer

b

Sodium ion (Na⁺)   Chloride ion (Cl⁻)

c

ionic compound (Na⁺Cl⁻)

salt crystals

d

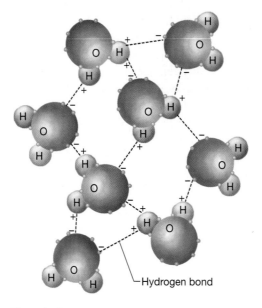

Figure 2.12
Hydrogen Bonding
The hydrogen bond, in this case between water molecules, is indicated by the dotted line. It exists because of the attraction between hydrogen atoms, with their partial positive charge, and the unshared electrons of the oxygen atom, with their partial negative charge.

To take a step back for a second, recall that bonding runs a gamut from the nonpolar covalent bonding (where electrons are shared equally) to the slightly charged polar covalent bonding (where they are shared somewhat unequally) to the charged ionic bonding (where electrons are gained or lost altogether). It's important to recognize that there is a *spectrum of polarity*, and that within it, some bonds are almost completely ionic (as with sodium chloride) while others are completely non-polar (as with methane).

A Third Form of Bonding: Hydrogen Bonding

We need to look at one more variant on bonding, called hydrogen bonding. Recall that in any water molecule, the stronger electronegativity of the oxygen atom pulls the electrons *shared* with the hydrogen atoms to-

ward the oxygen nucleus, giving the oxygen end of the molecule a partial negative charge and the hydrogen end of the molecule a partial positive charge. So what happens when you place several water molecules together? A positive hydrogen atom of one molecule is weakly attracted to the negative, *unshared* electrons of its oxygen neighbor. Thus is created the **hydrogen bond**, which links an already covalently bonded hydrogen atom with an electronegative atom (in this case with oxygen; see **Figure 2.12**). Hydrogen bonding is a linkage that, for our purposes, nearly always pairs hydrogen with either oxygen or nitrogen. These relatively weak bonds are important in linking the atoms of a single molecule to one another, but they are just as important in creating bonds *between* molecules, as in the example. The hydrogen bond is indicated by a dotted line. You will be seeing a lot of it in the molecules of life—in DNA, proteins, and elsewhere.

2.3 Some Qualities of Chemical Compounds

Molecules Have a Three-Dimensional Shape

We now need to make more explicit what has been noted only by implication in our diagrams of water molecules: that molecules and ionic compounds have a three-dimensional shape. It is useful to depict them as two-dimensional chains and rings and such, but in real life a molecule is as three-dimensional as a sculpture. A fair number of shapes is possible, even in simpler molecules. Atoms may be lined up in a row, or in triangles or pyramid shapes. As an example, look at another representation of the water and methane molecules (**Figure 2.13**). You can see that in water there is a very definite spatial configuration; its hydrogen atoms are splayed out from its oxygen atom at an angle of 104.5°.

Molecular Shape Is Very Important in Biology

Why does molecular shape matter? It is critical in enabling biological molecules to carry out the activities they do. This is so because molecular shape determines the capacity of molecules to latch onto or "bind" with one another. When, for example, you smell the

Figure 2.13
Three-Dimensional Representations of Molecules

a In the case of water there is an angle of 104.5° between hydrogen atoms.

b Methane is a molecule with an angle of 109.5° between hydrogen atoms.

a water (H_2O)

b methane (CH_4)

Figure 2.14
The Importance of Shape
Signaling molecules from the aroma of bread (the triangles) bind with specific receptors on the surface of the cell, setting in motion a cellular process. For this binding to take place, the molecules must fit together, and this fit is governed by the shape of each molecule.

aroma of fresh-baked bread, gas molecules wafting off the bread bind with receptor molecules in your nasal passages, thus sending a message to the brain about the presence of bread. It is the precise shape of the bread molecules and nasal receptor molecules that allows them to bind with one another. Look at **Figure 2.14** to see how this works, and in the bargain get an idea of how large some biological molecules are, relative to the molecules considered so far (**see Figure 2.15**).

Having learned a little about atoms and molecules, we now need to review a few final concepts that will aid us in understanding chemistry on a slightly larger scale.

Solutes, Solvents, and Solutions

Take a glass of water and pour a little salt in it. When you stir that up, the salt quickly disappears. It has not actually gone any-

Figure 2.15
Complex Binding
A computer model of some real-life molecular binding, in this case of a protein (the yellow atoms) binding to a length of DNA. This space-filling model provides some idea of the enormous number of atoms and the complicated shapes that make up some of the molecules employed by living things.

where, of course; it has simply mixed with the water. Now, if it has mixed uniformly, so that there are no lumps of salt here or there, you have created a **solution**: a homogeneous mixture of two or more kinds of molecules, atoms, or ions. The salt is what's being dissolved here, so it is the **solute**. The water is doing the dissolving, so it is the **solvent** (see **Figure 2.16**).

Molecules vary greatly in the degree to which they are *soluble* in different solvents. This is because, for something to act as a solvent, it must be able to break down the chemical bonds of the solute. Salt breaks down very quickly into its ionic components when placed in enough water, but don't stand around waiting for water to do the same thing to salad oil.

A general rule is that like dissolves like. Substances that are nonpolar dissolve best in nonpolar solvents; substances that are polar dissolve best in polar solvents. Salt and water are both polar, and the one dissolves in the other. Conversely, the ingredient in soap that actually breaks up nonpolar greases and dirt is a long, nonpolar molecular chain composed of hydrogen and carbon atoms.

On to Some Detail Regarding Water

In biology, you will often see casual references to a molecule being *soluble* in a certain way. And you will see this most commonly

Figure 2.16
A Solute Dissolved by a Solvent Makes a Solution
When we pour a small amount of table salt (an ionic compound composed of sodium and chloride ions) into water, the salt crystals dissolve into the water.

mentioned in two ways: as something being fat-soluble or water-soluble. Does it dissolve in fat or does it dissolve in water? Both things are important to life, but the importance of the latter—of water as a solvent—cannot be overstated. Water accounts for about 75–85 percent of a cell's weight, and most cells are surrounded by it. It is not difficult to see why water should be so prominent in this way. Life began in water, much life lives in it today, and our own ancestors made the transition from sea to land essentially by carrying their wet environment with them—inside themselves. So important to life is this simple, common molecule that we end with it here, only to take it up again in our next chapter in a more detailed way.

Chapter Review

Key Terms

atomic number 22

atomic weight 23

ball-and-stick model 27

chemical bonding 23

covalent bond 25

double bond 26

electrical charge 21

electron 20

electronegativity 27

element 22

free radical 28

hydrogen bond 30

Summary

2.1 The Nature of Matter: The Atom

- The fundamental unit of matter is the atom. The three most important constituent parts of an atom are protons, neutrons, and electrons. Protons and neutrons exist in the atom's nucleus, while electrons move around the nucleus, at some distance from it. Atoms have an electrical charge. Protons are positively charged, electrons are negatively charged, but neutrons carry no charge.

- An element is any substance that cannot be reduced to any simpler set of constituent substances through chemical means. Each element is defined by the number of protons in its nucleus.

- The number of neutrons in an atom can vary independently of the number of protons. Thus a single element can exist in various forms, called isotopes, depending on the number of neutrons it possesses.

2.2 Matter Is Transformed through Chemical Bonding

- Atoms can link to one another in the process of chemical bonding. Among the forms this bonding can take are covalent bonding, in which atoms share one or more electrons; and ionic bonding, in which atoms lose and accept electrons from each other.

- Chemical bonding comes about as atoms "seek" their lowest energy state. An atom achieves this state when it has a filled outer electron shell. Hydrogen and helium require two electrons in orbit around their nuclei to have filled outer shells, while all other elements require eight electrons to have filled outer shells.

- A molecule is a compound of a defined number of atoms in a defined spatial relationship. For example, two hydrogen atoms can link with one oxygen atom to form one water molecule.

- Atoms of different elements differ in their power to attract electrons. The term for measuring this power is electronegativity. Through electronegativity, a molecule can take on a polarity, meaning a difference in electrical charge at one end compared to the other. Covalent chemical bonds can be polar or nonpolar. A polar covalent bond exists when shared electrons are not being shared equally among atoms in a molecule, due to electronegativity differences.

- Two atoms will undergo a process of ionization when the electronegativity differences between them are great enough that one atom loses one or more electrons to the other. This process creates ions, meaning atoms whose number of electrons differs from their number of protons. The charge differences that result from ionization can produce an electrostatic attraction between ions. This attraction is an ionic bond. When atoms of two or more elements bond together ionically, the result is an ionic compound.

- Hydrogen bonding links a covalently bonded hydrogen atom with an electronegative atom. In water, a hydrogen atom of one water molecule will form a hydrogen bond with an unshared oxygen electron of a neighboring water molecule.

2.3 Some Qualities of Chemical Compounds

- Three-dimensional molecular shape is important in biology because this shape determines the capacity molecules have to bind with one another.

- A solution is a homogeneous mixture of two or more kinds of molecules, atoms, or ions. The compound being dissolved in solution is the solute; the compound doing the dissolving is the solvent.

- A general rule in chemistry is that like dissolves like: Substances that are nonpolar dissolve best in nonpolar solvents, while substances that are polar dissolve best in polar solvents.

Understanding the Basics

Multiple-Choice Questions

1. Carbon is an element with an atomic number of 6. Based on this information, which of the following statements is true? (More than one may be true.)
   a. Carbon can be broken down into simpler component substances.
   b. Carbon cannot be broken down into simpler component substances.
   c. Each carbon atom will always have 6 neutrons.
   d. Each carbon atom will always have 6 protons.
   e. Protons + electrons = 6.

2. Suppose that you are reviewing for a test, and a fellow student says that the equation for photosynthesis is $6CO_2 + H_2O \rightarrow C_6H_{12}O_6 + 6O_2$. How would you reply?
   a. They are right.
   b. The way this equation is written violates the law of the conservation of matter.
   c. The CO_2 is held together by ionic bonds.
   d. A hydrogen bond holds the two hydrogens to the oxygen in the water molecule.
   e. There are 12 carbons in sugar.

3. Neon used to be called an inert gas. Thus it
   a. easily forms perfect covalent bonds
   b. easily forms ionic bonds
   c. has a filled outer shell
   d. is polar
   e. all of these

4. Carbon and hydrogen have almost the same electronegativity. Thus
   a. They share electrons equally.
   b. Sometimes carbon takes electrons completely away from hydrogen.
   c. They share electrons, but unequally.
   d. Hydrogen is attracted to carbon, but does not bond with it.
   e. They have the same number of protons.

5. Fats are made up mostly of carbon and hydrogen, which have almost the same electronegativity. Based on this information, fats are made up mostly of
   a. polar covalent bonds
   b. hydrogen bonds
   c. ionic bonds
   d. nonpolar covalent bonds
   e. there is insufficient information to answer this question

6. You add sugar to your coffee, and the sugar dissolves. Thus the coffee is the _____ and the sugar is the _____.
   a. solute . . . solvent
   b. solvent . . . solute
   c. polar covalent bond . . . nonpolar covalent bond
   d. nonpolar covalent bond . . . polar covalent bond
   e. ionic bond . . . hydrogen bond

Brief Review

1. As with most elements, carbon comes in several forms, one of which is carbon-14. What are these forms called and how does one differ from the other?

2. Draw a line and label one end "complete + or − charge" and the other end "no charge" to indicate the charges on the molecules or ions after bonding has occurred. Along the line, indicate where polar covalent bonds, nonpolar covalent bonds, and ionic bonds should be placed.

7. The two strands of a DNA molecule are held together because large numbers of hydrogens that are covalently bonded to oxygen or nitrogen in one strand are weakly attracted to oxygens or nitrogens in the opposite strand. Therefore the two strands of DNA are held together by
   a. polar covalent bonds
   b. nonpolar covalent bonds
   c. ionic bonds
   d. inert bonds
   e. hydrogen bonds

8. While baking cookies for a friend, you're having a hard time keeping your roommate from eating them because the smell is driving her wild. She is able to smell the cookies because the molecules wafting from them
   a. have a shape that allows them to bind to receptors in her nose
   b. form permanent attractions to nerves in her nose by ionic bonding
   c. form temporary attractions to nerves in her nose by covalent bonding
   d. dissolve in her blood and travel to her brain
   e. travel along her nerves—causing changes as they go—until they reach the brain, where they are recognized

3. Why are free radicals so dangerous when they are produced in our bodies?

4. Compare the size of an atom with the size of its nucleus. Where are the electrons? In light of this, what occupies most of an atom?

5. Why are atoms unlikely to react when they have their outer shell filled with electrons?

Applying Your Knowledge

1. In the Middle Ages alchemists labored to turn common materials such as iron into precious metals such as gold. If you could journey back in time, how could you convince an alchemist that iron cannot be changed into gold? Remember that nothing was really known about atoms and elements in the Middle Ages.

2. Go to the Internet and find three references to free radicals. Are any of these related to selling a product designed to decrease the number of free radicals? Based on what you have learned in this chapter, do you think that taking these products is a good idea?

3. The next time you go to a large warehouse or discount store, look around. List as many reasons as you can to explain why it is important that molecules have a three-dimensional shape. How would sensations differ if we lived in a two-dimensional world?

Group Activity

With a group of your classmates, discuss each person's probable career choices. Then research the effects of chemistry on each career. Report on any environmental issues that might be important in each field. Prepare a brochure on the career choices that each student researched. Make the brochure available to other students through the career center at your school.

3

Water, pH, and Biological Molecules

Energy used:
Glucose can serve as an immediate source of energy.

Energy stored:
Or, glucose can be made into glycogen and stored in the liver and muscle for later use.

Energy from plants—use it now or store it for later on.
(page 52)

NUCLEOTIDES ARE THE BUILDING BLOCKS OF DNA

a

NUCLEOTIDE

nitrogenous base

sugar (deoxyribose)

P

O

G

C

O

P — phosphate group

P

O

T

A

O

P

The building blocks of DNA.
(page 59)

A portrait of the most famous molecule in all of biology.
(page 59)

Imagine that it's raining

outside as you read this, the rhythm of individual drops combining to create a sound that is like no other. It's a sound that most people find comforting. And isn't that the way with other sounds water makes when it's in motion? With ocean waves or a running stream? It may be that these things are soothing purely as *sounds*—something about their tone, or their steady pace. But isn't it possible that we are calmed by the sounds of rain or streams because they are the sounds of *water*? Rain has fallen on leaves for millennia, sending our ancestors—and now us—a timeless message: Water's here; you and your family will have enough to drink; food will be abundant; all is well.

3.1 The Importance of Water to Life

Human societies tend to come together precisely where water exists, of course. In places where it's plentiful, water seems less like a substance than an environment: People drink it, cook in it, bathe in it, wash wastes away in it, harness it for power, swim in it. Some 71 percent of Earth's surface is ocean water, and human bodies are about 66 percent water by weight, so that if we have, say, a 128-pound person, about 85 pounds of him or her will be water. If Earth amounts to a watery environment flecked by the landmasses we call continents, the human body amounts to a watery mass with significant proportions of other materials immersed in it.

This preponderance of water in living things gives rise to an important point. Recall that you have learned, in Chapter 2, what a solution is. Well, a specific kind of solution is one in which water is the solvent. This is an **aqueous solution**. Thus, when biologists talk about a given process taking place "in solution," what they mean, unless otherwise noted, is that it is occurring in *aqueous* solution.

Water Is a Major Player in Many of Life's Processes

Because water is the medium in which life's processes take place, it is of course a major

all-and-stick model of glucose

The most important energy
source for our bodies.
(page 45)

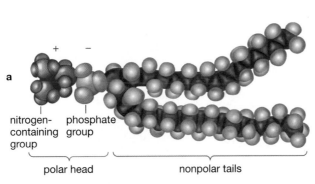

a

nitrogen-   phosphate
containing   group
group

polar head   nonpolar tails

A dual-natured molecule with head and tails.
(page 54)

A protein found over almost all of your body.
(page 57)

ANIMATION

Figure 3.1
Water's Power as a Solvent

a Both sodium ions (Na⁺) and chloride ions (Cl⁻), the component parts of this crystal of table salt, carry an electric charge. Thus, both are attracted to the charged regions of the water molecule—sodium to the negative charge of water's oxygen atom, and chloride to the positive charge of water's hydrogen atoms.

b Pulled from the crystal, and separated from each other by this attraction, sodium and chloride ions become surrounded by water molecules.

c This process of separating sodium and chloride ions repeats until both ions are evenly dispersed, making this an aqueous solution.

player in many of life's chemical reactions. One role that water plays is to break down the bonds of compounds that have been placed into it. (These compounds are the *solutes* to water's *solvent*.) This was reviewed in Chapter 2 in connection with table salt, or sodium chloride. Look at **Figure 3.1** for a more detailed view of water's solvent power. Attracted by the polar nature of the water molecule, the sodium and chloride ions separate from a crystal—and from each other. Each ion is then surrounded by several water molecules (Figure 3.1b). These units keep the sodium and chloride ions from getting back together. In other words, they keep the ions evenly dispersed throughout the water, which is what makes this a solution (Figure 3.1c). Water works as a solvent here because the ionic

compound sodium chloride carries an electrical charge. What *generally* makes water work as a solvent, however, is its ability to form hydrogen bonds with other molecules.

Water's Structure Gives It Many Unusual Properties

When we note water's ability to act as a solvent, we're actually not giving water its due. Water is not just *a* solvent: Over the range of substances, nothing can match it as a solvent. It can dissolve more compounds in greater amounts than can any other liquid.

Ice Floats Because It Is Less Dense than Water

But solvency-power is merely the beginning of water's abilities. It is a multi-talented per-

water
(solvent)

H
O
H

sodium
chloride
(solute)

Cl⁻ Na⁺

a   b   c

sodium and chloride ions dissolved in water

Figure 3.2
Under the Ice
Water is unusual in that its solid form (ice) is less dense than its liquid form. The fact that ice floats makes life possible in cold-weather aquatic environments. Here a giant jellyfish floats under the ice in the waters off Antarctica.

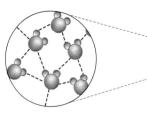

ICE

In ice, the maximum number of hydrogen bonds form, causing the molecules to be spread far apart.

WATER

In liquid water, hydrogen bonds constantly break and reform, enabling a more dense spacing than in ice.

former. And it achieves this status because it is . . . odd. Compared to other molecules, water is like some zany eccentric whose powers stem precisely from its eccentricity. Consider the fact that ice floats on water. This is so because the solid form of H_2O is less dense than the liquid form—a strange reversal of nature's normal pattern. Things work this way with H_2O because, when water molecules slow their motion in cooling, they are able to form the maximum number of hydrogen bonds with each other; this results in the water molecules being spaced *farther apart* when frozen. Thus, ice is less dense than water. This may seem like some minor, quirky quality but it actually has the effect of making possible life as we know it. Ice on the surface of water acts to insulate the water below, creating a warmer environment for organisms such as fish (**Figure 3.2**). If ice *sank*, on the other hand, the entire body of water would freeze solid at colder latitudes, creating an environment in which few living things could survive for long.

Water Has a Great Capacity to Absorb and Store Heat

Water serves as an insulator not only when it is frozen, but also when it is liquid or gas. This is so because water has what is known as a high **specific heat**: Compared to other substances, it takes a relatively large amount of energy to raise the temperature of water. Put a gram container of drinking alcohol (ethanol) side by side with one of water, heat them both, and it will take almost twice as much energy to raise the temperature of the water one degree Celsius as it will the alcohol. Having absorbed this much heat, however, water then has the capacity to *release* it when the environment around it is colder than the water itself. The result? Water acts as a great heat buffer for Earth. The oceans absorb tremendous amounts of radiant energy from the sun, only to release this heat when the temperature of the air above the ocean gets colder. Without this buffering, temperature on Earth would be less stable. People who have spent a day and a night in a desert can attest to this effect. The searing heat of the desert day radiates off the desert floor; but at night, with little water vapor in the air to capture this heat, the desert cools off dramatically. In the same way, our *internal* temperature can remain much more stable because the water that makes up so much of us is first able to absorb and then to release great amounts of heat. The sweat that we throw off in exercise has considerable cooling power because each drop of perspiration carries with it a great deal of heat.

Water derives these powers from its chemical structure, in particular from the hydrogen bonding described earlier. Weak and shifting though individual hydrogen bonds may be, collectively they have a great strength. Heat is the motion of molecules. To get molecules moving, though, chemical bonds must first be broken. Because water has a formidable set of hydrogen bonds, it takes a lot of energy to break them and get its molecules moving. This very same set of bonds gives water another notable characteristic: cohesion. With this quality, water can be drawn (by evaporation) in one continuous column from a plant's roots all the way up to its leaves and out into the air as water vapor.

Water's Cohesion Gives It Surface Tension

Water's cohesion imparts a special quality, called **surface tension**, in places where water meets air. Water molecules *below* the surface are equally attracted in all directions to other water molecules. *At* the surface, however, water molecules have no such attraction to the air above them—they are pulled down and to the side, but not up. This causes the "beading" that water droplets do on surfaces. More important for our purposes, surface water molecules pack together more closely than do interior molecules, allowing all kinds of small animals to move across the surface of water rather than sinking into it. Note the familiar water strider in **Figure 3.3**.

Having looked at the qualities of water, you may be tempted to think how *uncanny* it is that water has all these qualities that are conducive to life. But there actually is nothing surprising about this. Life on Earth began

Figure 3.3
Walking on Water

a Water's high surface tension enables small animals and insects, such as this pond skater, to walk on it.

b This same surface tension also causes water to bead up on waxy or oily surfaces, such as this goose feather.

a

b

in water; thus water has fundamentally conditioned life as we know it. Being surprised about water's life-enhancing qualities is like being surprised that a stage is a good place to put on a play.

What Water Cannot Do

With its great complexity, life requires molecules that *can't* be dissolved by water. Such is the case with nonpolar covalent molecules. Compounds made of hydrogen and carbon (**hydrocarbons**) are good examples of nonpolar molecules; Chapter 2 presented methane (CH_4) as a specific example of a nonpolar molecule. Petroleum products are more complex hydrocarbons than is methane, and you can see a vivid demonstration of their water-insolubility in oil spills (**Figure 3.4**). Oil doesn't dissolve in water because the oil carries no electrical charge that water can bond with; thus water has no way to separate one oil molecule from another.

Two Important Terms: Hydrophobic and Hydrophilic

The ability of molecules to form bonds with water has a couple of important names attached to it. Compounds that will interact with water—such as the sodium chloride considered earlier—are known as **hydrophilic** ("water-loving"), while those that do not interact with water are known as **hydrophobic** ("water-hating"). Both terms are misleading,

in that no substance has any known emotional relationship with water. Hydrophobic is particularly off the mark in that water does not repel hydrophobic molecules; rather, as water molecules bond with one another they exclude oil molecules, making them perpetual outsiders to the water.

The importance of hydrophobic molecules can be illustrated in part by the common milk carton. Why is the milk carton important? Because it can keep milk separate from everything else. We living organisms need some kind of "carton" that can separate the world outside of us from ourselves. Likewise, organisms have great use *within* themselves for compartments that can be sealed off to one degree or another. If water broke down every molecule of life it came in contact with, then it would break down all these divisions of living systems.

Molecules do not have to be completely hydrophilic or hydrophobic. Indeed, as you'll see, a number of important molecules have both hydrophilic and hydrophobic portions.

3.2 Acids and Bases Are Important to Life

When considering the question of aqueous solutions, an important concept is that of acids, bases, and the pH scale used to measure their levels.

We've all had experience with acids and bases, whether we've called them by these names or not. Acidic substances tend to be a little more familiar: lemon juice, vinegar, tomatoes. Substances that are strongly acidic have a well-deserved reputation for being dangerous; the unadorned term *acid* is often used to mean something that can sear human flesh. It might seem to follow that bases are benign, but ammonia is a strong base. The safe zone for living tissue in general lies with substances that are neither strongly acidic nor strongly basic.

Science has developed a way of measuring the degree to which something is acidic or basic—the pH scale. So widespread is pH usage that it pops up from time to time in television advertising ("It's pH-balanced!").

Acids Yield Hydrogen Ions in Solution; Bases Accept Them

The acronym pH comes from the French *pouvoir hydrogène* ("hydrogen power"), and

Figure 3.4
Oil and Water Do Not Mix
When there is an oil spill in the ocean, the oil stays concentrated, even as it spreads out, because oil and water do not form chemical bonds with each other. Here trawlers are using a boom to clean up after an oil spill in Great Britain.

from this comes a clue about what lies at the root of pH. An **acid** is any substance that *yields hydrogen ions* when put in solution. A **base**, conversely, is any substance that *accepts* hydrogen ions in solution.

How might this yielding or accepting come about? Suppose you put an acid—hydrochloric acid (HCl)—into some water. What happens is that HCl *dissociates* into its ionic components, H⁺ and Cl⁻. The HCl has therefore yielded a hydrogen ion (H⁺). Now, with a greater concentration of hydrogen ions in it, the water is more acidic than it was (**see Figure 3.5**).

Now, what about bases? There is a compound called sodium hydroxide (NaOH) that, when poured into water, dissociates into Na⁺ (sodium) and OH⁻ (hydroxide) ions. The place to look here is the OH⁻ ions. Negatively charged as they are, they would readily bond with positively charged H⁺ ions. In other words, they would *accept* H⁺ ions in solution, which is the definition of a base. Were OH⁻ ions to be put into a solution with H⁺ ions in it, they would accept the H⁺ ions, making the solution more basic—or, to look at it another way, *less acidic*. Thus, acids and bases are something like a teeter-totter: When one goes up, the other comes down. As you might have guessed, in the right proportions they can balance each other out perfectly. Look at Figure 3.5 to see how this would play out with the solutions you've looked at so far. Mixing them, the collection of H⁺ and OH⁻ ions now comes together in water, and for each pair of ions that interacts, the result is:

$$H^+ + OH^- \rightarrow H_2O$$

Right—water, which is neutral on the pH scale. The acid and the base have perfectly balanced one another out. The OH⁻ ion, generally referred to as the **hydroxide ion**, is important because compounds that yield them in quantity are strongly basic and can be used to move solutions from the acidic toward the basic.

Many Common Substances Can Be Ranked According to How Acidic or Basic They Are

Look at **Figure 3.6** on page 42 to get an idea of how acidic or basic some common substances are. Following from the notion of what pH

amounts to, it's clear that battery acid, for example, is strongly acidic because, when it dissociates in solution, it yields a large number of hydrogen ions. As you go up the scale, however, you run into weaker acids, meaning substances that yield fewer H⁺ ions in solution. By the time you get to seawater, you've arrived at substances that *accept* hydrogen ions.

The pH Scale Allows Us to Quantify How Acidic or Basic Compounds Are

The net effect of all this yielding and accepting of hydrogen ions is the *concentration* of H⁺ ions in solution, and it is through this that the notion of pH is *quantified*—has numbers attached to it—in the form of the **pH scale**. Look at Figure 3.6 again, this time in connection with the numbers that mark it off. You can see that zero on the scale is the most acidic while 14 is the most basic. It's important to note that the pH scale is *logarithmic*: A substance with a pH of 9 is *10* times as basic as a substance with a pH of 8 and 100 times as basic as a substance with a pH of 7.

**Figure 3.5
Hydrogen Ions and pH**

a Pure water is a "neutral" substance in terms of its pH levels.

b Hydrochloric acid (HCl), poured into the water, dissociates into H⁺ and Cl⁻ ions. With a higher concentration of H⁺ ions in it, the water moves toward the acidic end of the pH scale.

c An equal concentration of sodium hydroxide, poured into water, dissociates into Na⁺ and OH⁻ ions, moving the water toward the basic end of the scale.

d When the acid and base solutions are poured together, the OH⁻ ions from c accept the H⁺ ions from b, forming more water and bringing the solution back to a neutral pH.

pure water

ACID

BASE

HCl → H⁺ + Cl⁻   NaOH → Na⁺ + OH⁻

$$H_2O + H^+ + OH^- \rightarrow 2H_2O$$
acid   base   neutralized solution

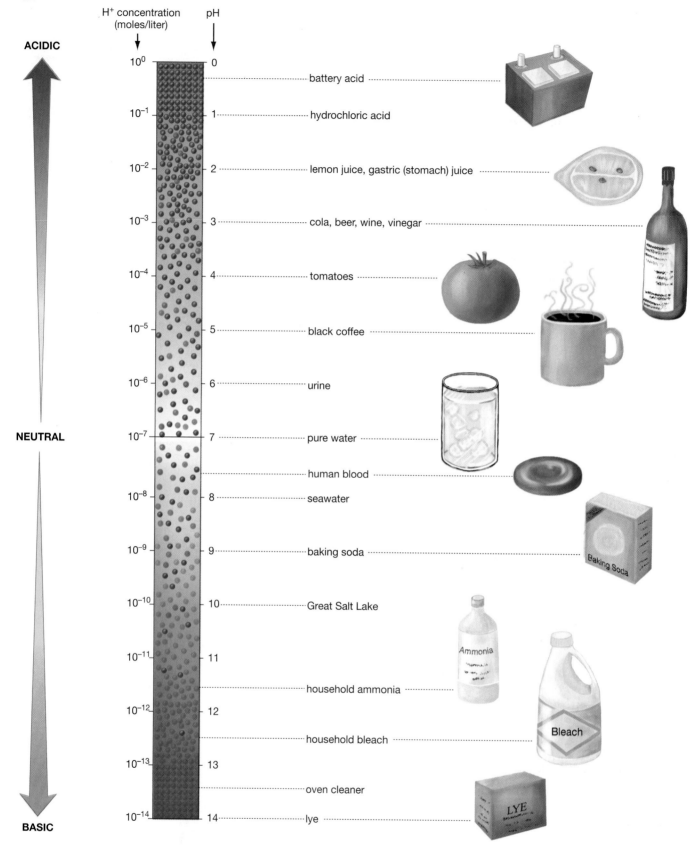

H+ concentration (moles/liter)

pH

ACIDIC

NEUTRAL

BASIC

10^0	0	battery acid
10^{-1}	1	hydrochloric acid
10^{-2}	2	lemon juice, gastric (stomach) juice
10^{-3}	3	cola, beer, wine, vinegar
10^{-4}	4	tomatoes
10^{-5}	5	black coffee
10^{-6}	6	urine
10^{-7}	7	pure water
		human blood
10^{-8}	8	seawater
10^{-9}	9	baking soda
10^{-10}	10	Great Salt Lake
10^{-11}	11	household ammonia
10^{-12}	12	household bleach
10^{-13}	13	oven cleaner
10^{-14}	14	lye

Figure 3.6
Common Substances and the pH Scale
Chemists use units called moles per liter to measure the concentration of substances in solution. The pH scale, derived from this framework, measures the concentration of hydrogen ions per liter of solution. The most acidic substances on the scale have the greatest concentration of hydrogen ions, while the most basic (or alkaline) substances have the least concentration of hydrogen ions. The scale is logarithmic, so that wine, for example, is 10 times as acidic as tomatoes, and 100 times as acidic as black coffee.

Some Terms Used When Dealing with pH

Here are some notes on pH terminology:

- As a solution becomes more basic, its pH *rises*. Thus the higher the pH, the more basic the solution; and the lower the pH, the more acidic the solution. Oven cleaner is spoken of as having a high pH, while lemon juice has a low pH.
- You may also recall that hydrogen is composed of one proton and one electron. Given this, when, in ionization, the hydrogen proton separates from its electron, the hydrogen ion amounts to nothing *but* a proton. It is also correct, therefore, to say that an acid is something that yields *protons* in solution, while a base is something that accepts protons. This is how you will often hear hydrogen ions talked about in biology.
- A solution that is basic is also referred to as an **alkaline** solution.

Why Does pH Matter?

So, why do we care about pH? The brief answer is: because living things are sensitive to its levels in many ways. In human beings, the dangers of insulin-dependent diabetes, of heart stoppages (cardiac arrests), and even of severe vomiting stem in part from the pH imbalances these conditions can bring about. (In all these cases, the danger stems from acidosis, meaning elevated acidity in the blood.)

In general, living things thrive at a pH that hovers around neutral: From about 6–8 is the usual range, and the pH of the cell is about 7. To put this another way, most living things just cannot tolerate extremes of pH. So sensitive are they to this condition that many have developed elaborate **buffering systems** to keep pH hovering at optimal levels. Buffers are generally weak acids or bases that work to neutralize any sudden infusion of acid or base by accepting or donating hydrogen ions.

Different parts of living things have different pH needs, however. The digestive structures in cells, called lysosomes, require a mildly acidic environment to get their work done. But human stomachs require a *very* acidic environment in order to function properly: Gastric juices in the stomach have a pH of between 1.6 and 1.8. pH matters on larger scales as well; acid rain has become a great problem for whole environments. (See "Acid Rain" on page 46).

3.3 Carbon Is a Central Element in Life

Carbon as a Starting Ingredient

You're crossing a bridge of sorts here. To this point, the discussion has largely been about water and hydrogen and various ions. From time to time, mention was made of the element *carbon* (**see Figure 3.7**), although only as one substance among many. But carbon is not that at all.

Life is based on carbon, or if you will, on carbon compounds immersed in water. You could think of carbon in life in the same way as you do flour in baking. How many times does a baking recipe begin: "Start with two cups of flour." With the molecules of life, the recipe often begins: "Start with a carbon ring," or "start with a chain of carbon atoms." But just as cinnamon or fruit will be folded into flour, thus imparting special qualities to the basic ingredient, so will other elements join carbon: A little addition of this here, a little of that on the end, and voilà! A complex molecule that serves a very specific purpose.

Carbon's Importance Stems from Its Bonding Capacity

How does carbon come by its great powers? Linkage. Recall from Chapter 2 that carbon has four outer-shell, or "valence," electrons, but that all elements (except hydrogen and helium) need *eight* outer-shell electrons for maximum stability. This means that carbon needs to link up with four more electrons to

Figure 3.7
Pure Carbon, Mostly Carbon
A diamond is pure carbon and the hardest natural material known. Surrounding the diamond is a lump of coal, which has a high carbon content, because it is made up mostly of the remains of carbon-rich ancient vegetation.

achieve stability. Moreover, the bonds that carbon creates are covalent—it is *sharing* electrons with other atoms—which means its linkages are more stable than those formed by elements that bond ionically. Given the natural forces that can buffet life about, such as ultraviolet radiation and heat, this stability is an important quality in first getting life going and then keeping it going. So important is carbon that it forms one whole branch of chemistry, **organic chemistry**, meaning the chemistry of compounds that include carbon.

How does carbon link up with itself or other atoms? You've already had a look at a model of a very simple carbon compound, methane (CH_4). Look at it again:

Note that the only elements present are hydrogen and carbon atoms. Now observe a slightly more complex hydrocarbon, the familiar gas propane (C_3H_8):

You can see that instead of being surrounded by four hydrogen atoms, the carbons here link up with other carbons, as well as with the hydrogens. This process of extension keeps going with still more complex hydrocarbons. For obvious reasons, the preceding configuration is known as a *straight-chain* carbon molecule. But carbon has more tricks up its sleeve than simple straight-line extensions. The next hydrocarbon up the line is butane (C_4H_{10}). It can be just another straight-chain extension, looking like this:

But it can also look like this:

These two forms of butane are known as **isomers**—molecules that are the same in their chemical formulas, but differ in the spatial arrangement of their elements.

Not content to branch only in this way, however, carbon can also form rings. Here is the structure of benzene (C_6H_6):

Note the three sets of double lines in the molecule. Recall that this means there are double *bonds* between these atoms; the atoms involved are sharing *two* pairs of electrons.

It just so happens that all the carbon molecules introduced so far are made of nothing but carbon and hydrogen. But this is the exception rather than the rule. Here are two representations of a molecule mentioned already—glucose, better known as blood sugar:

So here are some added oxygen atoms. Notice how the second model, below the first, with its heavy line down at the bottom, looks a little different than the benzene ring? Such a model is meant to give you a slightly more realistic picture of the actual arrangement of atoms when there is a carbon ring present. Think of the ring as lying at an angle to the paper, with the heavy line closer to you than the line in the back; the H and OH groups then lie above and below the plane of the ring, as shown. You may also notice that when things get this complicated, space starts getting a little tight for writing in the letters that stand for the various atoms that are a part of the molecule. Because most rings are formed predominately of carbon, it is common to dispense with the Cs when a structural formula is presented. Here is a third representation of the glucose model, written in this stripped-down form:

You can see that carbon is *assumed* to exist at each bond juncture in the ring; if some other element occupies one of these points, it is explicitly noted. Often the solitary Hs that are attached to carbons are left out as well.

Such models as these, as you'll recall from the last chapter, don't show the actual three-dimensional form of a molecule. To do so, ball-and-stick or space-filling models will do. Shown in **Figure 3.8** are examples of these models.

3.4 The Molecules of Life: Carbohydrates, Lipids, Proteins, and Nucleic Acids

The Building-Blocks Model of Organic Molecules

With this review of carbon structures under your belt, you're ready to begin looking at some of the classes of carbon-based molecules—the molecules of living things. You'll explore four groupings of these organic compounds: carbohydrates, lipids, proteins, and nucleic acids. As you get into this section, it's a great help to keep in mind that *complex* organic molecules often are made from *simpler* molecules. Many of the molecules you'll be reading about have a building-blocks quality to them: Take a simple sugar, or monosaccharide, such as glucose, put it together with another monosaccharide (fructose), and you have a larger *di*saccharide called sucrose (which is better known as table sugar). Put *many* monosaccharide units together, and you have a polysaccharide, such as starch or cellulose. The large units are called **polymers**, while the smaller building blocks of polymers are called **monomers**. Look at **Table 3.1** for examples of both.

Table 3.1 Monomers, Polymers	
If the Monomer is . . .	**The Polymer is . . .**
A monosaccharide (for example, glucose, fructose)	A polysaccharide (for example, starch, glycogen, cellulose)
An amino acid (for example, arginine, leucine)	A polypeptide or protein [for example, A- and B-chains of insulin (polypeptides) and insulin itself (protein)]
A nucleotide (sugar, phosphate, base in combination)	A nucleic acid (for example, DNA, RNAs)

Ball-and-stick model of glucose   Space-filling model of glucose

Figure 3.8
Ball-and-Stick and Space-Filling Models of Glucose
The simple sugar glucose is the single most important energy source for our bodies. Because it has several OH groups, it is highly hydrophilic and thus readily breaks down in water. Note how the carbon atoms (in black) form the core of the molecule, with oxygen (in red) and hydrogen (in white) at the periphery.

Acid Rain: When Water Is Trouble

If the concept of pH seems remote from the real world, consider this: Hundreds of lakes in the eastern and midwestern United States are dying and their condition has everything to do with pH. They have reached this state not by natural processes, however, but by civilization's impact on the environment.

This situation has been brought about by a phenomenon commonly called acid rain, but more properly known as acid deposition, since acid *snow* and a dry form of acid discharge are problems as well. In acid deposition, you can see an application of pH principles, only played out this time on an enormous scale. You've learned something about acids and bases and the importance of *internal* pH to living things. It makes sense, then, that when rainwater itself has a skewed pH, there is going to be widespread trouble. And this is indeed what has happened, not only in the United States, but in Canada, in Eastern Europe and England, in China—in short, worldwide, though industrialized nations are the most affected. In the 1980s, residents of the Scottish Highlands had the dubious distinction of being able to experience *black snow*, which was extremely acidic. In the pristine Adirondacks region of upstate New York some 200 lakes were, by the early 1990s, incapable of supporting fish. Scandinavian residents at that time might well have envied New York residents, however, because some 16,000 lakes in Norway and Sweden by then had no fish in them.

Acid precipitation comes primarily from two sources: the sulfur dioxide (SO_2) emissions of coal- and oil-burning factories (in particular power plants), and the nitric oxide (NO) and nitrogen dioxide (NO_2) emissions that come mostly from cars, with an assist from factories (**Figure 1**). When these compounds rise into the air, they come together with a culprit you've seen before: an oxygen radical, specifically the hydroxyl radical (OH). In so doing, they are converted to sulfuric acid and nitric acid. *These* compounds then combine with the rainwater in clouds. The result of this is acidic rain or snow.

Acid deposition can do its damage directly or indirectly. By simply making a lake more acidic, it begins to interfere with the metabolic processes of the animals who live there. Acid rain also has the effect, however, of leaching metals from the soil it passes through. Aluminum brought into lakes in this way damages the gills of fish, destroying their ability to absorb oxygen from the water. This same effect takes place in connection with land vegetation, where leached metals can kill plant roots outright or keep them from absorbing nutrients.

Acid deposition will have different effects on different environments, and here again you can see the basics of pH at work. Soils are not neutral in their pH; should acid rain fall on soil that is significantly basic (or "alkaline"), the buffering process you've read about will take place: Basic compounds in the soil will combine with the water and ac-

---

Figure 3.9
Carbohydrates in Foods
Breads, cereals, and pasta make up a significant proportion of our diets. These foods are all rich in carbohydrates, one of the four main types of biological molecules.

Carbohydrates: From Simple Sugars to Cellulose

Happily, for purposes of memory, the elements of carbohydrates are all hinted at in the name: **Carbohydrates** must contain carbon, oxygen, and hydrogen. In many instances, they will contain nothing *but* carbon, oxygen, and hydrogen. Furthermore, they usually contain exactly twice as much hydrogen as oxygen; with the carbohydrate glucose ($C_6H_{12}O_6$), there are twelve atoms of hydrogen and six of oxygen. Most people think of carbohydrates purely in terms of foods such as breads and pasta (**see Figure 3.9**); but as you'll see, carbohydrates have more roles than this in nature.

The building blocks of the carbohydrates are the **monosaccharides** mentioned earlier, which also are known as **simple sugars**. You've already seen several views of one of these molecules—glucose. Glucose has a use in and of itself: Much of the food we eat is broken down into it, at which cept hydrogen ions from the rain, thus resulting in water that is less acidic. Likewise, aquatic environments may contain basic ions that can neutralize incoming water.

The extent to which such environments possess basic ions is known as their *acid-neutralizing capacity*. Given what you've read about pH, it might be apparent that this capacity is not limitless but instead can dwindle as more acidic compounds are added. Indeed, soils can suffer from a *long-term* depletion of basic compounds, and this has happened in some eastern U.S. forests. A 1990 amendment to the federal Clean Air Act has resulted in a reduction in U.S. sulfur emissions, but this change has not brought about the improvement in eastern forests and lakes that many researchers had hoped for. At present it appears that returning these resources to a healthy state will be a long-term process.

a

b

Figure 1
How Acid Rain Forms
(a) Sulfur dioxide (SO_2) from coal- and oil-burning power plants rises into the air along with nitric oxide (NO) and nitrogen dioxide (NO_2), which come mostly from cars. These compounds combine with hydroxyl radicals (OH) in the atmosphere to produce sulfuric acid (H_2SO_4) and nitric acid (HNO_3), which combine with atmospheric water, making it acidic. **(b)** A stand of trees damaged by acid rain at the top of Mount Mitchell in North Carolina.

point it becomes our most important energy source. Glucose can also bond with other monosaccharides, however, to form more complex carbohydrates. Let's take a look at how this happens. If you look at **Figure 3.10**, you can see an example of two glucose molecules bonding to create the disaccharide maltose.

Several things are worth noting here. In maltose, as you can see, the link that joins the two glucose monomers is a single oxygen atom linked to carbons of each of the glucose units. To get this, two atoms of hydrogen and one atom of oxygen are split off (on the left side of the equation) from the original glucose molecules. What becomes of these three atoms? Look at the far right-hand side of the reaction and see the + H_2O. The product of this reaction is a molecule of maltose and a molecule of water.

Now note the arrows in the middle of the reaction. You've been used to seeing a single,

Figure 3.10
Carbohydrates Follow a "Building-Blocks" Model
In this example, two units of the monosaccharide (or simple sugar) glucose link to form the disaccharide maltose. In addition to maltose, the reaction yields water. The double arrows indicate that this reaction is reversible; a single maltose molecule can yield two glucose molecules.

glucose + glucose = maltose + water

CH_2OH $\quad$ CH_2OH $\quad$ CH_2OH $\quad$ CH_2OH $\quad$ + H_2O

rightward-pointing arrow; here the arrows go both ways. What this means is that this is a **reversible reaction**; it can go both ways. When maltose is placed in water, it can be split apart to yield two *glucose* molecules.

Kinds of Simple Carbohydrates

You have thus far seen examples of one of the most simple carbohydrates, the monosaccharide glucose, and one slightly more complex carbohydrate, a disaccharide called maltose. There are many kinds of mono- and disaccharides, however. Among the monosaccharides are, for example, fructose and deoxyribose. Among disaccharides, there are sucrose and lactose. At the risk of pointing out the obvious, note that all these sugars have *-ose* at the end of their name: If it's an *-ose*, it's a sugar (**see Figure 3.11**).

Complex Carbohydrates Are Made of Chains of Simple Carbohydrates

You've seen carbohydrates built up from monosaccharides to disaccharides. If you go another couple of bumps up, you get to the **polysaccharides**. The *poly* here means "many" sugars, and the term is apt. In the polysaccharide molecule cellulose, for example, there may be 10,000 glucose units linked up with one another. The basic unit here is that of the six-carbon monosaccharide glucose, $C_6H_{12}O_6$, from which chains of glucose units are built up. The complexity of these molecules gives them another name: complex carbohydrates. Four different types of complex carbohydrates interest us: starch, glycogen, cellulose, and chitin (**see Figure 3.12**).

Starch is familiar to most people, because it comes to us in the form of food. Potatoes, rice, carrots, corn; all these are starchy foods. Note that all of them come from plants; a closer look reveals that *in plants*, these starches serve as the main form of carbohydrate *storage*, sometimes as seeds (rice and wheat grains), or sometimes as roots (carrots or beets).

Glycogen does for animals what starch does for plants: It serves as the primary form in which carbohydrates are stored. For this reason, glycogen is sometimes called animal starch. The starches or sugars we eat are broken down, eventually into glucose, at which point some of the glucose may be used immediately. Some may not be needed right away, however, in which case it is moved into the muscle cells and liver to be stored as the more complex carbohydrate glycogen. The "carbohydrate loading" that swimmers and runners do before races (by eating lots of pasta, for example) could just as easily be called glycogen loading.

Cellulose is contained in the walls of the cells of many organisms and is the largest single component of the cell walls of plant cells. Despite this innocuous-sounding function, it would be fair to call cellulose *mighty cellulose*, because upon its function hinges so much of the structure of the natural world. It is easily the most abundant carbohydrate in the world: Trees, cotton, leaves, and grasses are largely made of it. When cellulose is enmeshed with a hardening compound called *lignin*, the result is a set of cell walls that can hold up giant redwood trees.

Because cellulose is so dense and rigid, it is not surprising that human beings and other mammals cannot digest it. This statement may make you wonder about grass-eating *cows*, but in fact cows have cellulose digested for them by special bacteria that reside in their digestive tract. Cellulose is important to humans in that it is our major source of insoluble fiber, which helps move foods through the digestive tract. Because cellulose exists in the cell walls of plants, we can get our fiber from such foods as whole grains and fresh fruits.

Chitin is a carbohydrate that forms the external skeleton or "exoskeleton" of the category

Figure 3.11
Sugars Come in Many Forms
Sucrose or table sugar comes to us from sugar cane or sugar beets, and glucose is found in corn syrup. Fructose comes to us in sweet fruits and in high-fructose corn syrup, which often is used to sweeten soft drinks.

of animals known as arthropods, a grouping so large it includes all insects, spiders, and "crustaceans," such as crabs. In all these animals, chitin plays a "structural" role similar to that of cellulose in plants: It gives shape and strength to the structure of the organism.

Lipids: Oils, Fats, Hormones, and the Outer Lining of Cells

The second class of biological molecules, **lipids**, turns out to be made of the same elements as carbohydrates—carbon, hydrogen, and oxygen—but lipids have much more hy-drogen, relative to oxygen, than do the carbo-hydrates. We're all familiar with some lipids; they exist as fats, as oils, as cholesterol, and as hormones, such as testosterone and estrogen. Unlike the other biological molecules you'll be studying, a lipid is not a polymer composed of component-part monomers; no single structural unit is common to all lipids. The one characteristic that pure lipids share is that they are insoluble in water. Remember an earlier discussion about the need life has to create internal containers? Well, lipids are able to serve this function because water

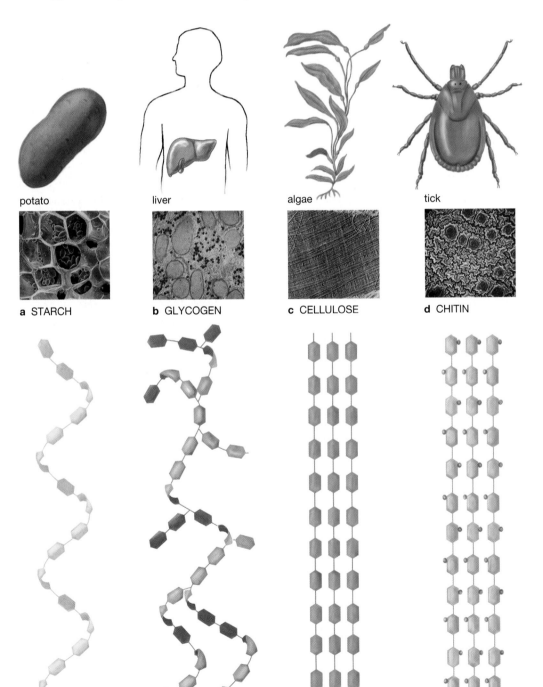

potato   liver   algae   tick

a STARCH   **b** GLYCOGEN   **c** CELLULOSE   **d** CHITIN

Figure 3.12
Four Examples of Complex Carbohydrates
All complex carbohydrates are composed of chains of glucose, but they differ in the details of their chemical structure.

a Starch serves as a form of carbohydrate storage in many plants. Here starch granules can be seen within the cells of a slice of raw potato.

b Glycogen serves as a form of carbohydrate storage, as is the case in this photo of glycogen globules in the liver.

c Cellulose, running as fibers through cell walls, provides structural support for plants. The photo is of sets of cellulose fibers running at right angles to one another in the cell wall of marine algae.

d Chitin provides structural support for some animals. The outer "skin" or cuticle of insects is composed mostly of chitin. The photo shows the exoskeleton of a tick.

can't dissolve them. In addition, lipids have considerable powers to store energy and to provide insulation.

One Class of Lipids Is the Glycerides

Glycerides, the most common kind of lipid, can be thought of as a molecule in two parts. The first is a "head" composed of a particular kind of alcohol, usually **glycerol**. Attached to the glycerol head is the second part of the molecule, **fatty acids**, each consisting of a long chain of carbon and hydrogen atoms punctuated at the end with an acid COOH group.

The structural formula for one of the fatty acids, stearic acid, which is found in animal fat, is shown in **Figure 3.13**. This monotonous chain of hydrogen and carbon is, as you might have guessed, the hydrocarbon portion of the fatty acid molecule. Fatty acids can have from 4 to 22 carbons linked up like this. On the left side, you can see the COOH group just mentioned (which is properly known as a *carboxyl group*).

Now look at the second part of the fat molecule, the alcohol known as glycerol:

The hallmark of alcohols is that they have an OH group, which you can see here. Bringing the OH portion of glycerol together with the carboxyl part of the fatty acid chain is what makes a glyceride. But it's obvious that the glycerol has *three* OH groups on the right. Thus there is, you might say, docking space on glycerol for three fatty acids, and in the synthesis of many glycerides, this linkage takes place: Three fatty acids link up with glycerol to form a **triglyceride**, which is an important form of lipid. **Figure 3.14** shows how it works schematically.

The Rs on the right end of the COOH group stand for whatever the hydrocarbon chain is of that particular fatty acid. To get a better idea of the shape of a completed triglyceride, look at the space-filling model in **Figure 3.15**. This particular triglyceride, called tristearin, has three stearic fatty acids stemming like tines on a fork from the glycerol. But this is only one possibility among many, and is exceptional, rather than usual, in that all three fatty acids are the same. Among the dozens of fatty acids that exist, several *different* kinds of fatty acids generally will hook up in glycerol's three "slots" to form a triglyceride. Actual fat products, such as butter, are composed of different proportions of these various fatty acids.

We've seen three fatty acids linking with glycerol to form *tri*glycerides, but monoglycerides (one fatty acid joined to glycerol) and diglycerides (two fatty acids joined with glycerol) can be formed as well. Triglycerides are the most important of the glycerides, however, because they constitute about 90 percent of the lipid

Figure 3.13
Structural Formula for Stearic Acid

Figure 3.14
Formation of a Triglyceride

glycerol + 3 fatty acids = triglyceride + water

Figure 3.15
The Triglyceride Tristearin
This lipid molecule is composed of three stearic fatty acids, stemming rightward from the glycerol OH "heads." Tristearin is found both in beef fat and in the cocoa butter that helps make up chocolate.

inspection, you can see that the palmitic acid has an unbroken line of single bonds linking its carbon atoms. Meanwhile, the oleic acid has one double bond between the carbons in its chain and the linoleic acid has two. Furthermore, note that these double bonds exist precisely where the kinks appear in these molecules.

Fatty Acids; from Solid to Liquid What these variations describe are the differences between a **saturated fatty acid** (no double bonds), a **monounsaturated fatty acid** (one double bond), and a **polyunsaturated fatty acid** (two or more double bonds). What a saturated fatty acid is saturated *with* is hydrogen atoms. If palmitic acid had a carbon double bond, it could take on more hydrogen atoms; but it doesn't, so it's saturated.

This may not sound like much of a difference, but it has several important consequences. First, at room temperature, as you move from saturated to unsaturated, you also move in general from fats in their solid form to fats in their liquid form, which we call **oils**. You can turn naturally occurring oils *into* fats by saturating them—by moving hydrogen into the oils in a process known as hydrogenation. How solid do you

weight in foods. In short, the substances we call fats usually are composed mostly of triglycerides.

Saturated and Unsaturated Fatty Acids: A Linkage with Solids and Liquids and with Health

Look now at **Figure 3.16**, which shows three different fatty acids: palmitic, oleic, and linoleic. On first glance, there is seemingly not much difference among the three. One thing that does jump out at you is the "kinks" in the oleic and linoleic acids compared to the straight-line palmitic acid. Upon closer

MediaLab Link
Diet and health, see p. 64.

Figure 3.16
Saturated and Unsaturated Fatty Acids
The degree to which fatty acid hydrocarbon chains are "saturated" with hydrogen atoms has consequences for both the form these lipids take and for human health.

a The hydrocarbon "tail" in palmitic acid is formed by an unbroken line of carbons, each with a single bond to the next.

b In oleic acid, a double bond exists at one point between two carbon atoms. An additional hydrogen atom could link to each of these carbon atoms instead, which would make this a saturated fatty acid—saturated with hydrogen atoms. As things stand, this is a monounsaturated fatty acid.

c The carbons in linoleic acid have double bonds in two locations, making this a polyunsaturated fatty acid. The "kinks" imparted by double carbon-bonds make unsaturated and polyunsaturated fatty acids more likely to be liquid oils, rather than solid fats, at room temperature.

a Palmitic acid

Saturated
(no double bonds)

b Oleic acid

Monounsaturated
(one double bond)

c Linoleic acid

Polyunsaturated
(more than one double bond)

want a lipid to be? Just decide, and saturate a given portion of its double bonds. This is one step in creating margarine from vegetable oils. Why does a linkage exist between the degree of saturation and a solid or liquid form? This has to do with the kinks in these molecules—or the lack of them. Being all of one straight-line form, the saturated fatty acids pack together tightly, like so many boards in a lumber yard. The unsaturated fatty acids lack this uniformity, however, and thus stick out at varying angles to one another, a relative disorder that generally makes them liquid at room temperature.

Saturated Fatty Acids and Health The distinction between saturated and unsaturated fatty acids also has another important consequence, this one related to human health. Think of it this way: To the degree that *fatty acids* are saturated, the *fats* they make up will be saturated—and saturated fats have been linked with heart disease. They are, in fact, more important than dietary cholesterol in determining how much cir-

culating or "blood" cholesterol a person will have. This is a complicated linkage, but its practical effect is that public health groups have advised people to limit their intake of saturated fats, which in their natural form are found primarily in animal products (meat, high-fat dairy products) and tropical oils (coconut and cocoa butter, which is used in making chocolate). Fats and oils aren't tempting to human beings because nature had it in for us, however. These lipids are a terrific source of energy, they're a great way to store energy, and they act as valuable insulators in cold weather. (Think of seals and polar bears, with their thick layers of fat.)

Energy Use and Energy Storage via Lipids and Carbohydrates

Lipids have something in common with carbohydrates in connection with energy *storage* on the one hand and energy *use* on the other. To be stored, lipids must be in their triglyceride form; to be used to provide energy, they must first be broken down into glycerol and fatty acids. Car-

Figure 3.17
Storage and Use of Carbohydrates and Lipids
Carbohydrates and lipids generally are *stored* in one form, but *used* in another.

a In plants, carbohydrates commonly are stored as starch, a complex carbohydrate that is a major component of such vegetables as carrots. Human beings can consume such a starch, after which it may be broken down into the smaller, simple carbohydrate glucose and used immediately—for exercise or other purposes—or stored in the liver and muscles as the larger, more complex carbohydrate glycogen.

b Cows use their fat for energy storage and insulation. The fat portion of the meat that humans consume may be used for energy, but only after being broken down into its glycerol and fatty acid components. Conversely, humans may convert the components back into triglycerides to be stored in fat cells, providing energy reserves for later.

a CARBOHYDRATES

Energy stored:
The starch in carrots stores energy.

Energy made available:
When we eat carrots, the starch is broken down into glucose.

Energy used:
Glucose can serve as an immediate source of energy.

Energy stored:
Or, glucose can be made into glycogen and stored in the liver and muscle for later use.

b FATS

Energy stored:
The fat in cows stores energy.

Energy made available:
When we eat meat, the fat is broken down into glycerol and fatty acids.

Energy used:
Glycerol and fatty acids can be used to provide energy right away.

Energy stored:
Or, glycerol and fatty acids can be converted into triglycerides and stored in fat cells for later use.

bohydrates, meanwhile, are stored as the complex carbohydrate glycogen; to be used for energy expenditure, they must be broken down to simple carbohydrates, often glucose. This process of alternately building up or breaking down molecules for storage or energy expenditure as needed is a major task of living things. If you look at **Figure 3.17**, you can see how carbohydrates and lipids are alternately stored and used in living things.

A Second Class of Lipids Is the Steroids

The class of molecules known as **steroids** represents a smaller kind of lipid, different from the glycerides. What all steroids have in common is a linked set of four carbon rings. What separates one steroid from another are the various side chains that can be attached to this four-ring structure (**see Figure 3.18**). When you see how different steroids are structurally from the triglycerides, you can understand why the monomers-to-polymers framework doesn't apply to lipids.

Among the most well-known steroids are cholesterol and two of the steroid hormones, testosterone and estrogen. Like fats in general, **cholesterol** has a bad reputation; but also like fats in general, it serves good purposes too. Compounds that come from cholesterol help break down fats in the first place. Cholesterol is part of the outer membrane of many cells; and it acts as a precursor for many steroids, among them the steroid hormones testosterone (a principal "male" hormone), and estrogen (a principal "female" hormone). The term *steroids* by itself undoubtedly rings a bell, because the phrase "on steroids" has come to mean artificially bulked-up or supercharged. In this common usage, *steroids* refers to manufactured drugs that are close chemical cousins of the muscle-building "male" steroid hormones (which actually exist in smaller amounts in females as well; see **Figure 3.19**).

a What all steroids have in common is a four-ring carbon unit:

b What makes individual steroids unique are the side chains that are attached to the rings:

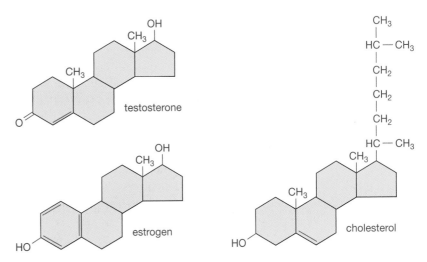

Figure 3.18
Structure of Steroids

a The basic unit of steroids, four interlocked carbon rings.

b Types of steroids, each differentiated from the other by the side chains that extend from the four-ring skeleton. Testosterone is a principal "male" hormone while estrogen is a principal "female" hormone. Both of these steroid hormones actually are found in both men and women, though in differing amounts. Cholesterol is also a prevalent and important steroid in both men and women. Although cholesterol has a bad reputation, it has several important functions—for example, breaking down fats.

a

b

Figure 3.19
Steroids, in Uses Natural and Unnatural

a Some steroid hormones such as estrogen and testosterone are important in natural processes, such as reproduction.

b Pharmaceutical steroids can be used in potentially harmful ways, for example, to build muscle mass. Pictured is a Chinese swimmer who, at a 1998 competition, tested positive for a banned substance that helps flush evidence of steroids from the body.

A Third Class of Lipids Is the Phospholipids

The final class of lipids, **phospholipids**, has something of the same makeup as triglycerides; but instead of three fatty acids stemming from the glycerol head, there now are only two (**see Figure 3.20**). Linking up with glycerol's third OH group is a **phosphate group**, meaning a phosphorus atom surrounded by four oxygen atoms. This change is extremely important because it gives phospholipids a dual nature: The fatty acid tails, being hydrocarbons, are hydrophobic, but the phosphate head is hydro*philic* because it is *charged*. In Figure 3.20, you can see the effect of this in solution. Imagine a phospholipid as a marker buoy in deep water. No matter how you push the hydrocarbon tail around, it's going to end up waving free *out* of the water, while the head is going to be submerged in it. You will learn more about these molecules later, when you get to cells; for now, just note that the material on the periphery of cells—the outer membrane of a cell—is largely made of phospholipids. Living things need the kind of partitions described earlier, and these partitions are composed to a significant extent of phospholipids.

Proteins

Living things must accomplish a great number of tasks just to get through a day, and the diverse biological molecules you've been looking at allow this to happen. You've seen carbohydrates do some things, and you've seen lipids do some more. But in the range of tasks that molecules accomplish, proteins reign supreme. Witness the fact that almost every chemical reaction that takes place in living things is hastened—or, in practical terms, *enabled*—by a particular kind of protein called an enzyme. These molecules function in nature like some vast group of tools, each one taking on a specific chemical task. Accordingly, more than a thousand enzymes are known to exist.

We might marvel at proteins solely because of what enzymes can do, but the amazing thing is that enzymes are only *one class* of proteins. Proteins also form the scaffolding, or structure, of a good deal of tissue; they're active in transporting molecules from one site to another; they allow muscles to contract and cells to move; some hormones are made from them. If you factor out water, they account for about half the weight of the average cell. In short, it's hard to overestimate the importance of these molecules. **Table 3.2** lists some of the different kinds of proteins.

Proteins Are Made from Chains of Amino Acids

Proteins are prime examples of the building-block type of molecule described earlier. The monomers in this case are called amino acids. String an arbitrary number of them together in a chain—some say 10, some say 30—and you have a **polypeptide**; when the polypeptide *folds up* in a specific three-dimensional manner, you have a **protein**. As a practical matter, proteins are likely to be made of hundreds of amino acids strung together and folded up. **Figure 3.21a** gives the fundamental structural unit for amino acids, followed by a couple of examples.

You can see carbon at the center of this unit, an amino group off to its left, and a carboxyl group to its right. (The name *amino acid* comes from the fact that there is an *amino* group and, in the carboxyl group, an *acid* group.) What differentiates one amino acid from another is the group of atoms that occupies the R or "side-chain" position. In **Figure 3.21b**, you can see examples of actual amino acids, tyrosine and glutamine, with their different occupants of the R position.

**Figure 3.20
A Dual-Natured Molecule**

a Phospholipids are comprised of two long fatty acid "tails" attached to a "head" containing both a phosphate group (with a negative charge) and a nitrogen-containing group (with a positive charge).

b Since the head is polarized, it can bond with water and thus will remain submerged in it; the tails, on the other hand, have no such bonding capability.

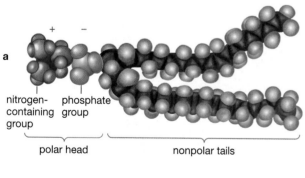

nitrogen-containing group | phosphate group

polar head | nonpolar tails

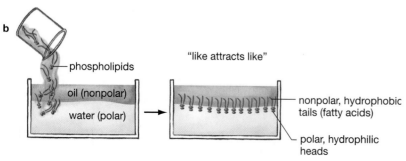

phospholipids

oil (nonpolar)
water (polar)

"like attracts like"

nonpolar, hydrophobic tails (fatty acids)

polar, hydrophilic heads

A Group of Only 20 Amino Acids Is the Basis for All Proteins in Living Things

Although only two examples are shown here, it is a group of *20* amino acids in total that is the basis for all the proteins that occur in living organisms. The thousands of proteins that exist can be made from a mere 20 amino acids, because these amino acids can be strung together in different *order*. Substitute an alanine here for a glutamine there and you've got a different protein. In this, amino acids commonly are compared to letters of the alphabet. In English, substituting one letter can take us from *bat* to *hat*. In the natural

a What all amino acids have in common is an amino group and a carboxyl group attached to a central carbon.

b What makes the 20 amino acids unique are the side-chains attached to the central carbon.

tyrosine

glutamine

Figure 3.21
Structure of Amino Acids

a The structural element that is common to all amino acids is an amino group and a carboxyl group (on the left and right), linked by a central carbon with a hydrogen attached to it. What makes one amino acid different from another is the side chain of atoms that occupies the R position.

b Two examples of actual amino acids; the differing occupants of the R position give us the amino acids tyrosine and glutamine.

Table 3.2
Types of Proteins

Type	Role	Examples
Enzymes	Quicken chemical reactions	Sucrase: Positions sucrose (table sugar) in such a way that it can be broken down into component parts of glucose and fructose
Hormones	Chemical messengers	Growth hormone: Stimulates growth of bones
Transport	Move other molecules	Hemoglobin: Transports oxygen through blood
Contractile	Movement	Myosin and actin: Allow muscles to contract
Protective	Healing; defense against invader	Fibrinogen: Stops bleeding Antibodies: Kill bacterial invaders
Structural	Mechanical support	Keratin: Hair Collagen: Cartilage
Storage	Stores nutrients	Ovalbumin: Egg white, used as nutrient for embryos
Toxins	Defense, predation	Bacterial diphtheria toxin
Communication	Cell signaling	Glycoprotein: Receptors on cell surface

What Is It?
See facing page.

Figure 3.22
Beginnings of a Protein
Amino acids join together to form polypeptide chains, which fold up to become proteins. The linking of amino acids yields water as a byproduct. In this figure, alanine (ala) first joins with glutamine (gln), which then is linked to isoleucine (ile). A typical protein would consist of hundreds of amino acids linked up sequentially.

world, 20 amino acids can be put together in different order to create a multitude of proteins, each with a different function. A list of the 20 primary amino acids appears in **Table 3.3**.

The stringing together of amino acids happens in a regular way; the carboxyl group of one amino acid joins to the amino group of another, with the loss of a water molecule resulting. Look at **Figure 3.22** to see how three amino acids come together.

Shape Is Critical to the Functioning of All Proteins

Now, recall that a protein is a chain of amino acids that has become *folded up* in a specific way. As the amino acids are being strung together in sequence, all the kinds of chemical forces discussed earlier begin to work on the chain; as a result, it begins to twist and turn and fold into a unique three-dimensional shape. And it turns out that in the functioning of proteins, this shape, or **protein conformation**, is utterly crucial. Here's an example of why. The hormone insulin, which is a protein, is released from a person's pancreas; it moves through the bloodstream and latches onto muscle cells; its presence then allows blood sugar to get into the muscles and provide energy. Now, how does insulin "latch onto" muscle cells? By *linking* itself with a molecule, called an insulin receptor, that lies on the surface of the muscle cell. But protein

and receptor must be precisely shaped for such linkage or "binding" to occur.

This binding of a protein hormone to surface receptor sounds important enough; but recall that in addition to being hormones, proteins are the chemical enablers called enzymes, and they're transport molecules, and so forth. In *all* these functions, shape is critical. The American architect Frank Lloyd Wright had a famous dictum about his designs: "Form follows function." With proteins, we can turn this around and say that function follows form (**Figure 3.23**).

There Are Four Levels of Protein Structure

So, what forms do proteins take? Well, the answer to this depends on what vantage point you adopt in looking at them. There are four levels of structure in proteins that determine their final shape. You can see all four levels in **Figure 3.24**. The first of these levels, the **primary structure** of a protein, is simply its sequence of amino acids. Everything about the final shape of a protein is dictated by this sequence. Electrochemical bonding and repulsion forces act on this structure, and the result is the folded-up protein.

Table 3.3 Amino Acids	
Name	**Abbreviation**
Alanine	ala
Arginine	arg
Asparagine	asn
Aspartic acid	asp
Cysteine	cys
Glutamic acid	glu
Glutamine	gln
Glycine	gly
Histidine	his
Isoleucine	ile
Leucine	leu
Lysine	lys
Methionine	met
Phenylalanine	phe
Proline	pro
Serine	ser
Threonine	thr
Tryptophan	trp
Tyrosine	tyr
Valine	val

The linkage of several amino acids...

...produces a polypeptide chain like this:

As it turns out, when these forces begin to operate on the amino acid sequence, a couple of common shapes begin to emerge in the **secondary structure** of proteins. The first of these common shapes, called the **alpha helix**, has a shape much like a corkscrew. The other common secondary structure in proteins, called the **beta pleated sheet**, is shaped something like the folds of an accordion. Proteins can be made almost entirely of alpha helices. This is the case with hair, nails, horns, and the like. Likewise, proteins can be made entirely of beta-pleated sheets. The most familiar example of this is

Figure 3.23
Made of Protein
A human hair, shown here emerging from a follicle in the skin, is composed mostly of the protein keratin.

FOUR LEVELS OF STRUCTURE IN PROTEINS

a

| ala | leu | ser | glu | glu | his | ala | gln | ile | ser | tyr | ala | ser | glu | glu |

amino acid sequence

Primary structure

What Is It?
Answer: A spider spinning silk for its web. The silk, which is magnified 500 times, is artificially colored purple.

b

pleated sheet

helix

random coil

Secondary structure

c

folded polypeptide chain

Tertiary structure

d

two or more polypeptide chains

Quaternary structure

Figure 3.24
Four Levels of Structure in Proteins

a The primary structure of any protein is simply its sequence of amino acids. This sequence determines everything else about the protein's final shape.

b Structural motifs, such as the corkscrew-like alpha helix, beta pleated sheet, and the less organized "random coils" are parts of many polypeptide chains, forming their secondary structure.

c These motifs may persist through a set of larger-scale turns that make up the tertiary structure of the molecule.

d Several polypeptide chains may be linked together in a given protein, in this case hemoglobin, with their configuration forming its quaternary structure.

silk, in which the beta sheets lie pancake-style on one another. Often, however, alpha helices and beta-pleated sheets form what we might think of as design motifs within a larger protein structure; they periodically give way to the less regular segments called *random coils*. The larger-scale twists and turns that the protein takes form its **tertiary structure**. The way in which *two or more* polypeptide chains come together to form a protein results in that protein's **quaternary structure**.

Proteins Can Come Undone

As noted, proteins have to maintain a precise conformation in order to function. As it turns out, however, proteins can *lose* their shape, and thus their functionality. A protein that has become deformed is a **denatured** protein. All of us have seen at least one example of this from start to finish. A protein in an egg, albumin, is clear in its starting state, but with the addition of heat in the frying pan, it goes from clear to opaque to white as the protein gets denatured. Alcohol works as a disinfectant on skin because it denatures the proteins of bacteria.

Lipoproteins and Glycoproteins

Some molecules in living things are hybrids, or combinations of the various types of molecules you've been looking at. **Lipoproteins**, as their name implies, are a combination of lipids and proteins; they are active in transporting fat molecules throughout the body via the bloodstream. They amount to a capsule of protein surrounding a globule of fat. Two kinds of lipoproteins have managed to enter public consciousness despite the handicap of having long names. These are high-density lipoproteins and low-density lipoproteins, also known as HDLs and LDLs. What makes them more or less dense is the ratio of protein to lipid in them; protein is more dense than lipid, so a high-density lipoprotein is one that contains a relatively greater amount of protein. LDLs have acquired a reputation as the villains of lipoproteins, because they carry cholesterol *to* outlying tissues including the coronary arteries of the heart, where this cholesterol may come to reside, thickening eventually into "plaques" that can block coronary arteries and bring about the blood stoppage that is

commonly known as a heart attack. HDLs, meanwhile, are regarded as the cavalry of lipoproteins; they carry cholesterol *away* from outlying cells to the liver; a high proportion of them in relation to cholesterol is predictive of keeping a healthy heart.

Glycoproteins are combinations of proteins and *carbohydrates*. Remember the discussion about insulin traveling through the bloodstream and latching onto a *receptor* on the surface of a muscle cell? Well, such receptors are usually glycoproteins, meaning mostly protein with a side-chain made of carbohydrate. A profusion of these receptors sits on cell surfaces like so many antennae, ready for a partner with just the right shape to come by and latch on. Some hormones themselves are glycoproteins, along with many other proteins released from cells.

Nucleotides and Nucleic Acids

Nucleotides are the last major type of biological molecule. They are important molecules in and of themselves and as building blocks for a couple of very important molecules.

As molecules in their own right, some nucleotides, called adenosine phosphates, serve as chemical carriers. When you study energy within cells, you'll find many references to a molecule called adenosine triphosphate, or ATP. As money is to shopping, so ATP is to getting things done in living organisms. In this text the most detailed discussion of nucleotides, however, comes in connection with their role as building blocks in two very large and important molecules, which are briefly considered next.

DNA Provides Information for the Structure of Proteins

You've learned that proteins perform a large number of biological functions, and that one class of proteins, the enzymes, has more than a thousand members in it. If you had a factory that turned out a thousand different kinds of tools, you would obviously need some direction on how each of these tools was to be manufactured: This part of the tool goes here first, and then that goes there, and so on. There is a molecule that in essence provides this kind of information for the construction of proteins. It is **DNA**, or **deoxyribonucleic acid**, and it is comprised of a huge number of nucleotides. In human beings, about 3 billion nucleotides are strung together to form our main molecule of DNA, and one copy of this molecule exists in each of our cells. A **gene** is generally a collection of DNA nucleotides that contains the instructions for an individual protein.

The information contained in the DNA molecule is much like the information contained in a cookbook, only what the DNA "recipes" are calling for are precisely ordered chains of amino acids, the building blocks of proteins. "Start with an alanine, then add a cysteine, then a tyrosine . . ." and so on for hundreds of steps, the DNA-encoded instruc-

tions will say, after which—through a series of steps—a protein becomes synthesized and then gets busy on some task. A player in this series of steps is another nucleic acid, **ribonucleic acid** or **RNA**, which is involved in ferrying the DNA-encoded instructions to the sites in the cell where proteins are put together.

The Structural Unit of DNA is the Nucleotide

Look at **Figure 3.25a** and you'll see the structure of the building blocks that make up DNA. Each nucleic acid **nucleotide** is a

NUCLEOTIDES ARE THE BUILDING BLOCKS OF DNA

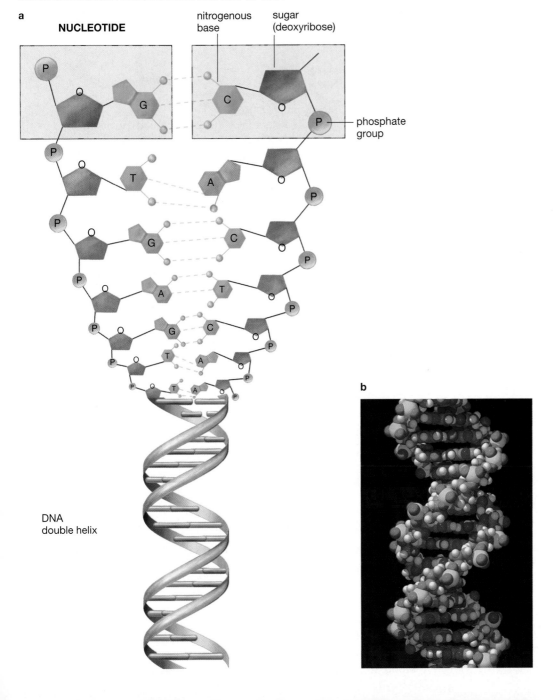

Figure 3.25
Nucleotides Are the Building Blocks of DNA

a The basic unit of the DNA molecule is the nucleotide, a molecule in three parts: a sugar, (deoxyribose), a phosphate group, and a nitrogen-containing base. A given nucleotide might contain any of four bases: Adenine (A), Guanine (G), Cytosine (C), or Thymine (T). The sugar and phosphate components of the nucleotides link up to form the outer "rails" of the DNA molecule, while the bases point toward the molecule's interior. *Two* chains of nucleotides are linked, via hydrogen bonds, to form DNA's double helix. Note that the hydrogen bond is between one base and another.

b A computer-generated space-filling model of DNA.

molecule in three parts: A *phosphate* group, a *sugar*, and a nitrogen-containing *base*. One nucleotide then attaches to another, forming a chain. *Two* of these chains then link together—as if a ladder, split down the middle, were coming together—forming the most famous molecule in all of biology, the DNA double helix.

On to Cells

Before getting to the story of this elegant DNA molecule, you need to learn a little bit about the territory in which it does its work. What you'll be looking at is a profusion of jostling, roiling, ceaselessly working chemical factories that make up all living things. These factories are called cells. Look at **Table 3.4**, and you'll find a summary of all the types of molecules that were reviewed in this chapter.

Table 3.4
Summary Table of Biological Molecules

Type of Molecule	Subgroups	Examples and Roles
Carbohydrates	Monosaccharides Disaccharides	Glucose: Energy source Sucrose: Energy source
	Polysaccharides	Glycogen: Storage form of glucose Starch: Carbohydrate storage in plants; used by animals in nutrition Cellulose: Plant cell walls, structure; fiber in animal digestion
	Chitin	External skeleton of arthropods
Lipids	Triglycerides 3 Fatty acids and glycerol	Fats, Oils (butter, corn oil): Food, energy, storage, insulation
	Fatty acids Components of Triglycerides	Stearic Acid: Food, energy sources
	Steroids Four-ring structure	Cholesterol: Fat digestion, hormone precursor, cell membrane component
	Phospholipids Polar head, nonpolar tails	Cell membrane structure
Proteins	Globular Soluble in water	Chemical: Enzymes, antibodies, receptors
	Fibrous Insoluble in water	Structural: Hair, cartilage
	Lipoproteins Protein-lipid molecule	HDLs, LDLs: Transport of lipids
	Glycoproteins Protein-sugar molecule	Cell surface receptors
Nucleotides	Adenosine phosphates	Adenosine triphosphate (ATP): Energy transfer
	Nucleic acids Sugar, phosphate group, base	DNA, RNAs: Contain information for and facilitate protein synthesis

Chapter Review

Summary

3.1 The Importance of Water to Life

- Water has several qualities that have strongly affected life on Earth. It is a powerful solvent, being able to dissolve more compounds in greater amounts than any other liquid. Because water's solid form (ice) is less dense than its liquid form, bodies of water in colder climates do not freeze solid in winter, which allows life to flourish under the ice.

- Water has a great capacity to absorb and retain heat; because of this, the oceans acts as heat buffers for the Earth, thus stabilizing Earth's temperature. Water has a high degree of cohesion, which allows water to be drawn up through plants, via evaporation, in one continuous column, from roots through leaves.

- Some compounds do not interact with water. Hydrocarbons such as petroleum are examples of such hydrophobic compounds. Water cannot break down hydrophobic compounds, which is why oil and water don't mix. Compounds that do interact with water carry an electric charge and are called hydrophilic compounds.

3.2 Acids and Bases Are Important to Life

- An acid is any substance that yields hydrogen ions when put in solution. A base is any substance that accepts hydrogen ions in solution. A base added to an acidic solution makes that solution less acidic, while an acid added to a basic solution makes that solution less basic.

- The concentration of hydrogen ions that a given solution has determines how basic or acidic that solution is, as measured on the pH scale. This scale runs from 0 to 14—0 being most acidic, 14 being most basic, and 7 being neutral. The pH scale is logarithmic; a substance with a pH of 9 is 10 times as basic as a substance with a pH of 8. Living things function best in a near-neutral pH, though some systems in living things have different pH requirements.

3.3 Carbon Is a Central Element in Life

- Carbon is a central element to life, because most biological molecules are built on a carbon framework. Carbon plays this central role because its outer shell has only four of the eight electrons necessary for maximum stability. Carbon atoms are thus able to form stable, covalent bonds with a wide variety of atoms, including other carbon atoms. The complexity of living things is facilitated by carbon's linkage capacity.

3.4 The Molecules of Life: Carbohydrates, Lipids, Proteins, and Nucleic Acids

- Carbohydrates are formed from the building blocks or monomers of simple sugars, such as glucose. These can be linked to form the larger carbohydrate polymers such as starch, glycogen, cellulose, and chitin. Carbohydrates generally are stored in polymer form in living things (glycogen in animals, for example) but broken down into monomers to be used (glucose in animals).

- There are several different varieties of lipids. Among the most important are the triglycerides, composed of a glyceride and three fatty acids. Most of the fats we consume are triglycerides. Another important variety of lipids is the steroids, which include cholesterol, and such hormones as testosterone and estrogen.

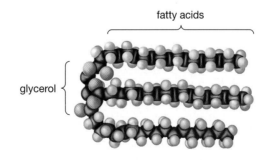

- Proteins are a diverse group of biological molecules composed of the monomers of amino acids. Important groups of proteins include enzymes, which hasten chemical reactions, and structural proteins, which make up such structures as hair. The primary structure of a protein is its amino

Key Terms

acid   41
alkaline   43
alpha helix   57
aqueous solution   37
base   41
beta pleated sheet   57
buffering system   43
carbohydrate   46
cellulose   48
chitin   48
cholesterol   53
denatured   58
deoxyribonucleic acid (DNA)   58
fatty acid   50
gene   59
glyceride   50
glycerol   50
glycogen   48
glycoprotein   58
hydrocarbon   40
hydrophilic   40
hydrophobic   40
hydroxide ion   41
isomer   44
lipid   49
lipoprotein   58
monomer   45
monosaccharide   46
monounsaturated fatty acid   51
nucleotide   59
oil   51
organic chemistry   44
pH scale   41
phosphate group   54
phospholipid   54
polymer   45
polypeptide   54
polysaccharide   48
polyunsaturated fatty acid   51
primary structure   56
protein   54
protein conformation   56
quaternary structure   58

acid sequence; this sequence determines how a protein folds up. The activities of proteins are determined by their final, folded shapes.

- Nucleic acids are polymers composed of nucleotides. The nucleic acid DNA is composed of nucleotides that contain a phosphate group, a sugar (deoxyribose), and one of four nitrogen-containing bases. DNA is a repository of genetic information. The sequence of its bases encodes the information for the production of the huge array of proteins produced by living things.

Understanding the Basics

Multiple-Choice Questions

1. Near an ocean or other large body of water, air temperatures do not vary as much with the seasons as they do in the middle of a continent. This tendency of water to resist changes in temperature is the result of water's
   a. high density
   b. low density
   c. being a good solvent
   d. low specific heat
   e. high specific heat

2. A frog survives a freezing-cold winter on the bottom of a pond because
   a. ice, which floats on water, insulates the water beneath it.
   b. the cells of a frog's body cannot freeze
   c. water has a low specific heat
   d. the surface tension of water protects the frog's body
   e. all of these are factors

3. Janine has dry skin, so she uses body oil every morning. The oil seals in some of the water on her skin, so that it doesn't get as dry. This is possible because oils:
   a. are hydrophilic
   b. are hydrophobic
   c. have a high specific heat
   d. are more dense than water
   e. are both nonpolar and hydrophobic

4. Some plants live in bogs in which the pH is about 2. Thus these plants are able to survive in a(n) _____ external environment.
   a. basic   d. neutral
   b. buffered   e. alkaline
   c. acidic

5. When you eat starch such as spaghetti, an enzyme in your mouth breaks it down to maltose. Eventually, the maltose enters your small intestine, where it is broken down to glucose, which you can absorb into your bloodstream. The starch is a _____, the maltose is a _____, and the glucose is a(n) _____.
   a. protein . . . dipeptide . . . amino acid
   b. monosaccharide . . . disaccharide . . . polysaccharide
   c. triglyceride . . . fatty acid . . . glycerol
   d. amino acid . . . dipeptide . . . protein
   e. polysaccharide . . . disaccharide . . . monosaccharide

6. Which of these is *not* an actual difference between saturated and unsaturated fats?
   a. Saturated fats are more likely to be solid at room temperature; unsaturated fats are more likely to be liquid.
   b. Saturated fats are a type of cholesterol or steroid, whereas unsaturated fats are a triglyceride.
   c. Saturated fats have fatty acids that pack closely together, owing to their straight-line construction.
   d. Saturated fats are composed of fatty acids that have no double bonds in their chemical structure. Unsaturated fats are composed of fatty acids that have one or more double bonds in their chemical structure.
   e. All of the above are differences between saturated and unsaturated fats.

7. The myoglobin protein, which carries oxygen in muscle cells, has only the first three levels of protein structure. In other words, it lacks a quaternary level. From this you can conclude that myoglobin
   a. is made of nucleic acids
   b. is made of only one polypeptide chain
   c. lacks hydrogen bonds
   d. is not helical or pleated
   e. is a fiber

8. You received your genetic information from your parents in the form of DNA. This DNA carried the instructions for making
   a. carbohydrates such as glycogen
   b. fatty acids
   c. phospholipids for making the membranes of your cells
   d. proteins
   e. all of the above

9. John is lactose intolerant. The -*ose* ending indicates that John cannot digest a certain
   a. sugar
   b. polysaccharide
   c. protein
   d. steroid
   e. enzyme

Brief Review

1. Describe two ways that water serves as a heat buffer.

2. Both low-density lipoproteins (LDLs) and high-density lipoproteins (HDLs) are involved with carrying fats through the bloodstream. If your LDL count is unusually high, should you be concerned? What if your HDL count is high? Why are they different?

3. List as many functions of proteins as possible. Why are proteins able to do so many different types of jobs? How does this affect the world we live in?

4. How does cohesion allow water to move through plants?

5. Early in 1999, scientists discovered that the hydrogen bond in water gets extra strength from the polar covalent bond that ties water molecules together. What effects should this have on our understanding of the properties of water? Look for information about this discovery on the Web.

Applying Your Knowledge

1. Would you buy a pH-balanced shampoo, or one that is not? Explain your answer.

2. Suppose that you are designing clothing to be worn at a space station on a very cold planet. Should you consider designing clothing that has a layer of water between two layers of plastic? Why or why not? What else might you need to consider?

3. Many species of beans produce a poisonous group of glycoproteins called lectins, which cause red blood cells to clump together (agglutinate) and cease to function. (Fortunately, cooking destroys most of them.) Using what you learned about the structure and function of glycoproteins, suggest how they might do this. What benefit do you think plants might get just by causing red blood cells to clump together?

Group Activity

With a group of your classmates, collect the following information in order to study acid rain in your area:

1. Research the acidity of precipitation in your area.

2. Find out the composition of soil and bedrock to determine whether acidity in your area is neutralized by the soil and thus has little effect.

3. Contact the local Department of Agriculture office to learn whether acidity has affected agriculture in your area.

4. Contact the Historic Preservation Commission or similar body to determine whether acid precipitation has affected buildings or monuments.

5. Find out whether your area has any special restrictions on automobile or power plant emissions.

6. Find out whether acid that is produced in your area is having an effect elsewhere.

When the data have been collected, write a formal proposal to an appropriate legislative body (town, city, state, national) suggesting what should be done in your area to limit acid precipitation.

MediaLab
You Are What You Eat:
Food and the Molecules of Life

Not only do we obtain all of the matter that makes up our body from the food and water we take in, but the quality of the foods we consume is critical to our overall health. To learn more, enter the *CD Activity* and learn to analyze food labels so you can interpret the energy sources and classes of macromolecules found in your diet. Using the *Web Investigation,* you can determine your specific daily Calorie requirements, and recommended dietary allowances (RDAs) for macromolecules, minerals, and vitamins. In the *Communicate Your Results* section, you will develop strategies for designing a healthy diet.

This *MediaLab* can be found in Chapter 3 on your CD-ROM and Companion Website (http://www.prenhall.com/krogh).

CD ACTIVITY

Obtaining enough Calories to perform our daily tasks is not hard for most Americans (although there is hunger in America). Indeed, many of us struggle with the problem of consuming too many Calories. Diets high in lipids (fats and oils) contribute to many health problems, including heart disease and cancer. How can you tell if your diet is healthy? In this *CD Activity,* you'll learn how to evaluate your diet using food labels.

Activity

1. *First you will determine the Calories (energy content) in carbohydrate, fat, and protein molecules, and learn to find Calorie content on a nutrition label.*

2. *Then you will use this knowledge in calculating serving sizes and determining what percentage of Calories come from fats, carbohydrates, and proteins, so that you can compare your results with the dietary levels suggested for a healthy diet.*

After learning to read a nutrition label, you'll be ready to tailor your diet to your specific energy requirements in the *Web Investigation.*

WEB INVESTIGATION

Investigation 1
Estimated time for completion = 5 minutes

Now that you have learned to read a food label and understand what you are eating, wouldn't it be nice to have a personal nutritionist determine the ideal diet for you? Select the Keyword **HEALTHY BODY CALCULATOR** on your CD or Website for this *MediaLab.* Then enter your personal description and submit it. You'll receive an analysis of your health status, including whether you are at a healthy weight and body fat percentage. You'll also receive a personalized daily guide for your food consumption.

Investigation 2

Estimated time for completion = 15 minutes, after keeping a 24-hour diet diary

Dieters often consider Calories when analyzing their diet, but forget to include healthy amounts of all food groups. To compare your recommended diet (from *Web Investigation 1*) with your actual diet, keep a 24-hour record of what you eat. When you are ready to analyze that diet, select the Keyword **DIET ANALYSIS TOOL** on your CD or Website for this *MediaLab*, and enter your personal demographic and diet information. The site will calculate the total Calories consumed (for fats, proteins, and carbohydrates) and compare the result to your specific requirements.

Investigation 3

Time for completion = 5 minutes

The United States has one of the fattest populations on Earth. Over 20 percent of U.S. adults can be classified as clinically obese (due largely to our inability to resist a barrage of good-tasting, inexpensive, high-fat foods. By adding

low-calorie fat substitutes, food manufacturers have responded to public pressure for foods with reduced fat content. How can you recognize a fat substitute, and are they good for you? Clip an ingredients label from a package of reduced or low-fat food, and make a list of any fat substitutes. To help you identify them, select the Keyword **FAT SUBSTITUTES** on your CD or Website for this *MediaLab*.

Now that you have analyzed your own diet, discuss with your peers what you have learned about creating a healthy diet.

COMMUNICATE YOUR RESULTS

Exercise 1

Estimated time for completion = 25 minutes

In the United States, many states now impose taxes on tobacco to discourage smoking and to help offset the health costs associated with smoking. However, heart disease, which has been linked to a high-fat diet, is the country's number one killer. In addition, more cancer cases in the United States are related to diet than to smoking. Compose a 250-word argument supporting or refuting the following plan: Taxing high-fat foods would be a great way to reduce the incidence of heart disease and help offset the cost of its treatment. The taxes could also be used to educate the public to the dangers of a high-fat diet, in the same way that taxes on cigarettes are used for antismoking ads.

Exercise 2

Estimated time for completion = 15 minutes

Compare the diet analysis that you did in *Web Investigation 2* with the diet of another student to determine ways you might help

each other create a more healthy diet. How are your diets related? What nutrients or food groups are you eating too much of, or too little of? Revise your day's diet to match RDA guidelines, especially with less than 30 percent of total Calories from fat, and post it on the textbook Website for other students to view.

Exercise 3

Estimated time for completion = 5 minutes

Make a list of all the fat substitutes that you found on your ingredient label from *Web Investigation 3*. Indicate what type of fat substitute you found (like the synthetic fat Olestra). Can you determine how many Calories you would save by eating this reduced-fat product instead of the high-fat one?

4

Life's Home
The Cell

Are those muffins I smell?
(page 68)

often multicellular

usually need oxygen to exist

membrane-bound organelles

Some cells need to breathe.
(page 69)

It's a small world.
(page 73)

Where proteins
shape up.
(page 76)

The United States Senate

has 100 members in it, and all of them do occasionally gather together at one time to consider given issues. But if you want to know how the Senate actually gets something *done*, you must look to the specialized working units of the Senate known as committees. Things work this way because the business of the nation is simply too complex for every senator to deal with every issue. Each senator must instead specialize in certain issues.

Specialization goes hand in hand with complexity, and this is as true of the natural world as it is of the political. So complex are the tasks that birds and trees and people carry out that if you want to know how any of these creatures actually gets something *done*, you must look again to specialized units, in this case the units known as cells.

4.1 Cells Are the Working Units of Life

Muffins are in the oven and you are in the living room. Gas molecules from the baking muffins waft into the living room, and some of them happen to make their way to your nose, there to travel a short distance to your upper nasal cavity and land on a set of ceaselessly moving hair-like projections called cilia. These actually are the extensions of some specialized nerve cells called olfactory receptors (**see Figure 4.1**, p. 68). If enough muffin molecules bind with enough of the cilia, an impulse is passed along (through other nerve cells) to trigger not only the *sensation* of smell, but the *association* of this smell with the memory of muffins past. How do we know muffins are in the oven? Through cells. How do we move our hands or read this page? Through cells. Life's working units are cells, and in our amazingly complex natural world, there is a great specialization in them.

Cells Bring Unity and Continuity to Life

And yet there is unity. Every form of life either is a cell, or is composed of cells. The one possible exception to this is viruses, but even they must commandeer the machinery of living cells in order to reproduce. There is unity, too, in the way cells come about: Every cell comes *from* a cell. Human beings are incapable of producing cells from scratch in the laboratory, and so far as we can tell, nature has fashioned cells from simple molecules only once—back when life on Earth got started. The fact that all cells come from cells means that each cell in your body is a link in a cellular chain that stretches back more than 3.5 billion years.

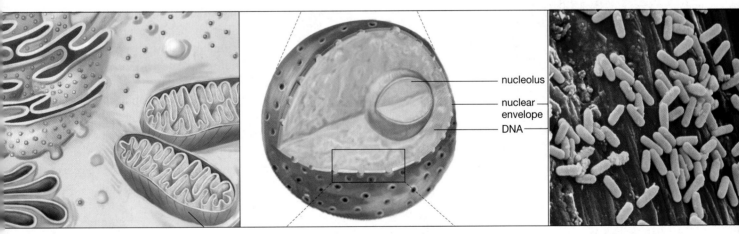

Inside the cell.
(page 71)

The center of attention.
(page 75)

nucleolus

nuclear envelope

DNA

Hidden life on a pin.
(page 72)

4.2 All Cells Are Either Prokaryotic or Eukaryotic

So, what are these tiny, working units we call cells? You've seen that cells can specialize and come in a variety of forms. It follows from this that there is no such thing as a "typical" cell. There are, however, certain *categories* of cells that are important, and the two most important of these are **prokaryotic cells** and **eukaryotic cells**. Every cell that exists is one or the other, and this simple either-or quality extends to the organisms that fall into these camps. All prokaryotic cells either are bacteria or another microscopic form of life known as archaea. Setting bacteria and archaea aside, *all other cells* are eukaryotic. This means all the cells in plants, in animals, in fungi, and in other creatures called protists that you'll get to in a later chapter. For a look at the person who first beheld the microworld of cells, see "First Sightings: Anton van Leeuwenhoek," on page 70.

Prokaryotic and Eukaryotic Differences

The name *eukaryote* comes from the Greek *eu*, meaning "true," and *karyon*, meaning "nucleus"; while *prokaryote* means "before nucleus." These terms describe the most critical distinction between the two cell types. Eukaryotes have a nucleus, bound within a thin membrane, that contains almost all their DNA. While the DNA of prokaryotes is localized (in a "nucleoid" region), it is not bound within a membrane (**see Figure 4.2**).

This distinction, though, only begins to describe the differences between these two cell types. However appealing it may be to think of two single-celled creatures—one a prokaryote, the other a eukaryote—as "alike," the distance between them as life forms is immense. Human beings and chimpanzees are nearly identical in comparison. Eukaryotes tend to be much larger than prokaryotes; indeed, thousands of bacteria could easily fit into an average eukaryotic cell (see "The Size of Cells" on page 72). Eukaryotes are quite often multicelled organisms, while prokaryotes are single-celled. And most eukaryotes are aerobic—they need oxygen to exist—whereas many prokaryotes can take oxygen or leave it and others actually are poisoned by it.

Compartmentalization in Eukaryotic Cells

Perhaps the most notable distinction between prokaryotes and eukaryotes, though, is that eukaryotes are *compartmentalized* to a far greater degree than are prokaryotes. The nucleus in eukaryotic cells is just one of their **organelles**, or "tiny organs"—internal compartments that are absent almost altogether in prokaryotes. For example, eukaryotes employ organelles called mitochondria that transform energy from food, while prokaryotes have no such structures.

What's the importance of this compartmentalization? Recall the observation made at the beginning of this chapter: Specialization and complexity go hand in hand. Eukaryotic cells are specialized *internally*, and this characteristic underlies their ability to specialize *as cells*. They have organelles, a much larger DNA "blueprint," and more proteins to go with it. The creatures made from eukaryotic cells can hear and see and fly. Prokaryotes, meanwhile, exist in a soundless, sightless, flightless world in which, for the most part, each cell is on its own. The complexity of animals and plants is based on the specialization of their component parts, or tissues, and these tissues are made of specialized cells.

This is not to say that prokaryotes are uniform, nor that they are unsuccessful. On

Figure 4.1
Cells Can Specialize
In more complex organisms, different cells carry out different functions. In the picture at left below, you can see one type of cell, a nerve cell, in this case located in the lining of the nose and surrounded by gray accessory cells. A closer look at one of these nerve cells, in the picture at right, shows a number of hair-like extensions, called cilia, protruding from it. When we smell muffins in the oven, gas molecules that waft off the muffins bind with the cilia.

cilia

nasal nerve cell

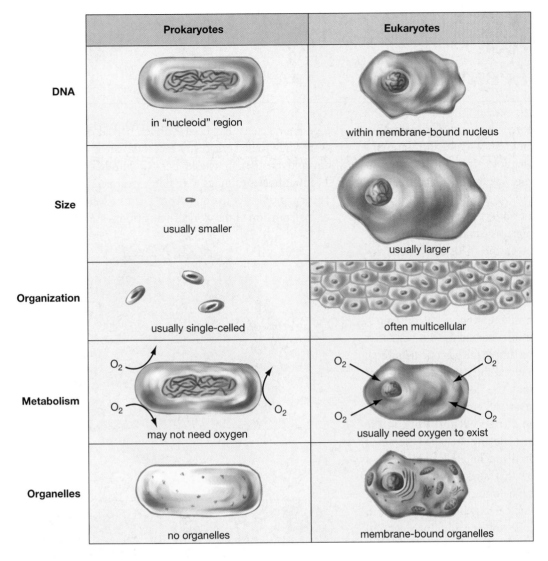

	Prokaryotes	Eukaryotes
DNA	in "nucleoid" region	within membrane-bound nucleus
Size	usually smaller	usually larger
Organization	usually single-celled	often multicellular
Metabolism	may not need oxygen	usually need oxygen to exist
Organelles	no organelles	membrane-bound organelles

Figure 4.2
Prokaryote and Eukaryote Cells Compared
A prokaryote cell is a self-contained organism, since the prokaryotes—bacteria and archaea—are essentially single-celled. Eukaryote organisms, which may be single- or multi-celled, include plants, animals, fungi, and protists.

the contrary, they were the first to exist on Earth—and they undoubtedly will be the last to go, if it comes to that. They are diverse, and extremely successful, if success is defined as living in a lot of places in huge numbers. As Lynn Margulis and Karlene Schwartz have observed, more bacteria are living in your mouth right now than the number of people who have ever existed. But prokaryotic cells are *limited* compared to the eukaryotic variety, and so are not the focus of this chapter. For the range of biological processes to be studied here, eukaryotes are the place to look.

4.3 The Eukaryotic Cell

Eukaryotic cells are made of many different components, but in overview these can be broken down into five parts (**see Figure 4.3**).

- The cell's **nucleus**, which contains most of its genetic material.

- Its other **organelles**, which lie *outside* the cell nucleus.
- The **cytosol**, a protein-rich, jelly-like fluid in which the cell's organelles outside the nucleus are immersed.
- The **cytoskeleton**, a kind of internal scaffolding that has several different kinds of units; some of these can be likened to tent poles, others to monorails.
- The outer boundary of the cell, the **plasma membrane**.

nucleus
cytoskeleton
cytosol
plasma membrane
other organelles

EUKARYOTIC CELL

Figure 4.3
All eukaryotic cells possess a nucleus, other membrane-bound organelles, jelly-like cytosol, a cytoskeleton, and an outer plasma membrane.

First Sightings: Anton van Leeuwenhoek

To the list of explorers that includes Columbus and Balboa we could, in a sense, add the name Anton van Leeuwenhoek. It was this unassuming Dutchman who was the great early voyager into another world, the micro-world.

It was not until the seventeenth century that human beings realized that such things as small as cells existed. Leeuwenhoek began to report in the 1670s on what he saw with the aid of a device that had been invented at the end of the 1500s—the microscope. (Note, though, that Leeuwenhoek actually was using devices that we would refer to today as magnifying glasses.) One of Leeuwenhoek's contemporaries, Englishman Robert Hooke, coined the term *cell* after viewing a slice of cork under a simple microscope (**Figure 1**), but Hooke's purpose was to reveal the detailed structure of familiar, small objects, such as the flea. Leeuwenhoek, by contrast, revealed the *existence* of creatures unimagined until his time. Moreover, he carried out this work in the most extraordinary fashion: Laboring alone in the small town of Delft with palm-sized magnifiers he himself had created, looking at anything that struck his fancy. (And many things struck his fancy; he once looked at exploding gunpowder under a microscope, nearly blinding himself in the process.) For 50 years, while working as shopkeeper and minor city official, this untrained amateur of boundless curiosity examined the micro-world and reported on it in letters he posted to the Royal Society in London.

Who could believe what he uncovered? How was it possible that there was a buzzing, blooming universe of "animalcules" (little animals) whose existence had been completely unsuspected? Prior to his work, no one thought that any creature smaller than a worm could exist within the human body. Yet, examining scrapings from his own mouth, Leeuwenhoek tells us:

I saw, with as great a wonderment as ever before, an inconceivably great number of little animalcules, and in so unbelievably small a quantity of the foresaid stuff, that those who didn't see it with their own eyes could scarce credit it.

The animalcules that Leeuwenhoek beheld over his career were single-celled organisms that today are known as bacteria and protists. It would be two hundred years before Leeuwenhoek's findings were fully integrated into a modern theory of cells. Yet the man from Delft had shown that only a small portion of the world's living things are visible things.

a

b

Figure 1

a What Leeuwenhoek Could See
Biologist Brian Ford used an actual 300-year-old Leeuwenhoek microscope to capture this image of spiny spores from a truffle. (×600)

b Leeuwenhoek's Microscopes Were Hand-Held and Paddle-Shaped
Leeuwenhoek revealing the micro-world to Queen Catherine of England, wife of King Charles II (1630–1685).

In any discussion of a cell, you are likely to hear the term **cytoplasm,** which simply means the *region* inside the plasma membrane but outside the nucleus. The cytoplasm is different from the cyto*sol*. If you removed all the structures of the cyto-plasm—meaning the organelles and the cytoskeleton—what would be left is the cytosol, which is mostly water. This does not mean that the cytosol is simply a passive medium for the other structures. But it is not an organized structure in the way the organelles are. Almost all the organelles you'll see are encased in their own membranes, just as the whole cell is encased in its plasma membrane. In the balance of this chapter, you'll explore all of these components except for the plasma membrane, which is so special it gets its own chapter.

The Animal Cell

Scientists sometimes make a convenient division of eukaryotic cells into two types: animal cells and plant cells. These cell types have more similarities than they do differences, but they are different *enough* that it will be helpful to look at them separately. You'll start by examining animal cells and then look at how plant cells differ from them.

Insofar as we can characterize that elusive creature, the "typical" animal cell (**see Figure 4.4**), it is roughly spherical, probably about 25 micrometers (μm) in diameter, surrounded by, and linked to, cells of similar type, and immersed in water.

**Figure 4.4
The Animal Cell**

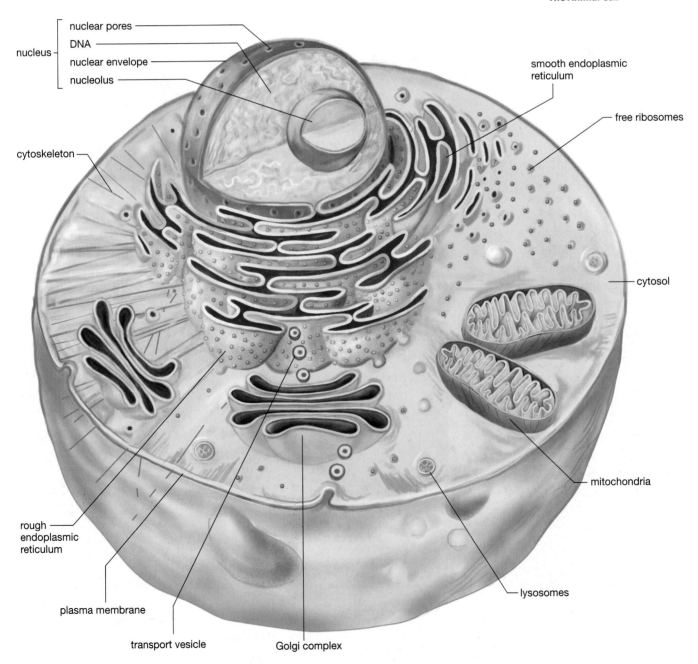

The Size of Cells

For several chapters, you've been going over atoms and ions and molecules and such—things small enough to be invisible to the unaided eye. For the most part, you've had to imagine what these things look like, simply because most of them are so small that we either have no pictures of them at all (as with electrons) or few clear pictures (as with atoms). If you flip through the pages of this chapter and those to come, however, you begin to see a fair number of actual photographs. They don't have the same quality as summer vacation snapshots, but they are recognizable as pictures, or more properly as **micrographs** (pictures taken with the aid of a microscope). Micrographs enable us to see surprising things; for example, hundreds of bacteria on the tip of a pin (**Figure 1**). So, with the cells that are introduced in this chapter, there has obviously been a bump up in size into a world that is more easily visible with the help of various kinds of technology.

How Small Are They?

In taking stock of the micro-world, two units of measure are particularly valuable. The smaller of them is the **nanometer**, abbreviated as nm; the larger is the **micrometer**, abbreviated μm. (The unfamiliar-looking first letter there is the Greek symbol for a small *m*. Scientists are not trying to be purposely obscure in using it; another unit of the metric system, the millimeter, lays claim to the mm abbreviation.) What these stand for is a billionth of a meter (nm) and a millionth of a meter (μm). A meter equals about 39.6 inches, or just over a yard, which gives you some starting sense of physical reality in understanding the rest of these sizes.

Now look at **Figure 2** to see what size various objects are. Atoms are down at the bottom of the scale, at about a tenth of a nanometer. Something less than a tenfold increase from there gets us to the size of the protein building-blocks, called amino acids, that you looked at last chapter. Another tenfold-plus increase and you've reached the upper limit on proteins.

You have to go better than 10 times larger than this, however, before you arrive at the size of something that is actually *living*, as opposed to something that is a component part of life. Because, as you've seen, life means cells, this means we're talking about the smallest cells in existence, which are bacteria measuring perhaps 200–300 nm. This extreme on the small side of cells has a counterpart on the large side with the single cells we call chicken or ostrich eggs and certain nerve cells that can stretch out to a meter in length. In general, however, we just cross into the micrometer range with the smaller bacterial cells: about 1–10 μm. The cell size for most plants and animals falls in a range that is 10 times larger than this, about 10–100 μm. At 10 μm, perhaps a billion cells could fit into the tip of your finger. And how about your whole body? One estimate is about 10 trillion.

So, Why So Small?

Having learned how small cells are, you might then well ask: *Why* are they so small? As noted, cells are small chemical factories, and just like any factory, they are constantly shipping things in and out. The size-limiting factor for cells is having enough surface area to export and import all that they need.

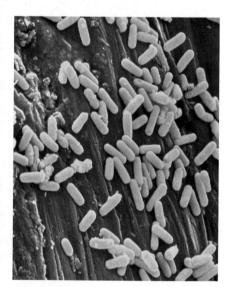

Figure 1
Hidden Life
Microscope enlargements of the tip of this pin show an abundance of life—in this case bacteria—thriving on an object that we normally think of as being devoid of living organisms. (a: ×85; b: ×425; c: ×2100)

This constraint comes about because of a fundamental mathematical principle: As the surface area of an object increases, its *volume* increases even more. Say you have a cube, 1 inch long on each of its sides. Its surface area is 6 square inches (length × width × number of sides), while its volume is one cubic inch (length × width × height). Now say you increase the side dimension to 8 inches. The surface area goes from 6 to 384, but the *volume* goes from 1 to 512. Where, at a one-inch dimension, there were *six* square inches of surface area for every cubic inch of volume, now there are only *three-quarters* of an inch of surface for every cubic inch of volume. Beyond a certain volume, then, a cell simply would not have enough surface area to "import and export" all the materials it needs to. This effectively sets an upper limit on how big cells can be and explains why most of the cells in an elephant are no bigger than those in an ant, though the elephant does have more cells than the ant. **Figure 3** shows you some size comparisons in the micro-world.

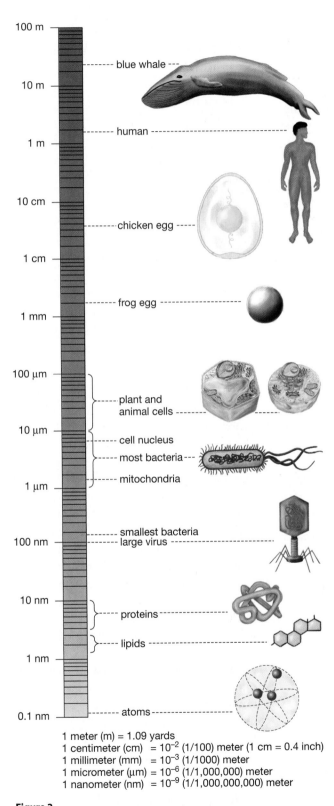

1 meter (m) = 1.09 yards
1 centimeter (cm) = 10^{-2} (1/100) meter (1 cm = 0.4 inch)
1 millimeter (mm) = 10^{-3} (1/1000) meter
1 micrometer (μm) = 10^{-6} (1/1,000,000) meter
1 nanometer (nm) = 10^{-9} (1/1,000,000,000) meter

Figure 2
Little and Big
The sizes of some selected objects in the natural world.

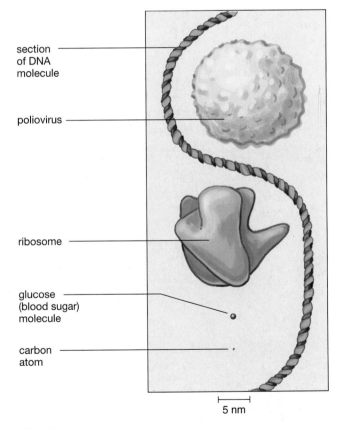

Figure 3
Small Is a Relative Thing
The sizes and shapes of five natural-world entities, each magnified a million times.

4.4 A Tour of the Animal Cell: Along the Protein Production Path

You are now going to take an extended tour of the animal cell; in this tour, it will often be helpful to think of a cell as a kind of living factory. Much of the first part of your trip will be spent tracing the way a cell puts together a protein for "export" outside of itself. **Figure 4.5** shows the path you'll be taking, from nucleus to the outer edge of the cell. Don't be bothered by the unfamiliar terms in Figure 4.5, because they'll all be explained in the text. The important thing is to have some sense of the path that protein production takes.

Figure 4.5
Path of Protein Production in Cells

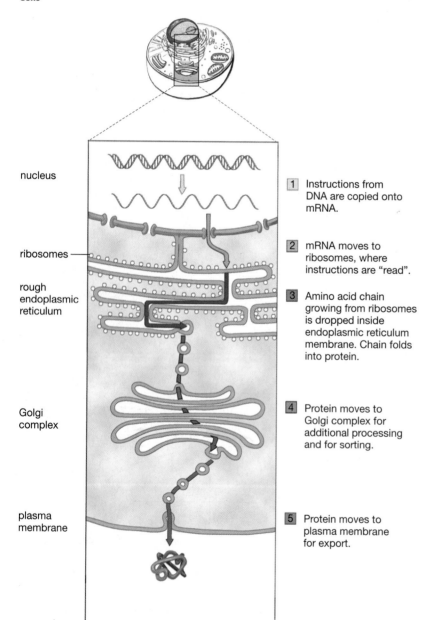

nucleus

1 Instructions from DNA are copied onto mRNA.

2 mRNA moves to ribosomes, where instructions are "read".

ribosomes

rough endoplasmic reticulum

3 Amino acid chain growing from ribosomes is dropped inside endoplasmic reticulum membrane. Chain folds into protein.

Golgi complex

4 Protein moves to Golgi complex for additional processing and for sorting.

plasma membrane

5 Protein moves to plasma membrane for export.

Beginning in the Control Center: The Nucleus

As noted in Chapter 3, proteins are critical working molecules in living things, and DNA contains the instructions for producing them. In the eukaryotic cell, DNA is largely confined within the cell **nucleus**, the outer boundary of which is a concentric, double membrane called the **nuclear envelope** (see **Figure 4.6**).

DNA itself is made of a long chain of building blocks, called nucleotides, which link together to form the famous double helix (see Chapter 3, pp. 57–59). There comes a point in the life of most cells when they divide, one cell becoming two, two becoming four, and so on. Because (with a few exceptions) all cells must possess the set of instructions that are contained in DNA, it follows that when a cell divides, its original complement of DNA must *duplicate*, so that each "daughter" cell can have its own DNA. The nucleus, then, is not just the site where DNA exists; it is the site where new DNA is put together, or "synthesized," for this duplication.

Messenger RNA

At the end of Chapter 3, you saw that the process of protein synthesis requires that DNA's instructions first get copied onto another long-chain molecule, RNA. This step is akin to having a cassette tape (of DNA) and then making a copy of it (onto RNA). Our RNA "tape" then moves out of the nucleus to continue the process of protein synthesis. As it turns out, RNA comes in several forms. The one that the DNA instructions are copied onto is called **messenger RNA** (mRNA), which is also synthesized in the nucleus. Given mRNA's function, it must have, of course, a way to *get out* of the nucleus; its exit points turn out to be thousands of **nuclear pores** that stud the surface of the nuclear envelope. Materials can go the other way through the nuclear pores as well: Proteins, energy molecules, and other "raw materials" pass from the cytoplasm into the nucleus by way of them.

mRNA Moves Out of the Nucleus

Imagine shrinking down in size so that the nucleus seems about as big as a house, with you standing outside it. What you would see in protein synthesis is lengths of mRNA, rapidly moving out through nuclear pores and dropping off into the cytoplasm. Several varieties of RNA actually come out like this, but for now, let's just follow the trail of the mRNA. You're about ready to leave the nucleus to continue your cell tour, but before you do, note two things. First, DNA contains information for making proteins, and the mRNA coming out of the nuclear pores amounts to a means of disseminating this information. Thus, it's not hard to conceptualize the nucleus as a control center for the cell. Beyond this, in looking at Figure 4.6, you've probably noticed that there is a rather imposing structure *within* the nucleus, called the **nucleolus**. For now, just *hold that thought*; the mRNA tapes have come out of the nu-

clear pores and are making a short trip to another part of the cell.

The Rough Endoplasmic Reticulum

If you look again at Figure 4.4, you can see that, though the nucleus cuts a roughly spherical figure out of the cell, there is, in essence, a folded-up continuation of the nuclear envelope on one side. This mass of membrane has a name that is a mouthful: the **rough endoplasmic reticulum**. It is rough because it is studded with structures you'll get to in a second called ribosomes; it is endo*plasmic* because it lies within (endo) the cyto*plasm*; and it is a reticulum because it is a network, which is what *reticulum* means in Latin. Understandably, it is nearly always referred to as the rough ER or RER.

nucleolus

nuclear envelope

DNA

DNA

mRNA

inner membrane

outer membrane

nuclear pore

Figure 4.6
The Cell's Nucleus
The DNA of eukaryotic cells is sequestered inside a compartment, the nucleus, which is lined by a double membrane known as the nuclear envelope. Compounds pass into and out of the nucleus through a series of microscopic channels called nuclear pores. The prominent spherical structure within the nucleus is the nucleolus, an area that specializes in the production of ribosomal RNA—the material that helps make up ribosomes. Protein production is dependent upon the information encoded in DNA's sequence of chemical building blocks. This information is copied onto a length of messenger RNA (mRNA), which then exits from the nucleus through a nuclear pore. (Micrograph: ×4400)

Embedded within the RER: Ribosomes

The **ribosomes** that are embedded in the rough ER are the destination for our mRNA tapes. Ribosomes are commonly described as the "workbenches" of protein synthesis, which is a fine metaphor; but following the notion of mRNA as a cassette tape, you might look at a ribosome as a kind of *playback head* on a cassette deck. What does such a head do in an actual deck? As a tape is run through it, it reads signals that have been laid down on it (as magnetic bits) and turns these into sound. Likewise, each ribosome acts as a site that an mRNA tape runs through, only the information on this tape results in the production of an *object* that grows from the ribosome: A chain of amino acids that folds up into the molecule we call a protein (**see Figure 4.7**).

Several Locations for Ribosomes

All of the mRNA that comes out of the nuclear pores goes to ribosomes, but only some of these ribosomes are embedded in the rough ER. A multitude also exist as **free ribosomes**, which is to say *free-standing*, in the cytosol. In general, those proteins that will reside in the cell's plasma membrane are synthesized in the ribosomes of the rough ER, as are proteins that will be exported out of the cell altogether (the **secretory proteins**). Meanwhile, most of the proteins that will be used inside the cell are made by free ribosomes. All told, a given cell in your body may contain 10 million ribosomes.

Back on the Protein Trail

You have seen ribosomes in the rough ER act as sites that the mRNA tapes migrate to, resulting in the completion of an amino acid chain. But such a chain is, in essence, an *unfinished* protein that needs to go though more processing before it can be exported. The first step in this processing leads only to the other side of the wall the ribosome is em-

Figure 4.7
Where Proteins First Take Shape: The Rough Endoplasmic Reticulum (RER)
Messenger RNA "tapes" move out from the nucleus to ribosomes embedded in RER, and there the tapes are "read" by the ribosomes. The result is an amino acid (or polypeptide) chain that drops into the cisternal space of the rough ER. There it will fold up into its protein shape and undergo processing (for example, with a side chain being added). The protein is then encased in a membrane-lined vesicle, ready for transport to the Golgi complex. (Micrograph: ×90,500)

rough endoplasmic reticulum

nuclear envelope

ribosomes

cisternae

cisternal spaces

mRNA

ribosome

amino acid chain

1. Amino acid chain produced by ribosomes moves into cisternal space.

2. Amino acid chain folds up making a protein.

3. Side chains added to protein.

4. Vesicle formed to house protein while in transport.

bedded in. As the rough ER ribosomes finish their work, the amino acid chains they have completed drop into *chambers* inside the rough ER.

If you look at Figure 4.7, you can see that the whole of the rough ER takes the shape of a set of flattened sacs (called **cisternae**). The membrane that the rough ER is composed of forms the periphery of these sacs. Inside are the **cisternal spaces**. As amino acid chains enter the cisternal spaces, they first fold up into their protein shapes, as you saw in Chapter 3. Beyond this, many proteins that are exported from cells have sugar side chains added to them—another thing that happens in the cisternal spaces. Once this work is done, the protein moves out of the rough ER for more processing before being exported.

A Pause for the Nucleolus

Before you continue on the path of protein processing, think back a bit, to the discussion about the nucleus, when you were asked to hold the thought about the large structure *inside* the nucleus, called the nucleolus. This is the point where its story can be told, because now you know what ribosomes are.

It turns out that *ribosomes* are mostly made of RNA. (They are made of a mixture of proteins and *ribo*nucleic acid.) But so great is the cell's need for ribosomes that a special section of the nucleus is devoted to their synthesis. This is the nucleolus. The *type* of RNA that's part of the ribosomes is, fittingly enough, called ribosomal RNA or rRNA, and it's one of the multiple varieties of RNA mentioned before. Ribosomes are brought to an unfinished state within the nucleolus, after which they pass through the nuclear pores and into the cytoplasm; when put together in final form there, they begin receiving mRNA tapes. The cell's traffic in ribosomes is considerable; with millions of them in existence, lots are going to be wearing out all the time. Perhaps a thousand need to be replaced every minute.

Elegant Transportation: Transport Vesicles

The proteins that have been processed within the rough ER need to move out of it and to their "downstream" destinations before being exported. But how do proteins move from one location to another within the cell? As noted, the rough ER and the nuclear membrane amount to one long, convoluted membrane. And recall in addition that most of the organelles in the cell are "membrane-bound," meaning they have membranes at their periphery. Each of these membranes has its own chemical structure, but collectively they have an amazing ability to work together: A piece of one membrane can *bud off*, as the term goes, carrying inside it some of our proteins-in-process. Moving through the cytosol, this tiny sphere of membrane can then *fuse* with another membrane-bound organelle, releasing its protein cargo in the process. This network of budding and fusing spheres is known as the **endomembrane system**. The spheres that move within it, carrying proteins and other molecules, are called **transport vesicles**.

This system gives cells a remarkable capability. One minute a piece of membrane may be an integral part of, say, the rough ER; the next it is separating off as a spheroid and moving through the cytosol, carrying proteins within. Most of the cell's structures are part of the endomembrane system.

It is this system that makes it possible for our proteins-in-process to move out of the rough ER. Note, though, that there are many different *kinds* of proteins being processed at any one time in the rough ER cisternae. It is as if the cellular factory has a lot of different assembly lines working at once. Almost all of the proteins under construction are, however, initially bound for the same place—the Golgi complex.

Downstream from the Rough ER: The Golgi Complex

Once a transport vesicle, bearing proteins, has budded off from the rough ER, it then moves through the cytosol to fuse with the membrane of another organelle, one first noticed by Italian biologist Camillo Golgi at the beginning of the twentieth century. The **Golgi complex** further processes proteins. Some side chains of sugar may be trimmed here, or phosphate groups may be added. But the Golgi complex does something else as well. Recall that, on the one hand, some proteins are bound for export or for use in the cell's membrane, but that, on the other, some proteins will remain within the cytosol. It follows that proteins have to be *sorted and shipped* appropriately, and the Golgi does just this, acting as a kind of distribution center. *How* it does this is something of a mystery. A given vesicle is likely to arrive at the Golgi bearing several sorts of proteins. For them to arrive at the correct final destination, they need to be "tagged" and routed somehow so that one can go here while another goes there.

The Golgi is similar to the ER in that it amounts to a series of cisternae, or connected membranous sacs with internal spaces (**see Figure 4.8**). A protein that arrives from the ER moves through these sacs by being packaged in successive varieties of the Golgi's own transport vesicles. It will exit the complex housed in a final transport vesicle that buds off from the Golgi's outside face.

From the Golgi to the Surface

For the kind of secretory protein you've been following, the journey that began with the transcription of DNA is almost over. Once a

vesicle buds off from the Golgi, all that remains is for it to make its way through the cytosol to the plasma membrane at the outer reaches of the cell; there, the vesicle fuses with the plasma membrane and the protein is ejected into the extracellular world. This last step, called **exocytosis**, is a process you'll be looking at next chapter. With it, one finished product of the cellular factory has rolled out the door.

4.5 Outside the Protein Production Path: Other Cell Structures

The Smooth Endoplasmic Reticulum

A functioning cell engages in more activities than the protein synthesis and shipment just reviewed. If you look back to Figure 4.4, you can see that there actually are *two* kinds of endoplasmic reticuli. The part of the ER membrane, farther out from the nucleus,

Figure 4.8
Processing and Routing: The Golgi Complex
Transport vesicles from the rough endoplasmic reticulum (RER) move to the Golgi complex, where they unload their protein contents by fusing with the Golgi membrane. The protein is then passed by other vesicles through the layers of disk-shaped Golgi cisternae, where editing of protein side chains may occur. The proteins are then sorted in vesicles and shipped to sites inside and outside the cell. The vesicles in the micrograph are the pink and purple spheres.

from RER

cisternae

cisternal space

vesicle

Golgi complex

to plasma membrane

to cytosol

for export out of cell

1. Side chains are edited (sugars may be trimmed, phosphate groups added).

2. Vesicle formed for protein transport.

that has no ribosomes is called the **smooth endoplasmic reticulum**, or smooth ER. It's "smooth" because it is not peppered with ribosomes, and this very quality means it is not a site of protein synthesis. In general, the smooth ER is the site of the synthesis of various kinds of lipids, and a site at which potentially harmful substances are detoxified. The tasks the smooth ER undertakes, however, will vary in accordance with cell type. The lipids we normally think of as "fats" (triglycerides) are synthesized and stored in the smooth ER of liver and fat cells, while the "steroid" lipids reviewed last chapter—testosterone and estrogen—are synthesized in the smooth ER of the ovaries and testes. The detoxification of potentially harmful substances, such as alcohol, takes place largely in the smooth ER of liver cells.

Tiny Acid Vats: Lysosomes and Cellular Recycling

In any factory, there must be some means by which old materials can be gotten rid of or recycled. Also, a factory has need of new materials coming in from the outside, and these have to undergo some processing before being used. A single organelle in the animal cell aids in these things, and it is called a **lysosome**. Several hundred of these membrane-bound organelles may exist in any given cell; you could think of them as sealed-off acid vats that take in large molecules, break them down, and then return the resulting smaller molecules to the cytosol. What they cannot return, they retain inside themselves or expel outside the cell. They carry out this work not only on molecules entering the cell from the outside (say, vesicles filled with food) but on materials that exist inside the cell—on worn-out organelle parts, for example (**see Figure 4.9**).

A given lysosome may be filled with scores of different enzymes that can break larger molecules into their component parts—an enzymatic array that allows each lysosome to break down whatever comes its way. The beauty of these digestive enzymes is that they do not become active until they are inside the lysosome. (They require an acidic pH, which the interior of the lysosome provides.) If things did not work this way, such enzymes would start to break down *valuable* cellular components from the moment of their synthesis.

A lysosome comes into existence like any other vesicle, budding off from the Golgi; it gets ahold of its macromolecule prey via the endomembrane system. A lysosome will fuse with the membrane surrounding a worn-out organelle part; proceeding to engulf it, it then goes to work breaking the organelle down. The small molecules that result then pass freely out of the lysosome and into the cytosol for reuse elsewhere. Thus there is recycling at the cellular level.

Cells carry out this kind of self-renewal at an amazing rate. Christian de Duve, who with his colleagues discovered lysosomes in the 1950s, has noted the effect of this activity on human brain cells. In an elderly person, he notes, such cells

Figure 4.9
Lysosomes are membrane-bound organelles that contain potent enzymes capable of digesting large molecules. When a lysosome fuses with a worn-out organelle, its enzymes break the organelle down so that the small resulting molecules can be returned to the cytosol and used elsewhere—a form of cellular recycling. Lysosomes can expel materials from the cell that they cannot digest. Lysosomes also digest small particles, such as food, that come from outside the cell.

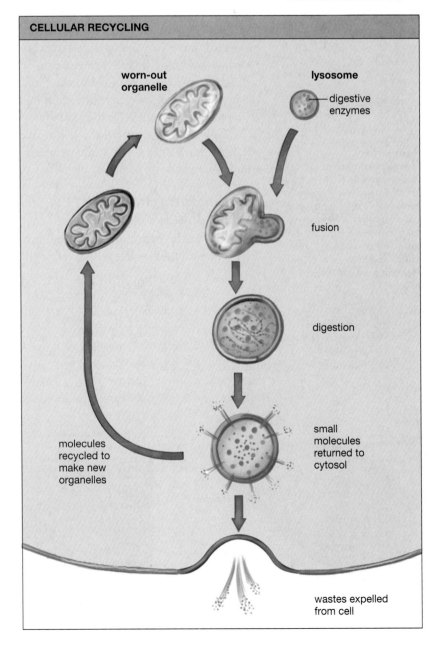

CELLULAR RECYCLING

worn-out organelle

lysosome

digestive enzymes

fusion

digestion

small molecules returned to cytosol

molecules recycled to make new organelles

wastes expelled from cell

have been there for decades. Yet most of their mitochondria, ribosomes, membranes, and other organelles are less than a month old. Over the years, the cells have destroyed and remade most of their constituent molecules from hundreds to thousands of times, some even more than 100,000 times.

So why do we grow old? Alas, we don't have any definitive answers to that question yet. One clue, however, is the fact that there are certain things even lysosomes cannot digest; and over time these substances begin to accumulate. In multicellular organisms, the lysosome's response to this is to hold on to this material, rather than releasing it outside the cell, since dumping it could harm extracellular structures. The result of this response, however, is a kind of permanent cellular bloat or constipation. Comic as this may seem, there is a serious side to it. The lysosomes of some people are deficient in certain enzymes; one manifestation of this can be Tay-Sachs disease, in which lysosomes are unable to digest great amounts of material and thus swell enormously, killing the cells that house them. The result is mental retardation and death at an early age.

Extracting Energy from Food: Mitochondria

Just as there is no such thing as a free lunch, there is no such thing as free mRNA transcrip-

tion, ribosomal action, or protein export. There is a price to be paid for all these things, and it is called energy expenditure. The fuel for this energy is contained in the food that cells ingest. But the energy in this food has to be converted into a molecular *form* that the cell can easily utilize. (Just as the energy in, say, coal needs to be converted by a power plant into a form that home appliances can easily use—electricity.) In eukaryotic cells, the place to look for most of this conversion is inside a group of organelles called **mitochondria**.

Figure 4.10 shows what a mitochondrion has in the way of structure: A continuous outer membrane enclosing an inner membrane that has a series of convolutions in it. The effect of these infoldings is to give mitochondria a larger internal surface area for carrying out their energy transformation activities.

Few details about mitochondria are included here, because much of Chapter 7 is devoted to them. Suffice it to say that to carry out their work, mitochondria need not only food, but *oxygen*. (Ever wonder why you need to breathe?) The *products* of mitochondrial activity, meanwhile, are carbon dioxide, water, and the energy-currency molecule adenosine triphosphate, or ATP, reviewed briefly last chapter.

Another thing to note about mitochondria is that they are probably the descendants of resident aliens. It is likely that they are the end products of bacterial cells that invaded other ancestral bacteria a billion or so years ago, only to end up living within them (see "The Stranger Within," on page 81).

4.6 The Cytoskeleton: Internal Scaffolding

When you see a set of eukaryotic cells in action, the word that comes to mind is hyperactive. Cells are not passive entities. Jostling, narrowing, expanding, moving about, capturing objects and bringing them in, expelling other objects: This is life as a bubbling cauldron of activity. It was once thought that cells did all these things with pretty much the equipment you've looked at so far, meaning that the Golgi complex, the ribosomes, the mitochondria, and other organelles were thought to be floating in an undifferentiated soup that was the cytosol. But advanced microscopy showed that there

Figure 4.10
Energy Transformers: Mitochondria
Just as a power plant converts the energy contained in coal into useful electrical energy, so do mitochondria convert the energy contained in food into a useful molecular form of energy for the cell—ATP.

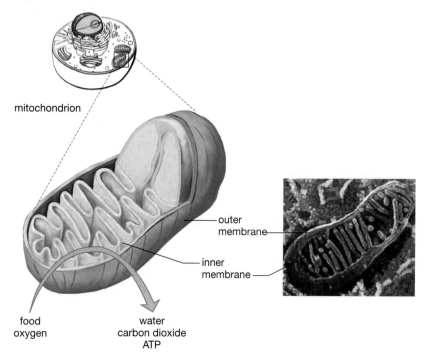

mitochondrion

outer membrane

inner membrane

food
oxygen

water
carbon dioxide
ATP

The Stranger Within:
Lynn Margulis and Endosymbiosis

If pressed to think about it, most people would probably agree that there's something slightly creepy about being made up of cells. Jostling and slithering as they are, dividing, working and dying within us without so much as asking permission, they give us the feeling that the unitary *self* that we so cherish actually amounts to an unruly collection of creatures within.

How sobering it is, then, to realize that we are composite beings in more ways than one. Each of our *cells* has within it the vestiges of yet other living things: the descendants of bacteria that long ago invaded our ancestors' cells, only to take up residence there. In plant and animal cells, the mitochondria that serve as cellular "powerhouses" are descendants of bacteria. Plant and algae cells have another set of bacterial descendants in them as well: the plastids, which include the chloroplasts that carry out photosynthesis.

The idea that these structures are descended from free-standing bacteria is called the endosymbiotic theory—*endo* for "within" the cell, and *symbiotic* for "symbiosis," meaning a situation in which two organisms not of the same species live in close association. Endosymbiosis is not a new idea; it was proposed as far back as the 19th century. It has, however, made quite a journey in recent years. Regarded as a crackpot notion as late as the 1960s, it came to be given some grudging credibility over time. By the mid-1980s it had made the transition from interesting hypothesis to established truth, in the sense that it was difficult to find a credible scientist who doubted it.

This is a kind of heroic journey, of course; and fittingly, one person is primarily responsible for it. She is Lynn Margulis, a professor of botany at the University of Massachusetts, Amherst. Resurrecting endosymbiosis in her 1965 Ph.D. thesis, Margulis went on to be its champion, at first taking it through derision and skepticism, and then seeing it through to vindication.

What makes us think that endosymbiosis really happened? First of all, mitochondria and plastids, which, remember, are *organelles* within eukaryotic cells, have many of the characteristics of free-standing *cells*, specifically free-standing bacterial cells. They have their own ribosomes and their own DNA, both of a bacterial type, and they reproduce through division, like bacterial cells, partly under their own genetic control. All of this makes mitochondria look suspiciously like once-independent cells that now find themselves living inside other cells.

Beyond this, we have the work done by a professor named Kwang Jeon, beginning in the 1970s at the University of Tennessee, Knoxville. Amoeba are a type of single-celled eukaryote; as it happened, a group of them Jeon had in his lab got infected by a species of rod-shaped bacteria. Most of the amoeba that were invaded in this way died; some, however, survived the attack, and the offspring of these hardy souls were nurtured by Jeon. Five years later, these amoeba were dividing at a normal rate *with* the bacteria living inside them (at a count of 40,000 bacteria per amoeba). Eventually, Jeon demonstrated that this strain of amoeba *required* the bacteria inside them in order to live. What began as an invasion ended as symbiosis, both parties functioning together, presumably to their greater good.

Margulis believes that, for mitochondria, the transition to endosymbiosis was completed about 1.4 billion years ago; prior to this, they existed as an aggressive form of oxygen-using bacteria. Seizing an opportunity, they invaded a form of bacteria that were fairly intolerant of oxygen. (In more recent versions of this theory, the bacteria invaded eukaryotic cells.) The arrangement the two species eventually came to was simple: The host provided food to the invader; and the oxygen-using invader, now domesticated, allowed the host to survive in an oxygenated world.

is, in fact, a tangled forest within the cytoplasm: Protein strands that give cells their shape, that anchor the structures you've been reading about, that act as monorails for particles moving within, and that allow the cells themselves to move. Taken as a whole, these protein strands are called the **cytoskeleton**.

Some of the cytoskeleton's parts are permanent and relatively static, but many are moving and some are assembled or disassembled very rapidly. The cytoskeleton usually is divided into three component parts. Ordered by size, going from smallest to largest in diameter, they are **microfilaments**, **intermediate filaments**, and **microtubules**

Microfilaments   **Intermediate filaments**   **Microtubules**

7 nm — Main function: changes in cell shape

10 nm — Main function: maintenance of cell shape

25 nm — Main functions: maintenance of cell shape, movement of organelles, cell mobility (cilia and flagella)

Figure 4.11
Structure and Movement:
The Cytoskeleton
Three types of fibers form the inner scaffolding or cytoskeleton of the cell: microfilaments, intermediate filaments, and microtubules.

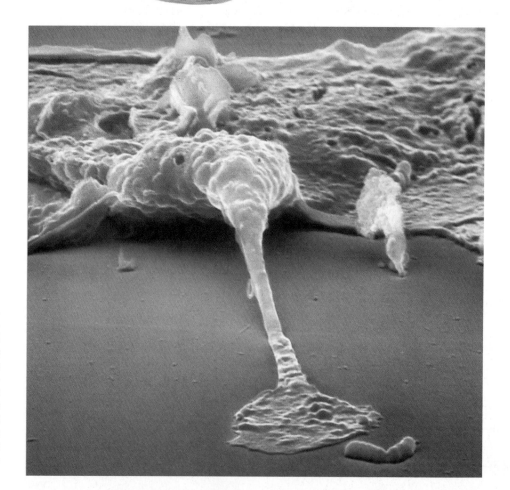

Figure 4.12
Microfilaments in Action
Certain cells can move or capture prey by sending out extensions of themselves called pseudopodia ("false feet"). It is the rapid construction of actin microfilaments—in the direction of the extension—that makes this possible. Here a type of blood-borne guard cell called a macrophage is about to use a pseudopodium it is constructing to capture a green bacterium.

(**see Figure 4.11**). Let's take a look at some of the characteristics of each.

Microfilaments

The most slender of our cytoskeletal fibers, microfilaments are made of the protein *actin*, which, by itself, exists as a support or "structural" filament in almost all eukaryotic cells. Actin microfilaments can also help cells move or capture prey, essentially by growing very rapidly at one end—in the direction of the movement or extension—while decomposing rapidly at the other end. Actin works with another protein, myosin, to bring about muscle contraction in muscle cells. You can see a vivid example of microfilament-aided extension in **Figure 4.12**.

Intermediate Filaments

These in-between-sized proteins are the most permanent of the cytoskeletal elements, perhaps coming closest to our everyday notion of what a skeleton is like. The positions of the nucleus and organelles within the cell are stabilized by intermediate filaments.

Microtubules

These largest of the cytoskeletal elements also play a structural role in the cell; in fact, theirs seems to be the preeminent structural role in the sense of determining the shape of the cell. But they take on several other tasks in addition. They serve, for example, as the monorails discussed earlier. Recall that protein-laden vesicles move from one organelle to another in the cell. These spheres are moving along the "rails" of microtubules, while sitting atop the "engine" of a so-called motor molecule (**Figure 4.13**).

Cell Extensions Made of Microtubules: Cilia and Flagella Microtubules also form the underlying structure for two kinds of cell extensions, **cilia** and **flagella** (see Figure 4.13). Cilia exist as a profusion of hair-like growths extending from cells. Their function is simple: Move back and forth very rapidly, perhaps 10 to 40 times per second. The purpose of this movement can be either to propel a cell or to move material *around* a cell. Cilia are extremely common among single-celled organisms and in some of the cells of simple animals (sponges, jellyfish). You saw an example of cilia in humans earlier in connection with our sense of smell; our lungs also are lined with cilia, whose job it is to sweep the lungs clean of whatever foreign matter has been inhaled. In this work, like most cilia, they all beat at once in the same direction, acting like rowers in a crew.

a transport monorails

transport vesicle

motor molecule

microtubule

b cilia

c flagellum

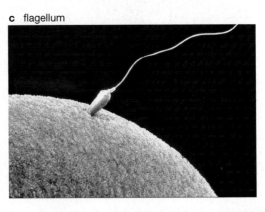

**Figure 4.13
Several Functions for Microtubules**

a They are the "rails" on which vesicles move through the cell, carried along by "motor molecules."

b They exist outside the cell in the form of cilia, which are profuse collections of hair-like projections that beat rapidly, forming currents that can propel a cell or move material around it.

c Microtubules are also found outside the cells in the form of flagella. The flagellum on this sperm cell is enabling it to seek entry into an egg.

Cilia grow from eukaryotic cells in great profusion, but it is a different story with flagella. Only a few generally will sprout from a given cell—indeed, there often is just a single tail-like flagellum. The function of flagella is cell movement. Only one kind of animal cell is flagellated, and it scarcely needs an introduction: A sperm is a single cell that whips its flagellum mightily in order to get to an unfertilized egg.

In Summary: Structures in the Animal Cell

In your tour of the cell so far, you've pictured a cell as a factory, one that synthesizes proteins in a "production line" that starts with DNA in the nucleus and then goes to the ribosomes (via mRNA), to the rough ER, to the Golgi complex, and finally to the protein's destination (the plasma membrane, export, etc.). In addition, you've seen that cells have other structures such as lysosomes for digestion and recycling, mitochondria for energy transformation, the smooth endoplasmic reticulum for detoxification and lipid synthesis, and the cytoskeleton for structure and movement. If you look at **Figure 4.14**, you can see, in metaphorical form, a "map" of these component parts within the cell. **Table 4.1** (on page 87 in the discussion of plant cells) lists cellular elements found in plants and animals, as well as some elements found only in plant cells.

4.7 The Plant Cell

As noted earlier, plant and animal cells have more similarities than they do differences. Among these similarities is that plant cells *do* most of the things that animal cells do (produce proteins, transform energy, etc.) A quick look at **Figure 4.15** will confirm for you how structurally similar plant and animal cells are as well. As you see, a plant cell has a nucleus (with a nucleolus), the smooth and rough ERs, a cytoskeleton—most of the things you've just gone over in animal cells. Indeed, there is only one structure present in the animal cells you've looked at that plant cells don't have: the lysosome. What jumps out at you when you look at plant cells is not what they lack compared to animal cells, but what they *have* that animal cells do not. As you can see in **Figure 4.16**, these additions are

- A thick cell wall
- A large structure called a central vacuole
- Structures called plastids, one important variety of which is the chloroplast

Figure 4.14
The Cell as a Factory
This comparison may help you to remember some roles of the different parts of the cell.

CONTROL CENTER (nucleus)

ASSEMBLY LINE (endoplasmic reticulum)

STRUCTURE (cytoskeleton)

WORKBENCHES (ribosomes)

DISTRIBUTION CENTER (Golgi complex)

POWERHOUSES (mitochondria)

CLEANING CREW (lysosomes)

SECURITY GATE (cell membrane)

Figure 4.15
Common Structures in Animal and Plant Cells

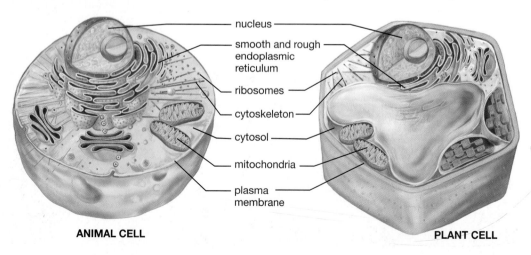

nucleus

smooth and rough endoplasmic reticulum

ribosomes

cytoskeleton

cytosol

mitochondria

plasma membrane

ANIMAL CELL

PLANT CELL

The Cell Wall

Plant cells have an outside layer that makes their plasma membrane, just inside it, look rather thin and frail by comparison. This is because it *is* thin and frail by comparison; the plasma membrane of a plant cell may be 0.01 µm thick, while the combined units of a cell wall may stretch to 7 µm or more. Cell walls are nearly always present in plant cells, whereas animal cells don't have them. You should note, however, that many organisms that are neither plant nor animal—bacteria, protists, and fungi—also have cell walls.

What do cell walls do? They provide plant cells with structural strength, put a limit on their absorption of water (as you'll see next chapter), and generally protect plants from harmful outside influences. Cell walls in plants can come in several forms, but all such forms will be composed to a considerable extent of a molecule you were introduced to last chapter: *cellulose*, a polysaccharide that is embedded within cell walls in the way reinforcing bars run through concrete. In some cell walls, cellulose will be joined by a compound called lignin, which imparts

Figure 4.16
The Plant Cell
The cell wall, central vacuole, and chloroplasts do not exist in animal cells, but the other components are common to both plant and animal cells.

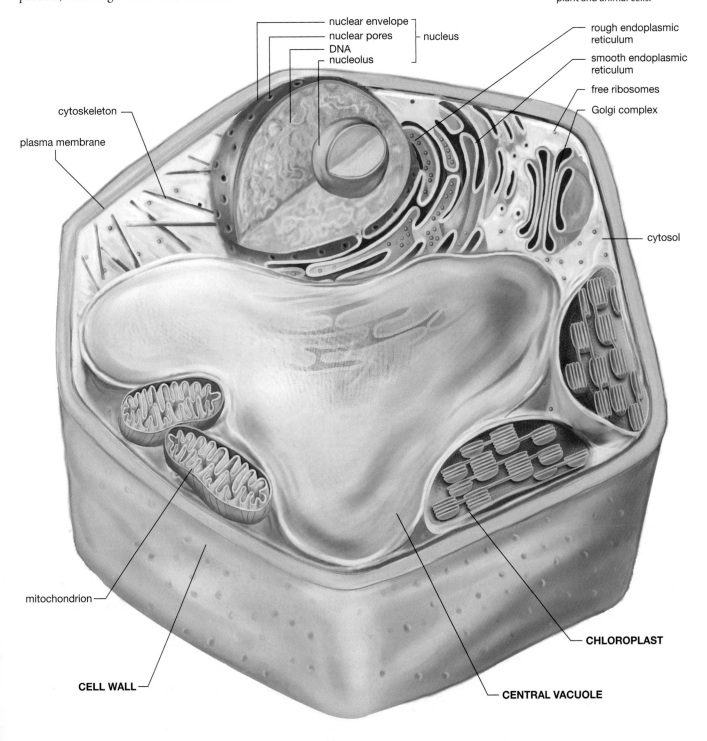

considerable structural strength. You can see a vivid demonstration of this in the material we know as wood, which is largely made of cell walls (see Figure 4.17a).

Cell walls can serve different functions over the life of an organism. Generally, they are the site of a good deal of metabolic activity, and thus should not be considered mere passive barriers. On the other hand, the outer portion of the barrier we call tree bark amounts mostly the cell walls of dead cells (see Figure 4.17b).

The Central Vacuole

Looking at the diagram of a plant cell, there is one structure, called a **central vacuole**, that is so prominent it appears to be a kind of organelle continent surrounded by a mere moat of cytosol. In a mature plant cell, one or two central vacuoles may comprise 90 percent of cell volume.

Although animal cells can have vacuoles, the imposing central vacuole in plants is different. For a start, it is composed mostly of water, which demonstrates just how watery plant cells are. A typical animal cell may be 70 to 85 percent water, but for plant cells the water proportion is likely to be 90 to 98 percent.

The watery milieu of the central vacuole contains hundreds of other substances. Many of these are nutrients, others are waste products; in addition, there are hydrogen ions, pumped in so that the cell's cytoplasm may be kept at a near-neutral pH. Given these materials, it may be obvious that the central vacuole does a lot of things: It stores nutrients, it is involved in metabolism, and it retains waste products or degrades them with digestive enzymes, like the lysosomes in animal cells. There is even an aesthetic side to the central vacuole: Many red and blue flowers owe their colors to the pigments that exist within it.

Table 4.1
Structures in Plant and Animal Cells

Name	Location	Function
Cytoskeleton	Cytoplasm	Maintains cell shape, facilitates cell movement and movement of materials within cell
Cytosol	Cytoplasm	Protein-rich fluid in which organelles and cytoskeleton are immersed
Golgi complex	Cytoplasm	Processing, sorting of proteins
Lysosomes (in animal cells only)	Cytoplasm	Digestion of imported materials and cell's own used materials
Mitochondria	Cytoplasm	Transform energy from food
Nucleolus	Nucleus	Synthesis of ribosomal RNA
Nucleus	Inside nuclear envelope	Site of most of cell's DNA
Ribosomes	Rough ER Freestanding in cytoplasm	Sites of protein synthesis
Rough endoplasmic reticulum	Cytoplasm	Protein processing
Smooth endoplasmic reticulum	Cytoplasm	Lipid synthesis, storage; detoxification of harmful substances
Vesicles	Cytoplasm	Transport of proteins and other cellular materials
Cell walls (in plant cells only)	Outside plasma membrane	Limit water uptake; maintain cell membrane shape, protect from outside influences
Central vacuole (in plant cells only)	Cytoplasm	Cell metabolism, pH balance, digestion, water maintenance
Plastids (in plant cells only)	Cytoplasm	Nutrient storage, pigmentation, photosynthesis (chloroplasts)

Plastids

Plastids are a diverse group of organelles that are found only in plants and algae. Some gather and store nutrients for plant cells; others are pigment-containing organelles that give us, for example, the red color of the tomato skin. The best known of the plastids, however, is the **chloroplasts**.

Photosynthesis: The Chloroplasts

The pigment chlorophyll, which runs through chloroplasts, is the central reason the living world is largely a green world. Color is the least of the reasons we should be grateful to chloroplast-containing plants, however, because we're indebted to them for almost all the food we eat and much of the oxygen we breathe. Chloroplasts are the sites of *photosynthesis*, in which sunlight is captured and used to produce food from nothing more than carbon dioxide, water, and a few minerals (**see Figure 4.18**). This may not seem like such a big deal until you try to name a food you eat that is not a plant or does not itself eat plants. A byproduct of photosynthesis is oxygen, whose significance to us can scarcely be overstated. From this, it may be obvious that plants could get along just fine without people, but the reverse is not true. A plantless world would soon be a personless world.

4.8 Cell Communication: Why Cells Need Not Be Islands

Most of what you've seen so far has made the cell seem like an isolated entity, but this is not the case at all. Single-celled organisms

a

b

Figure 4.17
Great Strength from Small Things

a Redwoods such as this one can reach enormous heights because of the strength of their wood, which is largely made of cell walls.

b Cell walls play important roles in living cells, but they also are valuable as the strong, remaining components of dead cells; here they help make up this redwood bark. (×240)

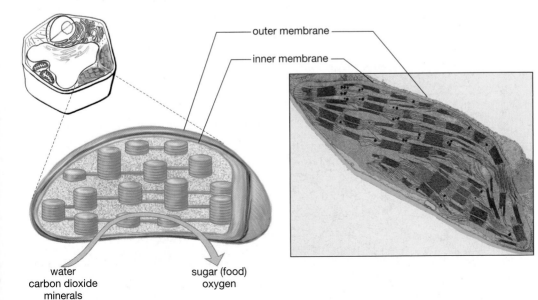

outer membrane

inner membrane

water
carbon dioxide
minerals

sugar (food)
oxygen

Figure 4.18
Food Source for the World
Chloroplasts, the tiny organelles that exist in plant cells, are sites of photosynthesis—the process that provides food for most of the living world. Using the starting materials of water, carbon dioxide, and a few minerals, plants use energy from sunlight to produce their own food. A double membrane lines the chloroplasts. (Micrograph: ×13,000)

can exist as separate entities, along with certain plant or animal cells (red blood cells, for example); but most plant and animal cells are linked together in organized assemblages referred to as **tissues**. Not surprisingly, these assemblages of cells, be they plant or animal, have the ability to communicate with one another.

Communication among Plant Cells

Having noted the thickness of something like the cell walls in plants, you might wonder how one plant cell could interact with another. Communication between plant cells takes place quite readily, however, through a series of tiny channels in the cell wall called **plasmodesmata** (singular, plasmodesma). The structure of these channels is such that the cytoplasm of one plant cell is continuous with that of another—so much so that the cytoplasm of an entire plant can be properly looked at as one continuous whole (**see Figure 4.19a**). The structure of plasmodesmata is more complex than that of a simple opening, but the basic idea is of a channel-like linkage between two plant cells.

Communication among Animal Cells

There are no plasmodesmata in animal cells, but there are three other kinds of **cell junctions**, one of which serves to facilitate cell communication. It is called a **gap junction**, and it consists of clusters of protein structures that shoot through the plasma membrane of a cell from one side to the other. When these structures line up in adjacent cells, the result is a channel for passage of small molecules and electrical signals (**see Figure 4.19b**).

On to the Periphery

Having looked at what is inside the cell, you've arrived at the cell's periphery, the plasma membrane, which is where you'll be staying for awhile. It may at first seem strange to devote a whole chapter to an outer boundary. How much attention would you pay, after all, to a factory's wall as opposed to its contents? The answer: A lot, if that wall could move, continually renew itself, and let some things in while keeping others out. Such is the case with the plasma membrane, a slender lining that manages to make one of the most fundamental distinctions on Earth: Inside, life goes on; outside it does not.

Figure 4.19
Cell Communication

a In plants, a series of tiny pores between plant cells, the plasmodesmata, allow for the movement of materials among cells. Thanks to the plasmodesmata channels, the cytoplasm of one cell is continuous with the cytoplasm of the next; the plant as a whole can be thought of as having a single complement of continuous cytoplasm.

b In animals, protein assemblies come into alignment with one another, forming communication channels between cells. A cluster of many such assemblies—perhaps several hundred—is called a gap junction.

a plant tissues

plasma membrane
cell walls
cytoplasm
plasmodesmata

b animal tissues

plasma membranes
cytoplasm
gap junction

Chapter Review

Summary

4.1 Cells Are the Working Units of Life

- With the possible exception of viruses, every form of life on Earth either is a cell or is composed of cells. All cells come into existence through the activity of other cells.

4.2 All Cells Are Either Prokaryotic or Eukaryotic

- All cells can be classified as prokaryotic or eukaryotic. All plants, animals, fungi, and protists either are single eukaryotic cells or are composed of eukaryotic cells. All prokaryotic cells are bacteria or archaea. Eukaryotic cells have most of their DNA contained in a membrane-lined nucleus, whereas prokaryotic cells do not have a nucleus. Eukaryotic cells also have more specialized structures, called organelles, than do prokaryotic cells. All prokaryotes are single-celled, whereas many eukaryotes are multicelled.

4.3 The Eukaryotic Cell

- There are five principal components to the eukaryotic animal cell: the nucleus, other organelles, the cytosol, the cytoskeleton, and the plasma membrane. Organelles are "tiny organs" within the cell that carry out specialized functions, such as energy transfer and material recycling. The cytosol is the fluid in which these organelles are immersed. (The cytoplasm is the region of the cell inside the plasma membrane but outside the nucleus.) The cytoskeleton is composed of several groups of proteins that give the cell support and facilitate transportation of cellular elements. The plasma membrane is the chemically active outer boundary of the animal cell. Plant cells have a plasma membrane that has, outside of it, a cell wall.

4.4 A Tour of the Animal Cell: Along the Protein Production Path

- Information for the construction of proteins is contained in the DNA located in the cell nucleus. This information is copied onto an informational "tape" of messenger RNA that departs the cell nucleus through nuclear pores and goes to the sites of protein synthesis, structures called ribosomes, many of which are embedded in a series of membrane sacs called the rough endoplasmic reticulum. Polypeptide chains that will become proteins are dropped from ribosomes into the internal spaces of the rough endoplasmic reticulum, where these chains undergo editing. Some ribosomes are not embedded in rough endoplasmic reticulum.

- Materials move from one structure to another in the cell via the endomembrane system, in which a piece of membrane, with proteins or other materials inside, can bud off from one organelle, move through the cell, and then fuse with another membrane-lined structure. Membrane-lined structures that carry cellular materials are called transport vesicles.

- Transport vesicles move proteins that are being produced from the rough endoplasmic reticulum to the Golgi complex, where the proteins are processed further and marked for shipment to appropriate cellular locations.

4.5 Outside the Protein Production Path: Other Cell Structures

- The smooth endoplasmic reticulum is a network of membranes that functions to synthesize lipids and to detoxify potentially harmful substances. Lysosomes are organelles that break down material coming into the cell (such as food) or worn-out cellular structures, returning the components of these structures to the cytoplasm for further use. Mitochondria function to extract energy from food and to transform this energy into a chemical form the cell can use.

4.6 The Cytoskeleton: Internal Scaffolding

- Cells have within them a web of protein strands, called a cytoskeleton, that provide the cell with structure, facilitate the movement of materials inside the cell, facilitate cell movement, and help cells capture prey.

- There are three principal types of cytoskeleton elements. Ordered by size, going from smallest to largest in diameter, they

Key Terms

cell junction 88

central vacuole 86

chloroplast 87

cilium 83

cisterna 77

cisternal space 77

cytoplasm 70

cytoskeleton 69

cytosol 69

endomembrane
  system 77

eukaryotic cell 68

exocytosis 78

flagellum 83

free ribosome 76

gap junction 88

Golgi complex 77

intermediate
  filament 81

lysosome 79

messenger RNA
  (mRNA) 74

microfilament 81

micrograph 72

micrometer 72

microtubule 81

mitochondrion 80

nanometer 72

nuclear envelope 74

nuclear pore 74

nucleolus 75

nucleus 69

organelle 68

plasma membrane 69

plasmodesma 88

prokaryotic cell 68

ribosome 76

rough endoplasmic
  reticulum 75

secretory protein 76

smooth endoplasmic
  reticulum 79

tissue 88

transport vesicle 77

are: microfilaments, intermediate filaments and microtubules. Microfilaments are made of the protein actin, and function in cell movement and the capture of prey. Actin works with the protein myosin to bring about muscle contraction. Intermediate filaments provide support and structure to the cell. Microtubules play a structural role in cells and facilitate the movement of materials inside the cell by serving as transport "rails."

- Cilia and flagella are extensions of cells composed of microtubules. Cilia extend from cells in great numbers, serving to move the cell or to move material around the cell. By contrast, one—or at most a few—flagella extend from cells that have them. The function of flagella is cell movement.

4.7 The Plant Cell

- Plant cells have almost all the structures found in animal cells. Plant cells also have a cell wall, a large central vacuole, and organelles called plastids, one variety of which is the chloroplasts that are the sites of photosynthesis. The cell wall gives the plant structural strength and helps regulate the intake and retention of water.

4.8 Cell Communication: Why Cells Need Not Be Islands

- Cells are able to communicate with each other through special structures: channels called plasmodesmata in plants and channels called gap junctions in animal cells.

Understanding the Basics

Multiple-Choice Questions

1. Jerome has strep throat, a bacterial infection. The cause of the infection is
   a. the growth of a virus
   b. the presence of archaea
   c. eukaryotic cells dividing in his throat
   d. organelles that take control of his organs—in this case, his throat
   e. prokaryotic cells

2. Where would you expect to find a cytoskeleton?
   a. primarily inside the nucleus
   b. as the internal structure of a mitochondrion
   c. holding the organelles together in bacterial cells
   d. throughout the cytosol
   e. as an outer coat on an insect

3. Suppose Dr. Hyde found a cell that had many mitochondria, a nucleus, a cell wall made of cellulose, and an endoplasmic reticulum, as well as many other parts. He might assume that he has found
   a. a plant cell
   b. an animal cell
   c. a bacterial cell
   d. one of these three types, but he will not know which type without further investigation
   e. either a plant cell or an animal cell, but not a bacterial cell

4. Cells in the pancreas manufacture large amounts of protein. Which of these would you expect to find a large amount or number of in pancreatic cells?
   a. DNA
   b. rough endoplasmic reticulum
   c. smooth endoplasmic reticulum
   d. chloroplasts
   e. plasma membrane

5. Suppose that a mutation caused a child to be unable to form intermediate filaments. Which of these problems would the child be most likely to develop?
   a. the inability to get energy out of food
   b. an inability to manufacture proteins
   c. a tendency for the nucleus and organelles in the cell to drift around inside the cell
   d. an inability of the skeletal muscles to contract
   e. an inability to move proteins from one part of the cell to another

6. Heart muscle cells have a number of gap junctions connecting them to the adjoining cells. From this you can conclude that heart muscle cells
   a. exchange nuclei very frequently
   b. have plasmodesmata
   c. move vacuoles from cell to cell
   d. communicate easily
   e. lack the ability to divide

7. Which is the correct ranking of these "small" things, from smallest to largest? (Use typical sizes.)
   a. animal cells < atoms < bacteria < proteins < amino acids
   b. atoms < amino acids < proteins < bacteria < animal cells
   c. atoms < proteins < amino acids < animal cells < bacteria
   d. bacteria < atoms < amino acids < proteins < animal cells
   e. bacteria < animal cells < atoms < amino acids < proteins

8. Cells need large amounts of ribosomal RNA to make proteins. The ribosomal RNA is made in specialized structures known as _____, which are found in _____.
   a. chloroplasts . . . cytosol
   b. ribosomia . . . cytosol
   c. endoplasmic reticula . . . nucleus
   d. nuclear envelopes . . . nucleus
   e. nucleoli . . . nucleus

Brief Review

1. Suppose your entire body were just one gigantic cell with one central nucleus and lots of organelles outside to perform the various functions. What problems can you envision with this system?

2. What are some of the differences between eukaryotic cells and prokaryotic cells?

3. Insulin, the hormone that controls sugar levels in the body, is a protein that is secreted from special cells in the pancreas. Trace the path from a piece of DNA in the nucleus that carries the code for insulin to the release of the hormone in the bloodstream.

4. What are cilia and flagella? What are some of the roles they play in the human body?

5. Consider what plants would be like if they had no plasmodesmata. Explain why this would be a problem.

6. What is a gap junction and what is its function?

Applying Your Knowledge

1. What would life on Earth be like if there were no plants?

2. In what ways is the cytoskeleton like bones?

Group Activity

Using the Web, the library, and textbooks, find as many roles as possible that prokaryotic cells play in the environment. Work in teams of three or four to prepare a website that illustrates and describes these roles. For example, one team might investigate the role of bacteria in breaking down dead organisms in the soil. That team would then make a Web page showing this role. One team should prepare the home page and link the other pages.

5

Life's Border
The Plasma Membrane

Moving out of the cell.
(page 105)

Journey through the plasma membrane.
(page 103)

An essential molecule.
(page 94)

As a professional football

quarterback, Boomer Esiason led the Cincinnati Bengals to the Super Bowl in 1989. His toughest challenge, however, has come not on the football field, but at home, where he and his wife have struggled to care for their son Gunnar, who was born in 1991 with the disease cystic fibrosis.

Gunnar's treatment has included the heartbreaking ritual of "percussive" therapy, which is used on young victims of the disease. They lie at an angle, head toward the ground, while their parents pound gently on their back, chest, and sides in an effort to dislodge the mucus that accumulates in their lungs. The lungs of healthy children are lined by mucus too, but in a layer that is thin and wet enough to be regularly swept away by the hairlike cilia that extend into lung passages like so many tiny brooms. In cystic fibrosis patients, the mucous layer is thicker and drier and becomes a site for repeated bacterial infections. In time, these infections can result in the lung passages being destroyed. The average life expectancy for a cystic fibrosis victim is 29 years.

5.1 The Importance of Activity at the Cell's Periphery

The difference between a healthy child and Gunnar Esiason comes down to this: Because of a faulty gene, the substance chloride cannot be transported in sufficient quantity from the inside of Gunnar's cells to the outside of them. Healthy individuals have a protein that acts as a channel for chloride—a kind of passageway that spans the cell's outer membrane. In contrast, the cells of cystic fibrosis patients have either defective chloride channels or none at all. The chemical imbalance caused by this lack of chloride brings on the mucus buildup.

As noted last chapter, life goes on only inside cells. The fact that the cell's outer, or "plasma," membrane is out on life's edge may prompt the thought that it's *merely* at the edge, as if the real action is taking place deep within the cell. But consider, as in cystic fibrosis, the effect of having a *defective* plasma membrane. The focus in this chapter is to learn more about the nature of this important border.

Portrait of the plasma membrane.
(page 101)

A cellular sandwich.
(page 94)

watery extracellular fluid

hydro*philic*

hydro*phobic*

hydro*philic*

watery cytosol

hydrophobic molecules pass through freely

hydrophilic molecules do not pass through freely

protein

Moving out of the cell.
(page 105)

5.2 Why Do We Need the Plasma Membrane?

The plasma membrane does perform one seemingly pedestrian task, but it is absolutely critical: *keeping the goods concentrated*. For life to get going on Earth, there had to be some way that biological molecules such as proteins and nucleotides could be collected in certain concentrations and the plasma membrane helped accomplish this. Not surprisingly, this task is related to a second function, which is keeping harmful materials *out*.

Cells are not isolated units, however. They need to have many substances moving in and out of them. This kind of controlled passage can be counted as the third of the major activities of the plasma membrane.

In a fourth function, the plasma membrane is critical in what is known as cell signaling. Almost all cells are constantly communicating with one another by means of chemical signals. Such communication has the ability to change the activity of the cells. (A signal might, for example, increase a cell's

Figure 5.1
Dual-Natured Lining
The essential building block of the cell's plasma membrane is the phospholipid molecule, which has both a hydrophilic "head" that bonds with water, and hydrophobic "tails" that do not. Two layers of phospholipids sandwich together to form the plasma membrane. The phospholipids' hydrophobic tails form the interior of the membrane, while their hydrophilic heads jut out toward the watery environments that exist both inside and outside of the cell. The bilayer forms a barrier to all but the smallest hydrophilic molecules, but hydrophobic molecules can pass through fairly freely.

polar head

nonpolar tails

a Phospholipid molecule

N +

P −

watery extracellular fluid

hydro*philic*

hydro*phobic*

hydro*philic*

watery cytosol

hydrophobic molecules pass through freely

hydrophilic molecules do not pass through freely

b Phospholipid bilayer

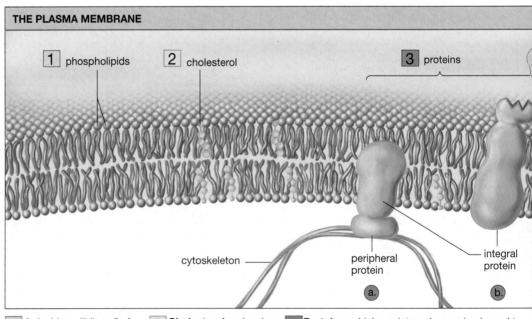

THE PLASMA MEMBRANE

1 phospholipids

2 cholesterol

3 proteins

cytoskeleton

peripheral protein

integral protein

a.

b.

1 A double or "bilayer" of **phospholipid molecules**, with their hydrophilic "heads" facing outward, toward the watery environment that lies both inside and outside the cell, and their hydrophobic "tails" pointing inward, toward each other.

2 **Cholesterol** molecules that act as a patching substance and that help the cell maintain an optimal level of fluidity.

3 **Proteins**, which are integral, meaning bound to the hydrophobic interior of the membrane, or peripheral, meaning not bound in this way. Membrane proteins serve four main functions:

a. **Structural support**, often when attached to parts of the cell's scaffolding, or "cytoskeleton."

b. **Recognition**. Binding sites on some proteins can serve to identify the cell to other cells, such as those of the immune system.

Figure 5.2
The Plasma Membrane

production of a protein.) The initial points of contact for cell signaling often are molecules that lie like so many antennae on the surface of the plasma membrane.

5.3 Four Components of the Plasma Membrane

What manner of material is the plasma membrane? First, it is very much like the membranes described in Chapter 4. Much of what follows about the plasma membrane could be said of all the membranes that are part of the membrane system—the membranes of the rough endoplasmic reticulum or the Golgi apparatus for example. It's necessary to take a more detailed look at membrane structure here, though. To do that, you'll need to reacquaint yourself with a molecule you've met before.

First Component: The Phospholipid Bilayer

Recall the discussion in Chapter 3 of phospholipids—molecules that have two long fatty-acid chains and a phosphate-bearing group (**see Figure 5.1**). The phosphate group is hydrophilic, or water-seeking, and the fatty-acid chains are hydrophobic, or water-evading. The result is a molecule that, once placed in water, has its phosphate "head" pointing in a different direction from its fatty-acid "tails." It behaves, in fact, as a buoy does when bobbing in deep water. No matter how you push the hydrophobic tails, they end up waving free out of the water, while the hydrophilic head stays submerged.

The form the plasma membrane takes is of two *layers* of phospholipids sandwiched together, the tails of each layer pointing inward, toward each other, and the heads pointing outward. This makes sense because what *lies* outward in both directions is a watery environment—attractive to phospholipid heads, but not to the tails. (Remember that the interior, or cytoplasm, of the cell is mostly water. This is usually the case with the extracellular fluid as well.) This basic structure of the plasma membrane is known as the **phospholipid bilayer**, which you can see in the context of the other membrane components in **Figure 5.2**.

This phospholipid composition makes the plasma membrane a very fluid structure.

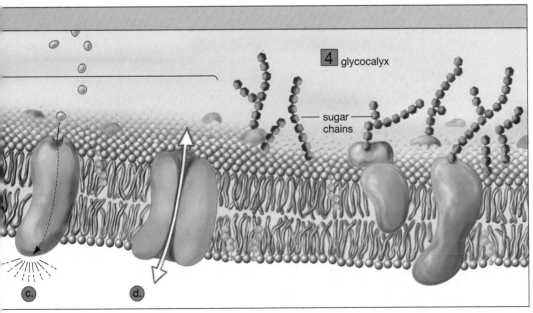

4 glycocalyx

— sugar chains —

4 The **glycocalyx**. Sugar chains that attach to communication or recognition proteins, serving as their binding sites. The glycocalyx can also lubricate cells and act as an adhesion layer for them.

c. **Communication**. Receptor proteins, protruding out from the plasma membrane, can be the point of contact for signals sent to the cell via traveling molecules, such as hormones.

d. **Transport**. Proteins can serve as channels through which materials can pass in and out of the cell.

The two layers of phospholipid tails mingle with each other in the membrane and are able to slide past one another, giving the membrane its loose nature. Recall, however, that these are two *hydrophobic* layers of tails. This gives the plasma membrane one of its most important characteristics: The only substances that can pass through it with any regularity are *other* hydrophobic substances or very small hydrophilic molecules, such as water. This stems from the rule of thumb about solvents in Chapter 3: like dissolves like. The bilayer is a hydrophobic substance—a *lipid* substance. Thus, various steroid hormones (which, remember, are lipids) gain fairly easy entry, as do fatty acids.

If the bilayer allows some things to pass through, however, it keeps many others out. Hydro*philic* substances—ions, polar molecules—do none of the dissolving that lipids do in the membrane and are thus not allowed in or out of the cell without help. Water, which is polar, gets a "pass" through the bilayer only because it is such a small molecule.

Second Component: Cholesterol

Molecules of the lipid material cholesterol nestle between phospholipid molecules throughout the plasma membrane, performing two functions. First, they act as a kind of patching substance on the bilayer, keeping some small molecules from getting through. Second, they help keep the membrane at an optimum level of fluidity.

Third Component: Proteins

Embedded within, or lying on, the membrane lipids is a third major group of membrane components: proteins, of which there are two major types, integral and peripheral. As usual, the names tell a good part of the story. **Integral proteins** may span the entire membrane, popping out on both sides of it, or they may merely extend partway into it. Either way, they are bound to its hydrophobic interior. **Peripheral proteins** lie on either side of the membrane, but are not bound to the hydrophobic interior. They are usually attached to integral proteins at the surface of the membrane.

Taken together, peripheral and integral proteins have a diverse range of *functions*. Here is a selected set of them.

Structural Support

Peripheral proteins that lie on the interior or "cytoplasmic" side of the membrane often are attached to elements of the cytoskeleton. Thus anchored, they help to stabilize various parts of the cell and play a part in giving animal cells their characteristic shape. (Why not plant cells? Because their shape is determined by the cell wall.)

Recognition

Much like military sentries, some cells need to know the answer to a critical question: Who goes there, friend or foe? The sentries are immune system cells, moving through the blood system and interacting with proteins that extend from the cell surface. The specific shape that **binding sites** on these proteins take can convey the message: Self here, pass on by. Conversely, a cell with a *foreign* set of binding sites on its proteins (generally an invader) elicits an immune system attack.

Communication

Cells communicate with one another in various ways. Signals can be sent from one cell to a neighboring cell, and communication over longer distances is possible through the chemical messengers known as hormones. The nervous system provides a means of fast communication throughout the body. In all these instances, signals are likely to be channeled through **receptor proteins** (often referred to simply as receptors). Many of these are integral proteins that protrude from the plasma membrane.

Each receptor has on it a binding site, shaped so that it can latch onto only a single communication molecule, or perhaps a group of closely related molecules. The usual outcome of such binding is a change in the cell's activity. For example, when a hormone binds with a receptor on the surface of a cell, the result can be a cascade of cellular reactions *inside* the cell that culminate in a protein being produced.

Transport

You have seen that most materials cannot simply pass through the cell's plasma membrane. Yet cells need many of these very materials. How do they get in? Different kinds of integral proteins take on a transport task, in ways that will be reviewed shortly.

Recall that life depends on the ability to make fine distinctions—to get more of this, but less of that; to respond to this, but not to that. The cell's ability to interact in very *particular* ways with its environment is greatly enhanced by proteins. Some of these say, in effect, "self here." Others say, "muscle cell here." Still others stand ready to facilitate the transport of a given ion—and only that ion—across the plasma membrane. Proteins have been likened before to a group of specialized tools; the proteins of the plasma membrane provide more cases in point.

Fourth Component: The Glycocalyx

If you look at Figure 5.2 and focus on the extracellular face of the plasma membrane, you can see, protruding from proteins and phospholipids alike, a number of small, branched extensions. These are carbohydrate or sugar chains that serve as the actual binding sites for many of the proteins mentioned earlier. Sugar chains also serve to lubricate cells, or to keep cells in place by acting as a sticky adhesion layer. Add to these some peripheral proteins, and you get an outer coat for the cell that goes by the name of **glycocalyx**, meaning "sugar coat."

The Fluid-Mosaic Membrane Model

Thus ends our review of the basic components of the plasma membrane. Taking a step back, what we have seen is a loose, lipid structure peppered with proteins and coated with sugars. A better image, however, might be of a *sea* of lipids that has proteins floating on it, because as it turns out, the plasma membrane is fluid enough that most of its elements are able to move sideways or "laterally" through it fairly freely. Indeed, membrane proteins frequently are compared to icebergs drifting through an ocean. (The proteins are numerous enough, however, that it's tempting to liken the cell surface to the Arctic Ocean at about the time the ice pack starts to break up.) When we overlay this quality of movement on the basic membrane structure just reviewed, the result is the modern view of the plasma membrane, the **fluid-mosaic model**: A mosaic of proteins moving within the fluid that is the phospholipid bilayer. For details on how this model was arrived at, see "How Did We Learn . . ." on page 100.

5.4 Moving Materials In and Out: Diffusions and Gradients

As we observed earlier, one of the main roles of the plasma membrane is to let in whatever's needed and keep out whatever's not. Having learned something about the membrane, you will now see how it carries out this task of passage and blockage.

It is necessary to begin, however, not with the cell, but with the more general notion of how it is that substances go from places where they are more concentrated to places where they are less concentrated. The way this works in liquids is exemplified in the placement of a little bit of dye into a container of water. We intuitively know what will happen here; a few drops of, say, red dye tumbles into the liquid, starts dispersing, and eventually results in a solution that is uniformly just a shade more reddish than plain water (**see Figure 5.3**, page 98). The question that will be addressed here is why this takes place.

Random Movement and Even Distribution

In the real world, several forces are at work in the distribution of dye in a liquid. But for purposes of explanation, here you'll look at just one of them. All molecules or ions are constantly in motion, and this motion is random. (The *degree* to which molecules are in motion defines their temperature; absolute zero is the point at which all molecular motion has ceased.) Liquid molecules are able to slide *past* one another, which increases their ability to move in comparison with solids. So, with random molecular motion and a liquid substance, the dye molecules are now in motion.

But how do these molecules come to be *evenly* distributed through the water? A rigorous explanation of this would require a tour of an area of science you'll be looking at in Chapter 6, thermodynamics. But to get a sense of why this distribution should happen, let us take another tack. Imagine the dye in water, only this time the process has been videotaped. The dye drops in and the tape rolls for just one tick of the molecular clock—just long enough for each molecule of the dye to make one of its motions. Then freeze-frame the action. Now, on the *second*

The Fluid-Mosaic Model of the Plasma Membrane

The richly detailed illustrations found in science textbooks do a great job of presenting information, but they also can be a little misleading. Given their clarity, such illustrations might leave students with the impression that many of the discoveries that scientists make would be apparent to anyone who simply looked in the right place with the right microscope.

Figure 1a presents a good example of the kind artist's rendering of the plasma membrane you've been seeing in this chapter, while **Figure 1b** presents an example of the kind of pictures (or "micrographs") we have of it. You'll probably agree that no one could have discerned the detail in 1a just by looking at 1b. How, then, do we know that the plasma membrane actually is structured as you see it in 1a? How do we know that it is largely lipid, and that proteins move about on the surface of it like so many icebergs? As it turns out, our view of the cell membrane was constructed piece by piece in an intellectual chain that stretched out for decades.

The idea that the surface of the cell is largely made of lipid material was set forth as early as 1855 by investigator Carl Nägeli. In 1899, Charles E. Overton came to the same conclusion by working with some cellular extensions of plant roots. What Overton found was that lipid-soluble substances moved quite easily into these "root hairs," while water-soluble substances did not. Remember now the adage about like dissolving like: Lipid substances would dissolve within *lipid materials*. Overton's conclusion was that the outside lining of the root cells was made of just such materials. From our vantage point—knowing about the phospholipid bilayer that makes up much of the membrane—we can see that he was right.

In 1925, two Dutch scientists, E. G. Gorter and F. Grendel, decided to take red blood cells and measure their lipid content. They knew from the work of another scientist that a given volume of phospholipids would cover a certain *area* when arrayed in a layer. What they found, however, was that the lipids they extracted from the blood cells spread out into an area *twice* as large as would be expected from this benchmark. This doubling, they concluded, came about because the cell membrane was doubled; it existed as the *bi*layer you've been reading about.

Following this advance, however, it was apparent that the plasma membrane had to be composed of more than just two layers of lipids pressed together. For one thing, it was clear that closely related types of hydrophilic molecules passed through the membrane at different rates. This was difficult to reconcile with a uniform phospholipid membrane, which would have let such substances pass through at nearly the same rate. Given such evidence, James Danielli, writing with Hugh Davson, proposed a view of the plasma membrane that ended up having a long run in science. With modifications over time, the Davson-Danielli model, as it was called, wore the mantle of being "generally accepted" by scientists from the time it was set forth in 1935 until the early 1970s. It proposed that a *layer of proteins* coated each side of the phospholipid bilayer.

By the 1950s, electron microscopes had become refined enough that they yielded a "railroad track" picture of the plasma membrane: two dark bands separated by a space in the middle (the view you see in **Figure 2**). Seduced, perhaps, by the comforts of what was "known" to be true, scientists took such pictures to be visual proof of

a

glycocalyx

proteins

cytoskeleton filaments

phospholipid

cholesterol

b

0.1μm

Figure 1
How Much Can a Picture Tell Us?
(a) An artist's rendering of the plasma membrane, and **(b)** a micrograph of it.

Davson-Danielli. The two dark lines were protein bands, it was thought, and the space in the middle was the phospholipid bilayer. We now know, however, that the "tracks" actually were electrons that bound with both membrane proteins and the charged portions of the phospholipid bilayer, thus creating the impression of a continuous track. In reality, there were no protein bands.

Through these years, it was also becoming apparent that there wasn't just one membrane in the cell, there were many: membranes encasing the mitochondria, membranes making up the Golgi apparatus, and so forth. In 1960, J. David Robertson set forth a proposal that saw *all* the cell's membranes as existing within the framework of Davson-Danielli. By this time the venerable model had been modified to include the notion of some proteins *spanning* the bilayer, from cytoplasm to outside face. This, then, was the generally accepted view of cell membranes in the early 1960s: Davson-Danielli-Robertson, which posited a phospholipid bilayer coated on each side with a protein layer, and shot through here and there with other proteins.

Figure 2
A Closeup of the Plasma Membrane
Electron micrograph of two adjoining cells, each of which has a plasma membrane with a "railroad track" appearance. The intercellular space is the space between the two cells. (×470,000)

Into this situation came a chemist, then at Yale, named Jonathan Singer. In 1962, Singer got to thinking about cell membranes and realized that Davson-Danielli-Robertson had a real problem: It would require hydro*phobic* parts of the supposed protein coat to be in contact with *water*. This was like assuming you could set one marble on top

---

Our view of the cell membrane was constructed piece by piece in an intellectual chain that stretched out for decades.

---

of another and expect it to stay there. By 1964, Singer was convinced not only that Davson-Danielli-Robertson had some shortcomings, but that he had a model that improved on it. It's one thing, however, to believe you're right, based on general chemical principles, and quite another to prove it.

"What happened was that I wasn't going to propose a model without some kind of experiments to go with it," he said in a telephone interview from his office at the University of California, San Diego. Working with a postdoctoral student of his, John Lenard, Singer published a scientific paper in 1966 that made the basic break with Davson-Danielli-Robertson. The view that emerged from this paper was that the membrane had a series of *individual* proteins that either were embedded in the bilayer or lay on it. By 1971, Singer was able to set forth a case for his model in great detail in a book chapter. The problem was that nobody paid much attention to it.

"This was a very thorough thing and it took up a lot of space," he says, "but ultimately I came to realize that nobody reads books. However, at that time, at least, if you wrote a review in *Science* [magazine] it would come to people's attention." So he set out to write just such a review. In the interval between the book chapter and the magazine article, however, one last piece of the membrane puzzle fell into place.

Recall that the modern view of the cell membrane is the fluid-mosaic model. It is *mosaic* because it posits discrete proteins within the phospholipid bilayer; but it is *fluid* because these proteins (and the phospholipids themselves) can *move* laterally through the membrane. In 1971, Singer had the mosaic part; by 1972, he had the fluid as well.

(continued)

The Fluid-Mosaic Model of the Plasma Membrane (continued)

"I came to Dr. Singer's lab in 1967," says Garth Nicolson, who was a graduate student with Singer in those days and now heads the Tumor Biology Department at the M. D. Anderson Cancer Center, University of Texas, in Houston. "I became very interested in the polarity of membranes . . . what I found when doing this work was that the distribution of components [on the surface of the cell] seemed to change dynamically." What Nicolson saw, in other words, was that the constituent parts of the membrane were shifting around on it.

Singer and Nicolson are agreed that a paper by L. D. Frye and Michael Edidin made Singer a believer in the rapid movement of materials across the cell surface. Their results, Singer says, "triggered the release of our inhibitions about proposing this kind of thing, which was, in a sense, the natural consequence of our model."

The upshot was a paper called "The Fluid Mosaic Model of the Structure of Cell Membranes," which appeared in *Science* magazine in February of 1972, with Singer and Nicolson as authors. It turned out to have great impact. Read any biology textbook today, and the account of the cell's plasma membrane is essentially the view set forth in this paper. The illustrations contained in textbooks likewise all look like a drawing in the *Science* article. Such was its effect that the fluid-mosaic model is just as likely to be referred to as the "Singer-Nicolson" model, even though, as Garth Nicolson is the first to say, "this is really [Singer's] model . . . he should get the credit for the membrane as far as I'm concerned."

Does this mean that the fluid-mosaic model has at last given us an accurate picture of the membrane? We might wonder about this because, after all, Davson-Danielli was generally accepted for more than 30 years. Singer believes, however, that fluid-mosaic does describe reality. "It's not going to change," he says. "Things rattle around for awhile, but then you get settled in on the right picture."

Osmosis and Plant Cells

It is in connection with the possibilities of solute imbalances that plants have a great advantage over animals—their cell walls. Remember from Chapter 4 that one function of the cell wall is to regulate water uptake. Well, now we can see why this is so valuable. Animal cells, which do not have a wall, can expand until they break when water comes in (**see Figure 5.6**). Plant cells, conversely, will expand only until their membranes push up against the cell wall with some force, setting up a pressure, or turgor, that keeps more water from coming in. Such tight quarters may sound uncomfortable, but this actually is an optimal condition for plants. A nice, crisp celery stick is one that has achieved this kind of *turgid* state, while a droopy stick that has lost this quality has cells that are *flaccid*; flowers in the latter condition are *wilted*.

MediaLab Link
Diffusion, see p. 112.

5.5 How Do Materials Get In and Out of the Cell?

This excursion into the land of diffusion and osmosis has prepared us to start looking at the ways materials actually move in and out of the cell, across the plasma membrane. The big picture here is that some molecules are able to cross with no more assistance than is provided by diffusion or osmosis. Other molecules require these forces and special protein channels to move through, and still others require channels and the expenditure of energy to get across. Any time energy is required for movement across the plasma membrane, **active transport** has occurred. When such energy does not need to be expended, **passive transport** is in operation (**see Figure 5.7**). Our review of transport begins with two varieties of passive transport.

Passive Transport

Simple Diffusion

Water molecules and such gases as oxygen and carbon dioxide need only move down their concentration gradients to pass into or out of the cell. Having been delivered by blood capillaries to an area just outside the cell, oxygen exists there in greater concentration than it does inside the cell. Moving down its concentration gradient, it dis-

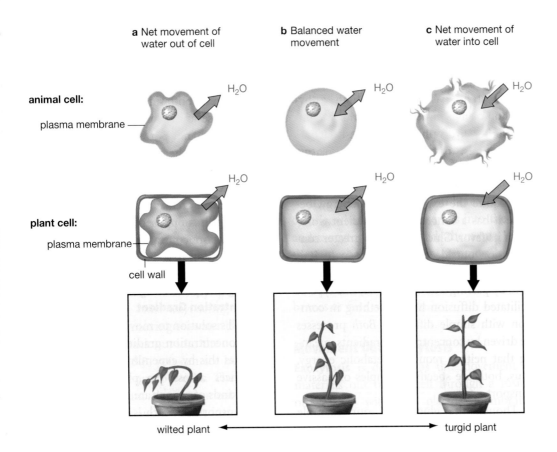

a Net movement of water out of cell

b Balanced water movement

c Net movement of water into cell

animal cell:

plasma membrane

plant cell:

plasma membrane

cell wall

wilted plant ← → turgid plant

Figure 5.6
Osmosis in Cells

a When solutes (such as salt) exist in greater concentration outside the cell than inside, water moves out of the cell by osmosis and the cell shrinks.

b When solute concentrations inside and outside the cell are balanced, there is a balance of water movement into and out of the cell.

c When solutes exist in greater concentration inside the cell than outside, water moves into the cell by osmosis. This influx may cause animal cells to burst, but plant cells are reinforced with cell walls and thus remain turgid—generally a healthy state for them.

Figure 5.7
Transport through the Plasma Membrane
Passive transport requires no expenditure of energy, but active transport does.

a In simple diffusion, materials move down their concentration gradient through the phospholipid bilayer.

b In facilitated diffusion, the passage of materials is aided both by a concentration gradient and by a transport protein.

c In active transport, molecules again move through a transport protein, but now energy must be expended to move them against their concentration gradient.

solves into the plasma membrane, travels through it, and emerges on the other side. This is an example of **simple diffusion**, meaning diffusion that does not require a special protein channel. Molecules that are larger than oxygen, but that are fat soluble, such as the steroid hormones mentioned before, also move into the cell in this manner. Carbon dioxide, which is formed *in the* cell as a result of cellular respiration, has a net movement *out* of the cell through simple diffusion.

Facilitated Diffusion: Help from Proteins

Water can traverse the lipid bilayer through simple diffusion, moving through *despite* the hydrogen bonding it does. It can pass through in this way (rather slowly) primarily because it's such a small molecule. Yet with water itself—and certainly with polar molecules larger than water—something else begins to be required for adequate amounts to pass through the plasma membrane. What's required are special channels—passageways formed by some of the integral proteins described earlier. With this, the method of passage is no longer *simple* diffusion; it is **facilitated diffusion**.

Here you can begin to see the protein specificity referred to earlier. The proteins that serve as these channels are called **transport proteins**, and each acts as a conduit for only one substance, or at most a small group of associated substances. The process of transport begins with the kind of binding explained earlier. A circulating molecule of glucose, for example, latches onto the binding site of a glucose transport protein. What happens next is not known with

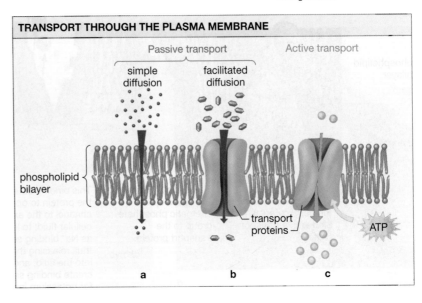

TRANSPORT THROUGH THE PLASMA MEMBRANE

Passive transport | Active transport

simple diffusion | facilitated diffusion

phospholipid bilayer

transport proteins

ATP

a | b | c

6. You get a cut on your finger and some bacteria enter. Your immune system cells kill off the invaders by eating them. This involves
   a. pinocytosis
   b. phagocytosis
   c. receptor-mediated endocytosis
   d. exocytosis
   e. active transport

7. According to the fluid-mosaic model of membrane structure, the membrane is made up of a
   a. nearly continuous base of proteins with phospholipids stuck in it
   b. A bilayer of phospholipds with proteins stuck in it that have fairly free lateral movement
   c. bilayer of protein with layers of phospholipids on the outside and inside
   d. bilayer of phospholipid with layers of proteins on the outside and inside
   e. bilayer of phospholipids sandwiched between two bilayers of proteins

8. The sodium-potassium pump uses _____ to move sodium ions _____ the cell and potassium ions _____ the cell.
   a. lipids . . . into . . . out of
   b. proteins . . . into . . . out of
   c. energy . . . out of . . . into
   d. cholesterol . . . out of . . . into
   e. binding . . . into . . . out of

9. Endocytosis and exocytosis are similar in that both involve the use of
   a. vesicles that carry materials into and out of the cell
   b. the Golgi apparatus to package materials
   c. diffusion to carry food and wastes across the plasma membrane
   d. starch to supply large amounts of energy
   e. rough ER to manufacture proteins

10. Osmosis refers to the movement of _____ a concentration gradient.
   a. energy down
   b. dissolved materials against
   c. water along
   d. dissolved materials along
   e. water against

Brief Review

1. Mentally picture a taste bud cell on your tongue. Assume that this particular taste bud responds to sweet tastes. Explain how the surface of the cell—its plasma membrane—might allow it to respond to sugar, but not to salt or other tastes.

2. You often hear how dangerous it is for athletes to take steroids to "bulk up" and how many problems steroids cause in various parts of the body. Why do these drugs affect all parts of the body—not just the muscles?

3. Explain why the phospholipid "heads" of the plasma membrane are always pointed toward the cytosol and extracellular fluid, whereas the "tails" are always oriented toward the middle of the membrane.

4. Most cells in your body do not respond to most hormones. For example, a hormone known as *erythropoietin* causes the bone marrow to make more red blood cells, but has little effect elsewhere in the body. Explain how the plasma membrane could control this effect.

5. As you saw in this chapter, all cellular membranes are essentially the same. However, the specifics given in the chapter relate primarily to the plasma membrane on the outside of the cell. What differences do you suppose might be true of membranes in the interior of the cell?

6. Place a number of marbles on one end of a cafeteria tray. Shake the tray gently a few times, keeping it level. What happens to the marbles? Start over and shake

faster the next time. What happens now? Explain how this is like diffusion. What is shaking the tray similar to in the process of diffusion?

7. The cells that line your stomach make it very acidic by pumping large numbers of H^+ ions from the inside of the cells into the stomach. They have to do this *against* a concentration gradient. What process do these cells use to do this pumping, and what do they need to carry it out?

Applying Your Knowledge

1. In the introduction to this chapter, you read about what happens in cystic fibrosis when a single protein on the surface of the cell cannot be made. Think about what would happen if any of the other three major components of the cell membrane (phospholipid bilayer, glycocalyx, cholesterol) could not be made. What would happen to the cell? What general conclusion can you then reach about the necessity of all parts of the cell membrane?

2. Because materials move from high concentration to low concentration by diffusion, and they move the same way by facilitated diffusion, why do cells need to make proteins to carry out facilitated diffusion? Isn't this just a waste of energy?

3. In 1998 scientists determined the structure of a channel protein that allows potassium ions to move across the plasma membrane. This channel works with the sodium-potassium pump to enable nerve cells to respond to changes in the environment. If you were a scientist, how might you use this information in understanding how cells communicate?

4. Recently, researchers discovered the structure of the protein that allows about 70 percent of all cold viruses to bind to the plasma membrane of a human cell, and enter to cause a cold. What benefit would there be to knowing the shape of that protein? How might you use this information, if you were working for a pharmaceutical company?

Group Activity

Divide into teams to further research the chemical nature and function of each of the four components of a membrane. Use this information to make a physical model of a plasma membrane. Contact a local high school or middle school biology teacher and arrange an opportunity to present the model and explain the components of the membrane to his or her students.

MediaLab
Balancing Your Imports and Exports: Membrane Transport

In this chapter, you read that without the efficient, rapid movement of substances across your membranes, you would not be able to move, absorb nutrients, or even see this page. The *CD Activity* illustrates the different types of membrane transport involved in all these processes. In the following *Web Investigation*, you will examine the importance of membranes for organisms from bacteria to humans, so that in the *Communicate Your Results* section you can analyze how manipulating membrane transport has been used to treat wounds, cystic fibrosis, or drug addiction.

This *MediaLab* can be found in Chapter 5 on your CD-ROM and Companion Website (http://www.prenhall.com/krogh).

CD ACTIVITY

Cells cannot live without exchanging substances with the environment. There are four different ways of transporting substances across membranes; each is illustrated through graphic animations in this *CD Activity*.

TRANSPORT THROUGH THE PLASMA MEMBRANE

Activity

1. First, you will discover how small substances may be able to diffuse directly across the lipid portion of the membrane in a process called **simple diffusion**.

2. Next, you will learn how **facilitated diffusion** is used to move lipid-intolerant substances through special protein channels into the cell.

3. Additionally, you will see how molecules are pumped across the membrane by carrier proteins in the process of **active transport**.

4. Finally, you will learn how other substances are moved in bulk within membrane "bubbles," a process called **endocytosis** or **exocytosis**.

After completing this activity, you should be familiar enough with these transport mechanisms to begin the *Web Investigation* into how they work together in cells.

WEB INVESTIGATION

Investigation 1
Estimated time for completion = 5 minutes

Cigarettes, chocolate, and cocaine: Have you ever wondered how they make us feel so good? Our brain contains millions of neurons that help us respond to the environment, coordinating sight, smell, hearing, touch, and even pleasure. All these neurons must communicate with each other using membrane transport. Most addictive drugs are able to elevate the natural neuron communication that occurs after pleasant sensations like a hug, or a bite of candy. Select the Keyword **ADDICTION** on your CD or Website to read more about nerve-cell communication. Once you have read about nerve cells, click on the next link, "drugs," at the bottom of the page to view how cocaine acts. After reading the article, select the Keyword **NERVES** to view an animation of one nerve cell communicating with another.

Investigation 2

Estimated time for completion = 5 minutes

Why do we wash our hands with soap, or cleanse a wound with alcohol or iodine? They are all antiseptics, substances that kill cells. What is it in soap or alcohol that would harm bacteria but not our own cells? Select the Keyword **ANTISEPTICS** on your CD or Website and read Table 2 in the linked article to find out how these antiseptics harm bacteria. Antiseptics are commonly found in many different household products, including cleaning supplies and mouthwashes. Select the Keyword **MOUTHWASH** to read an article listing the active ingredients in mouthwash. Then, clip a label from an antiseptic product and use these two Web sites to determine its active ingredients.

Investigation 3

Estimated time for completion = 10 minutes

At the opening of this chapter, you read about cystic fibrosis, a debilitating disease caused by a genetic disorder. The disease is caused by a defect in the gene that produces the protein CFTR, which forms channels in the cell membranes involved in the active transport of chloride ions. Because this disease greatly reduces life expectancy, you would expect few of these defective genes to be passed on to future generations. Yet, cystic fibrosis has been around for centuries. Could there possibly be an up side to having a defective gene like this? Select the Keyword **ADVANTAGE**, and read about evolution and the selective advantage of the cystic fibrosis allele.

COMMUNICATE YOUR RESULTS

Exercise 1

Estimated time for completion = 10 minutes

In *Web Investigation 1*, you saw how membrane transport operates to conduct nerve impulses. In 50 words, identify the types of membrane transport involved in passing nervous impulses, and describe their roles. These nerve impulses may be sending messages of pain (or pleasure in the case of addictive drugs) to your central nervous system. How does cocaine cause pleasurable sensations? Can you propose a way to interrupt neuron communications such as pain signals? Does this knowledge help to explain how painkillers and other drugs like cocaine or morphine work?

Exercise 2

Estimated time for completion = 5 minutes

What's the best way to cleanse your mouth, hands, or wounds of harmful bacteria? In *Web Investigation 2*, you read a list of antiseptics and saw that many of them kill bacteria by disrupting cell membranes. Create a list of all active ingredients from the label of your common household antiseptic that act by disrupting cell membranes.

Exercise 3

Estimated time for completion = 5 minutes

The article in *Web Investigation 3* explores the idea that if detrimental alleles like cystic fibrosis remain in the population, they must have some selective advantage. In 50 words, describe the selective advantage provided by the cystic fibrosis allele. Propose a hypothesis to explain this advantage, given what you know about the cause of cystic fibrosis.

Energy and Its Transformations

Most living creatures, such as this ant, derive their energy either from eating plants or from other creatures that consume plants (magnified 30 times).

6

Life's Mainspring
An Introduction to Energy

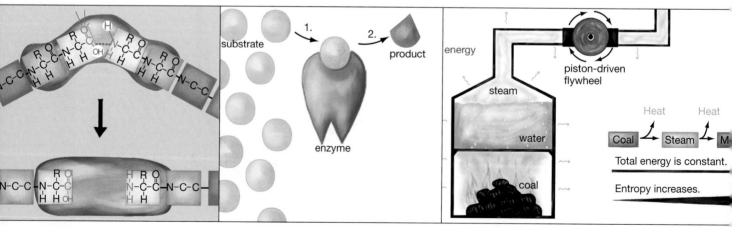

Enzymes at work.
(page 125)

Slowing a reaction.
(page 126)

Freeing energy
from bondage.
(page 119)

My eight-month-old daughter, Tessa, has a teddy bear with a music box in it that lets us listen to "Brahms's Lullaby." All that's required is that I wind up the key that attaches to the music box. After a few turns of the wrist, the melody cycles through several times, then slows for a last few bars, then stops. This process can be repeated as many times as I wind the music box. However—no surprise here—the music box never seems to wind itself. As we listen, a mechanism in the music box keeps the melody coming out at an even tempo. Even the slowing at the end seems just right for a lullaby, since the audience is a sleepy child.

6.1 Energy Is Central to Life

The forces that determine how Tessa's teddy bear works also turn out to condition life, since, in order to function, both the music box and life must be supplied with the same thing: energy. Just as the music from the teddy bear depends on energy from an "outside hand" (mine) to continue, so life on Earth depends on energy from an outside source: the sun, whose daily showering of energetic rays lets the green world of plants and algae flourish. These organisms then pass their bounty along to animals in the form of food to eat and oxygen to breathe.

Beyond this, the sun and the music box have another thing in common; their energy always runs downhill, from more concentrated to less. The spring in the music box spontaneously unwinds, but it does not rewind itself. The sun sends out light and heat, but this energy does not spontaneously come together to form another sun. The music from the music box eventually must come to a stop, and the sun will not shine forever. With all its grandeur, its energy is winding down, its brilliance dispersing. The difference is that once the sun's showering of energy comes to a stop, nothing will restart it.

While the sun's gift lasts, however, living things can transform its energy and use it to do things like sprout leaves and swim through oceans. Such complex activities make it imperative, though, that living things be able to *control* the energy they capture. The spring within the music box unwinds at a measured rate, spinning out the melody at just the right pace; similarly, living things have elaborate mechanisms in place that allow them to make the most out of the energy they receive.

A downhill reaction releases energy.
(page 121)

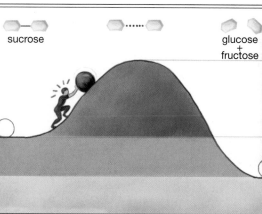

There must be an easier way—and there is.
(page 124)

Shaping up to fit in grooves and pockets.
(page 125)

Principles such as this are important enough to biology that this chapter and the two that follow are devoted to the subject of energy and life. In this chapter you'll learn some basics about energy, including information on how organisms employ substances called enzymes to lower the amount of energy needed to get chemical reactions going. Then, in Chapter 7, you'll see how living things "harvest" energy, meaning how they extract energy from food. Finally, in Chapter 8, you'll look at the ways that plants capture and transform the sun's energy in the process called photosynthesis.

6.2 What Is Energy?

Energy is commonly defined as the capacity to do work, but this definition merely passes the question mark along, to the word *work*. Given this, here's one possible definition of energy: The capacity to bring about movement against an opposing force. Speaking more generally yet, some observers have defined energy as the capacity to bring about change. The thing that makes energy so tricky as a concept is that although we can measure it with great precision, and we can experience its effects, we cannot grasp it or see it. We can see the water that drives a waterwheel, but who has seen the energy the water contains? (**See Figure 6.1.**)

The Forms of Energy

Energy can be thought of as coming in different forms, many of them familiar. Mechanical energy is captured in the wound-up spring of

Figure 6.1
Where's the Energy?
Water moving over a falls is plainly visible and the movement of this water releases energy. Yet the energy itself is invisible.

the musical teddy bear. Chemical energy is the energy held in the chemical bonds of a lump of coal. One other way to conceptualize energy is as potential and kinetic. **Potential energy** is stored energy: The rock perched precariously at the top of the hill; the charged ions kept on one side of a cell membrane. Conversely, **kinetic energy** is energy in motion, as with the rock tumbling down the hill or water driving a waterwheel.

The Study of Energy: Thermodynamics

Given how important energy is to life, it makes sense that there is a branch of biology, called *bioenergetics*, that links biology with a scientific discipline known as **thermodynamics**, the study of energy. Though thermodynamics is a mathematical and rather abstract discipline, research into it was prompted originally by a very down-to-earth goal—the desire to build a better steam engine. British inventors had harnessed steam power in the eighteenth century, and this discovery had the awesome impact of sparking the Industrial Revolution.

A Steam Engine Drives the Inquiry into Energy

The word *revolution* is fitting here because, before the advent of steam power, people used the stored energy in, say, wood or coal solely to produce *heat*. The steam engine was a device that allowed us to use heat to perform *work* on a scale that was previously impossible. Heat could be channeled, it was discovered, to drive an engine, which in turn could power any number of industrial processes.

With this development, new questions confronted scientists: How exactly did heat drive a steam engine? Was heat a substance that moved from one place to another? Was there a finite quantity of energy, such that it could all be used up? Research carried out in England, Germany, and France during the nineteenth century yielded the answers to these questions, ultimately resulting in the development of some insights known as the laws of thermodynamics.

The First Law of Thermodynamics: The Transformation of Energy

The **first law of thermodynamics** states that energy is never lost or gained, but is only transformed. The sun's energy is not used up by green plants; rather, some of it is *converted* by the plants into chemical form. Plants use the sun's energy to put together carbohydrates, whose *chemical bonds* contain some of the energy that previously existed in the sun's rays.

A critical point here is the qualification that *some of* the sun's energy takes on a chemical form in plants. As it turns out, a plant cannot convert *all* of the solar energy it receives into carbohydrates. Likewise, a steam engine actually converts only a fraction of the energy contained in coal into the movement of a piston. What happens to the energy that is not stored in carbohydrates or transformed into motion? It can't be destroyed; that's clear from the first law. What happens is that it is converted into *heat*. For every energy "transaction" that takes place, at least some of the original energy is converted into heat. This is true for every transaction in every energy system: cells burning food, cars burning gasoline, electric utilities burning coal, and your brain decoding this sentence.

The way this concept is usually phrased, however, is that some of the original energy is *lost* to heat. At first glance, this may seem like an unfair knock on heat, which, after all, has intrinsic value. Indeed, you may wonder about how this concept applies to the work of a steam engine. The piston is *driven* by heat that is produced by burning coal. Why, then, is energy "lost" to heat?

Well, consider what might be called the before-and-after of energy transfer. In the "before," a lump of coal is a very ordered object, containing carbon atoms that exist in a precise spatial relationship to one another. Thus bonded, these atoms are not free to move. However, bringing a flame to the lump of coal causes the coal's carbon atoms to react with oxygen. Chemical bonds are broken through this process, and the energy stored in these bonds is released as heat.

Now cut to the "after" of the energy transformation. Heat is the *random* motion of molecules. In the case of the steam engine, there are molecules in random motion—better known as hot air—inside and outside the apparatus. In comparison to the "before" of the transaction, it's clear that the energy contained in the *ordered, concentrated* chemical bonds of coal has been transformed into the *disordered, dispersed* energy of heat.

The Second Law of Thermodynamics: The Natural Tendency toward Disorder

And this gets us to the critical thing. Energy transformations will spontaneously run *only* from greater order to lesser order. Some of the heat produced in a steam engine is useful; it's driving a piston. Some of it is not, however—it's simply dissipated. But *all* of it amounts to a form of energy that is more dispersed than coal and that will quickly disperse further. Once this happens, there is no chance that it will spontaneously convert into anything *but* heat. We expect to see a lump of coal burn to ashes, and to feel the hot air that results disperse. But we do not expect to see the ashes reform into a lump of coal, nor to feel the hot air concentrate itself again.

In yielding energy, matter goes from a more-ordered state to a less-ordered state. The scientific principle that speaks to this is, in fact, the **second law of thermodynamics**: Energy transfer always results in a greater amount of disorder in the universe. In connection with this law there is a term you will hear—*entropy*, which is a measure of the amount of disorder in a system. The greater the entropy, the greater the disorder. **Figure 6.2** summarizes the thermodynamic principles of the first and second laws of thermodynamics.

The Consequences of Thermodynamics

All of this may seem like just so much—one might say—*hot air*. But few things so profoundly condition the universe we live in as these thermodynamic rules. Indeed, the relentless increase in entropy that the

Figure 6.2
The Transformations of Energy
In a steam engine, energy locked up in the chemical bonds of coal is transformed into heat energy and mechanical energy. There is no loss of energy in this process, but energy is transformed from a more-ordered, concentrated form (the chemical bonds of coal) to a less-ordered, more dispersed form (heat). Thus, the amount of disorder—or entropy—increases in the transaction.

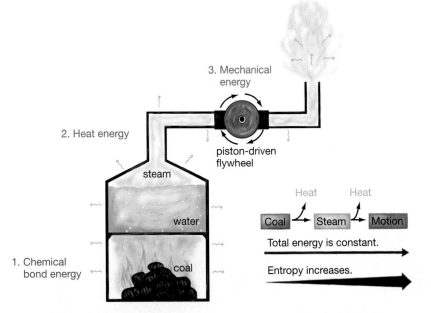

3. Mechanical energy

2. Heat energy

piston-driven flywheel

steam

water

1. Chemical bond energy

coal

Heat   Heat

Coal → Steam → Motion

Total energy is constant.

Entropy increases.

What Is It?
See p. 122.

second law entails may dictate nothing less than the fate of the universe.

Allow yourself, for a second, to think of the universe as consisting solely of the sun, the Earth, and the space between them. The sun is an ordered concentration of hydrogen atoms that are undergoing a nuclear reaction and thus releasing light and heat. Living things on Earth manage to capture buckets of this energy, primarily through the activities of plants. But every time one of these buckets is poured into another—every time light energy from the sun is transformed into carbohydrate energy in plants, for example—there is some spillage into heat. And once heat is generated, there is no spontaneous way it can make its way back into the orderly form of sunlight or carbohydrates. With every firing of a car's cylinder or contraction of a muscle, then, some amount of energy is transformed into heat.

Against this, life as we know it is made possible by ordered concentrations of energy. The sun sustains life, but dispersed heat does not. One day, our sun will have wound down like the spring in a music box, bringing to an end the energy source for life on Earth. One possible scenario for the universe as a whole is that *all* its stars eventually will darken, leaving a cold universe with little potential for life—indeed, with little activity of any sort. The laws of thermodynamics are powerful indeed.

6.3 How Is Energy Used by Living Things?

A sprouting plant is constantly building itself up; it is making larger, more complex molecules (proteins, polysaccharides) from smaller, simpler ones (amino acids, simple sugars). This growing degree of organization stands in contrast to the spontaneous course of things, which is breakdown and disorder. In the context of the universe as a whole, living things are contributors to its entropy. Any given biological activity—building a protein, say—generates heat and thus increases the disorder of the universe. But living things can bring about *local* increases in order (in themselves). It takes energy to do this, however, and on Earth the source of this energy ultimately is the sun. Living things can be quite efficient at converting energy from outside sources, as you can see by looking at **Table 6.1**.

Kinds of Work for Living Things

In considering *how* it is that living things use energy, there are really only a few kinds of work that they carry out. The most obvious of these is **mechanical work**, exemplified by the contraction of a muscle. There also is **transport work**; for example, moving sodium ions against their concentration gradient, as you saw in Chapter 5. Finally, there is **synthetic work**, which can be seen in the build-up of complex molecules (proteins) from simpler ones (amino acids).

Up and Down the Great Energy Hill

Synthetic work leads to the idea of an energy cycle in living things. A carbohydrate is a more complex thing than the simple sugars it is made of. It is a more *ordered* arrangement of atoms than is a group of individual simple sugars. As such, the second law of thermodynamics tells us, it should *take* energy to make such a substance from simple sugars, and this is indeed the case. Going in the opposite di-

---

Table 6.1
Relative Efficiencies of Several Energy Systems

The efficiency of several energy systems, as measured by the proportion of energy they receive relative to what they then make available to perform work. In measuring the efficiency of the car engine, the question is, how much of the energy contained in the chemical bonds of gasoline is converted by the car into the kinetic energy of wheel movement? In each system, most of the energy not available for work is lost to heat.

System	Efficiency
Conventional power plant	40%
Human respiration of glucose	37%
Automobile engine	25%

rection, the breakdown of a carbohydrate into simple sugars is an action favored by the second law, because it brings about *less* order. Such a process therefore should *release* energy, and this too is the case. This understanding leads to the concept of a great uphill and downhill cycle of energy in living things.

Downhill (Exergonic) and Uphill (Endergonic) Reactions

The breakdown of a carbohydrate is a specific example of a general kind of energy reaction. Such a breakdown is a form of **exergonic** energy exchange. An exergonic (meaning "energy out") chemical reaction is one that runs downhill, releasing energy; it is a reaction in which the starting set of molecules (the reactants) contain more energy than the final set of molecules (the products).

Conversely, an **endergonic** ("energy in") reaction is one that runs uphill; the products of such a reaction contain more energy than the reactants. Energy, in other words, is stored away in such a reaction. This is what happens when, for example, simple glucose molecules are brought together to form glycogen, which is a storage form of carbohydrates (**see Figure 6.3**).

Coupled Reactions

Now, as it happens, the two kinds of reactions are constantly *linked* in living things. Why should this be so? If you want to get something uphill—into a higher energy state—you have to power that process. And that means *using* some energy. (Think of getting a bicycle uphill.) Thus, endergonic reactions require exergonic reactions. When the two are brought together, it is known as a **coupled reaction**. You actually observed some of these reactions in Chapter 5, though they weren't labeled as such then. Remember how the sodium-potassium pump moved sodium ions out of the cell against their concentration gradient? (See pages 104–105.) For the sodium ions, this was an uphill reaction: They finished the reaction in a higher energy state than they started, because they were moved up their electrical and concentration gradients.

And how was this movement powered? Through a downhill reaction. All you observed in Chapter 5 was that an ATP molecule furnished the energy that powered the sodium pump. You have probably guessed by now, however, that what this really meant was the breaking down of a complex molecule into a simpler one, thus yielding the pump energy. The molecule being broken down was one that deserves a close consideration—ATP.

6.4 The Energy Molecule: ATP

Adenosine triphosphate. If you look at **Figure 6.4**, you can see that ATP is a nitrogen-containing molecule with three "phosphate groups" attached to it. Each group amounts to one phosphate atom and three oxygen atoms. Notice that each of the phosphate groups is negatively charged. This is the key to understanding why ATP serves as such an effective energy transfer molecule. Remember that like charges repel each other, and in

Figure 6.3
Energy Stored and Released
It takes energy to build up more complex molecules (in this case glycogen) from simpler molecules (in this case glucose). Such a build-up is thus an endergonic or uphill reaction. Conversely, energy is *released* in reactions in which more complex molecules are broken down into simpler ones. Such reactions are downhill or exergonic.

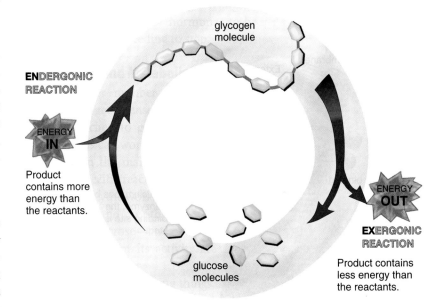

ENDERGONIC REACTION

ENERGY IN

Product contains more energy than the reactants.

glycogen molecule

ENERGY OUT

EXERGONIC REACTION

Product contains less energy than the reactants.

glucose molecules

Phosphate groups

adenine

Adenosine

ribose

ATP

ENDERGONIC REACTION

energy IN

energy OUT

EXERGONIC REACTION

P

ADP

Figure 6.4
Energy Release from Breakdown of ATP
ATP (adenosine triphosphate) is life's most important energy transfer molecule. It stores energy in the form of chemical bonds (shown in red) between its phosphate groups. When the bond between the second and outermost phosphate group is broken, the outermost phosphate separates from ATP and energy is released. This separation transforms ATP into adenosine *di*phosphate or ADP, which then goes on to pick up another phosphate group, becoming ATP again.

The substance that is being worked on by enzymes is known as a **substrate**. In our example, sucrose is the substrate for sucrase. The -*ase* ending you see in *sucrase* is common to most enzymes; as you can see, the substrate is identified in the enzyme's name. The sum of all the chemical reactions that a cell or larger organism carries out is known as its **metabolism**. To put things simply, enzymes are active in all facets of the metabolism of all living things.

6.6 Lowering the Activation Barrier through Enzymes

All this information about enzymes is well and good, you may be saying at this point, but what does it have to do with energy? The answer is that in carrying out their tasks, enzymes are in the business of lowering the amount of energy needed to get chemical reactions going. And this means that the reactions can get going faster.

An analogy may be helpful here. Consider the now-familiar rock perched at the top of a hill. Situated in this way, it has a good deal of potential energy. If it rolled down the hill, it would release this potential energy as kinetic energy. But here is the critical question: What would be required to get this rock going? It is *perched* at the top of the hill because it is *stable* at the top of the hill. To get it going would require *additional* energy in the form of, say, a push. This additional energy is referred to as the activation energy of this energy transaction. In living things, **activation energy** can be defined as the energy required to initiate a chemical reaction. Now consider the sucrose molecule. It is "perched" like the rock, in a sense, because it lies "uphill" from the glucose and fructose it can break into. But it is also stable like the rock. Left alone—sitting in a dry sugar bowl—it *stays sucrose*. This kind of stability is typical of most molecules in our everyday experience.

But the body needs the glucose and fructose that come from sucrose breakdown, and it can't wait days to get these molecules. How is it going to be supplied with them as needed? The elegant solution is enzymes. They lower the activation energy needed to carry out chemical reactions. If you look at **Figure 6.7**, you can see how this works. Imagine that you had to push a boulder up the hill in (a) to get it to roll down the other side. Now imagine that you only had to push a boulder up the hill in (b) to get the same result. In which situation would you be able to get your desired result *faster*?

How Do Enzymes Work?

How do enzymes carry out this task of accelerating chemical reactions? The short answer is that they bind to their substrates, and in so doing make these substances more vulnerable to chemical alteration. The amazing thing is that they do this without being permanently altered themselves. Enzymes are **catalysts**—substances that retain their original chemical composition while bringing about a change in a substrate. At the end of its cleaving of the sucrose molecule, sucrase has exactly the same chemical structure as it did before. It is thus free to pick up another sucrose molecule and split *it*. All this takes place in a flash: In one second, a given enzyme can carry out hundreds or even thousands of chemical transformations.

Figure 6.7
Enzymes Accelerate Chemical Reactions
How is sucrose split into glucose and fructose? Without an enzyme, the amount of energy necessary to activate this reaction is high. In the presence of the enzyme sucrase, however, a low activation energy is sufficient to get the process started. The energy released from the splitting of sucrose is the same in both cases.

a Without enzyme

b With enzyme

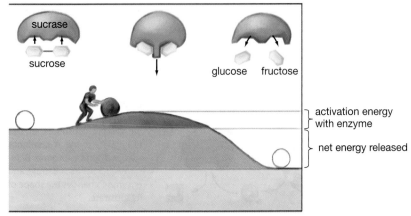

With only a handful of exceptions, all known enzymes are proteins. They are generally globular proteins whose shape includes a kind of pocket into which the substrate fits. If you look at **Figure 6.8**, you can see a space-filling model of one enzyme, called hexokinase. You can also see, buried there in the middle, its substrate, glucose—better known as blood sugar.

As with all proteins, enzymes are made up of amino acids, but only a few of the hundreds of amino acids in an enzyme are typically involved in actually binding with the substrate. Five or six would be common. The place in the enzyme where the action occurs is called the **active site**. There, in the pocket of the enzyme, is where the substrate is both bound and transformed. In some cases, the participants at the active site include, along with amino acids, one or more accessory molecules. One group of these molecules is known as **coenzymes**. If you've ever wondered what vitamins do, participation in enzyme binding is a big part of it. Once we have ingested them, many vitamins are transformed into coenzymes that allow enzymes to function.

A Case in Point: Chymotrypsin

The details of how enzymes carry out their work are very complex and vary according to enzyme. Let's consider in a very general way, however, the activity of a much-studied enzyme called chymotrypsin.

Chymotrypsin is delivered from the human pancreas to the small intestine, where it works with water to break down proteins we have ingested. Its function is to clip building-block amino acids one by one from within a protein chain. It does this by breaking the single bond that binds one amino acid to the next (**see Figure 6.9**). After having bound to part of a protein chain, chymotrypsin then interacts with it to create a transition-state molecule. In effect, it distorts the shape of the protein—and holds it in this new shape briefly—in such a way that the protein becomes vulnerable to bonding with ionized water molecules. This allows carbon and nitrogen atoms to latch onto new partners, and the protein chain is clipped. Chymotrypsin then returns to its original form and proceeds to a new reaction.

Figure 6.8
Shape Is Important in Enzymes
Substrates generally fit into small grooves or pockets in enzymes; the location at which the enzyme binds the substrate is known as the enzyme's active site. Pictured is a computer-generated model of a glucose molecule (in yellow) binding to the active site of an enzyme called hexokinase (in blue).

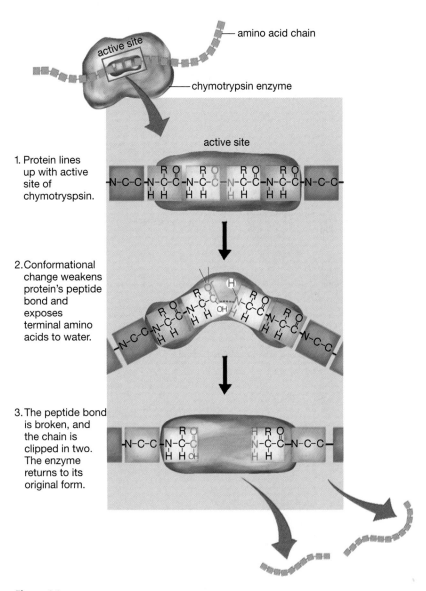

1. Protein lines up with active site of chymotryspsin.

2. Conformational change weakens protein's peptide bond and exposes terminal amino acids to water.

3. The peptide bond is broken, and the chain is clipped in two. The enzyme returns to its original form.

Figure 6.9
How the Enzyme Chymotrypsin Works
In this example, the enzyme chymotrypsin is facilitating the breakdown of a protein by changing the protein's shape. In the absence of chymotrypsin, this process would take a billion times longer.

Figure 6.10
When the Product Slows the Reaction
The frequency of chemical reactions can be controlled by negative feedback.

a In step 1, the substrate (yellow circles) binds to the enzyme. This transforms the substrate, in step 2, into the product (the orange wedges).

b The product now binds to a *different* site on the enzyme (step 3), causing a conformational change in the enzyme that prevents it from binding with any more substrate (step 4).

6.7 Regulating Enzymatic Activity

The activity of enzymes is regulated in several ways. The question here is, what factors influence the amount of "product" an enzyme turns out? One of these factors, pretty obviously, is the amount of substrate in the enzyme's vicinity; if there's no substrate to work on, no product can be turned out.

The work of enzymes can, however, be reduced in several other ways as well. For example, some molecules that are not the normal substrate for an enzyme can still bind to the enzyme at its active site anyway. An enzyme that is "occupied" in this way is not free to bind with substrate and thus is not able to carry out its usual catalytic activity.

Allosteric Regulation of Enzymes

Enzyme activity can also be lowered by binding that occurs at a location different from that of the active site. This binding brings into play the concept of *feedback* in a system.

a

substrate

1.

2.

product

enzyme

1. Substrate binds to enzyme.

2. Enzyme transforms substrate to product.

b

4.

3.

3. Product binds to a *different* site on the enzyme, causing the enzyme to change shape.

4. The new shape of the enzyme prevents it from binding to any more substrate.

Most of us have experienced the phenomenon known as **negative feedback** in connection with a common home thermostat. Falling temperature causes the thermostat to kick on the furnace. The *product* of this activity is hot air. When this air has warmed the room enough, the thermostat senses the heat and shuts the furnace down. Meanwhile, the product of an *enzyme's* work is an altered substrate. This chemical product, like heat feeding back to a thermostat, can diminish the activity of that enzyme. Earlier, you saw a model of an enzyme called hexokinase, which works on the substrate glucose. The product of this reaction is something called glucose 6-phosphate. It is to the enzyme as air is to a thermostat: It binds to hexokinase at a site other than the active site, and in so doing changes hexokinase's shape. This renders hexokinase unable to bind with its substrate, glucose, and its work is temporarily shut down. If you look at **Figure 6.10**, you can see how this process works schematically.

This process is known as **allosteric** regulation (*allo* means "other" in Greek, and *steric* means "shape"). The importance of such regulation is that enzymes are not simply fated to turn out product as long as there is substrate to work on. If a cell has too much of a substance, allosteric control can cut down on it; too little and its concentration can be increased. The product of a reaction can slow down the reaction itself; isn't that a neat trick?

On to Harvesting Energy from Food

So, to review for a second, you've seen that energy in living things can come packaged in food—and that the harvesting of this energy begins with food being broken down. The end result of energy harvesting is that the energy currency molecule, ATP, is produced and then used to power various processes. The breakdown of food and the synthesis of ATP occur in individual steps that are catalyzed by enzymes.

In Chapter 25, you'll look at what might be called the upstream part of this process—how we go from eating a potato, say, through digesting it and thus breaking it down into its constituent parts—in this case simple sugars. But for now, you'll be looking at the downstream part of this process: How does the body extract energy from food and use this energy to put together ATP? That is the subject of Chapter 7.

Chapter Review

Summary

6.1 Energy Is Central to Life

- Living things require a source of energy. The sun is the ultimate source of energy for most living things on Earth.

6.2 What Is Energy?

- Energy is the capacity to bring about movement against an opposing force.

- Two fundamental principles of energy are the first law of thermodynamics, which states that energy is never lost or gained, but is only transformed; and the second law of thermodynamics, which states that energy transfer will always result in a greater amount of disorder in the universe.

- In every energy transaction, some energy will be lost to the disordered form of heat. Entropy is a measure of the amount of disorder in a system; the greater the entropy, the greater the disorder.

- Living things can bring about local increases in order (in themselves), through their metabolic processes, but it takes energy to do this. For most living things, the source of this energy ultimately is the sun. Plants are able to grow by harvesting solar energy, and animals feed on plants.

6.3 How Is Energy Used by Living Things?

- Energy is stored away in endergonic (uphill) reactions in which the products contain more energy than the reactants. Conversely, energy is released in exergonic (downhill) reactions, which are reactions in which the reactants contain more energy than the products. The linkage of simple sugars to form a complex carbohydrate is an endergonic reaction; such a reaction requires an input of energy. The breakdown of a complex carbohydrate into simple sugars is an exergonic reaction; such a reaction releases energy.

6.4 The Energy Currency Molecule: ATP

- Adenosine triphosphate (ATP) is the most important energy transfer molecule in living things. It is the final product of the energy-harvesting process, in which energy is extracted from food. ATP drives reactions by donating the third of its three phosphate groups to those reactions, in the process becoming the two-phosphate molecule adenosine diphosphate (ADP). The energy supplied by food powers the process by which a third phosphate group is attached to ADP, making it ATP once again.

6.5 Efficient Energy Use in Living Things: Enzymes

- Enzymes are a class of proteins that greatly increase the rate of chemical reactions. Nearly every chemical process that takes place in living things is facilitated by an enzyme.

- The sum of all the chemical reactions that a cell or larger living thing carries out is its metabolism. Many activities in living things are controlled by metabolic pathways in which a series of interrelated steps is undertaken, each one of them facilitated by an enzyme.

6.6 Lowering the Activation Barrier through Enzymes

- Enzymes lower activation energy, meaning the energy required to initiate a chemical reaction. Enzymes are catalysts: They retain their original chemical composition while bringing about a change in the molecules they bind with, called their substrates.

6.7 Regulating Enzymatic Activity

- Enzyme activity can be controlled in several ways. One of these is allosteric regulation, in which the product of an enzyme-controlled reaction binds with the enzyme, thus reducing its activity.

Key Terms

Understanding the Basics

Multiple-Choice Questions

1. According to the first law of thermodynamics, energy
   a. is never lost or gained, but is only transformed
   b. always requires an ultimate source such as the sun
   c. can never be gained, but can be lost
   d. can never really be harnessed
   e. can never be transformed

2. Each time there is a chemical reaction, some energy is exchanged. According to the second law of thermodynamics, with each exchange
   a. The total amount of energy in the system remains unchanged.
   b. Energy cannot be exchanged in chemical reactions, but must come from the sun every time.
   c. A system goes from a more-ordered to a less-ordered state, and some energy becomes unavailable because it is lost as heat.
   d. Energy is gained for future use.
   e. Some energy is permanently and completely destroyed.

3. You are running a marathon and need lots of energy to keep your muscles contracting. ATP is broken down to ADP to supply the energy. Thus, the breaking down of ATP to ADP is
   a. caused by a coenzyme
   b. caused by a vitamin
   c. endergonic
   d. exergonic
   e. a, b, and c are all correct

4. ATP stores energy in the form of
   a. mechanical energy
   b. heat
   c. complex carbohydrates
   d. bond energy
   e. amino acids

5. People who are lactose intolerant lack a compound called lactase in their digestive tract. You know that this compound, lactase, is probably a(n)
   a. hormone, because it causes a change in the person
   b. coenzyme, because it helps the enzyme, lactose, do its job
   c. vitamin, because it is essential to health
   d. substrate, because it is involved with food
   e. enzyme, because it ends in -ase.

6. Glycolysis, a process that is discussed in the next chapter, is a metabolic pathway. From this we know that glycolysis
   a. consists of a series of steps
   b. uses a number of enzymes
   c. most likely is a large-scale activity
   d. a and c only
   e. a, b, and c

7. If you accidentally took a poison that prevented any ATP from being made, you would
   a. die
   b. have to use ADP for your direct source of energy
   c. have to use lipids for your direct source of energy
   d. have to use glucose for your direct source of energy
   e. have to use enzymes for your direct source of energy

8. Which of the following will lower the energy of activation of a reaction in a cell?
   a. increasing the temperature
   b. increasing the pressure
   c. using an enzyme
   d. changing the amount of the reactants
   e. supplying ATP

Brief Review

1. A plant leaf is more ordered and complex than is carbon dioxide and water. In photosynthesis plants take small, disordered carbon dioxide and water molecules and make complex, ordered sugars out of them. How is it possible for this to happen without violating the second law of thermodynamics?

2. The text notes that in the sodium-potassium pump, sodium ions are moved out of the cell against their concentration gradient and that ATP plays a part in this. Explain how this is a coupled reaction.

3. Where does the energy come from that is stored and released by ATP? In what way is ATP an energy currency molecule?

4. The average person thinks of enzymes only as something in their digestive tract that breaks down food. How would you explain the role of enzymes to your little sister, who has heard about digestion in grade school biology?

5. What is meant by allosteric regulation of an enzyme?

Applying Your Knowledge

1. In light of the laws of thermodynamics, explain why you get so much hotter when you run than when you are sitting still.

2. A box of macaroni, which is made almost entirely of starch, sits on a shelf. If you cook the macaroni and eat it, the starch starts to break down to sugars in the few seconds that the food is in your mouth. What helps to break down the starch? Would the box of macaroni eventually convert to sugar if you left it on the shelf?

Group Activity

An inevitable effect of the first and second laws of thermodynamics is that some of the products we use, whose manufacture consumes large amounts of energy—such as paper, machines, clothing, etc.—wear out and become less usable. The matter in them never decreases in amount, but as the energy becomes more and more disordered and dispersed, we are left with waste products that have to be discarded or recycled. As a class, or a group within the class, engage in a recycling project on campus. Advertise it among other students as a thermodynamic project. Use slogans such as "Fight Entropy! Recycle!" As you participate in this project, try to understand what energy changes must occur in order to make recycling work.

7

Vital Harvest
Deriving Energy from Food

Fruit of the vine.
(page 140)

A transition step in respiration.
(page 138)

Traveling electrons.
(page 141)

In 1986, the American cyclist Greg LeMond electrified the sporting world by becoming the first non-European to win the world's most prestigious bicycle race, the Tour de France. LeMond went on to win the race twice more, in 1989 and 1990, capturing his 1989 victory by coming from far behind to win on the race's last day.

By 1993, however, instead of winning races LeMond was dropping out of them. He would become exhausted, even sick, in competition; he could no longer climb hills. Through the course of the next year, he and his doctors did an intensive investigation on what might be ailing him. Eventually they came up with an answer: LeMond was suffering from a condition called mitochondrial myopathy. In essence, the mitochondria or "powerhouses" of his cells could not utilize oxygen efficiently. Many people who suffer from this illness have trouble making it from one side of a room to the other. In LeMond's case, the condition meant that he could no longer compete at the elite levels of cycling, and his racing career was over.

The existence of a condition such as mitochondrial myopathy drives home a paradox: No one needs to be reminded that we need to breathe in order to live, and most people are aware that oxygen is the most important thing that comes in with breath. That said, of the next 100 people you meet, how many could tell you what oxygen is *doing* to sustain life? Put another way, how many people could tell you why we need to breathe?

The short answer is that breathing and oxygen are in the energy transfer business. They are part of a system that allows us to extract, from food, energy that is then used to put together the "energy currency" molecule, ATP. As you'll recall from Chapter 6, the body uses ATP (adenosine triphosphate) to power activities that range from muscle contraction to thinking to cell repair. Living things need a certain amount of ATP to live, and organisms such as ourselves use oxygen to produce much of our ATP.

Oxygen is not required for all the "harvesting" we do of the energy contained in food. When you lift a box, the short burst of energy that is required comes largely from a so-called anaerobic transfer of energy, which you'll learn more about shortly. But even here, the essential *product* of energy extraction is ATP. Keep that in mind and you won't

Greg LeMond at the Tour de France in 1989

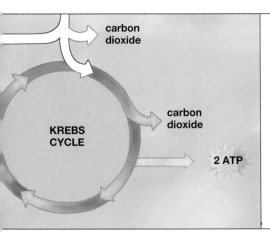

One stage in energy harvesting.
(page 135)

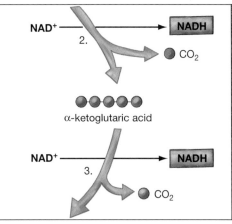

A closer look at energy harvesting.
(page 139)

One effect of oxidation: rust.
(p. 133)

get lost in the byways of energy transfer. How does the body produce enough ATP to allow us to read a book or climb a flight of stairs? In this chapter, you'll find out.

7.1 Energizing ATP: Adding a Phosphate Group to ADP

Given the importance of ATP to this story, let us begin with a brief review of how this molecule works. Recall from Chapter 6 that each ATP molecule has three phosphate groups attached to it and that ATP powers a given reaction by losing the outermost of these phosphate groups. In this process ATP is transformed into a molecule with *two* phosphate groups, adenosine diphosphate or ADP. To return to its more "energized" ATP state, it must have a third phosphate group attached again. This is, however, a trip *up* the energy hill, because the product (ATP) contains more energy than the reactant (ADP). Thus, getting that third phosphate group onto ADP requires energy. It is like preparing a spring-loaded mousetrap for action. To get the mousetrap bar in place requires energy, and the same is true of getting a third phosphate group in place on ATP (**see Figure 7.1**). Where does the energy come from? It comes from food; energy that is extracted from food powers the phosphate group up the energy hill—and literally onto ADP.

7.2 Electrons Fall Down the Energy Hill to Drive the Uphill Production of ATP

In tracking the extraction of energy from food, we will use as an example one particular molecule, glucose, to see how energy is harvested from it. Though the details here are complex, the essential story is simple. *Electrons* derived from glucose, which is high in energy, will be running downhill; they will be channeled off, a few at a time, and their downhill drop will power the *uphill* push needed to attach a phosphate group onto ADP. The glucose-derived electrons will be transferred to several intermediate molecules in their downhill journey, but the *final* molecule they will be received by at the bottom of the energy hill will be oxygen. We need to breathe because we need oxygen to serve as this final electron acceptor.

The Great Energy Conveyors: Redox Reactions

The basis for electron transfers down the energy hill is straightforward: Some substances more strongly attract electrons than do others. A substance that loses one or more electrons to another is said to have been **oxidized**. We hear this word all the time; when paint on an outdoor surface has become dulled, for example, or when metal rusts (**see Figure 7.2**). Meanwhile, the substance that *gains* electrons in this reaction is said to have undergone **reduction**. (This seems about as logical as saying a country lost a war by winning. One way to think about it is that because electrons carry a negative charge, any substance that gains electrons has had a reduction in its *positive charge*.)

In cells, oxidation and reduction never occur independently. If one substance is oxidized, another must be reduced. It's like a teeter-totter; if one side goes down, the other must come up. The combined operation is known as a reduction-oxidation reaction, or

Figure 7.1
Storing and Releasing Energy
Adenosine triphosphate (ATP) is the most important energy-storing molecule in our bodies. The energy it contains is used to power everything from muscle contraction to thinking.

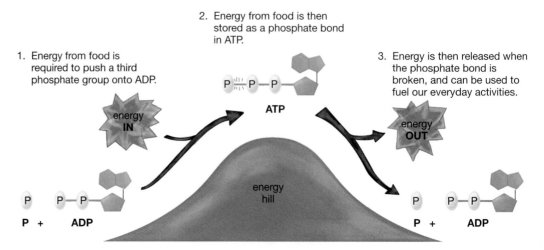

1. Energy from food is required to push a third phosphate group onto ADP.

2. Energy from food is then stored as a phosphate bond in ATP.

3. Energy is then released when the phosphate bond is broken, and can be used to fuel our everyday activities.

P P–P ATP

energy IN

energy hill

energy OUT

P + P P–P

P + ADP P P–P ADP

simply a **redox reaction**. Critically, the substance being oxidized in a redox reaction has its electrons traveling energetically *downhill*.

Many Molecules Can Oxidize Other Molecules

The term *oxidation* might give you the idea that *oxygen* must be involved in any redox reaction, but this is not the case. Any compound that serves to accept electrons from another is a so-called oxidizing agent. In living things a large number of molecules are involved in energy transfer, and each has a certain tendency to gain or lose electrons relative to the others.

This is how electrons can be passed down the energy hill: The starting "energetic" molecule of glucose is oxidized by another molecule, which in turn is oxidized by the *next* molecule down the hill. The whole thing might be thought of as a kind of downhill electron bucket brigade. Molecules that serve to shuttle electrons down the energy hill are known as **intermediate electron carriers**. The thing that makes their role a little complicated is that many of the electrons they accept are bound up originally in hydrogen atoms. If you can remember back to Chapter 3, a hydrogen atom amounts to nothing but one proton and one electron; in transferring a hydrogen atom, then, a molecule is transferring a single electron (bound to a proton), which means a redox reaction has taken place.

Redox through Intermediates: NAD

The most important electron carrier is a molecule known as **nicotinamide adenine dinucleotide**, or **NAD**, which can helpfully be thought of as a city cab. It can exist in two states: loaded with passengers or empty. And like a cab, it can switch very easily between those two states. The passengers that NAD picks up and drops off are electrons.

The "empty" state that NAD comes in is ionic: NAD^+. Remembering back to the discussion about ions, you can see that NAD^+ is positively charged, meaning that it has lost an electron. In a redox reaction, what NAD^+ does is pick up one hydrogen atom (an electron and a proton) and one solo electron (from a second hydrogen atom). The isolated electron that NAD^+ picks up turns it from positively charged to neutral ($NAD^+ \rightarrow NAD$); the whole hydrogen atom takes it from NAD to NADH. Keeping an eye on redox reactions here, NAD^+ has become NADH by oxidizing a substance—by accepting electrons from it.

So much for half of NAD's role: picking up passengers. *Now*, as NADH, it is loaded with these passengers. It can proceed down the energy hill to donate them to substances that have a greater potential to *accept* electrons than it does. Having dropped its passengers off with such a substance, it returns to being the empty NAD^+ and is ready for another pickup (**see Figure 7.3**). Through

Figure 7.2
An Effect of Oxidation
Rust has formed on a hook atop a corroded steel sheet.

Figure 7.3
The Electron Carrier NAD⁺
In its unloaded form (NAD^+) and its loaded form (NADH), this molecule is a critical player in energy transfer, picking up energetic electrons from food and transferring them to later stages of respiration.

1. NAD^+ within a cell, along with two hydrogen atoms that are part of the food that is supplying energy for the body.

2. NAD^+ is reduced to NAD by accepting an electron from a hydrogen atom. It also picks up another hydrogen atom to become NADH.

3. NADH carries the electrons to a later stage of respiration then drops them off, becoming oxidized to its original form, NAD^+.

this process, NADH transfers energy from one molecule to another.

How Does NAD Do Its Job?

The molecule that NAD^+ is oxidizing is the starting glucose molecule (actually derivatives of it). To carry out its oxidizing role, NAD^+ obviously needs to be brought together with the glucose derivatives. What have you been looking at that brings substances together in this way? Enzymes! One of the things you'll be seeing, therefore, is NAD^+ and glucose derivatives being brought together by enzymes, at which point NAD^+ accepts electrons from these derivatives, later to hand them off to another molecule. Electron transfer through intermediate molecules such as NAD^+ provides the energy for most of the ATP produced. Thus, when looking at the diagrams in this chapter that outline respiration, you'll see:

or:

What these drawings indicate is the electron carrier shifting between its empty state (NAD^+) and its loaded state (NADH), or vice versa.

7.3 The Three Stages of Cellular Respiration: Glycolysis, the Krebs Cycle, and the Electron Transport Chain

All the energy players are finally in place: glucose, redox reactions, electron carriers, enzymes, and the rest. In terms of a molecular formula, the big picture on respiration looks like this:

$$C_6H_{12}O_6 + 6O_2 + ADP \rightarrow 6CO_2 + 6H_2O + ATP$$

The starting molecule on the left ($C_6H_{12}O_6$) is glucose, which is storing chemical energy. You can also see on the left the oxygen ($6O_2$) that is needed as the final electron acceptor, and ADP. The *products* of this reaction, to the right of the arrow, are carbon dioxide ($6CO_2$) and water ($6H_2O$) as by-products, and energy in the form of ATP. How *much* ATP does the breakdown of one molecule of glucose yield? About 36 mole-

cules of ATP. Though many separate steps are involved in actually getting energy from glucose to ATP, respiration can be divided into three *sets* of steps. These are known as **glycolysis**, the **Krebs cycle**, and the **electron transport chain**.

Glycolysis Is the First Stage in Energy Harvesting

The division among these respiratory steps is, in one sense, a division of evolutionary history. For most eukaryotic creatures—ourselves, birds, trees—the first of these sets of steps, glycolysis, could be likened to a nineteenth-century steam engine working side by side with a state-of-the art electric turbine. Glycolysis does produce ATP, but its net production is only two ATP molecules per glucose molecule, which is pretty small in comparison to the 36 ATP molecules likely to come from all three sets of steps in oxygen-using or "aerobic" respiration. Human beings make use of the energy yielded in glycolysis, but we could not function for 3 minutes if it were our *sole* source of energy. The real import of glycolysis for us is that it's critical in supplying energy to us in certain situations—in a short burst of activity, for example—and it is a necessary precursor to the other stages of respiration. Meanwhile, for certain bacteria and one-celled eukaryotes, glycolysis can be the sole source of energy transfer.

Krebs and the Electron Transport Chain Were Later in Evolving and Are More Efficient

In this division of who uses what, you can probably begin to guess where glycolysis fits in. Scientists believe that it is a more ancient form of energy harvesting than the other two sets of steps. This is why it exists in all organisms. However, at some time in the evolutionary past, processes we now call the Krebs cycle and the electron transport chain were added. The electron transport chain requires one critical thing that glycolysis doesn't: the ability to use oxygen. The separation between glycolysis and the other two steps is also a physical separation, in that glycolysis takes place in the cell's cytoplasm while, for eukaryotes, the Krebs cycle and the electron transport chain take place within the cell structures noted earlier called mitochondria.

An Overview of the Three Stages

In overview, energy harvesting through these three phases goes like this: We get some ATP yield in glycolysis and the Krebs cycle, but for eukaryotes such as ourselves the *main* function of these phases is the transfer of electrons to the intermediate carriers such as NAD^+. Bearing their electron cargo, these carriers then move to the electron transport chain, where they are themselves oxidized (losing electrons). The movement of these electrons through the chain then yields about 32 of the approximately 36 molecules of ATP that are netted per molecule of glucose. If you look at **Figure 7.4**, you can see a schematic representation of the three phases of energy extraction. Taken as a whole, this three-stage harvesting of energy via oxidation is known as **cellular respiration**. Now let's see how the three phases of this process work in detail.

7.4 First Stage of Respiration: Glycolysis

Glycolysis means "sugar splitting," which is appropriate because our starting molecule is the simple sugar glucose. Briefly, here is what happens during glycolysis. First, this one molecule of glucose has to be prepared, in a sense, for energy release. ATP actually has to be *used*, rather than synthesized, in two of the first steps of glycolysis, so that the relatively stable glucose can be put into the form of a less-

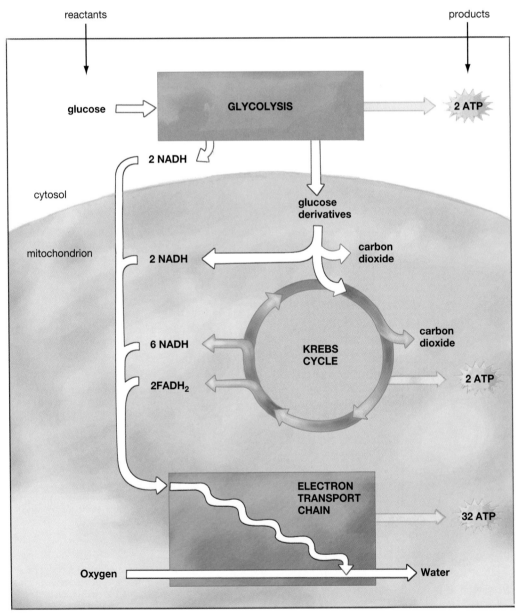

36 ATP maximum
per glucose molecule

Figure 7.4
Overview of Energy Harvesting
The energy derived from food (glucose in the example) is transferred to ATP in a process that has three major stages: glycolysis, the Krebs cycle, and the electron transport chain. A maximum of 36 molecules of ATP can be obtained from one molecule of glucose in these three sets of steps; glycolysis and the Krebs cycle yield 2 molecules of ATP each, and the electron transport chain yields a maximum of 32 ATP. Glycolysis and the Krebs cycle also yield electron carriers (NADH and $FADH_2$) that transport electrons to the electron transport chain. Oxygen is consumed in electron transport chain metabolism, while water and carbon dioxide are produced in it. Glycolysis takes place in the cytosol of the cell, but the Krebs cycle and electron transport chain take place in the cellular organelles called mitochondria.

stable sugar. This sugar is then split in half. The two molecules formed have three carbons each (whereas the starting glucose has six). Once this split is accomplished, the steps of glycolysis take place in *duplicate*: It is like taking one long piece of cloth, dividing it in two, and then proceeding to do the same things to both of the resulting pieces. Look now at **Figure 7.5** to see how glycolysis proceeds.

Here are the steps of glycolysis in detail.

The Steps of Glycolysis

1. Delivered by the bloodstream, glucose enters the cell and immediately has a phosphate group from ATP attached to it. This process, called phosphorylation, is facilitated by hexokinase, an enzyme introduced in Chapter 6. Because the

Figure 7.5
Summary of Glycolysis
In glycolysis, the single glucose molecule is transformed in a series of steps into two molecules of a substance called pyruvic acid. These two molecules then move on to the next stage of cellular respiration (the Krebs cycle). Meanwhile, glycolysis also produces two molecules of electron-carrying NADH, which move directly to the electron transport chain. Although two molecules of ATP are used up in the earlier stages of glycolysis, four more are produced in the later stages, for a net production of two ATP molecules per glucose molecule. The carbon atoms are represented by red circles, and the phosphate groups are represented by yellow circles.

SUMMARY OF GLYCOLYSIS

1. Glucose enters cell from bloodstream and has a phosphate group attached to it from ATP.

2. Glucose 6-phosphate is rearranged.

3. A second phosphate group is added from another ATP (total of 2 ATP molecules used so far).

4. Six-carbon sugar splits into two identical three-carbon sugars.

5. Each sugar molecule is oxidized by NAD^+, enabling the attachment of phosphate groups. Two NADH molecules result.

6. Each sugar molecule loses a phosphate group, releasing enough energy to push the phosphate onto ADP, producing one ATP molecule each.

7. Once again, the loss of a phosphate group from the sugar molecules fuels the production of ATP molecules.

Net production: **2 ATP** molecules produced for each 1 glucose molecule used phosphate is attached to the sixth carbon of glucose, it now goes under the name glucose 6-phosphate. Note that one molecule of ATP has been *used* in this step. The ATP ledger thus reads: minus-one.

2. Glucose 6-phosphate is rearranged to become a molecule called fructose 6-phosphate.

3. Fructose 6-phosphate has a second phosphate group added, so that it becomes a molecule called fructose 1,6-biphosphate. Another molecule of ATP has been used in getting this arrangement, however, which makes the ATP ledger now read: minus-two. But fructose 1,6-biphosphate now has phosphate groups on each end, which is important because it is about to be split.

4. Fructose 1,6-biphosphate becomes two molecules of glyceraldehyde 3-phosphate. We started with a six-carbon sugar, glucose. We now have two separate three-carbon sugars, each with phosphate groups attached. From here on out, glycolysis begins to proceed in duplicate. What happens to one of the glyceraldehyde molecules happens to the other.

5. At last, some energy from glucose. An enzyme brings together the glyceraldehyde 3-phosphate molecule, NAD$^+$, and a phosphate group. The outcome of this is that NAD$^+$ oxidizes glyceraldehyde 3-phosphate. Now NAD$^+$ in its new form, NADH, moves down the energy hill, occupied with one new hydrogen atom (a proton plus an electron) and one new solo electron. The oxidation it has carried out is energetic enough that it allows the phosphate group to become attached to the main molecule. Now called 1,3-diphosphoglyceric acid, it has two phosphates attached to it. Because everything in this step is happening in duplicate, this yields two NADH molecules.

6. ATP at last. Energy-enriched as it was, 1,3-diphosphoglyceric acid loses one of its phosphate groups downhill, thus becoming 3-phosphoglyceric acid. The fall is energetic enough to push this phosphate onto an ADP molecule and make it ATP. Because this is happening in duplicate, two molecules of ATP are produced; two ATP were used in steps 1 and 3, however, so that the ATP ledger now stands at zero.

7. Through a couple of reactions, 3-phosphoglyceric acid becomes a molecule called phosphoenolpyruvic acid, which generates more ATP when it transfers its phosphate group to ADP. As this takes place in duplicate, two more ATP molecules are created, which leaves the ATP ledger, at the end of glycolysis, at plus-two. The reaction turns phosphoenolpyruvic acid into pyruvic acid, the final form of the original glucose molecule in glycolysis.

Before continuing on to the other stages of energy harvesting, you may wish to read more about what might be called the consequences of glycolysis. For some organisms, this is the end of the harvesting line, as it is in certain instances even for oxygen-using organisms such as ourselves. You can read about these consequences in "When Energy Harvesting Ends at Glycolysis," on page 140. Glycolysis' role in providing quick bursts of energy for human beings means it fits into a special place in providing energy for human exercise; you can read about this in "Energy and Exercise" on page 142.

7.5 Second Stage of Respiration: The Krebs Cycle

Glycolysis has accomplished three valuable things in energy harvesting: It yielded two ATP molecules, it yielded two energized molecules of NADH, and it resulted in two molecules of pyruvic acid. These pyruvic acid molecules are the derivatives of the original glucose molecule. Now *they* are the molecules that will be oxidized—that will have electrons removed from them—in the next stage of energy harvesting. They serve this purpose because glycolysis by itself could not wring all the energy out of the starting glucose molecule. (After all, it yielded only two of the 36 molecules of ATP that eventually are harvested.)

After glycolysis, the second stage of respiration is called the **Krebs cycle**. This stage is named for the German and English biochemist Hans Krebs, who in the 1930s used the flight muscles of pigeons to find out how aerobic respiration works. Because the first product of the Krebs cycle is citric acid, the cycle is sometimes referred to as the **citric acid cycle**.

The ATP yield in this cycle is once again a paltry two molecules. The big harvest of ATP comes in the *final* stage of respiration, the electron transport chain (ETC). The Krebs cycle serves, however, to set the stage for this

MediaLab Link
Enzymes, see p. 148.

MediaLab Link
Glucose, see p. 148.

harvest by supplying energy-bearing electron carriers, as you will see.

Site of Action Moves from the Cytoplasm to the Mitochondria

Glycolysis yielded its two molecules of pyruvic acid in the cytoplasm, but the site of energy harvesting quickly shifts, with the pyruvic acid, to the new site of energy transfer, the mitochondria. These organelles are where all the action takes place from here on out, so now would be a good time to look at **Figure 7.6** and examine their structure. You can see that mitochondria have both an inner and an outer membrane. The Krebs cycle reactions take place to the interior of the inner membrane—in an area known as the inner compartment—while the reactions

of the electron transport chain take place *within* the inner membrane.

Between Glycolysis and Krebs, an Intermediate Step

There actually is a transition step in respiration between glycolysis and the Krebs cycle. In it, the three-carbon pyruvic acid molecule combines with a substance called coenzyme A, thus forming acetyl coenzyme A (acetyl CoA). One outcome of this reaction is that acetyl CoA enters the Krebs cycle. There are, however, two other products of this reaction. One is a carbon dioxide molecule that, like others you'll be seeing, eventually diffuses out of the cell and into the bloodstream. This is one source of the CO_2 that we exhale when we breathe. The other product is one more molecule of NADH, which finds its way to the ETC. This reaction takes place twice for each glucose molecule (**see Figure 7.7**).

cell

mitochondrion

Figure 7.7
Transition between Glycolysis and the Krebs Cycle
The pyruvic acid product of glycolysis does not enter directly into the Krebs cycle. Rather, it must first be transformed into acetyl coenzyme A. The consequences of this reaction are the production of CO_2, which dissolves into the bloodstream, and the production of an NADH molecule, which continues onto the electron transport chain. Because one molecule of glucose produces two molecules of pyruvic acid, two molecules of NADH are produced per glucose molecule in this transitional step.

Into Krebs: Why Is It a Cycle?

Acetyl CoA now enters the Krebs cycle. The reason this is a *cycle* becomes clear when you look at **Figure 7.8**. You can see that, in the first step of the cycle, Acetyl CoA combines with a substance called oxaloacetic acid to produce citric acid. If you look over at about 11 o'clock on the circle, however, you can see that the *last* step in the cycle is the *synthesis* of oxaloacetic acid. Thus, a substance that's necessary for this chain of events to take place is itself a product of the chain.

Stroll around the circle now to see how the Krebs cycle functions in a little more detail. In essence, what's happening is that, as the entering acetyl CoA molecule is transformed into these various substances, it is being oxidized by electron carriers, with the resulting electrons moving on to the ETC. ATP is also derived, and CO_2 is a product.

The only major player that's unfamiliar here is the FAD-$FADH_2$ that can be seen taking part in a redox reaction at about 7 o'clock. It is simply another electron carrier, similar to NAD^+/NADH.

In counting up the Krebs' yield of both ATP and electron carriers (NADHs, etc.), recall that *two turns* around this cycle result from the original glucose molecule (because glucose provided *two* molecules of the starting pyruvic acid). This means a total yield of 6 NADH, 2 $FADH_2$, and 2 ATP from Krebs for each glucose molecule.

Here are the steps of the Krebs cycle in detail.

The Steps of the Krebs Cycle

1. Acetyl CoA combines with the four-carbon oxaloacetic acid. The CoA fragment separates from this compound; the

SUMMARY OF THE KREBS CYCLE

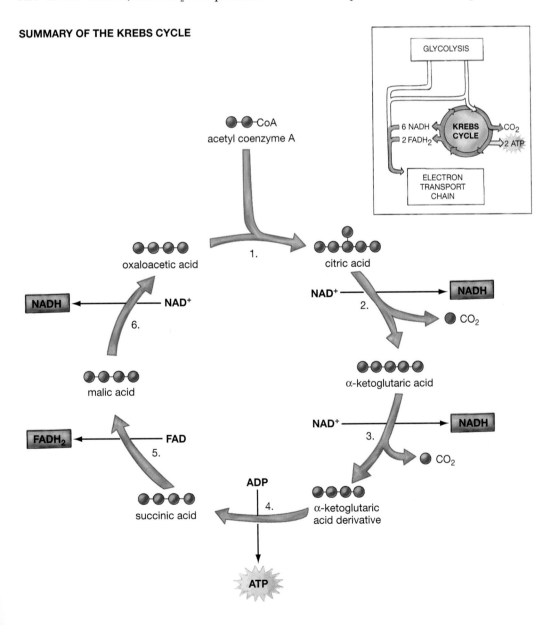

Figure 7.8
Summary of Krebs Cycle
The Krebs cycle is the major source of electrons that are transported to the electron transport chain by the electron carriers NADH and $FADH_2$. For each molecule of glucose, two molecules of acetyl coenzyme A enter the Krebs cycle. Through a series of reactions, a total of 6 NADH, 2 $FADH_2$, and 2 ATP are produced per glucose molecule.

When Energy Harvesting Ends at Glycolysis

You've seen that for many bacteria, and even sometimes for people, energy harvesting ends with the set of steps called glycolysis, rather than proceeding on through the Krebs cycle and electron transport chain. When this happens, organisms need a way to recycle their energy transfer molecules in a way that doesn't depend on oxygen. (The NADH they produced in glycolysis needs to lose its added electrons and become NAD$^+$ again, so that it can be reused in glycolysis.) The solution to this dilemma is a kind of electron dumping, the *products* of which turn out to be some of the most familiar substances in the world.

Glycolysis in Yeast

Yeast is a good example of how this works, because it is a busy single-celled fungus. Yeasts can live by glycolysis alone, or they can go through aerobic respiration. Say, then, that yeasts are working away on a sugar, going

---

For yeasts, alcohol is simply a waste by-product of the glycolysis they utilize for energy conversion.

---

through glycolysis, but doing so in an oxygenless environment. Recall that the final "substrate" product of glycolysis is two molecules of a substance called pyruvic acid. In yeasts, the pyruvic acid they end up with is converted to something called acetaldehyde, and *it* takes on the electrons from NADH, meaning the recycling problem is now solved. Having taken on NADH's electrons, though, acetaldehyde now is converted to something very familiar: ethanol, better known as drinking alcohol.

Human beings *put* yeasts in environments in which this will happen, of course, because this is how we make wine and beer. Imagine yeast cells inside a dark, airless wine cask, working away on burgundy grape juice, harvesting energy from the grape-juice sugars, but in so doing turning out pyruvic acid, which is turned into alcohol. Things are going along fine for these yeast cells, until

the alcohol content of the wine reaches a level (about 14 percent) at which they can no longer survive in it. This whole process is known as **alcoholic fermentation**.

For yeasts, alcohol is simply a waste by-product of the glycolysis they utilize for energy conversion. The same is true of the carbon dioxide that also results from their conversion of pyruvic acid. Here again, humans have found great use for the refuse of nature. Put the right yeast into dough, and its continuing fermentation produces the CO_2 that causes bread to rise and become "light," because of the air holes now in it. (And the alcohol? It evaporates.)

Fermentation in Animals

Alcoholic fermentation is the solution that fungi (and occasionally plants) employ to sustain glycolysis in the absence of oxygen. But animals take a different tack, because in them the product of glycolysis, pyruvic acid, accepts the electrons from NADH. In this process, pyruvic acid is turned into another substance: **lactic acid**. Thus, the kind of fermentation that animals (and certain bacteria) carry out is called **lactate fermentation**. Have you ever experienced the muscle "burn" that comes with, for example, climbing several flights of stairs? If so, you have experienced a buildup of lactic acid in your muscles.

Yeast Inside, Performing Glycolysis
This man is inspecting a gauge on a huge metal tank of fermenting wine. The alcohol in the wine is a by-product of the glycolysis carried out by yeast.

---

six-carbon citric acid that results is an energetic substrate that will now be oxidized.

2. A citric acid derivative loses a CO_2 molecule, thus becoming a five-carbon substrate that is oxidized in the NAD$^+$ → NADH reaction, with NADH going to

the ETC. Citric acid is now alpha-ketoglutaric acid.

3. Alpha-ketoglutaric acid loses a CO_2 molecule, and the resulting four-carbon molecule is oxidized by NAD$^+$.

4. A derivative of alpha-ketoglutaric acid is split, releasing enough energy to attach a

phosphate group to ADP. ATP is thus produced. Alpha-ketoglutaric acid has become succinic acid.

5. Succinic acid is oxidized by FAD, losing two hydrogen atoms (complete with electrons) to it. FAD is thus reduced to $FADH_2$ and moves to the ETC. In a series of steps, succinic acid is transformed into malic acid.

6. Malic acid is oxidized by NAD^+. This oxidation transforms malic acid into oxaloacetic acid, which is now ready to enter into another turn of the cycle.

7.6 Third Stage of Respiration: The Electron Transport Chain

Having moved through Krebs, you've reached the main event in cellular respiration: The production of 32 more ATP molecules through the work of the **electron transport chain (ETC)**. Glycolysis took place in the cytoplasm, and the Krebs cycle took place in the inner compartment of the mitochondria. Now, however, the action shifts to the mitochondrial inner membrane, the site of the electron transport chain (**see Figure 7.9**). The "links" in this chain are a series of molecules. You've been looking for some time now at how NADH and $FADH_2$ carry off the electrons derived from glycolysis and the Krebs cycle. This is the destination of these electron carriers and their cargo.

Upon reaching the mitochondrial inner membrane, the electron carriers donate electrons and hydrogen ions to the ETC. Once again, this is a trip down the energy hill, only *NADH* is now the molecule whose electrons exist at a higher energy level. Thus, when NADH runs into the appropriate protein in the ETC, NADH is oxidized, donating its electrons and proton to the ETC. This process then is simply repeated down the whole ETC, each carrier donating electrons

Figure 7.9
The Electron Transport Chain (ETC)
The movement of electrons through the ETC powers the process that provides the bulk of the ATP yield in respiration. The electrons carried by NADH and $FADH_2$ are released into the ETC and transported along its chain of molecules. The movement of electrons along the chain releases enough energy to power the pumping of hydrogen ions (H^+) across the membrane into the outer compartment of the mitochondrion. It is the subsequent energetic "fall" of the H^+ ions back into the inner compartment that drives the synthesis of ATP molecules by the enzyme ATP synthase.

Energy and Exercise

One minute we're sleeping, the next we're looking for our running shoes, and 10 minutes after that we're bounding down the road trying to cover three miles at a faster pace than yesterday. Coming to the steep hill at the end of our run, we go all out—charging up to the top of it, slowing down only after we start going downhill, breathless and spent. How can the human body cope with a range of energy demands that runs from sleeping to sprinting? Its secret is having a kind of energy trio at its service, with each member of the group specializing in delivering energy in particular situations.

The essence of any kind of exercise is the contraction of muscles, and the only energy molecule that can power the contraction of skeletal muscle is ATP. You have seen two processes by which the body puts together ATP. One is glycolysis, which does not use oxygen; the other is "aerobic" respiration, which does use oxygen. These are two members of the energy trio the body uses. Our cells also have a capacity to *store* small amounts of ATP, however, and they can also stockpile another molecule, phosphocreatine (PCr), that acts as a reservoir of phosphate groups that can be used to produce ATP. This stored ATP/PCr is the third member of the energy ensemble. Let's now see how all three players would work together in the course of a bike ride, starting out at a fairly brisk pace.

With the first turns of the pedals, bike riders bump up their body's energy demands tremendously. In their cycling, they could easily burn up 10 times the calories they would while in a resting state. (As a point of comparison, in doing housework a person typically burns up about three times the calories he or she does while at rest.)

So, how do muscle cells accommodate such large increases in their need for ATP? In the first few seconds, the small reservoirs of ATP/PCr take the lead role, with a proportionately smaller contribution from glycolysis and a smaller contribution yet from aerobic metabolism. These differences reflect the time it takes for each of these systems to get going. You've seen how many steps there are in glycolysis, and how many *more* steps there are in the oxygen-using aerobic respiration. By comparison, the output of ATP from stored ATP/PCr is instantaneous. Remember, though, that the ATP/PCr reservoir is small. A person who had to depend solely on it for a full-ahead sprint would have only enough energy to last about six seconds. On a longer bike ride, glycolysis and ATP/PCr are contributing roughly equal amounts of ATP within 10 seconds. Within 30 seconds, glycolysis has greatly eclipsed PCr as an energy provider.

At about this same 30-second mark, the third player, aerobic respiration, is supplying only about 30 percent of the ATP, but its contribution is growing fast. In **Figure 1**, you can see how the aerobic contribution to ATP yield grows over the course of a long, strenuous workout. At the start, oxygen-aided respiration is supplying less than 10 percent of our ATP, but 10 minutes into the workout it is delivering 85 percent.

You might think that this would be the final division of ATP contributions among the three players, but the

to the next electron carrier in line. A carrier is reduced by receiving these electrons; donating them to the next carrier, it then returns to an oxidized state. Each carrier thus alternates between oxidized and reduced states.

Visualizing the ETC

There are several ways to conceptualize the ETC. In Figure 7.9, you can see that one way is to view the ETC as three large protein complexes with two smaller mobile molecules that link them. When NADH arrives at the inner membrane, it bumps into the ETC's first carrier, which accepts electrons from NADH; this carrier then donates these electrons to the next carrier, and so on down the line to the last electron acceptor, which is oxygen.

Where's the ATP?

At this point you may be tapping your foot, waiting for this promised harvest of ATP to appear. So far, all that's taken place is a transfer of electrons in the ETC. Well, note what happens in the first ETC protein complex. The movement of electrons through it releases enough energy to power the movement of *hydrogen ions* (H^+ ions) through the complex, pushing them from the inner compartment into the outer compartment. By the time the electrons have completed their movement through the ETC, this pumping of H^+ ions will occur with two other complexes. Critically, these H^+ ions are being pumped *against* their concentration and electrical gradients (which were reviewed in Chapter 5). Put another way, the ions are being pumped

Of course, lots of sports have no aerobic component to them at all. The energy for a sport like football is supplied almost entirely by stored ATP/PCr. Even in a competition as long as a 400-meter sprint—about once around a regulation track—about 70 percent of the energy comes from anaerobic respiration, meaning a combination of ATP/PCr and glycolysis.

Duration of maximal exercise								
Seconds			Minutes					
10	30	60	2	4	10	30	60	120
Percent anaerobic								
90	80	70	50	35	15	5	2	1
Percent aerobic								
10	20	30	50	65	85	95	98	99

Figure 1
Different Contributions over Time
Relative contributions of anaerobic and aerobic respiration to exercise during the duration of a workout. At the 1-minute mark, aerobic respiration is supplying only 30 percent of the body's energy needs; at the 10-minute mark, it is supplying 85 percent. (Adapted from Astrand, P. O., and Rodahl, K. *Textbook of Work Physiology.* New York: Mc-Graw Hill Book Company, © 1977.)

amounts they supply can vary over time depending on the intensity of the activity. **Figure 2** shows how this would work in an actual professional-level bicycle race. Going along at a steady state in "pack" riding, you can see the overriding contribution of aerobic metabolism. In the final sprint, however, stores of ATP/PCr—replenished during the pack riding—are brought out to take the leading role, with the contribution of glycolysis increasing as well. In stretches of difficult hill climbing, meanwhile, glycolysis predominates.

Figure 2
Different Activities, Different Energy Stores
Energy harvesting processes used at different stages in a bicycle race. (Adapted from Kearney, J. T. "Training the Olympic Athletes." *Scientific American,* June 1996, p. 54.)

up the energy hill with energy supplied by the *downhill* fall of electrons through the ETC.

Here's where the ATP is produced at last. The H^+ ions that have been pumped into the outer compartment move *back down* their energy gradients, into the inner compartment, through a special enzyme called **ATP synthase**. This remarkable enzyme is driven by the H^+ ions flowing through it, which cause part of the enzyme to rotate (as fast as 100 revolutions per second). This spinning drives the synthesis of ATP from the constituent parts of ADP and phosphate.

This mechanism—the pumping of hydrogen ions powering the synthesis of ATP—was first proposed in 1961 by British biochemist Peter Mitchell, who won the Nobel Prize in 1978 for his contribution to bioener-

getics. Lest it pass by too quickly, *this* is the essence of aerobic respiration. The downhill drop of electrons derived from food has powered the uphill synthesis of the molecule that powers the vast majority of our activities.

Bountiful Harvest: ATP Accounting

In the ETC, each NADH molecule will be responsible for the production of three molecules of ATP. And, of course, every molecule of NADH that comes to the chain will result in this amount of ATP. And how many are coming? Ten: two produced in glycolysis, two from pyruvic acid conversion, and six from the Krebs cycle. The $FADH_2$ carrier that came from the Krebs cycle joins the ETC a little later than does NADH and as such produces only about two ATP molecules, compared to NADH's three. Given this, here's the final count on our energy harvest.

Stage	NADH	FADH$_2$	ATP Yield
Glycolysis	2		2
Pyruvic acid conversion	2		
Krebs cycle	6	2	2
Electron transport chain	10 →		30
		2 →	4
Loss, due to active transport:			–2
Total ATP:			36

What a mighty energizer is the ETC, eh? Think about how efficient this process is compared to glycolysis (34 ATP compared to 2). Imagine one car that got 34 miles to the gallon compared to another that got 2. The 36 total ATP yield actually is the amount that can be derived under optimal conditions. (The "loss, due to active transport" in the table refers to the expenditure of ATP required to move the NADH produced in glycolysis into the mitochondria.)

Finally, Oxygen Is Reduced, Producing Water

The story of the breakdown of glucose—the saga of *one molecule's* transformation—is nearly complete. The only unfinished business lies at the end of the ETC. As noted earlier, oxygen is the final acceptor of the working electrons. In the mitochondrial matrix, oxygen accepts two electrons and two H$^+$ ions (from the inner compartment). The result is H$_2$O: water. That's why the formula for respiration is written as:

$$ADP + C_6H_{12}O_6 + 6O_2 \rightarrow ATP + 6CO_2 + 6H_2O$$

ADP plus glucose plus oxygen *yields* ATP plus carbon dioxide plus water: That's the short story of aerobic respiration.

7.7 Other Foods, Other Respiratory Pathways

By looking at aerobic respiration as it applies to glucose, which was used as the starting molecule, you've been able to go through the whole respiratory chain. But it goes without saying that there are many kinds of nutrients besides glucose—for example, fats, proteins, and other sugars. All these can provide energy by being oxidized within the chain of reactions you've just gone over. However, none of them proceeds through the chain in exactly the same way as glucose. In addition, the body has more uses for foods than producing ATP. When an organism has enough energy, it does not *need* to convert more glucose to ATP; nor does it need to use fats, other sugars, or proteins for this purpose. At any given moment, an organism may need to go the other way: toward the *buildup* of proteins from amino acids or the synthesis of the fats we call triglycerides. Organisms are able to handle this variability because nutrients and their derivatives can be *channeled* through pathways in different directions in the respiratory chain in accordance with cell needs. The alternation of the breakdown of food molecules and the buildup of food storage molecules is one of the central functions that many living things carry out.

Alternate Respiratory Pathways: Fats as an Example

To give one example of how this channeling of nutrients works, consider the possible fates of triglycerides. As you learned in Chapter 4, triglycerides are fats made of a three-carbon glycerol "head" and three long hydrocarbon chains that look like tines on a fork. Imagine that a given cell needs energy from fat. The question is, how do *triglycerides* go through aerobic respiration, as opposed to glucose?

The first thing that happens is that a triglyceride molecule is split by enzymes into its constituent parts of fatty acids and glycerol. Glycerol is then converted into glyceraldehyde phosphate. This formidable name may not ring a bell, but if you look back to step 4 in Figure 7.5, you'll see that it is one of the *downstream* derivatives of glucose in glycolysis. To put this another way, glycerol does not first get converted to glucose and then march through *all* of the steps of glycolysis. Instead, it joins the glycolytic pathway several steps down from glucose and is then converted to pyruvic acid. Then, as pyruvic acid, it goes through the Krebs and so forth, producing ATP.

What Happens When Energy Is Not Needed?

So much for what happens when an organism needs energy from fats. But what happens to the triglyceride components when this need stops? Glycerol, which had been converted to glyceraldehyde phosphate and begun marching through glycolysis, can be converted to *glucose* in a process that, in its main steps, is glycolysis in reverse. The resulting glucose molecules can then be used to do any number of things. They might, for example, be put together to form the storage form of glucose, glycogen. You may be getting a sense that the metabolism is rather like a traffic circle: Molecules can enter at different points and leave at different points for varying destinations, depending on the needs of the organism. In **Figure 7.10**, you can see a summary of the way foods are broken down through various respiratory pathways.

On to Photosynthesis

This long walk through cellular respiration has illustrated how living things harvest energy from food. Recall, however, that this chapter began with the observation that we ultimately have one source to thank for this food: the sun's energy, trapped by plants in the chemical bonds of carbohydrates. This energy conversion takes place through a process that has a beautiful symmetry with the respiration you have just looked at. That process is called photosynthesis.

Figure 7.10
Many Respiratory Pathways
Glucose is not the only starting material for cellular respiration. Other carbohydrates, proteins, and fats can also be used as fuel for cellular respiration. These reactants enter the process at different stages.

Chapter Review

Summary

7.1 Energizing ATP: Adding a Phosphate Group to ADP

- The molecule adenosine triphosphate (ATP) supplies the energy for nearly all the activities of living things. For ATP to be produced, a third phosphate group must be added to adenosine diphosphate (ADP), a process that requires energy.

7.2 Electrons Fall Down the Energy Hill to Drive the Uphill Production of ATP

- An energetic fall of electrons derived from food powers the process by which the third phosphate group is attached to ADP, making it ATP.

- Electron transfer in the production of ATP works through redox reactions, meaning reactions in which one substance loses electrons to another substance. Electrons are carried between one part of the energy harvesting process and another by inter-mediate electron carriers, the most important of which is nicotinamide adenine dinucleotide or NAD^+.

7.3 The Three Stages of Cellular Respiration: Glycolysis, the Krebs Cycle, and the Electron Transport Chain

- In more complex organisms, the harvesting of energy from food takes place in three principal stages, glycolysis, the Krebs cycle, and the electron transport chain (ETC).

- Some organisms rely solely on glycolysis for energy harvesting. For more complex organisms, glycolysis is a primary process of energy extraction only in certain situations—when quick bursts of energy are required, for example—but it is a necessary precursor to the other two stages of energy harvesting. Together, the three stages of energy harvesting are known as cellular respiration.

- Glycolysis takes place in the cell's cytoplasm, while the Krebs cycle and the elec-

Key Terms

alcoholic fermentation 140

ATP synthase 143

cellular respiration 135

citric acid cycle 137

electron transport chain (ETC) 141

glycolysis 135

intermediate electron carrier 133

Krebs cycle 137

lactate fermentation 140

lactic acid 140

nicotinamide adenine dinucleotide (NAD) 133

oxidized 132

redox reaction 133

reduction 132

tron transport chain (ETC) take place in the cellular organelles called mitochondria. Glycolysis yields two net molecules of ATP per molecule of glucose, as does the Krebs cycle. The net yield in the electron transport chain is a maximum of 32 ATP molecules per molecule of glucose. In addition to ATP, Glycolysis and the Krebs cycle both yield electrons that are carried to the ETC via intermediate electron carriers.

7.4 First Stage of Respiration: Glycolysis

- The net energy yield of glycolysis is two molecules of NADH and two molecules of ATP per molecule of glucose.

7.5 Second Stage of Respiration: The Krebs Cycle

- The net energy yield of the Krebs cycle is six molecules of NADH. Two molecules of $FADH_2$, and two molecules of ATP per molecule of glucose.

7.6 Third Stage of Respiration: The Electron Transport Chain

- The ETC is a series of molecules that are located in the mitochondrial inner membrane. Upon reaching the ETC, the intermediate electron carriers NADH and $FADH_2$ are oxidized by molecules in the chain. Each carrier in the chain is then reduced by accepting electrons from the carrier that came before it. The last electron acceptor in the ETC is oxygen.

- The movement of electrons through the ETC releases enough energy to power the movement of hydrogen ions (H^+ ions) through the three ETC protein complexes, moving them from the mitochondria's inner compartment to its outer compartment. The movement of these ions down their energy gradients, back into the inner compartment through an enzyme called ATP synthase, drives the synthesis of ATP from ADP.

7.7 Other Foods, Other Respiratory Pathways

- Nutrients and their derivatives can be channeled through different pathways in cellular metabolism, in accordance with the needs of the organism. At any given moment, a cell may need to work on synthesizing organic molecules rather than breaking them down.

Understanding the Basics

Multiple-Choice Questions

1. You need energy to think, to keep your heart beating, to play a sport, to study this book. This energy is directly supplied by _____, which is produced in the process of respiration.
   a. enzymes
   b. ATP
   c. NAD$^+$
   d. vitamins
   e. proteins

2. Which of these statements best explains why most of the time your body does aerobic respiration rather than anaerobic?
   a. Aerobic respiration is simpler for cells to carry out.
   b. Anaerobic respiration requires more enzymes and thus more energy.
   c. Most cells lack the ability to make the enzymes needed to do anaerobic respiration.
   d. Anaerobic respiration produces less ATP than does aerobic.
   e. Anaerobic respiration does not require NAD$^+$.

3. Imagine a prokaryotic (bacterial) cell in an environment that did not contain oxygen but that did contain food. In such a situation, the cell could not
   a. extract energy from food
   b. perform glycolysis
   c. utilize the Krebs cycle
   d. utilize the electron transport chain
   e. utilize either the Krebs cycle or the electron transport chain

4. What happens when a cell in your body has all the ATP it needs for the moment?
   a. It stops producing ATP.
   b. It continues to produce ATP so that it is ready for the next emergency.
   c. It begins breaking down NAD$^+$ instead of making ATP.
   d. It forms glucose out of the extra ATP.
   e. It dies.

5. How many ATPs are produced in aerobic respiration?
   a. 2   c. 24   e. 75
   b. 8   d. 36

6. Where does the Krebs cycle occur in this model of a mitochondrion?

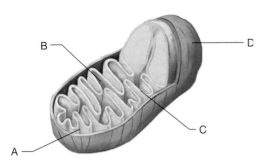

a. A
b. B
c. C
d. D
e. none of these

7. When NADH passes its electrons to the ETC, it is
a. electrolyzed
b. polarized
c. reduced
d. oxidized
e. deformed

8. In the first step of glycolysis, glucose enters the cell and immediately has a phosphate group from ATP attached to it. This process is called
a. phosphorylation
b. oxidation
c. photosynthesis
d. electron transport
e. the citric acid cycle

Brief Review

1. Picture yourself on a trampoline. Use the fact that you have to jump in order to start bouncing to explain how ATP can be formed as a result of electrons from glucose falling down an energy hill.

2. Explain why you should expect to have a lot of mitochondria in your muscle cells.

3. Since ATP is used for our cellular needs, why don't we just eat ATP?

4. A few desert animals are able to get by without ever drinking water. They use the water in their food to supply some of their need for water. They also use the water they produce in respiration. Where does that water come from?

5. FAD is derived from the B-vitamin, riboflavin. What is the function of FAD? What do you think happens if you fail to get enough riboflavin?

Applying Your Knowledge

1. When you eat foods that contain small amounts of the B-vitamin niacin, the vitamin is converted into NAD^+ in your body. Using what you know about the role of NAD^+, predict what will happen to people whose diets are deficient in niacin.

2. Why do you think that we use the same word—*respiration*—for breathing as we do for breaking down food to extract its energy?

3. Enzymes are now used for a host of processes outside of cells, including such things as cleaning clothes and hands. In the late 1990s researchers found an enzyme in spinach that will break down explosives such as TNT. In light of what you know about enzymes, how could this work? Think of other environmental problems that might be solved by using enzymes.

4. In Madeleine L'Engle's children's novel *A Wrinkle in Time*, the mitochondria in one of the characters start to die. Describe what would happen to a person if they lost their mitochondria, and explain why it would happen.

Group Activity

As a class, find out how the hereditary diseases known as metabolic disorders can cause people to be unable to use certain foods. Put this together into an information package and use it to organize a campus-wide fund-raiser to support research into the treatment of such diseases.

MediaLab
Dietary Fad or Miracle Drug?
Using Science to Understand Metabolism

You've heard the claims: "Burns fat while you sleep!" "Energy fuel capsules!" and even "Magic weight-loss pills!" Companies entice us to try some new product by exploiting our ignorance about the actual process of cellular metabolism—how cells extract energy from food. This *MediaLab* covers the basic concepts needed to understand how cells really use the energy from food. To learn more, enter the *CD Activity* for an animated review of metabolism. Once you have completed the *CD Activity*, try the *Web Investigations* to uncover how metabolism may be misrepresented in diet claims on the Internet. Finally, *Communicate Your Results* will help you practice the skills you'll need to critically appraise the material you read.

This *MediaLab* can be found in Chapter 7 on your CD-ROM and Companion Website (http://www.prenhall.com/krogh).

CD ACTIVITY

Americans are presented with such an abundance of food that they tend to equate the word *diet* with limiting the consumption of Calories, not as a necessary source of energy and nutrients. However, if you can see how your body uses the molecules in food, then you can begin to see why you need to consume certain foods. This *CD Activity* is designed to review metabolism.

ATP

Activity

1. *First, you will examine the role of the cellular energy currency, ATP.*

2. *Then you will investigate why we eat, by following the series of steps needed to convert the energy from food (carbohydrates, proteins, and fats) into ATP.*

3. *Finally, you will look at the role of enzymes in this process.*

After reviewing the concepts of metabolism in these processes, you'll be ready to test your knowledge in the next section by uncovering unfounded, impossible diet and exercise claims on the Internet.

WEB INVESTIGATION

Investigation 1
Estimated time for completion = 10 minutes

How do we know this diet stuff won't live up to its promises? There is an excellent Web page devoted to disseminating information on how to recognize fraudulent weight-loss products and programs. Select the Keyword **WEIGHT-LOSS FRAUD** on your CD or Website for this *MediaLab*. After you've read the criteria, perform a search of the Internet to locate examples of weight-loss Websites that make one of the promises described in the fraud article, like magic weight loss without dieting or exercise.

Investigation 2

Estimated time for completion = 5 minutes

Nutritional supplements are marketed on the principle that because we need enzymes, amino acids, vitamins, and other nutrients for these cellular functions, then it would be beneficial to add them as a supplement to improve our diets. Select the Keyword **ENZYME SUPPLEMENTS 1** or **2** on your CD or Website for this *MediaLab* to read two contradictory articles about enzyme supplements. Which one do you believe?

Investigation 3

Estimated time for completion = 5 minutes

Over 50 percent of adult Americans purchase dietary supplements to improve their health, but who tests the accuracy of the health claims made by supplement manufacturers? The responsibility for checking the safety of products and the truthfulness of label claims lies in the hands of consumers and manufacturers, not in an impartial government agency like the FDA. Select the Keyword **FDA GUIDE TO SUPPLEMENTS** on your CD or Website for this *MediaLab* to read the FDA's guide to dietary supplements.

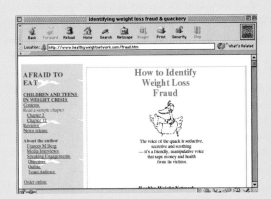

COMMUNICATE YOUR RESULTS

Determine the integrity of the Web sites from the *Web Investigation* section by applying rigorous questions in the following examples.

Exercise 1

Estimated time for completion = 5 minutes

How can you determine what to believe on the Internet, when Web sites include scientific terminology and definitions that may not be based on tested, controlled experiments? Consumers can be embarrassingly gullible when they believe incredible diet claims. Make a difference by posting information about one of these too-good-to-be-true sites (a message board is provided at the textbook's Companion Website). Describe in two or three sentences why you think the product falls under the examples listed in the diet fraud article.

Exercise 2

Estimated time for completion = 20 minutes

Many diet-supplement companies promote the idea that human health problems can be remedied through ingesting enzyme supplements. How would a scientist determine the credibility of these claims? Here is a list of guidelines:

- *Do the Web sites refer to studies published in reputable science journals with rigorous standards, or do they refer to testimonials and stories?*

- *Do the author's credentials lend credibility, or are they unrelated?*

- *Does the author stand to benefit financially by the sales of the product?*

Compose a 250-word argument describing the claims made at the two Web sites listed in *Web Investigation 2*, and using the scientific criteria just listed, explain which one is most credible.

Exercise 3

Estimated time for completion = 15 minutes

After reading the FDA report on supplements in *Web Investigation 3*, compose a letter to the FDA expressing whether you think the current system provides adequate protection to Internet consumers. Include your opinion on whether supplements should be subjected to the same rigorous clinical studies as drugs to determine their effectiveness, safety, possible interactions with other substances, and appropriate dosages. Should the FDA take more responsibility for removing potentially unsafe dietary supplements? Should the FDA take a greater role in enforcing prosecution of companies making false or erroneous health claims for its supplements?

Applying rigorous questions like these helps prepare you to critically analyze future material that you may find on the Internet. Internet material is not always filtered or checked for accuracy, so you'll need the skills practiced in this section to assess it on your own.

8

The Green World's Gift
Photosynthesis

A different path for photosynthesis.
(page 162)

Capturing the sun's energy.
(page 155)

An important cycle begins.
(page 158)

On a summer's evening,

out by a pond as the sun is going down, nature tunes up her symphony. The crickets bring their chirping to the fore, while frogs chime in with their own countermelody. Silently, something else is changing as the last light fades: The green world is shutting down. Microscopic pores on tree leaves and pond algae are closing up, ceasing to be openings for the carbon dioxide that flows in and the water that flows out during the day. The green world is alive at night, but the activities it carries out are not so much different from those a human listener might be undertaking while relaxing at water's edge. Both are building up tissue, storing away some foodstuffs while breaking down others, getting rid of the by-products that result from all these processes. The green world has a special talent, but sunlight must be falling for it to be put into practice.

8.1 Photosynthesis and Energy

The green world's special talent is called photosynthesis, and the wonder of it can perhaps best be appreciated by imagining what life would be like if we humans possessed it. The bone and muscle within us—and the energy to make them work—come from the foods we eat: carbohydrates and fats and proteins. Imagine, though, being able to flourish simply by spending time in the sun and having access to three things: water, small amounts of minerals, and carbon dioxide. In this condition we would not *eat* carbohydrates in the usual forms (bread, sugar, potatoes). Rather, we would transform the simple gas carbon dioxide *into* carbohydrates using nothing more than water, minerals, and sunlight. Carbohydrates produced in this way are as useful as any food; they can be stored away as starches, moved from one part of the body to another as simple sugars, or broken down to provide ATP for any energy needs. This, then, is what plants do: They *make their own food*. They use the energy provided by the sun to transform a simple gas, carbon dioxide, into an energy-rich carbohydrate. In this activity, they are joined by a variety of other photosynthesizing organisms (**see Figure 8.1** on page 152).

From Plants, a Great Bounty for Animals

The process of photosynthesis is not only good for plants; it's good for animals, since the entire animal kingdom ultimately depends upon photosynthesis for food. Try naming something you eat that isn't a plant or doesn't itself eat plants to survive, and you will come up with a very short list indeed.

Photosynthesis in action.
(page 157)

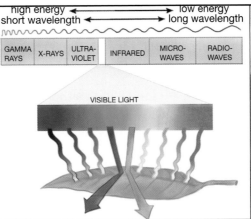

The electromagnetic spectrum.
(page 152)

Warm-weather photosynthesis.
(page 162)

151

a   b   c

Figure 8.1
Three Types of
Photosynthesizers
Photosynthesis is carried out not only by familiar plants, such as these sunflowers **(a)**, but by algae, such as this giant kelp **(b)**, and by some bacteria, such as these cyanobacteria **(c)**. (magnified ×1025)

Food, however, is only half the bounty that plants provide to animals. A by-product of photosynthesis—a kind of castoff from the work of plants—is oxygen. As part of photosynthesis, plants break water molecules apart. In doing so, they *use* electrons and protons from H_2O, but they *leave behind* oxygen molecules (O_2). This is the oxygen that exists in our atmosphere; it is the oxygen that we breathe in every minute of every day, or suffer the mortal consequences. Note also that one of the starting ingredients for photosynthesis is the carbon dioxide we breathe *out* with each breath, thus adding a small amount to the atmospheric store that exists. In the living world, one organism's by-product is another's life-sustaining substance.

With this, you can step back and view the largest picture of them all with respect to energy flow in living things. It is a great pathway in which energy comes from the sun and then, in photosynthesis, is stored in plants in the complex molecules we call carbohydrates. Thus stored—in such forms as grass leaves or wheat grains—these carbohydrates can be broken down and used, either by the plants

themselves or by the animals that eat them. One by-product of the breakdown of food is carbon dioxide, which keeps photosynthesis going. Meanwhile, one by-product of photosynthesis is oxygen, which keeps aerobic respiration going.

Up and Down the Energy Hill Again

Looked at another way, these two processes are simply trips up and down the energy hill that became so familiar in Chapter 7. In respiration, the trip is down the hill—from more stored energy (in food) to less, as the food is being broken down. Photosynthesis is a trip back up the energy hill. Here electrons are being removed from water, boosted to a more energetic state by the power of sunlight, and then brought together with an organic compound and carbon dioxide, resulting in an *energy-rich* carbohydrate (food). The story of how this is accomplished is the subject of much of the rest of this chapter.

8.2   The Components of Photosynthesis

What Kind of Light Drives Photosynthesis?

Think of photosynthesis as starting with the absorption of sunlight by leaves. *Absorption* here means just what it sounds like: Light is taken in by the leaves. Actually, leaves capture a *portion* of the light that falls on them. The sunlight that makes it to the earth's surface is composed of a spectrum of energetic rays (measured by their "wavelengths") that range from very short, ultraviolet rays, through visible light rays, to the longer and less-energetic infrared rays. (In **Figure 8.2**

Figure 8.2
The Electromagnetic Spectrum
Sunlight is composed of rays of many different wavelengths, but we can see only those in the visible light range (represented by the color spectrum). When the light of the sun shines on a green leaf, the sun's green rays are scattered, while frequencies in other ranges are absorbed; it is frequencies in the red and blue ranges that most strongly drive photosynthesis.

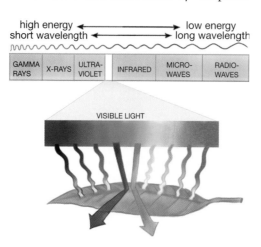

high energy ←→ low energy
short wavelength ←→ long wavelength

| GAMMA RAYS | X-RAYS | ULTRA-VIOLET | INFRARED | MICRO-WAVES | RADIO-WAVES |

VISIBLE LIGHT

you can see that these wavelengths are, in turn, part of a larger range of electromagnetic radiation.)

Photosynthesis is driven by part of the *visible light* spectrum—mainly by blue and red light of certain wavelengths. Tune a laser to emit blue light of a certain wavelength, shine this beam on a plant, and the plant will absorb a great deal of this light. On the other hand, tune the laser to emit *green* light, and the plant will scatter most of this light. This is why plants are green: They strongly scatter the green portion of the visible light spectrum.

Where in the Plant Does Photosynthesis Occur?

So, where is this light absorption going on within plants? To answer this question, you need to know something about the playing fields of photosynthesis, leaves.

The broad, flat leaves that are so common in nature have in essence a two-part structure: A *blade* (which we usually think of as the leaf itself) and a *petiole*, more commonly referred to as the leaf stem. If you look at the idealized leaf in **Figure 8.3**, you can see that the blade can be likened to a kind of cellular sandwich, with one layer of outer (or epider-

MediaLab Link
Light absorption by leaves, see p. 168.

**Figure 8.3
Site of Photosynthesis**

a The mesophyll cells of leaves contain many chloroplasts, which are the organelles in which photosynthesis is performed. Pigments that absorb solar energy are located on the highly folded thylakoid membranes which frequently pile into stacks (called grana) within the chloroplast. The interior water space of the thylakoids is called the thylakoid compartment.

b Top, a micrograph of a leaf cross section, with layers of mesophyll sandwiched between upper and lower epidermis layers. Bottom, a micrograph of a chloroplast. The groups of thin lines are the stacks of grana. The large white body is a starch body, produced through photosynthesis.

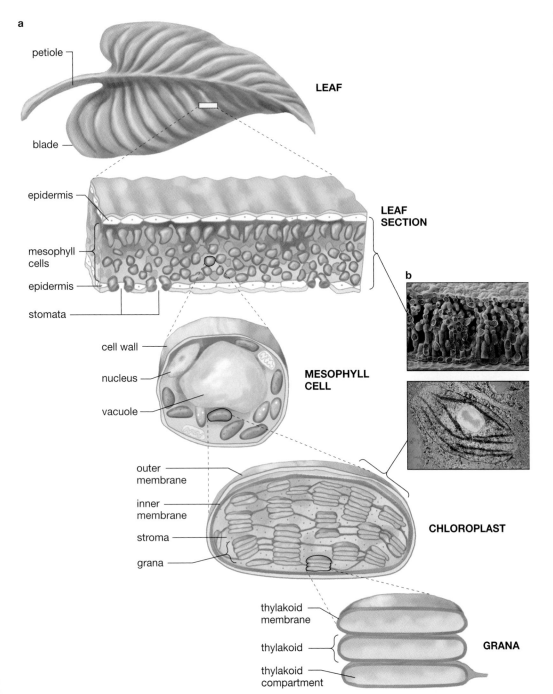

mal) cells at the top, one layer at the bottom, and layers of mesophyll cells in between.

The leaf's epidermal cells are dotted with numerous microscopic pores, called **stomata**; these are the openings mentioned at the beginning of the chapter that let carbon dioxide in and water out. There is no shortage of stomata for this work; a given square centimeter of plant leaf may contain from a thousand to a hundred thousand of them.

Photosynthesis Central: The Chloroplasts

A journey to the actual site of photosynthesis would take you to the leaf's layers of mesophyll cells, then inside the cells themselves, then into a group of organelles within the cells—the **chloroplasts**. Anywhere from one to several hundred of these oblong structures might exist within a given leaf cell, each capable of carrying out photosynthesis on its own.

Traveling into the *chloroplasts*, you'd come to the molecules, called pigments, that actually absorb sunlight. Anything, living or nonliving, that strongly absorbs certain wavelengths of sunlight is called a pigment. Chloroplasts contain a primary pigment, embedded in their membranes, called **chlorophyll *a***, which is aided by several substances known as **accessory pigments**. Working together, these pigments perform the task of absorbing the light wavelengths mentioned earlier.

Figure 8.3 shows you the detailed structure of a chloroplast. As has so often been the case with cell structures, what chloroplasts present are membranes within membranes. The chloroplast has an outer and an inner membrane at its periphery. Moving inside them, to the interior of the chloroplast, you see a network of membranes, called **thylakoids**. At frequent intervals, these thylakoids sprout vesicles that stack on top of one another like so many pancakes, creating structures called **grana**. Thylakoids, in turn, have an interior water space, called the **thylakoid compartment**. Thylakoids are immersed in the liquid material of the chloroplast, the **stroma**. All the things you'll be looking at take place where thylakoid compartment, thylakoid membrane, and stroma intersect.

There Are Two Essential Stages in Photosynthesis

Keeping these structures and concepts in mind, you can now begin tracing the steps of photosynthesis. It is helpful to divide photosynthesis into two main stages. In the first stage (the *photo* of photosynthesis), water is stripped of electrons by the power of sunlight, which also boosts these electrons to a higher energy level. From here, the electrons jump from one electron carrier to another, finally attaching to a mobile electron carrier ($NADP^+$) that will carry the electrons to the second set of reactions. In this second stage (the *synthesis* of photosynthesis), the electrons are brought together with carbon dioxide and a low-energy organic molecule. The attachment of the electrons to this compound produces a high-energy carbohydrate (food, in other words).

The processes in the first stage of photosynthesis obviously depend on sunlight, but those in the second stage do not (at least not directly). Given this division, scientists have named the first set of steps the **light-dependent reactions**, and the second set the **light-independent reactions**. You will be following these reactions as they occur in time, meaning from the collection of sunlight to the synthesis of carbohydrates.

The Working Unit of Photosynthesis Is Called a Photosystem

Absorption of sunlight takes place within *aggregations* of a few hundred pigment molecules. The majority of molecules in these units—most of the chlorophyll *a* and all of the accessory pigments—serve only as "antennae" that absorb energy from the sun and pass it on. At the center of the antennae system, however, is a molecular complex known as the **reaction center**, which contains a pair of special chlorophyll *a* molecules that transform solar energy into chemical energy.

You could think of photosynthesis as beginning when sunlight is absorbed by any of the hundreds of antennae pigment molecules in a photosystem unit. The energy that comes with this absorption is then passed on to the pair of chlorophyll *a* molecules in a reaction center. As a result, electrons from this pair are "moved" in a couple of ways, one physical and one metaphorical. Physically, these electrons jump to another electron carrier within the reaction center, a "primary electron acceptor." You can get a sense of how the reaction center and antennae pigments fit together to form a **photosystem** by looking at **Figure 8.4**.

You also could think of "distance traveled" in photosynthesis in a metaphorical way, however: as a movement up and down an energy hill. It is the energy from sunlight that is doing the pumping up the energy hill in photosynthesis.

Energy Transfer in Photosynthesis Works through Redox Reactions

How can electrons move up and down an energy hill? Recall from Chapter 7 that some substances have a tendency to *lose* electrons to other substances. Any time one substance loses (or "donates") electrons to another, it is said to have been oxidized. Meanwhile, the substance that gained electrons is said to have been reduced. These two reactions always happen together; if one substance is oxidized, another must be reduced. This kind of reduction-oxidation reaction has a shorter name: a *redox reaction*.

Many of the photosynthetic steps you will be seeing are redox reactions. In **Figure 8.5**, you can see where they lead in terms of movement up and down the energy hill. The

electrons are pumped up a couple of formidable energy gradients (in photosystems II and I) only to come partway back down them, as they "seek" their lowest energy state, releasing energy as they fall—just as a boulder releases energy by rolling downhill.

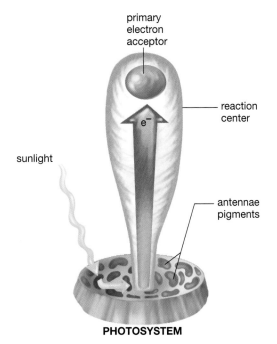

PHOTOSYSTEM

Figure 8.4
The Working Units of Photosynthesis
Photosystems are multipart units that bring together electrons derived from water and energy derived from the sun. Hundreds of antennae pigments absorb sunlight and transfer solar energy to the photosystem's reaction center. This energy gives a boost to an electron within the reaction center in two ways: It physically moves the electron to the center's primary electron acceptor, and it moves the electron up the energy hill to a more energetic state.

Figure 8.5
Collecting Solar Energy, Boosting Electrons
The solar energy gathered by photosystems II and I is used to energize electrons that exist within the photosystems' reaction centers. Energetically, the electrons are boosted up, and then move down, two energy hills. Physically, they first move to primary electron acceptors and then down electron transport chains until they are at last taken up by $NADP^+$ to form NADPH. The NADPH molecules then transfer the electrons to the light-independent reactions, where they are used to make sugars. When an electron is boosted in photosystem II, the resulting energy imbalance drives the process by which water molecules are split. Electrons derived from the water then move to the reaction center of the photosystem, where they are the next to be boosted by the sun's energy. The energetic fall of electrons down the electron transport chain between photosystems II and I provides the energy for the synthesis of ATP, which is used to power the light-independent reactions.

These are redox reactions, with each molecule in the chain undergoing successive oxidation and reduction reactions.

8.3 Stage 1: The Steps of the Light-Dependent Reactions

With these components and processes in mind, let's trace the steps of the light-dependent reactions (see Figure 8.5). The first step is that solar energy, collected by photosystem II's antennae molecules, arrives at the reaction center. This energy then gives a boost to an electron in the reaction center in the two ways noted before. The electron *physically* moves to another part of the reaction center protein complex, the primary electron acceptor. But it is also pumped up the energy hill. With this activity, the reaction center chlorophyll *a* has *lost* an electron. That loss leaves an energy "hole" in this chlorophyll, making it an oxidizing agent. With the energy provided by this imbalance, a special enzyme in the reaction center splits water molecules that lie within the thylakoid compartment. These water molecules are now being oxidized, which means they are *losing* electrons. The electrons travel to the reaction center, where they will be the next electrons in line for an energy boost.

A Chain of Redox Reactions and Another Boost from the Sun

By following Figure 8.5 in a general way, you can see what happens next to electrons that are boosted in photosystem II. After arriving at the primary electron acceptor, they fall back down the energy hill as they are transferred through a series of electron transfer molecules, each one oxidizing its predecessor. At the bottom of this hill they arrive at photosystem I, which includes a slightly different kind of reaction center. This center is also receiving solar energy, and uses it to boost electrons to a higher energy state. From this energetic state, electrons are transferred down the energy hill—until they are received by the electron carrier $NADP^+$, which is very much like the electron carrier NAD^+ reviewed in Chapter 7. In accepting electrons, $NADP^+$ becomes reduced to NADPH, an electron carrier that ferries the electrons into the second stage of photosynthesis: the light-independent reactions, which will yield the food that is the function of photosynthesis.

The Physical Movement of Electrons in Light-Dependent Reactions

In this movement up and down the energy "hill," the electrons have moved physically through the chloroplast. As **Figure 8.6** shows, they started out in the water of the thylakoid compartment, moved into and then through the thylakoid membrane—handed off at each step by the various electron transfer molecules—and then finally ended up in the stroma, attached to NADPH.

8.4 What Makes the Light-Dependent Reactions So Important?

The steps just reviewed are the essence of the light-dependent reactions. Lest you miss the forest for the trees, let's be clear about two momentous things that have taken place within these steps.

The Liberation of Oxygen from Water

The first notable event takes place near the start, when water is split. This reaction provides the traveling electrons whose path you've just followed. It also provides something else: the oxygen that today accounts for more than 20 percent of the Earth's atmosphere. *Here* is where we get the substance that is the breath of life itself for most species, all of it contributed by living things. Given its importance, it is no small irony that oxygen is, in another sense, a kind of green-world refuse; leftovers from the process by which plants strip water of the electrons they need to carry out photosynthesis.

The Transformation of Solar Energy to Chemical Energy

The second major feat that takes place during light-dependent reactions is the transformation of solar energy to chemical energy. When the energy of sunlight moves to a chlorophyll molecule in a reaction center, it boosts a chlorophyll electron (derived from water) from what is known as a ground state to an excited state. Thus far, this has been described as a movement up a metaphorical energy hill. In physical terms, however, this electron is moving farther out from the nucleus of an atom. To appreciate what's special about photosynthesis, it's necessary to note what the *common* fates are for such excited electrons. One fate is for the electrons to drop back down to the original (and more stable) state, in the process releasing as *heat* the energy they have absorbed. This is why black objects get hot on sunny days. Another possible fate is for falling electrons to release part of their energy as light in the process known as fluorescence.

In photosynthesis, however, the energized electrons are *transferred to a different molecule*—the initial electron acceptors of photosystems II and I. They don't fall back to their ground state, releasing relatively useless heat or light in the process. They are passed on in a redox reaction. This is the bridge to life as we know it. Without this step, the energy of the sun merely makes the Earth warmer. With it, the green world grows profusely; it captures some of the sun's energy and with it builds trunks and grains and stems. By one reckoning, photosynthesis produces up to 155 billion tons of material each year. This is an amazing amount—about 25 tons for every person on Earth. Electrons are taken from water, boosted to a higher energy state by the sun, and transferred to other, more stable molecules. The result is the food that sustains nearly all living things.

It's also worth noting that a primary function of the fall of electrons between photosystems II and I is a release of energy that is used for the production of ATP. This ATP is used to power the reactions that are coming up in the second stage of photosynthesis.

8.5 Stage 2 of Photosynthesis: The Light-Independent Reactions

Now, back to the central story. You are ready to see how carbon dioxide is incorporated into carbohydrates in the light-*independent* reactions. To begin, let's take stock of where the first set of reactions left off. The short answer is: in the stroma. That's where $NADP^+$ has become NADPH by accepting the energized electrons pouring out from photosystem I. There's also ATP in the stroma, freshly made from the light-dependent reactions.

Food Comes from a Cycle of Reactions: The Calvin Cycle

Even with the right ingredients, however, it takes more than one step to go from carbon dioxide to an energetic carbohydrate. It takes a set of steps that make up a cycle,

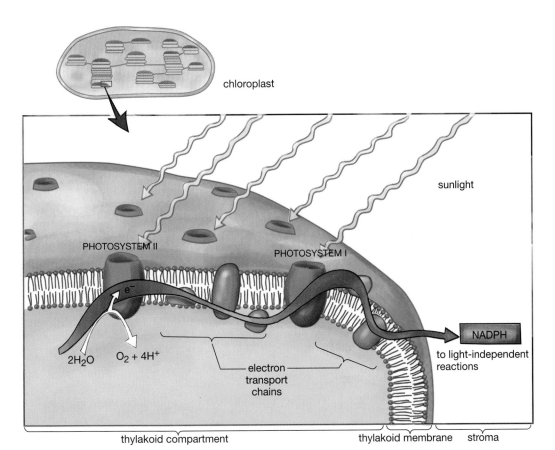

Figure 8.6
Light-Dependent Reactions
The light-dependent reactions take place in the thylakoid membranes within the chloroplasts. Electrons are donated by water molecules located in the thylakoid compartments. Powered by the sun's energy, these electrons are passed along the electron transport chain embedded in the thylakoid membrane and end up stored in NADPH in the stroma. An additional product of the splitting of water molecules is oxygen atoms, which quickly combine into the O_2 form. This is the atmospheric oxygen breathed in by organisms such as ourselves.

chloroplast

sunlight

PHOTOSYSTEM II

PHOTOSYSTEM I

e^-

NADPH

to light-independent reactions

$2H_2O$

$O_2 + 4H^+$

electron transport chains

thylakoid compartment

thylakoid membrane   stroma

commonly known as the Calvin cycle or the C₃ cycle. The highlights of this cycle are set forth in **Figure 8.7**. (The Calvin cycle is named after the chemist Melvin Calvin, whose work you can read about in "Plants Make Their Own Food, But How?" on page 160.)

The *synthesis* of photosynthesis is, in its first steps, a process of **fixation**—of a gas being incorporated into an organic molecule. Specifically, carbon dioxide is being fixed into carbohydrate.

Here are the steps in the **Calvin cycle**:

1. **Bringing together carbon dioxide and sugar.** The process begins with three molecules of CO_2 (coming in from the atmosphere) being joined to three molecules of a five-carbon sugar, RuBP. Given that one carbon of CO_2 is being added to a five-carbon molecule (and that it is being done three times), the result, of course, should be three *six-carbon* molecules. This is the case, but only for a vanishingly brief time. Each six-carbon molecule is so unstable that it is split apart immediately into two *three-carbon* molecules of a substance called 3-phosphoglyceric acid or 3-PGA. The three carbons here give the C₃ cycle its name.

Figure 8.7
Calvin Cycle

1. Carbon fixation. The enzyme rubisco brings together three molecules of CO_2 with three molecules of the five-carbon sugar RuBP; the three resulting six-carbon molecules are immediately split into six three-carbon molecules named 3-PGA (3-phosphoglyceric acid).

2. Energizing the carbohydrate. In two separate reactions, (a) Six ATP molecules react with six 3-PGA, in each case transferring a phosphate onto the 3-PGA. (b) The six 3-PGA derivatives oxidize (gain electrons from) six NADPH molecules; in so doing, they are transformed into the energy-rich carbohydrate G3P (glyceraldehyde 3-phosphate).

3. Exit of product. One molecule of G3P exits as the output of the Calvin cycle. This molecule, the product of photosynthesis, can be used for energy or transformed into materials that make up the plant.

4. Regeneration of RuBP. In several reactions, five molecules of G3P are transformed into three molecules of RuBP.

Overview: CO_2, ATP and electrons (contained in hydrogen atoms) from NADPH are input into the Calvin cycle, while a sugar (G3P) is its output.

SUMMARY OF THE CALVIN CYCLE

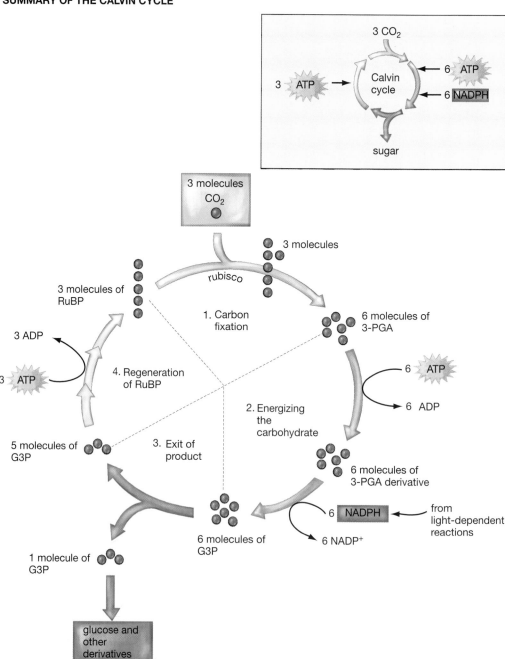

A key player in photosynthesis is the enzyme that brings together CO_2 and RuBP. This massive protein goes under several aliases, but the one used here is **rubisco**. You need to be clear about what it is doing, bringing together in its active site CO_2 and the sugar RuBP. Here is where the small, low-energy gas molecule CO_2 is joined to the ranks of organic molecules, specifically sugars.

2. **Energizing the sugar.** Carbon dioxide has now been added to a sugar; what remains is for this sugar to be energized. These things take place in the next two steps. First, ATP reacts with the molecules of 3-PGA, putting a phosphate group onto each of them. Then NADPH donates a pair of electrons to each of the resulting molecules, yielding six molecules of a substance called glyceraldehyde 3-phosphate or G3P.

3. **Our food has arrived.** There should now be a little fanfare, because with this step the food has arrived. Note that G3P was produced by receiving energetic electrons from NADPH—the electrons that came originally from water and that were boosted up the energy hill by sunlight. G3P thus has more potential energy than did its predecessor in the Calvin cycle. It is precisely this position—a greater distance up the energy hill—that makes *food*. Food can be broken down, thus providing materials or energy in accordance with the needs of a plant. G3P thus fits right into the chain of metabolism: the building up and breaking down of nutrient materials for usage, transfer, or storage. It is the product of photosynthesis.

G3P is, however, a product that is analogous to steel coming out of a factory: It can be turned into many things. Put two molecules of G3P together and you get the more familiar 6-carbon sugar, glucose. In turn, sugars can come together to form the large storage molecules known as starches, familiar to us in such forms as potatoes or wheat grains. Beyond this, sugar is used to make proteins, which are then used as structural components of the plant or as enzymes. Given all this, you may ask, what is the *ultimate* product of photosynthesis? The answer is *the whole plant*.

4. **Completing the Calvin cycle.** As you can see from Figure 8.7, the C_3 cycle still has some distance to go after the synthesis of G3P. Critically, it is only *one* of the six molecules of G3P that will exit as product, while the other five will continue through the cycle to be used in the regeneration of RuBP. This is a process that, as you can see, is powered by the expenditure of more ATP. **Figure 8.8** summarizes the process of photosynthesis.

8.6 Photorespiration: Undercutting Photosynthesis

You have now gone through the basic process of photosynthesis. It's necessary, however, to say something about what might be called a *glitch* in this process, and then note a mechanism that plants have evolved to deal with it.

The glitch is called **photorespiration**, and it concerns a deviation from the process you just went through. Recall that the Calvin cycle begins with the important enzyme, rubisco, binding with carbon dioxide and then combining it with a five-carbon sugar. As it turns out, however, rubisco frequently binds with *oxygen* instead of carbon dioxide. When this happens, no carbohydrate is produced, which means no plant growth results from the reaction.

The amazing thing is how often this takes place. In C_3 photosynthesis, rubisco may fix up to one molecule of O_2 for every three of CO_2 it incorporates. This is no trivial matter, because many food crops use the C_3 cycle; in these instances, photorespiration is undercutting food production.

Figure 8.8
Summary of Photosynthesis in the Chloroplasts of Plant Cells

In the light-dependent reactions, solar energy is converted to chemical energy in the thylakoids, and the chemical energy is stored in the form of ATP and NADPH. Water is required for this reaction, and oxygen is a by-product. The stored chemical energy is in turn used in the light-independent reactions (Calvin cycle) taking place in the stroma, in which sugar is made from carbon dioxide and a low-energy sugar, RuBP. The sugar can be used for food, or may become part of the plant's structure.

Plants Make Their Own Food, But How?

Science constantly is confronted with what are known as "black box" problems. In these problems, researchers know what goes into a process and they know what comes out of it, but what happens *in between* is a mystery. At the conclusion of World War II, plant science had a doozy of a black-box problem. It was clear that plants were starting with carbon dioxide from the air and ending with high-energy carbohydrates, but no one knew what came in between. What were the *intermediate* substances that carbon dioxide was being made part of on the route to food? Getting inside this black box turned out to require a three-part combination: insightful researchers, new scientific techniques, and old-fashioned hard work.

One of the peacetime benefits of World War II's atomic weapons research was that scientists could produce quantities of substances known as *radioactive isotopes*. As you saw in Chapter 2, isotopes result from variations in the number of neutrons an element has in its nucleus. Radioactivity, meanwhile, is the release of radiant energy from atoms undergoing rapid decay. The element carbon is everywhere in nature, and as it turns out, one of its isotopes, carbon-14, is radioactive. Carbon-14 occurs naturally, but the ability to *manufacture* this substance opened a host of possibilities for research in the mid-1940s.

Stepping up to the plate in 1946 to take advantage of these possibilities was a University of California, Berkeley, chemist named Melvin Calvin. Then just 35 years old, Calvin was encouraged by the physicist Ernest Lawrence of Berkeley to apply radiocarbon techniques to organic chemistry research. With carbon-14, Calvin realized, he could *tag* carbon dioxide to find out what it becomes part of during photosynthesis, much as a zoologist might tag a bear with a signaling device to find out where it goes during winter. The radiation that carbon-14 emits might be thought of as a signal whose constant message is: "Here I am."

The steps that Calvin and his colleagues followed were simple in conception but very difficult in execution. They first took some photosynthetic algae and put them in a flask that had light shining on it. Like any photosynthetic organism, these algae would, of course, be taking in carbon dioxide to carry out their work. In this case, however, the carbon dioxide in question contained carbon-14. The trick was to allow the algae to perform a little photosynthesis using this $^{14}CO_2$. Then the algae were plunged into boiling alcohol, which killed them. By the time the plunge was taken, the carbon dioxide would have made it to *some*

point in its transformation to carbohydrate; it would have been incorporated into one molecule in the series of intermediate molecules involved. Stopping photosynthesis at differing points—two seconds after it started, five seconds, and so on—meant stopping the progress of carbon dioxide as it marched through its changes.

All of this would have been for naught, however, without a way to figure out where the carbon was when the process was stopped. This is where the carbon-14 could do its job, but it needed to be joined to another technique that was also newly developed at that time: two-dimensional paper chromatography. If you look at **Figure 1a**, you can see how this technique works. Put a spot of ink on some filter paper and then stick one end of the paper in a solvent—water, for example. The water moves through the paper, carrying different components of the ink along with it at different rates. You can tell something about what the ink is made of by measuring how far its components move through the paper. You can tell still more by doing this in two dimensions: Letting the solvent move through the spot one way, then setting the paper down at a right angle in a *second* solvent and letting it move through in another direction (**Figure 1b**). This is what Calvin and his colleagues did, using as their "ink" the algae that had been killed following photosynthesis.

The paper chromatography left the algae cells drawn out in two directions by the solvents. But where was the original carbon in this material? Because it was radioactively *tagged* carbon, it was sending out its signal saying, "I'm here now." But only a special "receiver" could pick up this signal: photographic film (**Figure 1c**). Setting x-ray film tightly on top of the chromatography paper for long periods—usually about two weeks—caused the radiation from the carbon-14 to expose the film. Spots would show up saying, in effect, "carbon-14 is now here, in this concentration."

There is a wide gap, however, between seeing a black blotch on a piece of film and *identifying* that blotch as a specific substance. The difficulty the researchers encountered can be measured in time—10 years in this case. That's how long it took these workers to complete their investigation. During that time, Calvin and his principal colleagues—postdoctoral researcher Andrew Benson and graduate student James Bassham—worked extraordinarily long hours in deciphering what the downstream products of photosynthesis were and the order in which they were produced.

This long operation was a great success, however. Calvin's team dismantled the black box of carbon fixation, and the information they uncovered can be *used* to aid in such things as increased food production. The set of steps these workers elucidated has variously been referred to as the C_3 cycle, Calvin-Benson cycle, or the Calvin cycle. In 1961, Melvin Calvin received the ultimate in congratulations for a piece of scientific work when he was awarded the Nobel prize in chemistry for shining a light on photosynthesis.

a paper chromatography in first dimension

solvent A

Material of interest (algae for Calvin) is put on paper in solution form.

Solvent A moves through paper, separating out components of original material. (Components, shown here as green dots, would probably not be visible in experiments.)

b paper chromatography in second dimension

solvent B

Paper is rotated 90 degrees and put in second solvent.

Component compounds are separated in second dimension. Some compounds contain radioactively labeled carbon.

c autoradiography

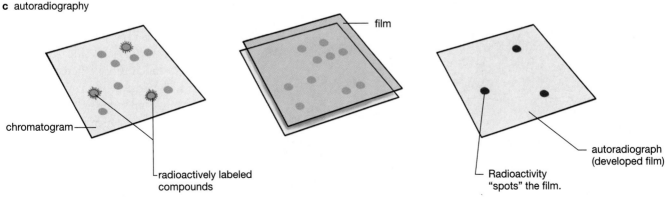

film

chromatogram

radioactively labeled compounds

Radioactivity "spots" the film.

autoradiograph (developed film)

Figure 1

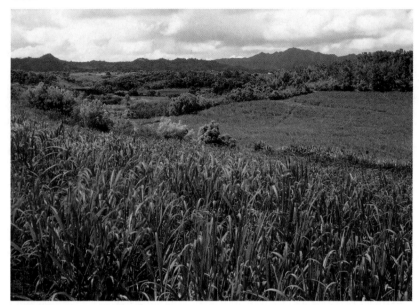

Figure 8.9
Warm-Weather Photosynthesis
Plants such as this sugarcane, shown here in Hawaii, practice C_4 photosynthesis.

Figure 8.10
The C_4 Pathway
The C_4 pathway is an adaptation of many warm-climate plants to the problem of photorespiration. Photorespiration occurs when the C_3 enzyme rubisco binds to oxygen rather than carbon dioxide (thereby stopping the production of sugar). C_4 plants contain a different enzyme in their mesophyll cells that binds to carbon dioxide, but not to oxygen. In this way, the carbon dioxide is escorted into the bundle-sheath cells, where rubisco can bind to it in the Calvin cycle and the production of sugar can proceed.

8.7 A Different Kind of Photosynthesis: The C_4 Pathway

Photorespiration is especially likely to take place in warm environments, because heat prompts the stomata on leaves to close in order to preserve water. The effect of closed stomata, however, is not only that water is kept *in*, but that CO_2 is kept *out*. With a deficit of CO_2, rubisco tends to bind more frequently with oxygen, prompting photorespiration—and plants that simply do not grow as much.

The adaptation that warm-climate plants developed to deal with this problem employs a special carbon-fixing "front-end" and then a carbon dioxide shuttle into the Calvin cycle. The so-called C_4 plants that utilize this system (**see Figure 8.9**) don't initially use rubisco to fix carbon-dioxide. Rather, they employ a different enzyme that does not bind with oxygen. The immediate product of this reaction is a four-carbon molecule (hence the C_4 cycle's name). The derivative of that molecule moves to a special group of cells and *releases* the CO_2, which will be used in the Calvin cycle. These special cells are wrapped around leaf veins and are hence named the **bundle-sheath cells** (**see Figure 8.10**). What the C_4 system provides—even at times when little CO_2 may be coming in through the stomata—is a relatively high concentration of CO_2 where it's needed: in the cells where the Calvin cycle takes place.

The C_4 Pathway Is Not Always Advantageous

The C_4 pathway is known to be employed in about 1,000 species of plants, most notably in some grasses, and in corn, sugarcane, and sorghum. But there are vastly more C_3 than C_4 plants. You might wonder why evolution hasn't weeded out C_3 plants in favor of the C_4 variety, since the latter seem to be so much more efficient at photosynthesis. The answer is that C_4 fixation does not confer an across-the-board advantage. It has a cost, which is the expenditure of ATP in shuttling carbon dioxide to the bundle-sheath cells. What this means is that the warm-weather advantage of C_4 fixation doesn't travel well; there are few C_4 plants in cold-weather climates.

8.8 Another Photosynthetic Variation: CAM Plants

When plants live in climates that are not just warm but *hot*, a large part of their survival comes down to retaining water. Photosynthesis, however, works against water retention, because as you have seen, when CO_2 can pass in, water can pass out.

The plants we call succulents (and some other plants) have a solution for this: Close the stomata during the day and open them at night. The plants then carry out C_4 metabolism at night, but only up to a point. They fix carbon dioxide into an initial four-carbon molecule and then stand pat. The CO_2 stays "banked" in them, awaiting the energy of the next day's sun, which will power the production of the ATP they need to carry out the range of photosynthetic operations (see **Figure 8.11**).

This process, called **CAM metabolism**, gets the job done for cactus, pineapple, orchid, mint family plants, and others. As you've no doubt guessed, CAM is an acronym; it stands for *crassulacean acid metabolism*. The succulent plant family *Crassulaceae* was the first group of plants in which CAM metabolism was discovered. The three methods of photosynthesis—C_3, C_4, and CAM—are summarized in **Figure 8.12**.

Closing Thoughts on Photosynthesis and Energy

Readers who have stayed the course on this account of photosynthesis now know at least one thing very well about it: How complicated it is, with its many components and long metabolic pathways. It's also probably clear by this point why scientists continue to study photosynthesis in such detail. It is the foundation of plant growth and upon plant growth hinges nothing less than the survival of all animals, including human beings. Without an understanding of this linkage, it's easy to see plants as a set of mute fixtures whose main contribution to human life is aesthetic. But with this knowledge, you can begin to see the central position that plants occupy in the interconnected web of life.

Over the last three chapters, you have learned about the endless back-and-forth of

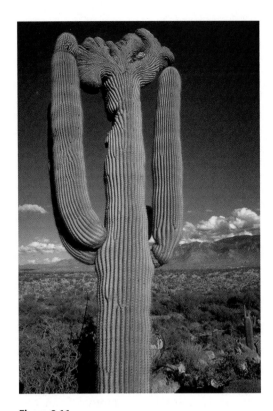

Figure 8.11
Hot-Weather Photosynthesis
Plants such as this saguaro cactus in Arizona practice CAM photosynthesis, thereby preserving precious water.

	C_3	C_4	CAM
Used by:	majority of plants	corn, sugarcane (warm environments)	cactus, pineapple, orchid (hot environments)
Benefits:	efficient use of ATP	less photorespiration	less water loss
Problems:	photorespiration	uses up more ATP	uses up more ATP; hard to "bank" enough CO_2

Figure 8.12
Three Modes of Photosynthesis

oxygen, carbon dioxide, and energy in the living world. "*Cycle*" has been an operant word in this long discussion, because the only one-way trip you've encountered has been the relentless "spillage" of energy from the sun down into heat. Looked at in a cynical way, Earth and its inhabitants constitute a kind of leaky holding tank for energy that comes from the sun. Looked at another way, however, the living world has been able, through photosynthesis, to take the sun's energy and build a remarkable edifice with it. Think of the forms and sheer mass of living material on Earth. One of the most amazing things about this structure is that we humans get to be both a part of it and witnesses to it.

Chapter Review

Key Terms

accessory pigment   154
bundle-sheath cell   162
Calvin cycle   158
CAM metabolism   163
chlorophyll *a*   154
chloroplast   154
fixation   158
granum   154
**light-dependent
    reaction**   154
**light-independent
    reaction**   154
photorespiration   159
photosystem   154
reaction center   154
rubisco   159
stoma   154
stroma   154
thylakoid   154
**thylakoid
    compartment**   154

Summary

8.1   Photosynthesis and Energy

- Photosynthesis is the process by which certain organisms produce their own carbohydrates, using carbon dioxide, water, minerals, and energy captured from the sun's rays.

- Photosynthesis has made possible life as we know it on Earth, because the organic material produced in photosynthesis is the source of food for most of Earth's living things. Photosynthesis also is responsible for the atmospheric oxygen used by many living things in respiration.

8.2   The Components of Photosynthesis

- In plants, photosynthesis takes place in the organelles called chloroplasts, which can exist in great abundance in the mesophyll cells of leaves. The energy for photosynthesis comes mostly from various blue and red wavelengths of visible sunlight that are absorbed by pigments in the chloroplasts. Plant leaves contain microscopic pores called stomata that can open and close, letting carbon dioxide in and oxygen and water vapor out.

- There are two primary stages to photosynthesis. In the first stage, electrons derived from water are energetically boosted by the power of sunlight. The electrons physically move in this process from special chlorophyll molecules through a series of electron carriers, ending up as part of the electron carrier NADPH, which carries the electrons to the second stage of photosynthesis. In this second stage, the electrons are brought together with carbon dioxide and a low-energy sugar to produce a high-energy carbohydrate.

8.3   Stage 1: The Steps of the Light-Dependent Reactions

- In its first stage, photosynthesis works through a pair of complexes, photosystems II and I. These photosystems are composed of antennae chlorophyll molecules that absorb and transmit solar energy; reaction centers that accept both this energy and electrons derived from water; and primary electron receptors (a part of the reaction centers), to which the electrons move after being energetically boosted.

- The energetic fall of electrons through an electron transfer chain between photosystems II and I yields the energy that produces the ATP used in the second stage of photosynthesis.

8.4   What Makes the Light-Dependent Reactions So Important?

- Two actions of great consequence take place in the light-dependent reactions. First, water is split, yielding not only the electrons that ultimately play a part in providing food, but liberating oxygen from water as well. This is the oxygen that organisms such as ourselves breathe in and use in extracting energy from food. Second, the electrons that are given an energy boost in the light-dependent reactions are transferred to a different molecule, the initial electron acceptor. This is one of the keys to the plant's ability to make its own food.

8.5 Stage 2 of Photosynthesis: The Light-Independent Reactions

- In the second stage of photosynthesis, through the process of the Calvin cycle, carbon dioxide from the atmosphere is brought together with a low-energy sugar, RuBP, by the enzyme rubisco. The resulting compound is energized with the addition of phosphate groups and electrons supplied by the first stage of photosynthesis. The result is the high-energy sugar G3P, which is the product of photosynthesis.
- G3P can be used for energy or for plant growth. Everything in the plant ultimately is derived from this carbohydrate.

8.6 Photorespiration: Undercutting Photosynthesis

- The enzyme rubisco frequently binds with oxygen, rather than carbon dioxide, thus undercutting productivity in photosynthesis.

8.7 A Different Kind of Photosynthesis: The C_4 Pathway

- Some plants in warm climates, where this problem is most pronounced, have evolved a C_4 method of photosynthesis. C_4 employs an enzyme that binds with carbon dioxide but not with oxygen. The carbon dioxide is then shuttled to special bundle-sheath cells in the plant and released, after which it moves into the Calvin cycle.

8.8 Another Photosynthetic Variation: CAM Plants

- Hot-weather plants such as cacti employ another form of photosynthesis, CAM metabolism. The plant's stomata open only at night, letting in carbon dioxide that is "banked" until sunrise, when the sun's rays will supply the energy needed to complete photosynthesis.

Understanding the Basics

Multiple-Choice Questions

1. Most of your house plants are green (if they are healthy) because they
   a. absorb green light
   b. take the energy out of white light and produce green light
   c. contain molecules that manufacture green light in the process of photosynthesis
   d. use green light in photosynthesis
   e. scatter green light

2. A college student asked his freshman biology teacher about the leaves on his plant, which were turning white. She told him he should
   a. be concerned because this means that the plant pigments are disappearing so the plant cannot photosynthesize
   b. be concerned because his plant was producing so much starch in photosynthesis that it would be unable to store any more
   c. not be concerned because plants use many colors of light for photosynthesis
   d. not be concerned because many parts (such as the roots) of plants are white, so plants photosynthesize just fine with white leaves
   e. be concerned because white pigments damage plants

3. A tulip tree turns yellow in the fall, because the accessory pigments show up when the chlorophyll has faded. These accessory pigments
   a. are formed to take over the function of chlorophyll in the fall
   b. are produced to protect the leaves in the fall
   c. are always present to help absorb light for photosynthesis
   d. help pass electrons in chemiosmosis in photosynthesis
   e. are found just outside the chloroplast to pass light into it

4. Suppose you become a biologist. On a bright summer day, while studying the rate of photosynthesis in a desert succulent, you discover that its stomata are closed—yet it is producing ATP. This plant must be doing
   a. photorespiration
   b. the C_3 cycle
   c. the C_4 pathway
   d. CAM metabolism
   e. the Krebs cycle

5. Most plant photosynthesis occurs in the _____ of leaves.
   a. upper epidermal cells
   b. lower epidermal cells
   c. guard cells
   d. stomata
   e. mesophyll cells

6. A leaf on a tree is photosynthesizing rapidly, capturing light energy through light-dependent reactions. Why do the cells of the tree also need light-*independent* reactions?
   a. to use up the excess energy
   b. to cause photorespiration to occur
   c. to make sugars so that the tree can use the energy it captures
   d. to release the energy to the atmosphere
   e. to form the woody portion of the tree

7. Oxygen is released to the atmosphere in the light-dependent reactions of photosynthesis when water is split to supply electrons to
   a. Photosystem I
   b. Photosystem II
   c. Calvin cycle
   d. C_4 pathway
   e. CAM

Brief Review

1. Because plants harness energy in photosynthesis, do they need cellular respiration as well? Why or why not?

2. Sunlight reaching Earth is composed of a spectrum of energetic rays. Describe the rays that make up this spectrum. Which portion of the spectrum drives photosynthesis?

3. How is a photosystem put together? What are its parts? Explain why these all need to work together for the photosystem to function.

4. Picture a world without photosynthesis. What would it be like? What would happen to plants, animals, fungi, etc.?

5. Which would you expect to grow more rapidly—a C_3 plant, a C_4 plant, or a CAM plant? Be careful; there may be more than one answer to this question. Explain your answer.

6. List the four steps in the Calvin cycle, and explain why they are necessary. Why should you assume that all these steps are necessary? Do plants or animals typically waste energy? Why or why not?

Applying Your Knowledge

1. In photosynthesis, plants make carbohydrates and then use them as a source of energy. Animals must eat plants to get carbohydrates and use them as a source of energy. Which process do you think is a more efficient use of the sun's energy? Why? What implications does that have for feeding the growing population of the world?

2. One of the hottest ideas in engineering today is the concept of building an energy-capturing system similar to photosynthesis. The system would use light energy to make artificial food (shades of the food generators on Star Trek). What two parts would such a system need, and how must these parts interact if they are to be like photosynthesis?

3. Rubisco has been called the most important protein on earth. What arguments can you make to support that hypothesis? Do you have any that would go against it?

4. One of the most dramatic environmental changes today is the increase in carbon dioxide levels in the atmosphere. We usually hear about it in connection with "global warming," but that's not the only issue. What effect would increased carbon dioxide have on C_3 and C_4 plants? Considering that a fairly large proportion of our major crops (corn, for example) and weeds (giant ragweed) are C_4 plants, what implications might this have? What actions do you think should be taken to prevent increases in carbon dioxide?

Group Activity

As a class, act out the light-dependent reactions of photosynthesis using ping-pong balls as electrons and students as the various kinds of molecules. What roles will you need? Be sure to have both photosystems represented, as well as the electron transfer chains. Think about how to represent the splitting of the water molecule.

MediaLab
Capturing Sunlight to Make Food: Photosynthesis

You've probably heard songs in which the singers attribute their life, the act of eating and breathing, to their romantic interest. Love is a powerful emotion, but you'd think at least one songwriter would give homage to the event that really does make our lives possible—the process of photosynthesis. Perhaps it wouldn't sell. The *CD Activity* reviews photosynthesis, so that in the following *Web Investigations* you are prepared to gather information on topics closely associated with it. In the *Communicate Your Results* section, you can interpret for yourself how important photosynthesis is in our lives.

This *MediaLab* can be found in Chapter 8 on your CD-ROM and Companion Website (http://www.prenhall.com/krogh).

CD ACTIVITY

The overall process of photosynthesis is often summarized in the following reaction:

water + carbon dioxide + light energy -> glucose + oxygen

As with most biological processes, photosynthesis does not occur in one step, but in a series of reactions. In this *CD Activity,* you'll review the three phases of photosynthesis, focusing on reactants and products.

Activity

1. First, you will learn about how light energy is absorbed by pigment molecules in plants.
2. Next, you will discover how light energy is converted to a stored chemical form for transport and later storage.
3. Finally, you will investigate the storage of energy in glucose.

After this review, you'll be prepared to apply, extend, and experience some of the consequences of the photosynthetic process in the following *Web Investigations*.

WEB INVESTIGATION

Investigation 1
Estimated time for completion = 10 minutes

If photosynthesis is to occur at all, the plant must be able to intercept light. The better a plant is in intercepting the light, the greater opportunities the plant has to produce food. Select the Keyword **LEAVES** on your CD or Website for this *MediaLab* to read an article describing how plants maximize the amount of light they can intercept. At the bottom of that page is a forward link to a discussion of leaf position in a canopy. Review this page. Why have so many plants developed broad, flattened leaves? What is heliotropism, and how does it make plants more efficient? What are sleep movements? Describe the model for leaf orientation in the upper, middle, and lower levels of a leaf canopy.

Investigation 2

Estimated time for completion = 10 minutes

Plants absorb light energy using special molecules called *photosynthetic pigments* (predominantly chlorophylls). These pigments absorb most of the visible light except for wavelengths in the green part of the spectrum. Consequently, plants appear green. But what happens in the fall when leaves change colors? Why are there so many different colors? And why are the colors more intense some years? Select the Keyword **COLORS** on your CD or Website for this *MediaLab* to access a site that will help you answer these questions.

Investigation 3

Estimated time for completion = 15 minutes

You've probably heard of global warming and the greenhouse effect, in which certain gases cause heat to be trapped in the Earth's atmosphere in the same way heat is trapped in a greenhouse. How hot will it be in your community in 100 years? Select the Keyword **HOTSPOT** on your CD or Website for this *MediaLab* to review predicted temperatures for the year 2100. One of the major greenhouse gases is carbon dioxide, which has been added to the atmosphere by the burning of fossil fuels. Of course, plants in photosynthesis trap CO_2 out of the atmosphere. Could plants help with global warming? Select the Keyword CO_2 to view changing CO_2 levels over 25 years in Barrow, Alaska; Mauna Loa, Hawaii; Samoa, and the South Pole. What might account for the yearly fluctuations in CO_2 levels? Where fluctuations are the greatest, is there a seasonal difference in plant activity?

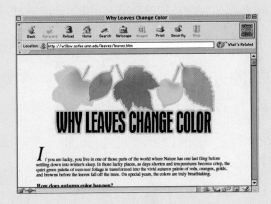

COMMUNICATE YOUR RESULTS

Exercise 1

Estimated time for completion = 15 minutes

Using the knowledge from the articles in *Web Investigation 1*, design a plant that can effectively intercept light while keeping the amount of plant tissues to a minimum. Here are some characteristics of this hypothetical plant to keep in mind: The plant grows in a seasonal environment, water conservation is important but not critical, and the plant typically has ten leaves located at five levels on the stem. Draw the leaves for this plant. Also draw a view of the plant from the side and a view of the plant from the top. Describe the decisions you made while creating this plant.

Exercise 2

Estimated time for completion = 15 minutes

If it's fall and the leaves are changing, make a collection of leaves from different species in your area. Otherwise, find pictures of fall leaves and clip or copy them to create a collection. Place these leaves on a poster or in a bound collection. Identify the species of plant as well as possible using available references. For each species, using the leaf's color as a guide, identify the dominant pigment present in the leaf. Is there a pattern of leaf color and plant type? Describe any patterns you observe.

Exercise 3

Estimated time for completion = 15 minutes

In *Web Investigation 3*, you saw an annual fluctuation in CO_2 levels. One possible explanation for the yearly cycle is that increased plant growth decreases atmospheric CO_2 levels during the growing season. How would you test this hypothesis? Prepare a short presentation or poster explaining the initial observation and hypothesis; and your test, observations, and conclusions. Include your conclusions on how photosynthesis could be used to fight global warming, and how deforestation would contribute to the problem.

How Life Goes On: Genetics

A common goldeneye
duck with young.

9

Introduction to Genetics; Mitosis and Cytokinesis

A closer look at the chromosome.
(page 179)

What does a cancer cell look like?
(page 178)

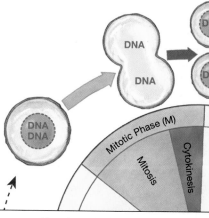

How does mitosis work?
(page 181)

Everyone knows how one

generation of living things gives rise to another, at least in a general sense. But the details of how this works are certainly not apparent just by looking at Earth's creatures. A brief sexual encounter takes place, and the next thing we know there are new kittens or chicks or babies. Only the encounter and the newborns are obvious; what comes in between is mysterious.

The problem, of course, is that most of the component parts of reproduction are hidden away and, in any event, so small that the process is completely invisible to the naked eye. Given this, it's no wonder that for centuries human beings wondered what it was that one generation passed on so that the next generation could be formed.

A similar kind of puzzlement existed over a related question: What is the "hand" that controls the development of a living thing as it goes from being a microscopic cell to being a human baby in full cry? For that matter, how is it that the adult body is able build muscle or to repair a minor wound? What, in short, is the hand that controls day-to-day physical functioning?

9.1 An Introduction to Genetics

The Key to Reproduction, Development, and Maintenance Is DNA

It took roughly a century of hard-won scientific advances for these questions to be answered. In the end, it turned out that a single substance is central to the reproduction of living things, their development from single cells, and their day-to-day functioning. This substance is deoxyribonucleic acid, a long, vanishingly thin molecule known everywhere today by its acronym, DNA.

DNA contains what is known as an organism's **genome**: the complete collection of that organism's genetic information, strung together in chemical units called genes that lie along DNA's famous double helix.

How is it that this one molecule can undertake all the tasks just described? To the question about what is it that humans *pass on* in reproduction, the answer is: about half

A frog egg undergoing cell division.
(page 184)

nuclear envelope forming

The mechanics of cell division.
(page 183)

amino acids

ribosome

mRNA

protein

Protein synthesis in action.
(page 175)

cell plate

Vesicles fuse together, forming a cell plate that grows toward the parent cell wall.

Plant cell division.
(page 184)

of a father's genome and about half of a mother's genome, both these halves coming together in a fertilized egg to produce a whole, new genome (**Figure 9.1**).

To the other questions about what controls the development and everyday functioning of living things, the answer is: the instructions provided by the genes along the genome.

DNA Contains Instructions for Protein Production

To grasp how genes can control the development and functioning of living things, note that there are logically two parts to such control. If genes are *giving* instructions, something needs to carry them out. As it turns out, the workers here are **proteins**; in tremendous diversity and number, they are the working products of the information contained in genes. Readers who went through Chapter 3 (on biological molecules) and Chapter 6 (on energy) should have a good grasp already of what proteins are capable of. As we have seen, there are proteins that form cartilage and hair; there are communication proteins that help form the receptors on cells; there are hormonal proteins and transport proteins. Then there are **enzymes**, the chemically active proteins that speed up—or in practical terms, *enable*—chemical reactions in living things. Without these protein molecules, almost nothing could get done in the body; with them, a dazzling set of chemical reactions takes place every second in every living thing.

How Do Genes Direct the Production of Proteins?

What most genes do, then, is contain the information for the production of proteins. In Chapter 14, you'll get the details of how protein production comes about, but for now let's just look at the essentials of the story. Imagine you are a telegraph operator in the Old West, and your "key" starts tapping frantically. You leap forward in your chair, listening for the sound of short and long bursts on the key, and you begin writing down the letters these sounds stand for. Here's the message you get:

.—.. —— —.—. —.— — ....— ..—. .

which means:

L   o   c   k   t   h   e   s   a   f   e

You walk over and do just that. There is thus a code here, Morse code, that has resulted in some action. In DNA there is likewise a code. But instead of a code specified in short or long sounds, it is a code specified in chemical substances. There are four of these substances, and they lie along the genome. They are the "bases" adenine, thymine, guanine, and cytosine, abbreviated A, T, G, C.

A particular sequence of these bases—usually thousands of letters long—contains information that can be acted upon, just as the Morse code did. Only the action here is the production of a particular protein. One series of A's, T's, C's, and G's contains the in-

Figure 9.1
Passing on Genetic Information
About half of each person's genome comes from the father, and half from the mother.

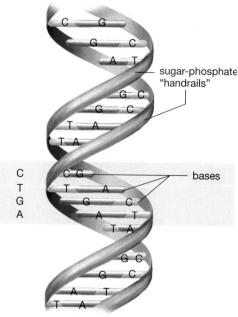

Figure 9.2
Information-Bearing DNA Molecule
The DNA molecule is composed partly of two handrails made of sugar and phosphate. The genetic information in the molecule is contained in the sequence of "bases" along one strand of the double helix. For example, the bases in the highlighted section of DNA are in the order CTGA.

sugar-phosphate "handrails"

bases

formation for the production of one protein, but a *different* sequence of A's, T's, G's, and C's specifies a *different* protein. These separate sequences of bases are separate *genes*.

You saw in Chapter 3 that proteins are composed of building blocks called amino acids. The base sequence that makes up a gene, then, is like a message that says, "Give me some of this amino acid, now this one, then this one . . ." and so on for hundreds of steps until a protein is created that folds up and gets busy on some task. This task can be one small part of *forming* a human being—as in embryonic development—or of *maintaining* one, as in an adult. It is by passing on the entire complement of this information that one generation gives rise to another.

The Architecture of DNA

It is the architecture of the DNA molecule that makes this protein production process possible. Look at the general scheme of it pictured in **Figure 9.2**. As you can see, DNA looks something like an open, spiral staircase, the handrails of which are a repeating series of sugar (deoxyribose) and phosphate molecules. Meanwhile, the messages you've just been reading about are contained in the steps of the staircase, in the order in which the bases A, T, C, and G are placed along the helix. As you can see, these bases extend inward from both of the DNA rails. The order in which any of the bases occurs *along the rails* is extremely varied. This order is what "encodes" the information for the production of proteins.

The Path of Protein Synthesis

Look now at **Figure 9.3** and see the path that is followed in protein synthesis. The DNA you have been reading about is contained in the nucleus of the cell. The first step is that a stretch of it unwinds there, and its message—the order of a string of bases—is copied onto a molecule called messenger RNA (mRNA). This length of mRNA, which could be thought of as an information tape being dubbed off a "master" DNA tape, then exits from the cell nucleus. Its destination is a molecular workbench in the cell's cytoplasm, a structure called a ribosome. It is here that both message (the mRNA tape) and raw materials (amino acids) come together to make the product (a protein). The mRNA tape is "read" within the ribosome, and as this happens a growing chain of amino acids is linked together in the ribo-

some in the order called for by the mRNA. When the chain is finished and folded up, a protein has come into existence.

Genetics as Information Management

In reflecting on this story of genes (and the field of **genetics**), the first thing to note is that it concerns the storage, duplication, and transfer of *information*. True, it is information encoded in *chemical form*, but that shouldn't bother us. Information comes in lots of forms; there are the familiar words on a page, but there are also bar graphs, smoke signals, or a baby's smile.

To grasp the fact of genetics as molecular information, think back to Chapter 6, where you looked at a particular protein, an enzyme called sucrase that helps break up table sugar (sucrose) into its components of fructose and glucose. In line with what you've seen so far, there is a gene that prompts the production of sucrase. This gene's message is copied onto mRNA, which then migrates to a ribosome in the cytoplasm. Sucrase is put together there, amino acid by amino acid, and, once synthesized, it folds up and gets busy clipping sucrose molecules.

Figure 9.3
The Path of Protein Synthesis
The information contained in a length of DNA is transcribed onto the "tape" of a length of messenger RNA (mRNA). The mRNA then exits the cell's nucleus and goes to a structure in the cell's cytoplasm called a ribosome. There, the mRNA tape is "read," and a string of amino acids is put together in the order specified by the tape. The result is a protein.

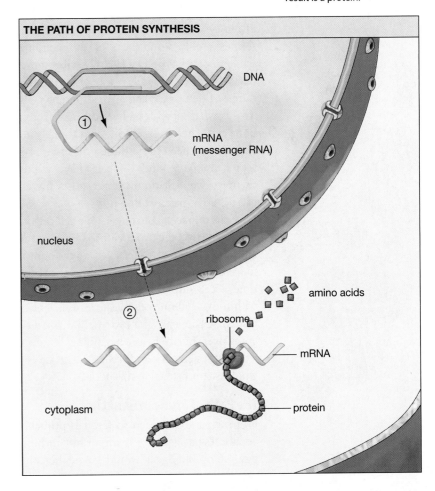

THE PATH OF PROTEIN SYNTHESIS

DNA

①

mRNA
(messenger RNA)

nucleus

②

amino acids

ribosome

mRNA

cytoplasm

protein

Now, consider the role of the gene in this process. It didn't do the sucrose clipping; it didn't even migrate to the cytoplasm to be involved in the synthesis of sucrase. In this case, its role was more like that of a . . . cookbook recipe. It specified some of this amino acid, then some of that amino acid, and eventually sucrase was produced.

From One Gene to a Collection

Now, in considering not just one gene, but rather the entire *collection* of genes in a living thing—its genome, in other words—it is easy to see that we are dealing with a vast *library* of information. The size of such a collection can vary, but to give you some idea, the human genome may include as many as 100,000 genes. We generally think that information comes to us after birth and from outside the body, but in fact we are born with a volume of information that has been amassed and edited over 3.5 billion years of evolution. It is this information that we bring to our environment from the moment of our conception.

The amazing thing about such collections is that there is not just one copy of them in living things; there may be trillions. Most cells within an organism contain a complete copy of that organism's genome. A given type of cell puts to use (or "transcribes") only some parts of this genome, while another type of cell transcribes other parts; but this is merely a way of saying that different genes are active in different cells. This is what separates a liver cell, say, from a bone cell. These cells do different things and, accordingly, need to synthesize different proteins. But *both* kinds of cells have within them the whole genomic library; they merely put different parts of it to use.

Cells duplicate, of course. One cell divides to become two, the two divide to become four, and so on. Given this, if each cell in the body is to have its own copy of the genome, then each time a cell duplicates, the genome that lies within that cell must duplicate as well. This is just what happens, in a process that you'll be going over shortly. Before beginning, however, let's take a step back and see how genetics as a whole is approached in this book.

The Path of Study in Genetics

From what you've seen so far, it is probably apparent that genetics is important in biology. Many biologists would go further and

MediaLab Link
Cancer and cell division, see p. 206.

argue that genetics lies at the *heart* of biology, because it is so central to what distinguishes the living world from the nonliving. The ability to reproduce is a distinguishing characteristic of life, and reproduction is the business of genetics. Beyond that, living things are notable for the degree of organization they possess. Genetics is, to a considerable extent, the means by which life's unimaginable complexity is managed; it is thus a key to the high level of organization we see in life.

Given genetics' importance, in this book you'll be looking at it from several angles. First, you will study DNA mostly as it comes packaged in units called chromosomes. Then you will learn how genetics functions in whole organisms, namely Gregor Mendel's famous pea plants. Only later will you return to DNA proper—to its replication and its protein-coding function. Finally, you will see how basic knowledge about genetics is being applied in the brave new world of genetic engineering. Let's begin the tour now, by seeing how life goes from one cell to many.

9.2 An Introduction to Cell Division

How does a baby grow, or a plant develop, or a wound heal? Always through cell division. As noted in Chapter 4, with the possible exception of viruses, life exists only inside cells. Further, we know that cells come only from other cells. And the *way* cells come from cells is by dividing. To understand the continuity in life, then, we need to have some understanding of how cells divide.

The sheer numbers involved in this process are mind-boggling; in your body, as many as four million cell divisions are taking place each second. It is likely that every one of these divisions serves some purpose in maintaining your body; yet there are instances in which cells divide too much. How does the disease we call cancer take hold? Always through the *unrestrained* division of cells. As you can see in "When Cell Division Runs Amok" on page 178, cancer is intimately related to basic cellular processes.

Cell division may be a somewhat misleading term if it is taken to mean a simple separation of cellular material, akin to a candy bar being split between two friends. Indeed, a splitting does occur in cell division; but certain parts of the cell must *duplicate* before this happens. Then the duplicated material is parceled out with fine precision, half of it going to one "daughter" cell and half to another.

What's being duplicated and divided is DNA. Because a cell's full complement of DNA contains such critical information, it would not do for a cell to be left with 50 or 75 percent of this information; rather, it needs the whole thing—no more and no less.

With this in mind, here is the big picture on cell division. First, there is a duplication of DNA; then there is the movement of two precisely matched quantities of it to opposite sides of the "parent" cell; finally there is the splitting of the parent cell into two daughter cells. The duplication of DNA is known as replication, the apportioning of it into two quantities is known as mitosis, and the splitting of the cellular material is known as cytokinesis (**Figure 9.4**). The goal for the remainder of this chapter is to learn a little bit about replication and a good deal more about mitosis and cytokinesis.

The Replication of DNA

So how does the first part of cell division work—the DNA replication? If you look at **Figure 9.5**, you can see a simplified representation of the DNA molecule shown in Figure 9.2. Notice that the strands of the double helix have started to unwind, which is an initial step in DNA replication (Figure 9.5, step 1). The two resulting single strands then serve as a template or pattern upon which a new strand is created (Figure 9.5, step 2). Earlier, DNA structure was compared to a spiral staircase. Now think of that staircase as splitting right down the middle of its steps (where the A and T or the G and C bases meet). Then a new *half*-staircase is created that bonds with one of the old DNA strands; a second half-staircase is likewise created that bonds with the *second* original strand (these are the green strands in Figure 9.5, step 2). This replication process, which doubles the cell's DNA content, takes place in advance of the cell dividing.

In Chapter 13, you will metaphorically walk along the rungs of the double helix to grasp the details of this doubling of DNA strands. Right now, however, you'll be considering DNA on a larger scale, as it comes *packaged*.

Figure 9.4
Overview of Cell Division

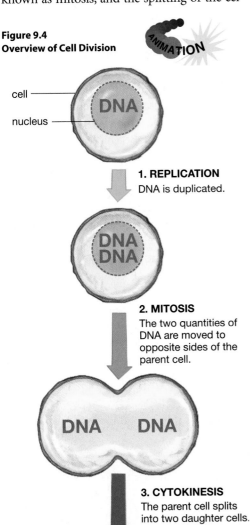

cell

nucleus

1. REPLICATION
DNA is duplicated.

2. MITOSIS
The two quantities of DNA are moved to opposite sides of the parent cell.

3. CYTOKINESIS
The parent cell splits into two daughter cells.

1. Original DNA molecule unwinds.

2. New DNA strands are synthesized from the two original strands.

Figure 9.5
DNA Replication

1. A DNA molecule unwinds.

2. Each of the single strands of the original molecule (in red) pairs with a new, complementary strand (in green). Thus, the quantity of DNA has effectively doubled.

When Cell Division Runs Amok: Cancer

In the main text, there is a review of the cell cycle. But what *controls* this cycle? This topic of is of compelling interest because the *unrestrained* division of cells goes by another name: cancer. As it turns out, much of the cancer research that goes on today is research on the control of the cell cycle.

What causes unrestrained cell growth? To use a common but apt metaphor, cells can come to this out-of-control state in one of two ways: They can get their accelerator stuck, or they can have their brakes fail. The control mechanisms that *induce* cell division can become hyperactive, or the mechanisms that *suppress* cell division can fail to perform. You may have heard a couple of terms used to describe the genetic components of such processes. There are normal genes that induce cell division, but which when mutated can cause cancer; these are the stuck-accelerator genes, called **oncogenes**. Then there are **tumor suppressor** genes, which in their mutated state can be like failed brakes.

Investigations in the 1980s and 1990s revealed that the master controllers of cell division are chemical complexes composed of two types of proteins: kinases and cyclins, which collectively are referred to as cyclin-dependent kinases or CDKs.

The details by which CDKs govern cell division are complex and need not concern us here. Suffice it to say that these compounds work together in a linked chain of protein activity. CDKs brought into production prompt some cell-division task—such as the breakdown of the nuclear envelope—after which they prompt the *next* step in cell division.

CDKs might appear to be the domain of basic cell biologists, while tumor suppressor genes might appear to be the domain of cancer researchers. But the work of these different types of researchers is increasingly coming together. In fact, cell and cancer scientists seem at times to be like two sets of miners who are burrowing in from opposite sides of a mountain, only to meet unexpectedly in the middle.

Consider, on the cancer research side, Utah scientist Mark Skolnick and his colleague Lisa Cannon-Albright. About 10 years ago, the two were looking at data from a group of families that had a high incidence of melanoma, which is an often deadly form of skin cancer. By 1992, the researchers found that these families had a chromosome that contained a susceptibility gene for melanoma. Skolnick and researcher Alexander Kamb then looked at melanoma cells and saw that they seemed to be missing one small part of this chromosome. This suggested that the culprit was a *gene* that was missing. When present and functioning normally, such a gene might *restrain* the growth of cells. It would be, in other words, one of the sets of brakes mentioned earlier—a tumor suppressor gene. This is exactly what Kamb and Skolnick believed they were on to. Accordingly, they set out to find it and describe it.

Meanwhile, digging in from the cell biology side of the mountain was one of the nation's experts on the cell cycle. In 1993 David Beach, at Cold Spring Harbor laboratory in Long Island, identified a gene that he called p16. As it turned out, p16 coded for a protein that inhibits cell growth. It does this by binding to, and thus suppressing, none other than a cyclin-dependent kinase (specifically one called cdk4). When genes are found and characterized, they are put into a kind of gene directory, which Beach did with p16.

Then, in early 1994, Kamb and Skolnick, looking for the missing gene, zeroed in on what they believed might be the culprit. Reading its sequence of T's, C's, G's and A's, they then looked up this sequence in the gene directory to see if it had been characterized before. It had, as a matter of fact—by Beach! Though Beach could not be sure at the time that it was a tumor suppressor gene, it turned out that the gene that caused *melanoma* by its dysfunction was none other than the p16 *growth-inhibition* gene that Beach had found. A malfunctioning p16 gene, then, is like a set of failed brakes—one that does not produce a protein that restrains cell division.

Researchers have also found the flip side of this: a gene, which codes for cyclin, that seems at times to simply "stay on." Because cyclins are required to make CDKs become active, a cyclin that was overproduced might well cause *excess* CDK activation and prompt excessive cell division. Here, in other words, is a stuck accelerator. It's worth noting that in general, for cancers to get started, cells have to experience both a stuck accelerator and failed brakes.

As may be apparent, the process of cancer initiation is very much bound up with the process of cell-cycle regulation. Only time will tell what gains in the fight against cancer will come from this knowledge.

Harmful Division
Pictured are skin cancer cells, taken from a tumor and growing in a laboratory. The red and yellow cell in the center is undergoing cell division. The DNA of the cells is stained blue. (×300)

9.3 DNA Is Packaged in Chromosomes

You have thus far conceptualized each cell's DNA content as spiraling out in *one* long double helix; but it does not, in fact, take that form. Rather, the DNA in each cell comes divided up and packaged into units called **chromosomes** (from the Greek for "colored bodies," a name stemming from the dye these structures absorbed when they were first seen in the nineteenth century). Different organisms have different numbers of chromosomes; human cells have 46, for example, while onion cells have 16. Chromosomes amount to DNA "packages" in several senses. They consist not only of DNA itself, but of a protein material around which DNA is wrapped. The resulting combination of DNA and protein is known as **chromatin**. Thus a number of chromosomes, composed of chromatin, exist in a cell's nucleus, detached but in close physical proximity to one another, and the *collection* of chromosomes makes up nearly the entire complement of a cell's DNA.

If you look at **Figure 9.6a**, you can see how to think about the double helix being packaged into chromosomes. **Figure 9.6b** shows you how to relate the "staircase-splitting" DNA replication just described to what this means at the chromosomal level. When we say that DNA is replicating, this is another way of saying that the chromosomes that DNA helps make up are *duplicating*. Once that has happened, the result is chromosomes in "duplicated state," which means individual chromosomes, each of which is made up of two sister **chromatids**. Each chromatid is one of the newly replicated DNA double helices, as you can see in Figure 9.6b. (It probably goes without saying that, despite the name, sister chromatids are neither siblings nor females.)

Matched Pairs of Chromosomes

Chromosomes are detached from one another, but that does not mean they are completely different from one another. In eukaryotes such as ourselves, in fact, chromosomes come in *pairs* that are close, but not exact

DNA wraps around protein to make chromatin

Chromatin folds up to make chromosomes

a   DNA   chromatin   duplicated chromosome   cell

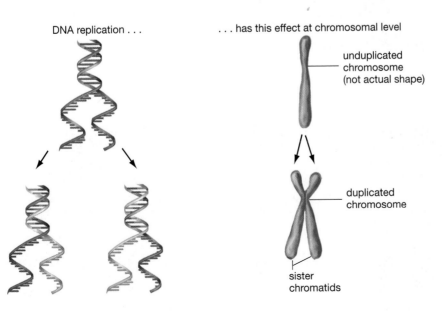

DNA replication . . .   . . . has this effect at chromosomal level

unduplicated chromosome (not actual shape)

duplicated chromosome

sister chromatids

b

Figure 9.6

a DNA Is Packaged in Units Called Chromosomes
The DNA molecule is bound up with proteins, with the resulting combination known as chromatin. This chromatin then makes up structures called chromosomes.

b DNA Replication at Two Levels
When DNA replicates, the result is two copies of the original DNA molecule. At the chromosomal level, the result of this replication is a single chromosome in duplicated state. The chromosome is composed of two sister chromatids—the two copies of the original DNA molecule. An unduplicated chromosome doesn't actually have the well-defined shape of the chromosome in the figure.

matches. The 46 chromosomes we have come to us as 23 chromosomes from *each parent*. Critically, with one exception that you'll soon get to, these are 23 matched pairs of chromosomes, each chromosome from the mother matching with one from the father.

Defining a "Matched Pair"

What is a "matched pair" of chromosomes? You have seen that DNA is an information-bearing molecule, and in this respect the genes that lie along the DNA helices can be compared to recipes. Now think of those recipes as being contained in *books*, with each book being a chromosome. For actual books, a matched set would be a pair whose *informational content* was the same, or nearly the same. Thus it is for matched chromosomes: Their informational content—their set of genes—is very much alike. We have a chromosome 1 that we inherit from our mother, and it contains a set of genes very similar to those that lie on the chromosome

1 we inherit from our *father*. The same is true for the maternal and paternal copies of chromosome 2, chromosome 3, and so on. Each of us, then, has 23 pairs of **homologous chromosomes**, "homologous" here meaning the same in shape and function.

However, homologous chromosomes are not *exactly* alike. A given gene from our father may help code for red hair, for example, while its matched gene, on the maternal chromosome, helps code for *blonde* hair. This difference in gene content is small, however, in relation to the similarity between the two. Both chromosomes have genes that code for hair color, as well as for thousands of other traits, and as such are said to be homologous.

X and Y Chromosomes

The one exception to the matched-pairs rule in human chromosomes is the so-called sex chromosomes of males. Human females have the 23 pairs of homologous chromosomes mentioned before; 22 of these are *autosomes*, or nonsex chromosomes, and one is a pair of X chromosomes that, when present, means that a growing embryo will be a female. Males, on the other hand, have the 22 pairs of homologous autosomes and one X chromosome; but then they also have one Y chromosome, which has a gene on it that confers the *male* sex on an embryo.

This whole scheme is laid out for you in **Figure 9.7**. There you can see a **karyotype**, or pictorial arrangement of a full set of human chromosomes. There are 46 chromosomes in all, 22 of them matched pairs, and, in this case, one pair having one X chromosome and one Y chromosome, meaning this is a male.

The chromosomes in Figure 9.7 are in the duplicated state noted earlier, each of them

Figure 9.7
A Karyotype Displays a Full Set of Chromosomes

One member of each chromosome pair comes from the individual's father and the other member from the mother. Each paired set of chromosomes is said to be "homologous," meaning the same in shape and function. (The two chromosomes over the number 1 are a homologous pair, the two over number 2, and so forth.) Homologous chromosomes are not exactly alike, however; the genes on them may differ somewhat in the effects they produce. Because this karyotype set is from a human male, there are 22 pairs of matched chromosomes and then one X and one Y chromosome. All the chromosomes are in the duplicated state.

Figure 9.8
DNA Can Be Arranged in Two Ways

a Replication occurs when the chromosomes DNA comes packed in are in a relatively formless state.

b Mitosis occurs after DNA condenses into the easily discernable shapes of duplicated chromosomes. The micrograph is of color-enhanced duplicated human chromosomes.

a DNA in uncondensed form

b DNA condensed into chromosomes

being composed of two sister chromatids. For the sake of clarity, it would be nice to see a picture of some *un*duplicated chromosomes, but we have no pictures that are the counterpart to Figure 9.7. This is so because chromosomes take on an easily discernable shape only when they duplicate prior to cell division. As cell division approaches, the DNA-protein complex changes its form; it tightens up, condensing mightily to produce well-defined chromosomes (**Figure 9.8**). After cell division is complete, the cell's chromosomes return to their relatively formless state.

Why this change in form? This condensing before cell division has the same effect as you taking all your scattered belongings and packing them into boxes just before moving from one apartment to another. Remember that DNA is packing up to leave as well, in this case for life in a successor cell. Were it not to tighten into its duplicated chromosomal form, its elongated fibers would get tangled up in the move.

Chromosome Duplication as a Part of Cell Division

With the chromosomes duplicated, you are now ready to look at the process by which they will split up. Before following this path, though, recognize that this separation occurs within the larger process that we are following: the division of the cell as a whole. As you saw in Chapter 5, there are many more things inside a cell than its complement of chromosomes. There are mitochondria, lysosomes, ribosomes, and so forth, and the cytosol in which they are immersed. Together, these things make up the cell's cytoplasm—the material *outside* the nucleus, as opposed to chromosomes, which lie *inside* the nucleus. With cell division, about half of the cytoplasmic material goes to one daughter cell and half to the other. It is helpful, then, to conceptualize cell division as having two separable components: **mitosis**, which is the division of the cell's chromosomes, and **cytokinesis**, which is the division of its cytoplasm.

The Cell Cycle

If you look at **Figure 9.9**, you can see how these two processes are linked over time. The figure illustrates the **cell cycle**: the repeating pattern of growth, genetic duplication, and finally division that holds true for many kinds of cells. There are two main phases in this cycle. First there is **interphase**, in which

the cell simultaneously carries out its work, grows, and—in preparation for division—duplicates its chromosomes. Second, there is **mitotic phase** (or M phase), in which the duplicated chromosomes separate properly and then the cell as a whole splits in two.

These two larger phases are in turn subdivided into smaller phases. In interphase, first there is G_1, standing for "gap-one." What goes on here are normal cell operations and cell growth. The cell then enters the *S* or "synthesis" phase, which is the synthesis of *DNA*, resulting in the duplication of the chromosomes. When this ends, interphase's G_2 period begins, during which there is more cytoplasmic growth and a preparation for cell division. Because DNA synthesis is stopped in G_2 as well as G_1, the word *gap* in G_1 and G_2 can be thought of as a gap in DNA synthesis.

The length of the cell cycle varies greatly from one type of cell to another. In a typical animal cell, the total cell cycle length is about 24 hours. Within this cycle, mitotic phase takes up only about 30 minutes. Within interphase, an animal cell spends roughly 12 hours in G_1, 6 hours in S, and 6 hours in G_2.

Figure 9.9
The Cell Cycle
The three main stages of cell division—DNA replication, mitosis, and cytokinesis—can be seen here in the context of a complete cell cycle. This cycle traditionally is divided into two main phases, interphase and mitotic (or "M") phase, which are in turn divided into other phases.

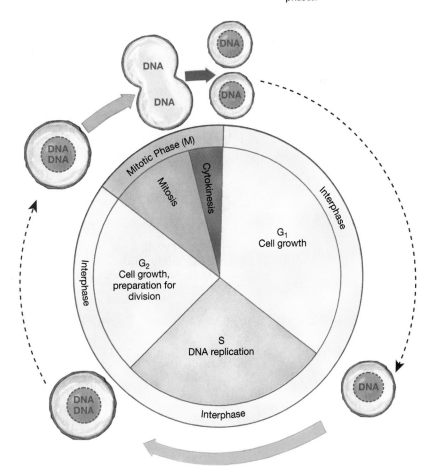

9.4 Mitosis and Cytokinesis

It is the second overarching phase of the cell cycle, mitotic phase, that is the focus for much of the rest of this chapter. Now the concern is not with how a cell carries out its general functions, but with how it *divides*. Mitotic phase includes both mitosis—during which time the cell's newly duplicated chromosomes condense, align themselves, and then separate—and cytokinesis, in which a cell's cytoplasm divides and one cell becomes two. Starting with mitosis, you'll look at its four phases in order (**see Figure 9.10**).

The Phases of Mitosis

Prophase

The beginning of mitosis marks the end of the cell's interphase. The starting state is that the cell's DNA, which doubled back in S phase, has just begun to pack itself into well-defined chromosomes. When we can finally *see* such chromosomes, we can say that mitosis has begun, with prophase. Now the packing job continues; before it is done, the DNA will have coiled and condensed into one eight-thousandth of its S-phase length. When this is over, there are 46 well-defined chromosomes in the nucleus in duplicated state (meaning they are composed of 92 chromatids). Meanwhile, the nuclear envelope—the double membrane surrounding the nucleus—begins to break up.

While this is going on, big changes are taking place outside the nucleus. Recall from Chapter 4 the protein structures called **microtubules** that are part of the cell's cytoskeleton. In some of their roles, microtubules can be compared to tent poles that can shorten or lengthen as need be. Throughout mitosis, microtubules stretch the cell as a whole, and physically move the cell's chromosomes around.

In an interphase cell there exists, just outside the nucleus, a microtubule organizing center, a **centrosome**. This centrosome doubles, so that now there are two microtubule organizing centers (see Figure 9.10). Now, in prophase, these two centrosomes start to move apart. Where are they going? To the poles. From here on out, it is convenient to think of mitosis and cytokinesis in terms of a global metaphor: the centrosomes migrate to the cellular poles, while the chromosomes first align, and then separate, along a cellular

Figure 9.10
Mitosis and Cytokinesis

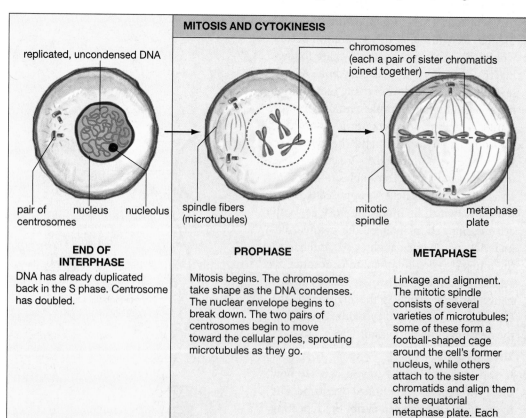

MITOSIS AND CYTOKINESIS

replicated, uncondensed DNA

chromosomes
(each a pair of sister chromatids
joined together)

pair of centrosomes | nucleus | nucleolus

spindle fibers
(microtubules)

mitotic spindle | metaphase plate

END OF INTERPHASE

DNA has already duplicated back in the S phase. Centrosome has doubled.

PROPHASE

Mitosis begins. The chromosomes take shape as the DNA condenses. The nuclear envelope begins to break down. The two pairs of centrosomes begin to move toward the cellular poles, sprouting microtubules as they go.

METAPHASE

Linkage and alignment. The mitotic spindle consists of several varieties of microtubules; some of these form a football-shaped cage around the cell's former nucleus, while others attach to the sister chromatids and align them at the equatorial metaphase plate. Each chromatid now faces the pole opposite that of its sister chromatid.

equator called the **metaphase plate**. As the centrosomes move apart, they begin sprouting microtubules in all directions. By the time they have reached their respective poles, they will have several types of microtubules stemming from them. One variety of them forms a football-shaped cage around the nuclear material, while a second variety attaches to the chromosomes themselves. Taken together, the microtubules active in cell division are known as the **mitotic spindle**.

Metaphase

By the time metaphase begins, the nuclear envelope has disappeared completely, and the microtubules that were growing toward the chromosomes now *attach* to them. With this, what had been random becomes orderly. Through a lively back-and-forth movement, the chromosomes eventually are *aligned* at the equator: each pair of chromatids is now moved to the metaphase plate by action of the microtubules. Just as important, each chromatid now faces the pole *opposite* that of its sister chromatid, and each chromatid is *attached* to its respective pole by perhaps 20 microtubules.

Anaphase

At last the genetic material divides. As you may have guessed, this is a parting of sisters. It is the sister chromatids that are pulled apart, each now becoming a full-fledged chromosome. All 46 chromatid pairs divide at the same time, and each member of a chromatid pair is pulled toward its respective pole by the disassembly of its microtubules.

Telophase

Telophase represents a return to things as they were before mitosis started. The newly independent chromosomes, having arrived at their respective poles, now unwind and lose their clearly defined shape. New nuclear membranes are forming. When this work is complete, there are two finished daughter nuclei lying in one elongating cell. Even as this is going on, though, something else is taking place that will result in this one cell becoming two.

Cytokinesis

Cytokinesis actually began back in anaphase and is well underway by the time of telophase. It works through the tightening of a cellular

What Is It?
See p. 185.

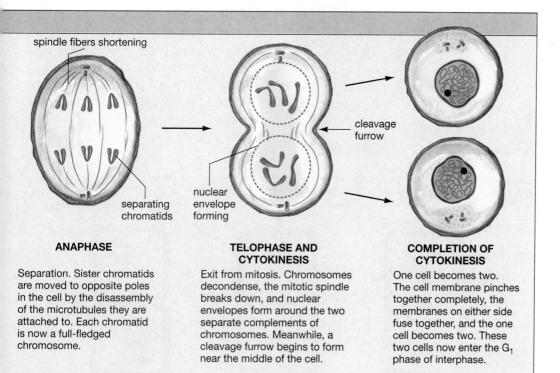

ANAPHASE

Separation. Sister chromatids are moved to opposite poles in the cell by the disassembly of the microtubules they are attached to. Each chromatid is now a full-fledged chromosome.

TELOPHASE AND CYTOKINESIS

Exit from mitosis. Chromosomes decondense, the mitotic spindle breaks down, and nuclear envelopes form around the two separate complements of chromosomes. Meanwhile, a cleavage furrow begins to form near the middle of the cell.

COMPLETION OF CYTOKINESIS

One cell becomes two. The cell membrane pinches together completely, the membranes on either side fuse together, and the one cell becomes two. These two cells now enter the G_1 phase of interphase.

What Is It?
See p. 186.

Figure 9.11
Cytokinesis in Animals
Cytokinesis in animal cells begins with a cleavage furrow, shown here in a dividing frog egg. (×85)

Figure 9.12
Cytokinesis in Plants
The cell walls of plants are too rigid to form cleavage furrows, as animal cells do when dividing. Therefore, plant cells use a different strategy in cytokinesis. They build a new cell wall down the middle of the parent cell to separate the two daughter cells from the inside out.

waistband that is composed of two sets of protein filaments working together. These materials form a **contractile ring** that narrows along the cellular equator (**see Figure 9.11**). A **cleavage furrow** results from the ring's contraction; consequently, the fibers in the mitotic spindle are pushed closer and closer together, eventually forming one thick pole that is destined to break. The dividing cell now assumes an hourglass shape; as the contractile ring continues to pinch in, one cell becomes two by means of something you looked at in Chapter 4: membrane fusion. The membranes on each half of the hourglass circle toward each other and then fuse.

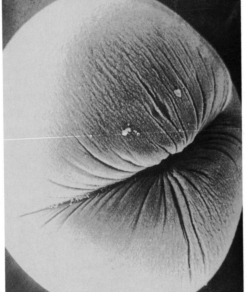

With this, the two cells become separate. Mitosis and cytokinesis are over, and the two daughter cells slip back into the relative quiet of interphase.

9.5 Variations in Cell Division

Plant Cells

For the most part, plant cells carry out the steps of mitosis just as animal cells do. In the splitting of cytokinesis, however, plant cells must deal with something that animal cells don't have to: the cell wall noted in Chapter 4. The way animal cells carry out cytokinesis—pinching the plasma membrane inward via the contractile ring—wouldn't work with plant cells, because their cell wall lies *outside* the plasma membrane.

The plant cell's solution to cytokinesis is to grow a new cell wall and plasma membrane that run roughly down the middle of the parent cell. If you look at **Figure 9.12**, you can see how this works. A series of membrane-lined vesicles begins to accumulate near the metaphase plate; inside these vesicles is a complex sugar that is a precursor to the cell wall. The vesicles then fuse together, forming a single, flat **cell plate** that grows toward the parent cell wall. Its membrane material then fuses with the original plasma membrane, while the cell-wall material within is released to fuse with the original cell wall. The result is two distinct daughter cells, complete with cell walls.

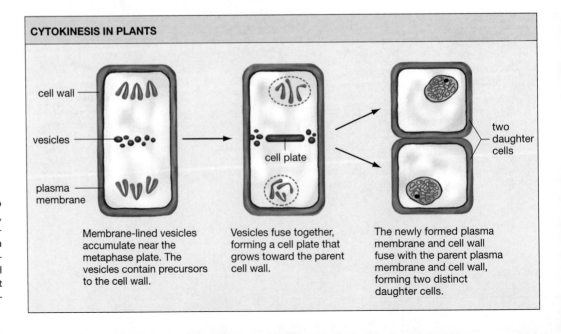

CYTOKINESIS IN PLANTS

cell wall

vesicles

plasma membrane

cell plate

two daughter cells

Membrane-lined vesicles accumulate near the metaphase plate. The vesicles contain precursors to the cell wall.

Vesicles fuse together, forming a cell plate that grows toward the parent cell wall.

The newly formed plasma membrane and cell wall fuse with the parent plasma membrane and cell wall, forming two distinct daughter cells.

Prokaryotes

It's also important to recognize that some cells have fundamentally different ways of replicating. The cells of prokaryotes—the bacteria and archaea you looked at in Chapter 4—are a case in point.

If you look at **Figure 9.13**, you can see how cell division works among the prokaryotes. Bacteria are single-celled and have a single, circular chromosome. Bacterial cells do not, however, have their chromosome sequestered inside a nucleus, as with eukaryotic cells, for the simple reason that they *have* no nucleus. Instead, their lone chromosome is attached to their plasma membrane.

A bacterial cell begins its division by duplicating its chromosome, utilizing the same DNA unwinding and replication described earlier. As the daughter chromosomes are completed, however, they attach to *different sites* on the plasma membrane. The cell then pinches in between the two points of attachment of the chromosomes. As parts of the bacterial membrane and cell wall pinch together, the pinched ends close off, and two cells come to exist where one did previously (**see Figure 9.14**). This process is obviously very different from the eukaryotic mitosis and cytokinesis you looked at. As such, prokaryotic cell division goes under an entirely different name: **binary fission**.

What Is It?
Answer: Chromosomes moving on mitotic spindle during metaphase of mitosis. (×600)

Figure 9.13
Binary Fission in Bacteria
The key to bacterial cell division is that the bacterial chromosome attaches to the cell's plasma membrane. Both daughter chromosomes attach to the plasma membrane, but in different locations. The cell then pinches in between these two locations, eventually leaving one chromosome in each daughter cell.

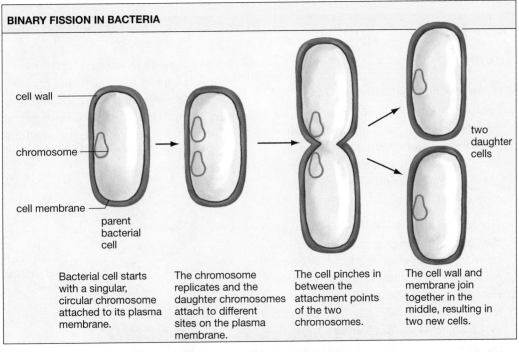

BINARY FISSION IN BACTERIA

cell wall

chromosome

cell membrane

parent bacterial cell

two daughter cells

Bacterial cell starts with a singular, circular chromosome attached to its plasma membrane.

The chromosome replicates and the daughter chromosomes attach to different sites on the plasma membrane.

The cell pinches in between the attachment points of the two chromosomes.

The cell wall and membrane join together in the middle, resulting in two new cells.

Figure 9.14
A Micrograph of an *E. coli* Bacterium Completing Cell Division
These bacteria exist in great quantity in the human digestive tract, generally harmlessly. (×30,000)

Variations in the Frequency of Cell Division

There are also variations in cell division among different types of eukaryotic cells. Not all cells divide throughout their existence. Most human brain cells are formed in the first three months of embryonic existence and then live for decades, with very few of them ever dividing again. Much the same is true of the leaf cells in plants; they divide only when the leaf is very small, then grow for a time, and then function at this mature size for as long as the leaf lives. At the other extreme, "stem" cells located in human bone marrow never *stop* dividing as they produce blood cells.

On to Meiosis

The cell division reviewed in this chapter concerns "somatic" cells—the kind that form blood, bone, muscle, nerves and many other sorts of tissue. Given a distribution this wide, you might well ask: What kind of cells are *not* somatic? The answer is only one kind: the kind that forms the basis for each succeeding generation of living things.

Chapter Review

Key Terms

binary fission   185

cell cycle   181

cell plate   184

centrosome   182

chromatid   179

chromatin   179

chromosome   179

cleavage furrow   184

contractile ring   184

cytokinesis   181

enzyme   174

genetics   175

genome   173

homologous chromosomes   180

interphase   181

karyotype   180

metaphase plate   183

microtubule   182

mitosis   181

mitotic phase   181

mitotic spindle   183

oncogene   178

protein   174

tumor suppressor gene   178

Summary

9.1   An Introduction to Genetics

- DNA is an information-bearing molecule that plays a critical role in the reproduction, development, and everyday functioning of living things. DNA contains the information for the production of proteins, which carry out an array of tasks in living things.

- The information in DNA is encoded in chemical units called genes. Each gene is composed of a series of chemical bases, with the order of these bases determining what protein the gene codes for. There are four DNA bases: adenine (A), thymine (T), guanine (G), and cytosine (C). A typical gene is composed of thousands of bases sequenced in a given order.

- Protein synthesis begins with the information in a sequence of DNA bases being copied onto a molecule called messenger RNA (mRNA). This molecule moves out of the nucleus to a structure in the cell's cytoplasm called a ribosome. There, the mRNA "tape" is "read," and amino acids brought to the ribosome are put together in the sequence called for in the mRNA tape. The result is a chain of amino acids that folds into a protein.

- Most of the cells in an organism contain a complete copy of that organism's genome, meaning its collection of genetic information. Before cells divide, their genome must first be copied and the resulting copies apportioned evenly into what will become two daughter cells.

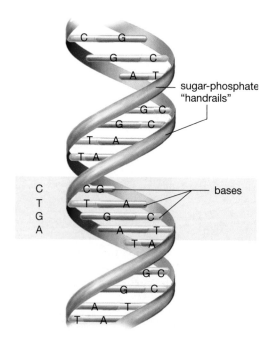

9.2 An Introduction to Cell Division

• Cell division includes the copying of DNA (known as replication), the apportioning of it into two quantities (mitosis), and the splitting of cellular material into two (cytokinesis).

9.3 DNA Is Packaged in Chromosomes

• DNA comes packaged in units called chromosomes. These chromosomes are composed of DNA and its associated proteins—a combined chemical complex called chromatin. Chromosomes exist in an unduplicated state until such time as DNA replicates, prior to cell division. DNA replication results in chromosomes that are in duplicated state, meaning one chromosome composed of two identical sister chromatids.

• Chromosomes in human beings (and many other species) come in matched pairs, with one member of each pair inherited from the mother, and the other member of each pair inherited from the father. Such homologous chromosomes have closely matched sets of genes on them, though many of these genes are not identical. A given paternal chromosome may have genes that code, for example, for slightly different hair or skin color than the counterpart genes on the homologous maternal chromosome. Human beings have 46 chromosomes—22 matched pairs and either a matched pair of X chromosomes (in females) or an X and a Y chromosome (in males).

• Cell division fits into the larger framework of the cell cycle, meaning a repeating pattern of growth, genetic replication, and cell division. The cell cycle has two main phases. One is interphase, in which the cell carries out its work, grows, and duplicates its chromosomes in preparation for division. The second is mitotic phase, in which the duplicated chromosomes separate and the cell splits in two.

9.4 Mitosis and Cytokinesis

• There are four stages in mitosis: prophase, metaphase, anaphase, and telophase. The essence of the process is that duplicated chromosomes line up along the midpoint of the parent cell (the metaphase plate) with the sister chromatids that make up each duplicated chromosome on opposite sides of the plate. The sister chromatids are then moved apart, to opposite poles of the parent cell. Once cell division is complete, sister chromatids will reside in separate daughter cells, with each sister chromatid now functioning as a full-fledged chromosome.

• Cytokinesis works through a ring of protein filaments that tightens at the middle of a dividing cell. Membranes on the portions of the cell being pinched together then fuse, resulting in two daughter cells.

9.5 Variations in Cell Division

• Because of their cell walls, plant cells must carry out cytokinesis differently from animal cells. The plant's solution is to grow new cell walls and plasma membranes along the metaphase plate, thus dividing the parent cell into two daughter cells. Prokaryotes such as bacteria double their single chromosome, with the two resulting chromosomes attaching to different sites on the plasma membrane. The cell then pinches in, between these two sites, in carrying out cell division. This process is called binary fission.

Understanding the Basics

Multiple-Choice Questions

1. A protein that accelerates many chemical reactions in the body is
   a. an enzyme   d. DNA
   b. a histone   e. glucose
   c. a gene

2. The four bases used in the DNA code are
   a. adenine, thymine, guanine and cytosol
   b. adenine, thyroxine, glucose and cytosine
   c. adrenaline, thymine, glucosamine and uracil
   d. adenine, thymine, guanine and cytosine
   e. none of the above

3. The "handrails" of the DNA structural staircase are composed of
   a. adenine, paired with thymine
   b. fatty acids and sugars
   c. phosphate groups and nitrogen-containing bases (ATCG)
   d. deoxyribose and phosphate groups
   e. sugars and proteins

4. The human genome contains
   a. 23 chromosomes
   b. 44 chromosomes
   c. about 100,000 genes
   d. about 500,000 genes
   e. about 1,000,000 genes

5. A(n) _____ is the cellular structure that mediates the formation of proteins.
   a. nucleus
   b. ribosome
   c. mitochondrion
   d. smooth endoplasmic reticulum
   e. amino acid

6. Initiation of prophase is indicated by
   a. the ability to see well-defined chromosomes
   b. the presence of a cleavage furrow
   c. alignment of condensed chromosomes along the metaphase plate
   d. duplication of chromosomes
   e. separation of sister chromatids

7. A _____ is one set of protein filaments used to cause cell division during cytokinesis.
   a. dividing fiber
   b. metaphase plate
   c. spindle
   d. contractile ring
   e. centriole

8. The length of the cell cycle is
   a. 16 hours   d. 2 hours
   b. 24 hours   e. 12 hours
   c. 30 minutes

9. By far, the longest phase of the cell cycle is
   a. prophase   d. cytokinesis
   b. interphase   e. telophase
   c. metaphase

10. An arrangement of a full set of chromosomes by size, shape, and banding pattern is called a(n)
    a. chromosome grid
    b. genome
    c. genetic "picture"
    d. autosome alignment
    e. karyotype

11. If a cell is not dividing, but instead is in the process of normal operation and growth, what phase of the cell cycle would it be in?
    a. reproductive phase
    b. synthesis (S) phase
    c. gap-one (G_1) phase
    d. gap-two (G_2) phase
    e. either c or d would be correct

Brief Review

1. The drawing shows a cell in a phase of mitosis. What phase is shown? How many chromatids are present in this stage? What is the total number of chromosomes each daughter cell will have?

2. How do mitosis and cytokinesis differ?

3. Prokaryote cell division is called _____ _____.

4. Name the five phases of the cell cycle and give the importance of each, describing the major events occurring in each phase.

5. Given this half of the DNA "spiral staircase," what would the other half be?

|-A

|-A

|-C

|-T

|-A

|-G

6. What additional cell structure must plant cells deal with when they are undergoing cytokinesis (as compared to animal cells)? How do plant cells compensate for this difference?

7. What does CDK stand for? What is the role of the CDK complex?

8. How does the prokaryote chromosome structure differ from the eukaryote chromosome structure?

9. The protein synthesis pathway—DNA → mRNA → protein—is a very important concept in biology.
   a. What are the information-bearing "building blocks" of DNA?
   b. Where is DNA located in the cell?
   c. What does the "m" in mRNA stand for?
   d. Using this knowledge, what is the purpose of mRNA, and what is its cellular pathway?
   e. What are the "building blocks" of proteins?
   f. Where does protein formation occur?

Applying Your Knowledge

1. If a human cell fails to undergo cytokinesis, how many chromosomes will be in the final cell?

2. A gene is known to contain 15 percent guanine. What percentages of each of the other three nucleotide bases are contained in the same gene?

3. The total number of chromosomes in a cell is six. Draw the cell in each of the four phases of mitosis. How many chromosomes and chromatids are present in the first phase?

4. How do cancer cells and healthy human cells differ in rate of replication? What are two types of genes in a cancer cell that may contribute to this difference?

5. If each cell contains the entire genome of an organism, what prevents a human stomach epithelial cell from behaving as a hair follicle?

Group Activity

As a group, take a few minutes to investigate recent genetic developments in a popular magazine or national newspaper. What are some of the ethical considerations associated with these technical advances?

10

Making Reproductive Cells
Meiosis

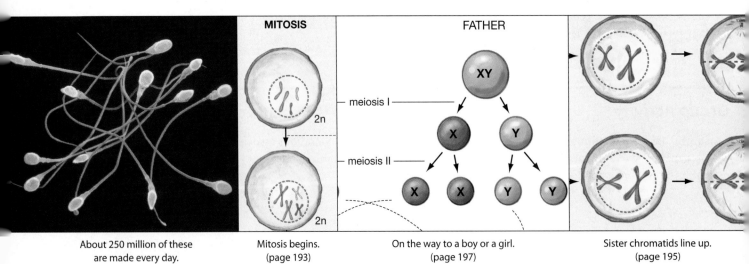

About 250 million of these
are made every day.
(page 200)

MITOSIS

Mitosis begins.
(page 193)

FATHER

meiosis I

meiosis II

On the way to a boy or a girl.
(page 197)

Sister chromatids line up.
(page 195)

A real-life couple—let's call them Jack Fennington and Jill Kent—combine their last names when they get married and thus become Jack and Jill Fennington-Kent. Now, what would happen if the Fennington-Kent's children were to continue in this tradition? Their daughter Susie might marry, say, Ralph Reeson-Dodd, which would make her Susie Fennington-Kent-Reeson-Dodd. If *her* daughter, Carol, were to keep this up, she might be fated to become Carol Fennington-Kent-Reeson-Dodd-Smith-Lee-Minderbinder-Green, and so on.

This growing chain of names illustrates a kind of problem that living things have managed to address in carrying out reproduction. "Reproduction" in this context does not mean the mitotic cell division you looked at in Chapter 9. Reproduction here means *sexual* reproduction, in which specialized reproductive cells come together to produce offspring. The cells described in Chapter 9—the ones that undergo mitosis—are known as *somatic* cells, which in animals means all the cells in the organism except for one type. What type is this? Reproductive cells; cells called *gametes*, one variety of which goes by the names egg and sperm.

Eggs and sperm are not commonly thought of as cells, but that's just what they are, complete with plasma membranes, nuclei, chromosomes, and all the rest (though sperm eventually lose much of this material). Eggs and sperm are *special* cells, however, in that they are not destined to divide like somatic cells and thus make more eggs and sperm. Rather, they will *fuse*—sperm fertilizing egg—to make a zygote, which grows into a whole organism. It is sperm and egg that take part in sexual reproduction and that thus bear some relation to the Fennington-Kent problem. Here's how it plays out in humans. Human somatic cells, as you have seen, have 23 pairs of chromosomes or 46 chromosomes in all. If an egg and sperm each brought 46 chromosomes to their *union*, the result would be a zygote with 92 chromosomes. The *next* generation down would presumably have 184 chromosomes, the next after that 368, and so on. This would be about as functional as having Fennington-Kent-Reeson-Dodd-Smith-Lee-Minderbinder-Green as a last name.

End of two journeys.
(page 200)

Chromosomes link as they condense.
(page 194)

Meiosis begins.
(page 193)

Changing partners.
(page 194)

10.1 An Overview of Meiosis

A Chromosome Reduction before Union of Egg and Sperm

How do chromosomes avoid the problem of doubling with each generation? In sexual reproduction, chromosome union is preceded by chromosome *reduction*. The reduction comes in the cells that give rise to sperm and egg. When these cells divide, the result is sperm or egg cells that have only *half* the usual, somatic number of chromosomes. Human sperm or egg, in other words, have only 23 chromosomes in them. (Not the 23 *pairs* of chromosomes that somatic cells have, mind you; 23 chromosomes.) Each 23-chromosome sperm can then unite with a 23-chromosome egg to produce a *46-chromosome* zygote that develops into a new human being. In each generation, then, there is first a halving of chromosome number (when egg and sperm cells are produced), followed by a coming together of these two halves (when sperm and egg unite). This basic pattern holds true for sexual reproduction in all eukaryotes.

Some Helpful Terms

Now here are a few terms that will be helpful. The kind of cell division that results in the halving of chromosome number is called *meiosis*, in contrast with *mitosis*, which you looked at in Chapter 9. In mitosis, there is no halving of chromosome number; just a doubling of chromosomes and then a division of these chromosomes into each of two daughter cells.

In animals, cells that go through meiosis—thus getting the half-number of chromosomes—are always reproductive cells, the term for which is **gametes**, as we've seen. Such reduced-chromosome cells are said to be in the **haploid** state, the term *haploid* literally meaning "single vessel." When egg and sperm unite, however, it marks a return to the **diploid**, or "double vessel" state of cellular existence, meaning 46 chromosomes in human beings. **Meiosis** can thus be defined as a process in which a single diploid cell divides to produce haploid reproductive cells. Diploid cells are also sometimes referred to as **2n** cells (the "n" here standing for a doubled chromosome *number*) while haploid

cells are said to be **1n**. Thus, eggs and sperm are 1n, while the diploid cells that give rise to them are 2n.

10.2 The Steps in Meiosis

To see how this chromosomal halving takes place, it's desirable to go over the process of meiosis. **Figure 10.1** shows meiosis, in a stripped-down form, as compared to the mitosis described in Chapter 9. As you can see, meiosis includes one chromosome duplication followed by two cellular divisions. This is the key to how the chromosome numbers work out. In mitosis, there was one chromosome duplication and one subsequent division of the doubled chromosomes (into two cells). In meiosis there is one duplication followed by *two* divisions (into four cells). The first of the meiotic divisions is a separation not of chroma*tids*, as in mitosis, but of homologous chromo*somes*. The second meiotic division is then the separation of the chromatids that make up these homologous chromosomes. (You may remember from Chapter 9 that homologous chromosomes are chromosomes that are the same in shape and function. We have a chromosome 1 that we inherit from our mother, and it has a set of genes on it that are very similar to those on the chromosome 1 we inherit from our father; so it is for chromosomes 2, 3, 4, and so on.)

Let's walk through the steps of meiosis now; these details will reveal a process that not only divides up genetic material, but provides for a great deal of the diversity that exists among living things. The steps of meiosis are illustrated in **Figure 10.2**.

The process starts with a doubling (that is, a replication) of DNA, as you saw in Chapter 9. DNA doubling in meiosis means the same thing as it did in mitosis: All 46 chromosomes in a human cell replicate, yielding 46 chromosomes in duplicated state (meaning they are composed of 92 chromatids).

Meiosis I

Prophase I

The steps of meiosis are separated into two multistep stages, called meiosis I and meiosis II. The big picture here is that in meiosis I, homologous chromosomes are positioned close together, on opposite sides of the cellular equator called the metaphase plate. Then these chromosomes will be pulled apart into different daughter cells. In meiosis II, the chroma*tids* of these now-separated chromosomes will, in turn, separate into different daughter cells.

Meiosis I begins, in prophase I, with the same appearance of 46 identifiable chromosomes that you observed in mitosis. The first big difference between meiosis and mitosis also appears in prophase I. It is that, in meiosis, *homologous chromosomes pair up*. In mitosis, 46 individual chromosomes condensed, but did not pair up in any way. In meiosis, however, each pair of homologous chromosomes link up as they condense. Maternal chromosome 5, for example, intertwines with *pa*ternal chromosome 5, maternal chromosome 6 with paternal 6, and so on. When two homologous chromosomes have paired up in this way, they are called a **tetrad** (Figure 10.2b).

A critical bit of part-swapping then takes place between the non-sister chromatids of the paired chromosomes. This process is called *recombination*, and you'll be looking at it in detail later. Once this recombination has finished, the homologous chromosomes begin to unwind from one another, though they remain overlapped.

Metaphase I

Still existing as tetrads, the homologous chromosomes are moved by microtubules to the metaphase plate; in this step of meiosis, the maternal member of a given pair lies on one side of the plate and the paternal member on the other. A critical point about this, however, is that the alignment adopted by any one pair of chromosomes bears no relation to the alignment adopted by any other pair. It may be, for example, that paternal chromosome 5 will line up on what we might call side A of the metaphase plate; if

**Figure 10.1
Meiosis Compared to Mitosis**

1 Both meiosis and mitosis begin with diploid cells, meaning cells that contain paired or homologous sets of chromosomes. Two sets of homologous chromosomes are shown in both the mitosis and meiosis figures. Though homologous chromosomes are similar, they are not identical; thus one homologous pair in each figure is represented by the larger set of chromosomes (one blue, one green), while the other homologous pair is represented by the smaller set of chromosomes (one yellow, one red).

2 In both mitosis and meiosis, the chromosomes duplicate. Each chromosome is now composed of two sister chromatids.

3 In mitosis, the chromosomes line up on the metaphase plate, one sister chromatid on each side of the plate. In meiosis, meanwhile, homologous chromosomes—not sister chromatids—line up on opposite sides of the metaphase plate.

4 In mitosis, the sister chromatids separate, each of them now becoming full-fledged chromosomes, after which these chromosomes will come to reside in separate cells. In meiosis, the homologous pairs of chromosomes separate, one member of each homologous pair going to one cell, the other member to the other cell. Because each of these cells now has only a single set of chromosomes, each is in the haploid state. Now these single chromosomes line up on the metaphase plate, with their sister chromatids on opposite sides of the plate.

5 In the final steps of meiosis, the sister chromatids of each chromosome then separate and the cells divide again, yielding four haploid cells.

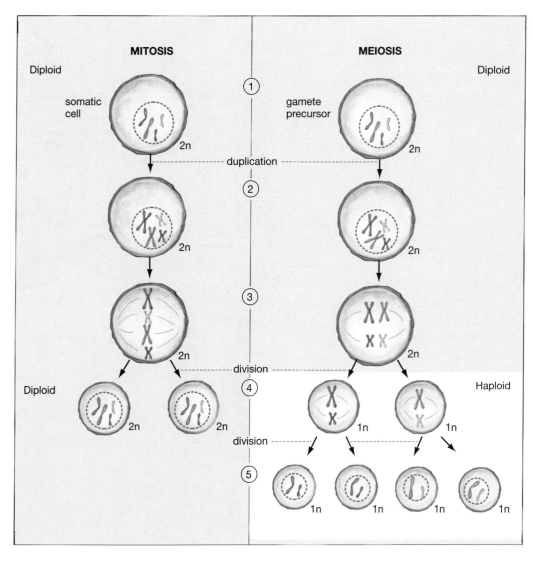

so, then maternal chromosome 5 ends up on side B. Shift to chromosome 6, however, and things could just as easily be reversed: The *maternal* chromosome might end up on side A, and the paternal on side B. It is thus a throw of the dice as to which side of the plate a given chromosome lines up on. More important, it is this chance event that determines which *cell* a chromosome joins, be-

cause each *side* of the plate becomes a separate *cell* once cell division is finished. As you will see, this randomness is critical in the shuffling of genetic material.

Anaphase I

The paired chromosomes now begin to move away from each other, toward their respective poles, via the same microtubular

Figure 10.2

a The Steps of Meiosis

b Crossing Over

c Independent Assortment

MEIOSIS I

Diploid

END OF INTERPHASE

DNA has already duplicated

PROPHASE I

Homologous chromosomes link as they condense, forming tetrads.

Crossing over occurs.

METAPHASE I

Microtubules move homologous chromosomes to metaphase plate.

Independent assortment occurs.

ANAPHASE I

Microtubules separate homologous chromosomes (sister chromatids remain together).

Two very important sources of genetic variation

CROSSING OVER

Exchange of parts of non-sister chromatids.

duplicated maternal chromosome

duplicated paternal chromosome

tetrad

sister chromatids

non-sister chromatids

means that operated in mitosis. Though homologous chromosomes have now separated, the sister chromatids that make them up are still together.

Telophase I

With chromosome movement toward the poles completed, the original cell now undergoes cytokinesis, dividing into two completely separate daughter cells. With this, there are two haploid cells, whereas in the beginning there was one diploid cell.

Meiosis II

Prophase II and Metaphase II

There is little pause for interphase between meiosis I and meiosis II; little time for the

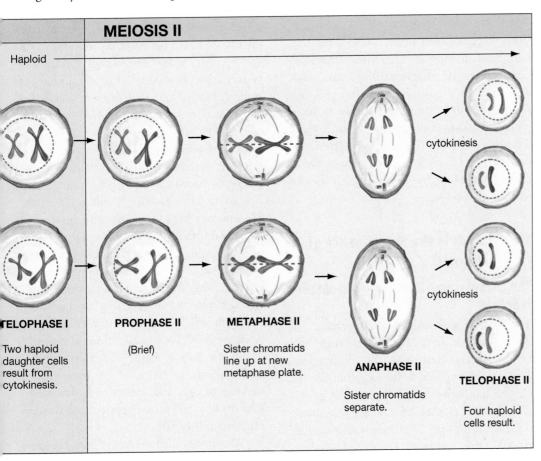

MEIOSIS II

Haploid

TELOPHASE I

Two haploid daughter cells result from cytokinesis.

PROPHASE II

(Brief)

METAPHASE II

Sister chromatids line up at new metaphase plate.

cytokinesis

ANAPHASE II

Sister chromatids separate.

cytokinesis

TELOPHASE II

Four haploid cells result.

INDEPENDENT ASSORTMENT

Random alignment of maternal/paternal chromosomes at the metaphase plate.

METAPHASE I   **METAPHASE II**   **TELOPHASE II**

Homologous chromosomes lined up this way in this meiosis ...

... but they could have lined up this way, yielding a different outcome.

daughter cells to regroup and carry on regular metabolic activities. What happens instead is another division—or set of divisions, since there are now two cells. Each of these cells now has 23 duplicated chromosomes in it. After a very brief prophase, these chromosomes line up at a new metaphase plate, with sister chromatids on opposite sides of each plate.

Anaphase II and Telophase II

What happens next is the same thing that happened in mitosis: The *sister chromatids* now separate, moving toward opposite poles, and with this assuming the role of full-fledged chromosomes. Once this is completed, cytokinesis occurs; where once there were two cells, now there are four. The *difference* between this process and mitosis is that each of these cells has 23 chromosomes in it instead of 46.

10.3 What Is the Significance of Meiosis?

The Chromosome Duplication Problem Is Solved

You have now examined the mechanics of meiosis (Figure 10.2a). But what are the effects of this process? First of all, the "Fennington-Kent" problem noted at the beginning of the chapter has been solved. There will be no 92-chromosome zygotes, because egg and sperm will each bring only 23 chromosomes to their meeting, thus yielding a combined 46 chromosomes in human somatic cells. Just as notable, however, are the possibilities for diversity that meiosis brings about.

Meiosis Ensures Genetic Diversity in Two Ways

The Independent Assortment of Chromosomes

One of the ways that meiosis ensures diversity is a crucial event that took place in metaphase I with the random alignment or **independent assortment** of homologous chromosomes at the metaphase plate (see Figure 10.2c). To grasp the importance of this event, remember that one member of each pair of homologous chromosomes is inherited from the father, while the other member of the pair is inherited from the mother. If we wonder how traits

MediaLab Link
Inheriting a defective gene, see p. 206.

from mother and father can get "mixed up" in offspring, independent assortment is a large part of the answer.

Independent assortment has to do with the possibilities that follow from the way chromosomes line up at the metaphase plate. If human beings had only two chromosomes in each cell that went through meiosis, there would be only two possible "states" of line-up at the plate, which might be called "maternal on side A" of the plate and "maternal on side B" (with the paternal homologue on the opposite side). The actual number of possibilities, however, is this number of states (2) *raised to the power of* the number of chromosome pairs. Thus, if three chromosome pairs existed, the number of possible outcomes would go to 2^3, or 8. Now get your calculator ready. With the 23 pairs of chromosomes human beings have, the possibilities go to 2^{23}, or about 8 million. General Motors may be able to mix and match engines and upholstery to come up with scores of different models of a given car, but in any given human meiosis, chromosome pairs can line up along the metaphase plate in 8 million different ways—meaning any given egg or sperm will be one of about 8 million possible "models." To get a basic idea of how this works out, look at Figure 10.2c to see what it would mean, in terms of the final genetic makeup of eggs and sperm, to have a variation on how the homologous chromosomes lined up at the plate.

We will be returning to the importance of this process in later chapters. For now, remember that there are *genes* on these chromosomes, and that the genes on any given paternal chromosome may differ somewhat from those on its matched maternal chromosome. Which chromosome the offspring gets thus makes a difference—in eye color, height, or any of thousands of other characteristics. Given this, each of us is very much affected by the way paternal and maternal chromosomes happened to line up in two very special meioses: those that produced the egg cell and the sperm cell that gave rise to us.

Genetic Diversity Is Also Ensured through Recombination

Independent assortment produces genetic diversity as the result of the placement of *whole chromosomes* on one side of the metaphase plate or the other. Recall, however, that, back in prophase I of meiosis, the homologous chromosome pairs intertwined for a time and then engaged in something called **recombination** (or **crossing over**). In this, there is an exchange of *parts* of chromosomes prior to their lining up at the metaphase plate.

Look at Figure 10.2b to see how this important process works. In the actual process, there is a physical breaking of non-sister chromatids and then a "reunion" of the loose ends onto new chromatid partners. With this, what had been a "maternal" chromosome now has a portion of the paternal chromosome within it, and vice versa.

The linkage between recombination and genetic diversity probably is apparent; the ability of reciprocal lengths of DNA to be exchanged between chromosomes provides one more means by which the genetic deck can be reshuffled prior to the formation of gametes. When we add to this process the independent assortment of chromosomes along the metaphase plate, it's easy to see why separate children from the same parents can come out looking so different—from each other as well as from the parents themselves (**see Figure 10.3**). Overall, then, meiosis generates *diversity*, while its counterpart, mitosis, does not. Mitosis makes genetically exact copies of cells; it *retains* the qualities that cells have from one generation to the next. Meiosis mixes genetic elements each time it produces reproductive cells and thus brings about genetic variation in succeeding generations.

Meiosis and Sex Outcome

Let us think now about what meiosis means in terms of passing on *sex* to offspring. In humans there is one exception to the rule that chromosomes come in homologous pairs. Human females do indeed have 23 pairs of matched chromosomes, including 22 pairs of "autosomes" and one matched pair of **sex chromosomes**, meaning the chromosomes that determine what sex an individual will be. The sex chromosomes in females are called X chromosomes, and each female possesses two of them. In males, conversely, there are 22 autosomes, one X sex chromosome, and then one Y sex chromosome. It is this Y chromosome that confers the male sex.

In meiosis I in a female, the female's two X chromosomes, being homologous, line up together at the metaphase plate. Then these chromosomes separate, each of them going to different cells (**see Figure 10.4**). This same

Figure 10.3
Ensuring Variety
The shuffling of genetic material that occurs during meiosis is the primary reason that children look different from their parents, and from each other.

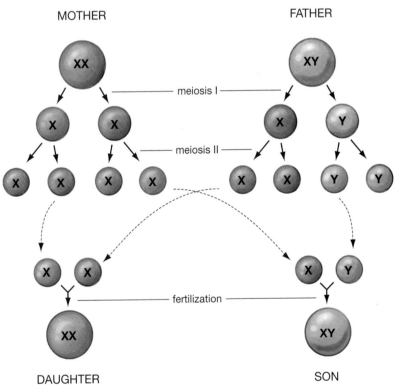

Figure 10.4
Sex Outcome in Human Reproduction
Early in meiosis I, the mother's two X chromosomes line up at the metaphase plate; meanwhile, in the father's meiosis, his X and Y chromosomes line up. The X-X and X-Y pairs then separate into different cells. In meiosis II, the chromatids that made up the duplicated chromosomes separate, yielding individual eggs and sperm. In fertilization, should a Y-bearing sperm from the male reach the egg first, the child will be a boy; should the male's X-bearing sperm reach the egg first, the child will be a girl.

The Revealing Y Chromosome

A debate that existed in the United States for nearly 200 years was apparently settled in 1998 when genetic tests revealed a high probability that Thomas Jefferson, the third president of the United States, fathered at least one child by a slave he owned named Sally Hemings. Historians had argued the case for years on the basis of fragmentary evidence, but it was genetic testing of the modern-day descendents of Hemings and Jefferson's paternal uncle that all but ended the debate.

What scientists examined in the case was the descendants' DNA—in every case, DNA extracted from their Y chromosomes. Why just the Y chromosomes? These chromosomes turn out to be something of an oddity in the human chromosome collection and it was this very quality that made them useful in the Jefferson-Hemings investigation.

The Y chromosome is unusual because it is the only human chromosome that exists in one sex but not the other—in men, but not in women. Females, as we've seen, have two X chromosomes, but males have one X chromosome and one Y. The Y is also unusual in that it is a pip-squeak of a chromosome, having in it perhaps 20 genes in all, compared to the thousands of genes that typically exist in other chromosomes. Finally, the Y is different in that it does almost no part-swapping in meiosis. As explained in the main text, in the early stages of meiosis, homologous chromosomes pair up and exchange reciprocal parts in the process called *crossing over* or *recombination*. The two X chromosomes that a woman has will come together in meiosis and do just this. The Y chromosome a man has, however, engages in only the slightest bit of recombination with his X chromosome. Nature has kept the X and Y separate in this way in order to keep the sexes separate. Imagine that through crossing over, the single gene on the Y chromosome that initiates male development ended up on the female X chromosome. The whole male-female distinction would begin to break down.

The Y's abstention from recombination means, however, that the DNA "book" that it contains is not having pages inserted in it, and removed from it, every time a new generation is produced. (Think what recombination

> *Nature has kept the X and Y chromosomes separate in this way in order to keep the sexes separate.*

is doing: removing one part of a chromosome while inserting a part of a chromosome from another person.) In historical genetic testing, as was done in connection with Jefferson and Hemings, this absence of recombination is critical. With the Y chromosome, the genetic record—the sequence of As, Ts, Cs, and Gs—is not getting scrambled in every generation.

The scientists who did the Jefferson and Hemings tests looked for similar repeating sequences of Y-chromosome DNA in 19 descendants of the Hemings and Jefferson families—all of them males, of course—and found such closely matching sequences that there is little doubt these families are related. The only remaining question is whether it could have been one of Jefferson's relatives—rather than Jefferson himself—who fathered Hemings' child.

---

thing happens with males; only with them, one resulting cell gets an X chromosome while the other gets the male's Y chromosome.

Sex determination is then simple. Each of the eggs produced by a female bears a single X chromosome. If this egg is fertilized by a *sperm* bearing an X chromosome, the resulting child will be a female. If, however, the egg is fertilized by a sperm bearing a Y chromosome, the child will be a male.

Because females don't have Y chromosomes to pass on, it follows that the Y chromosome that any male has had to come from his father. Meanwhile, the single X chromosome that any male has had to come from his mother. Females, conversely, carry one X chromosome from their mother and one from their father. (For details on how Y chromosomes can be used in genetic investigations, see "The Revealing Y Chromosome" above.)

10.4 Gamete Formation in Humans

You've seen that, through meiosis, a single diploid cell gives rise to four haploid gametes. But what are the diploid cells that go through meiosis? As you might expect, there is a whole sequence of actions involved in the process by which egg and sperm are produced—a process known as *gamete formation*. The starting female cells are known as **oogonia**, while the male cells are **spermatogonia**. These are diploid cells that give rise to two other sets of diploid cells, the **primary oocytes** and **primary spermatocytes**. *These* then are the cells that undergo meiosis, yielding haploid sperm and egg cells.

Sperm Formation

Spermatogonia exist in the male testes and are capable either of giving rise to primary spermatocytes or of generating more spermatogonia. It is this latter, self-generating ability that qualifies spermatogonia as "stem" cells and that allows males to *keep* producing sperm throughout their lives, beginning at puberty. These sperm-producing "factories," in other words, generate not only sperm but more factories as well (**see Figure 10.5a**).

The primary spermatocytes that come from spermatogonia go through meiosis I and thus produce the haploid **secondary spermatocytes**. Meiosis II then ensues, and the result is four haploid **spermatids**. These are not yet the familiar, tadpole-like sperm; those come only after spermatids develop further, a process that takes about 3 weeks. In this process, a given sperm cell must fully develop its tail-like flagellum. As it turns out, these "tails"—the only such structure in the human body—are composed of none other than the microtubules you have been seeing so much of in cell division. Moving the flagellum in a whip-like motion, microtubules serve to propel the sperm on their journey (**see Figure 10.6**). Sperm also need to "travel light," to which end developing sperm get rid of nearly all the cellular organelles reviewed back in Chapter 4: lysosomes, the Golgi complex, and so forth. A mature sperm thus

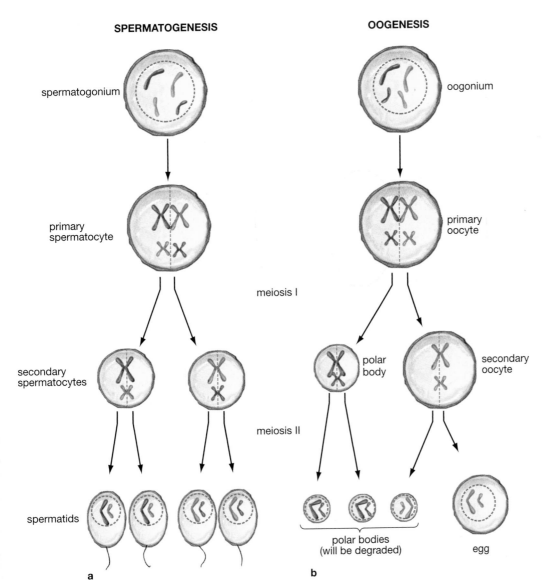

Figure 10.5
Sperm and Egg Formation in Humans

a In sperm formation (spermatogenesis), diploid cells called *spermatogonia* produce primary spermatocytes. These are the diploid cells that go through meiosis, yielding haploid secondary spermatocytes. These spermatocytes then go through meiosis II, yielding four haploid spermatids that will develop into mature sperm cells.

b In egg formation (oogenesis), cells called *oogonia*, produced before the birth of the female, develop into primary oocytes. These diploid cells will remain in meiosis I until they mature in the female ovary, beginning at puberty. (Only one oocyte per month, on average, will complete this maturation process.) Oocytes that mature will enter meiosis II, but their development will remain arrested there until they are fertilized by sperm. An unequal meiotic division of cellular material leads to the production of three polar bodies from the original oocyte and one well-endowed egg. The egg can go on to be fertilized, but the polar bodies will be degraded.

amounts to little more than a haploid set of DNA-containing chromosomes in front, a collection of "engines" (mitochondria) in the middle, and a propeller in the back (the flagellum). About 250 million sperm are made each day and about that number will be released with each ejaculation. Sperm that are not released will age and eventually are destroyed by the body's immune system.

Egg Formation

The need of male sperm to become stripped-down DNA packages stands in contrast to the needs of female ova or eggs, which must have rich resources of nutrients and building materials in order to sustain life once they have been fertilized. This difference in function is reflected in size: The volume of an egg is 200,000 times the volume of a sperm (**see Figure 10.7** to get an idea of this difference). Accordingly, eggs are created in a process that maximizes cell content, at least for the fraction of eggs that become viable.

Egg formation begins with the cells called the *oogonia* of the female; these may sound like the counterpart to the male sper-

matogonia but actually are very different. Whereas some spermatogonia are self-generating stem cells, oogonia are produced in large numbers only up to the seventh month of embryonic life, and none are created after birth. Of the millions produced, most will die while the female is still an embryo; but some survive to become primary oocytes and thus, as diploid cells, enter meiosis I (see Figure 10.5b). Amazingly enough, meiosis I is where these oocytes *stay*—for years. After puberty, an average of one oocyte per month is selected to complete maturation in the ovary, and thus enters into meiosis II. Even this meiosis is arrested, however, until this oocyte is fertilized by a sperm. It is the union with a sperm that prompts the completion of meiosis II. Thus, an oocyte that came into being when a female was an embryo might remain suspended in the cell cycle for years or even decades. Even then, it is not *all* primary oocytes that move through the cycle, but only a selected few; out of the millions initially created, only a few hundred will mature to ovulation during a woman's lifetime.

One Egg, Several Polar Bodies

The oocytes that do mature go through the rest of meiosis I and then meiosis II, as described earlier. However, in this progression—from one diploid oocyte to four haploid eggs—almost all the cytoplasm will be shunted to *one* of the four eggs, the better to build up its stores of nutrients and other cytoplasmic material. This happens in two rounds: meiosis I and meiosis II. We have thus far thought of the metaphase plate as existing in the middle of a cell. In egg formation, however, a primary oocyte's metaphase plate is effectively shifted to one side of the cell, meaning that when cytokinesis comes, one daughter cell gets almost all the cytoplasm while the second daughter cell gets almost none. This same thing happens in meiosis II. The result? One richly endowed haploid ovum and three—or sometimes only two—very small **polar bodies** (Figure 10.5b). (Only two polar bodies and the ovum will be the end result if the polar body produced in meiosis I does not continue through meiosis II, which often happens.) The egg has a chance at taking part in producing offspring, but the polar bodies are bound for oblivion rather than a

Figure 10.6
Mobile Cells, Bearing Chromosomes
Human sperm, colored to show detail.

Figure 10.7
Big Difference in Size
A human egg surrounded by much smaller human sperm.

shot at immortality; they will eventually degrade into their constituent substances.

How do certain oocytes get "selected" to mature out of prophase I? This occurs in connection with an intricate interplay that involves their surrounding tissue, hormones, and the happenstance of timing. Of the relatively few oocytes that do mature, in general only *one* will fully develop and be expelled from the ovary for a journey through the uterine tubes, where it can be fertilized by sperm, and this happens only once every 28 days on average. This number of days may ring a bell; the process of oocyte maturation is directly linked to the process of menstruation. You'll be going over all of this in Chapter 27.

10.5 Life Cycles: Humans and Other Organisms

The process of the formation of the human gametes called egg and sperm is followed, as you have seen, by the union of these gametes, which creates a fertilized egg. This egg (or zygote) then develops into a whole human being, mostly through the mitotic division of cells. But even as mitosis goes on, another group of egg and sperm is being readied in the new human being through the process of meiosis. This process amounts to a **life cycle**: the repeating series of steps that occur in the reproduction of an organism. You have looked at the human life cycle, but it's important to recognize that it amounts to only one among many.

Not All Reproduction Is Sexual

Human reproduction is sexual reproduction: the union of two reproductive cells to create a new organism. But in many other species there is **asexual reproduction** as well. Bacteria are single-celled organisms that possess a single, circular chromosome. (This latter fact makes them haploid by definition.) You saw in Chapter 9 that bacteria divide through the process of binary fission, one bacterium becoming two in the course of its cell cycle (**Figure 10.8**). As it turns out, that's it for bacteria as far as reproduction goes. For them, in other words, the cell cycle *is* the life cycle. There is no true diploid state or fusion of two cells; just an exact replication of DNA and a splitting of cytoplasm. There is asexual reproduction among eukaryotes as well; many single-celled eukaryotes reproduce through simple cell division, though there are lots of variations on this theme.

Asexual reproduction is also seen as we move through the branching evolutionary tree. Anyone who has ever seen a "cutting" taken from a plant knows that a snipped-off stem or branch will, when properly planted, sprout its own roots and grow into an independent plant, complete with the ability to take part in sexual reproduction. This process is called **vegetative reproduction**, and it has a counterpart in the animal world. Cut the arm and part of a central disk from a sea star, for example, and, through **regeneration**, this severed limb will grow into a whole, multi-armed sea star (**see Figure 10.9**).

Figure 10.8
Reproduction in Bacteria
There is no fusion of eggs and sperm in bacteria, just a division of one cell into two. Pictured is an *E. coli* bacterium undergoing this cell division, referred to as binary fission.

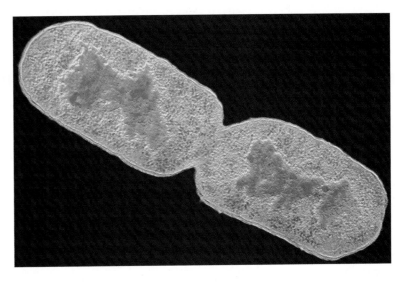

Figure 10.9
Regenerating the Whole from a Part
A sea star off the coast of Papua New Guinea regenerating a body and other arms from a piece of one arm.

Note that all these forms of asexual reproduction produce offspring that are genetically identical to their parents. There is no mixing and matching of chromosomes from different parents in these asexual processes, just a cloning of one genetically identical individual from its parent.

Variations in Sexual Reproduction

Within the framework of sexual reproduction, as you can imagine, there are many variations. For example, separate gametes need not come from separate organisms. Some animals, such as tapeworms, are hermaphrodites that can fertilize themselves. Most plants contain both male and female parts, and some plants "self-fertilize"—their sperm can fertilize their eggs—as with a pea plant you will be looking at in Chapter 11.

On to Patterns of Inheritance

The tour of cell division over the last couple of chapters has yielded not only a look at how one cell becomes two, but at how chromosomes operate within this process. As you have seen, it is the DNA in these chromosomes that is the bearer of the "qualities" or "traits" that are passed on from one generation to the next. Given the paired nature of chromosomes and their precisely ordered activity during meiosis, it stands to reason that there would be some predictability or pattern to the passing on of traits. This is the case, as it turns out. And the person who first recognized this order was a monk who worked in nineteenth-century Europe in an obscurity that was richly undeserved.

Chapter Review

Key Terms

1n 192

2n 192

asexual reproduction 201

crossing over 197

diploid 192

gamete 192

haploid 192

independent assortment 196

life cycle 201

meiosis 192

oogonium 199

polar body 200

primary oocyte 199

primary spermatocyte 199

recombination 197

regeneration 201

secondary spermatocyte 199

sex chromosome 197

spermatid 199

spermatogonium 199

tetrad 193

vegetative reproduction 201

Summary

10.1 An Overview of Meiosis

- In more complex organisms, nearly all cells have paired sets of chromosomes, meaning these cells are diploid. Meiosis is the process by which a single diploid cell divides to produce four haploid cells—cells that contain a single set of chromosomes.

- The haploid cells produced through meiosis are called gametes. Female gametes are eggs; male gametes are sperm. They are the reproductive cells of more complex organisms.

- When the haploid sperm and haploid egg fuse, a diploid fertilized egg (or zygote) is produced, setting into development a new generation of organism.

10.2 The Steps in Meiosis

- In meiosis, there is one round of chromosome replication, followed by two rounds of cell division. There are two primary stages to meiosis, meiosis I and meiosis II. In meiosis I, chromosome duplication is followed by a pairing of homologous chromosomes with one another. These chromosome pairs are then separated, in the first round of cell division, into sepa-

rate daughter cells. In meiosis II, the chromatids of the duplicated chromosomes are separated, in the second round of cell division, into four daughter cells.

10.3 What Is the Significance of Meiosis?

- Meiosis generates diversity by ensuring that the gametes it gives rise to will differ from one another genetically. In this, it is unlike regular cell division, or mitosis, which produces daughter cells that are exact genetic copies of parent cells.

- Meiosis generates genetic diversity in two ways. First, in prophase I of meiosis, homologous chromosomes pair with each other and, in the process called recombination or crossing over, exchange corresponding chromosomal segments with one another. Second, in metaphase I of meiosis, there is a random alignment or independent assortment of maternal and paternal chromosomes on either side of the metaphase plate. This chance alignment determines which daughter cell each chromosome will end up in.

- Human females have 23 matched pairs of chromosomes—22 autosomes and two X chromosomes. Human males have 22 autosomes, one X chromosome, and one Y chromosome. Each egg that a female produces has a single X chromosome in it. Each sperm that a male produces has either an X or a Y chromosome within it. If a sperm with a Y chromosome fertilizes an egg, the offspring will be male. If a sperm with an X chromosome fertilizes the egg, the offspring will be female.

10.4 Gamete Formation in Humans

- The human male diploid cells called primary spermatocytes go through meiosis, producing, by the end of meiosis II, spermatids that develop into mature sperm. Human egg formation begins with oogonia cells, all of which are produced prior to the birth of the female. These give rise to primary oocytes, whose development is arrested in meiosis I until they are released in ovulation and fertilized.

10.5 Life Cycles: Humans and Other Organisms

- Not all reproduction is sexual reproduction. Bacteria reproduce through binary fission, meaning a chromosomal replication followed by cell division. Plants can engage in vegetative reproduction, and other organisms, such as worms and sea stars, in regeneration.

Understanding the Basics

Multiple-Choice Questions

1. In reference to chromosomes, meiosis involves
   a. no duplications, 1 reduction
   b. 1 duplication, 2 reductions
   c. 2 duplications, 1 reductions
   d. 2 duplications, 2 reductions
   e. none of the above

2. At prophase of meiosis I, the number of chromatids for a 2n = 8 cell is equal to
   a. 16
   b. 8
   c. 4
   d. 2
   e. twice that of mitosis prophase

3. Independent assortment occurs during
   a. prophase I
   b. anaphase II
   c. metaphase I
   d. metaphase II
   e. both c and d

4. Which of the following statements is *not* true regarding egg formation in humans?
   a. Oogonia are not produced until the eleventh or twelfth year.
   b. At the completion of division, one egg and three polar bodies appear.
   c. In both meiosis I and II, the metaphase plate is shifted to one side of the cell.
   d. Oocyte maturation occurs in approximately a 28-day cycle.
   e. Each egg contains 23 chromosomes.

5. What are the two sources of genetic diversity in meiosis?
   a. binary fission and regeneration
   b. recombination and independent assortment
   c. oogenesis and recombination
   d. spermatogenesis and oogenesis
   e. vegetative reproduction and regeneration

6. An egg has approximately
   a. 200,000 times the volume of a sperm
   b. 100,000 times the volume of a sperm
   c. 50,000 times the volume of a sperm
   d. 25,000 times the volume of a sperm
   e. 0.5 times the volume of a sperm

7. The cytoplasmic content of sperm is very limited compared to that of an egg. The major ingredients of sperm cell are
   a. a nucleus and Golgi complex
   b. ribosomes and mRNA
   c. a haploid set of chromosomes and endoplasmic reticulum
   d. mitochondria and ribosomes
   e. a haploid set of chromosomes and mitochondria

8. In prophase I of meiosis,
   a. maternal chromosome 1 joins with maternal chromosome 2
   b. maternal chromosomes exchange genetic material with other maternal chromosomes

c. paternal chromosomes exchange genetic material with other paternal chromosomes

d. tetrads form, joining homologous chromosomes (for example, maternal chromosome 1 with paternal chromosome 1)

e. both c and d

9. At the completion of meiosis I, each human cell contains
   a. 23 chromosomes, 46 chromatids
   b. 46 chromosomes, 46 chromatids
   c. 46 chromosomes, 92 chromatids
   d. 23 chromosomes, 23 chromatids
   e. none of the above

10. In sexual reproduction, there is first a(n) _____ of the normal number of chromosomes in somatic cells, followed by a _____ of male and female reproductive cells.
    a. multiplication, division
    b. doubling, decrease
    c. maintenance, halving
    d. tripling, splitting
    e. halving, union

Brief Review

1. Does chromosome reduction take place during meiosis I or II?

2. Is asexual reproduction confined to the plant world?

3. Why are men able to keep producing sperm throughout their lifetime, from puberty forward?

4. How does meiotic cytokinesis in the production of eggs differ from mitotic cytokinesis?

5. Define and distinguish between somatic cells and gametes.

6. An organism has a 2n value of 4. Name the phases of mitosis or meiosis shown in the following figures.

7. Draw the following cells at the phase indicated:
   a. n = 3, metaphase of meiosis II
   b. 2n = 8, prophase of meiosis I

8. Which anaphase of meiosis is most like anaphase in mitosis? Why?

9. Why is an egg so big relative to a sperm?

10. Define the following terms:
    a. gamete
    b. crossing over (or recombination)
    c. asexual reproduction
    c. haploid

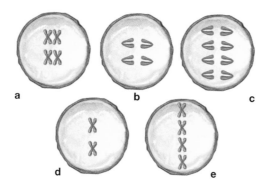

205

Chapter 10
Meiosis

Applying Your Knowledge

1. Why would crossing over *not* be of significant value during prophase II of meiosis?

2. Ultimately, is it the paternal or maternal gamete that delivers the sex-determining gene? Why?

3. Does asexual reproduction or sexual reproduction allow for the greatest genetic variation? Why?

4. Draw one progeny cell in the following phases and indicate the number of chromosomes and chromatids present.

	drawing of cell	number of chromosomes	number of chromatids
2n = 4 during metaphase of mitosis			
2n = 6 during anaphase of meiosis II			
2n = 4 following telophase of meiosis I			

Group Activity

Many students have heard of the cloned sheep "Dolly." Briefly, scientists "created" Dolly by fusing a donor somatic cell (with genetic material) to an enucleated recipient egg cell (no genetic material—no nucleus). Why did the genetic material need to be removed from the egg before the somatic cell genome was introduced? As a group, take some time to investigate the most recent advances in animal cloning. How do the new techniques differ from that used to create Dolly, and why do they improve the chances of successful cloning?

MediaLab

Too Much Division: The Cell Cycle and Cancer

This *MediaLab* is designed to stimulate interest in cell division by allowing you to examine the role of the cell cycle in cancer. To learn more, enter the *CD Activity* for an animated review of cell division. Once you have completed this section, you will be guided through a series of *Web Investigations*, where you can utilize information on the Internet to examine your own personal risk of contracting cancer, taking into account such factors as inheritance and exposure to environmental carcinogens. Finally, in *Communicate Your Results*, you can evaluate the relevance of these factors.

This *MediaLab* can be found in Chapter 10 on your CD-ROM and Companion Website (http://www.prenhall.com/krogh).

CD ACTIVITY

One cell dividing to create two is crucial for growth, for wound repair, and for creating the next generation of organisms. But there are major differences between cell division for growth and cell division for sexual reproduction. You will be looking at these differences in an animated tutorial.

Activity

1. *First, you will review the cycle in which our body uses existing cells to refresh and replace cells when we grow, or when a wound heals.*

2. *Next, you will compare this cell division to the type of cell division our body uses to make specialized sex cells carrying only half the genes from one parent, so they can be combined with the second half from the other parent to create a child.*

3. *Finally, you will see how cancer is caused when cell division becomes uncontrolled.*

After reviewing the concepts of cell division, you can evaluate your fears and misconceptions about cancer in the *Web Investigations* and apply your knowledge of cell division to achieve a better understanding of cancer.

WEB INVESTIGATION

Investigation 1
Estimated time for completion = 15 minutes

Cells can become cancerous (accelerated and unregulated growth) because they have a mutation in one of the genes that normally controls the rate at which a cell travels through the cell cycle. Scientists believe that inheriting a defective gene may predispose a person toward developing some specific cancers. What proportion of cancers are inherited?

Actually, a very small number. In families with these defective genes, the same type of cancer is seen in two or more close relatives, with an earlier-than-usual onset. Select the Keyword **INHERITED CANCERS** on your CD or Website for this *MediaLab* to view a list of inherited cancers. Then, select Keyword **PERSONAL HEALTH FORM** to complete a personal health history form, which will likely relieve you of your concern about inheriting these cancers.

Investigation 2

Estimated time for completion = 5 minutes

Daily media warnings expound on evidence that chemicals, drugs, and diet can increase our chances of acquiring cancer. But how can you determine which risk factors are serious, and which may be rare or insignificant? Select the Keyword **CANCER RISK FACTORS** on your CD or Website for this *MediaLab* to read an article comparing the relative risk factors in contracting cancer from natural and human-made compounds.

Investigation 3

Estimated time for completion = 5 minutes

Americans worry that cancer rates are rising because we are increasingly exposed to chemicals or other factors that increase cancer (carcinogens). Select the Keyword **NATIONAL CANCER INSTITUTE** on your CD or Website for this *MediaLab* to view research on suspected carcinogens, and read one of the articles posted at this site.

COMMUNICATE YOUR RESULTS

How well can scientists predict who will get cancer, when both heredity and the environment are such important factors? In the following exercises, you'll evaluate your own hereditary and environmental cancer risk.

Exercise 1

Estimated time for completion = 5 minutes

Now that you have filled out your personal health history form from *Web Investigation 1*, prepare a 50-word document indicating what type of cancer is found in your family. If it has an earlier-than-normal onset and is not found in two or more close relatives, it is probably not inherited.

Exercise 2

Estimated time for completion = 30 minutes

How can you evaluate the significance of news stories on such substances as chlorinated water or artificial sweeteners as potential cancer causes? After completing *Web Investigation 2*, compose a 250-word essay describing which statement you think is most correct:

1. *Synthetic chemicals manufactured by humans are responsible for more cancers (are more carcinogenic) than are natural compounds like sunlight and natural foods.*

2. *Chemicals found in natural foods and sunlight are more carcinogenic than are synthetic chemicals.*

3. *Both naturally occurring and synthetic chemicals are equally responsible for causing cancer.*

Exercise 3

Estimated time for completion = 15 minutes

Web Investigation 3 introduced you to cancer risk factors evaluated by the National Cancer Institute. Form a group of your peers, and have each member present one of these factors (for example, artificial sweeteners) to the group. In your presentation, be sure to inform the group whether they should be concerned about getting cancer from this factor. Also tell them about what tests could be done to relieve this concern.

11

The First Geneticist
Mendel and His Discoveries

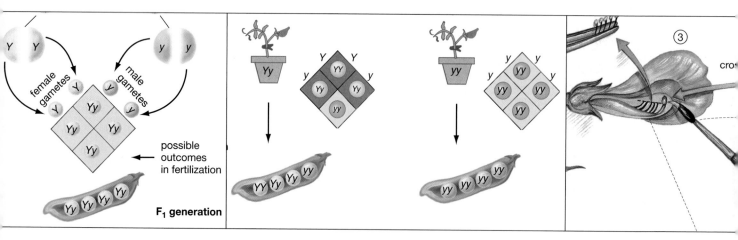

Green and yellow peas.
(page 214)

More pea experiments.
(page 215)

Cross-pollination.
(page 211)

In the Czech Republic, in the oldest part of the city of Brno, stands the former Monastery of St. Thomas. Now a research center, St. Thomas has a small, fenced-in garden that sits just outside its historic living quarters. At one end of the garden stands an ivy-covered tablet that is inscribed with a few simple words: "Prelate Gregor Mendel made his experiments for his law here." Visitors to the garden can cast their eyes upward and see the rooms that Gregor Johann Mendel lived in when he was carrying out this work in the years between 1856 and 1863 (**Figure 11.1**). Mendel labored during this time on a species of common peas, *Pisum sativum*, using tweezers and an artist's paintbrush as his tools. Like many scientific discoverers, Mendel looked at ordinary objects; and as with many scientists, his work was tedious and exacting. Nevertheless, the distance from what Mendel did to what Mendel learned is the distance from the ordinary to the historic.

Coming to adulthood in the mid-nineteenth century, Mendel knew nothing of the elements of genetics we have reviewed so far. Chromosomes and the genes that lie along them had not even been discovered when he was working with his peas; little knowledge existed of meiosis or mitosis, to say nothing of DNA. And as it happened, Mendel himself played no direct role in discovering any of these things. Yet this unassuming monk, the son of eastern European peasant farmers, generally is accorded the title of father of genetics. What contribution earned him this honor?

**Figure 11.1
Austrian Monk and Naturalist
Gregor Mendel**

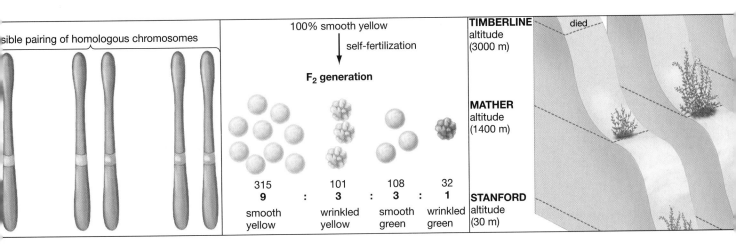

Alleles on chromosomes.
(page 217)

From a smooth yellow pea
to four phenotypes.
(page 218)

Genes and environment.
(page 222)

11.1 Mendel and the Black Box

Mendel's achievement was to comprehend what was going on inside what might be called the "black box" of genetics without ever being able to look inside that box himself. Science is filled with so-called black-box problems, in which researchers know what goes *into* a given process and what comes *out*. It is what is going on in between—in the black box—that is a mystery. In the case of genetics, what lies inside the black box is DNA and chromosomes and meiosis, and so forth; all the material *parts* of genetics, in other words, and the way they work together. Because of the timing of his birth, Mendel had no knowledge of any of these things. Until Mendel, however, nobody had looked carefully at even the starting and ending points of this black-box problem: at what went in and what came out. Mendel's original pea plants represented the "input" side to the black box of genetics, while the offspring he got from breeding these plants represented the output. By looking very carefully at generations of parents and offspring—at both sides of the box, in a sense—Mendel was able to infer something about what had to be going on within. As you will see, his main inferences were correct: (1) That the basic units of genetics are material elements that come in pairs; (2) that these elements (today called *genes*) can retain their character through many generations; and (3) that gene pairs *separate* during the formation of gametes.

These insights may sound familiar, because all of them were approached from another direction in Chapters 9 and 10. Here are the lessons you went over then:

- That genes are material elements—lengths of DNA
- That in human beings, genes come in pairs, residing in pairs of chromosomes
- That chromosomes make copies of themselves, thus giving the genes that lie along them the ability to be passed on intact through generations
- And that in meiosis, homologous chromosomes line up next to each other and then *separate*, with each member of a pair ending up in a different egg or sperm cell

By observing generations of pea plants and applying mathematics to his observations, Mendel inferred that something like this had to be happening in reproduction. Because Mendel was the first to perceive a set of principles that govern inheritance, we date our knowledge of genetics from him.

11.2 The Experimental Subjects: *Pisum sativum*

Beginning work in his monastery garden after a period of study in Vienna, Mendel managed to pick, in *Pisum sativum*, a nearly perfect species on which to carry out his experiments. If you look at **Figure 11.2**, you can see something of the anatomy and life cycle of this garden pea. Note that what we think of as peas in a pod are seeds in this plant's ovary.

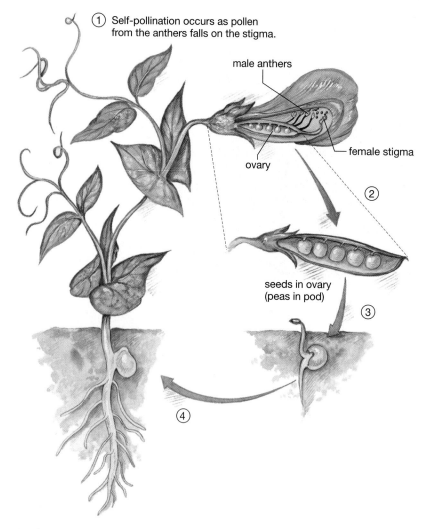

① Self-pollination occurs as pollen from the anthers falls on the stigma.

male anthers

female stigma

ovary

②

seeds in ovary
(peas in pod)

③

④

Figure 11.2
Anatomy and Life-Cycle of the Pea Plant

1 Sperm-bearing pollen, which exists in the plant's anthers, lands on the plant's stigma.

2 This results in the fertilization of the plant's eggs, which are housed in the plant's ovary. These eggs will develop into seeds in the ovary (peas in a pod). Each seed is fertilized separately.

3 After being planted, each seed has the potential to grow into a separate plant.

4 The seedlings develop into mature seed plants, capable of producing their own offspring.

Each of these seeds begins as an unfertilized egg, just as a human baby begins as a maternal egg that is unfertilized. Sperm-bearing pollen, landing on the plant's stigma, then set in motion the fertilization of the eggs.

Importantly, *Pisum* plants can *self*-pollinate. The anthers of a given flower release pollen grains that land on that flower's stigma. Each of the seeds that develops from the resulting fertilizations can then be planted in the ground and give rise to a new generation of plant. The way to think of the seeds in a pod is as multiple offspring from separate fertilizations—one pollen grain fertilizes this seed, another pollen grain fertilizes another. This is why, as you'll see, seeds that are in the same pod can have different characteristics.

Though the pea plants can self-pollinate, Mendel could also **cross-pollinate** the plants at will—he could have one plant pollinate another—by going to work with his tweezers and paintbrushes, as shown in **Figure 11.3**.

In directing pollination, Mendel could control for certain attributes in his pea plants—qualities now referred to as **characters**. If you look at **Table 11.1** on page 212,

you can see that these included seven qualities in all—such things as stem length, seed color, and seed shape. Note that each of these characters comes in two varieties; there are *yellow* or *green* seeds, for example, and purple or white flowers. Such variations are known as character **traits**. You'll notice that each of these variations is referred to as being either a "dominant" or a "recessive" trait. You'll learn the meaning of this shortly.

Phenotype and Genotype

Let us now put these traits in another context. Taken together, traits represent what are called **phenotypes**. Broadly speaking, a phenotype is a physical function, bodily characteristic, or action of an organism. In the context of Mendel, phenotype means the pea plant's visible physical characteristics. Purple flowers are one phenotype, white flowers another; yellow seeds are one phenotype, green seeds another. Meanwhile, any phenotype is in significant part determined by an organism's underlying **genotype**, meaning the genes that lie along its chromosomes. You could think of phenotype as the observable outcome of genotype.

Figure 11.3
How to Cross-Pollinate Pea Plants

1 Before fertilization occurs, peel back the closed petals of a pea plant (in this case, one that came from a line that yielded yellow peas). Then pull out the pollen-bearing stamens with tweezers so that self-fertilization is no longer possible.

2 Next, gather pollen from another plant by dabbing its anthers with a paintbrush.

3 Finally, rub these pollen grains onto the stigma of the first plant. The results of the cross-pollination can be observed when the fertilized eggs mature into seeds in the ovary, meaning peas in a pod. The resulting seeds are yellow in this case, because yellow is dominant over green.

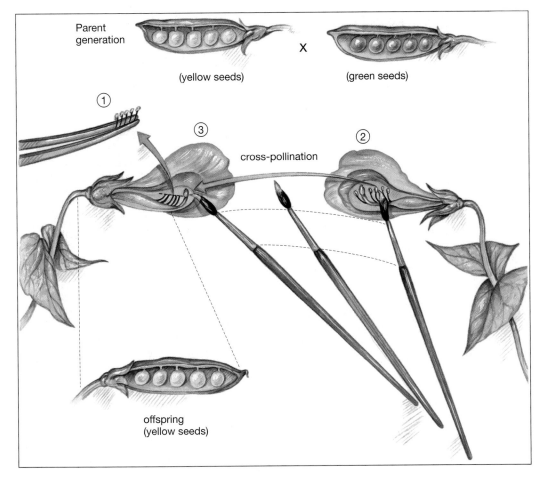

Parent generation

X

(yellow seeds)   (green seeds)

① ③ ②

cross-pollination

offspring
(yellow seeds)

11.3 Starting the Experiments: Yellow and Green Peas

Parental, F₁, and F₂ Generations

Mendel began by making sure that his starting plants "bred true" for the phenotypes under study. That is, he assured himself that, for example, all of his purple-flowered plants would produce nothing but generations of purple offspring. The **parental generation** in such experiments is often referred to in shorthand as the **P** generation. Meanwhile, the offspring of the parental generation are known as the **first filial generation**, the word *filial* indicating a son or a daughter. The shorthand for first filial is F_1. The F_x form can be used for any succeeding generation.

You will be seeing a lot of the second filial, or F_2 generation, for example.

So how did Mendel proceed with his plants? Let's start with just one example. Mendel took, for his P generation, some plants from a line that bred true for yellow seeds and other plants from a line that bred true for green seeds. Then he took pollen from one variety of the plants and fertilized the other variety, as seen in Figure 11.3. The plants fertilized in this way then produced their own seeds—every one of which was yellow. (Remember that these seeds represent the new generation—the F_1 generation.) Getting all yellow seeds was an interesting result in itself. Because each seed in a pea-plant pod is fertilized separately, it *can* be the case that a given pod will contain both yellow and green seeds. (Other pea characters can be variable within a pod as well, as you can see in **Figure 11.4**.) Yet all the seeds in this generation were yellow.

Having viewed these results, Mendel then let the F_1 generation *self*-pollinate. What he got from this, in the F_2 generation, was 6,022 yellow seeds and 2,001 green seeds.

The Power of Counting

With this counting, Mendel had taken a giant step forward. Why? Experimenters before Mendel had gotten the kind of results he had. What they did not do, however, was undertake a careful *counting* of their results and then analyze the results in terms of proportions.

Looking at the specific results, two things stand out. First, green seeds disappeared in F_1, but came back in F_2. Recall that there was not a green seed to be found in the F_1 generation pods, but in F_2, there are 2,001 of them. Second, green seeds came back in F_2 as a *specific proportion* of the seeds as a whole. The F_2 generation isn't divided 50/50, half-yellow and half-green; instead, there are roughly 3 yellow seeds for every 1 green seed—or to put it another way, a 3:1 ratio of yellow to green seeds. As it turned out, Mendel got the same result with each of the seven characters he was studying, as you can see in **Table 11.2**. Note that the 3:1 ratio is a proportion of dominant to recessive traits—more yellow seeds than green, more purple flowers than white. It was *solely* the dominant traits that appeared in the F_1 generation, but in the F_2 generation the dominants are simply appearing in greater *proportion* than the recessives.

Table 11.1
Pea-Plant Characters Studied by Mendel

Character Studied	Dominant Trait	Recessive Trait
Stem Length	Tall	Dwarf
Seed Shape	Smooth	Wrinkled
Seed Color	Yellow	Green
Pod Shape	Inflated	Wrinkled
Pod Color	Green	Yellow
Flower Color	Purple	White
Flower Position	On stem	At tip

Table 11.2
Ratios of Dominant to Recessive in Mendel's Plants

Dominant Trait	Recessive Trait	Ratio of Dominant to Recessive in F_2 Generation
Tall stem	Dwarf stem	2.84:1 (787 tall plants, 277 dwarfs)
Smooth seed	Wrinkled seed	2.96:1 (5,474 smooth, 1,850 wrinkled)
Yellow seed	Green seed	3.01:1 (6,022 yellow, 2,001 green)
Inflated pod	Wrinkled pod	2.95:1 (882 inflated, 299 wrinkled)
Green pod	Yellow pod	2.82:1 (428 green, 152 yellow)
Purple flower	White flower	3.14:1 (705 purple, 224 white)
Flower on stem	Flower at tip	3.14:1 (651 along stem, 207 at tip)
	Average ratio, all traits:	3:1

Interpreting the F_1 and F_2 Results

No "Blending" in Inheritance

What to make of these results? Mendel saw, for one thing, that inheritance for his peas was not a matter of the "blending" of their characteristics. For example, the flower-colors purple and white were retained as just that. No intermediate phenotypes—say, light blue flowers—resulted from the cross in any generation. His finding in this regard ran contrary to the notion, popular in his time, that two given traits would blend into a homogenous third entity as coffee and milk will blend together.

Dominant and Recessive Elements Come in Pairs

Beyond this finding, it was apparent that, dominance notwithstanding, plants could retain the *potential* for recessive phenotypes, even though those phenotypes might not be expressed in a given generation. Mendel's F_1 plants had no green seeds, but by F_2 green seeds were back. It was reasonable to assume, therefore, that the yellow-seed F_1 plants retained a green-seed *element*, which got expressed only in the F_2 generation. Finally, because his phenotypes came in pairs, it was reasonable for Mendel to hypothesize that the elements likewise came in pairs.

If you think Mendel's *pairs of elements* sound suspiciously like the "pairs of genes" on homologous chromosomes we've seen before, you are right. However, it's time to start referring to these "matched pairs of genes" in scientific terminology. It is more accurate to think of matched pairs of genes as alternative forms of a single gene. The proper name for an alternative form of a gene is an **allele**. Thus, the pea plant had a single gene for seed color that came in two alleles, one of which coded for yellow seeds, the other of which coded for green seeds. These alleles resided on separate chromosomes.

11.4 Another Generation for Mendel

Having seen recessive phenotypes reappear in F_2, Mendel decided to extend his breeding. In his next experiment, he took a set of F_2 generation seeds—all of them yellow—planted them and let the plants they grew into self-pollinate. Then he looked at the color of the seeds produced by these plants. What he found was that, of 519 plants grown, 166 yielded only yellow seeds, while

Figure 11.4
Variation within a Pea Pod
Since each garden pea is fertilized separately, individual peas within a pod can have different character traits. Note that some of these peas have a smooth texture, while others are wrinkled.

**Figure 11.5
a Mendel's F₁ Crosses**

1 Mendel started out by cross-breeding plants that for generations had yielded either all yellow seeds or all green seeds. In the example pictured, female gametes are being provided by a plant that has the dominant, yellow alleles (*YY*), while the male gametes are being provided by a plant has the recessive, green alleles (*yy*).

2 The cells of the pea plants that give rise to gametes start to go through meiosis.

3 The two alleles for pea color, which lie on separate chromosomes, separate in meiosis, yielding gametes that each bear a single allele for seed color. In the case of the female, each of these gametes bears a *Y* allele; in the case of the male, each bears a *y* allele.

4 The Punnett square shows the possible combinations that can result when the male and female gametes come together in the moment of fertilization. (If you have trouble reading the Punnett square, see Figure 11.5b.) The single possible outcome in this fertilization is a mixed genotype, *Yy*.

5 Because *Y* (yellow) is dominant over *y* (green), the result is that all the offspring in the F₁ generation are yellow, because they all contain a *Y* allele.

b How to Read a Punnett Square
This Punnett square concerns pea-plant flower color, which is either purple (dominant) or white (recessive). Two of the four possible combinations are shown.

1 A *p* gamete from the male combines with a *p* gamete from the female to produce an offspring of *pp* genotype (and white color).

2 A *p* gamete from the male combines with a *P* gamete from the female to produce an offspring of *Pp* genotype (and purple color).

the remaining 353 plants had a mixture of yellow and green seeds, in the familiar 3:1 ratio. When he similarly took F₂ green seeds, planted them, and let them self-pollinate, however, these plants produced nothing but green seeds.

What could have caused these results? The underlying elements Mendel had hit upon, meaning the alleles for the color gene. When he planted his yellow seeds, some of them yielded all-yellow offspring, but others yielded mixed (green and yellow) offspring. He reasoned that there must be *two kinds* of yellow plants: Those that are "pure" yellow (the 166 that yielded solely yellow seeds) and those that were "mixed" yellow (the 353 that produced both yellow and green seeds). Pure yellow had *nothing but* yellow alleles within, and would produce nothing but yellow seeds when self-pollinated. Meanwhile, the mixed yellow would have one yellow and one *green*

a MENDEL'S F₁ CROSSES

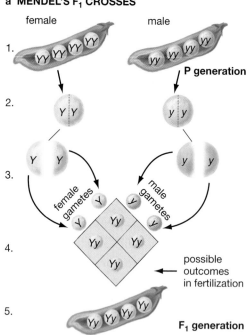

b HOW TO READ A PUNNETT SQUARE

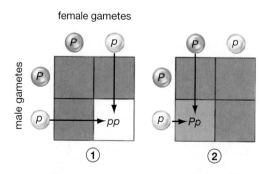

allele within, and thus would produce pods that contained seeds of both colors. Finally, the green seeds had nothing but green alleles and thus produced nothing but green seeds.

Mendel's Generations in Pictures

The F₁ Generation
Let's make these points more clear by reviewing all three generations of experiments schematically. It will be helpful to introduce a convention here. Let us refer, as Mendel did, to dominant and recessive gene types by uppercase and lowercase letters. Thus a "pure" yellow-seeded plant, having *two* yellow alleles, would be symbolized as *YY*. Meanwhile, pure green-seeded plants are *yy*, while mixed seeds are *Yy*. Let us say, just as an example, that female gametes are being supplied by a plant that is *YY*, while male gametes are coming from a plant that is *yy*. Each of these plants will contribute *one member* of its gene pair to its respective gametes—the *YY* female contributing a *Y* gamete, and the *yy* male contributing a *y* gamete. When these gametes fuse in the moment of fertilization, the result will be a *Yy* hybrid in the F₁ generation. If you look at **Figure 11.5**, you will be introduced to a time-honored way to represent such outcomes, the Punnett square, and see Mendel's F₁ results symbolized as well.

Note that, because *Y* is dominant, all the seeds in the F₁ offspring pods will have a yellow *phenotype*, even though every one of these seeds contains a mixed *genotype*. (Every one is *Yy*.) It only takes one *Y* allele for a seed to be yellow; for a seed to be green, then, it must have two *yy* alleles.

The F₂ Generation
Next come the F₁ crosses that yielded the F₂ generation. The starting point here is the F₁ seeds, which are all of the mixed *Yy* type, as you can see in **Figure 11.6**. Meiosis then occurs in the gamete precursors, which results in a separation of these alleles: Half the gametes now contain *Y* genes and the other half *y*. You can see from the figure why this cross can now give us back the green-seed phenotype, on average in a 1:3 proportion with yellow seeds. Note that there are twice as many mixed-genotype seeds (*Yy*) as either variety of pure seed (*YY* or *yy*). Important mathematical prin-

ciples underlie this outcome. You can read more about them in "Proportions and Their Causes" on page 216.

The F₃ Generation

Finally, in Figure 11.6, you can see why Mendel got the results he did when he went from F₂ to F₃. Recall that when Mendel planted his *yellow* F₂ seeds, he got 519 plants, of which 166 produced nothing but yellow seeds. Now you can see why: These were not only phenotypically yellow; they were "pure" yellow in terms of genotype (*YY*). Also, he got 353 plants that produced both yellow and green seeds. This is because the seeds these plants grew from were of mixed genotype

(*Yy*); in reproduction, their offspring would be of both pure (*YY*, *yy*) and mixed (*Yy*) genotype. Finally, when Mendel planted *green* F₂ seeds, all the plants that resulted from these seeds bred true for green seeds, because all of them began as *yy*. **Figure 11.7** illustrates how the three genotypes discussed here yield only the two phenotypes of yellow or green.

The Law of Segregation

For inheritance to work out this way, Mendel saw what has already been touched on in this chapter: That, though plant cells may contain *two* copies of a gene relating to a given phenotype, these copies must *separate* when a plant makes gametes. How else could two *Yy*

MediaLab Link
Punnett square, see p. 228.

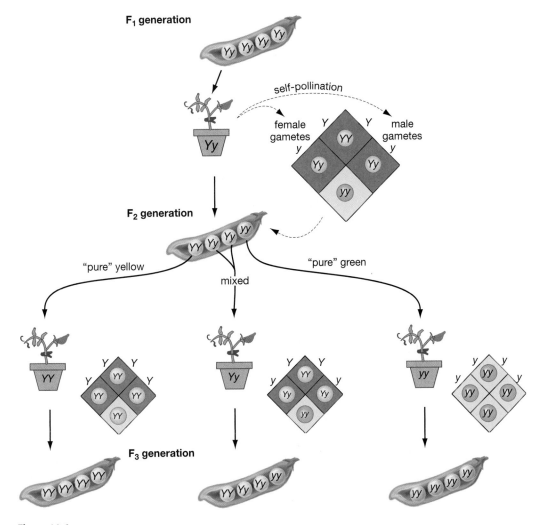

Figure 11.6
From the F₁ to the F₃ Generation
F₁ to F₂: The starting point is the F₁ generation, a set of seeds that all have the *Yy* genotype. These seeds are planted and the plants go through meiosis, yielding the gametes shown in the Punnett square. When these gametes come together in self-fertilization, the possibilities include *YY* and *yy* combinations, as well as the *Yy* combination seen in the F₁ generation. The existence of *yy* individuals is the reason green seeds reappear in the F₂ generation. Because *Y* is dominant, the green phenotype could not appear in seeds that had even a single *Y* allele. **F₂ to F₃:** With three starting genotypes (*YY*, *Yy*, *yy*) the F₂ generation yields plants that have these three genotypes, though there are more plants of "mixed" genotype than of either "pure" genotype.

Proportions and Their Causes: The Rules of Multiplication and Addition

To be a good songwriter, a person needs to be both a good musical composer and a good lyricist. Of all the people who decide to take up songwriting, say that 1 in 10 is a good composer and that 1 in 10 is a good lyricist. Now, what is the probability that a person who takes up songwriting will be *both* a good composer and a good lyricist? The answer to this question turns out to have relevance to all kinds of questions in our world, including the results that Gregor Mendel got.

When Mendel produced his F_3 generation, he got many more plants that had a mixed seed-color (yellow and green) than plants whose seeds were either all-yellow or all-green. Here's a Punnett square showing the results of the meiosis that produced this outcome:

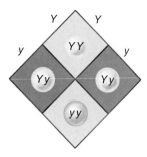

The Punnett square provides an intuitive visual sense of why things turned out the way they did. There were twice as many mixed seeds (Yy) as either type of "pure" seeds (YY or yy). It's even more helpful, however, to understand a principle that underlies this outcome. A simple coin-tossing example can illustrate what is at work. Suppose that you are going to throw two coins in the air—a nickel and a dime. What is the likelihood that *both* coins will come up heads? There is only one way to get this result:

1. Nickel = heads; dime = heads.

Now consider tossing the two coins and specifying a *mixed* result. There are *two* ways that this condition can be satisfied:

1. Nickel = heads; dime = tails
2. Dime = heads; nickel = tails

The seeds are in the same situation. What is the likelihood of getting a "pure" yellow seed? As you can see by looking again at the Punnett square, there is only one way

to get this: female Y, male Y. Now consider specifying a mixed result. There are two ways to get this: female y, male Y; and female Y, male y.

The Rule of Multiplication

It's necessary, though, to look at this concept in a more formal way, in order to state a principle that can generalize to any situation. What are the odds of tossing up one

---

What are the odds of tossing up two coins and having both come up heads?

---

coin and getting heads? Fifty-fifty, right? This means a 50 percent chance, which equals 0.5. Now what are the odds of tossing up two coins and having both come up heads? A principle called the **rule of multiplication** comes into play here. It states that the probability of any *two* events happening is the product of their respective probabilities. In this case, the probability for each head coming up is 0.5, so the equation is $0.5 \times 0.5 = .25$, or a 25 percent probability of two heads.

The Rule of Addition

Now what is the probability of getting one coin that comes up heads and one that comes up tails—when either of them can be heads or tails? When an outcome can occur in two or more *different* ways, as is the case here, the probability of this happening is the *sum* of the respective probabilities. This principle is known as the **rule of addition**. The probability of nickel = heads, dime = tails is the same 0.25 probability you saw in the rule of multiplication. But the probability of nickel = tails, dime = heads is also 0.25. The probability that *either* of these outcomes will take place is thus $0.25 + 0.25 = 0.50$, or 50 percent.

All this provides a way to figure the songwriter probability—or any set of independent probabilities. Remember the stipulation that 1 in 10 aspiring songwriters is a good composer and 1 in 10 is a good lyricist. Thus, under the rule of multiplication, $0.10 \times 0.10 = 0.01$, meaning 1 in 100 will be good at both composing and lyric writing.

parents ever give rise to *yy* (or *YY*) progeny, unless the *Yy* elements could first separate (in the parents) and then recombine in different ways (in the offspring)? Thus did Mendel derive his insight, sometimes called Mendel's First Law or the **Law of Segregation**: That organisms have two genetic elements that separate when gametes are formed. As noted, the physical basis for this law, which Mendel knew nothing of, is the separation of homologous chromosomes during meiosis.

Homozygous and Heterozygous Traits

It's time to add a couple more terms to the concepts you have been considering. There is scientific terminology for the genotypically "pure" and "mixed" organisms noted earlier. An organism that has two identical alleles of a gene for a given character is said to be **homozygous** for that character (as with *YY* or *yy*). An organism that has differing alleles for a character is said to be **heterozygous** for that character (as with *Yy*). You often see homozygous used in combination with the terms *dominant* and *recessive*. For example, a *yy* plant is a **homozygous recessive**, while a *YY* is a **homozygous dominant** (see Figure 11.8).

11.5 Crosses Involving Two Characters

Thus far, the subject has been how pea plants come to differ in *one* of their characters, seed color. When people breed plants for a single difference such as this, looking to see how the offspring will come out, the procedure is known as a **monohybrid cross**. Mendel, however, went on to ask: What happens if you breed plants for *two* characters? What happens, in other words, if you undertake what are known as **dihybrid crosses**?

Crosses for Seed Color *and* Seed Shape

One dihybrid cross that Mendel performed involved, for its first character, the yellow and green seeds you've become so familiar with. In addition, this cross involved a sec-

ond character of these seeds: their shape, which can be smooth or wrinkled. It's clear that yellow color is dominant to green and, as Table 11.1 shows, smooth seed shape is dominant to wrinkled. We have been denoting the alleles for seed color as *Y* for yellow and *y* for green; let us now denote the alleles for seed shape as *S* for smooth and *s* for wrinkled.

You can see, in **Figure 11.9a** on page 218, what the phenotypic outcomes were for these crosses between the P and F_2 generations. In the F_1 generation, all the seeds were smooth and yellow. When the F_1s were self-pollinated, however, the F_2 phenotypes that resulted came out as: 315 smooth yellow seeds; 108 smooth green seeds; 101 wrinkled yellow seeds; and 32 wrinkled green seeds. These numbers work out to 9:3:3:1 ratio, meaning 9 parts smooth yellow; 3 parts smooth green; 3 parts wrinkled yellow; and 1 part wrinkled green.

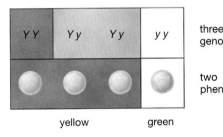

three
genotypes

two
phenotypes

yellow   green

Figure 11.7
Three Genotypes, Two Phenotypes
The two alleles for seed color (*Y* = yellow and *y* = green) can result in three genotypes (*YY*, *Yy*, *yy*), but these can yield only two phenotypes (yellow and green).

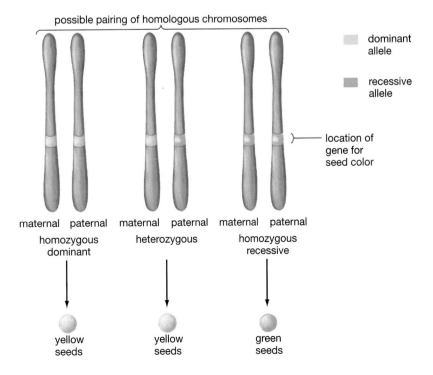

possible pairing of homologous chromosomes

dominant allele

recessive allele

location of gene for seed color

maternal paternal   maternal paternal   maternal paternal
homozygous   heterozygous   homozygous
dominant   recessive

yellow   yellow   green
seeds   seeds   seeds

Figure 11.8
Chromosomes and Phenotypes
The figure shows how alleles on chromosomes yielded the pea-color phenotypes that Mendel observed.

A Hidden, Underlying Ratio

For Mendel, these imposing figures represented another opportunity to perceive something about how the sealed box of genetics operated. He realized, early on, that this ratio might have been a composite that hid an underlying reality for each of the *single* characters he was studying. Think for a second, as Mendel did, of only one of these characters, the *color* of the seeds. Here the result is:

315 (smooth) yellow seeds	108 (smooth) green seeds
<u>101</u> (wrinkled) yellow	<u>32</u> (wrinkled) green
416 yellow seeds	140 green seeds

This equals a yellow:green ratio of about 3:1. In other words, the familiar 3:1 ratio in the F_2 generation still holds. It held as well when Mendel looked only at seed *shape*. This suggested something to Mendel that turned out to be another of his major insights: That *characters*—in this case, seed shape and seed color—are transmitted *independently* of one another. When these two characters were being crossed at the same time, the results for each were the same as when they were crossed as single characters in the earlier experiments. Thus, one character's transmission did not appear to affect the other's.

Understanding the 9:3:3:1 Ratio

Still, what's the basis of the formidable 9:3:3:1 ratio that Mendel got in his dihybrid cross, with its mixture of wrinkled yellows, smooth greens, and so forth? This is simply a more complex example of the kind of outcomes you saw earlier with the aid of the Punnett square. If you look at the square in **Figure 11.9b** and start adding internal squares *by phenotype*, Mendel's 9:3:3:1 ratio starts making sense. You can see, for example, where the 1 in the ratio comes from: There is only 1 part wrinkled green seeds, because it is only the *sy* (male) and *sy* (female) combination that yields this phenotype. Likewise, you will get three parts wrinkled yellow seeds by adding up the three possible male and female gamete combinations that could bring this about.

The Law of Independent Assortment

This outcome could only come about, however, if Mendel's fundamental insight about dihybrid crosses was correct: That characters were being transmitted independently of one another. If one character had *affected* another's transmission, these ratios would have been very different. This insight of Mendel's

Figure 11.9
Phenotype Ratios in a Dihybrid Cross

a In one of his dihybrid crosses, Mendel cross-bred plants that had smooth yellow seeds with those that had green wrinkled seeds. The result of this was a generation of plants that all had smooth yellow seeds. When these plants self-fertilized, the result was an F_2 generation that had the phenotypes shown in a 9:3:3:1 ratio.

b The Punnett square demonstrates why Mendel got the 9:3:3:1 phenotypic ratio in his dihybrid cross. Nine combinations yield smooth yellow seeds, 3 yield smooth green seeds, 3 yield wrinkled yellow seeds, while only 1 results in a wrinkled green seed.

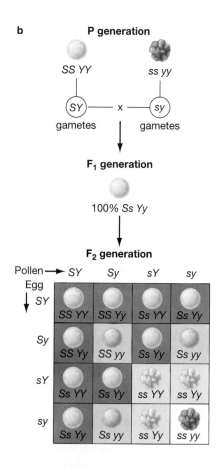

is now known as Mendel's Second Law or the **Law of Independent Assortment**. It states that during gamete formation, gene pairs assort independently of one another.

Independent Assortment and Chromosomes

Now recall that the underlying *physical* basis for this law was set forth in Chapter 10: that, in meiosis, pairs of homologous chromosomes *assort independently* from one another at the metaphase plate. It may be, for example, that paternal chromosome 5 will line up on side A of the metaphase plate; if so, then maternal chromosome 5 ends up on side B. For chromosome 6, however, things could just as easily be reversed: The *maternal* chromosome might end up on side A, and the paternal on side B (see Figure 10.2c, page 195.) The genes for Mendel's seed color and seed shape existed on separate chromosomes, and hence assorted independently at the metaphase plate, meaning they were *passed on* independently to future generations.

11.6   Reception of Mendel's Ideas

When Mendel finished his experiments, he delivered two lectures on them to a local scientific society in 1865. It would be nice to report that the society members' jaws dropped open once Mendel let them in on how inheritance worked in the living world, but no such thing happened. His findings were ultimately published in a scientific journal that was distributed in Germany, Austria, the United States, and England. Mendel even took it upon himself to get 40 reprints of his paper, thereafter sending some out to various scientists in an effort to spark some exchange on his findings. All to no avail. His work sank nearly without a trace, finally to be rediscovered in 1900, 16 years after his death. Today, the consensus within the scientific community is that nobody cared about Mendel's findings in his own time simply because nobody grasped their significance (see "Why So Unrecognized?" page 220).

This early reception notwithstanding, Mendel's insights have stood the test of time. For certain kinds of phenotypes, his rules of inheritance are extremely reliable. To this day, biologists will begin an observation by noting that "Mendelian rules tell us that . . ." or "Mendelian inheritance operates here." Even

allowing that he had intellectual predecessors, the fact is that Mendel did not just *add* to the discipline of genetics; he founded it.

11.7   Variations on Mendel's Insights

Incomplete Dominance

Mendel knew as well as anyone that the rules he discovered did not apply in all instances of inheritance. In his peas, crossing white- and purple-flowered plants resulted in an F_1 generation in which all flowers were purple. In some other species, however, the results are different. Crossing a red-flowered *snapdragon* with a white-flowered snapdragon produces neither red- nor white-flowered F_1 snapdragons, but *pink* snapdragons (**see Figure 11.10**).

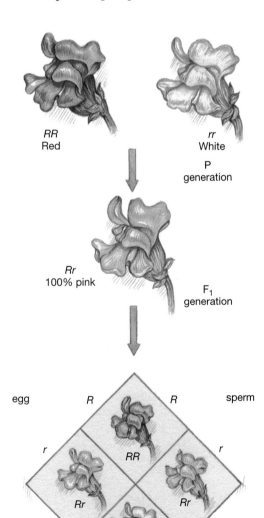

**Figure 11.10
How Red and White Yield Pink: Incomplete Dominance in Snapdragons**
A single red allele (*R*) yields only enough pigment to produce a flower that is pink—the only phenotype in the F_1 generation. In the F_2 generation, however, red alleles combine (*RR*) to produce the red-flower phenotype.

RR
Red

rr
White

P generation

Rr
100% pink

F_1 generation

egg   *R*   *R*   sperm

r   RR   *r*

Rr   Rr

rr

F_2 generation

1   :   2   :   1
red   pink   white

Why So Unrecognized?

"If fame belonged to me, I could not escape her—if she did not, the longest day would pass me on the chase," wrote Emily Dickinson in 1862, in seeming ambivalence about becoming a published poet. Dickinson's words came in a letter to Thomas Higginson, who had recently advised her against seeking publication of her verses on grounds that they were not yet strong enough. Higginson was the second, and as it turned out, last person to disparage Dickinson's efforts to publish her poetry. After his advice was offered, Dickinson kept her own company as a writer; by the time she died, only a close circle around her had seen her work.

Half a world away, Dickinson had a counterpart of sorts in her contemporary, Gregor Mendel (born eight years before her, and deceased two years before). Like Dickinson, Mendel had something special to offer the world. But it was an offer made to a world that was not prepared to listen. Mendel died in obscurity in 1884, his work appreciated by, so far as we can tell, no one but himself. Indeed, 34 years were to pass between publication of his research results and recognition of their significance. Yet from the time his work was rediscovered in 1900, Mendel has been treated with great respect, such that today he is generally recognized as the founder of genetics.

How, then, could Mendel's work have gone unrecognized in his own time? A couple of easy but unsatisfying answers are that he labored in a scientific backwater (in the hinterland of the Austrian empire) and that his results were published in an obscure scientific journal.

Mendel appears to have been, in essence, a victim of his own originality.

Even granting this, a good number of scientists still had access to his report, if not by reading it in the journal that published it, then by receiving one of the reprints of it that Mendel himself mailed out. If fame didn't follow from such exposure, why not simple recognition by at least one person?

The answer appears to be fairly straightforward: Mendel wasn't appreciated in his lifetime by anyone because he wasn't comprehended in his lifetime by anyone.

At first, this might seem to provide evidence that inheritance can indeed work through *blending* of genetic traits—a notion that Mendel's work had quashed. Blending is not taking place, however. When breeding is continued through the F_2 generation, red and white flowers come back, along with the pink, in a familiar ratio: 1 part red, 2 parts pink, 1 part white. This 1:2:1 ratio demonstrates that red and white alleles have not irretrievably blended into pink, but rather have come together in F_1, only to separate out again in F_2. Still, how can the pink F_1 snapdragons be explained?

Genes Code for Proteins

First, think about what genes do: They contain information regarding the production of proteins. In the case of colors, such proteins can bring about the formation of pigments. This is the case with the snapdragons; they have a gene for color, and one of the alleles of this gene brings about the production of red pigment. Meanwhile, the second allele of this gene is nonfunctional—it brings about the production of no pigment at all. Two of the red alleles, then, yield a red color, while one red allele produces only enough pigment to yield a pink color. Neither the red nor the white allele, therefore, is completely dominant, and each is thus said to be **incompletely dominant**.

11.8 Lessons from Blood Types

The notion of genes as protein-producing entities can help us steer through another variation on Mendelian inheritance, this one having a great importance for variety in the living world.

Codominance: A and B Types of Red-Blood-Cell Proteins

When people speak of blood "types," what they mean is types of proteins that cover the surface of red blood cells. These proteins

Indeed, intense exposure to his ideas didn't seem to make any difference in this regard. Consider that he carried on a seven-year correspondence regarding his work with a famous botanist, Carl Nägeli, during which time Nägeli, like everyone else, seemed to miss the significance of Mendel's experiments entirely. Assuming that nineteenth-century scientists were as bright as those in the twentieth century, how could this have happened?

First, Mendel himself did not make much of the principles of heredity he uncovered. In his long paper, he punctuates a great deal of detail about his experiments with some understated formulations of general principles. In itself, this is not surprising. Scientists often let data "speak for themselves," resting assured that their colleagues will not miss a major point, however subtly it may be presented. Most scientists can do this, however, because they are confident that their colleagues will be *looking* for major points in their papers. In Mendel's case, nobody was looking for points about how heredity worked because scarcely anybody but Mendel had a concept of pure heredity. As Daniel L. Hartl and Vitezslav Orel have observed, in the minds of nineteenth-century scientists, heredity was merely a part of development—the process by which a fertilized egg becomes a fully formed organism. Mendel, meanwhile, thought of heredity as operating under its own independent set of rules, after which he went on to say what those rules were. Mendel thus delivered an answer to a question that nobody had asked: How does heredity work?

Other barriers existed to understanding as well. Mendel used mathematics to derive his insights on genetics, an approach that was novel in the biology of his day. Beyond this, Charles Darwin's *Origin of Species* had been published in 1859, an event that sent scientists scurrying to look for "continuous," or subtle variations between organisms—a phenomenon that they thought might shed some light on evolution. Conversely, Mendel's paper, coming seven years later, talked about pairs of seemingly unchangeable genetic elements that gave rise to such features as smooth or wrinkled seeds. It was difficult to fit such elements into a scheme of gradual evolutionary transformation.

Looking over all these factors, Mendel appears to have been, in essence, a victim of his own originality. (This puts him again in the company of Emily Dickinson, whose language was too idiosyncratic for her early readers.) There is a saying that it doesn't pay to be more than 10 minutes ahead of your time. Gregor Mendel had the misfortune of being about 34 years ahead of his.

---

come in many different varieties, the two most important of which are designated A and B.

Blood types are completely under genetic control, with a single gene that lies on chromosome 9 determining what blood type a person will have—A, B, AB, or O. As is almost always the case, this gene exists in duplicate, with one copy lying on the chromosome 9 inherited from the mother, and the other lying on the chromosome 9 inherited from the father. Thus, there exists the possibility for *alleles*, or gene variants.

Now, a single person could have two alleles that code for type A protein, or a person could have both alleles coding for type B molecules. In the first case a person would have type A blood, and in the second type B blood. However, a person could have one allele that codes for the type A molecule *and* one that codes for type B, in which case that person would have type AB blood. Finally, a person could have two alleles that are inactive—that code for neither surface protein— and would thus have type O blood.

Getting Both Types of Surface Proteins

What's new in this case is, first, another aspect of the idea of dominance in alleles. In the blood types, *both* the A and B alleles produce proteins (for the cell surface molecules). People who are heterozygous for these alleles thus do not get a phenotype that, like pink flowers, lies halfway between A and B. They get *both* the A and B proteins on the surface of their blood cells. Alleles that have this kind of independent effect are said to be **codominant**.

How to Think of Dominance

All this gets us to an important point: how to conceptualize dominance. In common speech the word *dominance* is used all the time, to mean "Bringing another under submission by means of superior power." Alleles, however, are not locked in struggle with one another. Alleles either call up the production of proteins or they don't. Such proteins (or lack of them) can create dominant phenotypes (smooth peas), intermediate phenotypes (pink snapdragons), or codominant phenotypes (AB blood).

Multiple Alleles

ABO blood types have one more lesson for us. In Mendel's seed shapes, there were two allelic variants: smooth and wrinkled. Now, in blood types, there are three: A, B, and O. As you know, no one person can have more than *two* alleles for a given gene: A person may be, say, A and B; but he or she cannot be A, B, and O. It is thus only in a *population* of humans that the full range of ABO alleles can be found. In short, alleles can come not just in two, but in many variants. When three or more alleles of the same gene exist in a population, they are known as **multiple alleles**, and they are extremely important. Multiple alleles are the reason that human beings don't really have "black" or "white" or "red" skin, but instead have skin that comes in a range of hues in which one

color shades imperceptibly into another. They are the reason that trees of a given species come in a range of heights, or that the beaks of birds come in a range of lengths. In short, it is multiple alleles that are responsible for much of the diversity seen in living things.

11.9 One Gene, Several Effects: Pleiotropy

In the genes considered so far, there has been a very clear-cut relationship between gene and phenotype: one gene, one character. A given gene, in other words, has controlled for, say, seed shape or flower color; but not for seed shape *and* flower color. This kind of one-gene, one-character relationship actually is fairly rare, however. The usual case is that a given gene will have many effects, for the simple reason that the various processes of living things are so interrelated.

The situation of one gene having many effects is called **pleiotropy**. Many examples of this phenomenon exist; one of the more interesting, perhaps, is a debilitation known as *fragile-X syndrome*, which is the most common cause of inherited mental retardation. The "X" in its name refers to the X chromosome. "Fragile" comes from the break on the long arm of the X chromosome that can be seen in people who suffer from the syndrome. Fragile-X is now thought to be caused by a defect in a single gene, dubbed *FMR-1* by the team of researchers who discovered it.

The upshot of this defect is that fragile-X victims lack a functional *FMR-1* gene, which is to say their *FMR-1* does not appear to prompt production of any protein. One clear result of this is mental retardation; fragile-X sufferers commonly have an IQ of about 40–70 (as opposed to the norm of 100). For our purposes, fragile-X also has effects that seemingly stand far afield from mental retardation. Fragile-X sufferers often have an abnormally long face; large, protuberant ears; and (when male) large testicles. This broad range of effects, seemingly stemming from the inactivation of a single gene, contains a message: Genes normally work together in an interrelated web, rather than functioning under a one-gene → one-phenotype model.

**Figure 11.11
Environment and Genes Interact**
The *Potentilla glandulosa* plant grew in nature at the three elevations shown. Cuttings were then taken from the plants and grown in experimental plots at all three elevations. Thus, for example, the Stanford *Potentilla*, which grew naturally at 30 meters (100 feet), was planted at 1,400 meters (4,600 feet) and 3,000 meters (10,000 feet) as well as at its native elevation. The results show how environment and genes interact to produce the physical traits or phenotypes of organisms. Note, for example, that the Stanford and Mather plants grew best at their native elevations, while the Timberline plant did not.

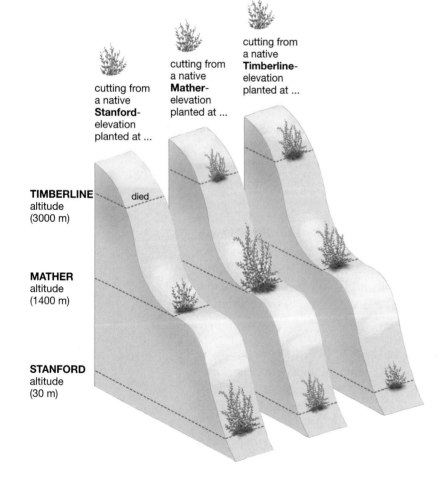

cutting from a native **Stanford**-elevation planted at ...

cutting from a native **Mather**-elevation planted at ...

cutting from a native **Timberline**-elevation planted at ...

TIMBERLINE altitude (3000 m)   died

MATHER altitude (1400 m)

STANFORD altitude (30 m)

11.10 Genes and Environment

In nature there is another force, quite separate from genotypic complexity, that works to create additional phenotypic variation; this is the effect of environment.

If you look at **Figure 11.11**, you will see a species of plant that is native to California, *Potentilla glandulosa*. Varieties of this plant grew in the wild at three elevations: the Stanford plant at 30 meters (100 feet), the Mather plant at 1,400 meters (4,600 feet), and the Timberline plant at 3,000 meters (10,000 feet). Cuttings from each of these plants were then replanted in experimental plots, both at the native elevation and the other two elevations. Thus, genetically identical versions of the Timberline plant, for example, were grown at 30 meters and 1,400 meters (as well as at its native 3,000 meters). You can see for yourself what happened: The Timberline cutting actually grew larger at 1,400 meters, rather than its native 3,000 meters; but it grew least well at 30 meters.

Now, the same genes existed in the Timberline plant no matter where it was grown, because all the plants were taken from cuttings. Yet this plant did much better in one location as opposed to another. The message here is one of genetic limitation: Genotype specifies only so much about phenotype.

Protein products that bring about a relatively tall plant at 3,000 meters may bring about a relatively short one at 30 meters.

We can put a little finer point on this. Look at the Potentillas again. Is it fair to say that any of the three varieties contains a "tall" gene? Grown at its native height, the Mather is the tallest plant overall. But how can the idea of a "tall" Mather gene be squared with the plant's short stature at either the Timberline or Stanford elevations? It can't. The plant's stature depended partly on environment. A light bulb may come with a wattage rating that holds true whether the bulb is burning in Denver or Death Valley, but genes do not possess such unvarying strengths or potentials. Imagine a light bulb whose brightness *changed* incrementally with each locale it operated in, and you begin to get the idea about genes.

On to the Chromosome

This completes the survey of Mendel's work and the variations on his ideas of inheritance. As you have seen, Mendel's work lay dormant for some decades in the nineteenth century. What sent scientists scurrying back to his findings were experiments on those tiny entities you have looked at in detail before: chromosomes, the subject of the next chapter.

Chapter Review

Summary

11.1 Mendel and the Black Box

- Gregor Mendel was the first person to comprehend some of the most basic principles of genetics. He reached these understandings in the mid-nineteenth century, working in what is now the Czech Republic and using as his experimental subjects a species of garden pea, *Pisum sativum*.

11.2 The Experimental Subjects: *Pisum sativum*

- Mendel looked at seven characters in his plants—such attributes as seed color and texture. In his plants, each of these characters came in two varieties or traits, one of

them dominant, the other recessive. His experiments involved breeding pea plants, starting with plants that had a given set of traits and then observing which of those traits showed up in succeeding generations.

- The physical functioning, bodily characteristics, or actions of an organism are its phenotype. In Mendel's plants, purple flowers were one phenotype and white flowers were another. Phenotypes in any organism are in significant part determined by that organism's genotype, meaning its genetic makeup. Mendel realized that the phenotypes in his plants were being controlled by what we would today call their genotypes.

Key Terms

allele 213

character 211

codominant 221

cross-pollinate 211

dihybrid cross 217

first filial generation (F₁) 212

genotype 211

heterozygous 217

homozygous 217

homozygous dominant 217

homozygous recessive 217

incompletely dominant 220

11.3   Starting the Experiments: Yellow and Green Peas

- Mendel realized that it was possible for organisms to have identical phenotypes—for all his pea plants to have yellow seeds, for example—and yet to have differing underlying genotypes.

- One of Mendel's central insights was that the basic units of genetics are material elements that, in his pea plants, came in pairs. These elements, today called *genes*, come in alternative forms called *alleles*. One member of an allele pair resides on one chromosome, while the other allele resides on a second chromosome that is homologous to the first.

- Another of Mendel's insights was that genes retain their character through many generations, rather than being "blended" together. Genes that coded for yellow pea color, for example, were retained in their existing form over many generations.

11.4   Another Generation for Mendel

- A third insight of Mendel's was that alleles separate prior to the formation of gametes. Though Mendel did not know it, the physical basis for this is that the alleles he was observing resided on homologous chromosomes, which always separate in meiosis.

11.5   Crosses Involving Two Characters

- Mendel observed that the genes for the different characters he studied were passed on independently of one another. This was so because the genes for these characters resided on separate chromosomes. The physical basis for what he found is the independent assortment of chromosomes during meiosis.

11.6   Reception of Mendel's Ideas

- Gregor Mendel published his work, but the significance of it was never recognized in his lifetime. It was only rediscovered 16 years after his death, in 1900.

11.7   Variations on Mendel's Insights

- Not all inheritance works through the principles Mendel perceived in his peas. Incomplete dominance operates when neither allele for a given gene is completely dominant, with the result that heterozygous genotypes can yield an intermediate phenotype (such as pink snapdragons).

11.8   Lessons from Blood Types

- An organism that has two identical alleles of a gene for a given trait is said to be homozygous for that trait. An organism that has differing alleles for a trait is said to be heterozygous for that trait. Human beings and many other species can have no more than two alleles for a given gene, each allele residing on a separate, homologous chromosome. Many allelic variants of a gene can, however, exist in a population. Such multiple alleles are responsible for much of the diversity seen in living things.

11.9   One Gene, Several Effects: Pleiotropy

- Genes work in an interrelated fashion, such that a single gene is likely to have multiple effects. Pleiotropy is a situation in which one gene has many effects.

11.10   Genes and Environment

- The effects of genes can vary greatly in accordance with the environment in which the genes are expressed. An organism's genotype and environment interact to produce that organism's phenotype.

Understanding the Basics

Multiple-Choice Questions

1. The first filial generation, or F_1, results when
   a. two P generation organisms are crossed
   b. the progeny of a P generation is crossed
   c. a P generation is crossed with a F_2 organism
   d. an F_3 organism is crossed with an F_2 organism
   e. both a and d

2. When Mendel crossed pure-bred yellow peas with pure-bred green peas, the F_1 generation contained
   a. all yellow peas
   b. all green peas
   c. green and yellow peas
   d. green, wrinkled peas
   e. both b and d

3. A very important characteristic of the garden peas used by Mendel was that
   a. each trait examined had more than two different varieties
   b. all yellow peas had the same genotype
   c. the F_1 generation always resulted in a 1:1 phenotypic ratio
   d. all crosses resulted in a 1:3 genotypic ratio
   e. the plants had the ability to self-fertilize

4. If a cross is made between a pure breeding green, round (*yyRR*) plant, and a pure breeding yellow, wrinkled (*YYrr*) plant, what is the result?
   a. all green, wrinkled peas
   b. all yellow, wrinkled peas
   c. all yellow, round peas
   d. yellow, wrinkled; green wrinkled
   e. none of the above

5. If a cross has the following genotype (*YYRr* × *yyRr*), the correct term for the type of cross is:
   a. recessive cross
   b. dominant cross
   c. monohybrid cross
   d. dihybrid cross
   e. codominant cross

6. Mendel's first law states:
   a. During the formation of gametes, gene pairs demonstrate independent assortment.
   b. During the formation of gametes, gene pairs segregate from each other.
   c. The probability of any two events happening is the product of their respective probabilities.
   d. If an organism is pure-breeding for any trait, it will display incomplete dominance.
   e. Mendel did not write any laws regarding genetics.

7. A trait whose phenotype in the heterozygous condition is an intermediate of the two homozygous conditions (that is, heterozygous pink flowers, homozygous red and white flowers) demonstrates:
   a. pleiotropy
   b. incomplete dominance
   c. intermediate dominance
   d. codominance
   e. multiple alleles

8. Mendel's second law states:
   a. That no organism can have more than two alleles for a given character.
   b. That no two organisms can ever be genetically identical.
   c. That organisms will tend to be heterozygous for most traits, rather than homozygous.
   d. That the way one gene pair assorts during gamete formation will affect the way other gene pairs assort.
   e. That gene pairs assort independently of one another during gamete formation.

9. The tool introduced in the early 1900s which diagrammatically represents the genotype frequencies of a genetic cross is called:
   a. a Mendelian square
   b. a genotype grid
   c. a phenotype plate
   d. a Punnett square
   e. none of the above

10. Which of the following is true regarding the effects of genes and environment on phenotypic variation?
    a. In the native environment, gene expression will always be greatest.
    b. Gene expression is not affected by a change in environment.
    c. Phenotypes are based only on environmental influences.
    d. Environment may easily affect genotype, not phenotype.
    e. Genetically identical organisms may be phenotypically affected in different environments.

Brief Review

1. What are the differences between:
   a. phenotype and genotype?
   b. dominant and recessive?
   c. codominance and incomplete dominance?

2. True or False: When Mendel conducted his genetic research, neither DNA nor chromosomes had been discovered.

3. What genetic principle is demonstrated in the fragile-X syndrome? Explain how this principle differs from the relationship between genes and physical outcomes that Mendel observed in his pea plants.

4. In a case of Mendelian genetics where one gene affects one trait, how many different genotypes may represent a dominant phenotype (for example, yellow pea color)?

5. If a heterozygote yellow pea plant is crossed with a homozygous green pea plant, what percentage of the progeny peas will be yellow?

6. What term is now given to Mendel's "element of inheritance," also referred to in the "particulate theory of inheritance"?

7. What two laws are credited to Mendel and his research?

8. What Mendelian crosses would result in the following phenotypic ratios?
   a. 1:1   b. 3:1   c. 1:1:1:1   d. 9:3:3:1

Applying Your Knowledge

1. What is meant by the term *allele*? Why can no more than two alleles of any gene exist simultaneously in a person?

2. A researcher is studying the traits of a recently identified plant species. She is particularly interested in two characteristics: leaf shape (round or oval), and flower color (red or blue). After crossing two true-breeding plant lines, she self-pollinates the F_1 generation plants, and observes the following results in the F_2 generation:

   Round leaves, Blue flowers: 110 plants
   Round leaves, Red flowers: 29 plants
   Oval leaves, Blue flowers: 317 plants
   Oval leaves, Red flowers: 105 plants

   Using these data, which two traits are dominant? What were the genotypes of the parents?

3. A purple flowering plant (*PP*) is crossed to a white flowering plant (*pp*).
   a. Using a Punnett square, diagram the cross. What are the genotypic and phenotypic outcomes?
   b. Two F_1 individuals from the previous cross are crossed to each other. Draw a new Punnett square, diagramming the new outcome. What is the ratio of purple flowers to white flowers? How many genotypes are there? In what frequencies do they occur?

4. A pure breeding, purple flowered, yellow pea plant (*PPYY*) is crossed to a white flowered, green pea plant (*ppyy*). The F_1 generation is allowed to self-fertilize. What are the phenotypic and genotypic ratios of the F_2 plants?

5. Using what you have learned in this chapter, and the following genetic cross, predict the phenotypic expression of any F_1 mouse. (Assume capital abbreviations represent dominant phenotypes.)

   G = gray, g = brown
   L = long hair, l = short hair
   C = curly tail, c = straight tail
   W = whiskers, w = no whiskers
   E = big ears, e = small ears

   a. Cross: *GGLLCCwwee* × *ggllCCwwEE*
      What is the phenotype of the F_1 generation?
   b. If the F_1 generation is crossed, what is the probability that the offspring will have the following genotype: *GgLlCCwwee*?

6. ABO blood types: In many paternity cases the first test involves a blood typing, since it is an easily detected genetic trait.

   a. In such a paternity case, a man denies fathering a child of blood type O. He is blood type AB, while the mother is blood type O. Based on this test, could he have fathered the type O child? Why or why not?

   b. If the man were type A, and the mother type O, would the conclusion be any different? Why or why not?

7. Stands of aspen trees often are a series of genetically identical individuals, with each succeeding tree growing from the severed shoot of another tree. Using what you've learned of genetics in this chapter, would you expect one aspen tree in a stand to differ greatly from another in terms of phenotype? Would you expect each to look exactly like the next in terms of phenotype?

8. In the years just prior to the time Mendel was formulating his laws of genetics, the Englishman Charles Darwin published a book on the subject of evolution. Darwin hypothesized that the way new organisms evolved was that a given organism that had a trait that favored its survival—say a slightly longer beak in a bird—would live to have more offspring than other such birds. Thus this trait would be passed on to more birds in a succeeding generation than would the trait of a shorter beak. Such changes would then build up over time until one species evolved from another. Do you see anything in the genetic rules that Mendel uncovered that tends to support Darwin's ideas about evolution?

Group Activity

Get together with a group of classmates, choose a trait easily visible in your families (for example, hair color, eye color), and construct a pedigree diagramming the inheritance of that trait in each of your families. Then compare the results you got with those obtained by your classmates.

MediaLab

Where Did I Get This Nose?
Understanding Mendelian Genetics

Can you roll your tongue? Is there a cleft in your chin? What is the color of your eyes? These and countless other characteristics that make you a unique individual were determined by the traits you inherited from your parents. In this *MediaLab*, you'll firmly establish your understanding of the rules governing the inheritance of simple genetic traits. In the *CD Activity*, you will work through these rules using some of Gregor Mendel's original genetic studies of peas. In the *Web Investigation*, you can apply these rules in breeding simulations for flies—and your own family—and make predictions about the inheritance of these traits in the *Communicate Your Results* section.

This *MediaLab* can be found in Chapter 11 on your CD-ROM and Companion Website (http://www.prenhall.com/krogh).

CD ACTIVITY

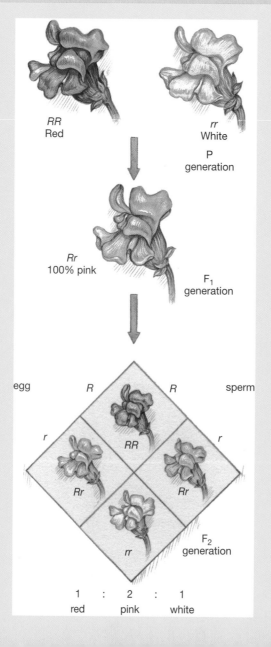

RR
Red

rr
White

P
generation

Rr
100% pink

F₁
generation

egg   *R*   *R*   sperm

r   *RR*   *r*

Rr   *Rr*

rr

F₂
generation

1   :   2   :   1
red   pink   white

It is obvious to most people that physical traits are inherited in families. But the rules that govern them were a mystery until a nineteenth-century Augustinian friar, Gregor Mendel, discovered the pattern behind inheritance of traits, by conducting breeding experiments on garden peas. With this *CD Activity,* you can review Mendel's experiments in order to lay the foundation for applying these rules to human genetics.

Activity

1. *First, you will examine how different versions of a single trait (like white versus purple flower color) are passed from parents to offspring.*

2. *Then you will learn to use a Punnett square, a chart that shows all possible combinations of alleles in the offspring of any parents, allowing you to predict the chances of inheriting one or more different traits.*

WEB INVESTIGATION

Investigation 1

Estimated time for completion = 5 minutes

Select the Keyword **VIRTUAL FLY** on your CD or Website for this *MediaLab*, to begin this investigation. First select the traits of male and female fruit flies, breed them, and observe the types of offspring they produce. Leave all of the radio buttons at the site selected for the default (wild type) setting for both male and female, except for the Wing Size Mutations. For that, select "apterous" for the male trait. Leave the female as "wild type." Scroll to the bottom and click on the "Mate Designed Flies" button. What is the phenotype of the offspring F_1 flies? Select a male and female F_1 offspring, and then breed them by clicking on the "Mate Selected Flies" button. What are the phenotypic ratios of the F_2? Do they fit an expected Mendelian ratio? Chose another of the traits and repeat. What phenotypic ratios do you observe?

Investigation 2

Estimated time for completion = 15 minutes

In human genetics studies, scientists cannot set up breeding experiments, for obvious reasons. Instead, they use pedigree analysis, investigating family histories to find existing crosses and inferring the nature of the trait's inheritance. Select the Keyword **HEMOPHILIA** on your CD or Website to view a pedigree of one of the better-documented traits (inheritance of hemophilia in the British royal family). Using this chart as a guide, form a group and use one of your own families, or a famous family, to create a personal pedigree for an inherited human trait. Examples of traits you can use are earlobes that are attached or free, presence or absence of a widow's peak, cleft chin, tongue-rolling, or any other trait that might be inherited in your family.

Investigation 3

Estimated time for completion = 10 minutes

Understanding the inheritance of Mendelian traits is just the beginning of understanding how genes mold our identity. Our appearance, health, even behavior is dictated in large part by our genes. Currently underway is a multi-national project, called the Human Genome Project, which is intended to uncover all of the genes located on all the human chromosomes. Select the Keyword **GENOME** on your CD or Website to view a current map of all the human genes that have been located thus far. Select a chromosome and read as much information as you can about a human disease-causing gene that has been located there. What is the name of the gene, and what disease does it cause? Is there a genetic test available for the gene? Has identifying the gene helped create any new therapies?

Now that you have practiced some of Mendel's laws in the preceding investigations, apply and extend your abilities by addressing some genetics problems.

COMMUNICATE YOUR RESULTS

Exercise 1

Estimated time for completion = 5 minutes

Having worked with Virtual Flylab in *Web Investition 1*, you should be at a point where you can determine the nature of each trait's inheritance. Construct a chart indicating whether the two traits you investigated were dominant or recessive, sex-linked or not. A good indication of sex linkage is when the proportion of males and females of the various types differ. Be prepared to discuss your findings with other students and to justify your conclusions.

Exercise 2

Estimated time for completion = 5 minutes

Your group created a human pedigree chart like the one for hemophilia in *Web Investigation 2*. Use your chart to determine how the trait is inherited. Is it a dominant or recessive trait? Is it an X-linked or an autosomal trait? List all the factors that convinced you, including any specific crosses that were definitive. Compose a list of guidelines for identifying dominant/recessive traits and X-linked or autosomal traits in human pedigrees.

Exercise 3

Estimated time for completion = 5 minutes

Prepare a 50-word description of the gene you investigated in *Web Investigation 3* to inform your classmates about the current state of human genetics research.

12

Chromosomes and Inheritance

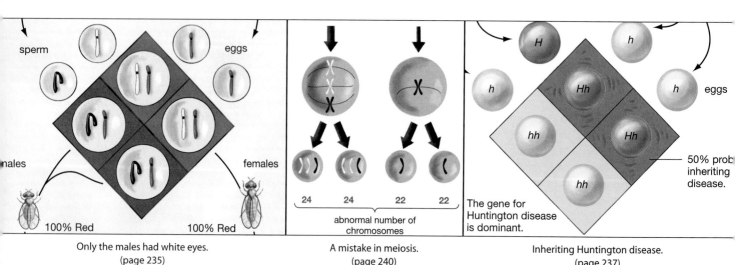

Only the males had white eyes.
(page 235)

A mistake in meiosis.
(page 240)

Inheriting Huntington disease.
(page 237)

The tragically common

condition known as Down syndrome is almost always caused by the inheritance of an extra chromosome. But how can a parent who is healthy pass on a harmful extra chromosome to a son or daughter? The gene that causes sickle-cell anemia seems to also offer some protection against malaria. How can it have both effects? The chromosomal makeup of women offers them a kind of protection against being hemophiliacs. Why should this be so?

From these questions, it may be apparent that chromosomes and the genes on them are centrally involved in issues of human health. You saw last chapter that it was chromosomal activity that lay at the root of several of the genetic principles uncovered by Gregor Mendel. Mendel's "pairs of elements," which separate in reproduction, had as their physical basis the pairs of homologous chromosomes—one inherited from the male, one from the female—that first pair up and then separate from one another in meiosis. The differing traits that Mendel investigated so thoroughly, such as yellow or green peas in a pod, had as their physical basis the pairs of **alleles** or differing forms of a gene that lie on homologous chromosomes.

But there are aspects of chromosomal functioning that Mendel's work scarcely touched on. For example, in Chapter 10 you saw that a critical step in meiosis is the **crossing over** or **recombination** that occurs early in it. In this process, homologous chromosomes first intertwine and then swap pieces of themselves.

Then there is the issue of sex chromosomes. As noted in Chapter 10, human females have 22 pairs of autosomes and then two sex chromosomes—their X chromosomes, which are the chromosomes that make them female. Meanwhile, human males have 22 pairs of autosomes, one X chromosome, and one Y chromosome, the latter of which confers the male sex. But apart from conferring sex, what part do these special chromosomes play in heredity?

Finally, chromosomes in general can be key players in some of the diseases and afflictions that affect human beings, such as Down syndrome.

These issues frame the three broad subjects covered in this chapter: (1) how the work that Mendel began was extended by later researchers, (2) what special role sex chromosomes play in human heredity, and

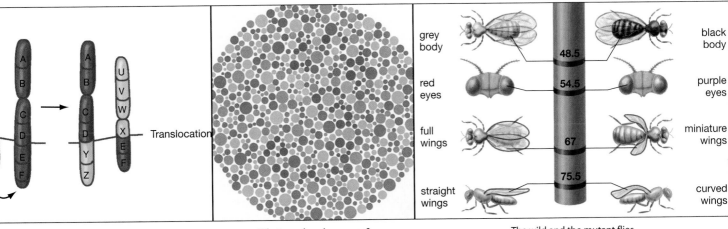

Translocation.
(page 244)

What number do you see?
(page 232)

The wild and the mutant flies.
(page 236)

grey body — black body — 48.5

red eyes — purple eyes — 54.5

full wings — miniature wings — 67

straight wings — curved wings — 75.5

MediaLab Link
Hemophilia, see p. 228.

(3) what human conditions can result from chromosomes that don't function properly. For an account of the work that followed Mendel's research, see "Thomas Hunt Morgan" on page 234. Let us start here with sex and heredity.

12.1 Sex-Linked Inheritance in Humans

Minor cuts are simply an irritation to most of us, but for people suffering from the condition known as hemophilia, such cuts can be life-threatening. Hemophilia is a failure of blood to clot properly. A group of proteins interact to make blood clot, but about 80 percent of hemophiliacs lack a functioning version of just one of these proteins (called Factor VIII). It is *genes* that contain the information for the production of proteins, of course, so at root hemophilia is a genetic disease. In this respect, it turns out to have something in common with two other afflictions: a disease called Duchenne muscular dystrophy, and a far less serious condition, red-green color blindness.

What do all these disorders have in common? They are all known as X-linked conditions, and they all claim more male victims than female. Let's see why.

X Chromosome: Male Vulnerability, Female Protection

It is the chromosomal makeup of males that makes them more susceptible to these afflictions. Remember that men have but a single X chromosome, while women have two X chromosomes. As it happens, genes for blood clotting, color vision, and Duchenne muscular dystrophy *lie* on the X chromosome. Thus, a male will be red-green color blind if only one of his chromosomes—his lone X chromosome—carries a nonfunctional allele for it. A female, meanwhile, can carry one nonfunctional allele for color vision, but be protected by a "good" allele along her *second* X chromosome.

How Color Vision Works

How does such protection work? Think, once again, of genes as entities that code for proteins. In the case of color vision, within the eye there are molecules called pigments, which absorb different colors of light. In humans, proteins make these pigments. A gene that codes for blue pigment lies on chromosome 3, while the genes for both red and green pigments lie very close to one another on the X chromosome. A red-green color-blind person, then, is one in whom these X-located genes fail to code for the proper proteins. The result, in the most severe cases, is a person who cannot distinguish red from green (**Figure 12.1**).

Color Blindness as a Recessive Condition

All it takes, however, is one set of functioning red and green alleles for a person *not* to be color blind. Red-green color blindness is thus a **recessive condition**: It will not exist in the presence of functional alleles. Given this, think about the interesting way that color blindness is passed along. Say there is a mother who is not color blind herself, but

Figure 12.1
Typical Test for Red-Green Color Blindness
If you do not see a number, you may have this recessive trait.

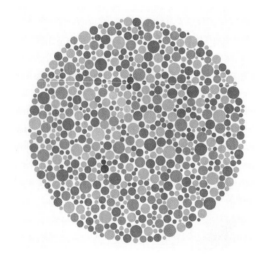

Figure 12.2
The Value of Having Two X Chromosomes
The mother in the figure is not herself color blind but has one nonfunctional set of red-green alleles, which she passes on to one son and one daughter. The daughter is not color blind because she has inherited a second X chromosome from her father—one that has functional red-green alleles. Meanwhile the son is color blind because his only X chromosome is the flawed one he inherited from his mother. (The other son, shown in the square's leftmost cell, is not color blind because he has inherited his mother's functional alleles.)

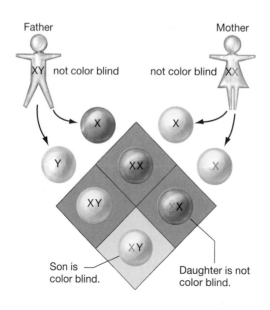

Father — XY not color blind

Mother — not color blind XX

X

X

Y

XX

X

XY

X

XY

Son is color blind.

Daughter is not color blind.

who is heterozygous for the trait. That is, she has functional color alleles on one of her X chromosomes but nonfunctional alleles on her other X chromosome. Should her son happen to inherit this second X chromosome, he will be color blind, because the X chromosome he got from his mother is his *only* X chromosome (**Figure 12.2**). Meanwhile, a daughter who inherited this chromosome would likely be protected by her second X chromosome—the one she got from her father. Not surprisingly, then, more males are color blind than females: About 8 percent of the male population has some degree of color blindness, while for females the figure is about half a percent.

12.2 Autosomal Genetic Disorders

X and Y chromosomes are, of course, only two of the chromosomes in the human collection, and any chromosome can have a malfunctioning gene on it. The chromosomes other than the X and Y chromosomes are called *autosomes*, as you've seen. A recessive dysfunction related to an autosome is thus known as an **autosomal recessive disorder.** (For a list of all the disorders you'll be looking at and more, see **Table 12.1**.)

Sickle-Cell Anemia

A well-known example of an autosomal recessive disorder is sickle-cell anemia, which affects populations derived from several areas on the

Table 12.1
Selected Examples of Human Genetic Disorders

Type	Name of Condition	Effects
X-linked recessive disorders	Hemophilia	Faulty blood clotting
	Duchenne muscular dystrophy	Wasting of muscles
	Red-green color blindness	Inability to distinguish red from green
Autosomal recessive disorders	Albinism	No pigmentation in skin
	Sickle-cell anemia	Decreased oxygen to brain and muscles
	Cystic fibrosis	Impaired lung function, lung infections
	Phenylketonuria	Mental retardation
	Tay-Sachs disease	Nervous system degeneration in infants
	Werner syndrome	Premature aging
Autosomal dominant disorders	Polydactyly	Extra fingers or toes
	Campodactyly	Inability to straighten little finger
	Huntington disease	Brain tissue degeneration
Aberrations in chromosome number	Down syndrome	Mental retardation, shortened life span
	Turner syndrome	Sterility, short stature
	Kleinfelter syndrome	Dysfunctional testicles, feminized features
Aberrations in chromosome structure	Cri-du-chat syndrome	Mental retardation, malformed larynx
	Fragile-X syndrome	Mental retardation, facial deformities

Thomas Hunt Morgan: Using Fruit Flies to Look More Deeply into Genetics

Gregor Mendel used pea plants in his work, but generations of researchers after him have employed a common fruit fly, *Drosophila melanogaster*, in trying to unravel the secrets of genetics. One of the earliest *Drosophila* workers was a Kentuckian named Thomas Hunt Morgan, who greatly deepened our understanding of heredity with work he began about 1908 in his lab at Columbia University.

The tiny *Drosophila* fly has a lot going for it as an experimental subject, including a short generation time. Whereas Mendel got one generation of pea plants a year, Morgan got one generation of *Drosophila* every 12 days or so. In practical terms, this meant that Morgan's group had to wait at most a month to see the results of their experiments.

Lessons from a Mutation: White-Eyed Flies

Morgan's path to discovery began with a chance event. One day he looked into one of the empty milk bottles in which he kept his flies and saw something strange: Among numerous normal or "wild-type" *Drosophila* with red eyes was a single male with white eyes (**see Figure 1**). Morgan correctly assumed that the variation he saw had come about because of a spontaneous genetic change or "mutation" in the fly. He then crossed his white-eyed male with females that he knew bred true for red eyes, and got all red-eyed flies in the F_1 generation. No surprise there; it simply looked as though red eyes were the same type of dominant trait that Mendel had seen in, for example, his yellow peas. Indeed, when Morgan went on to breed the F_1 generation—those that had mixed red and white alleles—he got a 3:1 red-to-white ratio in the F_2 generation. The strange thing was, however, that every white-eyed fly was a *male*. Why should that be?

As it turns out, sex chromosomes exist in *Drosophila* as they do in human beings: Females have two X chromosomes, while males have one X and one Y. Doing further experiments on these flies, Morgan eventually decided that the gene for eye color had to lie on a *particular* chromosome: the X chromosome. Why? Any fly that was white-eyed had to have only white-eyed alleles (because white was recessive). Normally, of course, this would mean *two* white-eyed alleles, but there was an exception to this: Males could be white-eyed if they had but a single white allele—because they have only a *single X chromosome*. You can look at **Figure 2** to see how this plays out schematically. The importance of Morgan's insight was not that he had grasped something about eye color in the fly. It was that he had linked a particular trait to a particular chromosome, which was a first.

a

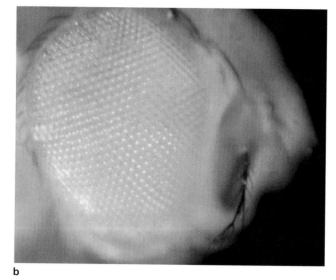
b

Figure 1
a The Compound Eye of a Red-Eyed *Drosophila*
b The Compound Eye of a White-Eyed *Drosophila*

When Traits Travel Together: Linkage in Genetic Transmission

Continuing with *Drosophila* breeding work, Morgan and his colleagues soon were finding all kinds of mutations in the flies. There was, for example, a mutation for "miniature" wings, as opposed to full ones. This trait followed the rules described earlier for the white-eyed mutation and accordingly was deemed to be an X-linked characteristic. Not surprisingly, the white-eyed and miniature-winged mutation tended to be transmitted *together* when

> *The importance of Morgan's insight was not that he had grasped something about eye color in the fly. It was that he had linked a particular trait to a particular chromosome, which was a first.*

both of these mutations were bred for, because the genes for both of them lay on the same chromosome. Morgan and his colleagues eventually were to find, however, not just one but several clusters of traits that tended to be transmitted together, thus giving rise to the concept of linked transmission in genetics.

Linked Genes Can Separate: Crossing Over

The question was, though, did genes on the same chromosome *always* travel together? Surprisingly, the answer was no. For example, miniature wings would usually be passed on together with white eyes, but not always. Sometimes one trait would appear in an offspring but the other would not. Because genes for these traits were on one chromosome, how did they get separated from one another?

A paper published in Europe in 1909 gave Morgan a clue: Viewing cells under a microscope, F. A. Janssens had observed chromosomes entwining with each other during an early phase of meiosis. Given this and his own data, Morgan made a creative leap: He suggested that chromo-

(continued)

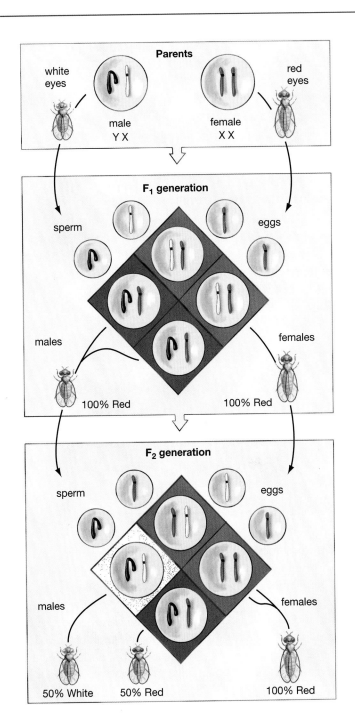

Figure 2
Why All Morgan's White-Eyed F$_2$s Were Males
The F$_2$ female flies that inherited one white-eyed X-chromosome allele also inherited a second *red-eyed* X allele, making them red-eyed. However, the *males* inheriting the white-eyed allele had no second X chromosome. Instead, they inherited a Y chromosome—which has no genes on it for eye color—leaving them white-eyed.

Thomas Hunt Morgan: Using Fruit Flies to Look More Deeply into Genetics *(continued)*

somes can swap parts with each other during their meiotic intertwining. Sound at all familiar? Remember, in the Chapter 10 review of meiosis the discussion of the phenomenon known as **crossing over** or **recombination**? Here, with Morgan, is the insight that led to this now-established tenet of science. To see how crossing over makes a difference in inheritance, look at Figure 10.2b on page 194.

Making the First Chromosomal Maps

Morgan's group went on to correctly infer that the closer two genes lay on a chromosome, the less likely they were to separate during crossing over. (Imagine having a stick with two lines on it; the closer the two lines are, the less likely they would be to separate should the stick be broken.) By tracking the *frequency* with which genes separated, they were able to begin making the first maps of

---

> *Morgan's group went on to correctly infer that the closer two genes lay on a chromosome, the less likely they were to separate during crossing over.*

---

chromosomes. You can see one of these maps in **Figure 3**, showing the sequence of some of the genes on *Drosophila*'s chromosome II. The distance between each gene, measured in "map units," does not show the *physical* distance between the genes along the chromosome, but rather their relative distance from one another, as indicated by the frequency of their separation in meiosis. Different kinds of gene mapping remain a staple of genetics to this day.

In his work, Morgan and his colleagues managed to confirm the essence of Mendel's theories of inheritance, even as they ventured into areas that lay beyond Mendel's principles, such as gene linkage and crossing over. Appropriately enough, Morgan's efforts won him the Nobel prize in 1934.

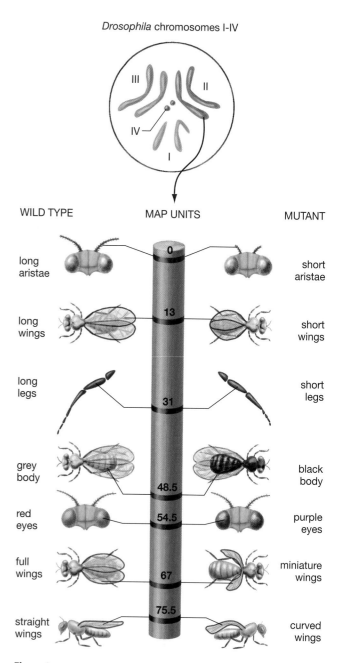

Figure 3
Partial Map of Chromosome II in *Drosophila*
The "map units" mark off not physical distance but rather relative separation of the genes, calibrated by the frequency with which they separated from one another in meiosis.

globe, including Africa. In the United States it is, of course, most widely known as a disease affecting African Americans. The "sickle" in the name comes from the curved shape that is taken on by the red blood cells of its victims. Red blood cells carry oxygen to all parts of the body; in their normal shape they look a little like doughnuts with incomplete holes. When they take on a sickle shape, however (**Figure 12.3**), red cells clog up capillaries, thus resulting in decreased oxygen supplies to brain and muscle, a condition that is usually fatal in the long run.

The question is, what causes a red blood cell to take on this lethal, sickled form? There is a protein, called *hemoglobin*, that carries the oxygen within red blood cells. The vast majority of people in the world have one form of hemoglobin, called *hemoglobin A*; but sickle-cell anemia sufferers have another form of this protein, *hemoglobin S*, which coalesces into crystals that distort the cell.

The pattern of inheritance for sickle-cell anemia is very simple. A person with one allele producing hemoglobin S does not suffer the condition to any serious extent, but is a **carrier** for it—someone who can pass it on (as with the mother who is heterozygous for color blindness). Both mother and father must be at least single-allele carriers—both must be heterozygous for the condition, in other words—in order for an offspring to inherit *two* hemoglobin S alleles and thus suffer from the disease.

Figure 12.3
Healthy Cells, Sickled Cells

a Normal red blood cells.

b Sickled red blood cells indicative of sickle-cell anemia.

a

b

Thus, the laws governing inheritance of this disorder are simply those of Mendel's monohybrid cross. You can look at the Punnett square in **Figure 12.4a** to see how this works out.

a

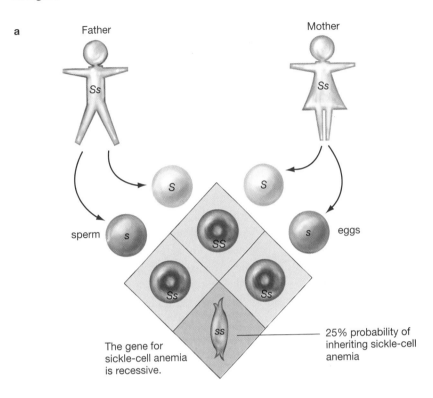

The gene for sickle-cell anemia is recessive.

25% probability of inheriting sickle-cell anemia

b

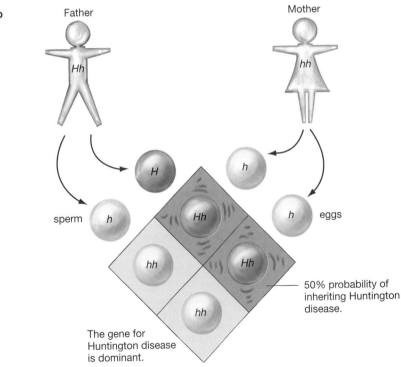

The gene for Huntington disease is dominant.

50% probability of inheriting Huntington disease.

Figure 12.4

a Sickle-cell anemia is a recessive autosomal disorder; both the mother and father must carry at least one allele for the trait in order for a son or daughter to be a sickle-cell victim. When this is the case, there is a 25 percent chance that any given offspring will inherit the condition.

b Conversely, in Huntington disease, if only a single parent has a Huntington allele there is a 50 percent chance that a son or daughter will inherit the condition.

Malaria Protection: Hemoglobin S as a Useful Protein

Sickle-cell anemia offers another lesson in connection with the notion of hemoglobin S as a "faulty" protein. It is in fact quite a functional protein in certain circumstances; namely, when it appears heterozygously (with hemoglobin A) in people who live in geographical regions where malaria is common. The severest form of malaria is caused by a single-celled parasite that is transmitted into human beings by the *Anopheles* mosquito. Traveling through the bloodstream, these parasites invade red blood cells, in the process destroying them. For reasons we still don't understand, having hemoglobin S in one's system makes red blood cells resistant to invasion. Thus, it is likely that hemoglobin S became as widespread as it is because of its value in resisting malaria. The price of this resistance, however, was that some offspring would be not heterozygous for hemoglobin S, but would be homozygous for it, meaning they would have sickle-cell anemia.

Dominant Disorders

Though sickle-cell anemia is an autosomal disorder and red-green color blindness a sex-linked disorder, what both have in common is that they are recessive conditions: A person with even a single properly functioning allele will not suffer from them. However, there are also **dominant disorders** in which a single faulty allele can *cause* damage, even when a second, "good" allele exists. There is, for example, an **autosomal dominant** condition called Huntington disease, affecting about 30,000 Americans, that results in both mental impairment and uncontrollable spastic movements called *chorea*. Its most famous victim was perhaps the American folksinger Woody Guthrie, who, like most Huntington sufferers, did not begin to show symptoms of the disease until well into adulthood—meaning *after* the time he had children, who then may have had the disease passed along to them. You can look at the Punnett square in **Figure 12.4b** to see how the inheritance pattern of an autosomal dominant disease, such as Huntington, differs from that of an autosomal recessive illness. Because all a parent has to do is pass on a *single* Huntington allele for a son or daughter to suffer from the condition, the chances of

MediaLab Link
Inherited traits, pedigrees, see p. 228.

any given offspring getting the disease from a single affected parent are one out of two.

Tracking Traits with Pedigrees

Confronted with a given medical condition that is running through a family, or a larger community, scientists sometimes find it helpful to construct a kind of family tree of the condition, creating a diagram called a **pedigree**. Such a diagram does not just give a family history of the disease, however: it can be used to ascertain whether the condition is dominant or recessive, which can help establish probabilities for *future* inheritance of the condition, something that can be very helpful for couples thinking of having a child.

If you look at **Figure 12.5**, you can see a simple pedigree for albinism, a lack of skin pigmentation, which is known to be an autosomal recessive condition. In the figure, you can see some of the standard symbols used in pedigrees. A circle is used for a female and a square for a male. Parents are indicated by a horizontal line connecting a male and a female, while a vertical line between the parents leads to a lower row that denotes the parents' offspring. This second row—a horizontal line of siblings—has an order to it: oldest child on the left, youngest on the right. A circle or square that is filled in indicates a family member who has the condition (in this case albinism), while a symbol that is half filled in indicates a person known to be heterozygous for a recessive condition. One of the strengths of a pedigree is that it can sometimes tell researchers which persons are heterozygous carriers of a recessive condition. Remember that a carrier does not display *symptoms* of the condition. So how can a pedigree reveal this?

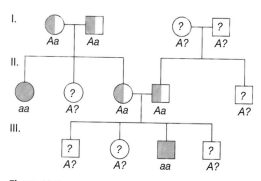

Figure 12.5
A Hypothetical Pedigree for Albinism through Three Generations

Take a look at Figure 12.5. The only thing that would be apparent from looking at any of the people in the pedigree is that two of them—a female in generation II and a male in generation III—had the condition of albinism. But a little knowledge of Mendelian genetics also allows some other deductions. If the condition had been dominant, then it would have manifested itself in at least one of the parents on the left in generation I. Because this was not the case, it is fair to deduce that this is a recessive condition. The fact that it is recessive, however, means that *both* parents in generation I had to be carriers for the allele—both had to be heterozygous for the condition, which is why their symbols can be half shaded in. Things are less clear with the parents on the right in generation I. One of their sons did not manifest the condition himself, but went on to have a son who did. From this, we know that the son in generation II had to have been heterozygous for the condition. But from which parent did this son get his albinism allele? Either, or both, of his parents in generation I could have been heterozygous for the condition and yet between them only have passed along a single albinism allele to him.

This mixture of certainty and uncertainty leads to the genotype labeling you see in the figure, with *A* representing the dominant "normal" allele and *a* representing the recessive albinism allele. We know, for example, that both parents on the left in generation I had to be *Aa*, but all we can say for sure about the parents on the right in the same generation is that each of them had to have at least one *A* allele.

12.3 Aberrations in Chromosomal Sets: Polyploidy

All of the maladies you've looked at so far have resulted from dysfunctional *genes* that exist among a standard array of chromosomes. In humans this array is our 22 pairs of autosomes and either an XX combination (in females) or an XY (in males). Meanwhile, Mendel's peas had 7 pairs of chromosomes, while the *Drosophila melanogaster* fly has 4 pairs. Whatever the *number* of chromosomes, note the similarity in all these species: Their chromosomes come in pairs, or to put it another way, they all have two *sets* of chromosomes (meaning they are **diploid**).

There are many situations, however, in which organisms don't end up with the two sets of chromosomes that are standard for their species. The state of having more than two sets of chromosomes is called **polyploidy**.

Valuable in Plants, a Disaster in Humans

Polyploidy is a calamity for human beings, but in plants it can result in perfectly robust organisms—so much so that plant polyploidy has come about countless times through evolution. You will look more closely at its effects during the review of evolution in Chapter 18 (see pages 352–353.) For now, simply note that changes in chromosome set number are not necessarily a bad thing in all species. For human beings, however, polyploidy is such an unmitigated disaster that perhaps 1 percent of human embryos with the condition will survive to birth—and none of these babies lives long. Concerns about chromosome number in living persons, then, center not on addition or deletion of whole sets of chromosomes, but rather on the gain or loss of *individual* chromosomes.

12.4 Incorrect Chromosome Number: Aneuploidy

The condition called **aneuploidy** is one in which an organism has either more or fewer chromosomes than normally exist in a full set for that organism. So common is aneuploidy in humans that at least 5 percent of all human pregnancies are thought to be affected by it; of the aneuploid embryos, the vast majority will spontaneously abort. It is possible, however, for embryos with the wrong number of chromosomes to become fetuses and then babies. It is possible, in other words, for human beings to end up with the wrong number of chromosomes in their diploid set, meaning the wrong number of chromosomes in every cell in their bodies.

A Common Cause of Aneuploidy: Nondisjunction

A common cause of human aneuploidy is a phenomenon known as **nondisjunction**, which simply means a failure of homologous chromosomes or sister chromatids to separate during meiosis, the result being that some daughter cells can end up with one chromosome too many, while others can end up with one chromosome too few.

Looking at **Figure 12.6**, you can see that nondisjunction can occur either in meiosis I or in meiosis II. In meiosis I, two homologous chromosomes, on the opposite side of the metaphase plate, can be pulled to the *same* side of a dividing cell, producing daughter cells with imbalanced numbers of chromosomes. Conversely, nondisjunction can take place in meiosis II, by means of sister chromatids going to the same daughter cell after failing to "disjoin" (hence the cumbersome term *nondisjunction*). Such actions then produce an egg or sperm that has 24 or 22 chromosomes instead of the standard haploid number of 23; after union with a normal egg or sperm, this results in a zygote that has either 47 or 45 chromosomes, instead of the standard 46.

Down Syndrome

It is the gain of an autosome that brings about the most commonly noticed form of aneuploidy in human beings. **Down syndrome** is, in some 95 percent of cases, a condition in which victims have *three* copies of chromosome 21, rather than the standard two. What usually causes this is nondisjunction in egg formation: Female eggs develop in such a way that they contain one too many copies of chromosome 21. However, about 10 percent of the nondisjunctions that lead to Down syndrome take place in the development of sperm. Down syndrome is seen in about 0.1 percent of all live births, and results in an array of effects: smallish, oval heads, IQs that are well below normal, infertility in males, short stature and reduced life span in both sexes. The condition is all the more heartbreaking because Down syndrome children seem to be extraordinarily warm and loving.

Since the late 1960s, it has been possible to test developing fetuses for Down syndrome. Today, several methods are commonly employed, as you can see in "Testing for Genetic Trouble" on page 242. It is well known that when women pass the age of about 35, their risk of giving birth to a Down syndrome child increases dramatically. It seems not as well known that even at maternal age 40, the odds of conceiving such a child are less than 1 in 100 (**Figure 12.7**). Scientists are not certain why the mother's age should figure so prominently in Down syndrome. One hypothesis is that older mothers are more likely than younger mothers to carry a Down syndrome fetus to term rather than spontaneously aborting it. We also

Figure 12.6
A Mistake in Meiosis Brings About an Abnormal Chromosome Count
Nondisjunction can occur either in meiosis I or meiosis II, when either chromosomes or chromatids fail to separate properly. When this occurs in meiosis I, 100 percent of the resulting gametes will be dysfunctional; when it takes place in meiosis II, only 50 percent will be dysfunctional.

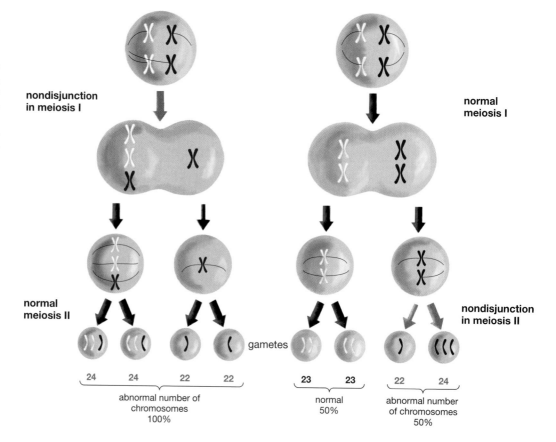

know that as a woman ages, her developing eggs are more likely to make "mistakes" in meiosis, such mistakes being, as you've seen, the usual root cause of Down syndrome.

Abnormal Numbers of Sex Chromosomes

There are also numerous kinds of *sex* chromosome additions or deletions that living persons can have, though these usually result in deformities or debilitations. An example of such a condition is Turner syndrome, which produces people who are phenotypically female, but who have only one X chromosome. Such women have, then, only 45 chromosomes, rather than the usual 46. Their state is sometimes referred to as XO, the "O" signifying the missing X chromosome. The absence of a second sex chromosome causes a range of afflictions: Turner syndrome victims' ovaries don't develop properly (which causes sterility), and they are generally short and retarded and often have brown spots (called *nevi*) over their bodies.

Turner syndrome results from a loss of a chromosome, but all manner of sex chromosome *additions* can also take place. There are, for example, XXY men who, while phenotypically male in most respects, tend to have a number of feminine features: some breast development, a more feminine figure, and lack of facial hair. When coupled with other characteristics (such as tall stature and dysfunctional testicles), the result is a condition called Klinefelter syndrome.

12.5   Structural Aberrations in Chromosomes

Aberrations can occur *within* a given chromosome, sometimes resulting from interactions *between* chromosomes. A frequent cause of such change is that pieces of chromosomes can break off from the main chromosomal body. Such a chromosomal fragment may then be lost to further genetic activity, or it may rejoin a chromosome—either the one it came from or another—often to harmful effect. These structural aberrations may come about spontaneously, but they may also be caused by exposure to such factors as radiation, viruses, and chemicals.

Deletions

A chromosomal **deletion** occurs when a chromosome fragment breaks off and then does not rejoin any chromosome. Such an event might take place during meiosis in a parent whose own somatic cell chromosomes were perfectly normal. In such an instance, a healthy parent could pass along a complement of 23 chromosomes, one chromosome of which would be missing a segment. If the deleted piece were large enough, the zygote would likely not survive. You can see how critical even a portion of a chromosome is by looking at children who suffer from a rare condition, called *cri-du-chat* ("cry of the cat") syndrome, that results from the deletion of the far end of the "short arm" of chromosome 5. If you look at **Figure 12.8**, you can see a

Figure 12.7
Down Syndrome: Increasing Risk with Age

a An adolescent girl with Down syndrome works with her teacher.

b The risk of giving birth to a Down syndrome child increases dramatically past maternal age 35.

b

Mother's Age	Chance of Giving Birth to a Child with Down Syndrome
20	1 in 1925
25	1 in 1205
30	1 in 885
35	1 in 365
40	1 in 110
45	1 in 32

a

b

a

Figure 12.8
Cri-du-Chat Syndrome

a A five-year-old boy affected by cri-du-chat syndrome. Note the small head and low-set ears.

b Karyotype of a person with cri-du-chat syndrome.

Testing for Genetic Trouble

"Is the baby OK?" In a life full of questions, it's possible that none carries more weight than this one. In most cases the answer will be reassuring, but in some it will be devastating. Small wonder, then, that scientists constantly are trying to perfect ways to know in advance about whether a newborn will be healthy. A new fetal testing technique now in development holds out the promise of being as effective as current techniques, but much safer to boot.

You have seen that a large number of debilitating human conditions have genetic causes. Further, you know that nearly every cell in a person's body contains a complete copy of that person's genome, or set of genes. Therefore, to check on someone's genetic well-being, all that is necessary is to have access to a small collection of that person's cells. And such cells exist, of course, in fetuses as well as in newborn babies and adults. Thus, it's not hard to see what the procedure would be for prenatal genetic testing: gather fetal cells and examine their DNA and chromosomes.

Given that such testing is now commonplace, it's worth pointing out that the first general genetic screening method, **amniocentesis**, became available only in the late 1960s. Indeed, it was only in 1956 that scientists finally arrived at the correct *number* of human chromosomes. This is obviously a piece of information that had to be in place before a technique like amniocentesis could perform one of its main tasks: checking for an *improper* number of chromosomes in a developing fetus.

If you look at **Figure 1**, you can see how amniocentesis is carried out. A physician, using an ultrasound image of mother and fetus to "steer," inserts a needle through the mother's abdomen and into the amniotic fluid in which the fetus is suspended. The physician then draws up a small amount of amniotic fluid, which contains epidermal (skin) cells that the fetus has sloughed off as it has developed. Some tests are conducted right away on the fluid itself; other tests require isolated fetal cells, which are needed in relatively large quantity. To get the latter, technicians put the amniotic fluid into test tubes that are then centrifuged, which separates the cells from the fluid around them. The cells are then cultured, or grown to large numbers in the laboratory, after which their genetic material is examined. DNA itself can be tested for abnormalities, while the number of chromosomes the fetus has can be determined by creation of the **karyotype**, or picture of the set of fetal chromosomes.

An alternative to amniocentesis is **chorionic villus sampling** (CVS), which was introduced in the early 1980s.

In it, cells derived from the embryo are suctioned from the "villi," or extensions of the chorionic membrane that surrounds the placenta.

Though amniocentesis and CVS are powerful diagnostic tools, they are not everything we might hope for in a prenatal test. For one thing, there is the question of timing. Traditional amniocentesis is done 14 to 16 weeks into a pregnancy. Once the procedure is completed, it then

---

CVS is, along with amniocentesis, an invasive procedure: It involves inserting medical instruments into the delicate surroundings of the fetus.

---

takes an additional 10 to 14 days to culture the cells for DNA testing and karyotyping. As a result, more than four months of a nine-month pregnancy may elapse before parents can learn about the health of their unborn. The major advantage of CVS over amniocentesis is that it is faster: It usually is performed between 9 and 12 weeks after conception and most of its results are available within 24 hours.

Despite this advantage CVS is, along with amniocentesis, an invasive procedure: It involves inserting medical instruments into the delicate surroundings of the fetus. In the United States, the risk of fetal loss through CVS is about 1 percent of procedures performed, while the figure for amniocentesis is about half that. Beyond this, both amniocentesis and CVS are expensive, in large part because they are labor-intensive.

Given these limitations, amniocentesis and CVS are not routine screening procedures. In general, they are performed only when some genetic risk factors exist in a pregnancy, the most common of these being a mother who is past the age of 35. Yet babies with genetic afflictions are most often born to couples who were thought to be at low risk: Testing for Down syndrome only among older women reveals fewer than a quarter of the cases that exist, simply because most babies are born to younger women. What's clearly needed is a genetic test that *can* become routine—one that is inexpensive, quick, noninvasive, and that can be performed early in the pregnancy. Scientists are now hopeful that they may have hit on a technique that addresses all these issues.

This technique is new enough that it doesn't yet have a short name (like amniocentesis). Rather, it is referred to by the unwieldy title of *fetal cells derived from maternal circulation*. Its starting point is the now-established fact that fetal cells are present in small numbers in the *mother*'s bloodstream fairly early in pregnancy. The possibility this has raised is of drawing a few ounces of blood from a mother's arm, isolating fetal cells from the blood, and then testing the genetic material in the cells for defects.

At present, one question about the technique is whether it can reliably yield the desired fetal cells from among an ocean of maternal cells in a cost-effective way; in 6 fluid ounces of blood, there may be 100 billion maternal red blood cells but only 100 to 1,000 fetal red blood cells. Nevertheless, it may be that analyzing fetal cells from maternal circulation will be the thing that transforms genetic testing from a specialized procedure into one that is routine.

Figure 1
Two Means of Fetal Genetic Testing and Their Results
Amniocentesis and chorionic villus sampling compared.

karyotype—a visual image of a chromosome set—that shows you this deletion. Children born with the cri-du-chat condition exhibit a host of maladies, among them mental retardation and an improperly constructed larynx that early in life produces sounds akin to those of a cat. Some cri-du-chat victims, however, learn to communicate verbally and to care for themselves to some extent.

Inversions and Translocations

When a chromosome fragment rejoins the chromosome it came from, it may do so with its orientation "flipped," so that the fragment's chemical sequence is out of order—an **inversion** (see Figure 12.9). A **translocation** can occur when two chromosomes that are *not* homologous exchange pieces, leaving both with improper gene sequences, which can have phenotypic effects.

Duplications

There is, of course, a perfectly normal process of chromosomal part swapping—crossing over—but consider what might happen if two homologous chromosomes exchanged *unequal* pieces of themselves in crossing over: One would lose genetic material (effectively a deletion) while the other would gain it; because these are homologous (meaning paired) chromosomes, the fragment added to the latter chromosome would *duplicate* some of the material it already has. Duplications can take place in several different ways, and they are not always harmful. Nevertheless, some of our most devastating diseases are caused by genetic duplications. The Huntington disease noted earlier in this chapter has as its root cause an abnormal number of "repeats" of given genetic sequences.

On to DNA

For several chapters, you've been looking at genetics from the point of view of the genetic packages called *chromosomes*. Important as chromosomes are, it's been apparent that real genetic control lies with the units that help make up chromosomes: the sequences of A's, T's, C's and G's we call *genes*. As you've seen, it is genes that make pea plant flowers purple or white—and that give a person Huntington disease or not. It's time to explore just what these genes are and how they operate.

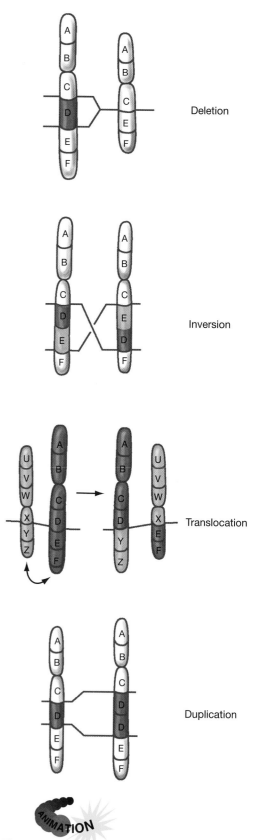

Figure 12.9
Chromosomal Structural Changes

Chapter Review

Summary

12.1 Sex-Linked Inheritance in Humans

- Certain human conditions, such as hemophilia and Duchenne muscular dystrophy, are called **X-linked conditions**, because they stem from nonfunctional genes located on the X chromosome. Men are more likely than women to suffer from these conditions because men have only a single X chromosome. A woman with, for example, a nonfunctional blood-clotting allele on one of her X chromosomes often will be protected from hemophilia by a functional allele that lies on her second X chromosome.

- Hemophilia and Duchenne muscular dystrophy are examples of recessive genetic conditions, meaning conditions that will not exist in the presence of functional alleles.

- Given the nature of recessive genetic conditions, persons can carry one allele for a condition and yet not suffer from the condition themselves. Such persons, referred to as **carriers** of the trait, are heterozygous for it—the alleles they have for the trait differ, one of them being functional, the other being nonfunctional.

12.2 Autosomal Genetic Disorders

- Sickle-cell anemia is an example of a recessive autosomal disorder. It is autosomal in that the genetic defect that brings it about does not involve either the X or Y chromosome. It is recessive in that persons must be homozygous for the sickle-cell allele to suffer from the condition—they must have two alleles that code for the same sickle-cell hemoglobin protein. This protein offers some protection against malaria, however, so that persons who are heterozygous for hemoglobin proteins—who have different alleles coding for different proteins—do not suffer from sickle-cell anemia and are afforded some protection against malaria.

- Some genetic disorders are referred to as **dominant disorders**, meaning those in which a single allele can bring about the condition, regardless of whether a person also has a normal allele.

12.3 Aberrations in Chromosomal Sets: Polyploidy

- Human beings and many other species have diploid or paired sets of chromosomes. In human beings, this means 46 chromosomes in all: 22 pairs of autosomes and either an XX chromosome pair (for females) or an XY pair (for males). The state of having more than two sets of chromosomes is called **polyploidy**. Many plants are polyploid, but the condition cannot be tolerated by human beings.

12.4 Incorrect Chromosome Number: Aneuploidy

- Aneuploidy is a condition in which an organism has either more or fewer chromosomes than normally exist in a full set for that organism. One example of aneuploidy in human beings is Down syndrome, whose victims generally have a third copy of chromosome 21.

- A common cause of aneuploidy is nondisjunction, in which homologous chromosomes or sister chromatids fail to separate correctly in meiosis, leading to eggs or sperm that have one too many or one too few chromosomes.

12.5 Structural Aberrations in Chromosomes

- Aberrations can occur within chromosomes, including deletions, inversions and translocations, and sequence duplications.

Key Terms

allele 231

amniocentesis 242

aneuploidy 239

autosomal dominant 238

autosomal recessive disorder 233

carrier 237

chorionic villus sampling 242

crossing over 231

deletion 241

diploid 239

dominant disorder 238

Down syndrome 240

inversion 244

karyotype 242

nondisjunction 239

pedigree 238

polyploidy 239

recessive condition 232

recombination 231

translocation 244

Understanding the Basics

Multiple-Choice Questions

1. Examples of sex-linked diseases include
   a. Sickle-cell anemia, red-green color blindness, and polyploidy
   b. Duchenne muscular dystrophy, Down syndrome, and red-green color blindness.
   c. Hemophilia, red-green color blindness, and Duchenne muscular dystrophy.
   d. Huntington disease, hemophilia, and Klinefelter syndrome.
   e. Down syndrome, cri-du-chat, and hemophilia

2. Sickle-cell anemia patients have a different form of _____, which leads to _____.
   a. hemoglobin, increased muscle activity
   b. a sex chromosome, sterility
   c. pigment, color blindness
   d. skin cell, nevi
   e. hemoglobin, decreased oxygen transport

3. The individuals who are most protected against malaria are those who
   a. are heterozygous for sickle-cell anemia (one hemoglobin A, one hemoglobin S)
   b. are homozygous for sickle-cell anemia (two hemoglobin S)
   c. are hemophiliacs
   d. are homozygous for hemoglobin A
   e. both a and c

4. Inheritance of a dominant autosomal disorder differs from inheritance of an autosomal recessive disorder in that:
   a. A dominant disorder may be passed on only if both parents are affected.
   b. A dominant disorder is evident only if the offspring is homozygous for the allele.
   c. A dominant disorder is more often seen in females.
   d. A dominant disorder may be passed on even if only one parent is affected.
   e. both b and d

5. An individual having 44 autosomes and one X chromosome would be classified as:
   a. polyploid   d. Turner
   b. aneuploid   e. both b and d
   c. Klinefelter

6. A human embryo with 69 chromosomes would
   a. die in the womb or shortly after birth
   b. be considered polyploid
   c. have fewer problems than a plant with the same condition
   d. both a and b
   e. all of the above

7. Which of the following statements is *not* true regarding Down syndrome?
   a. Affected individuals usually have three copies of chromosome 21.
   b. The condition is often caused by non-disjunction in egg formation.
   c. A woman over 35 increases her chance of having a Down syndrome child to 1 out of every 40 births.
   d. Approximately 0.1 percent of all live births are children with Down syndrome.
   e. The physical characteristics associated with the disorder include mental retardation, male infertility, short stature and reduced life span.

8. Inherited non-disjunction can occur in
   a. meiosis I or mitosis
   b. mitosis or meiosis II
   c. only meiosis II
   d. only meiosis I
   e. either meiosis I or II

9. A defect in crossing over may lead to which of the following?
   a. polyploidy
   b. duplication of chromosome region
   c. translocation of chromosome region
   d. b and c
   e. a and b

10. Which of the following diseases is not evident until well into adulthood?
    a. Huntington disease
    b. Down syndrome
    c. cri-du-chat
    d. Turner syndrome
    e. none of the above

Brief Review

1. For the diseases listed below, indicate the mode of inheritance, using the following list: autosomal recessive, X-linked, chromosomal deletion, aneuploidy, autosomal dominant, polyploidy.
   a. Huntington disease
   b. cri-du-chat
   c. Turner syndrome
   d. Down syndrome
   e. red-green color blindness
   f. sickle-cell anemia
   g. Klinefelter syndrome
   h. hemophilia
   i. Duchenne muscular dystrophy

2. In sex-linked diseases, which gender is more frequently affected: male or female? Why?

3. Why can people with red-green color blindness still see other colors, such as blue?

4. Compare and contrast the four chromosome structural aberrations discussed in this chapter: deletions, inversions, translocations, and duplications.

5. Given the following situations, answer accordingly:
   a. A human cell with 47 chromosomes has probably undergone _____.
   b. A cell duplicates DNA, but fails to undergo cytokinesis. This would lead to _____.

6. Why would a person with Klinefelter syndrome exhibit both male and female features?

7. Certain genotypes may be advantageous in specific situations. Give an example of a heterozygous genotype that is advantageous over either homozygous genotype. (Hint: especially in mosquito-infested areas.) Why is this the case?

Applying Your Knowledge

1. A red-green color-blind man marries a woman who is a carrier for red-green color blindness. (a) What percentage of their sons could be expected to be color blind? (b) What percentage of their daughters?

2. A couple has a son with Duchenne muscular dystrophy. It is known that external environment and exposure to harsh chemicals may lead to birth defects. The child's father works at a local chemical plant, and suspecting it played a part in the condition of his son, sues the company for damages. You are called in to testify as an expert in genetics. How would you interpret the data? Does the man have a valid case? Why or why not?

3. Genetics techniques are making it more feasible for couples to have healthy children. Some tests are currently in use for specific selection during in vitro fertilization, such as disease screening and sex preferences. When do you think these techniques are justified? (Sex determination for sex-linked disease prevention vs. choosing sex for "family balancing" only.) Where do we draw the line between cure and convenience?

4. A couple comes to you for genetic counseling, worried that if they have a child, it will inherit sickle-cell anemia. The mother has sickle-cell anemia, but the father is known not even to be a carrier for it. (That is, he has two alleles that code for hemoglobin A, rather than the sickle-cell-causing hemoglobin S.) Their question to you: What are the odds that any of our children would inherit sickle-cell anemia?

Group Activity

A new race is discovered on the planet of Zor. It is known that there are both male and female genders, and also that they have a total of three chromosomes—two autosomes and one sex chromosome. When a male and female Zoranian conceive, they always give birth to twins, one male and one female. Observing that all births are split-gender twins, and only one sex chromosome is present, you investigate sex determination. What is one possible explanation?

Genetics Problems

Note: In several of the problems that follow, you probably will need to construct Punnett squares to get an answer.

1. **a.** For an X-linked recessive condition such as red-green color blindness, what is the chance of having an affected son under the following two scenarios?

   Scenario I: The mother is a carrier and the father is normal.
   Scenario II: The father is affected and the mother has no family history of the disease.

   b. Using X for the dominant color-producing allele on the X chromosome, x for the recessive (color-blind) allele, and Y for the male chromosome, what are the genotypes of the parents in the two scenarios?
   c. In scenario I, if the parents have a daughter, what are the chances that she will be a carrier (that is, heterozygous for the condition)? What are the chances of a daughter being a carrier in scenario II?

2. Coat color in Labrador retrievers is governed by two genes: B/b and E/e.

   B = black pigment
   b = brown pigment
   E = allows pigment to be deposited
   e = does not allow pigment to be deposited

   The B allele is dominant to the b, and E is dominant to e. The ee genotype specifies a golden lab regardless of what B/b allele combination it carries. This is because the black or brown pigment cannot be deposited in a proper manner. If a brown lab is mated to a brown lab and they have some golden puppies.

   a. What are the genotypes of the parents?
   b. What is the chance that these parents will be able to generate black lab puppies? Brown labs? Golden labs?
   c. In a litter of 12 puppies, how many theoretically will be of each color based on your answer in (b)?

3. A couple, Jan and Barry Smith, see a genetic counselor because they are concerned about the possibility that their unborn son will have Marfan syndrome, a connective tissue disorder that can lead to serious problems with blood circulation. Both of their families have a history of Marfan syndrome, with Jan's brother, and Barry's mother, father, and sister being affected. Jan and Barry are unaffected. No other information is known about the family history. Assume that Marfan syndrome is controlled by a single gene on an autosome, and answer the following questions.

   a. Draw a pedigree that includes all individuals mentioned above. Label Jan and Barry Smith with stars.
   b. Is this disease dominant or recessive? Why?
   c. What is the genotype of each of the following individuals? (Give all possibilities.)

   Barry's mom, Barry's sister, Barry, Jan

   d. What are the chances that Barry and Jan will have a child with Marfan syndrome?

4. A mother with blood type A and a father blood type B have a child that is blood type O. Write the genotypes of all three individuals using the following alleles: A, B, and O.

5. A man in his fifties is showing signs of Huntington disease, which has an autosomal dominant pattern of inheritance. Two of his daughters, both in their thirties, are worried that they will get the disease, and that they have already passed the gene to their children. The daughters get tested for the disease-causing allele. One daughter, Debbie, is free of the allele. The other daughter, June, carries one copy of the allele.
   a. What is the genotype of each daughter? (*H* and *h* are the alleles.)
   b. What is the chance that each daughter will get Huntington disease?
   c. What is the chance that each daughter has passed the gene to her children?

6. The following couples plan to marry and want to know the chance of having a child with the following diseases. You are a genetic counselor who needs to answer their questions.
   a. two carrier parents for cystic fibrosis (autosomal recessive)
   b. one parent with Marfan syndrome and the other parent with no family history for the syndrome (autosomal dominant)
   c. an unaffected father and carrier mother for hemophilia (X-linked recessive)
   d. two parents with achrondroplasia (an autosomal dominant bone disorder that can result in dwarfism)

7. a. What are the possible gametes from the following persons:

   Person I: *Aa*
   Person II: *aa*
   Person III: aaBb
   Person IV: *AaBb*

   b. What are the genotypes and frequencies of the offspring produced if you cross person I with person II and person III with person IV?

8. Determine the most likely mode of inheritance for the following pedigrees. What is the genotype of the affected individuals in each case?

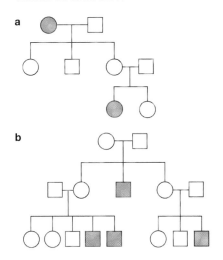

9. A curious genetics student wants to reproduce Mendel's pea plant results. She mates a true breeding tall plant (*TT*) to a short plant (*tt*). She self-crosses the F_1 plants and obtains 200 F_2 plants. How many of these will be tall, short, or intermediate in height? (Assume height is controlled by a single gene and that tall is dominant to short).

13

DNA Structure and Replication

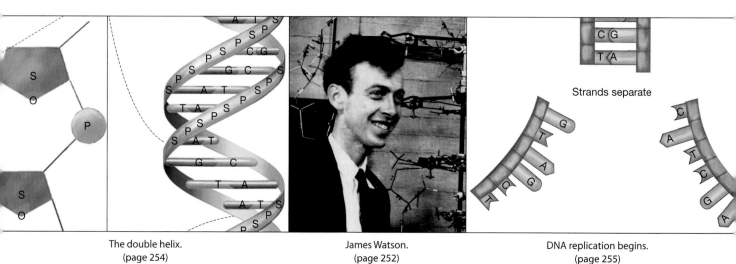

The double helix.
(page 254)

James Watson.
(page 252)

DNA replication begins.
(page 255)

"In research the front line is almost always in a fog," wrote Francis Crick in his 1988 memoir, *What Mad Pursuit*. Crick has firsthand knowledge of this. In 1953, with James Watson, he discovered the structure of DNA, an effort that brought him to the front line of research, which is where he has remained ever since. Crick's book makes clear what a tremendous difference there is between learning something about nature and discovering something about it. In learning, dozens of voices stand ready to instruct; there are books, lectures, videos, CDs—a world of well-organized information. The person who seeks to discover something about nature, meanwhile, is confronted with silence. Nature goes on: Cells divide, chromosomes condense, birds migrate. But *why* do these things operate the way they do? All the researcher has as a guide are the things themselves, working away. The trick is to devise some test (called an *experiment*) that can make nature's routine operations yield up information. In this process of teasing out truth, knowledge is gained in very small increments. Scientists are like the first people down a darkened maze of tunnels; they must make their way exceedingly slowly, feeling each square inch of wallspace as they go, after which they can only leave lights on *behind* them (in the form of their scientific papers, books, and so forth).

13.1 What Do Genes Do, and What Are They Made of?

To get a feel for the process of discovery, consider the state of genetics early in the twentieth century. By 1920, it had been demonstrated beyond any doubt that genetic information resided on chromosomes. Within a decade or so, by looking at some abnormally large fly chromosomes, scientists could observe in great detail such chromosomal processes as crossing over. By viewing the banding patterns on these chromosomes, they could even identify the rough location of some genes.

Yet what was a gene? What was the physical nature of this unit of heredity, and how did it work? No one knew. Because genes are much smaller than the chromosomes on which they resided, there was no hope of simply viewing one under a microscope. In observing chromosomes, scientists were only "looking" at genes in the way that any of us is

One kind of mutation.
(page 258)

Francis Crick.
(page 252)

A famous experiment.
(page 257)

"looking" at lunar rocks by glancing up at the moon. Through the 1920s and 1930s, then, genes could be described only in vague, functional terms, as in: A gene is an entity that lies along a chromosome and brings about a phenotypic trait in an organism (for example, the green or yellow peas of Mendel's experiments).

Things were to clear up somewhat in the ensuing years. The seminal achievement in genetics in the late 1930s and early 1940s was a more concrete description of what genes do: They bring about the production of proteins. (See "Getting Clear about What Genes Do" on page 256.) The central achievement in genetics from the mid-1940s to the early 1950s was strong evidence indicating what genes are composed of: deoxyribonucleic acid, or DNA for short.

DNA Structure and the Rise of Molecular Biology

By the early 1950s, a key question became *how* DNA carried out its genetic function. To find out, scientists had to piece together its exact chemical structure. This investigation turned out to be a watershed event in biology. Just as opening a mechanical watch and observing how its parts fit together would allow a person to understand how the watch works, so deciphering the structure of DNA allowed biologists to understand how genetics works at its most fundamental level. How could a section of DNA specify a protein? How could genetic information be copied? The structure of the DNA molecule sug-

gested answers to these questions, as you'll see in this chapter.

Apart from what this investigation uncovered, the inquiry itself stood as a symbol of a new era in biology. When Gregor Mendel did his experiments with peas, he was looking at whole organisms. When T. H. Morgan was doing his work with *Drosophila*, he was interested in whole chromosomes, which he knew to be composed of several types of molecules. In the early 1950s, however, the search turned to the constituent parts of a single molecule. What atoms of oxygen, hydrogen, and carbon were in DNA, and how were they arranged? Biological research of this sort has grown ever more important in the decades since the 1950s. It is today known as **molecular biology**: The investigation of life at the level of the constituent parts of its individual molecules.

13.2 Watson and Crick: The Double Helix

James Watson and Francis Crick may not be instantly recognized scientific names in the way that, say, Albert Einstein or Louis Pasteur are, but there's a certain public awareness that these two researchers did something important in connection with DNA (**Figure 13.1**). What they did was present to the world the structure of DNA in 1953: Atom-by-atom, bond-by-bond, this is how DNA fits together, they said in unveiling DNA's now-famous configuration, the double helix.

The two were a seemingly unlikely pair to make an epochal scientific discovery. Watson, an American, was a 23-year-old who was scarcely more than a year out of graduate school, and Crick, an Englishman, was a 35-year-old just then working on his doctorate when the two met in the fall of 1951 at Cambridge University in England. Ending up together by coincidence at the same laboratory, they realized in short order their mutual interest in the structure of DNA and, to the neglect of projects they were supposed to be working on, they began several rounds of model building and brainstorming that resulted in their breakthrough.

Watson and Crick were greatly aided in their investigation by the work of others. Though the DNA molecule was too small to be seen by even the most powerful microscopes of the time, something about its

Figure 13.1
Young and Famous
James Watson, on the left, and Francis Crick, with a model of the DNA double helix, shortly after they published their paper on the molecule's structure.

structure could be inferred from a technique called *x-ray crystallography*. In this process, a purified form of a molecule (a crystal) is "grown" to a considerable mass, and pieces of it are bombarded with x-rays; the way these rays scatter upon impact then reveals something about the structure of the molecule.

If you look at **Figure 13.2**, you can see the results of some x-ray crystallography. It may be obvious that it takes a highly trained observer to be able to perceive anything about the structure of a molecule from such an image. Fortunately for Watson and Crick, such a person was working just up the road from them. She was Rosalind Franklin, a researcher at King's College in London and one of the handful of individuals then skilled in performing x-ray crystallography on DNA. She and her colleague, Maurice Wilkins, were themselves working on the structure of DNA at the time, as were other researchers in Amercia. Thus did Watson, at least, regard the search for DNA structure to be a race between several teams, a fact that concentrated his efforts wonderfully. In 1962 Watson, Crick, and Wilkins were awarded the Nobel Prize in Medicine or Physiology for their work on DNA. Rosalind Franklin died in 1958 at the age of 37. Nobel Prizes are not awarded posthumously; it's unknown what would have happened had she lived.

Let's start now to look at DNA's structure, which will put you in a position to appreciate the achievement of Watson, Crick, and their fellow researchers.

13.3 The Components of DNA and Their Arrangement

If you look at **Figure 13.3a**, you can see the component parts of DNA. First, there is a phosphate group, and then a sugar called *deoxyribose*. By looking at Figure 13.3b and focusing on the red-colored molecules on the left, you can see how sugar and phosphate fit together to form a kind of chain, with its links going: sugar-phosphate-sugar-phosphate (symbolized in the figure by S and P). If you look now at Figure 13.3c and think of the entire DNA double helix as a staircase, the two "handrails" on it are formed by these sugar-phosphate chains.

There is then a third component to DNA, the "steps" that lie in between its handrails.

These steps are composed of DNA's "bases": adenine, guanine, thymine, or cytosine—A, G, T, and C for short. If you look at Figure 13.3b, you can see how these bases are linked to the DNA handrails.

As the figure shows, each of the bases amounts to a *half*-stair-step, if you will. Each extends inward from one handrail of the double helix and is then joined to a complementary base extending inward from the *other* handrail of the DNA molecule. (The bases are linked via hydrogen bonds, symbolized by the blue dotted lines in Figure 13.3b.)

The figure shows some examples of an A base being paired with a T base, and one of a C base being paired to a G base. This turns out to be one of the fundamental rules about DNA's structure. If you viewed a billion DNA **base pairs** such as this, you would find the same thing: A always pairing with T, and G always pairing with C, across the helix.

a

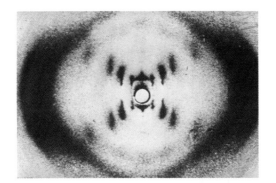

b

Figure 13.2

a DNA Investigator
Rosalind Franklin, whose work in x-ray crystallography was important in revealing the structure of the DNA molecule.

b Imaging DNA
One of Franklin's x-ray crystallography images of DNA. The "cross" formed of dark spots indicated that the molecule had a helical structure.

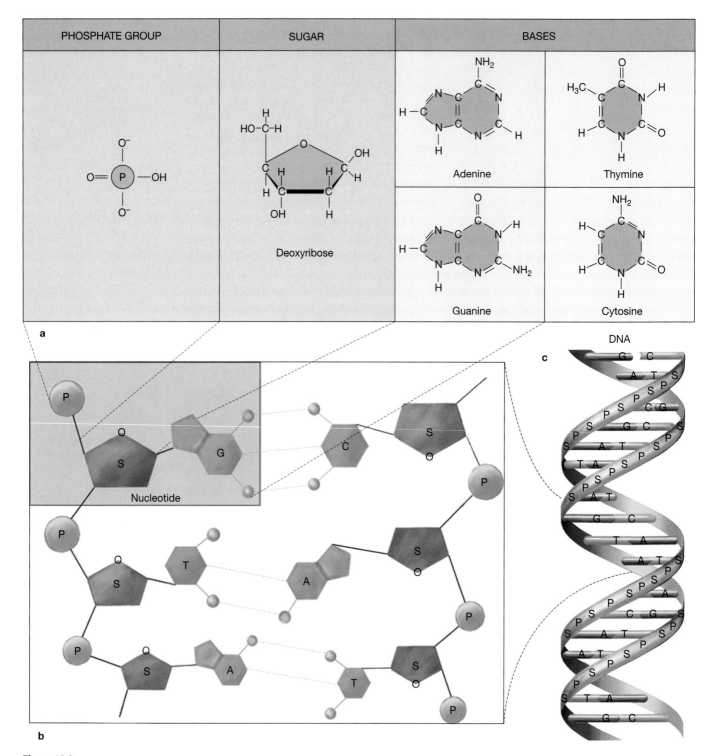

PHOSPHATE GROUP	SUGAR	BASES

Deoxyribose

Adenine

Thymine

Guanine

Cytosine

a

Nucleotide

b

DNA

c

Figure 13.3
Component Molecules of DNA

a The DNA molecule is composed of three types of component molecules: phosphate groups, the sugar deoxyribose, and the bases adenine, thymine, guanine, and cytosine (A, T, G, and C).

b The building blocks of DNA are units called nucleotides, each of which is composed of one phosphate group, one sugar, and one of the four bases—in this example, G. Across the strands of the helix, G always pairs with C, and T with A.

c Sugar and phosphate link together to form the "handrails" of the double-helix "staircase," while the bases form the "stairsteps," each base extending across the helix to link with a complementary base extending from the other side.

The Structure of DNA Gives Away the Secret of Replication

It was DNA's very structure that suggested the answer to one of the great questions of genetics: How is genetic information passed on? As you've seen, a full complement of our DNA—a complete copy of our genome—is contained in nearly every cell in our body. Yet cells divide, and the *daughter* cells contain exactly the same genetic information as the parent cell. This means that genetic information is passed on by means of *DNA being copied*, with one copy of the genome ending up in each daughter cell. But how?

The structure Watson and Crick suggested a way. We've observed that A must always pair with T, and G with C, across the two strands of the double helix. This rule, Watson and Crick saw, meant that each single strand of DNA could serve as a *template* for the synthesis of a new single strand (**see Figure 13.4**). Each A on an old strand would specify the place for a T on the new, each G on the old a place for C on the new, and so forth. All that was required was that the two old strands separate—splitting the stair steps right down the middle—and that new strands be synthesized that were complementary to the old. As it turned out, this was exactly how things worked, as you'll see shortly.

The Structure of DNA Gives Away the Secret of Protein Production

The second thing the structure suggested was a partial answer to another great question of genetics: How the molecule of heredity could be versatile enough to specify the dazzling array of proteins that all living things produce. The handrails of DNA's double helix are monotonous—phosphate, sugar, phosphate, sugar. But the bases can be laid out along these handrails in an extremely varied manner. A had to pair with T and G with C *across* the helix, but *along* the handrails, the bases could come in any order. Look at Figure 13.4 and see the order this hypothetical group of bases comes in; starting from the bottom right, A-G-C-T-A-C. Strung together by the thousands, these bases can specify a particular protein, just as in Morse code a series of short and long clicks can specify a particular word.

The Building Blocks of DNA Replication

You have already looked briefly at the basic process by which DNA is copied or "replicated," but you also need to note some of the details of this process. The basic steps here are straightforward. The essential building block of DNA is the unit called the **nucleotide** that can be seen in the box at the upper left of Figure 13.3b: one sugar and one phosphate, joined to one of the four bases (in this case G). In replication, as you can see in Figure 13.4, the double helix unwinds and the nucleotides on each of its single strands are paired with free-floating nucleotides that

Figure 13.4
DNA Replication
The result of DNA replication is two identical strands of DNA, whereas the process began with one.

DNA to be replicated

Strands separate

Each strand now serves as a template for the synthesis of a separate DNA molecule as free nucleotides base-pair with complementary nucleotides on the existing strands.

This results in two identical strands of DNA.

Order of bases encodes information for protein production.

Getting Clear about What Genes Do: Beadle and Tatum

In the 1930s, biologists were confused not only about what genes were made of (DNA) but also about what it was that genes did. We now know that a large part of what they do is contain the information for the production of proteins, including the reaction-hastening proteins called *enzymes*. But it took work over a period of years by Stanford University researchers George Beadle and Edward Tatum to make this clear. Their experimental subject was a red fungus, the bread mold *Neurospora*.

Following the Growth of a Fungus

Research scientists are always looking for ways to control the confusing conditions of nature in order to tease out the rules it plays by. Beadle and Tatum's control mechanism was, first, to *induce* genetic mutations in *Neurospora* spores by bombarding them with x-rays. These mutated spores were then transferred one at a time into test tubes, where the second element of control came in. Natural or "wild type" *Neurospora* can manufacture most of what they need to live, given just a few basic biological molecules in their surroundings. But could the descendants of the *mutated* group of *Neurospora* do this? Could they be put in a so-called minimal test tube medium—a liquid environment that lacked the biological molecules the fungus normally has access to—and still multiply into a thriving colony? The answer for many of the mutant strains was no. The mutations these organisms had un-

dergone had rendered them unable to produce some substance that they needed to thrive. By being selective about *adding* substances to the minimal media, Beadle and Tatum could tell what the missing substance was. (If they added the right thing, the colony would grow, as you can see in **Figure 1**.) For some of the mutant strains, the missing substance turned out to be the amino acid arginine.

We now know that arginine is produced in one of the metabolic pathways reviewed in Chapter 6. In such a pathway, a precursor compound is modified several times

> *In the 1930s, biologists were confused not only about what genes were made of but also about what it was that genes did.*

in a chain, with a different enzyme facilitating each step of the operation. By selectively supplying different strains of mutant spores with arginine's *precursor* substances, Beadle and Tatum could discern the step at which the arginine metabolic pathway had been blocked. One strain required only precursor A to produce arginine, for example, while another strain required precursors A and B.

In the tradition of T. H. Morgan, the researchers linked these blockages to genes on three separate *Neu-*

line up in a new complementary strand. Because *both* strands of the original double helix are being paired with new strands, the end product is two double helices, where before there was one.

DNA Replication: Something Old, Something New

It's worth drawing a little finer point on one aspect of this process: Each resulting double helix is a combination of the old and the new. Each represents one "parental" strand of DNA and one newly synthesized complementary strand. This is conceptually important because it is how life builds upon itself. You can see this illustrated in **Figure 13.5**.

Eventually, the two double helices are fully formed. At that point they part company, during mitosis, with each double helix moving into a separate cell and thereafter serving as an independently functioning segment of DNA. (At the chromosomal level, newly replicated helices of DNA are the sister chromatids you've seen so much of in the last few chapters.)

However simple this process may sound in overview, its details are enormously complicated. As you might imagine, such a process could not proceed without enzymes to catalyze it; the most important group of them at work here are collectively known as **DNA polymerases**. They move along the double helix, bringing free-standing nurospora chromosomes. The summary result then was clear. A given *Neurospora* strain had become a mutant by having its genes altered through radiation. As a result, it no longer carried out the transformation of, say, precursor A to precursor B. It took an *enzyme* to facilitate this transformation. What genes were doing, then, was bringing about the production of enzymes.

The One-Gene, One-Enzyme Hypothesis

Eventually, a famous label was applied to these experimental results: The **one-gene, one-enzyme hypothesis**. What each gene does is call up production of one enzyme, which undertakes a particular metabolic task. Though this notion is now dated, the essential insight was not only correct, it was critical. With it, scientists at last understood what genes were doing. For this work, which they completed in 1941, Beadle and Tatum shared in the Nobel prize in 1958.

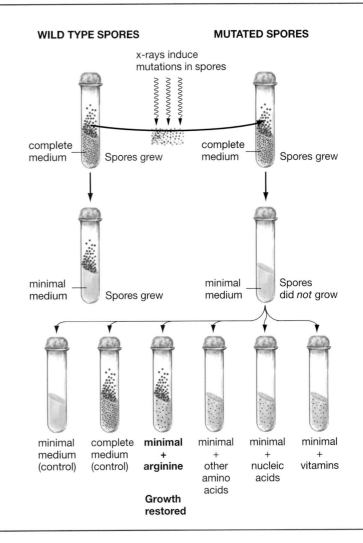

WILD TYPE SPORES   **MUTATED SPORES**

x-rays induce mutations in spores

complete medium — Spores grew

complete medium — Spores grew

minimal medium — Spores grew

minimal medium — Spores did *not* grow

minimal medium (control)

complete medium (control)

minimal + arginine

Growth restored

minimal + other amino acids

minimal + nucleic acids

minimal + vitamins

Figure 1
Beadle and Tatum Experiment with Mutated Spores
Using spores of the fungus *Neurospora*, Beadle and Tatum found that some mutated versions of the spores could not grow in a medium that lacked a basic set of organic molecules. By selectively *adding* molecules to this "minimal medium," they demonstrated that the mutation was preventing the spores from producing certain compounds, among them the amino acid arginine. By tracing the metabolic pathway leading to arginine, Beadle and Tatum realized that the *Neurospora* genes that had been made dysfunctional were coding for enzymes that facilitated the steps leading from the precursors of arginine to arginine itself.

---

cleotides into line with the complementary bases of the parental nucleotides. Another group of enzymes—the **DNA ligases**—is then involved in bonding sugar and phosphate in replicating DNA.

Editing Out Mistakes

The base pairing that goes on in replication happens hundreds, then thousands, then millions of times in a given stretch of DNA: Free-standing A's are brought into line with complementary T's, while C's are aligned with G's. Given the numbers of base pairs involved, the amazing thing is how few mismatches there are by the time the process is completed. In *E. coli* bacteria, the error rate—the rate at which the *wrong* bases have

OLD

NEW

Figure 13.5
How Life Builds on Itself
Each newly synthesized DNA molecule is a combination of the old and the new. An existing DNA molecule unwinds, and each of the resulting strands (the old) serves as a template for a complementary strand that will be formed through base pairing (the new).

been brought together—might be one in 10 million by the end of replication. Yet *during* replication, such a mistake might be made once in every 100,000 bases. Obviously, to start with one number of mistakes but to end up with far fewer, the cell's genetic machinery has to be capable of correcting its errors.

This happens partly through the services of the versatile DNA polymerases, which have the ability to perform a kind of DNA editing: They remove a mismatched nucleotide and replace it with a proper one. Interestingly enough, this happens through a kind of backspacing. Normally the DNA polymerases are moving along a DNA chain, lining up base pairs. When an error is detected, however, they stop, move "backward," remove the incorrect nucleotide, put in the correct one, and then move forward again.

13.4 Mutations: Another Name for a Permanent Change in DNA Structure

MediaLab Link
Cancer, see p. 206.

With this consideration of alterations in DNA structure, you have arrived at an important concept: that of a **mutation**, which can be defined as a permanent alteration in a cell's DNA base sequences. DNA's makeup can be altered in many ways. A slight change in the chemical form of a base might, for example, cause a G to link up across the helix with a T (instead of with its normal partner, C). Conversely, a sequence of base pairs *along* the handrails might be inserted in error. (A C-G base pair might be put in, for example, where an A-T normally resides.) (**See Figure 13.6** for these examples.) Permanent mistakes such as these are called **point mutations**, meaning a mutation at a single location in the genome. Such a mutation stands in contrast to the kind of whole-chromosome aberrations presented in Chapter 12, though these also qualify as mutations under some definitions.

An Example of Mutations Passed Along in a Line of Cells: Cancer

Point mutations are passed along in a line of cells—from parent cell to daughter cells as the original DNA mistake is copied over and over in subsequent DNA replications. Most such changes have no noticeable effect on an organism. Occasionally, however, the result is disastrous. Almost all cancers get started in just this way; cancers are a line of cells that have undergone a special kind of mutation—one that causes the affected cells to proliferate wildly.

The reason cancers aren't more common is that multiple mutations are required to produce malignant groups of cells. In colon cancer, for example, from four to seven mutations are required for a malignant tumor to be produced. This is one reason that cancer rates generally rise with age; it takes time for all of the necessary mutation events to occur. In addition, these mutations often are the result of years of environmental exposure—to cigarette smoke, to the sun, and so forth.

Heritable Mutations: Those That Occur in Germ-Line Cells

Most mutations come about in the body's **somatic cells**, which is to say cells that do not become eggs or sperm. Conversely, some mutations arise in **germ-line cells**, meaning the cells that do become eggs or sperm. The important point here is that germ-line cell mutations are *heritable*—they can be passed on from one generation to the next. Many generations ago, the DNA in some germ-line

Mutation: Incorrect base-pairing

Normal DNA

Mutation: Incorrect sequence of base pairs

— DNA with point mutations —

Figure 13.6
The Nature of Mutations
Shown are two examples of mistakes that can be made during DNA replication. Such "point mutations" come about fairly frequently, but are usually corrected by DNA polymerases during the process of replication.

cells underwent mutations that resulted in the conditions we now call sickle-cell anemia and Huntington disease, for example.

Think of the difference between these mutations and the mutations that result in cancer. In, say, pancreatic cancer, a group of cells in the pancreas undergoes a series of mutations that causes them to grow out of control. However harmful this line of cells may be, it is separate from the line of cells that gives rise to eggs or sperm. In short, these cancerous mutations cannot be passed on. What can be inherited are genes that *dispose* a person to cancer. Often a person inherits one mutated gene (meaning allele) for a given cell-growth process, but also inherits a second good gene. It is then a mutation to this second gene that gets the malignant process going. Up to 10 percent of cancers are thought to be caused by genetic predisposition.

As observed earlier, DNA replication has a very low error rate. After learning about the kinds of effects mutations can have on organisms, you can understand why: Life as we know it could not exist with a *high* error rate. The cause of sickle-cell anemia turns out to be the substitution of a *single* T for an A in the gene that codes for protein hemoglobin in red blood cells. (This substitution produces a form of hemoglobin that causes blood cells to assume the potentially harmful sickle shape, as you can see in Figure 12.3 on page 237). Given this, you can see why genetic error rates have to be low, at least for some DNA sequences. How could a human being or a bird or a tree survive a high error rate in the DNA that codes for proteins?

The Value of Mistakes: Evolutionary Adaptation

We should also not be surprised that some genetic mistakes go uncorrected, for the simple reason that mutations are a critical part of evolutionary adaptation in organisms. Mutations are the only means by which species can add completely new genetic information to their genome, in the form of new alleles (meaning variant forms of genes). Organisms can combine *existing* alleles, in myriad ways. Think of meiosis, with its chromosomal part swapping (crossing over) and reshuffling of chromosomes (independent assortment). Valuable as these processes are, no amount of genetic recombination could have produced, for example, the eyes that some living things possess. To go from no eyes to eyes, there had to have been some mutations along the line—some accidental reorderings of DNA sequences such that entirely new proteins were produced.

To put perspective on this, most mutations do nothing to an organism, and when they do have some effect, it is generally harmful. Yet mutations are occasionally useful. In the big picture, they are *vital* to populations of organisms, given the struggle of living things to get along in changing environments. You'll be looking at this topic again, but for now, isn't it interesting to ponder the fact that the living world adapts partly through its mistakes?

On to How Genetic Information Is Put to Use

Back in Chapter 9, you saw that since antiquity, human beings have speculated about how one generation of living things can give rise to another—about *how* it is that life goes on. The questions posed by the Greeks, by scientists in the Renaissance, by Mendel and Morgan and others were finally answered in the 1950s and 1960s, when scientists came to understand DNA replication at the molecular level. The something-old, something-new quality to this replication—parental DNA strands serving as templates for new strands—was the detailed answer to the ancient question of how the qualities of living things can be passed down through generations.

Splendid as this function is, it is only one of the two great tasks carried out by our genetic machinery. You have just reviewed the process by which genetic information is replicated; now it's necessary to look at the process by which this information is used. How can a stretch of DNA bring about the production of a protein? You'll see in the chapter coming up.

Chapter Review

Key Terms

base pair 253

DNA ligase 257

DNA polymerase 256

germ-line cell 258

molecular biology 252

mutation 258

nucleotide 255

Summary

13.1 What Do Genes Do, and What Are They Made of?

- James Watson and Francis Crick discovered the chemical structure of DNA in 1953. This event ushered in a new era in biology because it allowed researchers to understand some of the most fundamental processes in genetics.

DNA to be replicated

Strands separate

Each strand now serves as a template for the synthesis of a separate DNA molecule as free nucleotides base-pair with complementary nucleotides on the existing strands.

This results in two identical strands of DNA.

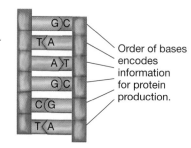

Order of bases encodes information for protein production.

- In trying to decipher the structure of DNA, Watson and Crick were performing work in molecular biology. This is the investigation of life at the level of the constituent parts of its individual molecules. Molecular biology has grown greatly in importance since the 1950s.

13.2 Watson and Crick: The Double Helix

- Watson and Crick met in the early 1950s at Cambridge University in England and set about to decipher the structure of DNA. Their research was aided by the work of others, including Rosalind Franklin, who was using x-ray crystallography to learn about DNA's structure.

13.3 The Components of DNA and Their Arrangement

- The DNA molecule is composed of building blocks called nucleotides, each of which consists of one sugar (deoxyribose), one phosphate group, and one of four bases: adenine, guanine, thymine, or cytosine (A, G, T, or C). The sugar and phosphate groups are linked together in a chain that forms the "handrails" of the DNA double-helix. Bases then extend inward from the handrails, with base pairs joined to each other in the middle by hydrogen bonds. In this base-pairing, A always pairs with T, while G always pairs with C.

- DNA is copied by means of each strand of DNA serving as a template for the synthesis of a new, complementary strand. The DNA double helix first divides down the middle. Each A on an original strand then specifies a place for a T in a new strand, each G specifies a place for a C on the new strand, and so forth.

- Each double helix produced in replication is a combination of one parental strand of DNA and one newly synthesized complementary strand. This is how life builds upon itself. A group of enzymes known as DNA polymerases are central to DNA replication; they move along the double helix, bringing new nucleotides into line with the complementary base pairs of the parental nucleotides. Another group of enzymes, the DNA ligases, is involved in bonding sugar and phosphate in replicating DNA.

- DNA can encode the information for the huge number of proteins utilized by living things because the sequence of bases along DNA's handrails can be laid out in an extremely varied manner. A collection of bases in one order encodes the information for one protein, while a different sequence of basis encodes the information for a different protein.

13.4 Mutations: Another Name for a Permanent Change in DNA Structure

- The error-rate in DNA replication is very low, owing partly to the fact that DNA polymerases are able to correct mistakes in base pairs. When such mistakes are made and then not corrected, the result is a mutation, meaning a permanent alteration in a cell's DNA base sequences.

- Most mutations have no effect on an organism; when they do have an effect, it is generally negative. Cancers result from a line of cells that have undergone types of mutations that cause them to proliferate wildly.

- Some mutations come about in the body's germ-line cells, meaning cells that become eggs or sperm. Such mutations are heritable: They can be passed on from one generation to another. The genes for sickle-cell anemia and Huntington disease, which exist in germ-line cells, are mutated forms of normal genes.

- Mutations are a part of the evolutionary adaptation of organisms. They are the only means by which species can add completely new genetic information to their genomes, in the form of new alleles.

Understanding the Basics

Multiple-Choice Questions

1. The structure of DNA was determined in _____ by Watson and Crick.
   a. 1951   d. 1958
   b. 1953   e. 1962
   c. 1957

2. The type of biological study that was exemplified through the discovery of DNA structure is called
   a. detailed biology
   b. component biology
   c. microscopic biology
   d. elemental biology
   e. molecular biology

3. The components of the DNA handrails are _____ and _____.
   a. nucleotides, phosphates
   b. sugars, nucleotides
   c. sugars, bases
   d. sugars, phosphates
   e. none of the above

4. Which of the following combinations is an example of a nucleotide?
   a. sugar + phosphate + adenine
   b. sugar + phosphate + guanine

   c. sugar + phosphate + thymine
   d. sugar + phosphate + cytosine
   e. all of the above

5. Which of the following DNA sequences contains a mutation?
   a. TCAA
      AGTT
   b. CAGC
      GTCG
   c. GATA
      CTCT
   d. AATT
      TTAA
   e. GACG
      CTGC

6. In DNA replication, the parent strand of DNA is called the _____ strand, while the daughter strand is called the _____ strand.
   a. template, complementary
   b. pattern, mirror
   c. template, opposite
   d. complementary, template
   e. opposite, mirror

7. Though the DNA structure could not be detected by a microscope, Rosalind Franklin aided in its discovery by using
   a. electron microscopy
   b. x-ray crystallography
   c. mass spectrometry
   d. high-pressure liquid chromatography
   e. a combination of b and c

8. DNA polymerases work to
   a. strengthen base–base bonds
   b. align new nucleotides to form daughter strand
   c. edit mismatched nucleotides
   d. a and c only
   e. b and c only

9. Which of the following processes or substances is believed to have contributed the most to evolutionary adaptation?
   a. "normal" crossing over
   b. DNA ligase
   c. cytokinesis
   d. mutations
   e. chromosome condensation

10. How are the bases of the DNA double helix linked together?
   a. phosphate bonds
   b. hydrogen bonds
   c. carbon–carbon bonds
   d. sugar bonds
   e. basic bonds

Brief Review

1. How does the author's phrase "something old, something new" describe the method of DNA replication employed by the cell?

2. Give one example from this chapter, and one from Chapter 12, of cellular functions that may lead to mutations.

3. True or False: DNA polymerases work only in a forward direction.

4. True or False: All mutations are detrimental.

5. Why can such a simple set of materials (bases A, T, C, and G) encode for every protein of our body?

6. Give an example of a point mutation that produces an allele that results in a disease.

Applying Your Knowledge

1. Given the following DNA sequence, determine the complementary strand that would be added by DNA polymerase:
   ATTGCATGATAGCC

2. Why does a germline mutation carry greater potential significance than a somatic mutation?

3. Why does "normal crossing over" not contribute significantly to species changes?

4. Why is it important that DNA replication have a low error rate?

5. How do we know that DNA replicates only once in every cell cycle, and not many times?

6. In light of the functions of DNA polymerases described in this chapter, how many of the following scenarios could be attributed to faulty polymerase function?
   a. failure of the chromosomes to separate at anaphase
   b. failure of the hydrogen bonds to strengthen
   c. failure of the sugar-phosphate bonds to form
   d. failure of the bases to align with their complement bases
   e. failure of the complementary DNA strand to undergo editing

Group Activity

This chapter describes the challenge of understanding and "looking" at chromosomes without powerful microscopes—as somewhat like our trying to understand and "look" at lunar rocks just by looking at the moon. Brainstorm with a group of your classmates on ways to describe lunar rocks in as much detail as possible even though you could never travel to the moon yourself.

14

How Proteins Are Made

Genetic Transcription, Translation, and Regulation

The triplet code.
(page 268)

A ribosome.
(page 273)

Genetic regulation in action.
(page 278)

Cracking the code.
(page 271)

The scientific advance that

was reviewed in Chapter 13, the discovery of the structure of DNA in 1953, bears some comparison to the birth of Napoleon Bonaparte: Both events had momentous consequences, but not for a while. Only slowly did the discoveries come that, in time, were seen as hinging on James Watson and Francis Crick's work. So significant were these advances, however, that the years between 1953 and 1966 are now regarded by some scholars as a kind of golden age of genetics; a time when basic findings about this field came fast and furious, such that by the mid-1960s scientists had at last grasped the essentials of how the genetic machinery works.

These discoveries were made in connection with both of the two great tasks of genetics: On the one hand, the copying or "replication" of DNA that you reviewed in Chapter 13; on the other, the means by which DNA's genetic information brings about the production of proteins. The first task has to do with how genetic information is preserved; the second with how genetic information is utilized. It is the latter task that is the subject of this chapter.

14.1 The Structure of Proteins

Because you'll be dealing a good deal with proteins here, now may be a good time to review some basic information about them. As you saw in Chapter 3, proteins fit into the "building blocks" model of biological molecules. The blocks in this case are amino acids. String a number of these together and you have a **polypeptide** chain, which then folds up in a specific three-dimensional manner, resulting in a protein. Proteins are likely to be made of hundreds of amino acids strung together and folded up. It is the three-dimensional shape of proteins that gives them their working ability—generally speaking, the ability to bind with other molecules in very specific ways.

Synthesizing Many Proteins from 20 Amino Acids

Though there are tens of thousands of different proteins, all of them are put together from a starting set of a mere 20 amino acids. If you wonder how such diversity can proceed from such simplicity, think of the English language, which has thousands of words, but only 26 letters in its alphabet. It is the

One step in a chain.
(page 274)

Mass production.
(page 275)

Transferring information.
(page 267)

order in which letters occur that determines whether, for example, "cat" or "act" is spelled out; just so, it is the order of amino acids that determines what protein is synthesized.

If you look at **Figure 14.1a**, you can see the chemical structure of two free-standing amino acids, glycine (gly) and isoleucine (ile). If you look at Figure **14.1b**, you can see that these two amino acids are, in fact, the first two that occur in one of the polypeptide chains in the unusually small protein insulin, which helps move blood sugar into cells. **Figure 14.1c** then shows a three-dimensional representation of insulin—the "folded up" form this protein assumes before beginning its work. A list of all 20 primary amino acids and their three-letter abbreviations can be found in **Table 14.1**.

For our purposes, the question is: How do the chains of amino acids that make up a protein come into being? How do gly, ile, val, and so forth come to be strung together in a specific order to create this protein? You

know from what you've studied so far that genes "order up" proteins, and that genes amount to a series of chemical "bases"—the As, Ts, Cs, and Gs along a DNA strand. How, then, does *this* series of DNA bases come to specify *that* series of amino acids?

14.2 Protein Synthesis in Overview: Transcription and Translation

In overview, this process can be described fairly simply, as those of you who have read Chapter 9 know. The DNA just referred to is contained in the nucleus of the cell. The first step is that a stretch of it unwinds there, and its message—the order of a string of its bases along one of its strands—is copied onto a molecule called messenger RNA (mRNA). This length of mRNA, which could be thought of as an information tape being dubbed off a "master" DNA tape, then exits from the cell nucleus (**Figure 14.2**).

Figure 14.1
The Structure of Proteins

a The building blocks of proteins are amino acids such as glycine and isoleucine, which differ only in their side-chain composition (white background).

b These amino acids are strung together to form polypeptide chains.

c These chains function as proteins only when they are folded into their proper three-dimensional shape, as shown here.

a

glycine (gly)   isoleucine (ile)

b

H_3N^+—| gly | ile | val | glu | gln | cys | cys | ala | ser | val | cys | ser | leu | tyr | gln | leu | glu | asn | tyr | cys | asn |—C⟨O, O⁻

c

The destination of this mRNA is a molecular workbench in the cell's cytoplasm, a structure called a *ribosome*. It is there that both message (the mRNA tape) and raw materials (amino acids) come together to make the product (a protein). As the mRNA tape is "read" within the ribosome, a growing chain of amino acids are linked together in the ribosome in the order the mRNA calls for; when the chain is finished and folded up, a protein has come into existence. And how do amino acids get to the ribosomes? They are brought there by a *second* type of RNA, transfer RNA (tRNA).

The first part of this process, in which DNA's information is copied onto mRNA, is called **transcription**. The second part of this process, in which a polypeptide chain is produced from the mRNA tape's instructions, is called **translation**. These two terms describe the two main stages in protein synthesis. Let's look at both of them now, starting with transcription.

The Nature of Transcription

An Important Player: RNA

From what's reviewed so far, you can see that a key player in transcription (and translation) is RNA, whose full name is ribonucleic acid. If you look at **Figure 14.3** on page 268, you can see just how similar RNA is to DNA. For one thing, RNA has the sugar and phosphate components you saw in DNA last chapter. Then, in both molecules, this two-part structure is joined to a third element, a base.

Among several differences between DNA and RNA, the most important, for our purposes, has to do with a substitution in bases. Recall that any given DNA nucleotide has one of four bases: adenine, guanine, cytosine or thymine (A, G, C, or T). RNA utilizes the first three of these, but then substitutes uracil (U) for the thymine (T) found in DNA.

Table 14.1
Amino Acid Abbreviations

Amino Acids	Abbreviations
alanine	ala
arginine	arg
asparagine	asn
aspartic acid	asp
cysteine	cys
glutamine	gln
glutamic acid	glu
glycine	gly
histidine	his
isoleucine	ile
leucine	leu
lysine	lys
methionine	met
phenylalanine	phe
proline	pro
serine	ser
threonine	thr
tryptophan	trp
tyrosine	tyr
valine	val

Figure 14.2
The Two Major Stages of Protein Synthesis
In transcription, a section of DNA unwinds and nucleotides on it base-pair with nucleotides on a length of messenger RNA. This mRNA sequence then leaves the cell nucleus, bound for a ribosome in the cell's cytoplasm, where translation takes place. Amino acids are also brought to the ribosome by transfer RNA molecules. The messenger RNA tape is then "read" within the ribosome, resulting in a growing chain of amino acids being linked together in the order specified by the mRNA tape. When the chain is finished and folded up, a protein has come into existence.

Passing on the Message: Base Pairing Again

Given the chemical similarity between DNA and RNA, it's not hard to see how DNA's genetic message can be passed on to messenger RNA: Base pairing is at work again. Recall from Chapter 13 how DNA is replicated. The double helix is unwound, after which bases (the As, Ts, Cs, and Gs) along the now-single DNA strands pair up with *complementary* DNA bases that are part of free-floating nucleotides. Every T on a single DNA strand pairs with a free-floating A, and every C links up with a G, thus yielding a complementary DNA *chain*. Because of RNA's similarity to DNA, the bases RNA has can *also* form base pairs with

DNA. The twist to this RNA-DNA base pairing is that each *A* on a DNA strand links up with a *U* on the RNA strand, instead of the T that would be A's partner in DNA-to-DNA base pairing. Thus, complementary base pairing is the means by which a DNA message is transferred to RNA. The length of mRNA that results from this process is called a **transcript**.

A Triplet Code

Thus far, you have been looking at what might be called the flow of genetic information (from DNA to mRNA to ribosomes). A separate, difficult issue that scientists faced in the 1950s and 1960s, however, came to be known as "the coding problem" in protein synthesis. Allowing that DNA bases are coding for amino acids, one primary question that arose was how *many* DNA bases it takes to code for each amino acid in a protein. It seemed clear that the answer was not one: Because there are only four DNA bases (A, T, C, and G), if each coded for its own amino acid, only four amino acids could have been incorporated into proteins, as opposed to the 20 that actually are. Nor could the answer be two, because the number of possible amino acids this number could yield up was only 16. Thinking that nature probably would work as economically as possible, the South African-born biochemist Sydney Brenner suggested that DNA worked in a **triplet code**, meaning that each *three* DNA bases specified a single amino acid.

Brenner turned out to be right. As **Figure 14.4** shows, each three bases in a DNA sequence pairs with three mRNA bases, but each group of three mRNA bases then codes for a *single* amino acid. Each coding triplet of mRNA bases is known, appropriately enough, as a **codon**.

The Second Stage in Protein Synthesis: Translation

The completion of transcription in protein synthesis leaves an mRNA transcript arriving at the ribosomes—but with no protein synthesized from it yet. A protein does come together, however, in the second stage in synthesis, **translation**: The process by which information encoded in a sequence of mRNA bases is transferred to a polypeptide chain. Ribosomes are the molecular workbenches at which the two ends of this protein-synthesizing process come together. On the one hand, the genetic message comes to the ribosomes in the form of the messenger RNA tape. On the other, amino acids

Figure 14.3
RNA and DNA Compared
Each building-block nucleotide of both DNA and RNA is composed of a phosphate group, a sugar—ribose in RNA, deoxyribose in DNA—and one of four bases. RNA and DNA both utilize the bases adenine, guanine, and cytosine; but RNA utilizes the base uracil instead of the thymine that DNA employs.

RNA nucleotide

DNA nucleotide

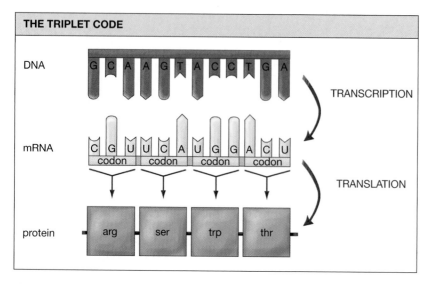

Figure 14.4
Triplet Code
Each triplet of DNA bases codes for a triplet of mRNA bases (a codon), but it takes a complete codon to code for a single amino acid.

must be brought to the ribosome to be strung together in the order specified by the tape. As you've seen, the molecule that brings the amino acids to ribosomes is a second form of RNA, **transfer RNA** (or **tRNA**).

Aptly Named: Transfer RNA

If you look now at **Figure 14.5**, you can see why this molecule is aptly placed within the translation phase of protein synthesis. When, in common discourse, we think of a *translator*, we think of someone who can communicate in *two* languages. Transfer RNA effectively does this. One end of each tRNA molecule links with a specific amino acid, which it finds floating free in the cytoplasm; transferring this amino acid to the ribosome, this tRNA molecule then employs its *opposite* end to form base pairs with *nucleic* acids—a codon on the mRNA tape that is being "read" inside the ribosome. Thus tRNA is the translator between the molecular languages of amino acids and nucleic acids.

Though there's many a detail to go, with this you've arrived at the essence of protein synthesis. Each tRNA carries a specific amino acid to a ribosome. There, three bases in the tRNA molecule form base pairs with three bases (a codon) in the mRNA transcript. As a *succession* of mRNA codons is read in the ribosome, a corresponding succession of amino acids is brought to the ribosome. In this process, these amino acids are linked together end to end to form a growing polypeptide chain—in the order specified by the mRNA transcript. And, of course, this transcript was in turn specified originally by the DNA sequence back in the cell's nucleus. Thus you can see the truth of a saying James Watson once coined regarding protein synthesis: "DNA makes RNA makes protein."

14.3 The Importance of the Genetic Code

Very shortly, you'll look at this process of protein synthesis in greater detail. Before that, however, it will be helpful to learn a little more about something only briefly touched on—the genetic code.

Cracking the Code

Earlier, it was noted that the genetic code is a triplet code, each three messenger RNA bases

① tRNA and amino acids float freely in cytoplasm.

② tRNA links to an amino acid and transfers it to the ribosome.

④ A polypeptide chain is produced.

③ tRNA links to the appropriate mRNA codon at the ribosome.

ribosome

mRNA

Figure 14.5
Bridging Molecule
The four steps in producing a polypeptide chain. Transfer RNA (tRNA) molecules link up with amino acids on the one hand and mRNA codons on the other, thus forming a chemical link between the two kinds of molecules in protein synthesis.

Cracking the Genetic Code

Children often are amused by giving common words special meanings. "If I say, 'flower,' that means to stand up," one will say to another. With this code in mind, the second child dutifully pops right up when the magic word is spoken. In the early 1960s, scientists were like observers in such a game. This one, however, involved genetics, and the whole point was to decipher a genetic code that was being employed. "If I say to you, 'the codon is UUA,'" the riddle went, "what amino acid gets added to a protein chain?"

At the time, this was a vexing question indeed. One of the difficulties in answering it was that researchers knew that a working cell is like a molecular Times Square: It is jammed with the traffic of many proteins working away or undergoing synthesis. Marshall Nirenberg and Heinrich Matthaei, scientists at the National Institutes of Health in the early 1960s, knew they had to work *outside* of this great complexity in order to follow the path of a *single* RNA triplet and see what amino acid it led to. Thus, they utilized what is known as a "cell-free" system, meaning they watched amino acid translation play out in test tubes rather than cells.

You've seen that, within cells, stretches of messenger RNA are synthesized by means of free-standing RNA nucleotides forming base pairs with lengths of DNA. Working outside of cells, Nirenberg and Matthaei synthesized an *artificial* mRNA chain by using a special enzyme that can string together RNA nucleotides. Critically, it can string together *whatever* nucleotide chain researchers wish, depending on what bases the researchers start with.

Running through the list, the researchers hit pay dirt with the amino acid phenylalanine.

For example, putting this enzyme together with the basic building blocks of RNA, but no base except uracil (U), scientists got an *mRNA chain* that had no bases but uracil in it (UUUU . . .).

So then the question became: What amino acid did the base triplet UUU code for? To find out, Nirenberg and Matthaei took their mRNA and put it into a test tube containing ribosomes, all 20 amino acids, and other molecules. Thus, this mix had in it all the molecular machinery necessary for protein translation, but it had only *one* mRNA tape to work with—the UUU mRNA transcript the researchers had made. It was therefore reasonable to

specifying a single amino acid. Scientists knew this by the 1960s; but the question that then confronted them was: If you have a given three bases, *which* amino acid do they specify? In Morse code, the sounds .__. (short-long-short) stand for the letter P. If we know that a given codon has the base sequence UCC (uracil, cytosine, cytosine) what *amino acid* does this code for? If you look at "Cracking the Genetic Code" at the top of this page, you can get an idea of how this code was cracked. **Figure 14.6** shows the genetic code it in its entirety. Now a couple of notes about the nature and importance of the genetic code.

A Redundant Code

One of the notable things about the genetic code is that not every mRNA triplet in it codes for an amino acid. Note, in Figure 14.6, that three mRNA codons specify "stop" codes that bring an end to the synthesis of a polypeptide chain. One triplet (AUG) specifies a special amino acid, methionine (met), which serves as the starting amino acid for most polypeptide chains. Note also that the genetic code is redundant. The acid phenylalanine (phe), for example, is coded for not only by the UUU triplet, but by UUC as well. Indeed, almost all the amino acids are coded for by more than one mRNA codon. Leucine (leu) is coded for by no fewer than six. The **genetic code** can be defined as the inventory of linkages between specific nucleotide triplets and amino acids.

The Genetic Code and Life's Unity

You have read a good deal in this book about the diversity and unity of life. The genetic code is a sterling example of the unity. With only a few exceptions, the genetic code is universal in all living things. This means that the base triplet CAC, for example, codes for the expect that this system would turn out polypeptide chains that contained only one amino acid—whichever amino acid was coded for by the UUU triplet.

The trick was to look at all the amino acids in the mix by means of radioactively labeling *one* of them (out of the 20) with each experiment (**see Figure 1**). Such labeling effectively puts a kind of tag on a given amino acid, while other techniques allowed the researchers to tell whether such an amino acid had remained free-standing or had been incorporated into a polypeptide chain. Running through the list, the researchers hit pay dirt with the amino acid phenylalanine. It clearly was part of a polypeptide chain, and the only way it could have been incorporated in this way was through the activity of mRNA. Because the only mRNA triplet in the system was UUU, this meant that UUU coded for phenylalanine.

Several other techniques had to be worked out in the ensuing years (by several researchers) to decipher the entire genetic code; all of these techniques, however, relied on the cell-free systems and synthetic mRNAs that Nirenberg and Matthaei employed. By 1968, the entire code had been cracked. In that same year, Marshall Nirenberg shared the Nobel Prize with researchers Robert Holley and Har Gobind Khorana for undertaking this critical piece of molecular deciphering.

Figure 1
Cracking the Code
In their work, Nirenberg and Matthaei radioactively labeled, in each experiment, one amino acid out of the total group of 20. Other techniques allowed them to tell whether such an amino acid remained free-standing or had been incorporated into a polypeptide chain. When they labeled phenylalanine, the researchers found that it had been incorporated into a chain. Because the only mRNA triplet in their system was UUU, this told them that UUU had to code for phenylalanine.

amino acid histidine, whether this coding is going on in a bacterium or in a human being.

Why is this important? First, it is evidence that all life on Earth is derived from a single ancestor. How else can we explain all of life's diverse organisms sharing this very specific code? Such a complex molecular linkage—this triplet equals that amino acid—is extremely unlikely to have evolved *more* than once. What seems likely is that it came to be employed in an ancient common ancestor, after which it was passed on to all of this organism's descendants—every creature that has subsequently lived on Earth. Note, then, that this code has been passed on from one generation to the next, and from one evolving species to the next, for billions of years. We humans share in an informational linkage that stretches back billions of years and that runs through all the contemporary living world.

		Second Base				
		U	**C**	**A**	**G**	
First Base	**U**	UUU UUC } phe / UUA UUG } leu	UCU UCC UCA UCG } ser	UAU UAC } tyr / UAA **Stop** UAG **Stop**	UGU UGC } cys / UGA **Stop** / UGG trp	U C A G
	C	CUU CUC CUA CUG } leu	CCU CCC CCA CCG } pro	CAU CAC } his / CAA CAG } gln	CGU CGC CGA CGG } arg	U C A G
	A	AUU AUC } ile / AUA / AUG met **(start)**	ACU ACC ACA ACG } thr	AAU AAC } asn / AAA AAG } lys	AGU AGC } ser / AGA AGG } arg	U C A G
	G	GUU GUC GUA GUG } val	GCU GCC GCA GCG } ala	GAU GAC } asp / GAA GAG } glu	GGU GGC GGA GGG } gly	U C A G

Third Base

Figure 14.6
The Genetic Code Dictionary
If we know what a given mRNA codon is, how can we find out what amino acid it codes for? This genetic code dictionary offers a way. In Figure 14.4, you saw that the codon CGU coded for the amino acid arginine (arg). Looking that up here, C is the first base (go to the C row along the "first base" line), G is the second base (go to the G column under the "second base" line) and U is the third (go to the codon parallel with the U in the "third base" line).

Apart from this, the universality of the genetic code has another very practical consequence. It means that human genes can function within such organisms as bacteria—and vice versa. This has both good and bad consequences for human beings. First the bad: Viruses that cause diseases ranging from colds to AIDS can "hijack" the human cellular machinery for their own purposes, precisely because their genes function in human cells. (The human cellular machinery will put together proteins whether these proteins are called for by human or viral DNA.) Now the good: Using "biotechnology" processes you'll be learning about in Chapter 15, human beings are now able to use viruses and bacteria to manufacture all kinds of products, including medicines such as human insulin and human growth hormone. In these cases, human genes are being put to work inside microorganisms.

14.4 A Closer Look at Protein Synthesis

Thus far, you've reviewed in outline the means by which DNA codes for proteins. Now this process is described in somewhat greater detail.

Protein Synthesis Begins: Transcription

Gene transcription begins with a section of the DNA double helix unwinding, after which one of the now-single DNA strands serves as a template for creation of a messenger RNA strand. It will come as no surprise to learn that an enzyme is critically involved in this process. **RNA polymerase**, as this enzyme is known, undertakes both of these initial tasks: unwinding the DNA sequence and stringing together a chain of RNA nucleotides that is complementary to it.

The import of this process is not that RNA is linked to DNA, but that the RNA nucleotides are also linked one to another,

yielding an RNA *chain*—the messenger RNA "tape"—synthesized in the order specified by the DNA bases.

Messenger RNA Processing

It turns out that the eukaryotic mRNA tape is not functional from the moment it is synthesized; instead, it must undergo some editing before it is a finished product. What form does this mRNA editing take? After some initial steps, the most important part of it is an operation that bears comparison to film editing. When directors put movies together, they first assemble a "rough cut" of the footage they've shot, meaning all the scenes they want included, sequenced in their proper order. Then they go to work with "scissors and paste." In looking at the work, a director may decide that a given scene is useless in its entirety, in which case it is snipped out. Then the film segment that comes before this scene and the film segment that comes after it are *spliced together*.

Introns and Exons

This is roughly what happens with eukaryotic mRNA. As you can see in **Figure 14.7**, special molecular splicing complexes cut out sections of the mRNA transcript that are known as **introns**. Then the remaining portions of the mRNA tape are spliced together. This edited form of the transcript eventually will contain no introns at all. What remains are **exons**: portions of the mRNA transcript that code for amino acids. From this, it follows that introns are *noncoding* portions of the original mRNA tape; they do not contain information for the sequencing of amino acids in a protein. (The intron and exon names are derived from their function—or lack of it. Introns are *in*tervening sequences of mRNA or DNA, while exons are *ex*pressed as proteins.) The obvious question that follows from this is: Why are introns included in the DNA chain in the first place? As you can see in "Making Sense of 'Nonsense' DNA" on p. 276, clues are steadily emerging about the function of these "junk" sequences.

Translation in Detail

With editing concluded, the finished mRNA transcript moves out into the cell's cytoplasm, where it is destined to link up with a

Figure 14.7
Editing Out Genetic Material
Special enzymes remove the noncoding regions of mRNA (introns) and splice the coding regions (exons) back together to create an edited transcript.

EXON 1   INTRON   EXON 2

mRNA transcript

enzymes cut out the intron

enzyme

edited mRNA transcript

ribosome for the second phase of protein synthesis. As you've seen, another key player must also make its way to these ribosomal sites for this work to be completed: transfer RNA molecules, each one of which comes with an amino acid in tow. Ribosomes act in one sense as a "reading head," through which the mRNA tape will pass.

The Structure of Ribosomes

If you look at **Figure 14.8**, you can begin to get an idea of the structure of ribosomes. Note that they are comprised of two "sub-units"—one larger than the other—both made of a mixture of proteins and yet another type of RNA, **ribosomal RNA** (or **rRNA**). The two types of subunits generally float apart from another in the cytoplasm until prompted to come together by the process of translation. You can see that when the subunits have been joined, there exist two binding sites (which it is convenient to think of as slots) one of them a "P" site, the other an "A" site. You'll be looking at their roles shortly.

The Structure of Transfer RNA

Figure 14.9 provides a more detailed look at the transfer RNA molecule. The role of tRNA as a "translator" between the molecular languages of nucleic acids (RNA) and amino acids was noted earlier. Figure 14.9 gives you an idea of how this one molecule can serve as this kind of bridge. One end of the molecule has three bases on it, called an **anticodon**, that can form a base pair with the appropriate codon in an mRNA tape. At the *opposite* end of the tRNA molecule (the 12 o'clock position in the figure) there is an attachment site for an *amino acid*. Each tRNA molecule has specificity for a particular amino acid, as well as a particular codon, so that the correct amino acid sequence will result from a given mRNA sequence.

The Steps of Translation

Now translation can begin. An mRNA transcript has left the nucleus and is ready to begin binding with a ribosome. Meanwhile, nearby tRNA molecules have bound with their appropriate amino acids. For purposes of simplicity, we'll follow the process as it occurs in prokaryotes.

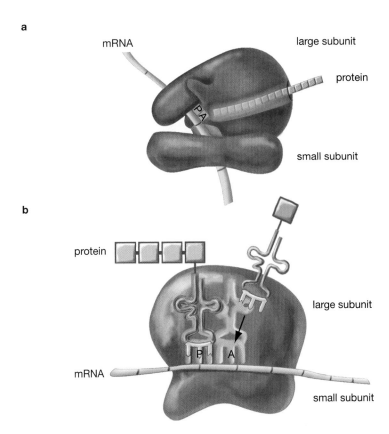

Figure 14.8
The Structure of Ribosomes

a Ribosomes are composed of two subunits that come together during translation.

b A simplified cross section of the ribosome illustrates the "P" and "A" sites where tRNA molecules bind during translation.

Figure 14.9
The Two-Dimensional Structure of Transfer RNA

Figure 14.10
The Steps of Translation

1. A messenger RNA transcript binds to the small subunit of a ribosome as the first transfer RNA is arriving. The mRNA codon AUG is the "start" sequence for most polypeptide chains. The tRNA, with its methionine (met) amino acid attached, then binds to this AUG codon.

2. The large ribosomal subunit joins the ribosome, as a second tRNA arrives, bearing a leucine (leu) amino acid. The second tRNA binds to the mRNA chain, within the ribosome's A site.

3. A bond is formed between the newly arrived leu amino acid, and the met amino acid, thus forming a polypeptide chain. The tRNA that had been in the P site leaves.

4. The ribosome in effect moves down the mRNA chain, one codon to the right, thus switching the second tRNA (bearing the polypeptide chain) into the P site and bringing a new codon into the A site. The A site is then available for a new tRNA, and the process of elongation continues.

mRNA Binds to Ribosome; First tRNA Arrives First, the mRNA tape arrives at the ribosome and binds to the ribosome's small subunit (**see Figure 14.10**). The mRNA codon AUG is the usual "start" codon for a polypeptide chain. Next, a tRNA molecule with the appropriate anticodon sequence (UAC) binds to this AUG codon. This tRNA arrives bearing its appropriate amino acid, which is methionine (met). Following this, the large ribosomal subunit becomes part of the ribosome, providing the ribosome's A and P binding sites.

Polypeptide Chain Is Elongated Now the polypeptide chain will have more amino acids added to it. As you can see, this chain elongation begins with a second incoming tRNA molecule binding to an mRNA codon in the A site. Because it happens to be a CUG codon, a tRNA with a GAC anticodon binds to it. This tRNA comes bearing the amino acid leucine (leu).

The met amino acid attached to the tRNA in the P site now bonds with the leu amino acid attached to the tRNA in the A site. In this process, the bond is broken between met and its original tRNA. A polypeptide chain is now attached to the tRNA in the A site, and, with this, the original tRNA is released from the P site. Now a kind of molecular musical chairs ensues: The ribosome can be thought of as moving down the mRNA line, three base pairs to the right. This action has two effects: The tRNA that had been in the A is relocated to the P site; and a new mRNA codon moves into the now-vacated A site. The rest of elongation amounts to a repetition of this process: A new incoming tRNA binds with the new codon in the A site, the growing polypeptide chain bonds with that amino acid, and so on.

Termination of the Growing Chain Recall that when you looked at the genetic code, there were three codons that acted as stop signals for polypeptide synthesis. Any time one of these "termination" codons moves into the ribosome's A site, it doesn't bind with an incoming tRNA, but instead brings about a severing of the linkage between the P-site tRNA and the polypeptide chain. Indeed, the whole polypeptide apparatus comes apart at this point, with the polypeptide chain being released to fold up and be processed as a protein. Translation has been completed.

Speed of the Process; Movement through Several Ribosomes How fast does this process go? Approximately five amino acids are added to a growing polypeptide chain every second. This figure is somewhat misleading, however, in that mRNA tapes often are "read" not by one ribosome, but by many. As you can see in **Figure 14.11**, several ribosomes—perhaps scores of them—might move, in effect, over a given mRNA transcript, with identical polypeptide chains then growing out of each ribosome.

14.5 Genetic Regulation

At this point, you have a basic understanding of what genes are: of how they call forth proteins, of how they make copies of themselves, and so forth. Even with this, however, some significant questions need to be addressed. If you think about it, in the entire discussion of genetics over the past few chapters, you have seen no reason for a given gene to turn on or off—to begin transcription or to end it. Scientists know that this does happen; that the cells in our pancreas, for example, turn out the protein insulin *intermittently*—in

**Figure 14.11
Mass Production**

a An mRNA transcript can be translated by many ribosomes at once, resulting in the production of many copies of the same protein.

b A micrograph of this process in operation. The figure shows two mRNA strands with ribosomes spaced along their length. In the upper strand, translation is underway and polypeptides can be seen emerging from the ribosomes.

Making Sense of "Nonsense" DNA

I magine that you pick up a book, a novel, that begins: "It was the best of times, it was the pzknlku ljkh uiop nk;lkj bhji ;lkjhjoiugh qpmzxd kjhb olkjb worst of times." Reading on, you see that things go on like this for hundreds of pages, right to the end. Baffled? So were scientists when, in 1977, they discovered that eukaryotes such as ourselves have genetic "books" that are filled with passages of such seeming nonsense.

You've already seen that the "sense" side to an organism's "genome," or store of genetic material, is the information contained in that organism's DNA. Such information comes in the form of As, Ts, Cs, and Gs that code for proteins. If a genome were nothing *but* sense, one "coding" triplet of these bases would follow another in succession, with only some housekeeping sequences in between them. What scientists began perceiving in the late 1970s, however, is that it is the *coding* sections of the genome that are few and far between. In human beings, for example, a mere three percent of the genome is thought to code for protein, which means that 97 percent of the human genome is noncoding. Some portions of the noncoding DNA are the housekeeping or "regulatory" sequences of DNA noted in the text, while other portions play a part in keeping chromosomes intact. Some DNA, however, has no known function at all and thus, years ago, was given a name that has stuck: **junk DNA**.

Such "junk" can be thought of as existing in two locations. First, there are the **introns** mentioned earlier, which are noncoding nucleotide strings *within* gene sequences. Introns bear some comparison to the nonsense strings in the first paragraph, in that they interrupt related strings of bases that *do* code for a given protein. It would not be unusual for a gene to contain half a dozen introns, with these intervening sections comprising most of the bases in the sequence.

Junk DNA can also lie, however, *in between* segments that code for genes. Some of this junk can be stupendously monotonous; there exist, for example, some 30,000 to 50,000 short segments of DNA in the human genome that consist of nothing more than a couple of bases repeated over and over, as in: CACACACACA . . .

This latter sequence, a molecular broken record, would certainly seem to fit the definition of junk, if by that we mean something that is so useless that it might as well be thrown away. And yet, research in recent years has cast doubt on whether any sequence in the genome is insignificant in this way. It is true that many sequences do not code for protein, but that is not to say that they serve no purpose at all.

Consider, for example, a sequence in the genomes of primates that was long thought to be the essence of junk. Known as the *Alu repeat*, this sequence is about 280 base pairs long, which is not particularly extensive in the con-

response to what we've eaten—rather than ceaselessly transcribing the insulin gene. So, what are the start and stop signals for this protein-synthesizing machinery? For that matter, why is it that only *pancreatic* cells turn out insulin when almost all other cells in the body possess copies of the insulin gene? To come to an understanding of these things, you need to learn something about *genetic regulation*.

DNA Is the Cookbook, Not the Cook

We have thus far thought of the "genome" or genetic inventory of a living thing as a kind of cookbook, containing recipes (DNA sequences) that lead to products (proteins) that undertake a wide variety of tasks (hastening chemical reactions, serving as structural components of tissue, and so on). It is tempting to go from these ideas to thinking that DNA's the boss, at least in terms of all

the metabolic processes that lie outside our conscious control. After all, it's giving "orders" for the body to put together these all-important proteins, isn't it?

Well, sticking with the metaphor, a cookbook does not give orders; it merely contains information, which an *agent* (called a cook) then carries out. It becomes clear that DNA is not the cook when you consider that by itself, DNA can't synthesize anything. Think back to the "transcription" by which a DNA sequence is copied onto an mRNA tape. The DNA double helix does not unwind itself, nor does it bring mRNA nucleotides into alignment with its own bases, nor does it snip the completed mRNA transcript, nor does it effect any of the other steps you looked at. Instead, all these tasks are carried out by *proteins*. (Recall that the protein complex known as RNA polymerase is central in carrying out the steps of transcription.)

text of the 3-billion base-pair human genome, until you consider that about 500,000 *copies* of Alu are believed to be dispersed throughout the human chromosomes. If so, this sequence alone may comprise some 4–5 percent of our DNA.

The Alu repeat is thought to have been *inserted* into the primate genome tens of millions of years ago in much

---

Some DNA has no known function at all and thus, years ago, was given a name that has stuck: junk DNA.

---

the same way that a lot of junk DNA gets there: by a process in which a virus inserts copies of its own genome into the genome of an organism it invades. Under this view, the Alu sequence is a kind of transient that came and stayed. It did no particular harm to its "host" genome, but did no good for it either; it simply made more and more copies of itself, which continued to be inserted into the primate genome. Thus, the Alu was thought of as falling into a particular category of junk DNA, **selfish DNA**, meaning DNA whose only function is to replicate itself.

In 1994, wondering if the Alu repeat might have some significance beyond this, Roy Britten of the California Institute of Technology looked at the base-pair order of thousands of Alu repeats in primate DNA. Britten found that these sequences have been highly "conserved," meaning they have retained their original structure over the millions of years. Further, such change that has occurred in the Alu sequence has clearly been "suppressed," as Britten put it, in certain *parts* of the Alu sequence—those that are likely to serve as binding sites for proteins that help turn primate genes on and off. It is difficult to explain why nature would have selectively retained these parts of the Alu sequence unless they were doing something for the primate hosts. Though scientists still don't know what that something is, it appears that the Alu repeat is not a genetic squatter; rather, it has been conscripted into duty in primates. Such a binding-site role has clearly been demonstrated for segments of noncoding DNA other than the Alu repeat.

In the view of some scientists, evidence such as this increasingly calls into question the whole notion of junk DNA. Some genome sequences may indeed turn out to be useless (except to themselves), and other "inserted" genetic sequences clearly can be harmful to hosts. But more and more we seem to be finding that if a sequence is in the genome, it probably is performing some function for the organism that houses it.

DNA is indeed the cookbook, but it is powerless to prepare its own recipes. A critical insight in genetic regulation is that, through chemical binding, information comes *to* the double helix (from other molecules), as well as from the double helix.

A Model System in Gene Regulation: The Operon

Genetic regulation can occur at any of the steps in the protein synthesis process. Here, however, the focus is on just one example of genetic regulation, one that takes place in the transcription stage of protein synthesis. You'll look at what's known as a "model" regulatory system in a model organism. The system is called the *operon*, and the organism is the *Escherichia coli* bacterium. *E. coli* has long been a model organism because it is simple, it can easily be experimented upon, and it reproduces in as little as 20 minutes.

Jacques Monod and François Jacob's Experiments

The sterling qualities of *E. coli* as a research subject made it an ideal choice for two French researchers, Jacques Monod and François Jacob, who in the 1950s wanted to learn the means by which genes turn on and off in living creatures. By then it was clear that in most cases, genes didn't simply *stay* on. The activity of at least some genes had to be *inducible*, meaning triggered by conditions in the organism's environment. One means of inducement seemed clear; it had been known for many years that bacteria would synthesize certain enzymes only if the substance these enzymes worked on (their "substrate") was present in the bacterial cell. Accordingly, Monod and Jacob focused on a group of enzymes that allow *E. coli* first to obtain the sugar lactose and then to break it down into two simple sugars. This latter step is a clipping operation in which lactose (commonly known as *milk sugar*) is broken into its component simple sugars, galactose and glucose.

Three genes are involved in lactose metabolism in *E. coli*, two of which are important for our purposes. As you can see in **Figure 14.12**a, there is a *y* gene that codes for a so-called *permease* enzyme, which transports lactose into the cell in the first place; then there is a *z* gene, which codes for the enzyme beta-galactosidase (or β-galactosidase), which does the lactose clipping mentioned before. (A third, *a*, gene produces a related enzyme; but this need not concern us, other than to note why it sits next to the other two.)

Activity Induced by Lactose

The genes involved in lactose metabolism were known to be inducible, and the **inducer**—the substance that prompted their activity—was known to be lactose. Put simply, in the presence of lactose, these genes turned on. Yet what was the control mechanism here? The experiments of Jacob and Monod eventually led them to propose that *E. coli*'s lactose metabolism is governed by an elegant multipart genetic system, which they dubbed the **operon**.

In the lac operon, the *z*, *y*, and *a* genes are transcribed as a unit onto a single messenger RNA transcript. "Upstream" from the *z* gene is a segment of DNA known as the **promoter** sequence. It is the actual binding site for the

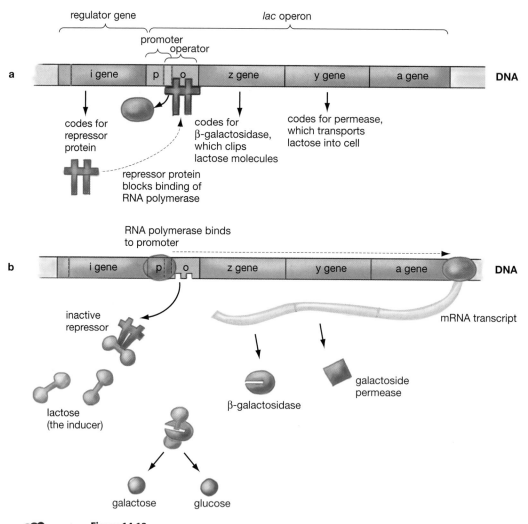

Figure 14.12
Genetic Regulation in Action
The lac operon enables the production of lactose-processing enzymes, but only when they are needed.

a In the absence of lactose, the repressor protein binds to the operator and inhibits the transcription of lactose-processing enzymes.

b When lactose is present, the repressor is inhibited from binding with the operator, thus allowing transcription and hence production of the lactose-processing enzymes.

RNA polymerase—the enzyme that brings about transcription by going down the line, pairing RNA nucleotides with their DNA counterparts. Note, however, that *in between* the promoter site and the *z* gene sequence is a region that Jacob and Monod called the *operator*. This sequence of DNA can exist in one of two states. In one, it is bound to a protein called a *repressor*, which effectively *blocks* the binding of RNA polymerase to the promoter, thus shutting down transcription. In the other state, it is unbound—free of the repressor—in which case transcription can proceed. Because the repressor is a protein, it is itself the product of a gene (the *i* gene) that lies just a little farther yet upstream.

Now, the critical question is: What causes the repressor either to bind with the operator (thus halting transcription) or to stay clear of it (thus allowing transcription)? The answer is the absence or presence of lactose. If you look at Figure 14.12b, you can see how this plays out. The critical factor is that the repressor protein has *two* binding sites: One of them binds to the operator, as already noted, but the other binds to *lactose*. When this lactose binding takes place, a repressor that is bound to an operator changes shape; like a clothespin that's been pinched, it opens up and is released from the operator site. The result? RNA polymerase is no longer blocked, meaning it can bind to the promoter site and begin transcription of the *z, y,* and *a* genes; they produce their enzymes, and lactose metabolism proceeds apace. When all the lactose has been broken down, however, the repressor molecules return to their original shape, meaning the shape that can bind to the operator, thereby shutting down gene transcription.

Thus did Jacob and Monod elucidate a detailed mechanism of gene regulation. Within this concept, they confirmed the existence of genetic sequences whose sole functions are *regulatory*: The operator, the regulatory *i* gene, and the promoter site. Such regulatory sequences stand in distinction to the genes that code for end-product proteins (*z, y,* and *a*). As a piece of scientific research, Jacob and Monod's work on the operon was so stellar that, with André Lwoff, they were awarded the Nobel Prize in 1965. For our purposes, the lac operon exemplifies the general concept of gene regulation in transcrip-tion. The proteins that are produced through DNA's instructions *feed back* on the DNA molecule itself, thus helping to control the production of *other* proteins.

14.6 The Magnitude of the Metabolic Operation

Before leaving the subject of protein synthesis, let's now consider the metabolic operation that is managed under genetic regulation and then something about the gene itself.

The Number of Proteins Utilized

The human genome is thought to code for somewhere between 50,000 and 100,000 proteins. Any one human cell, however, might produce as few as 5,000 of these proteins or perhaps as many as 20,000. Five thousand proteins seems, in any event, to be the minimum number found in all eukaryotic cells. A quantity this large may seem surprising, but think of what a cell must do. Immature red blood cells called *erythroblasts* are unique in their production of the protein hemoglobin; but like most cells, erythroblasts must be generalists as well as specialists. Almost all cells need proteins for ribosomes, membranes, a cytoskeleton—the machinery, in short, that makes cells work—and all cells must respond to their environment as well. These factors coupled with the special tasks that eukaryotic cells take on mean that most of them must contain a huge number of proteins. Some of these proteins are produced almost continuously, while others are "inducible," as with the lactose-clipping β-galactosidase in *E. coli*.

The Size of the Genome

Not surprisingly, to code for all these proteins, eukaryotic cells must also have an immense genome. As noted, the human genome is about 3 billion base pairs long. To make this large number a little less abstract, consider the following exercise. If we took the base sequence of this genome and simply arrayed the single-letter symbols for the bases on a printed page, going like this:

AATCCGTTTGGAGAAACGCGCCCTATT
GGCAGCAAGGCTCTCGGGTCGTCAACG
CGTATTAAACATATTTCAAGGCTCTA . . .

MediaLab Link
Human Genome Project, see p. 228.

it would take about 1,000 telephone books, each of them 1,000 pages long, merely to record it all. We're talking, then, about an unbroken series of these base symbols that goes on for a million pages. That is one measure of the size of the human genome. You can get an idea of the size of several different genomes if you look at **Table 14.2**.

14.7   What Is a Gene?

Finally, given the material presented in this chapter, you are ready for a more accurate definition of the most basic unit in genetics, the gene. Thus far a gene has been viewed as a length of DNA that codes for a protein. But you saw in the section on translation that *ribosomes* are made partly of RNA. Just like messenger RNA, this ribosomal RNA must be coded for by DNA. A segment of DNA unwinds and then forms base pairs with RNA nucleotides to produce an RNA sequence. Only this is a *ribosomal* RNA sequence that doesn't code for anything; it simply migrates to the cytoplasm to become part of a ribosome. So great is the cell's need for ribosomes that there is an entire section of the nucleus (called the nucleolus) whose DNA is constantly being transcribed for rRNA. Given this, here's a more accurate definition of a gene. A **gene** is a segment of DNA that brings about the transcription of a segment of RNA. Note that under this definition, a gene includes both the DNA regulatory sequences and the DNA segments that actually form base pairs with RNA.

On to Biotechnology

You have seen in this chapter that metabolic functioning is governed by a complex of molecules, only one of which is DNA. Yet to say that something is simply one part of a chain is not to say that it is not powerful. DNA may "only" be the cookbook, but think of the profound effect this single function stands to have. Following the metaphor, think how different a pie would be if an editing error in a recipe resulted in the insertion of the word *cinnamon* where *sugar* should be (and if our cook had to insert the cinnamon). Now imagine the change for the better if you could *change the recipe* to include sugar instead of cinnamon. Scientists have gotten very good in the last 20 years at figuring out one part of the genetic network: the chain from DNA to protein. The ability to *manipulate* DNA's instructions lies at the heart of what is commonly called biotechnology, and it is this exciting, promising area of biology that is the subject of Chapter 15.

Table 14.2
Sizes of Different Genomes

Different genetic sets. The genome sizes of a virus and six organisms, measured in number of base pairs.

Virus/Organism	Genome Size in Base Pairs
T2 virus	200,000
E. coli bacterium	4 million
Yeast	20 million
Fruit fly	200 million
Chicken	2 billion
Human	3 billion
Peas	9 billion

Chapter Review

Key Terms

Summary

14.1   The Structure of Proteins

- Proteins are composed of building blocks called *amino acids*. A string of amino acids is called a *polypeptide chain*; once such a chain has folded into its working three-dimensional shape, it is a *protein*. Though there are tens of thousands of different proteins, all of them are put together from a starting set of 20 amino acids. It is the order in which the amino acids are linked in a polypeptide chain that determines which protein will be produced.

14.2 Protein Synthesis in Overview: Transcription and Translation

- There are two principal stages in protein synthesis. The first stage is transcription, in which the information encoded in DNA is copied onto a length of messenger RNA, which then moves from the cell nucleus to structures in the cytoplasm called *ribosomes*. The second stage is translation, in which a polypeptide chain is produced in accordance with the instructions encoded in the mRNA sequence.

- The information in DNA is transferred to messenger RNA through complementary base pairing. The length of mRNA that results from this base pairing is called a *transcript*.

- Each three coding bases of DNA pairs with three mRNA bases, but each group of three mRNA bases then codes for a single amino acid. Each triplet of mRNA bases that codes for an amino acid is called a *codon*.

- Transfer RNA serves as a bridging molecule, because it can bond with both mRNA and amino acids. A given tRNA molecule bonds with a specific amino acid in the cell's cytoplasm and then transfers that amino acid to a ribosome in which an mRNA transcript is being "read." When the tRNA arrives, another portion of it, called an *anticodon*, bonds with the appropriate codon in the messenger RNA chain.

- Specific mRNA nucleotide triplets code for specific amino acids or for "start" and "stop" sequences in protein synthesis. The entire inventory of these linkages is known as the *genetic code*.

14.3 The Importance of the Genetic Code

- The near-universality of the genetic code in living things is evidence for a single, common ancestor for all life that now exists on Earth. A common genetic code is the reason viruses can utilize the genetic machinery of other living things, but it is also the basis for today's biotechnology industry.

14.4 A Closer Look at Protein Synthesis

- Many of the sequences of DNA in eukaryotes do not code for proteins. These intervening sequences or *introns* are spliced out of mRNA transcripts before the transcripts leave the cell nucleus. The protein-coding portions of the genome are called *exons*. Other sequences of DNA are purely regulatory, serving only to affect transcription. Still other sequences have no known function.

14.5 Genetic Regulation

- Protein synthesis can be facilitated or impeded at many points. This process is called *genetic regulation*. The operon system reviewed in the text concerned the regulation of mRNA transcription. In such regulation, the proteins that are produced through DNA's instructions feed back on the DNA molecule itself, thus helping to control the production of other proteins.

14.6 The Magnitude of the Metabolic Operation

- The human genome is thought to code for somewhere between 50,000 and 100,000 proteins. Any one human cell, however, might produce as few as 5,000 of these proteins or perhaps as many as 20,000. With the special tasks that eukaryotic cells take on, most of them must contain a huge number of proteins. Not surprisingly, to code for all these proteins, eukaryotic cells must also have an immense genome.

14.7 What Is a Gene?

- A gene can be defined as a segment of DNA that brings about the transcription of a segment of RNA.

Understanding the Basics

Multiple-Choice Questions

1. The working ability of a protein is found in its
   a. polypeptide sequence
   b. ability to encode DNA polymerase
   c. two-dimensional structure
   d. three-dimensional structure
   e. both b and c

2. How does DNA differ from RNA?
   a. DNA uses deoxyribose sugar RNA uses ribose sugar
   b. DNA is a double-stranded molecule RNA is single-stranded
   c. DNA uses the bases A, U, C, G; RNA uses the bases A, T, C, G
   d. a and b
   e. all of the above

3. The "t" of tRNA stands for:
   a. translator   d. teleporting
   b. transfer   e. tracking
   c. transcribing

4. What mRNA sequence signals the start of a sequence to be translated?
   a. ATG   d. UAG
   b. UGA   e. UAA
   c. AUG

5. What happens during mRNA editing
   a. Introns are removed in the nucleus prior to export.
   b. Exons are removed in the nucleus prior to export.
   c. Introns are removed in the cytoplasm prior to translation.
   d. Exons are removed in the cytoplasm prior to translation.
   e. Introns are removed in the cytoplasm following translation.

6. Which of the following statements is true regarding ribosomes?
   a. They are composed of two separate units of rRNA combined with protein.
   b. They are used to translate the mRNA sequence into a protein composed of amino acids.
   c. They have two different sites for tRNA binding: P, where tRNA exits the ribosome and A, where new tRNA attaches to the ribosome.
   d. a and b only
   e. all of the above

7. Why can transfer RNA be referred to as a "bridging" molecule?
   a. It forms a chemical bridge between DNA and messenger RNA.
   b. It creates the linkage between the base pairs of the double helix.
   c. It links species through evolution.
   d. It links ribosomes together.
   e. It forms chemical bonds with both amino acids and messenger RNA.

8. The minimum number of proteins found in all eukaryotic cells is believed to be
   a. 100,000   d. 5,000
   b. 50,000   e. 2,000
   c. 20,000

9. Feedback was employed in the example of genetic regulation in the text. What was feeding back onto what?
   a. Proteins produced through the information encoded in DNA went on to regulate the transcription of that same DNA.
   b. Ribosomes produced through the information encoded in DNA went on to serve as the sites of protein synthesis.
   c. Messenger RNA strands synthesized as a complement to DNA went on to reduce DNA transcription.
   d. Transfer RNA molecules limited translation at the ribosomes.
   e. Transfer RNA molecules sped up translation in mitochondria.

10. The genetic code is the same in:
    a. All organisms except bacteria
    b. Nearly all organisms
    c. Both organisms and crystals
    d. All organisms except plants
    e. All organisms except humans

Brief Review

1. What two great tasks are carried out by our genetic machinery?

2. What are the three types of RNA discussed in this chapter? What do their abbreviations stand for? What are their functions?

3. What is the difference between transcription and translation?

4. What is one detrimental effect of the universal nature of the genetic code?

5. What is the difference between introns and exons?

6. In the process of gene regulation, define the following terms:
   a. inducer
   b. operon
   c. promoter
   d. operator
   e. repressor

7. How many DNA bases does it take to code for an RNA codon? How many amino acids does an RNA codon code for?

8. In what sense does each human being share in an ancient informational linkage?

Applying Your Knowledge

1. Why does the activity of at least some genes need to be inducible?

2. Given the following sequence of DNA, what would the complementary mRNA sequence be?

   TACCCGTATACGATCATGGTCAA-GTCGTAC

3. Given the following sequence of mRNA, what would the resulting amino acid sequence be?

   AUGAAACGGGGACCAAUGGAUA-ACUAA

4. In a hypothetical situation, you have identified a new species that uses six bases, instead of four, in its genetic code. It has also been discovered that the proteins this species forms are made up of 220 amino acids. How many bases need to be included in each codon in order to ensure that each amino acid has a separate and distinct encoding sequence?

5. In Chapter 13, point mutations were introduced as sources of error in DNA replication. Shown below are five mRNA sequences, including one "normal" sequence and four different point mutations. What effect will each of these point mutations have on the amino acid sequence? What does this indicate about the significance of the location and type of point mutation that occurs? (Remember that an mRNA initiates an amino acid sequence with a start codon and ends the sequence with a stop codon.)

   Normal mRNA: AUG AAA CAU GCA CUA AUG UAA CCU
   Point mutation 1: AUG UAA CAU GCA CUA AUG UAA CCU
   Point mutation 2: AUG AAA CAU CCA CUA AUG UAA CCU
   Point mutation 3: AUG AAA CAU GCA CUA AUG AAA CCU
   Point mutation 4: AUG AAG CAU GCA CUA AUG UAA CCU

6. You hear that scientists have discovered an animal-like organism living at the bottom of the ocean that not only looks different from anything we've seen, but that employs a different genetic code. (It uses DNA and the standard 20 amino acids, but in a completely different linkage to one another.) What are some hypotheses that could account for this?

Group Activity

When an RNA polymerase reads a strand of DNA, it is known that the RNA polymerase transcribes only one of the two strands. Additionally, in any DNA sequence the RNA polymerase may read in either direction, and it may start in one of three "reading frames" on each strand. That is, because there are three bases in each codon, a sequence of six bases may code four different codons (codons are in bold):

    CGATTCG → reading frame 1: **CGA TCG**
                  reading frame 2: C **GAT** CG
                  reading frame 3: CG **ATC** G

Sequences of these codons that begin with a start codon, and continue for a significant length without encountering a stop codon are called *open reading frames,* or ORFs.

Using this knowledge, in the following DNA sequence, which of the two strands is more likely to be transcribed? Why? What is the mRNA sequence and the amino acid sequence that results? Compare your answers with those of your classmates.

Strand 1: CTACCAGGAATGATATAAGTGTAAG
Strand 2: GATGGTCCT TACTATAT TCACAT TC

15

The Future Isn't
What It Used to Be
Biotechnology

Hello, Dolly!
(page 289)

Restriction enzymes.
(page 287)

From bacteria to a protein.
(page 288)

In our long walk through

the subject of genetics, you've had many occasions to think about genes: What is a gene, and what does it do? How does it function, and how can its functioning be disrupted? But now here's a very different question: What is a gene *worth*?

It's a bracing sort of question, because its underlying premise seems to be that a gene can be bought and sold. But this is, in a sense, the state of affairs today; through the patent process, what someone can lay claim to is *knowledge* about what a gene does. That knowledge can then be sold, or more likely licensed to a company that can do something with it.

Within this framework, in 1995, a Southern California biotechnology company, Amgen, Inc., decided what a gene called ob was worth to it: $20 million. Amgen paid that amount to Rockefeller University, whose scientists had deciphered the base-pair sequence of the *ob* gene and figured out what *ob* (for "obese") codes for: a protein hormone, called leptin, that is produced in fat cells and that acts to regulate the appetite and metabolic rate of mice (**see Figure 15.1**). Amgen paid so much for *ob* not because it

was interested in controlling the weight of mice, of course. It was interested in products that might control the weight of *people*, the idea being that *ob* may function in human beings as it does in mice.

If this is true, Amgen and Rockefeller may have a lucrative future ahead of them. Leptin subsequently was shown to have dramatic effects on mice: Grossly overweight mice, lacking a functioning *ob* gene, dropped 40 percent of their body weight within a month when given daily leptin injections. By 1998,

Figure 15.1
Slimmer through Genetics
Both of these mice have a defect in a gene called *ob* (for obese), but the mouse on the right received daily injections of the protein the *ob* gene codes for, leptin. Mice given the leptin ate less and burned more calories.

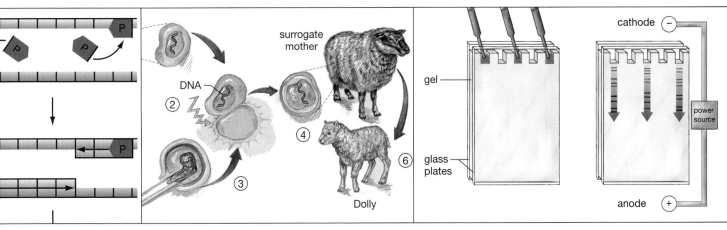

A copying machine.
(page 291)

How it was done.
(page 290)

Visualizing DNA.
(page 292)

results from human studies were being reported and the news was encouraging; overweight subjects who were given the highest doses of the protein lost an average of about 16 pounds over a period of six months.

Even so, we might ask: $20 million—just as an initial payment—for the right to develop products from an unproven protein? Well, consider the potential. Americans alone spend some $36 *billion* a year trying to keep slim; if leptin can indeed become a magic bullet for obesity, Amgen may be seen one day as having gotten a rock-bottom bargain. Then again, leptin could be a bust, but that's the nature of risk out on the biotech frontier.

There are other measures of worth besides money, of course. Consider the case of a Virginia man who, on the basis of visual identification, was convicted of rape in 1985. Ten years later, two crusading New York attorneys were able to obtain a vaginal swab taken after the crime was committed and, by means of DNA testing, prove that the accused man could not have been the rapist. In this case, a stretch of DNA was worth an innocent man's freedom. Hundreds of innocent people have been freed through this process in recent years.

The power of what is usually called "biotechnology" can be seen in many areas beyond these: in the ability to let abnormally short children grow to a more normal height; in the hope, as yet unrealized, that we may be able to cure such genetic diseases as cystic fibrosis and muscular dystrophy; in the potential to grow more hardy, disease-resistant crops to help feed a hungry world.

All of these things are the fruits of a particular sort of labor: The research in basic genetics that has been going on since the time of Mendel. If you wonder what use we have for knowledge about such things as DNA sequences and mRNA transcripts, you need look no further than an innocent man's freedom or the satisfying height of an adolescent treated with genetically engineered human growth hormone.

As you'll see, however, biotechnology raises a host of troubling questions as well: questions of manipulating an organism's genome for human ends; questions of safety; and questions of what limits there should be on the uses of genetically engineered products.

15.1 What Is Biotechnology?

In a strict sense, **biotechnology** can be defined as the use of living organisms to create products or to facilitate processes. A brief look at one real-world biotech product can show us what this notion can mean in practice.

The human growth hormone (HGH) noted earlier normally is secreted by the human pituitary gland and is critical to the growth of children. A faulty pituitary gland can greatly reduce the amount of HGH children have in their system, leaving them abnormally short. For years, the only way to get HGH for such children was to laboriously extract it from the pituitaries of dead human beings, a practice that not only yielded too little HGH to go around but also turned out to be unsafe.

Enter biotechnology, which in the mid-1980s produced synthetic HGH in the following way. Using collections of human cells, the gene for HGH was isolated from the human genome, snipped out of it, and inserted into the *E. coli* bacterium. Each bacterium that took on the HGH gene thereupon began transcribing and then translating this gene, which is to say turning out a small quantity of HGH. These bacteria were then grown in vats by the billions. The result? Collectible quantities of HGH, clinically indistinguishable from that produced in human pituitary glands, manufactured by a biotech firm and shipped to pharmacies worldwide.

This account may make genetic engineering sound like a straightforward process, but consider the knowledge that has to be in place for a process such as HGH synthesis to go forward. How do you cut DNA to get a human gene out of the human genome? How do you then get this gene working away inside the bacterial mini-factories? Let's now take a look at some of the basics of biotechnology, so that you'll be in a better position to consider ways this kind of work stands to affect humankind. Imagine as a starting point that, as with HGH, the goal is to produce quantities of a hypothetical human protein through the use of a living organism.

15.2 Some Tools of Biotechnology

Restriction Enzymes

In the early 1970s it became possible for scientists to cut genomes at particular places, with the discovery of **restriction enzymes**.

These proteins occur naturally in bacteria, where they serve the purpose of cutting up foreign DNA, such as that of invading viruses. In isolating restriction enzymes, scientists found that many of them had a wonderful property: They didn't just cut DNA randomly; they cut it at very specific places. Here is how it works with an actual restriction enzyme called *Bam*HI. The two strands of DNA's double helix are complementary, as we've seen, and they also turn out to run in opposite directions. Thus, the sequence GGATCC would look like **Figure 15.2a** if we were viewing both strands of the helix.

Now, the *Bam*HI restriction enzyme will move along the double helix, leaving the DNA alone *until* it comes to this series of six bases, known as its **recognition sequence**, and here it will make a cut, always between adjacent G nucleotides (**see Figure 15.2b**).

When another *Bam*HI molecule encounters another GGATCC sequence, it too will make a cut, which effectively will be like making a second cut in a piece of rope, giving us rope *fragment* (**see Figure 15.2c**). *Bam*HI's recognition sequence may be GGATCC, but another restriction enzyme will have a *different* recognition sequence and will likely make its cut between a different pair of bases. Indeed, nearly 1,000 different restriction enzymes have been identified so far, which cut in hundreds of different places. The fact that they make cuts at so many specific locations has given scientists a terrific ability to cut up DNA in myriad ways.

Note that with *Bam*HI, each of the resulting DNA fragments has one strand that protrudes. Restriction enzymes that make this kind of cut are particularly valuable, for they produce "sticky ends" of DNA, so named because they have the potential to *stick to* other complementary DNA sequences. In the fragment on the left, for example, the protruding sequence CTAG could now easily form a base pair with *any* piece of DNA whose sequence is the complementary GATC. So great is the need for restriction enzymes that they can now be ordered from biochemical suppliers, much as a person might order a set of socket wrenches from a hardware store.

Another Tool of Biotech: Plasmids

From the overview of manufacturing human growth hormone, recall that the human gene for HGH is inserted into *E. coli* bacteria, which then start turning out quantities of this protein. The question is: How did this human gene get into a bacterium? Several methods of transfer are available, but for now let us focus on a specific kind of DNA delivery vehicle. As it turns out, bacteria have small DNA-bearing units that lie *outside* their single chromosome. These are the **plasmids**, extra-chromosomal rings of bacterial DNA that can be as little as 1,000 base pairs in length (**see Figure 15.3**). Plasmids can replicate independently of the bacterial chromosome; but just as important for biotech's purposes, they can *move into* bacterial cells.

How do plasmids do this? Bacteria are capable of taking up DNA from their surroundings, and they then proceed to incorporate this DNA into their own genetic machinery. Appropriately enough, this process is known as **transformation**: a cell's incorporation of genetic material from outside its boundary. Some bacterial cells are naturally adept at transformation, while others, such as *E. coli*, can be induced to perform it by means of chemical treatment. Critically, plasmid DNA can be taken in via transformation.

MediaLab Link
Plasmids, see p. 304.

Figure 15.2
The Work of Restriction Enzymes

a A portion of a complementary strand of DNA has the highlighted recognition sequence GGATCC.

b A restriction enzyme moves along the DNA strand until it reaches the recognition sequence and makes a cut between adjacent G nucleotides.

c A second restriction enzyme makes another cut in the strand at the same recognition sequence, resulting in a DNA fragment.

Figure 15.3
Transfer Agent
Plasmids are small rings of bacterial DNA that are not a part of the bacterial chromosome. They can exist outside bacterial cells and then be taken up by these cells. This artificially colored micrograph shows a type of plasmid, from *E. coli* bacteria, that is commonly used in genetic engineering.

Using Biotech's Tools: Getting Human Genes into Plasmids

At this point, you know about a couple of tools in the biotech tool kit: restriction enzymes and the transformation process involving plasmids. Let's now see how they work together. As you can see in **Figure 15.4**, the process starts with a gene of interest in the human genome. The first step is to use restriction enzymes on this human DNA. Knowing, say, the starting and ending sequence of the gene of interest, a restriction enzyme is selected that allows the genome to be cut into a manageable fragment including this gene, preferably in a sticky-ended form.

Here's where the beauty of restriction enzymes really comes into play. If the same restriction enzyme is now used on the DNA of isolated *plasmids*, the result is *complementary* sticky ends of plasmid and human DNA. In other words, these segments of human and plasmid DNA, through sticky-ended base pairing, fit together like puzzle pieces.

When the DNA fragments are mixed with the "cut" plasmids, this is just what happens: Human and plasmid DNA form base pairs, and the human DNA is incorporated into the plasmid circle. With this, a segment of DNA that was once part of the human genome (the fragment that includes the selected gene) has now been *re-combined* with a different stretch of DNA (the plasmid sequence). Thus arises the term **recombinant DNA**, defined as two or more segments of DNA that have been combined by humans into a sequence that does not exist in nature.

**Figure 15.4
How to Use Bacteria to Produce a Needed Human Protein**

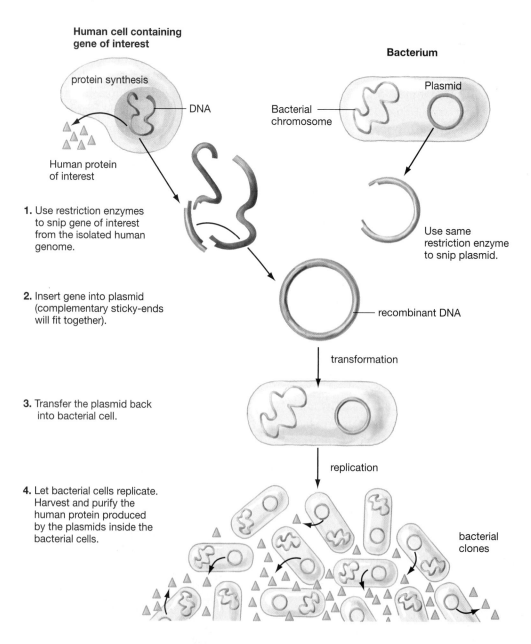

Human cell containing gene of interest

protein synthesis

DNA

Human protein of interest

Bacterium

Plasmid

Bacterial chromosome

1. Use restriction enzymes to snip gene of interest from the isolated human genome.

Use same restriction enzyme to snip plasmid.

2. Insert gene into plasmid (complementary sticky-ends will fit together).

recombinant DNA

transformation

3. Transfer the plasmid back into bacterial cell.

replication

4. Let bacterial cells replicate. Harvest and purify the human protein produced by the plasmids inside the bacterial cells.

bacterial clones

Getting the Plasmids Back inside Cells, Turning out Protein

To this point, what has been done is to produce a collection of independent plasmids having a human gene as part of their makeup. Remember, though, the goal is to turn out quantities of whatever protein the human gene is coding for. To do this requires a vast quantity of plasmids working away, which means working away *back inside* bacterial cells, and it's here that transformation comes into play. If the plasmids are put into a medium containing compatible bacterial cells, a few of these cells will take up plasmids through transformation, after which the plasmids start *replicating* along with the bacterial cells themselves. As one bacterial cell becomes two, two become four, and so on, the plasmids are replicating away as well, and as the cell count reaches into the billions, collectible quantities of the protein begin to be turned out via instructions from the human gene inserted into the plasmid DNA.

A Plasmid Is One Kind of Cloning Vector

In this example, plasmids were the vehicle that served first to take on some foreign DNA and then to ferry it into working bacterial cells. Note, however, that plasmids are only one vehicle among several that can be used. Collectively, such vehicles are known as **cloning vectors**, "vector" simply meaning a transfer agent. Next to plasmids, the most common cloning vector is a type of virus known as a **bacteriophage**.

15.3 Cloning and the Wider World of Biotechnology

The word *cloning* in the preceding paragraph may seem at once familiar and foreign. Thanks to movies and recent real-life events, this word has taken on some sinister implications, as if biological cloning were inherently a Frankenstein-like procedure. But it need be no more threatening than the process just reviewed. To **clone** simply means to make an exact genetic copy of. A clone is one of these exact genetic copies. In the example you've been looking at, what's being cloned is a gene, copies of which were made in order to obtain quantities of the protein it codes for. Once a sufficient number of these clones exist, the trick becomes to extract this pro-

tein and then purify it, so that what is shipped to pharmacies is a substance that bears no traces of the living cells that produced it, other than the protein itself.

This said, the range of what can be cloned has expanded enormously in the past few years. The potential for such cloning appears almost limitless—something that has inspired both great optimism and great concern.

Biotechnology has now joined with various reproductive technologies to produce clones not just of genes or cells, but of whole, complex living things. The most famous example is Dolly the sheep, cloned by researcher Ian Wilmut and his colleagues in Scotland in 1997 (see **Figure 15.5**). This is **reproductive cloning**. Powerful in its own right, it gains added potential when combined with the basic biotechnology processes you've been reviewing.

How Dolly Was Cloned

Dolly the sheep is a clone as that term is defined: She is, to a first approximation, an exact genetic replica—in this case, of another sheep.

MediaLab Link
Cloning, see p. 304.

Figure 15.5
Revolutionary Sheep
Dolly, the first mammal ever cloned from an adult mammal, is shown here as a young sheep with her surrogate mother. Note that Dolly is white-faced, like the sheep she was cloned from, while her surrogate mother is black-faced.

Here's how she was produced (**see Figure 15.6**): A cell was taken from the udder of a six-year-old adult sheep and then grown in culture in the laboratory, meaning that the original cell divided into many "daughter" cells. While this was going on, researchers took an *egg* from a second sheep and removed its nucleus, meaning they removed all its nuclear DNA. Then the udder cell (which had DNA) was placed next to the egg cell (which did not), and a small electric current was applied to the egg. This had two effects: It caused the two cells to fuse into one, and it mimicked the stimulation normally provided when a *sperm* cell fuses with an egg. With this, the udder cell DNA began to be reprogrammed. As a result of this reprogramming, the fused cell started to develop as an embryo. (Though the egg cell had its DNA removed, it still contained all kinds of egg-cell proteins whose normal function is to trigger development of an embryo. It was these factors still in the egg that reprogrammed the donor-cell DNA.) After the embryo had developed to a certain point, it was implanted in a third sheep, who served as a surrogate mother. The result, 21 weeks later, was a perfectly normal lamb, Dolly, who has now given birth to a healthy lamb of her own.

Every cell in Dolly has DNA in its nucleus that is an exact copy of the DNA in the six-year-old donor sheep. Thus Dolly is a clone of that sheep. There was no second parent contributing chromosomes to Dolly, no mixing of genetic material at the moment of conception—just a copying of one individual from another.

Reproductive Cloning and Recombinant DNA

Ian Wilmut and his colleagues were not so much interested in cloning a sheep as they were in coupling the power of reproductive cloning to the techniques of *recombinant DNA* that you looked at earlier. Indeed, the workers at Wilmut's research institute then went on to produce Polly, who is not only a clone, but a clone who has had a human gene inserted into her genome. People who suffer from the disease hemophilia B are missing a blood-clotting protein called *factor IX*. Using the same kinds of DNA-cutting techniques described in this chapter, scientists snipped the gene that codes for factor IX out of some human DNA and inserted it into a sheep donor cell. This cell was then fused with a denucleated egg, just as with Dolly, and the result was Polly. Now, here's the import of this: Polly secretes the human factor IX in her milk. She is expressing a needed human protein. You might say: Hasn't our protein "factory" now simply become a sheep instead of a bacterium? This is true, but because, compared to a bacterium, the sheep's genetic machinery is more like a human being's, some of the proteins the sheep produces are in a more useful chemical form than those produced by bacteria.

This only scratches the surface of the potential for such "transgenic" clones, however. To give but one other example, thousands of people die each year because their liver or heart or kidneys failed and they could not find a compatible organ donor. The essential problem is that the human immune system will attack any organ that it recognizes as "foreign," and of course most donated organs are recognized in that way. Imagine, though, taking a pig donor cell and removing from its DNA the genes that code for the proteins perceived as foreign by the human immune system. Now fuse this cell with a denucleated egg, and the egg grows into a pig.

Figure 15.6
How Dolly Was Cloned

1. A cell was taken from the udder of a six-year-old white sheep and then allowed to divide many times in the laboratory. Meanwhile an egg was taken from a second black sheep.

2. One of the resulting udder cells was selected to be the "donor" cell for the cloning. Meanwhile, using a slender tube called a *micropipette,* the DNA was sucked out of the egg.

3. The donor cell and egg were put next to each other, and an electric current was applied to the egg cell.

4. This caused the two cells to fuse and prompted an activation that reprogrammed the donor-cell DNA. This caused the fused cell to start developing as an embryo.

5. After some incubation, the embryo was implanted in a third sheep, who served as the surrogate mother.

6. This mother gave birth to Dolly, who grew into a healthy adult.

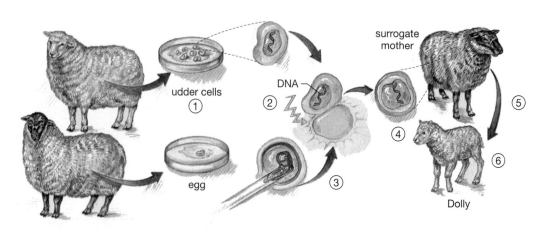

This is a pig, however, whose organs can be "harvested" for transplantation into human beings, because its organs will not elicit a strong human immune response. You'll read more about the ethical ramifications of such work later in the chapter.

15.4 Getting Quantities of a DNA Segment Quickly: PCR

Thus far, you've seen a couple ways of turning out human proteins through the use of recombinant DNA. Sometimes, however, the goal is not to get proteins, but to "amplify" DNA—to get many copies of a segment of DNA when the starting sample of DNA is small. The technique for doing so is called the **polymerase chain reaction**, or PCR for short (**see Figure 15.7**).

The essence of this molecular copying machine is very simple: Heat a starting quantity of DNA until the two strands of its double helix separate, resulting in two single strands of DNA. Then add a collection of individual DNA nucleotides, and DNA polymerase (which, remember, goes down the line on single-stranded DNA affixing nucleotides to its available bases). Also add some DNA "primer" sequences—short sequences of DNA that act as signals to DNA polymerase saying "start base pairing here." As the mixture cools, primers will attach to one of the ends of each of the now-separate strands of DNA. Then the DNA polymerases go down the line linking nucleotides, thus producing strands that are complementary to the original strands. The result is *two* double-stranded lengths of DNA, both identical to the original strand. In short, the DNA sample has been doubled. Then the entire process just described is repeated. By the time 10 cycles have passed, the DNA has been "amplified" a thousand times; with 20 cycles there is a million-fold amplification.

This technique is useful any time researchers want to quickly get a quantity of DNA, but it is particularly useful when the starting sample of DNA is very small. Crime scenes may contain tiny amounts of human blood or semen, for example, but larger amounts are needed for testing.

PCR was invented in 1983 by Kary Mullis, then a researcher at a California biotech firm. So important has the process become to science that in 1993 Mullis was awarded the Nobel Prize in Chemistry for his achievement.

15.5 Visualizing DNA Sequences

In working with DNA, scientists need a way to compare one segment of DNA to another. Does a given segment of DNA divide into three segments or five when a particular restriction enzyme is used on it? What is the exact sequence of bases *within* any of these segments? This question refers to DNA "sequencing," whose importance becomes apparent when you recall that, among the thousands of bases that make up the human hemoglobin gene, it is the substitution of *one base* (a T for an A) that brings about the disease sickle-cell anemia. Scientists can both sequence DNA and judge the sizes of larger DNA fragments through a process of visualization that is a kind of scientific "bar coding."

How can something as small as a segment of DNA be visualized? The key is that

**Figure 15.7
DNA Copying Machine**
The polymerase chain reaction (PCR) makes copies of a given length of DNA very quickly.

1. The section of DNA a researcher is interested in is heated, causing the two strands of the double helix to separate.

2. As the mixture cools, short DNA sequences called *primers*, which can form base pairs with the ends of the DNA strands, are added to the mix. The primers form base pairs with their respective strands.

3. DNA polymerase goes down the line, synthesizing complementary DNA strands.

4. The end result is a doubling of the original length of DNA. Then the process is repeated many times.

DNA carries a negative electric charge. As such, it is attracted to a positive charge. This fact helps scientists stage a kind of sprinting match of DNA fragments in order to separate them by size.

Running DNA through a Gel

There is a material, called a polyacrylamide gel, that in some ways is like the food gelatins we're used to. A quantity of this gel has running through it a series of microscopic spaces (or "pores"). Given this, polyacrylamide gel acts like a sifter: A substance such as DNA can pass through it, but only with some coaxing and at a controlled rate.

The starting length of DNA might be a whole genome's worth, but more likely will be some subset of a genome; whatever the case, the process begins with using a given restriction enzyme to chop up a starting sample—from a collection of cells—into a quantity of fragments. Chopping up these segments with the same restriction enzyme results in a huge number of identical copies of fragment 1, a huge number of copies of fragment 2, and so on. As you know, restriction enzymes don't make evenly spaced cuts throughout a length of DNA; they cut at specific recognition sequences. Thus fragment 1 will be a different size than fragment 2, which in turn will be a different size than fragment 3, and so on.

The question is, how do researchers learn the size of each fragment? The answer comes from the DNA sprint through the agarose gel. As you can see in **Figure 15.8**, a tray is used that has two critical features. First, it has electrodes at each end—one negative and one positive—and, second, it has a series of indentations or "wells," which lie on the same end of the tray as the negative electrode.

The process begins by pouring gel into the tray. Then the tray is placed in a solution that conducts electricity. Then the collection of DNA fragments is poured into the wells, which lie on the end of the tray *opposite* that of the positively charged electrode. In electricity, opposites attract, and the negatively charged DNA fragments are thus drawn to the positively charged end of the tray. But: They must make it through the gel to get there. Given this obstacle, it is the *smallest* of the fragments that will travel the farthest in a given amount of time, while the largest of the fragments will travel the least amount of distance.

Learning the Size of the Fragments

What this means is that the fragments are separated out by size. A fragment that is, say, one thousand nucleotides long will make it farther through the gel than a fragment that is two thousand nucleotides long. Researchers can *see* where each fragment has ended up in the tray because, after putting the DNA fragments in the wells, they pour in a dye that stains the DNA. The result is a series of bands within the gel (similar to supermarket bar codes), with each band representing a huge collection of DNA fragments of about the same size. The

Figure 15.8
Visualizing Lengths of DNA
In gel electrophoresis, negatively charged DNA fragments travel through the porous gel toward the positive end of the tray when a charge is applied to the gel. Large fragments, however, do not travel as far in a given amount of time as small fragments. Thus the fragments separate out by size. The first "stripe" at the top of a lane represents a collection of fragments of a given size, the second stripe a second collection of fragments of a smaller size, and so on.

size of each fragment is learned by putting, in one "lane" of the tray, a series of so-called reference fragments, meaning fragments of known size. If one of the reference fragments was 10,000 nucleotides long, and if one of the sample's fragments traveled far enough through the gel to line up perfectly with it, the researchers know that their sample fragment is roughly 10,000 nucleotides long as well.

This technique of sizing DNA fragments is called **gel electrophoresis**, a name that makes sense considering the gel that's used, the electricity, and the fact that *phoresis* is a Greek word meaning "to be carried."

Sequencing Requires a Different Operation

Characterizing a stretch of DNA by the number and size of its fragments is not, however, the same as sequencing. Knowing the size of a given fragment is not the same as knowing the sequence of As, Ts, Cs, and Gs *within* that fragment. The details of sequencing need not concern us here. Suffice it to say that there are two primary "manual" methods of sequencing small segments of DNA, both of which use gel electrophoresis in a way that produces fragments that differ in size by one nucleotide. The bands that result, corresponding to individual As, Ts, Cs, and Gs, can be "read" by a researcher, thus yielding the sequence of the DNA segment. The sequencing of larger stretches of DNA is now facilitated by automated sequencing machines, which analyze bases that have been "tagged" with fluorescent dyes (**Figure 15.9**).

Having learned how separation of fragments is done, you might well ask, to what end? Electrophoresis is used routinely in all kinds of scientific research. But to focus on one very specific use of it, imagine that two of the lanes of DNA fragments in Figure 15.8 came from two different suspects in a violent crime, while the third lane came from the person who committed the crime—say, from blood spilled by the perpetrator at the crime scene. If a suspect's bands *line up* with those of the perpetrator, this can be evidence that the suspect *is* the perpetrator. You can learn about this application of gel electrophoresis by looking at "DNA in the Courtroom" on page 294.

15.6 The Human Genome Project

Imagine that you go to a library and ask for a particular book on the history of Russia, only to be told that the library has no idea whether it owns a copy of the book. The librarian is pretty sure that the library has books on Russian history, but really doesn't know where they are in the library's stacks. As you start walking around this library, you realize that there has been some orderly cataloguing of a few collections, but they are few and far between. The essential message here is lots of books, but little knowledge about where they are in relation to one another or about what information each book might contain.

This imaginary library bears comparison to the human genome and our knowledge of it. If you are a scientist confronted with a genetic issue, traditionally you have been confronted with a *genetic* library (the human genome) that has lots of books in it (genes), but little cataloguing of these books and little knowledge of the information each book contains. This situation is, however, changing rapidly.

In a real library, you might go to its electronic catalogue and type in: Huckleberry Finn, or even just Huckle*, after which you could find the book itself and look through it, word by word and letter by letter. If you had a counterpart *genetic* catalogue, all you would need to type in would be, say, a few nucleotides from your gene's sequence. With

MediaLab Link
Human Genome Project, see p. 228.

Figure 15.9
Sequencing by Machine
Large-scale DNA sequencing is carried out today with automated DNA sequencing machines, such as the one shown here.

DNA in the Courtroom

A rape has been committed, but no fingerprints were left at the scene and there are no witnesses, apart from the victim. A suspect has been identified by the victim, but what physical evidence is there to corroborate her testimony? How can we tell whether the suspect is in fact the criminal?

In the mid 1980s, British geneticist Alec Jeffreys realized that the individual patterns each of us carries in our DNA might serve as just this kind of physical evidence. As a result of his work, courts in the United States have, since 1988, accepted so-called **forensic DNA typing**, better known as "genetic fingerprinting," in criminal cases.

The essence of this work is to get two sets of physical samples: first, a sample left by the perpetrator at the crime scene in such forms as blood or semen; and second, a blood sample from the suspect. By comparing the DNA patterns of both samples, odds can be established as to whether the suspect *is* the perpetrator. (Conversely, it would be possible to use a *victim*'s blood sample and test it against, for example, blood found on the suspect's clothing.)

This is evidence that can *be* used both to convict and acquit. If the DNA patterns of suspect and perpetrator do not match up, a prosecution often ends its case right there, the innocence of the suspect having been established. (If the evidence reveals the presence of someone *else*'s semen, how could the suspect have been involved?)

So, what are these individual DNA patterns that each of us carries? Human genomes are filled with short, repeated stretches of DNA called *variable-number tandem repeats*, or VNTRs. One of these might be the nucleotide sequence GGAGG repeated over and over. Such repeats don't code for any protein and as such, nature has put no premium on their remaining as they are. Over evolutionary time they have mutated—added or lost nucleotides—at very high rates. This gives them their usefulness in forensics: At a given location in the genome, one

---

This is evidence that can be used both to convict and acquit.

---

person might have 15 copies of a tandem repeat, while another person will have 30. Thus VNTR segments differ in *length*, and as you saw in the main text, different lengths of DNA can be visualized.

Forensic work begins by taking both a crime-scene DNA sample and a sample of the suspect's DNA. Now DNA from both samples is chopped up with a restriction enzyme, and run through gel electrophoresis, thus separating DNA fragments in the samples by length. What ultimately results are two patterns of DNA fragments: The pattern from the suspect and the pattern from the perpetrator. It is then a matter of comparing the two (**see Figure 1**). In crime cases, it is common to look not just at one length of DNA, but at several. The FBI looks at 13 DNA sequences, each about 100 to 600 bases long.

that information, the genetic catalogue would direct you to the "book" you're after, which is all the information that exists about the gene: whether such an entity seems to exist at all; and if so, its location on a particular chromosome, its complete nucleotide sequence, and anything else scientists have learned about it.

There is a project underway that has begun creating nothing less than this kind of repository of genetic information. Called the **Human Genome Project**, its goal is to locate and catalogue every bit of genetic information in the human genome.

By this point in your study of genetics, it may go without saying that this is a massive undertaking. Simply cataloguing 3 billion base pairs of information would be difficult enough. What scientists have to do, however, is discern the text in the "books" in the first place: They have to sequence every nucleotide in every human gene and all the DNA segments in between them.

In the late 1980s, science policymakers in the United States decided that a project of this nature was (1) worth doing and (2) so large that it would require an effort organized by the federal government. Thus was born, in October 1990, the Human Genome Project (HGP), whose budget in the United States was about $300 million in 1998. Though the United States is the acknowledged leader in HGP research, it is by no means the only country contributing to it; an international

Once this lab work is done, statistics plays a part in establishing guilt. Say the suspect's first VNTR pattern matched the pattern found in the crime-scene sample. Looking through the population at large, what fraction of people can be expected to *share* this pattern? The common way of establishing this has been to analyze samples from blood banks. If a given VNTR pattern is found in 1 out of 25 samples, it's assumed that 1 person in 25 has this pattern. This is not the kind of statistic on which to convict a person of rape, but say you are looking at five DNA segments. Imagine that the matches between suspect and perpetrator continued with each of these segments, and that the odds of sharing in each of the remaining four were 1 in 50, 1 in 30, 1 in 50 and 1 in 25. Using the rule of multiplication, the odds of a person other than the suspect sharing in all *five* patterns can thus be calculated:

$$1/25 \times 1/50 \times 1/30 \times 1/50 \times 1/25 = 1/46,875,000$$

This simple calculation, then, seems to have established that only one person in 46.8 million could be expected to have the same five VNTR patterns as the suspect. Thus the claim, "sure the samples match, but it could have been someone else," seems to evaporate.

DNA fingerprinting has proved enormously useful in recent years in areas other than crime. It was this technique that in 1998 established that American President Thomas Jefferson fathered at least one child by a slave he owned, named Sally Hemings. The DNA samples being compared in this instance came from descendants of Hemings and of Jefferson's paternal uncle.

Figure 1
Telltale Stains
A defendant in this real-life murder case claimed that blood stains found on his clothing (see the rows under "shirt") came from his own blood rather than from the blood of a young woman who was stabbed to death. The DNA evidence said otherwise. Look at the pattern in the row under V (for victim) and see how closely it matches the pattern in the shirt rows. The probability that the blood on the defendant's clothing came from anyone but the victim was 1 in 33 billion. (Courtesy of Cellmark Diagnostics, Inc., Germantown, Maryland.)

body called the Human Genome Organization (HUGO) is integrating the HGP efforts of at least 40 nations around the world. Initial planning called for the HGP to take 15 years to complete, thus being finished in 2005. In 1998, however, the two U.S. agencies funding the project announced a plan for generating a "working draft" of the human genome DNA sequence by 2001 and providing a finished sequence by 2003.

Limitations of the Human Genome Project

Impressive as this effort is, be aware that sequencing the human genome is not the same thing as possessing complete knowledge of human genetics. Indeed, what the HGP will

provide is a kind of basic reference work for human genetics—a quick means of *getting to* more complete knowledge.

An analogy may be helpful here. If you were interested in learning more about everyone you had gone to a summer camp with years ago, and you happened to run across a list of names and addresses of these people, that would be enormously *helpful* information. But it wouldn't begin to tell you what these people are doing now, whom they interact with, or any number of other things you might want to know. Just so with the genome: It is entirely possible to know that a gene exists in a certain place in the genome, what its sequence is, and what its protein product is, and yet not have a clue about what this protein *does*, or what other proteins it may incidentally interact with. The Human Genome Project aims to get genetic names and addresses together; coming to a complete understanding of all the genes in this "library" is the work of the twenty-first century.

15.7 Uses of Biotechnology

High-Tech Pharmaceuticals

You have already looked at some real-world applications of biotechnology: recombinant human growth hormone, reproductive cloning, and so forth. Let us now examine a few other ways in which biotechnology is changing our world. If you look at **Table 15.1,** you can see one straightforward example. We now have a growing list of **biopharmaceuticals**, meaning drugs, such as human growth hormone, that have been produced through recombinant DNA technology.

Human Gene Transfer

Looked at one way, the bacteria that produce quantities of, say, recombinant insulin are a kind of middleman we'd just as soon dispense with. What many genetic disease victims suffer from, after all, is a lack of a work-ing gene. Might it be that we could insert working genes directly into the cells of persons lacking them, after which these people could produce their *own* proteins? This is the goal of **gene therapy**, also known as **human gene transfer**, which to date has inspired considerable hope and effort but no clear-cut successes.

The idea behind gene therapy is simple: Find a vector, such as a virus, and strip it of genes that would cause any significant harmful effects to humans. Then splice the missing human gene into this virus, and get the virus to the cells in the human body where the human gene would be useful. The virus then infects these human cells, inserting its recombinant DNA into them. These infected cells then start producing the needed human protein. In the best of cases, the cells entered by the vector would be so-called stem cells, which *give rise* to the cells producing the needed protein. In such an instance, each new daughter cell would carry a copy of the inserted gene, and the person treated would thus receive long-term—perhaps lifelong—therapeutic benefits.

Scientists have thus far encountered difficulties with several of these steps. First, it is hard to get vectors such as viruses to enter

Table 15.1
Selected Biopharmaceuticals Approved by the U.S. Food and Drug Administration

Drug	Approved in	Used for
Trastuzumab	1998	One type of metastatic breast cancer
Recombinant interferon beta 1-B	1993	Management of one type of multiple sclerosis
DNase	1993	Management of cystic fibrosis
Proleukin, IL-2	1992	Kidney cancer
Recombinant antihemophiliac factor	1992	Hemophilia A
Recombinant alteplase	1990	Treatment of heart attack
Adenosine deaminase	1990	Immunodeficiency in children
Recombivax HB	1986	Hepatitis B prevention vaccine
Somatrem	1985	Human growth hormone deficiency
Recombinant human insulin	1982	Diabetes

enough human cells in the right locations; second, when such infection does occur it sometimes triggers a human immune system response that can undercut the therapy and cause other problems for the patient; third, it is difficult to get recombinant lines of cells to *keep* producing the needed protein over succeeding generations. Many researchers remain enthusiastic about the prospects for gene therapy, however, believing that it is only a matter of time until we learn how to adjust vector and host interactions to get a successful, long-lasting transfer.

Biotechnology and Food Production

Biotechnology has a broad potential to reshape food production, because it is capable of making improvements in so many different areas. One current product is genetically engineered bovine growth hormone, produced in quantity by the methods you've reviewed for cloning and host-cell reproduction. This hormone has been shown to boost milk production in cows by up to 25 percent. Meanwhile, researchers have already transplanted into Pacific salmon a gene that codes for growth hormone. Normal salmon take three to four years to mature to eating weight, but these experimental, genetically engineered salmon arrived at this weight in 10 to 12 months (**Figure 15.10**).

Genetically altered crop seeds first came to market in the United States in 1994; four years later, some 45 million acres of U.S. farmland had been planted with these recombinant seeds, most of them producing corn, soybeans, cotton, or potatoes.

The genetic alterations to these crops mostly fall into two categories. First, there are genes spliced into plant genomes that allow the plants to resist human-made herbicides, meaning weed-killers. Second, there are genes spliced into plants that allow the plants to resist insect pests (**Figure 15.11**). In the latter category, there is an example that showcases both the promise and potential dangers of biotechnology.

There is a bacterium, found naturally in the soil and called *Bacillus thuringiensis*, that produces proteins that are toxic to a number of insects. Collectively, these proteins constitute a natural insecticide, known as Bt, that has been sprayed on crops for years, mostly by "organic" farmers, meaning those who do

Figure 15.10
Super Salmon
A gene for growth hormone was inserted into the large Pacific salmon shown in the upper part of this photo. The fish in the lower part of the photo, which were not treated, averaged less than one-tenth the weight of the treated fish.

not use human-made pesticides or herbicides. Enter biotechnology firms, which in the 1990s spliced Bt genes into crop plants, with the result that these plants—corn, cotton, and potatoes—now produce their *own* insecticide. The result in some ways is an environmentalist's dream: Cotton plants that normally would get 10 to 15 pesticide applications per growing season may not be sprayed at all now, but will produce the same yields.

Yet there is another side to this. Those few insects that survive in the Bt-enhanced fields are likely to build up resistance to the natural toxin much faster than they would have before, with only periodic Bt *sprayings*. This raises the prospect of a valued natural insecticide losing its effectiveness against

Figure 15.11
Built-In Resistance to Pests
The potato plants in the center row had genes spliced into them from the bacterium *Bacillus thuringiensis*. These genes code for proteins that function as a pesticide. The center-row plants were thus able to produce their own pesticide, which protected them from predators. Meanwhile, the plants in the rows to the left and right had no such protection and were devastated by the Colorado potato beetle.

MediaLab Link
Genetically altered crops, see
p. 304.

insects over the long run. So alarming is this possibility that several organizations, including the U.S. Environmental Protection Agency, have called on farmers to set aside crop "refuges," meaning plots of crops that do not have the Bt genes spliced into them. The idea is that the few Bt-resistant bugs will, through mating, continue to share genes with the many nonresistant bugs in the refuges, thus dramatically slowing the buildup of resistance to Bt.

Applied Genetics as a Learning Tool

Sequences and the Tree of Life

Applied genetics goes beyond helping us *do* things; it is proving immensely helpful in allowing us to *learn* things. For example, long-held ideas about evolution have been either confirmed or rejected in recent years by looking at how a common gene or protein differs among organisms. The underlying assumption here is simple: Genes mutate over time, and the protein sequences that result from gene sequences also change. The longer it is since any two species have diverged from a common ancestor, the greater the sequence differences we can expect in genes or proteins that the species share.

Sure enough, when scientists have looked at a protein such as hemoglobin among species as diverse as humans, gorillas, and frogs, they have found almost no amino acid sequence difference between humans and gorillas, but a great deal of difference between either of these species and frogs. When scientists begin comparing sequence differences among closely related species—say, a particular variety of Antarctic fishes—they can begin to confirm or deny ideas about the evolutionary relationships that these species have to one another.

Deciphering the Dead Sea Scrolls

In another example (among dozens that could be cited), the historically important Dead Sea Scrolls, found in the middle-eastern desert in the 1940s, today include not only whole, deciphered scrolls, but about 10,000 tiny fragments thought to have come from some 800 different scrolls. Owing to their physical decay, these fragments cannot be fit together like pieces in a jigsaw puzzle. How, then, can they be combined and the writing they contain understood? The answer is DNA testing: The scrolls were written on the skins of animals, which after 2,000 years still contain enough DNA to yield experimental samples when the polymerase chain reaction (PCR) technique, reviewed earlier, is employed. Given this, scientists can begin to sort fragments first by the *species* of animal skin they were written on and then by *individuals* within these species, the idea being that the skin from a given animal is likely to contain a significant portion of an individual "entry" in the scrolls.

15.8 Ethical Questions in Biotechnology

When the Human Genome Project was started in the United States, policy makers decided that 3 percent of its budget would be used to fund research into the ethical ramifications of applied genetics. Subsequently, the proportion of HGP funds devoted to ethics was bumped up to 5 percent, making this expenditure "the largest investment in bioethics in the history of the world," in the words of Francis Collins, the director of the National Institutes of Health's Human Genome Program. It is unusual, to say the least, for a science project to have an ethical study component built into it, but the action seemed appropriate given applied genetics' unprecedented potential for manipulating life forms. Here is a brief look at a few of the applied genetics questions that have been raised.

Is It Inherently Unethical for Humans to Modify Themselves or Other Organisms?

One argument against this viewpoint is that the human race has been modifying itself for centuries through such means as better nutrition, vaccinations, dentures, and plastic surgery. Furthermore, humans have been modifying animals for this long through breeding (which is how something like a Dachshund has come about).

There is, however, an obvious difference between the insertion of dentures and the insertion of recombinant DNA into an organism. That difference raises a real ethical question: How far should we go in modifying organisms? In 1995, researchers at the University of Basel activated, in *Drosophila* fly embryos, a gene that brings about eye development. The twist was that they activated this gene in embryonic regions of the flies destined to become legs and antennae. The result was that the flies grew *eyes* on their legs and antennae. This does not mean that scientists are capable of modifying organisms in whatever way they wish. It does suggest, however, that we have not yet begun to learn the limits of what is possible. One observer has suggested a question that puts the ethical portion of this subject in sharp relief: It might be economical to produce an egg-laying chicken that had no eyes and no legs. If you could do such a thing, would you?

Then, of course, there are the possibilities raised by the cloning of Dolly, chief among them the prospect of cloning human beings. The question for the future is whether societies will decline to carry out a procedure that may become technologically feasible and for which there may be some demand. Imagine a couple in which one member is infertile but who nevertheless wish to have a child who possesses only their genes—or actually the genes of the single partner who is cloned.

Will Biotechnology Produce New, Harmful Organisms?

This practical question was debated most fiercely in the 1970s and seemed to be largely laid to rest then, but has returned in the 1990s. There are concerns about introduced genes moving from one species to another, particularly among plants. There are also worries that whether, by accident or by malicious design, scientists might create recombinant organisms that could be harmful to humans or to the environment, perhaps by generating new plant or animal diseases. The risks of "creating monsters" appear to be low, but that does not mean they are nonexistent.

Are Biotech Diagnoses Running Far Ahead of Biotech Treatments?

There is no question that this is the case. For example, there is a terrible affliction known as Huntington disease, which usually doesn't begin to cause problems for its victims until they are about 35. It is inevitably fatal, but its manifestations prior to death include spastic writhing and dementia. In recent years, it has become possible to test persons with a family history of the illness to tell whether they carry the single copy of the Huntington gene necessary to cause it. At present, however, we can do nothing to *treat* anyone who has this disease. Thus the dilemma: If you're at risk, do you want to know about something you are powerless to do anything about? A 20-year-old who carries the defective Huntington gene might well have 20 years of symptom-free life ahead; is it better to live those years in uncertainty, or in the certainty of a grim fate? If the answer is to live in uncertainty, does that mean not having children, or does it mean living in uncertainty about their fate as well?

Will Biotechnology Lead to "Genetic Discrimination?"

Given our new-found ability to test for genetic diseases, experts believe that early in the twenty-first century, physicians will easily be able to conduct sophisticated "panels" of genetic tests on adults, children, or fetuses to diagnose their susceptibility to a host of diseases. If so, the question is: Who gets to share in this information? Employers? Insurance companies? Persons seeking life insurance can generally be required to be tested for such diseases as AIDS. What's the difference between this and the requirement for submitting genetic screening information? If the answer is to forbid insurance companies from even asking for such information, imagine the effect it would have. Individuals at high risk—who *could* possess this information—would flock to sign up for life- and health-insurance policies, a turn of events that could demolish the assumptions of shared risk under which the insurance industry operates.

On to Evolution

In touring the strands of DNA's double helix, you've had many occasions to see how this microscopic molecule can profoundly affect our macroscopic world. DNA replicates, it mutates, it is passed on from one generation to the next. And some organisms will have more *success* in passing on their DNA—in reproducing, in other words—than will others. This latter fact has been critical in producing the huge *number* of forms of life that Earth houses, from bacteria to bats to trees. Over billions of years, genetics has interacted with Earth's myriad environments to produce the grand story of life. That story is called evolution, and it is the subject of the next four chapters.

Chapter Review

Key Terms

bacteriophage   289

biopharmaceutical   296

biotechnology   286

clone   289

cloning vector   289

forensic DNA typing   294

gel electrophoresis   293

gene therapy   296

human gene transfer   296

Human Genome Project   294

plasmid   287

polymerase chain reaction   291

recognition sequence   287

recombinant DNA   288

reproductive cloning   289

restriction enzyme   286

transformation   287

Summary

15.1 What Is Biotechnology?

- Biotechnology is the use of living organisms to create products or to facilitate processes. Much of its power is derived from the manipulation of segments of DNA.

15.2 Some Tools of Biotechnology

- Restriction enzymes are proteins derived from bacteria that can cut DNA in specific places. Plasmids are small, extrachromosomal rings of bacterial DNA that can exist outside of bacterial cells and that can move into these cells through the process of transformation.

- Human genes can be inserted into plasmid rings. Scientists use the same restriction enzyme on both the human DNA of interest and the plasmids. Complementary "sticky ends" of the fragmented human and plasmid DNA will then bond together, thus splicing the human DNA into the plasmid. This produces recombinant DNA: two or more segments of DNA that have been combined by humans into a sequence that does not exist in nature.

- Once plasmids have had human DNA spliced into them, the plasmids can then be taken up into bacterial cells through transformation. As these cells replicate, producing many cells, the plasmid DNA inside them replicates as well. These plasmids are producing the protein coded for by the human DNA that has been spliced into them. The result is a quantity of the human protein of interest.

- Plasmids are one type of cloning vector, meaning an agent that, in cloning, functions in the transfer of genetic material. Viruses known as *bacteriophages* are another common cloning vector.

15.3 Cloning and the Wider World of Biotechnology

- A clone is a genetically identical copy of a biological entity. Genes can be cloned, as can cells. Reproductive cloning is the process of making clones of whole, complex organisms. Dolly the sheep is a reproductive clone.

- Reproductive cloning can work in tandem with various recombinant DNA processes. Reproductive clones may produce proteins or even organs of interest.

15.4 Getting Quantities of a DNA Segment Quickly: PCR

- The polymerase chain reaction (PCR) is a technique for quickly producing many copies of a segment of DNA. It is useful in situations in which a large amount of DNA is needed for some sort of analysis, yet the starting quantity of DNA is small.

15.5 Visualizing DNA Sequences

- Gel electrophoresis allows scientists to visualize the sizes of DNA segments and aids in the sequencing of lengths of DNA. Sequencing is the process by which scientists learn the exact order of bases in a length of DNA. Through use of electrical attraction, DNA fragments will move through a gel, with shorter fragments moving farther in a given period of time. Thus the fragments separate out by size.

- In sequencing, gel electrophoresis is used to separate out individual DNA bases, whose order can then be "read" from the gel. Large-scale sequencing is facilitated today by sequencing machines.

15.6 The Human Genome Project

- The Human Genome Project is an international undertaking whose goal is the identification and cataloguing of all 3 billion base pairs in the human genome.

- The Human Genome Project will provide a means for researchers to know what DNA sequences are in the human genome. It will not by itself provide knowledge of the function of genes or other DNA sequences.

15.7 Uses of Biotechnology

- Biotechnology has both commercial and academic applications. In the commercial realm, it has been used to produce medicines, disease-resistant crops, and chemical compounds. The goal of human gene transfer is the introduction of genes into the human body that will produce needed proteins.

- In the academic realm, biotechnology has been used in the construction of evolutionary trees and to organize historical artifacts.

15.8 Ethical Considerations in Biotechnology

- Some ethical concerns connected to biotechnology are assertions that it is unethical for humans to genetically modify themselves or other organisms; that biotechnology may produce harmful new organisms; that biotechnology has delivered more in the way of medical diagnosis than treatment; and that biotechnology could lead to genetic discrimination.

Understanding the Basics

Multiple-Choice Questions

1. How is biotechnology defined?
   a. the technical advances made by researchers in biology over the last decade
   b. the study of technical biology
   c. the use of living organisms in creating new products and in facilitating existing processes
   d. the process of cloning genes
   e. the use of commercialization in biological research

2. Which sequence of events would be followed in the process of harvesting a protein of interest?
   a. Replicate vectors in host cells; isolate gene that codes for protein; insert gene into vector; harvest protein of interest; transform the vector into bacteria.
   b. Transform vector into bacteria; isolate gene that codes for protein; replicate vectors in host cells; insert gene into vector; harvest protein of interest.
   c. Harvest protein of interest; isolate gene that codes for protein; insert gene into vector; replicate vectors in host cells; transform the vector into bacteria.
   d. Isolate gene that codes for protein; insert gene into vector; replicate vectors in host cells; transform vector into bacteria; harvest protein of interest.
   e. Isolate gene that codes for protein; insert gene into vector; transform vector into bacteria; replicate vectors in host cells.

3. The process by which a cell takes up genetic material from outside its cell membrane is called
   a. genetic absorption
   b. transformation
   c. translocation
   d. transposition
   e. pinocytosis

4. If a criminal investigator wanted to amplify a DNA sample from a crime scene, the technique he or she would use would most likely be
   a. polymerase chain reaction
   b. genomic library creation
   c. gel electrophoresis
   d. DNA sequencing
   e. restriction fragment length polymorphisms (RFLPs)

5. Biotechnology has been used to accomplish all of the following, except
   a. artificially produced human growth hormone
   b. artificially produced insulin
   c. whole organism cloning
   d. a genetically engineered human
   e. DNA fingerprinting techniques

6. Which of the following would be an example of recombinant DNA?
   a. human growth hormone gene introduced into a bacterial plasmid
   b. injection of artificially produced insulin into a human
   c. swapping of bacterial and plant genes in a plant system
   d. a and c
   e. all of the above

7. Dolly the sheep was the result of
   a. applying an electric current to an egg and sperm in a laboratory
   b. taking an embryo from one sheep and implanting it in another
   c. taking a mammary cell from an adult sheep and fusing it with a sheep egg that had its DNA removed
   d. inserting new genes into an existing mammary cell
   e. removing the DNA from a mammary cell of an adult sheep and fusing the resulting cell with a sheep egg

8. The goal of the Human Genome Project is
   a. To insert disease-resistance genes into all vulnerable parts of the human genome
   b. To allow us to develop a genetic fingerprint for every human being
   c. To learn the basis of all genetic diseases
   d. To produce a catalogue of the protein coded for by each human gene
   e. To locate and catalogue all the genetic information in the human genome

9. To date, most genetically engineered food crops have been altered to
   a. produce needed human proteins
   b. resist human herbicides
   c. resist insect pests
   d. both a and b
   e. both b and c

10. What is a clone?
    a. an exact genetic copy of a gene, cell or larger living thing
    b. a cell that can take up plasmids
    c. a segment of DNA of a specified length
    d. an offspring produced by test-tube fertilization
    e. a segment of DNA that can prime the polymerase chain reaction

Brief Review

1. Why is it useful to have organisms such as bacteria turning out human proteins?

2. In gel electrophoresis, how do the DNA fragments found at one end of a "lane" differ from the fragments found at the other?

3. Why is the polymerase chain reaction (PCR) so valuable?

4. The Human Genome Project has been regarded as one of the greatest advances in genetics to date. Though this is very likely true, the project has some limitations that also must be recognized. What are some of these limitations?

5. Describe the basics of human gene transfer (gene therapy).

6. Why are restriction enzymes that cut with sticky ends useful? If the following piece of DNA were cut with *Bam*HI, what would the fragment sequences be?
   ATCGGATCCTCCG
   TAGCCTAGGAGGC

7. Why did Dolly the sheep end up looking like the sheep who "donated" the udder cell, rather than the sheep who donated the egg?

Applying Your Knowledge

1. Why are restriction enzymes important in recombinant DNA technology?

2. How is it possible for a single genetically engineered food crop to be both welcomed by environmentalists and feared by them?

3. If an organism were genetically cloned, would its experience of life also be a clone of the first organism? What other factors may be influential, other than genetic makeup? How, for example, might a clone of Abraham Lincoln experience life

differently from the original man in the 1800s?

4. You have recently been identified as having Huntington disease, which is a late-onset, autosomal-dominant, lethal, and currently untreatable disorder. As troubling as this information is to you, how do you deal with this information in relation to your family? Do you inform your siblings? What about your children? Do you also have them tested for the disease? Do you refrain from having any additional children?

5. One of the concerns mentioned in the chapter regarding the use of genetic technology centers around privacy issues—coined "genetic discrimination." Once the ability to screen the human genome for diseases becomes more detailed and readily available, who do you think should undergo testing? Would you be tested if you had the choice? Once such extensive genetic information is available about someone, who do you feel has the right to such knowledge? The government? Insurance agencies? Health-care providers? Extended-family members?

Why or why not? If for example, an insurance company provided the funds necessary for a genetic test, have they "bought" the right to this information? How might the genetic revolution influence the current health-care/insurance system?

6. An extension of biotechnology techniques employed in this chapter leads to even more mind-boggling possibilities. What if, beyond gene therapy treatments, whole organisms could be "engineered" with specific desirable traits in mind? The technology for "genetic engineering" may not be as far off as people may think. Do you feel that genetic engineering is justified? Why or why not? What about genetic engineering as a preventative measure for families more prone to a genetic disorder? Would engineering be acceptable to prevent multiple sclerosis? Huntington's? Red-green color blindness? What about using genetic engineering to control eye color or weight tendencies? Who decides what is allowable or preferable, and how does society prevent the creation of a separate "superior" race?

Group Activity

In the last few chapters, you have learned a lot about the history of genetics, the most recent advances, and hopes for using genetic knowledge in the future. When it was published, this text contained up-to-date information regarding genetic research. However, because the field advances so rapidly, there are sure to have been some new developments since this text was published. In your class, divide into small groups and research one of the following questions. Then, present your findings to the class to provide a more recent overview of genetic advances.

1. What is the progress of the Human Genome Project, and when is its completion expected? (Or, is the genome already completed?) What changes in genetic technology have been introduced as a result of the HGP? Benefits? Drawbacks?

2. When Dolly (see page 289) was originally cloned, there was some concern that because the DNA introduced to "create" her was "used," she would experience a larger number of health difficulties. How has (was) her health been? Is this what you would have predicted? Why or why not?

3. What are the latest advances in biopharmaceuticals? In gene therapy?

4. What are the latest governmental regulations (if any) on genetic technology? Do you agree with these regulations? Why or why not? Do you think these regulations are adequate to control the research of nongovernmental agencies?

5. What would you predict for the future of genetic research over the next 10 years? What are some potential benefits/problems with such advances?

MediaLab
Holding Out Promise or Peril? Biotechnology

Biotechnology promises to create new pharmaceuticals (even replacement organs), improve livestock yields, and even reduce the need for pesticides and herbicides in agriculture. However, many people question whether scientists are proceeding at a rate outdistancing careful, cautious evaluation. Using genetically manipulated organisms for food has met great resistance in Europe, where it is called Frankenfarming—a reference to Mary Shelley's Dr. Frankenstein and the famous monster he created by combining human corpses. To learn more, enter the *CD Activity* for an animated review of the major techniques used to swap DNA between species. Don't be intimidated by the technology; the animations will unveil the simplicity of the techniques. In the *Web Investigation* and *Communicate Your Results* sections, you will evaluate and discuss different perspectives on the value of biotechnology—its promise, or perhaps its threat.

This *MediaLab* can be found in Chapter 15 on your CD-ROM and Companion Website (http://www.prenhall.com/krogh).

CD ACTIVITY

The process of manipulating DNA has little of the mystique displayed in science fiction. It is more akin to the mundane process of cooking.

This *CD Activity* will lead you through some common techniques of biotechnology.

Activity

1. *First, you will learn about restriction enzyme digestion and gel electrophoresis to create and organize DNA fragments.*

2. *Next, you will learn about how plasmids help introduce foreign DNA into bacterial hosts.*

3. *Finally, you will investigate how DNA is introduced into other species for cloning to make identical copies of that genetically engineered species without the aid of sexual reproduction and meiosis to shuffle genes.*

Although the techniques are trivial, the result is not. Scientists are tinkering with the exact blueprint of an entire organism, of which only a small fraction is understood. Many people fear that scientists may make unforeseen changes that could irrevocably alter our lives.

WEB INVESTIGATION

Investigation 1

Estimated time for completion = 5 minutes, followed by keeping a diet diary for 24 hours

Currently, 20 percent of corn, 40 percent of soybeans, and 30 percent of cotton grown in the United States are the product of genetic engineering, adding DNA from foreign species to create plants more resistant to insect pests or herbicides. Could you be using some of these foods? Select the Keyword **GE FOODS** from your CD or Website for this *MediaLab* to read a list of genetically engineered agricultural products that are currently marketed and sold. Then select the Keyword **GE DIET** to see if you are eating any of these products. Keep a diet diary for one day to see how many genetically engineered foods you may be eating.

Investigation 2

Estimated time for completion = 15 minutes

Today, biotech food is one of the most explosive environmental issues in Europe, where protesters have destroyed crops they call Frankenplants, and insisted that food labels contain warnings if they contain genetically engineered products. Meanwhile, most Americans are already serving these foods for dinner. This may be the first time you have realized that fact, because the FDA has ruled that these genetically altered plants do not need to carry special labels warning consumers. Select the Keywords **GE PROTESTS** and **FDA GE POLICY** from your CD or Website for this *MediaLab* to read about both sides of the issue.

Investigation 3

Estimated time for completion = 5 minutes

Genetic engineering has already demonstrated its potential for great good, by alleviating diseases in countless hemophiliacs, diabetics, cancer sufferers, and immune-compromised children (see Table 15.1). These diseases are treated by pharmaceuticals manufactured in transgenic organisms, for example, pigs containing the human gene for human hemoglobin to be used in blood transfusions. Select the Keyword **CHOCOLATE COW** on your CD or Website for this *MediaLab* to test your knowledge of the techniques used to create transgenic organisms by creating cows that make chocolate milk. Then select the Keyword **TRANSGENIC HEMOGLOBIN** to visit a more serious site describing the creation of transgenic pigs for blood donation.

COMMUNICATE YOUR RESULTS

Now that you have been introduced to the level of genetically engineered products you come into contact with daily, use the following questions to analyze and communicate your feelings on the subject of biotechnology.

Exercise 1

Estimated time for completion = 5 minutes

You created a record of your diet for 24 hours. What percentage of your diet could be from genetically engineered crops? How do you feel about it?

Exercise 2

Estimated time for completion = 20 minutes

The promise of lowering the use of pesticides and herbicides in growing genetically engineered food would appear to be a dream come true for organic farmers and environmental groups. However, environmental groups have organized a letter-writing campaign to force the FDA to inform consumers about genetically engineered food on ingredient labels. Do you think that food labels should include information that identifies ingredients derived from genetically engineered foods? Form a group and compose a letter to the FDA supporting your position.

Exercise 3

Estimated time for completion = 30 minutes

What are some ethical concerns about using animals to replace human organs or to provide human medications? Write an essay defending or refuting this statement: *It is not only ethically appropriate, but it would be unethical not to pursue creating transgenic organisms to provide blood substitutes as a way to save human lives.* Consider these facts: Recombinant hemoglobin from livestock would cost less than human blood and would be free from HIV. Currently only 5 percent of Americans are blood donors; at this rate, it is estimated that by 2030 we will have a shortfall of 4 million units of blood. Over 60 million pigs are slaughtered for food each year, and our nation's donor blood needs could be entirely met by 1.2 million pigs.

305

Life's Organizing Principle: Evolution and the Diversity of Life

A fossilized
ammonite.

16

An Introduction to Evolution
Charles Darwin, Evolutionary Thought, and the Evidence for Evolution

Fantastic voyage.
(page 312)

Island of discovery.
(page 314)

Standing in a coastal

redwood grove in California, we see trees stretching hundreds of feet into the air and wonder: Why are they so tall? Watching a nature program on television, we observe the exquisite plumage of the male peacock, but are puzzled that such a *cumbersome* display could exist on a bird. Leafing through a book, we discover that the color vision of honey bees is most sensitive to the range of colors that exist in the very flowering plants they pollinate. We wonder: Can this be an accident?

For as long as humans have existed, people have asked questions such as these. Yet for most of human history, even the most informed individuals had no way to make *sense* of the variations that nature presented. Why does the redwood have its height or the peacock its plumage? Through the middle of the nineteenth century, the answer was likely to be either a shrug of the shoulders or a statement such as: "The creator gave unique forms to living things for reasons we can't understand."

Jumping ahead to our own time, modern-day biologists would give different answers to these questions. Their replies would be something like this:

- Redwoods (and trees in general) are so tall because, over millions of years, they have been in competition with one another for the resource of sunlight. Individual trees that had the genetic capacity to grow tall got the sunlight, flourished, and thus left relatively many offspring, which made for taller trees in successive generations.

- Extravagant plumage exists on male peacocks because it is attractive to *female* peacocks when they are choosing mating partners. Such finery is indeed cumbersome and carries a price (in lack of mobility, for example), but this price is outweighed by the value of the plumage. Over time, the peacock ancestors with more attractive plumage mated more often and thus sired more offspring who tended to have more attractive plumage; this made for successive generations of peacocks with more elaborate plumage.

- It is not an accident that honey bee eyesight is most sensitive to the colors that exist in the flowering plants they pollinate. Both bees and flowering plants flourish because of their interactions with one another—bees for the food they derive, flowers for the pollination. It is likely that

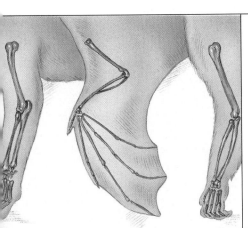

How is a cat like a bat?
(page 321)

Tracking evolution.
(page 322)

Charles Darwin.
(page 311)

flower color, or bee vision, or both, were *modified* over time to maximize this interaction.

Comparing contemporary answers such as these to that single nineteenth-century answer, you might well ask: What is it that came *in between*, so that we no longer must simply stand in puzzled wonderment before the diversity of nature?

16.1 Evolution and Its Core Principles

Common Descent with Modification

What intervened was a set of intellectual breakthroughs in the mid-nineteenth century that today go under the heading of the theory of **evolution**. Two principles lie at the core of this theory. One of them has been labeled **common descent with modification**. It holds that particular groups, or species, of living things can undergo modification in successive generations, with such change sometimes resulting in the formation of new, separate species—one species separating into two, these two then separating further. If you had a videotape of this "branching" of species and then ran it *backward*, what you'd eventually see is a reduction of all these branches into one, meaning that all living things on Earth ultimately are descended from a single, ancient ancestor.

Natural Selection

MediaLab Link
Natural selection, see p. 344.

Descent with modification is then joined to a second principle, that of **natural selection**, which states that the "fit" of an organism with its environment selects those traits that will be passed on with greater frequency from one generation to the next. In the redwood example, trees that grew taller had a better fit with their environment than trees that didn't, in the sense that the taller trees got more sunlight. Accordingly, these trees left more offspring. Thus, the quality of tallness was *selected* for transmission to future generations. As you will see, there are other forces that shape evolution, but none is as important as natural selection.

The Importance of Evolution as a Concept

These two principles are not just important to biology; they are central to it. If biology were a house, these ideas would be the mor-

tar that binds its stones together. Why? Because all life on Earth has been shaped by these factors. This is easy to see once you think about any given organism. Take your pick: a fish, a tree, a human being? *All* of them evolved from ancestors, all of them ultimately evolved from a common ancestor, and natural selection was the primary force that shaped their evolution.

Given this, the theory of evolution provides a means for us to understand nature in all its buzzing, blooming complexity. It allows us to understand not just things that are familiar to us, such as the forms of the redwoods and peacocks, but all manner of *new* natural phenomena that come our way. When physicians observe that strains of infectious bacteria are becoming resistant to antibiotics, they don't have to ask: How could this happen? General evolutionary principles tell them that organisms evolve; that bacteria are capable of evolving very quickly (given their capacity to produce many generations a day); and that antibiotic resistance is simply an evolutionary adaptation.

Evolution Affects Human Perspectives Regarding Life

The theory of evolution has an importance beyond the domain of biology, however; its impact extends into the realm of basic human assumptions about the world and the place of people in it. Once the truth of the theory was recognized, it meant that human beings could no longer view themselves as something *separate* from all other living things. Common descent with modification tells us that we are descended from ancestors, just like other organisms. Thus, we sit not on a pedestal, viewing the rest of nature beneath us, but rather on one tiny branch of an immense evolutionary tree.

Beyond this, evolution meant the end of the idea of a *fixed* living world, in which, for example, birds have always been birds and whales have always been whales. Birds are the descendants of dinosaurs, while whales are the descendants of medium-sized mammals that walked on land. Life-forms only *appear* to be fixed to us, because human life is so short relative to the time frames in which species diverge. Even within the span of a single human life, however, we can literally see species evolving, as **Figure 16.1** illustrates.

Finally in evolution, human beings are confronted with the fact that, through natural selection, life has been shaped by a force that has no mind, no goals, and no morals. Like a river that digs a canyon, natural selection shapes, but it does not design; it has no more "intentions" than does the wind. Trees that received more sunlight left more offspring and thus trees grew taller over time. This is not a "decision" on the part of nature; it is an outcome of impersonal force. The interesting thing is that something that has no inherent direction has channeled living things in so many.

If evolution has no consciousness, it follows that it can have no morals. This force that has so powerfully shaped the natural world is neither cruel nor kind, but simply indifferent.

16.2 Charles Darwin and the Theory of Evolution

Darwin's Contributions

How did such a far-reaching theory come into existence? Those who have read through the unit on genetics in this book know that the basic principles of that field were developed over the course of about a century by a large number of scientists. By contrast, a single person, the nineteenth-century British naturalist Charles Darwin, is credited with bringing together the essentials of the theory of evolution (**see Figure 16.2**). Darwin's contribution was twofold. First, he developed existing ideas about descent with modification while providing a large body of evidence in support of them. Second, he was the first person to perceive that natural selection is the primary force behind evolution.

To be sure, the study of evolution has been refined and expanded since Darwin. And it should be noted that, before Darwin had ever published anything on his theory, a contemporary of his, fellow Englishman Alfred Russel Wallace, independently arrived at the insight that natural selection could drive the origin of species in evolution. (This makes Wallace the co-discoverer of

Figure 16.2
Scientist of Great Insight
Charles Darwin, late in his life. (Painting by John Collier, 1883. London, National Portrait Gallery. © Archiv/Photo Researchers.)

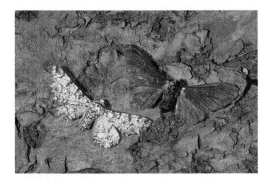

Figure 16.1
Evolution in Action
In the England of the mid-nineteenth century, the peppered moth *Biston betularia* was seen only in a light-colored form. As the century progressed, however, the Industrial Revolution did too, with its increased air pollution. By century's end, more than 90 percent of all peppered moths near British industrial areas had the darker coloration seen in the moth on the right, while moths in rural areas remained unchanged. Then, with Britain's clean-air legislation of the 1950s, the lighter form of the moth seen on the left began to reappear and today is dominant throughout Britain. Why the change? The soot of the Industrial Revolution blackened the habitat of the moth; thus lighter moths, standing out against a darker background, were more visible to birds that preyed on them. Therefore the dark moths came to predominate—through natural selection. Britain's reduction in air pollution then reversed this process so that the lighter moths survived in greater numbers. This same sequence of events has now been documented for a peppered moth population in the Detroit, Michigan area.

this principle.) Yet, there is near-universal agreement that it is Charles Darwin who deserves primary credit for providing us with the core ideas that exist in evolutionary biology even today.

Figure 16.3
What the _Beagle_ Looked Like
A reconstruction of the HMS _Beagle_, sailing off the coast of Tierra del Fuego in South America.

Darwin's Journey of Discovery

Darwin was born on February 12, 1809 in the country town of Shrewsbury, England. He was the son of a prosperous physician, Robert Darwin, and his wife, Susannah Wedgwood Darwin, who died when Charles was eight. Young Charles seemed destined to follow in his father's footsteps as a doctor, being sent away to the University of Edinburgh at age 16 for medical training. But he found medical school boring, and his medical career came to a halt when, in the days before anesthesia, he found it unbearable to watch surgery being performed on children. His father then decided that he should study for the ministry. At the age of 20, Darwin set off for Cambridge University to spend three years that he later recalled as "the most joyous in my happy life." Darwin's happiness came in part from the fact that theology at Cambridge took a backseat to what had been his true passion since childhood: the study of nature. From his early years, he had collected rock, animal, and plant specimens and was an avid reader of nature books. His studies at Cambridge did yield a divinity degree, but they also gave Darwin a solid background in what we would today call life science and Earth science.

Figure 16.4
Journey into History
The main mission of HMS _Beagle_'s 5-year voyage of 1831–1836 was to chart some of the commercially promising waters of South America. Charles Darwin served on board the ship as a resident naturalist and companion to ship's captain Robert FitzRoy. Darwin observed nature and collected specimens throughout the ship's journey, but for purposes of the theory of evolution, the ship's most important stops came in 1835 on the Galapagos Islands, west of South America. Making many stops, the ship had a complex itinerary; for example, it went twice to the Falkland Islands off the southeast coast of South America, once in 1833 and again in 1834.

An Offer to Join the Voyage of the *Beagle*

Darwin's training, and the contacts he made at school, came together in one of the most fateful first-job offers ever extended to a recent college graduate. One of Darwin's Cambridge professors arranged to have him be the resident naturalist aboard the HMS *Beagle*, a ship that was to undertake a survey of coastal areas around the world (**see Figure 16.3**). If you look at **Figure 16.4**, you can trace the journey of the *Beagle*. In addition to numerous stops on the east coast of South America (the ship's primary survey site), the *Beagle* also stopped briefly at the remote Galapagos Islands, about 970 kilometers or 600 miles west of Ecuador, a fact you'll learn more about later.

Thus did Darwin spend time on a research vessel, five years in all, beginning in England two days after Christmas 1831, and ending back there in October 1836. Just 22 when he left, he was prone to seasickness; he had to share a 10-by-15-foot room with two other officers; he was not a traveler by nature; and the journey was dangerous (three of the *Beagle*'s officers died of illness during it). Yet Darwin was happy because of the work he was doing: looking, listening, collecting, and thinking about it all. Here he is writing about a part of coastal Brazil:

> A most paradoxical mixture of sound and silence pervades the shady parts of the wood. The noise from the insects is so loud that it may be heard even in a vessel anchored several hundred feet from the shore; yet within the recesses of the forest a universal silence appears to reign. To a person fond of natural history such a day as this brings with it a deeper pleasure than he can ever hope to experience again.

16.3 Evolutionary Thinking before Darwin

Charles Lyell and Geology

Darwin beheld these sights and sounds at a time when change was in the air with respect to ideas about life on Earth. The single book he took with him when he boarded the *Beagle* was the first volume of Charles Lyell's *Principles of Geology*, published in 1830 and bearing a message that even Lyell's fellow scientists found hard to accept: That geological forces *still operating* could account for the changes geologists could see in the Earth's surface. (As Darwin put it, "that long lines of inland cliffs had been formed, the great valleys excavated, by the agencies which we still see at work.") Under this view, Earth had not been put into final form at a moment of creation, but rather was steadily undergoing change; if such a thing were possible for the Earth itself, why not for the creatures that lived on it? (**See Figure 16.5.**)

Jean-Baptiste de Lamarck and Evolution

An idea along this line *had* been proposed, by the French naturalist Jean-Baptiste de Lamarck, in a book published in 1809. Lamarck believed that organisms changed form over generations through what has been termed the inheritance of acquired characteristics. Ducks or frogs did not originally have webbed feet, he said, but in the act of swimming they stretched out their toes in order to move more rapidly through the water. In time, membranes grew between their toes, effectively becoming webbing; critically, this characteristic was passed along to their offspring (**see Figure 16.6** on page 314). Over time, he believed, an animal would acquire enough changes that one

**Figure 16.5
Geologic Strata and What They Contain**

a Several strata of sedimentary rock, with fossils embedded within.

b A fossilized sea urchin found in Great Britain, dating from at least 65 million years ago.

a

b

species would diverge into two, with this branching extending all the way to human beings. In his work, Lamarck lent support to a *means* of evolution that we know today is false. (Animals don't pass along traits in accordance with their behavior.) Yet note what he got right: That organisms can evolve; that one kind of organism can be ancestral to a different kind of organism.

Georges Cuvier and Extinction

Another French scientist, Georges Cuvier, examining the fossil-laden rocks of the Paris Basin early in the nineteenth century, provided conclusive evidence of the extinction of species on Earth. This was a radical notion at the time, because many Christians believed that the creator would never allow one of his creatures to perish. (One amateur fossil collector who believed this was Thomas Jefferson.) This much about extinc-

tion Cuvier got right. However, seeing in his rock layers seeming "breaks" in the sequence of animal forms—a layer of simple forms would be followed by a layer of more complex forms as he went from older to newer layers—he held that there had been a series of catastrophes, such as floods, that wiped out life in given areas, after which the creator had carried out *new* acts of creation, bringing more complex life-forms into being with each act.

If you were to survey a broad swath of scientific thought leading up to Darwin, you'd repeatedly find what you've seen with Lamarck and Cuvier: a given scientist getting things *partly* right and partly wrong (which is almost always the way in science). The result was a rich mix of scientific findings and fanciful speculation that existed in Darwin's world as he boarded the *Beagle*. He himself believed at the time that species were fixed entities; that they did not change over time.

16.4 Darwin's Insights Following the *Beagle's* Voyage

Perceiving Common Descent with Modification

Darwin observed and collected wherever he went, but the most important stop on his journey took place nearly four years into it,

Figure 16.6
The Duck's Feet
Jean-Baptiste de Lamarck proposed that new species evolve through a branching evolution, which is correct. He also proposed, however, that one of the forces driving this evolution is the inheritance of acquired characteristics, which does not take place. Under Lamarck's view, common at the time, a duck got its webbed feet because of activities it carried out during its lifetime, and this change was then passed on to its offspring.

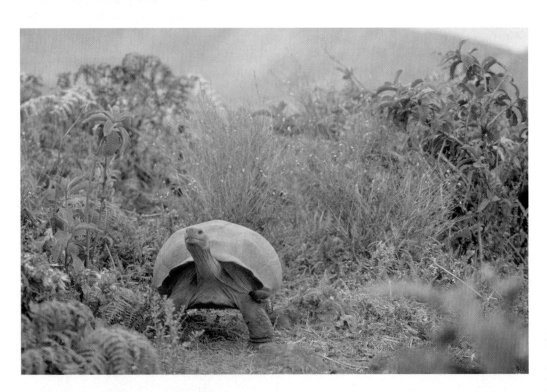

Figure 16.7
The Galapagos Islands
Galapagos is Spanish for *tortoise*. A part of Ecuador, the islands are still an important site of evolutionary research. Here a giant tortoise moves through the Alcedo Volcano area on one of the Galapagos Islands, Isabela.

when the *Beagle* stopped for a scant five weeks at the remote series of volcanic outcroppings called the Galapagos Islands (**see Figure 16.7**). There, in the dry landscape, amidst broken pieces of black lava, he saw strange iguanas and turtles and mockingbirds that varied from one island to another; he shot and preserved several of these mockingbird "varieties" for transport back home, along with a collection of small birds that he took to be blackbirds, wrens, warblers, and finches.

Docking in England in October of 1836, Darwin soon donated a good deal of his *Beagle* collection to the Zoological Society of London. If there was a single most important flash of insight for him regarding descent with modification, it came when one of the Society's bird experts gave him an initial report on the birds he had collected on the Galapagos. Three of the mockingbirds were not just the "varieties" that Darwin had thought; they were separate species, as judged by the standards of the day. Beyond that, the small birds he had believed to be blackbirds, finches, wrens, and warblers were all finches—separate species of finches, each of them found nowhere but the Galapagos (**see Figure 16.8**). With this, Darwin began to see it: The Galapagos finches were related to an ancestral species that could be found on the mainland of South America, hundreds of miles to the east. Members of that ancestral species had come by air to the Galapagos and then, fanning out to separate islands, had diverged over time

into separate species. Thus it was with Galapagos tortoises and iguanas and cactus plants as well; they had common mainland ancestors but, on these islands, had diverged into separate species. Darwin began to perceive the infinite branching that exists in evolution.

Perceiving Natural Selection

But what drives this branching? In England and set on a career as gentleman naturalist, Darwin married and settled in London for a time before moving to a small village south of London, where he would live out his days (**see Figure 16.9**). Two years after his homecoming, inspiration on evolution's guidance mechanism came to him not from looking at nature or thinking about the life sciences, but from reading a book on human population and food supply, *An Essay on the Principle of Population* by T. R. Malthus. As Darwin later wrote in his autobiography:

> I happened to read for amusement Malthus on Population, and being well prepared to appreciate the struggle for existence which everywhere goes on . . . it at once struck me that under these circumstances favourable variations would tend to be preserved and unfavourable ones to be destroyed.

Figure 16.8
Darwin's Finch
Geospiza fuliginosa feeds on sedge seeds in the Galapagos Islands.

Figure 16.9
Where the *Origin of Species* Was Written
In this study in his house in rural England, Charles Darwin conducted scientific research and wrote his ground-breaking work, *On the Origin of Species by Means of Natural Selection*.

Thus did natural selection occur to him as the driving force behind evolution. You might think that, with two major insights in place—common descent with modification and natural selection—Darwin would have rushed to inform the world of them. In fact, more than 20 years were to elapse between his reading of Malthus and the 1859 publication of his great book, *On the Origin of Species by Means of Natural Selection*. In between, though incapacitated much of the time with a mysterious illness, Darwin published on geology, bred pigeons, and spent eight years studying the variations in barnacles—activities that each had relevance to his theory, and that yet constituted a kind of holding pattern for him as he contemplated informing the world of a theory that he knew would be controversial. Darwin finally got to work on what he thought would be his "big species book" in 1856.

16.5 Alfred Russel Wallace

Two years after Darwin began this labor, half a world away and unbeknownst to him, a fellow English naturalist lay in a malaria-induced delirium on an Indonesian island. Alfred Russel Wallace made a living collecting bird and butterfly specimens from the then-exotic lands of South America and Southeast Asia, selling his finds to museums and collectors. He had thought long and hard (and even published) about how species originate. Now, shivering with malarial fever in a hut in the heat of the tropics, with the question on his mind again, Wallace came upon the very insight that Darwin had had 20 years earlier: Natural selection is the force that shapes evolution. Recovering from his illness, Wallace wrote out his ideas over the next few days and sent them to a scientific hero of his in England—Charles Dar-

win! He asked Darwin to read his manuscript and submit it to a journal if Darwin thought it worthy.

Darwin was stunned, as he faced the prospect of another man being the first to bring to the world an insight he thought was his alone. Nevertheless, he would not allow himself to be underhanded with the younger scientist. He informed some of his scientist friends about his plight, and (without informing Wallace) they arranged for both Wallace's paper and some of Darwin's letters, sketching out his ideas, to be presented at a meeting of a scientific society in London on July 1, 1858. The readings before the scientific society turned out to have little immediate effect, but they did prod Darwin into working on what would become *The Origin of Species*, published some 16 months later. This event had a great effect. Indeed, it set off a thunderclap whose reverberations can still be heard today.

16.6 Descent with Modification Is Accepted

Sparking both scorn and praise, Darwin's ideas were fiercely debated in the years after 1859. At least within scientific circles, however, it did not take long for common descent with modification to be accepted. Fifteen years or so after *The Origin of Species* was published, almost all naturalists had become convinced of it, and it's not hard to see why: With evolution as a framework, so many things that had previously seemed curious, or even bizarre, now made sense. Years before *Origin* was published, scientists had known that, at a certain point in their *embryonic* development, species as diverse as fish, chickens, and humans all have structures known as pharyngeal slits (**see Figure 16.10**). In fish, these structures go on to be *gill* slits; in humans they develop into the eustachian tubes, among other things. The question was: Why would such different animals share this common structure in embryonic life? With evolution it became clear. All of them shared a common vertebrate *ancestor*, who had the slits. The vertebrate ancestral line had branched out into various species, yet elements of the common ancestor persisted in these different species in their embryonic stage.

In bringing together countless "loose ends" such as this, evolution became the

Figure 16.10
Different Embryos, Same Structure
Pharyngeal slits, color-coded here, exist in the embryos of these five vertebrate animals: **(a)** sea lamprey, **(b)** pond turtle, **(c)** chicken, **(d)** domestic cat, and **(e)** human being. The common structure is evidence that all four evolved from a common ancestor. (Adapted from M. K. Richardson, 1997.)

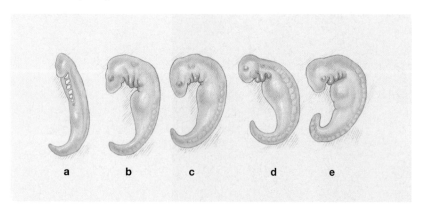

a   b   c   d   e

mortar that unified the study of the living world. (I like to think of biologists from this period as readers who start going over a detective novel a *second* time, saying, "Of course!" and "Why didn't I see that?") For another example of evolution's explanatory power, see "Can Darwinian Theory Make Us Healthier?" on page 318.) Given this, scientists had little trouble accepting the *fact* of evolution—that is, the occurrence of descent with modification. But the primary force that Darwin said was driving evolution, natural selection, was not embraced so quickly. Indeed, the twentieth century would be almost halfway over before it was generally accepted.

16.7 Darwin Doubted: The Controversy over Natural Selection

How Can Differences Be Passed On?

The essential stumbling block for natural selection was this: It asserted that there had to be traits that could *vary* in ways that would confer reproductive advantages on given individuals, and that these variations could be passed on from one generation to the next. With the redwood trees considered earlier, the trait was height, and the variation that conferred a reproductive advantage was *more* height. Thus, in a given environment, added height represented a *difference* between some members of a species and other members of that species. After many generations, enough of these differences would have accumulated that evolutionary branching would have taken place—two species would exist where one did previously. But in the middle of the nineteenth century, no one could imagine how small differences could reliably be passed down over many generations. In short, what scientists of the time did not understand was genetics.

In the genetics section of this book, you saw that a character such as height in trees is likely to be governed by many genes working in tandem, each of them bearing information that, when expressed as a protein, has some effect on tree height. Such "height genes" are shuffled in countless ways when egg and sperm are formed, and when they fuse in fertilization. The result is that gene combinations that yield particularly tall trees will come together in certain instances. In addition, outright mutations—alterations in genetic information—can take place, manifesting themselves as new physical traits (one of them being a taller tree). Most important for our purposes, whatever the genetic information is, it *persists*; it is passed on largely intact from one generation to another in the small packets we call genes.

Darwin could not have imagined a better mechanism for carrying out natural selection as he had envisioned it. The problem was that neither he nor anyone else in his time knew that things worked like this. Most nineteenth-century scientists believed that inheritance worked through a "blending" process akin to mixing different-colored paints. If you blend red and blue paint, for example, you get something that lies halfway between them, purple. Under such a system, differences in characters would not be preserved over time; they would be averaged into oblivion. Blending would work to make groups of organisms the *same*, not to allow differences among them. Given this, what differences would there be for nature to select *from*?

Coming to an Understanding of Genetics

The vindication of natural selection required, first, a demonstration (by Gregor Mendel) that inheritance doesn't work by blending. Rather, it works by the transmission of the stable genetic units we now call genes. Then, in the early part of the twentieth century, came the demonstration that genes could account for very small physical differences that arise within a given population. Trees come in a *range* of heights, with the differences in this range governed by the combinations of individual genes described earlier. Genes are the mechanism for the process Darwin envisioned: units of inheritance, persisting over generations, that can bring about small physical differences.

Darwin Triumphant: The Modern Synthesis

Advances in genetics such as this were later joined to other types of evolutionary research, such that in the period roughly from 1937 to 1950 there took place what is known as the **modern synthesis**: The convergence of several lines of biological research into a unified evolutionary theory. Taxonomists out in the field reported on how species are distributed through-

Can Darwinian Theory Make Us Healthier?

Why do some pregnant women have "morning sickness"? Why do we like fats so much? Should we try to bring down the temperature of anyone who has a fever? As noted in this chapter, evolution is central to biology because it has such great explanatory power. In the 1990s, researchers began trying to harness that power to combat human disease and discomfort. The result is an emerging field called Darwinian medicine.

Two of the field's founders, Randolph M. Nesse (a medical doctor) and George C. Williams (an evolutionary biologist) note that their discipline doesn't seek to explain the direct causes of disease; rather, it aims to examine the physiological traits that make humans *vulnerable* to disease, thereby providing insight on how to treat it. The direct cause of morning sickness could perhaps be found at the level of molecules that are produced in some women early in pregnancy. But this does not tell us why so many women should be vulnerable to such a malady in the midst of something as common as pregnancy. More important, it does not tell us whether production of these molecules may be doing something *useful* as well as discomforting.

Darwinian medicine has a hypothesis about morning sickness. Put forth by the independent scientist Margie

Profet, it is that morning sickness is an evolutionary *adaptation*; it is a means of minimizing the potential for exposure of a growing fetus to harmful substances. Nauseated women are less likely to consume strong-tasting substances; few of these are likely to be harmful to a fetus, of

---

It has taken only a few thousand years for humans to go from being hunters and gatherers to being computer programmers, but evolution cannot move this fast.

---

course, but in something as important as reproduction, nature is likely to be operating under what Nesse and Williams call the "smoke-detector principle." In our homes, a smoke detector would be set off by an actual fire, of course, but it might also be set off by a burned piece of toast. The price of avoiding something as catastrophic as a fire is the existence of a super-sensitive alarm system. Just so, the female body may have a super-sensitive "alarm" on

out the world. Mathematical geneticists provided guidance on how evolution *could* work, given the shifting around of genes among organisms. Paleontologists studied evolution in the fossil record. Critically, these different types of scientists were involved in a process of "mutual education" with one another; the findings of geneticists, for example, were now known to taxonomists—and vice versa. The upshot was a greatly deepened understanding of how evolution works, providing conclusive evidence for Darwin's idea about the importance of natural selection. The architect of evolutionary theory had been vindicated—nearly 100 years after publishing his great work.

16.8 Opposition to the Theory of Evolution

The False Notion of a Scientific Controversy

So, why are Darwin's essential insights so widely believed to be correct? You'll now review some of the evidence for these insights,

but first a word about why this is desirable. When you studied cells in this book, you did not review the "evidence for cells." Nor did you go over the "evidence for energy transfer" when studying cellular respiration, nor the "evidence for lipids" when studying biological molecules. Evolution has an unusual status among major biological subdisciplines in that, almost alone among them, its findings are regularly challenged as being unproven or simply wrong. For the average person, who can't be bothered with the details of such a controversy, these attacks make it appear that the theory of evolution is not a body of knowledge, solidly grounded in evidence, but rather a kind of scientific guess about the history of life on Earth.

What Is a Theory?

Several factors are critical in allowing the continued appearance of a "scientific debate" on the validity of the theory of evolution, when in fact none exists. The first of these has to do with a simple misunderstanding re-during early pregnancy to guard against the possibility of fetal damage. The price of avoiding catastrophe in pregnancy is an intermittent aversion to some foods, or to eating at all. Profet has made a first test of this thesis by examining the outcomes of a number of pregnancies. What she found was that women with more nausea were less likely to have a miscarriage.

With a result such as this, Darwinian medicine begins to assume some potential importance in the treatment of disease. Suppose that subsequent tests bear out Profet's hypothesis. We would then have to ask whether morning sickness should be "treated" at all, as it has been in the past. Much the same thing could be said of fevers—at least up to a certain level—because there is increasing evidence from humans and other animal species that higher body temperatures are a means of fighting off microbial invaders.

This is not to say that every baffling affliction that humans have is adaptive. Another hypothesis in Darwinian medicine is that near-sightedness may in part be a consequence of the modern practice of constantly looking at objects that are close to us, such as books, computer screens—something that our evolutionary ancestors did not do. No one claims that near-sightedness serves any positive function; the idea is to combat it more sensibly by understanding its deeper cause.

Darwinian medicine holds that a consistent culprit in human physical affliction is the difference between the pace of cultural evolution on the one hand and physiological evolution on the other. It has taken only a few thousand years for humans to go from being hunters and gatherers to being computer programmers, but evolution cannot move this fast. In near-sightedness, the evolution of our eyes hasn't had time to catch up with the change we've undertaken in how we spend our time. In another area, if we look at a typical "stone-age" diet, it turns out to be fairly close to the ideal diet recommended by doctors today: Low in saturated fats, high in fiber, high in vegetables and fruits. It was only with the rise of civilization that humans began consuming the large quantities of saturated fat that are a cause of heart disease. Yet fats taste good to us, because our bodies have not evolved much from the time when fats were nutritional "prizes," that packed lots of energy into small quantities of food. In this area, as in many others, the assumption of Darwinian medicine is that to understand our present condition, we need to look at our past.

garding terminology. You saw in Chapter 1 that to the average person, the word *theory*, as in "theory of evolution" implies that the whole subject amounts to speculation. In science, however, a theory is a general set of principles, supported by evidence, that explains some aspect of the natural world. Accordingly, we have Isaac Newton's theory of gravitation, Albert Einstein's general theory of relativity, and many others. There are lots of principles that go into making up the theory of evolution, some of them on surer footing than others. Over time, however, some of these principles have achieved the status of established fact, so that we are as sure of them as we are of the fact that the Earth is round. Among these principles are the fact that evolution has indeed taken place and the fact that this has occurred over billions of years.

The Nature of Historical Evidence

A second factor that provides an opening for the opponents of evolution is the nature of the evidence for it. If you were called upon to provide the "evidence for cells" referred to earlier, a look through a microscope at some actual cells might be enough to stop this "debate" in its tracks. Conversely, one of evolution's most important manifestations—radical transformations of life-forms—can never be observed in this way, because these transformations take place over vast expanses of time. Asking to "see" the equivalent of a dinosaur evolving into a bird before you'll believe it is like asking to see the European colonization of North America before you'll acknowledge it.

Evolution is taking place right now, all around us, but the *evidence* for evolution is *historical* evidence to a degree that is not true of, say, the study of genetics or of photosynthesis. Any ancient historical record is fragmentary, and evolution's historical record is ancient indeed. The fossils that scientists analyze are what remain from tens of millions of years of weathering and decay; such an incomplete record leaves room for a great deal of interpretation among scientists—far more

What Is It?
See p. 320.

What Is It?
Answer: Colored x-ray of a fossil ammonite showing the spiral shell with its many chambers. Ammonites lived from the Devonian period (approximately 400 million years ago) to the end of the Cretaceous period (approximately 64 million years ago).

than is the case in purely experimental science. This interpretation has to do, however, with the *details* of evolution and not with its core principles. It has to do with what group descended from what other; with the rate at which evolution has proceeded or with the role that pure chance has played in evolution. It does not have to do with *whether* evolution has occurred.

16.9 The Evidence for Evolution

So, what makes scientists so sure about the core principles of evolution? In any search for truth, we are more comfortable when lines of evidence are internally consistent and then go on to agree with *other* lines of evidence. The evidence for evolution satisfies both criteria; if we were to find even a single glaring inconsistency within or between lines of evidence, the whole body of evolutionary theory would be called into question; but no such inconsistencies have turned up yet, and scientists have been looking for them for about 140 years.

Radiometric Dating

One claim of evolutionary theory, as you have seen, is that evolution proceeds at a leisurely pace, with billions of years having elapsed between the appearance of life and the present. Yet how do we know that Earth isn't, say, 46,000 years old, as opposed to the 4.6 billion that scientists believe it to be? The bedrock upon which we have determined the age of the Earth and its organisms is **radiometric dating**. As volcanic rocks are formed, they incorporate into themselves various elements that are in their surroundings. Some of these elements are radioactive, meaning they emit energetic rays or particles and "decay" in the process.

With such decay, one element can be transformed into another, the most famous example being uranium-238, which be-

comes lead-206 though a long series of transformations. The critical thing is that this transformation proceeds at a *uniform rate*; it is as steady as the most accurate clock imaginable. It takes 4.5 billion years for half a given amount of uranium-238 to decay into lead-206 (hence the term **half-life**). Therefore, comparing the *proportion* of uranium-238 to lead-206 in a rock sample provides a date for the rock with a fair amount of precision (because it takes a given amount of time for a sample of given size to yield the proportion of lead we find). Other radioactive elements decay at different rates and are incorporated into different kinds of materials; given the range of such "radiometric clocks," scientists have been able to date objects from nearly the formation of the Earth to the present, though there are some gaps in the picture.

The Placement of Fossils

Today, one line of evidence for evolution is the similarity of fossil type by sedimentary layers. Looking at the same geologic layers of sediment worldwide, scientists find similar types of fossils, with a general movement toward more complex organisms as they go up through the newer strata. In Chapter 1, it was noted that scientific claims must be *falsifiable*; they must be open to being proved false upon the discovery of new evidence. The fossil record presents a falsifiable claim. For example, creatures called *trilobites* (**see Figure 16.11**) are one of the earliest examples of the animal grouping called *arthropods*. Trilobites had a long run, existing in the ancient oceans from about 500 million years ago until some 245 million years ago, when they became extinct. By contrast, the evolutionary lineage that humans are part of, the primates, *began* about 60 million years ago. If scientists were to find, in a single fossil bed, fossils of trilobites existing side by side with those of early primates, our whole notion of evolutionary sequences would be called into question. No such incompatible pairing has happened, however, and the strong betting is it won't. Given this, the line of fossil evidence is internally consistent, but there's more; when we compare fossil placement with the dates we get from radiometric dating, we get excellent agreement *between* these two lines of evidence. We don't find trilobites embedded in

Figure 16.11
Ancient Organism, Now Extinct
Shown is a fossilized trilobite, a kind of organism that became extinct long before the animals known as primates came to exist. This fossilized trilobite dates from about 370 million years ago and was found in present-day Morocco.

sediments that turn out to be 60 million years old.

Comparative Morphology and Embryology

Morphology is the study of the physical *forms* that organisms take, while **embryology** is the study of how animals *develop*, from fertilization to birth. As it turns out, the evidence provided for evolution by comparative embryology was touched on earlier, in noting the pharyngeal slits that exist in the embryos of creatures as diverse as fish and humans.

In comparative morphology, some classic evidence for evolution is seen in the similar forelimb structures found in a very diverse group of mammals—in a whale, a cat, a bat, and a gorilla, as seen in **Figure 16.12**. Such features are said to be **homologous**, meaning the same in structure owing to inheritance from a common ancestor. Look at what exists in each case: one upper bone, joined to two intermediate bones, joined to five digits. Evolutionary biologists postulate that the four mammals evolved from a common ancestor, adapting this 1-2-5 structure over time in accordance with their varying needs.

Evidence from Molecular Biology

In the genetics unit you saw that every living thing on Earth employs DNA and utilizes an almost identical "genetic code" (*this* triplet of DNA bases specifying *that* amino acid). At the very least, this means that there is a unity running through all earthly life; it is also consistent with the idea that all life on Earth ultimately had a single starting point—the single common ancestor mentioned earlier.

In recent years, molecular biology has also provided another check on what scientists have long believed to be true about evolutionary relationships—about which organisms are more closely related and about how long it's been since they've shared a common ancestor. We know that homologous genes may function in an array of many different organisms. We also know that genes *mutate* very slowly over time, meaning there are changes in the sequence of their DNA building blocks, known as *nucleotide bases*. If the *rate* of this change is constant, then base changes are like a molecular clock that's ticking; with the passage of a given amount of time, you get a set number of base mutations. Given this, the

longer it has been since two organisms shared a common ancestor, the greater the number of *differences* we should see in the base sequence of any gene they share.

This hypothesis was put to the test in connection with a gene that codes for an enzyme called *cytochrome C oxidase*, which exists in organisms as different as humans, moths, and yeasts. Going into this experiment, evolutionary theory predicted that there should be more base-pair differences between the cytochrome C genes of, say, a human and a moth than between a human and a duck, because all the *other* evidence we had told us that humans and moths are more distantly related than are humans and ducks. If you look at **Figure 16.13** (page 322), you can see the story this work told. There are 17 sequence differences between a human and a duck, but 36 between a human and a moth. Indeed, all the differences between the species pictured fall into line with evolutionary theory. Thus we have another confirmation for evolution *between* lines of evidence—between DNA sequencing *and* the fossil record *and* radiometric dating *and* comparative morphology.

Experimental Evidence

Finally, much evidence for evolution has been provided in recent years by experiment and observation. At first glance, this may seem rather improbable because as you've seen, evolutionary change can often be perceived only over long stretches of time. Scientists have devised clever ways to catch evolution in the act, however.

Consider the experiments of John Endler, who believed that the male guppies he was

Figure 16.12
Four Animals, One Forelimb Structure

Whales, cats, bats, and gorillas are all descendants of a common ancestor. As a result, the bones in the forelimbs of these diverse organisms are very similar despite wide differences in function. Four sets of homologous bones are color-coded for comparison.

Whale   Cat   Bat   Gorilla

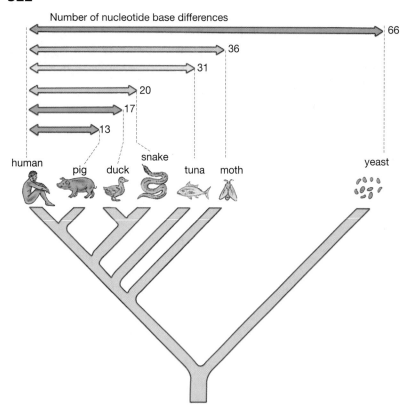

Figure 16.13
Using Molecules to Track Evolution
Diverse organisms—such as yeast, moths, and pigs—all have genes that code for an enzyme called *cytochrome C oxidase*. These organisms inherited cytochrome C genes from a common ancestor many millions of years ago. Over time, however, the cytochrome C genes have undergone mutations that have altered the sequence of their DNA "building blocks," called *nucleotide bases*. The longer it has been since any two species shared a common ancestor, the more differences there should be in their cytochrome nucleotide bases. There are 13 differences between the nucleotides found in human cytochrome C genes and those found in pigs, but there are 66 differences between the human and yeast genes. Data from Whitfield, Philip. 1993. From *So Simple a Beginning: The Book of Evolution*. (New York: Macmillan: Maxwell Macmillan International)

studying were being pulled in two directions by natural selection. On the one hand, those who were larger and had brighter coloration were chosen more often by females for mating. On the other, these very characteristics made the males more vulnerable to predators. What Endler did was to put some guppies in a predator-free environment. In only a few generations, the males evolved brighter coloration and larger tails. When Endler then reintroduced predators into this population, things went the other way; over several generations, the males evolved smaller tails and less brilliant colors. Both of these outcomes are precisely what evolutionary theory would predict. In both instances the guppies evolved in a direction that would maximize their reproductive success—more mates with the brighter colors in the predator-free environment, and a longer life (and hence more reproduction) with the drab coloration they evolved in the predator-laden environment.

On to How Evolution Works

Given the abundance of evidence for evolution, biologists long ago stopped asking the question of whether it occurred. The really interesting questions for decades have been: Through what means has evolution proceeded? At what pace? In what direction, if any? These are the questions this text will address in the next three chapters.

Chapter Review

Key Terms

common descent with modification 310

embryology 321

evolution 310

half-life 320

homologous 321

modern synthesis 317

morphology 321

natural selection 310

radiometric dating 320

Summary

16.1 Evolution and Its Core Principles

- In the theory of evolution, descent with modification describes the process by which species of living things can undergo modification over time, with such change sometimes resulting in the formation of new, separate species. All species on Earth have descended from other species and a single, common ancestor lies at the base of the evolutionary tree.

- Natural selection has been the most important force in shaping evolution. Through natural selection, the "fit" of an organism with its environment will select those traits that will be passed on with greater frequency

from one generation to the next. Groups of living things evolve in the direction of traits that increase survival and reproduction.

- The theory of evolution has an importance beyond the domain of biology. Through it, human beings have become aware that (1) they are descended from other varieties of living things, and (2) the organisms that populate the living world are not fixed entities, but instead are constantly undergoing change.

16.2 Charles Darwin and the Theory of Evolution

- Charles Darwin deserves primary credit for the theory of evolution. He developed existing ideas about descent with modification while providing a large body of evidence in support of them. And he was the first person to perceive that natural selection is the primary force behind evolution.

- Darwin's insights were inspired by the research he carried out during a five-year voyage he took around the world on the ship HMS *Beagle*, beginning in 1831.

16.3 Evolutionary Thinking before Darwin

- Some of Darwin's ideas can be traced to the work of Charles Lyell, Jean-Baptiste de Lamarck, and Georges Cuvier, who respectively noted the dynamic geological nature of the Earth, the possibility of descent with modification, and the extinction of species on Earth.

16.4 Darwin's Insights Following the *Beagle*'s Voyage

- Darwin understood descent with modification for several years before he comprehended that natural selection was the most important force driving it. It was his reading of a work by Malthus that sparked his realization about natural selection.

16.5 Alfred Russel Wallace

- English naturalist Alfred Russel Wallace is the co-discoverer of the principle that evolution is shaped by natural selection.

16.6 Descent with Modification Is Accepted

- Descent with modification was accepted by most scientists not long after publication of Darwin's *On the Origin of Species By Means of Natural Selection* in 1859. Scientists accepted it because it explained so many facets of the living world.

16.7 Darwin Doubted: The Controversy over Natural Selection

- The importance of natural selection was not generally accepted until the middle of the twentieth century. Its acceptance hinged on a modern synthesis in the theory of evolution that brought together lines of evidence from genetics, the fossil record, and the distribution of organisms throughout the world.

16.8 Opposition to the Theory of Evolution

- Even today the theory of evolution is regularly challenged as being unproven or simply wrong. One factor leading to the appearance of a "scientific debate" is confusion over the meaning of the word "theory." Though the average person may equate "theory" with uncertain speculation, in science, a theory is a general set of principles, supported by evidence, that explains some aspect of the natural world.

16.9 The Evidence for Evolution

- Five lines of evidence are consistent with the theory of evolution. First, radiometric dating indicates that the earth is greater than 4 billion years old. Second, the placement of fossils is consistent with the theory of evolution and with radiometric dating. Third, the theory of evolution explains the common occurrence of homologous physical structures in different organisms. Fourth, the theory of evolution is consistent with variations found in the DNA, RNA, and protein sequences of various organisms. Fifth, experimental demonstrations of evolution have been carried out in the laboratory and in nature.

Understanding the Basics

Multiple-Choice Questions

1. In science, a theory is
   a. an untested hypothesis
   b. a statement of a natural law that explains observed facts
   c. speculation
   d. an observation
   e. an idea that has been thoroughly tested

2. Which of the following are central ideas in the theory of evolution by natural selection (circle all that apply):
   a. Organisms vary in characteristics that affect their survival and reproduction.
   b. Descent with modification occurs over generations.
   c. Evolution has a goal.

d. Individuals with characteristics that allow them to survive and reproduce better than others are favored by natural selection.

e. Characteristics that are acquired during the life of an individual contribute to evolution.

3. Which of the following observations provide evidence for evolution? (Circle all that apply.)

a. Dogs and trilobites are never found in the same fossil beds.

b. Athletic training can produce an increase in muscle mass.

c. Extinctions can be documented in the fossil record.

d. All organisms use a common genetic code.

e. Plant and animal breeders can produce new breeds of crops and animals through artificial selection.

4. During the formulation of his theory of evolution by natural selection, Darwin brought together ideas and results from several disciplines. Match the person with the phrase that describes his work.

a. Lyell	Catastrophic extinction and new creations explain fossil record
b. Wallace	Inheritance of acquired characters
c. Lamarck	Natural selection is differential survival or reproduction
d. Cuvier	Geological forces observable today caused changes in Earth

5. Some of the following structures are homologous with each other. Which ones are they, and which ones don't belong?

a. whale flipper
b. insect leg
c. bat wing
d. horse leg
e. octopus tentacles

6. The discovery of _____ provided the evidence needed for natural selection to be widely accepted as the mechanism of evolution (select one):

a. methods for radiometric dating
b. extinctions in the fossil record
c. genes as the cause of disease

d. genes as the units of particulate inheritance
e. fossil bacteria

7. Important implications of the theory of evolution by natural selection include (select all that apply):

a. All organisms, including humans, are descended from a common ancestor.

b. The biological world is constantly evolving.

c. Natural selection leads to evolutionary progress and the appearance on Earth of superior organisms.

d. The characteristics of organisms are molded by an impersonal force without a goal.

e. Humans are just a small part of the tree of life.

8. John Endler performed an experiment involving guppies in which he demonstrated that predators can cause natural selection and lead to evolution in prey populations. Which of the following are true statements about the results of his experiments (select all that are correct):

a. Males with longer tails and brighter coloration were more frequently eaten by predator fish.

b. Females preferred to mate with dull-colored males.

c. When predator fish were added to a population that had not previously had predators, males with dull coloration were favored by natural selection.

d. When predators were removed, the average brightness of male tails in the population increased.

e. When predators were removed, the average brightness of females increased.

9. When chemical insecticides were first used in the 1940s, very low doses of insecticides were effective in killing most insects. Now, many pest insects are resistant to insecticides and are not killed unless pesticide concentrations are very high, and sometimes not even then. Agriculturalists are working to control pests with reduced pesticide use. Given your knowledge of how evolutionary processes lead to the evolution of resistance, which of the following might help to reduce the

speed at which resistance to a new chemical might evolve?

a. Apply chemicals only when absolutely necessary to avoid economic damage.

b. Apply chemicals as soon as a single pest insect is seen in a given field.

c. Use plant varieties that are resistant to insect attack when possible.

d. Try to shift planting to a time of year when the pest insect is not normally present.

e. Apply chemicals on a regular schedule, just to be sure.

Brief Review

1. What is homology, and why does it provide evidence for evolution?

2. How does the evidence presented in Figure 16.13 support the conclusion that humans are more closely related to pigs than to yeast?

3. Describe two examples in which concurrence of different lines of evidence provides solid evidence for evolution.

4. What is one observation that Darwin made in the Galapagos that influenced his thinking about evolution, and why was it important? Would he have been as likely to formulate his theory had he not had the opportunity to sail on the *Beagle*? Why or why not?

Applying Your Knowledge

1. Using evolutionary principles, explain why large ears might be expected to evolve in a terrestrial herbivore such as a rabbit or deer, but not in an aquatic mammal such as a seal.

2. Artificial selection is a process people use to decide which of a group of organisms will be the parents of the next generation, based on whether they possess particular desirable characteristics. Darwin used the differences that breeders had produced among breeds of pigeons as evidence for evolution. Describe how, using artificial selection, you would go about breeding a type of pigeon with a pronounced tuft of feathers on the head.

3. Explain the evolutionary steps by which bacteria may become resistant to antibiotics, using the core requirements for evolution by natural selection.

4. If all organisms had not descended from a common ancestor, and did not possess many common genes and mechanisms of development, would we be able to perform valid medical research using mice and in some cases, *Drosophila*? Why or why not?

Group Activity

Break up into groups that each read either of the following two papers. Have groups form with half of the students having read one article and the other half the second article. In a group, have the students discuss the evolutionary issues involved in antibiotic resistance and in control of crop pests. What are some of the similarities and differences?

Levy, Stuart, "The challenge of antibiotic resistance," *Scientific American* v278, n3, March 1998.

Gould, Fred, "The evolutionary potential of crop pests," *American Scientist* v79, n6, Nov.–Dec. 1991.

17

The Means of Evolution
Microevolution

A significant change.
(page 330)

Big beak, little beak.
(page 339)

Seashells by the seashore.
(page 338)

When people think of the

word *evolution*, what generally comes to mind is the grand sweep of evolution: The story of how microorganisms were the first life-forms; of how dinosaurs came and went; of how human beings arose in the last second of an evolutionary year. It is quite a story, to be sure, but to fully appreciate it we need to look first not at these *outcomes* of evolution, but rather at evolution's *mechanisms*. How is it that generations of organisms become modified over time? That's the subject of this chapter. When you understand it, you'll be in a position to comprehend evolution on its grander scale.

17.1 What Is It That Evolves?

In approaching the topic of how organisms become modified over generations, the first question is: What is it that evolves? If you think about it, it's pretty clear that it is not individual organisms. A tree may inherit a mix of genes that makes it slightly taller than other trees, but if this tree is considered in isolation, it is simply one slightly taller tree. If it were to die without leaving any off-spring, nothing could be said to have evolved, because no persistent quality (added height) has been passed on to any group of organisms.

In Chapter 16 you looked at the idea of **species**, which can briefly be defined as a group of organisms who may successfully interbreed with one another in nature, but who don't successfully interbreed with members of other such groups. You might think that it is species—such as horses or American elm trees—that evolve. Indeed, scientists often speak this way when they refer, for example, to an amphibious mammal having evolved into today's whales. But this is a kind of shorthand whose inaccuracy becomes apparent when species are considered as they live in the real world.

Populations Are the Smallest Units That Can Evolve

Think of a hypothetical species of frogs living in, say, equatorial Africa in a single expanse of tropical forest. Suppose that a drought persists for years, drying up the forest such that this single lush range is now broken up into two ranges separated by an

iving population
different allele
ency.

This different allele frequency is reflected in today's population.

The bottleneck effect.
(page 332)

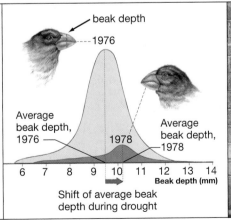

beak depth

1976

Average beak depth, 1976

Average beak depth, 1978

1978

6 7 8 9 10 11 12 13 14
Beak depth (mm)

Shift of average beak depth during drought

Surviving a drought.
(page 337)

Island of the color-blind.
(page 333)

expanse of barren terrain (**see Figure 17.1**). When the separation occurs, there is still but a single species of frog; but that species is now divided into two *populations* that are geographically isolated and hence no longer interbreeding. Each population now faces the natural selection pressures of its *own* environment—environments that may differ greatly even though they are nearly adjacent. One area may get more sun, while the other has a larger population of frog predators, for example. Thus, each population stands to be modified individually over time—to evolve. What is it that evolves? The essential unit that does so is a **population**, which can be defined as members of a species that occupy the same geographic region.

The question then becomes, *how* do populations evolve? The population of frogs that has more predators may, over many generations, evolve a coloration that makes them less visible to predators. But how does this slightly different coloration come about? Through genes. A given frog may be unable to change its spots, but *generations* of frogs can have theirs changed through the shuffling, addition, and deletion of genetic material.

Genes Are the Raw Material of Evolution

Recall from Chapter 11 that the genetic makeup of any organism is its **genotype**. Many genes are likely to be involved in producing a trait such as coloration. In sexually reproducing organisms, *each* of these coloration genes will likely come in two copies or **alleles**, with offspring inheriting one allele from their father and one from their mother (**see Figure 17.2**). Though both alleles help code for coloration, one may result in slightly lighter or darker coloration than the other. Genes don't just come in two allelic variants, however; they can come in many. In many species, no one organism can possess more than two alleles of a given gene, but a *population* of organisms can possess many allelic variants of the same gene (which is one reason human beings don't just come in one or two heights, but in a continuous *range* of heights.)

Figure 17.2
Genetic Basis of Evolution
There are many genes that produce a trait such as body coloration, and each gene often has many alleles or variants. Each individual in a population, however, can possess only two alleles for each gene, one allele inherited from its mother and one inherited from its father. The two frogs in the figure both have maternal and paternal copies of chromosome 3 that house genes for coloration. The chromosomes of the two frogs will differ, however, in the allelic variants they have of these genes. For example, the frog with dark coloration may possess alleles a_1 and a_2 of a chromosome-3 gene that codes for coloration, whereas the light-colored frog may possess alleles a_2 and a_4 of this same gene.

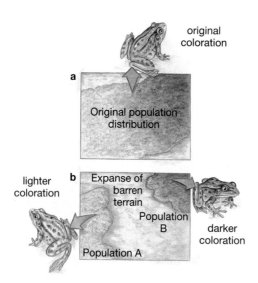

Figure 17.1
Evolution within Populations

a A hypothetical species of frog lives and interbreeds in one expanse of tropical forest.

b After several years of drought, the forest has been divided by an expanse of barren terrain. The single frog population has thus been divided into two populations, separated by the barren terrain. The two frog populations now have different environmental pressures, such as the kinds of predators each faces, and they no longer interbreed; after many generations their coloration diverges as they adapt to the different pressures.

When the concept of a population is put together with that of genes and their alleles, the result is the concept of the **gene pool**: all the alleles that exist in a population. This is the raw material that evolution works with. If evolution were a card game, the gene pool would be its deck of cards. Individual cards (alleles) are endlessly shuffled and dealt into different "hands" (the genotypes of individuals), with the strength of any given hand dependent upon what game is being played. (Survival in a drought? Survival against a new set of predators?)

17.2 Evolution as a Change in the Frequency of Alleles

Thinking of genes in terms of a gene pool gives us a new perspective on what evolution is at root: a change in the frequency of alleles in a population. This may sound a little abstract until you consider the frog coloration example. A frog inherits, from its mother and father, a coloration that allows it to evade predators slightly more successfully than other frogs of the same generation. It thus lives longer and leaves more offspring than the other frogs. For simplicity's sake, let's assume that a single combination of alleles gives this frog its advantageous coloration. Relative to other combinations, these alleles have thus *increased* in frequency in the frog population, because there are now relatively more of this frog's offspring in the population. In looking at any example of evolution in a population over time, you would find this kind of change in allele frequency as its basis, assuming a stable environment.

The change of allele frequencies in populations over time is called **microevolution**, and it is essentially the subject of this chapter. Why the *micro* in microevolution? Because evolution *within* a population is evolution at its smallest scale. Taking a step back, the large-scale *patterns* produced by microevolution become visible, as with the evolution of, say, mammals from reptiles. This larger-scale evolution is called **macroevolution**.

So, what causes the frequency of alleles to change in a population? In the frog-coloration example, a familiar force was at work: natural selection. This particular coloration was selected for passage to more frogs by its "fit" with the frog's environment. The coloration the frog had was selected from among others in the population because of its value in helping the frog produce more offspring; thus the relative abundance of alleles for this coloration increased in the population.

17.3 Five Agents of Microevolution

Natural selection is not the only force that can change allele frequencies, however. There are five "agents" of microevolution that can alter allele frequencies in populations. These are mutation, gene flow, genetic drift, nonrandom mating, and natural selection. You can see these forces summarized in **Table 17.1**. Each of them is now considered in turn.

Mutations: Alterations in the Makeup of DNA

As you saw in the genetics unit, a mutation is any permanent alteration in an organism's DNA. Some of these alterations are heritable, meaning they can be passed on to future generations. Mutations can be as small as a change in a single base pair in the DNA chain

Table 17.1
Agents of Change: Five Forces That Can Bring about Change in Allele Frequencies in a Population

Agent	Description
Mutation	Alteration in an organism's DNA; generally has no effect or a harmful effect. But beneficial or "adaptive" mutations are indispensable to evolution.
Gene Flow	The movement of alleles from one population to another. Occurs when individuals move between populations or when one population of a species joins another, assuming the second population has different allele frequencies than the first.
Genetic Drift	The chance alteration of gene frequencies in a small population. Can occur when populations are reduced to small numbers (the bottleneck effect) or when a few individuals from a population migrate to a new, isolated location and start a new population (the founder effect).
Nonrandom Mating	Occurs when one member of a population is not equally likely to mate with any other member. Includes sexual selection, in which members of a population choose mates based on the traits the mates exhibit.
Natural Selection	Some individuals will be more successful than others in surviving and hence reproducing, owing to traits that give them a better "fit" with their environment. The alleles of those who reproduce more will increase in frequency in a population.

MediaLab Link
Silversword plant, see p. 478.

(a point mutation) or as large as the addition or deletion of a whole chromosome or parts of it. Whatever the case, a mutation is a change in the informational set an organism possesses (**see Figure 17.3**). Looked at one way, it is a change in one or more alleles.

The rate of mutation in most organisms is very low; in human beings, a given gene may undergo a mutation as little as once in every 100,000 times a "gamete" (sperm or eggs) is formed. And of the mutations that do arise, very few are beneficial or "adaptive"; most do nothing, and many are harmful to organisms. As such, mutation is not likely to account for much of the change in allele frequency that is observed in any population. This is because not many mutations occur in the first place and, of those that do, a significant number are never passed along.

But a few mutations do occur that are adaptive. These genetic alterations are something like creative thinkers in a society: They are rare but very important. Such mutations are the only means by which *new* genetic information comes into being—by which new proteins are produced that can modify the form or capabilities of an organism. The evolution of eyes or wings had to involve mutations. No amount of shuffling of *existing* genes could get the living world from no-eyes to eyes. Of course, no mutation can bring about a feature such as eyes in a single step; such changes are the result of many mutations, followed by rounds of genetic shuffling and natural selection, generally over millions of years.

Gene Flow: When One Population Joins Another

Allele frequencies in a population can also change with the mating that can occur after the arrival of members from a *different* population. This is the second microevolutionary agent, **gene flow**, meaning the movement of genes from one population to another. Such movement takes place through **migration**, which is the movement of individuals from one population into the territory of another. Some populations of a species may truly be isolated, such as those on remote islands, but migration and the gene flow that goes with it are the rule rather than the exception in nature. It may seem at first glance that migration would be limited to animal species, but this isn't so. Mature plants may not move, but plant seeds and pollen do; they are carried to often-distant locations by wind and animals (**see Figure 17.4**). Of course, for a migrating population to alter allele frequencies of another population, its gene pool must be different from that of the population it is joining.

Genetic Drift: The Instability of Small Populations

To an extent that may surprise you, evolution turns out to be a matter of chance; in the allele-frequency changes that come about with the third microevolutionary agent, genetic drift, you can almost see the dice rolling.

Imagine a hypothetical population of 10,000 individuals and an allele in the gene pool that is carried by one out of ten of them, meaning that 1,000 individuals carry it. Now imagine that some disease sweeps over the population, killing half of it. Say that this allele had nothing to do with the disease, so the illness might be expected to decimate the allele carriers in rough accordance with their proportion in the population. If this were the case, 5,000 individuals in the population would survive, and 1 in 10 or about 500 of them could be expected to be carriers of this allele. Let us say, however, that

Figure 17.3
Basis of New Genetic Information
A mutation is any permanent alteration in an organism's DNA. Examples of mutations include **(a)** point mutations, in which the nucleotide sequence is incorrect, and **(b)** deletions, in which part of a chromosome is missing.

NORMAL   POINT MUTATION

a

correct nucleotide sequence

incorrect nucleotide sequence

Point mutation

NORMAL   DELETION

Deletion

b

complete chromosome 5

incomplete chromosome 5

just by chance, *550* of the allele carriers were killed, thus leaving the surviving population of 5,000 with only 450 allele carriers. In that scenario, the frequency of the allele in this population would drop from 10 percent to 9 percent (**see Figure 17.5a**).

Now the critical step. Imagine the same allele, with the same 1-in-10 frequency, only now in a population of *10*. There is now but a single carrier of the allele, and that individual may not be one of the 5 members of the population to survive the disease. In this case, the frequency of this allele drops from 10 percent to zero (**see Figure 17.5b**). It can be replaced only by a mutation (which is extremely unlikely) or by migration from another population. Assuming that neither happens, no matter how this population grows in the future, in genetic terms it will be a different population than the original, in that it lost this allele. The allele might be helpful or harmful, but its adaptive value doesn't matter. It has been eliminated, strictly through chance. This is an example of **genetic drift**: The chance alteration of allele frequencies in a small population. It is true that some genetic drift has taken place in the larger population, but look at how small the effect is: The allele simply went from a 10 percent frequency to a 9 percent, with no loss of allele. Chance events can have much greater effects on small populations than on large ones.

Two scenarios, common in evolutionary history, produce the small populations that are a prerequisite for genetic drift. Populations can be greatly reduced in number through such means as disease or natural catastrophe; or a small subset of a population can migrate elsewhere and start a new population. Let's have a look at both these scenarios.

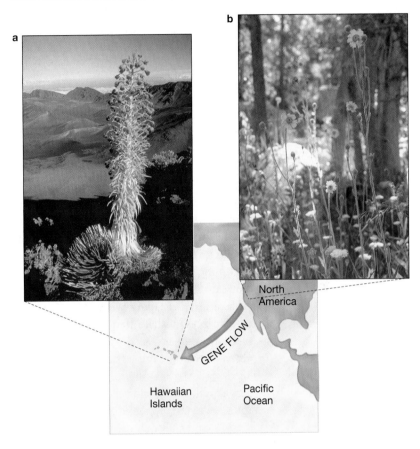

Figure 17.4
Plant Migration
Brought about by volcanic eruptions, the Hawaiian Islands have always been surrounded by the Pacific Ocean. Therefore all the plant species that exist on the islands today are descended from species that were introduced to the islands through one means or another—human activity, wind currents, water currents, or animal dispersal of seeds and pollen. **(a)** Hawaiian silverswords are derived from **(b)** a lineage of California plants commonly known as tarweeds.

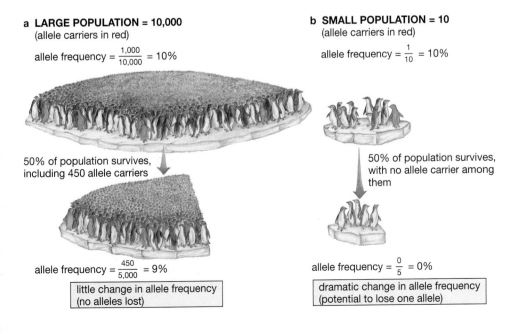

a LARGE POPULATION = 10,000
(allele carriers in red)

allele frequency = $\frac{1,000}{10,000}$ = 10%

50% of population survives, including 450 allele carriers

allele frequency = $\frac{450}{5,000}$ = 9%

little change in allele frequency (no alleles lost)

b SMALL POPULATION = 10
(allele carriers in red)

allele frequency = $\frac{1}{10}$ = 10%

50% of population survives, with no allele carrier among them

allele frequency = $\frac{0}{5}$ = 0%

dramatic change in allele frequency (potential to lose one allele)

Figure 17.5
Genetic Drift

a A hypothetical population of 10,000 exists, in which 1 individual in 10 carries a given allele. The population loses half its members to a disease, including 550 individuals who carried the allele. The frequency of the allele in the population thus drops from 10 percent to 9 percent.

b A population of 10 with the same allele frequency likewise loses half its members to a disease. Because the one member of the population who carried the allele is not a survivor, the frequency of the allele in the population drops from 10 percent to zero.

The Bottleneck Effect and Genetic Drift

A phenomenon called the **bottleneck effect** can be defined as a change in allele frequencies due to chance during a sharp reduction in a population's size. Real populations, or even species, can go through dramatic reductions in numbers. For example, Northern Elephant Seals, which can be found off the Pacific coast of North America, were prized for the oil their blubber yielded and thus were hunted so heavily that by the 1890s, fewer than 50 animals remained. Thanks to species protection measures, the seals' numbers have rebounded somewhat in recent decades. But genetically, all the members of this species are very similar today, because they all descended from the few seals who made it through the nineteenth-century bottleneck. What occurs in these reductions is a "sampling" of the original population with the "sample" being those who survived the devastation (**see Figure 17.6**).

The reason allele frequencies change in such an event has to do with the nature of probability and sample size. Imagine a box filled with M&M's candies, with equal numbers of red, green, and yellow M&M's inside. You close your eyes and grab a handful of M&M's and pull out 12. With such a small sample, you might get, say, six reds, four greens, and two yellows, rather than the four-of-each-kind that would be expected from probability. If you pulled out *120* M&M's, conversely, your reds, greens, and yellows would be much more likely to approach the 40-of-each-kind that would be expected. In just such a way is a small sample of *alleles* likely to yield a gene pool that's different from the distribution found in the larger population.

The Founder Effect and Genetic Drift

Genetic drift can also result from the **founder effect**, which is simply a way of stating that when a small subpopulation *migrates* to a new area to start a new population, it is likely to bring with it only a portion of the original population's gene pool. This is another kind of "sampling" of the gene pool, in other words, only in this case the cause is migration of a few individuals rather than the survival of only a few. This sample of the gene pool now becomes the founding gene pool of a new population. As such, it can have a great effect; whatever genes exist in it become the genetic set that is passed on to all future generations, as long as this population stays isolated.

The power of the founder effect can be seen most clearly when a founding population brings with it, or develops, the alleles for rare genetic diseases. For example, on a Micronesian island called Pingelap, about 10 percent of the human population is totally color-blind because of a genetic condition so rare that it is found almost nowhere else in the world (**see Figure 17.7**). Such a high incidence-rate makes it almost certain that the gene for the disease had to appear by mutation in a member of the Pingelap population at a point in which this population was very small. After this, it continued to appear in the population at a high *rate*, no matter how large the population grew.

From this and the earlier examples, you may have perceived by now that there is an inherent value in genetic diversity; you can

Figure 17.6
The Bottleneck Effect
The Northern Elephant Seals were hunted so heavily by humans that in the 1890s, fewer than 50 animals remained. The population has grown from the few survivors (represented here by the three brown seals), but the resulting genetic diversity of this population is very low. (Seal coloration for illustrative purposes only.)

"bottleneck" only allows a few individuals through

Original population, original allele frequency.

Hunting of seals in late 1800s greatly reduced population size.

Surviving population had different allele frequency.

This different allele frequency is reflected in today's population.

read about this in "Lessons from the Cocker Spaniel," on page 334.

Nonrandom Mating: When Mating Is Uneven Across a Population

You've looked at mutation and genetic drift as agents of change with respect to allele frequencies in a population. Now let's look at a third agent, nonrandom mating. Consider an imaginary population of eight animals, four males and four females, who pair off, mating for life and producing 16 offspring. Now imagine this same population of eight, but with *one* of its males mating with all four females while the other three males don't mate at all. It may be obvious that the allele frequencies of this population stand to differ depending on which of these two scenarios takes place, since, in the second scenario, a single male's alleles will be much more heavily represented in succeeding generations. This is an example of **nonrandom mating**, which is simply mating in which a given member of a population is not equally likely to mate with any other given member.

Some forms of nonrandom mating do not directly affect allele frequencies. Such is the case with **assortative mating**, which occurs when males and females that share a particular characteristic tend to mate with one another. Humans practice assortative mating in that short individuals tend to pick short mates, while tall individuals tend to pick tall mates. Such mating tends to bring similar alleles together ("short" with short, etc.) but it does not directly alter the *prevalence* of alleles in a population.

Nonrandom mating does affect allele frequency, however, when some members of a population mate *more* than others, as with the eight-animal example. In practice, this is mating based on **phenotype**, which you may recall is any observable trait in an organism, including differences in appearance and behavior. It is the *appearance* of particularly nice plumage on a male peacock that makes female peacocks choose it for mating, rather than another male, while it is the *behavior* of superior fighting in a given male sage grouse that causes a succession of female grouse to mate with it rather than nearby competitors (**see** Figure 17.8). Such **sexual selection** has something in common with natural selection, in that both processes result in some in-

dividuals passing along more of their genes to future generations than others. But while natural selection has to do with differential *survival* (and hence reproduction), sexual selection has to do with differential *mating* (and hence reproduction).

Natural Selection: Evolution's Adaptive Mechanism

You've already gone over a good deal about natural selection in this review of evolution, but it's time now to look at it as the last agent of change in microevolution. First, what is meant by **natural selection**? Here is a very clear description of it, written by the biologist Julian Huxley almost 60 years ago.

MediaLab Link
Natural selection, see p. 344.

Figure 17.7
Founder Effect
About 10 percent of the people who live on the island of Pingelap in Micronesia are color-blind due to an extremely rare genetic condition. Pictured is an island family whose members all suffer from the condition. Note how they squint into the sun. Sensitivity to bright light is another result of the affliction.

Figure 17.8
Sexual Selection
Individuals in some species choose their mates based on appearance or behavior. Female sage grouses prefer to mate with males who have demonstrated superior fighting ability. Here, two males prepare to fight on the breeding ground.

Lessons from the Cocker Spaniel: The Price of Inbreeding

"Marry afar" is advice given by the pygmy men of Africa, by which they mean, "marry a woman who lives far from your home." The pygmies point to the hunting rights a man acquires in distant territory by marrying a woman from a remote region, but there is another benefit to this practice as well: It helps ensure genetic diversity among the pygmies. When pygmies from remote locations marry, it cuts down on the chances of "inbreeding"—in which close relatives have children together—which can have harmful effects not only in humans, but in any species.

What are these dangers? Well, consider what has happened in the United States with "purebred" dogs (**Figure 1**), which is to say dogs that over many generations have been bred solely with members of their own breed (cocker spaniels mating only with cocker spaniels, etc.). In the mid 1990s it was estimated that up to one-fourth of all U.S. purebred dogs had some sort of serious genetic defect— ranging from improper joint formation to heart defects, to deafness. We also have a good deal of experimental evidence from other animals on the effect of inbreeding. In one instance, a group of white-footed mice that had been inbred and then let out into the natural environment survived at only 56 percent the rate of a group of genetically diverse mice. As already noted, negative biological outcomes can appear in humans too; for that reason, many Western societies have made it illegal for first cousins to marry.

Why does inbreeding cause such trouble? As you've seen, every organism has a whole series of genes, almost all of which come in two copies, or alleles, one copy inherited from the father and one from the mother. In all organisms, a very small proportion of these alleles is potentially harmful or even lethal. But in most instances such genes

In the mid 1990s it was estimated that up to one-fourth of all U.S. purebred dogs had some sort of serious genetic defect.

are recessive. *One* defective allele is not enough to cause trouble, because the second differs from it and provides sufficient information for the organism to function properly. Alleles that differ in this way are said to be heterozygous, whereas alleles that are identical are homozygous.

So how do these biological facts play out in inbreeding? Imagine two seals, brother and sister, that have mated together because their population has been so reduced. Imagine that their mother had a recessive allele that brings about a heart defect, and that this allele occurs in only one out of every hundred seals. There is a one-in-two chance that the mother will pass this allele on to any

Since there is a struggle for existence among individuals, and since these individuals are not all alike, some of the variations among them will be advantageous in the struggle for survival, others unfavorable. Consequently, a higher proportion of individuals with favorable variations will on the average survive, a higher proportion of those with unfavorable variations will die or fail to reproduce themselves.

By this process, the traits of those who are more successful in reproducing will be-

come more widespread in a population—as the alleles that bring about these traits increase in frequency from one generation to the next.

A key concept here is that of **adaptation**, meaning a modification in the structure or behavior of organisms over generations in response to environmental change. Environmental change may come to a population (through streams drying up, for example), or a population may come to environmental change (through migration). Either way, natural selection is the force that pushes populations to adapt to new conditions. The frog population noted at the outset of the chapter one of her offspring, and let's say it happened with the male and female here: Both brother and sister have inherited the defective allele. When *they* mate, the chance that either of them will pass along the defective allele to their offspring is likewise one out of two. More important, however, the chance that *both* of them will pass the allele along is one out of four. If, on the other hand, the brother had mated with an unrelated member of the population, the chances of this union passing along two copies of this defective allele are one in *four hundred.** Why the difference? Because the unrelated female had only a one-in-a-hundred chance of having the allele in the first place, whereas the sister had a one-in-two chance of getting it from her mother.

The essence of this lesson, as you might be able to see, is that inbreeding has brought together identical alleles—the recessive homozygous alleles that are required for many genetic diseases. Random breeding, meanwhile, tends to keep homozygous alleles apart. It tends to preserve genetic diversity, to put it another way.

---

*Remembering the rule of multiplication from Chapter 12, to calculate the odds of passing on two defective alleles in the random mating example, take the son's chances of passing along the allele (1 in 2, or 50 percent, or 0.5) × 0.01 (the chance of his female partner *having* the allele) × 0.5 (the chance of her passing the allele along). This works out to 0.0025, which equals 1 in 400. For the brother-and-sister mating, the figures are 0.5 (the chance of the brother passing it along) × 0.5 (the chance of the sister passing it along), which equals 0.25—or 1 in 4.

You might think from this that there are no uses for genetic *similarity*, but consider what human beings have been doing for centuries with dog breeding. Cocker spaniels did not come about by accident. Dogs that had cocker spaniel features were bred together for many generations, ultimately giving us today's dog. It was the *novel features* of the spaniel that people wanted, and novel features are often the result of homozygous recessives coming together. We don't value the seals' heart condition, but looked at one way it too is a novel feature, made possible by homozygous recessives pairing, perhaps like the long ears of a cocker spaniel. But there is a price to be paid for getting the features we want, and it is the diseases, early deaths, and other afflictions that go with inbreeding.

Figure 1
Result of Generations of Controlled Breeding

---

adapted to a new predator, for example, by taking on a darker coloration.

Among the five agents of microevolution, natural selection is the only one that consistently works to adapt organisms to their environment. Genetic drift is random; it could as easily work against an adaptive trait as for it. Although mutations can have an adaptive effect, they more often have no effect or even a negative effect.

Natural selection is, however, constantly working to modify organisms in accordance with the environment around them. As such, it is generally regarded as the most important agent in having shaped the natural world—in having given zebras their stripes and flowers their fragrance.

17.4 What Is Evolutionary Fitness?

Even the strongest supporters of natural selection as a shaping force would not maintain that it is some perfect mechanism working to produce perfect organisms. The concept of **fitness** is helpful here, if only to clear up some misconceptions. To a biologist, fitness means the success of an organism in passing on its genes to offspring *relative* to other members of its population in a particular environment. An organism cannot be deemed to be "fit" even if it has 1,000 offspring. It can only have *more or less* fitness relative to other members of its population (who might have 900 or 1,100 offspring).

This concept then gets us to the notion of "survival of the fittest," and the misunderstandings that arise from it. The phrase can be taken to imply the existence of superior beings; that is, organisms that are simply "better" than their counterparts, with images of being faster, more muscular, or smarter coming to mind. In fact, however, evolutionary fitness tells us nothing about organisms being *generally* superior, and it certainly tells us nothing about the value of any particular capacity, be it brawn or brain. All it tells us about are organisms who are better than others in their population at passing along their genes in a given environment at a given point in time. An accurate phrase, as others have pointed out, would not be "survival of the fittest," but rather "survival of those who fit—for now." Let's look at a real-life example of natural selection in action to see why this is true.

Galapagos Finches: The Studies of Peter and Rosemary Grant

When Charles Darwin stopped at the Galapagos Islands in 1835, one of the animal varieties he collected were various species of finches. Over the years, biologists kept coming back to "Darwin's finches" because of the very qualities Darwin found in them: They seemed to present a textbook case of evolution, with their 13 species having evolved from a single ancestral species on the South American mainland. Yet for more than 100 years after Darwin, it was a puzzle for scientists to figure out how Darwin's posited mechanism of evolution, natural selection, could have been at work with the finches. This changed beginning in the 1970s, when the husband-and-wife team of Peter and Rosemary Grant began a painstakingly detailed study of the birds.

Natural selection in the finches came into sharp focus in 1977, when a tiny Galapagos island, Daphne Major, suffered a severe drought. Rain that normally begins in January and lasts through July scarcely came at all that year. This was a disaster for the island's two species of finches; in January 1977 there were 1,300 of them, but by December the number had plunged to fewer than 300. Daphne's medium-sized ground finch, *Geospiza fortis*, lost 85 percent of its population in this calamity. The staple of this bird's diet is plant seeds. When times get tough, as in the drought, the size and shape of *G. fortis* beaks—their beak "morphology"—begins to define what one bird can eat as opposed to another (**see Figure 17.9**). In *G. fortis*, larger body size and deeper beaks turned out to make all the difference between life and death in the drought of 1977. Measuring the beaks of *G. fortis* who survived the drought, the Grant team found they were larger than the beaks of the population before the drought by an average of some 6 percent. This was a difference of about half a millimeter, or roughly two-hundredths of an inch; by such a difference were the survivors able to get into large, tough seeds and make it through the catastrophe, eventually to reproduce (**see Figure 17.10**).

This is natural selection in action, but there is more to be learned from the Grant study, which is to say *evolution* made visible. The Grant team knew that beak depth had a high "heritability" in the finches, meaning that beak depth is largely under genetic control. As it turned out, the *offspring* of the drought survivors had beaks that were 4 percent to 5 percent deeper than the average of the population before the drought. In other words, the drought had preferentially preserved those alleles from the starting population that brought about deeper beaks, and the result was a population that evolved in this direction.

Figure 17.9
A Product of Evolution on the Galapagos
A male of the species of finch *Geospiza fortis*, which is native to the Galapagos Islands.

But the Grant study yielded one more lesson. In 1984–85 there was pressure in the opposite direction: Few large seeds and an abundance of small seeds provided an advantage to *smaller* birds, and it was they who survived this event in disproportionate numbers.

Lessons from *G. fortis*

So, where is the "fittest" bird in all this? There isn't any. Evolution among the finches was not marching toward some generally superior bird. Different traits were simply favored under different environmental conditions. Secondly, there is no evolutionary movement toward combativeness or general intelligence here. Survival had to do with size—and not necessarily *larger* size at that. Looking around in nature, it's true that some showcase species, such as lions and mountain gorillas, gain success in reproduction by being aggressive. And it's true that our own species owes such success as it has had to intelligence. But in most instances, it is not brawn or brain that make the difference; it is something as seemingly benign as beak depth and its fit with the environment.

Finally, consider how imperfect natural selection is at the genetic level. Suppose the smaller *G. fortis* that disproportionately died off in 1977 also disproportionately carried an allele that would have aided just slightly in, say, long-distance flying. In the long run this might have been an adaptive trait, but it wouldn't matter. The flight allele would have been *reduced* in frequency in the population (as the smaller birds died off) because flight distance didn't matter in 1977, beak depth did. Evolution operates on the phenotypes of whole organisms, not individual genes. As such, it does not work to spread *all* adaptive traits more broadly; the destiny of each trait is tied to the constellation of traits the organism possesses. Genes are "team players," in other words, that can only do as well as the team they came in on.

17.5 Three Modes of Natural Selection

In what directions can natural selection push evolution? As noted earlier, a character such as human height is under the control of many different genes and is thus **polygenic**. Likewise, it has been noted that such a character is "continuously variable." There are not one or two or three human heights, but an innumerable number of them in a range. When natural selection operates on characters that are polygenic and continuously variable, it can proceed in any of three ways. The essential question here is: Does natural selection favor what is average in a given character, or what is extreme? (**See Figure 17.11.**)

Figure 17.11
Three Modes of Natural Selection
The bell curves represent the number of individuals in a population with a certain characteristic, in this case color intensity. Under stabilizing selection, the individuals that possess *extreme* values of a character—here the lightest and darkest colors—are selected against and die, or fail to reproduce. Over several generations, more and more members of the population move toward the average color of the population. Under directional selection, one of the extremes produces a better fit with the environment, meaning that individuals at the other extreme are selected against. Over several generations, the population thus moves in a direction—toward a darker coloration. Under disruptive selection, individuals with average coloration are selected against and die. Over many generations, half of the population becomes lighter, while the other half becomes darker.

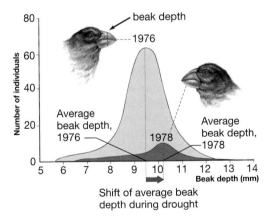

Figure 17.10
Who Survives in a Drought?
A large percentage of the population of *Geospiza fortis* died on a Galapagos Island, Daphne Major, during a drought in 1977. Peter Grant observed in 1978 that the individuals who survived the drought had a greater average beak depth than the average individuals surveyed before the drought, in 1976. Individuals with the larger beaks were better able to crack open the large, tough seeds that were available during the drought. The offspring of the survivors likewise had larger average beak size than did the population before the drought. Thus, evolution through natural selection was observed in just a few years on the island.

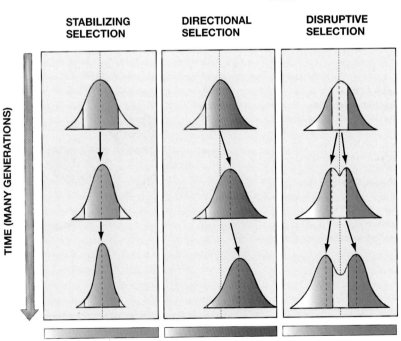

STABILIZING SELECTION **DIRECTIONAL SELECTION** **DISRUPTIVE SELECTION**

TIME (MANY GENERATIONS)

RANGE OF A PARTICULAR CHARACTERISTIC (SUCH AS COLOR)

Stabilizing Selection

In **stabilizing selection**, intermediate forms of a given character are favored over extreme forms. A clear example of this is human birth weights. If you look at **Figure 17.12**, you can see, first, the weights that human babies tend to be. Notice that there are not only relatively few 3-pound babies, but relatively few 9-pound babies as well; a great proportion of birth weights fall at the average or "mean" of a little less than 7 pounds. Now look at the infant-mortality curve. Infant deaths are highest at both extremes of birth weight; low-birth-weight babies *and* high-birth-weight babies are more at risk than are average-birth-weight babies (though low birth weight poses the greater risk). Put another way, the children most likely to survive (and reproduce) are those carrying alleles for intermediate birth weights. Thus, natural selection is working to make intermediate weights even more common; it is not working to move birth weights toward the extremes of higher or lower weights.

Directional Selection

When natural selection moves a character toward *one* of its extremes, **directional selec-** tion is in operation—the mode in which we most commonly think of evolution operating. You saw directional selection operating in Chapter 16, in connection with the moth *Biston betularia* as it got darker and then lighter again in response to the environmental changes around it. If you look at **Figure 17.13**, you can see an example from a much longer stretch of evolutionary time.

Disruptive Selection

When natural selection moves a character toward *both* of its extremes, the result is **disruptive selection**, which appears to occur much less frequently in nature than the other two modes of natural selection. This mode of selection is visible in the beaks of yet another kind of finch. A species of these birds found in West Africa (*Pyrenestes ostrinus*) has a beak that comes in only two sizes. Thomas Bates Smith, who has studied the birds, has observed that if human height followed this pattern, there would be some Americans who are 4 to 5 feet tall, and some who are 6 to 7 feet tall, but no one who is 5 to 6 feet tall (**see Figure 17.14**).

The environmental condition that leads to this mode of selection is, once again, diet. When food gets scarce, large-billed birds specialize in cracking a very hard seed, while small-billed birds begin feeding on several soft varieties of seed, with each type of bird being able to outcompete the other for its special variety. Given how these birds have evolved, it's probably safe to assume that a bird with an intermediate-sized bill would get less food than one with a bill of either ex-

Figure 17.12
Human Birth Weights and Infant Mortality
Note that infant deaths are more prevalent at the upper and lower extremes of infant birth weights.

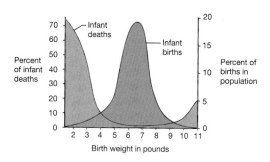

Figure 17.13
Directional Selection:
Evolving toward Smoothness
Over a period of 10 million years, beginning about 415 million years ago, directional selection operated on these brachiopods, such that their shells became progressively smoother over time. (Modified from T. Dobzhansky et al., *Evolution*. New York: Freeman, 1977, p. 329.)

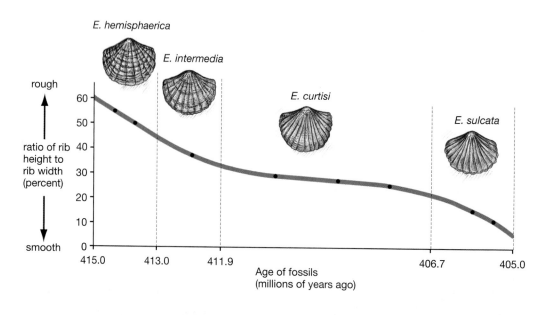

treme type. You might think that what really exists here are two separate species of bird, but large- and small-billed birds mate. Bill size seems to be under the control of a single genetic factor, so that bills are able to come out either large or small. This genotype was presumably shaped by natural selection over generations, such that any alleles for intermediate-sized bills were weeded out.

On to the Origin of Species

Stabilizing selection does what it says: It stabilizes given traits of a population, thereby keeping it a single entity. However, both disruptive and directional selection can serve as the basis for speciation—for bringing about the transformation of a single species into one or more *different* species. How is it, though, that such stabilization and speciation work? How do we classify the huge number of species that are the outcome of evolution's myriad branchings? These are the subjects of the next chapter.

Figure 17.14
Evolution That Favors Extremes in Two Directions
Finches of the species *Pyrenestes ostrinus*, found in West Africa. The birds of this species have beaks that come in two distinct sizes—large and small.

Chapter Review

Summary

17.1 What Is It That Evolves?

- A population is all the members of a species living in a defined geographical area. Natural selection acts on individuals, but it is populations that evolve.

17.2 Evolution as a Change in the Frequency of Alleles

- Genes may be found in variant forms, called *alleles*; in most species, no individual will possess more than two alleles for a given gene, but a population may possess many such allelic variants. The sum total of alleles in a population is referred to as that population's *gene pool*.

- The basis of evolution is a change in the frequency of alleles within a population and the appearance of new alleles through mutation. Evolution at this level is called *microevolution*.

17.3 Five Agents of Microevolution

- Five evolutionary forces can result in changes in allele frequencies within a population. These agents of microevolution are mutation, gene flow, genetic drift, nonrandom mating, and natural selection.

- Mutation happens fairly infrequently, and most mutations either have no effect or are harmful; yet adaptive mutations are vital to evolution in that they are the only means by which entirely new genetic information comes into being.

- Gene flow, the movement of genes from one population to another, takes place through migration, meaning the movement of individuals from one population into the territory of another.

- Genetic drift is the chance alteration of allele frequencies in a small population. Two scenarios can produce the small populations that are a prerequisite for genetic

Key Terms

adaptation 334

allele 328

assortative mating 333

bottleneck effect 332

directional selection 338

disruptive selection 338

fitness 335

founder effect 332

gene flow 330

gene pool 329

genetic drift 331

genotype 328

macroevolution 329

microevolution 329

migration 330

natural selection 333

nonrandom mating 333

phenotype 333

polygenic 337

drift. The first of these is the *bottleneck effect*, defined as a change in allele frequencies due to chance during a sharp reduction in a population's size. The second is the *founder effect*, the fact that when a small subpopulation migrates to a new area to start a new population, it is likely to bring with it only a portion of the original population's gene pool.

acts to adapt organisms to their environments. As such, it is generally regarded as the most powerful force in evolution.

17.4 What Is Evolutionary Fitness?

- The phrase "survival of the fittest" is misleading, because it implies that evolution works to produce generally superior beings who would be successful competitors in

a LARGE POPULATION = 10,000
(allele carriers in red)

allele frequency $= \frac{1,000}{10,000} = 10\%$

b SMALL POPULATION = 10
(allele carriers in red)

allele frequency $= \frac{1}{10} = 10\%$

50% of population survives, including 450 allele carriers

50% of population survives, with no allele carrier among them

allele frequency $= \frac{450}{5,000} = 9\%$

little change in allele frequency (no alleles lost)

allele frequency $= \frac{0}{5} = 0\%$

dramatic change in allele frequency (potential to lose one allele)

- Nonrandom mating is mating in which a given member of a population is not equally likely to mate with any other member. Sexual selection is a form of nonrandom mating that can directly affect the frequency of alleles in a gene pool. It occurs when differences in reproductive success arise from differential success in mating.

- Some individuals will be more successful than others in surviving and hence reproducing, owing to traits that give them a better "fit" with their environment. This phenomenon is known as *natural selection*. Natural selection is the only agent of microevolution that consistently

any environment. Evolutionary fitness, however, has to do only with the relative reproductive success of individuals in a given environment at a given point in time.

17.5 Three Modes of Natural Selection

- Natural selection has three modes: stabilizing selection, directional selection, and disruptive selection. Stabilizing selection moves a given character in a population toward intermediate forms; directional selection moves a given character toward one of its extreme forms; and disruptional selection moves a given character toward two extreme forms.

Understanding the Basics

Multiple-Choice Questions

1. Every individual within a typical population of plants or animals has _____ copies of each gene, which may be the same or different allelic variants. A population may have _____ allelic variants for a given gene.
   - **a.** 1, 2
   - **b.** 2, 4
   - **c.** 2, many
   - **d.** 1, many
   - **e.** 2, 2

2. Match the terms with their meanings:
   - **a.** gene pool — the form of a gene at a particular locus
   - **b.** allele — the set of all alleles in a population
   - **c.** allele frequency — the genetic makeup of an organism
   - **d.** genotype — exchange of genes between populations
   - **e.** gene flow — the proportion of a given form of a gene in a population

3. Which of the following statements is true of microevolution (select all correct statements)?
   - **a.** It is a change in allele frequency within populations.
   - **b.** It is caused by inheritance of characters acquired during the life of an organism.
   - **c.** It is seen when some allelic variants cause individuals to have increased survival or reproduction.
   - **d.** It always leads to adaptation.
   - **e.** It is caused only by natural selection.

4. Agents of change in allele frequency in populations include (select all that apply)
   - **a.** genetic drift
   - **b.** meiosis
   - **c.** mutation
   - **d.** segregation
   - **e.** natural selection

5. Agents that consistently produce adaptive evolution include (select all that apply)
   - **a.** genetic drift
   - **b.** mutation
   - **c.** meiosis
   - **d.** natural selection
   - **e.** gene flow

6. Mutations are (select all that apply)
   - **a.** generally beneficial
   - **b.** changes in the genetic material that may be inherited
   - **c.** always caused by changes in single base pairs of DNA
   - **d.** the ultimate source of new genetic variation
   - **e.** prone to occur with great frequency

7. In stabilizing selection, _____ individuals have the highest fitness:
   - **a.** large
   - **b.** intermediate
   - **c.** bright
   - **d.** extreme
   - **e.** small

8. Match the agent of evolution to its description:
   - **a.** mutation — chance alterations of allele frequencies in a population
   - **b.** gene flow — differential survival and reproduction
   - **c.** natural selection — movement of alleles between populations by migration
   - **d.** genetic drift — genetic drift due to a few colonizing genotypes
   - **e.** founder effect — changes in the genetic material

9. When an insect population becomes resistant to insecticides, this is an example of a response to _____ selection?
   - **a.** disruptive
   - **b.** stabilizing
   - **c.** sexual
   - **d.** directional
   - **e.** artificial

10. Finches in the Galapagos experienced _____ selection on beak depth following a drought, while the African finch *P. ostranus* experiences _____ selection on beak size because they eat either very small, soft seeds or much larger, hard seeds.
    - **a.** stabilizing, directional
    - **b.** sexual, disruptive
    - **c.** directional, disruptive
    - **d.** disruptive, directional
    - **e.** stabilizing, disruptive

11. A large lizard population (1,000 individuals) on the coast contains mostly individuals that are plain brown, but a few have white spots. Coloration is genetically determined. One day during a storm, two spotted lizards hitch a ride on a piece of driftwood to a nearby island, where they join a population of 100 lizards, some of which are spotted and some of which are plain brown. The island's lizard population allele frequency has thus changed due to
   a. natural selection
   b. mutation
   c. gene flow
   d. non-random mating
   e. inbreeding

Brief Review

1. Explain the statement, "Individuals are selected, populations evolve."

2. What is a gene pool and why do changes in gene pools lie at the root of evolution?

3. What are the five causes of allele frequency changes (microevolution) and how does each work?

4. Why is evolutionary fitness always a measure of relative fitness in a population?

5. Why is natural selection the only agent of evolution that can produce adaptation?

6. Is the evolutionary fitness of an individual expected to be the same or different in different environments? How did the Grant's study of beak depth of finches after the drought and then again in 1984–1985 provide a real example of this?

Applying Your Knowledge

1. Cheetahs have long legs relative to other large cats. However, leg length in a cheetah population is more likely to be under stabilizing rather than directional selection. Why?

2. If a compound is known to be a mutagen, people generally try to avoid it. Why is that so, even though we know that mutations are the raw material for adaptive evolution?

3. Explain why genetic drift may be important when captive populations of animals or plants are started with just a few individuals.

4. Two moth populations of the same species utilize different host plants. One rests on leaves and has evolved a green color that allows it to escape predation by blending in with the leaves. The other population rests on tree trunks, and is brown in color. The colors of these moths are genetically determined by different alleles of the same gene. What would be some consequences of migration of moths leading to gene flow between these two populations?

5. The text notes that natural selection is the force that pushes populations to adapt to environmental change. Is natural selection still going on in human populations? Why or why not?

Group Activity

1. Many large zoos have captive breeding programs in which animals (or their sperm) are exchanged for breeding purposes. Contact a zoo or search the Internet to find details of how a breeding program for a particular type of organism is formulated. Then come back to the group and share what you've found.

2. Imagine that your group is in charge of developing a plan for restocking populations of a particular species of trout in a western river that has been depleted by fishing. You have the ability to start a fish hatchery to rear any fish that you may wish to use in your program. Describe how you might take the following issues into account in devising your plan:

   a. The population of the wild trout in the river is genetically unique.
   b. This species of trout can be raised in hatcheries, but with difficulty. It would be easier to raise a different trout species.
   c. Trout of different species readily breed with one another in nature.

   How could you avoid genetic drift and/or inbreeding in your hatchery population?

   Search on the Internet or in the library to read about real cases of trout conservation plans. Critique one or more of these according to evolutionary principles.

MediaLab
Are Bacteria Winning the War?
Natural Selection in Action

The last time your doctor prescribed an antibiotic for you, was it for a common cold or a real bacterial infection? If you had a bacterial infection, the first antibiotic might not have worked. Why not? You may have been infected with a strain of bacteria resistant to that antibiotic. The rise and spread of drug-resistant bacteria are consequences of natural selection—evolution in action. The *CD Activity* will help you review three different ways populations can be molded by natural selection. The following *Web Investigation* lets you examine computer simulations of how the environment contributes to natural selection and introduces the serious threat of evolving antibiotic-resistant bacteria. In the *Communicate Your Results* section, you can look for causes and solutions to this new problem.

This *MediaLab* can be found in Chapter 17 on your CD-ROM and Companion Website (http://www.prenhall.com/krogh).

CD ACTIVITY

In any population, organisms vary. In a group of bacteria, for example, there may be some rare antibiotic-resistant members. When the environment changes, let's say you take a prescribed antibiotic, and natural selection acts. All the sensitive bacteria die, leaving the antibiotic-resistant bacteria to thrive, thus changing the makeup of the population. This is just one of three types of natural selection that are animated in this activity.

Activity

1. First, you will learn how selecting for organisms with one extreme trait, like antibiotic resistance, can cause directional selection.

2. Next, you will discover how stabilizing selection favors most of the members of a population and works against the extremes in a stable environment.

3. Finally, you will learn how disruptive selection works against the average value, producing a two-humped curve, with both extremes being favored.

Use the Internet in the following *Web Investigations* to view modern examples of natural selection, including antibiotic resistance in bacteria, and see how natural selection affects us all.

STABILIZING SELECTION | DIRECTIONAL SELECTION | DISRUPTIVE SELECTION

TIME (MANY GENERATIONS)

RANGE OF A PARTICULAR CHARACTERISTIC (SUCH AS COLOR)

WEB INVESTIGATION

Investigation 1
Estimated time for completion = 15 minutes

Review the types of natural selection by selecting the Keyword **NATURAL SELECTION** on your CD or Website for this *MediaLab*. Pay particular attention to the examples given. What type of selection leads to the development of a population consisting of two or more distinctive forms? Can you think of a current example? Search the Internet to find another example of one of these types of selection.

Investigation 2

Estimated time for completion = 5 minutes

Select the Keyword **EVOLVE IT** on your CD or Website for this *MediaLab*. You'll access a site that lets you choose hypothetical environmental conditions on two islands, and then use a computer simulation to chart the evolution of finches with different variations of beak size. At the site, click on "Proceed to evolution laboratory." Click through the list of option radio buttons at the left of the screen. Leave them all at the original settings, but increase the precipitation of Darwin Island to 85cm annual rainfall. Observe the kinds of seeds that are produced on the two islands. Read the information in the right sidebar to predict what kind of finches should survive best under these different conditions. Click the "Run Experiment" button. Did the results match your predictions?

Investigation 3

Estimated time for completion = 10 minutes

Why should people care if antibiotic resistance occurs in bacteria? Read an on-line article by

selecting the Keyword **ANTIBIOTIC RESISTANCE** on your CD or Website for this *MediaLab*. Note the biological causes of antibiotic (or antimicrobial) resistance, and the problems predicted. Two factors seem to be operating in this situation: antibiotic resistance by the bacteria in the context of natural selection, and the overuse (and misuse) of antibiotics by people.

Now that you have seen some real-life examples, do some of the following exercises. They'll give you an understanding of the factors influencing natural selection that are caused by overuse of antibiotics.

COMMUNICATE YOUR RESULTS

Exercise 1

Estimated time for completion = 5 minutes

In *Web Investigation 1*, you searched the Internet to find current examples of natural selection in action. Prepare a short presentation of your example to add to a list created by other students. Include the type of natural selection, the organism involved, and what environmental influences are most likely responsible for this evolution.

Exercise 2

Estimated time for completion = 5 minutes

The Galapagos Islands are famous because of Darwin's observations about the finch populations. Recently, researchers have observed that finch populations are still changing—average bill size in one species has increased after a recent drought. These finches are adapted to utilize various seed as a food source, as you observed in the simulation in *Web Investigation 2*. In 50 words or less, specu-

late on why drought would favor larger bill sizes. What type of natural selection has occurred? What type of natural selection occurs when a population of small-beaked finches experiences rainy conditions?

Exercise 3

Estimated time for completion = 5 minutes

What will happen when antibiotics are rendered useless against more and more antibiotic-resistant bacteria? What should we be doing to stop this trend? Using information from *Web Investigation 3*, compile a list of bacterial species that have developed antibiotic resistance, and indicate when. How has natural selection been a factor in the development and spread of antibiotic resistance? Write a list of common guidelines that you could use in explaining this problem to your friends and family. Include suggestions for stopping the spread of antibiotic resistance—and the consequences if we do not.

18

The Outcomes of Evolution
Macroevolution

Not a horse, not a donkey. What is it?
(page 354)

Jerky and smooth speciation.
(page 357)

Are birds reptiles?
(page 362)

How many types of living

things are there on Earth? How many varieties of life-forms are there that we can recognize as being fundamentally different from one another? It may surprise you to learn that we haven't the foggiest idea. The eminent naturalist Edward O. Wilson has pointed out that the number of species could be around 10 million, or it could be upwards of 100 million. Scientists simply have been unable to catalogue the vast diversity that exists on our planet.

Even if the 10 million figure is closer to the truth, isn't this number staggering? This seems particularly true given the starting point for it. A single type of organism arose some 3.6 billion years ago, branched into two types of organisms, and then the process continued—branches forming on branches—until at least 10 million different types of living things came to exist on Earth. Now here's the real kicker: These are just the *survivors*. The fossil record indicates that more than 99 percent of all species that have ever lived on Earth are now extinct. Branching indeed.

The question to be answered in this chapter is: How does this branching work?

How do we go from the microevolutionary mechanisms explored in Chapter 17 to the actual divergence of one species into two? And how do we classify the creatures that result from this branching?

18.1 What Is a Species?

There is a basic unit of living things you'll be reading about in this chapter, the species, that has a great importance to biology. If we allow that some fish and birds are endangered, that some bacteria are harmful to human beings while others are helpful, that some varieties of rice carry disease-resistance genes while others do not, then it follows that there must be some means of distinguishing one of these groupings from another (**see Figure 18.1**). The whole notion of knowing something about the natural world begins to break down if we can't say *which* grouping it is that is endangered or harmful.

Over the centuries, human beings have devised various ways of defining the groups now called species. But in science today, the most commonly accepted definition stems from what is known as the **biological species concept**, which uses breeding behavior to

A new generation of wheat.
(page 353)

How is a bat like a gorilla.
(page 360)

Endangered or not?
(page 348)

make classifications. We actually looked at a version of the biological species definition in Chapter 17. Here it is again, as formulated by the evolutionary biologist Ernst Mayr:

> Species are groups of actually or potentially interbreeding natural populations which are reproductively isolated from other such groups.

Note that the breeding behavior Mayr talks about can be real or potential. Two populations of finch may be separated from one another by geography, but if, upon being reunited in the wild, they began breeding again, they are a single species. Note also that Mayr stipulates that species are groups of *natural* populations. This is important because breeding may take place in captivity that would not in nature. No one doubts that lions and tigers are separate species; yet they will mate in zoos, producing little tiglons or ligers (depending on whether the father was a tiger or a lion). In natural surroundings, however, they apparently have never interbred, even when their ranges overlapped centuries ago.

You might think that, with the biological species concept in hand, scientists would be able to study any organism and, by discern-

ing its breeding behavior, pronounce it to be a member of this or that species. Nature is so vast and varied, however, that this doesn't always work. We can't look at the breeding activity of bacteria, for example, because they don't *have* any breeding behavior; they multiply instead by simple cell division. (Microbiologists define bacterial groupings by sequencing their DNA or RNA and looking for the degree to which these sequences are the same.) Then there are separate species that sometimes interbreed in nature, producing so-called hybrid offspring as a result. Such mixing between species turns out to happen more in the plant world than the animal, but it does take place in both. If species are supposed to be "reproductively isolated" from one another, what are we to make of these crossings? (Mayr points out that it is *populations* in his definition that are reproductively isolated, not individuals, and that whole populations do tend to stay within their species confines.)

Despite these difficulties, the biological species concept provides a useful way of defining the basic unit of Earth's living things. Bacteria notwithstanding, most species carry out sexual reproduction, and relatively few species outside the plant world

Figure 18.1
Separate Entities
The concept of a species can have very practical effects.

a This bird is a northern spotted owl (*Strix occidentalis*), which is an endangered species, protected by federal law.

b This bird is a barred owl (*Strix varia*), which is not endangered.

a

b

regularly produce hybrids. Moreover, this is a species concept that is rooted in a critical behavior of organisms themselves—mating, which controls the flow of genes. And as you saw in Chapter 17, it is the change in the genetic makeup of a population (a change in its allele frequencies) that lies at the root of evolution.

18.2 How Do New Species Arise?

Two Modes of Speciation: Cladogenesis and Anagenesis

Having defined a species, let's now see how one variety of them can be transformed into another in the process called **speciation**. It is worth noting at the outset, however, that evolution does not always entail a multiplication of species. As you've observed, a single species can diverge into two species, the "parent" species continuing, while another branches off from it. This form of speciation, known as branching evolution or **cladogenesis**, is the central subject of this chapter (**see Figure 18.2**). It's also possible, however, for an entire species to undergo change without any branching taking place. If the change is extensive enough, a new species will evolve—one that presumably would be reproductively isolated from its ancestral species. This type of evolution is known as nonbranching evolution, or **anagenesis**.

Speciation Occurs When Populations Cease to Interbreed

In the branching evolution scientists are most concerned with, the central question is: What brings it about? The answer has to do with the flow of genes reviewed in Chapter 17. As you saw then, evolution within a population means a change in that population's allele frequencies. Now imagine *two* populations of a single species of bird, with one being separated from the other—say, by one of them having flown to a nearby location. To the extent that they continue to breed with one another, with individuals moving between the locations, each population will *share* in whatever allele frequency changes are going on with the other population. Hence the two populations will evolve *together*, remaining a single species. Now imagine that the migration stops between the two populations. Each population continues to

undergo allele frequency changes, but it no longer shares these changes with the other population. Alterations in form and behavior may accompany allele changes, and these alterations may pile up over time. Coloration or bill lengths may change; feeding habits may be transformed. After enough time, should the two populations find themselves geographically reunited, they may no longer freely interbreed. At that point, they are separate species. Speciation has occurred.

The critical change here came when the two populations quit interbreeding. For scientists, the key question thus becomes: Why would this happen? What could drastically reduce interbreeding, and hence gene flow, between two populations of the same species?

The Role of Geographic Separation: Allopatric Speciation

In the preceding example, a very clear factor operated in reducing gene flow between the bird populations: They were separated geographically. And geographical separation turns out to be the most important starting point in speciation.

Populations can become separated in lots of ways. On a large scale, glaciers can move into new territory, cutting a previously undivided population into two. Rivers can change course, with the result that what was a single population on one side of the river may now be two populations on both sides of it. On a smaller scale, a pond may partially dry up, leaving a strip of exposed land between what are now two ponds with separate populations. Such environmental changes are not the only ways that populations can be separated, however. Part of a population might

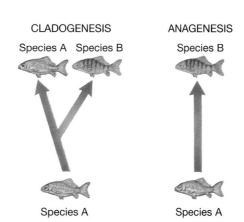

Figure 18.2
Modes of Speciation
Speciation can occur either by the branching of one lineage into others (cladogenesis) or by gradual change along a single lineage (anagenesis).

migrate to a remote area, as did the bird population, and in time be cut off from its larger population. For a real-life example of how migration can bring about speciation, **see Figure 18.3**.

When geographical barriers divide a population and the resulting populations then go on to become separate species, what has occurred is **allopatric speciation** (*Allopatric* literally means "of other countries"; **see Figure 18.4**). Look along the banks on either side of the Rio Juruá in western Brazil, and you will find small monkeys, called tamarins, that differ genetically from one an-

other in accordance with how wide the river is at any given point; where it is widest, the members of this species do not interbreed, while at the narrow headwaters they do. Are the nonbreeding populations on their way to becoming separate species? Perhaps, but only if gene flow is drastically reduced between them, probably for a very long time.

Reproductive Isolating Mechanisms Are Central to Speciation

While geographical separation is the most important factor in getting speciation going, it cannot bring about speciation *by itself*. Following geographical separation, two populations of the same species must then undergo physical or behavioral changes that will *keep* them from interbreeding, should they ever be reunited. Allopatric speciation thus operates through a one-two process: First the geographic separation, then the development of differing characteristics in the two resulting populations—characteristics that will *isolate* them from each other in terms of reproduction.

Thus arises the concept of **reproductive isolating mechanisms**, which can be defined as any factor that, in nature, prevents interbreeding between individuals of the same species or of closely related species. Geographic separation is itself a reproductive isolating mechanism, in that it is a factor that prevents interbreeding. (Populations that are physically separated cannot interbreed.) But, because the mountains or rivers that are the actual barriers to interbreeding are *extrinsic* to the organisms in question, geographic separation is called an **extrinsic isolating mechanism**. For allopatric speciation to take place, what also must occur is the second in the one-two series of events: The evolution of *internal* characteristics that keep organisms from interbreeding. Such factors are referred to as **intrinsic isolating mechanisms**: Differences in form or behavior that prevent interbreeding between individuals of the same species or of closely related species.

What are these intrinsic mechanisms? The following are the most important ones. An easy way to remember them is to think about what sequence of events would be required for fertile offspring to be produced in *any* sexually reproducing organism. First, organisms that live in the same area must en-

1. The original population started in the north and migrated southward.

2. The population split to the east and west of the central valley. Then two populations began to evolve independently.

3. Evolution of eastern population

3. Evolution of western population

Central Valley

4. The east and west populations came back together in Southern California, but could no longer interbreed (or produced infertile hybrid offspring).

Figure 18.3
Speciation in Action
Millions of years ago, the salamander *Ensatina eschscholtzii* began migrating southward from the Pacific Northwest. When it reached the Central Valley of California—an uninhabitable territory for it—populations branched west (to the coastal range) and east (to the foothills of the Sierra Nevada mountains). Over time, the populations took on different colorations as they moved southward. By the time the two populations were united in Southern California, they differed enough, genetically and physically, that they either did not interbreed or produced infertile hybrid offspring when they did. (Salamanders are falsely colored in drawing for illustrative purposes.)

counter one another; if they don't, the "ecological isolation" mechanism is in place. If they do encounter one another, they must then mate in the same *time-frame*; if they don't, then temporal isolation is in place. This then continues on down through the list (**see Table 18.1**).

Six Intrinsic Reproductive Isolating Mechanisms

Ecological Isolation

Two closely related species of animals may overlap in their ranges and yet feed, mate, and grow in separate areas, which are called *habitats*. If they use different habitats, this means they may rarely meet up. If so, gene flow will be greatly restricted between them. Lions and tigers *can* interbreed, but they never have in nature, even when their ranges overlapped in the past. One reason for this is their largely separate habitats: Lions prefer the open grasslands, tigers the deep forests.

Temporal Isolation

Even if two populations share the same habitat, if they do not mate within the same time frame, gene flow will be limited between them. Two populations of the same species of flower may begin releasing pollen at slightly different times of the year. Should their reproductive periods cease to overlap altogether, gene flow would be cut off between them.

Behavioral Isolation

Even if populations are in contact and breed at the same time, they must choose to mate with one another for interbreeding to occur. Such choice, it turns out, is often based on specific courtship and mating displays, which can be thought of as passwords be-

Table 18.1
Reproductive Isolating Mechanisms

EXTRINSIC ISOLATING MECHANISM	**GEOGRAPHIC ISOLATION** Individuals of two populations cannot interbreed if they live in different places (the first step in allopatric speciation).
	ECOLOGICAL ISOLATION Even if they live in the same place, they can't mate if they don't come in contact with one another.
INTRINSIC ISOLATING MECHANISMS	**TEMPORAL ISOLATION** Even if they come in contact, they can't mate if they breed at different times.
	BEHAVIORAL ISOLATION Even if they breed at the same time, they will not mate if they are not attracted to one another.
	MECHANICAL ISOLATION Even if they attract one another, they cannot mate if they are not physically compatible.
	GAMETE ISOLATION Even if they are physically compatible, an embryo will not form if the egg and sperm do not fuse properly.
	HYBRID INVIABILITY or INFERTILITY Even if fertilization occurs successfully, the offspring may not survive, or if it survives, may not reproduce (e.g.,mule).

tween members of the same species. Birds must hear the proper song, spiders must perform the proper dance, and fiddler crabs must wave their claws in the proper way for mating to occur.

Mechanical Isolation

Reproductive organs may come to differ in size or shape or some other feature, such that organisms of the same or closely related

Figure 18.4
Geographical Separation—Leading to Speciation?
These two varieties of squirrel had a common ancestor that at one time lived in a single range of territory. Then the Grand Canyon was carved out of land in Northern Arizona, leaving populations of this species separated from one another, about 10,000 years ago. In the area of the Grand Canyon, **(a)** the Abert squirrel lives only on the Canyon's south rim, while **(b)** the Kaibab squirrel lives on the north rim. It's unclear whether these two varieties of squirrel today would be reproductively isolated if reunited—that is, whether they are now separate species—but the geographical separation they have experienced is the first step on the way to allopatric speciation.

a

b

New Species through Genetic Accidents: Polyploidy

We have generally thought of speciation as something that takes place over many generations. Yet plants (and some animals) have a means of speciating in a single generation. It is called *polyploidy*, and it is very important in the plant world; more than 100,000 species of flowering plants in existence today are thought to have been brought about through it. Here is one version of how it works. From the genetics unit, you may recall that human beings have 23 pairs of chromosomes, *Drosophila* flies four pairs and Mendel's peas seven. Whatever the *number* of chromosomes, the commonality among all these species is that their chromosomes come in *pairs*, which is the general rule for species that reproduce sexually.

As you've noted, plants are more adept than animals at producing "hybrids," with the gametes from two *separate* species coming together to create an offspring. Often these offspring are sterile, however, because when it comes time for them to produce *their* gametes (the counterparts to human sperm or eggs), the sets of chromosomes they inherited from their different parental species may not "pair up" correctly in meiosis, owing to differences in chromosome number or structure.

Now comes the accident. Suppose that, back as a single-celled zygote, the hybrid offspring carried out the usual practice of doubling its chromosome number in preparing for its first cell division. Now, however, the cell fails to actually divide; whereas it was supposed to put half its complement of chromosomes into one daughter cell and half into another, it doesn't do this. This single cell now contains *twice* the usual number of chromosomes. It then proceeds to undergo regular cell division, meaning that every cell in the plant that follows will have this doubled number of chromosomes. Critically, this plant will have doubled both its *sets* of chromo-

somes—the set it got from parental species A and the set it got from species B. If you double a set, by definition every member of that set now has a partner to pair with in gamete formation. Thus the roadblock to a hybrid producing offspring has been removed; the chromosomes can all pair up.

This pairing up yields gametes (eggs and sperm), which can then come together and fuse thanks to another capability of plants. Recall that many plants can *self-fertilize*. A single plant contains male gametes that can fertilize that same plant's female gametes. Thus the plant

---

Many plants have employed a means of speciating in a single generation.

---

with the doubled set of chromosomes can fertilize itself, theoretically beginning an unending line of fertile offspring. This line of organisms is "reproductively isolated" from either of its parental species because it has a different number of chromosomes than either parental species. Were gametes from it to be "crossed" with gametes from either parent, sterile offspring would be the likely result. With reproductive isolation, we have a *separate species*; and with self-fertilization, we have a species that can perpetuate itself.

A multiplication of the normal two sets of chromosomes to some other set number is known as polyploidy; here we have speciation by **polyploidy**, which is one type of sympatric speciation. The importance of this in the plant world is immense. Many of our most important food crops are polyploid, including oats, wheat, cotton,

---

species can no longer mate. Different species of alpine butterfly look very similar, but their genital organs are different enough that one species cannot mate with another.

Gametic Isolation

Even if mating occurs, offspring may not result if there are incompatibilities between sperm and egg or between sperm and the fe-

male reproductive tract. In plants, the sperm borne by pollen may be unable to reach the egg lying within the plant's ovary. In animals, sperm may be killed by the chemical nature of a given reproductive tract, or may be unable to bind with receptors on the egg.

There is one form of gametic isolation, called *polyploidy*, that is very important in plants; indeed, at least 50 percent of all flow-potatoes, and coffee. Indeed, the type of polyploid speciation we've looked at—which begins with a hybrid offspring—often produces bigger, healthier plants. As a result, breeders have developed ways to artificially induce polyploidy.

Many of our most important food crops are polyploid, including oats, wheat, cotton, potatoes, and coffee.

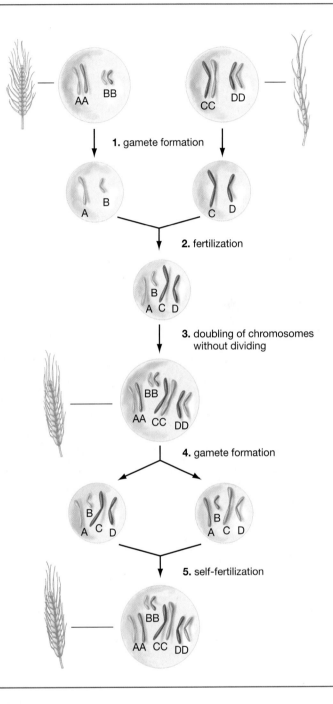

Polyploidy in Wheat

Two different species of wheat exist in nature, with slightly different "genomes" or complements of DNA.

1 Gametes (egg and sperm) are formed in the different species.

2 These gametes fuse, in fertilization, to form a zygote, a single cell that will develop into a new plant. Such a mixed-species or "hybrid" zygote generally develops into a sterile plant, because its chromosomes cannot pair up correctly when the plant produces its own gametes during meiosis. In this case, however, the zygote

3 doubles its chromosomes in preparation for cell division, but then fails to divide. With this doubling, each chromosome now has a compatible homologous chromosome to pair with during meiosis.

4 Gamete formation thus takes place in the plant and these gametes from the same plant then

5 fuse, because this is a self-fertilizing plant. With this,

6 a new generation of wheat plant has been produced—one that is a different species from either parent generation, because each of the parent species has two pairs of chromosomes, while this new hybrid species has four pairs.

ering plant species are thought to have come about through it. You can read about it in "New Species through Genetic Accidents," above.

Hybrid Inviability or Infertility

Even if offspring result, they may develop poorly—they may be stunted or malformed in some way—or they may be infertile, meaning unable to bear offspring of their own. A well-known example of such infertility is the mule, which is the infertile offspring of a female horse and a male donkey (**see Figure 18.5** on page 354).

Sympatric Speciation

Thus far, you have looked at the development of intrinsic reproductive isolating mechanisms strictly as a second step in speciation—one that follows a geographic separation of populations. As it happens, however, intrinsic reproductive isolating mechanisms can develop between two populations in the *absence* of any geographic separation of them. If these isolating mechanisms reduce interbreeding between the populations sufficiently, speciation can occur. What occurs in this situation is not allopatric speciation, however; it is **sympatric speciation**, which can be defined as any speciation that does not involve geographic separation (*Sympatric* literally means "of the same country").

Sympatric speciation has been a contentious subject in biology for years. One form of sympatric speciation is the polyploidy reviewed in "New Species through Genetic Accidents." But, setting it aside, most biologists thought for years that if sympatric speciation took place at all, it was of trivial importance. By the mid 1990s, however, new evidence had convinced some of the most prominent researchers in evolutionary biology that sympatric speciation does have significance in the broader picture of speciation.

Sympatric Speciation in a Fruit Fly

To see how sympatric speciation works, consider one of its best-studied examples, a species of fruit fly named *Rhagoletis pomonella*. Prior to the European colonization of North America, *R. pomonella* existed solely on the small, red fruit of hawthorn trees (**see Figure 18.6**). The Europeans brought *apple* trees with them, however, and by 1862 some *R. pomonella* had moved over to them. It seemed to at least one mid-19th century observer, however, that with the introduction of apples there had arisen separate *varieties* of the flies—what might be called apple *R. pomonella* and hawthorn *R. pomonella*—with each variety courting, mating, and laying eggs almost exclusively on its own type of tree. A modern researcher named Guy Bush led the way, beginning in the early 1960s, in investigating whether this was the case, and the answer turned out to be yes. The two varieties of flies are separated from each other in all these ways. In one study undertaken on them, only 6 percent of the apple and hawthorn flies interbred with one another. The hawthorn and apple *R. pomonella* are not separate species yet, but they certainly give indication of being in transition to that status.

For our purposes, the first thing to note is that this separation has not come about because of *geographic* division; the apple and hawthorn trees that the two varieties of flies live on may scarcely be separated in space at all. Given this, how did this single species move toward becoming two *separate* species? Bush offers a likely scenario, based on a critical difference between the hosts the two species live on. Apples tend to ripen in August and September while hawthorn fruit ripens in September and October. In the summer, all fruit flies emerge as adults after wintering underground as larvae, after which

Figure 18.5
Mules Are Infertile Hybrids
Even though mules cannot themselves reproduce, humans frequently cross horses and donkeys to produce mules in order to take advantage of their exceptional strength and endurance. Chromosomal incompatibilities between horses and donkeys leave the mules sterile.

Figure 18.6
A Species Undergoing Sympatric Speciation?
The fruit fly species *Rhagoletis pomonella*, pictured here on the skin of a green apple.

they fly to their host tree to mate and lay eggs in the fruit.

Bush believes that about 150 years ago (when hawthorns were hosts to *all* the flies) some individuals in a population of *R. pomonella* experienced either a mutation or perhaps a chance combination of rare existing alleles. In either event, this change did two things: It caused these flies to emerge slightly *earlier* from their underground state than did most flies; and it drew these flies to the smell of *apples* as well as hawthorns. Because apples mature slightly earlier than the hawthorn fruit, these flies had a suitable host waiting for them. More important, these flies would have bred *among themselves* to a high degree; recall that the other flies were emerging later, and the adult fly only lives for about a month. Thus, the variant alleles were passed on to a selected population the next generation. Today, the apple *R. pomonella* flies indisputably do emerge earlier than the hawthorn flies. This is one of the things that ensures reproductive isolation between the two groups; their periods of mating don't fully overlap.

Such a lack of overlapping mating periods may sound familiar because it is one of the intrinsic reproductive isolating mechanisms you looked at earlier. (It is temporal isolation.) You also can see another ecological isolation in operation with these flies. Though the two types of flies live in the same area, they meet up with relatively little frequency, because they have different habitats (hawthorn vs. apple trees). In short, these populations have developed intrinsic reproductive isolating mechanisms without ever having been separated from one another geographically. They are headed toward speciation, in other words, but in this case it is sympatric speciation.

18.3 When Is Speciation Likely to Occur?

The horseshoe crab (**see Figure 18.7**) is not really a crab at all, but instead is distantly related to land-dwelling arthropods such as spiders and scorpions. (It lacks the antennae that real crabs possess, but it *has* the pincers around the mouth that all spiders and scorpions do.) Horseshoe crabs have been around in something like their modern form for more than 300 million years. They exist today in a scant four or five species, each one of which is pretty much like the others. The crabs live in the shallow oceans off North America and Asia, pushing through sand or mud to feed on everything from algae to small-bodied invertebrate animals.

Now consider the Galapagos (or Darwin's) finches you looked at in Chapter 17. There are *thirteen* of these species on the small Galapagos archipelago, all of whom are derived from a single species of South American finch that arrived on one of the islands perhaps 100,000 years ago.

Think of the difference between horseshoe crabs and Darwin's finches. The former have remained almost unchanged for more than 300 million years throughout the world, while the latter diverged into 13 different species in the last 100,000 years in a small cluster of islands off South America.

Specialists and Generalists

Why the great differences in the rate of speciation here? Two general principles are at work. First, as Niles Eldredge has pointed out, horseshoe crabs are *generalists*: Their diet is extremely diverse; they will eat plants and small animals but will also scavenge for debris on the ocean bottom. By contrast, you saw in Chapter 17 how *specialized* some

**Figure 18.7
Different Groups of Organisms Undergo Different Rates of Speciation**

a Because horseshoe crabs look essentially the same today as they looked 300 million years ago, some scientists have called them "living fossils." Compare the modern organism, left, with the fossil on the right, which is at least 145 million years old.

b By contrast, the Galapagos finches have diversified tremendously, forming 13 species in as little as 100,000 years.

a

b

of the Galapagos finches are in their feeding behavior, particularly when food has become scarce because of drought conditions, as happens on the islands. In such times, species might exist on a single variety of plant seed. Species that are tied in this way to a particular food or environmental condition must adapt in connection with changes in them or face extinction. Think of how quickly the bills of the *Geospiza fortis* population on Daphne Major evolved toward greater depth when having a deeper bill meant the difference between life and death (see Chapter 17, page 336). By definition, this kind of adaptation means change—and change is what speciation is all about. By contrast, the horseshoe crab shifts from one food to another, depending on what is available, not adapting greatly in response to changes in any one food source.

New Environments: Adaptive Radiation

The second lesson offered by horseshoe crabs and finches concerns the kinds of environments that induce speciation. The Galapagos islands were formed by volcanic eruptions that brought the islands above the ocean's surface only about 5 million years ago. At that time, these volcanic outcroppings gave new meaning to the word *barren*. For a brief time at least the islands were utterly sterile, with no life on them. Very quickly, however, life did come to the islands, with bacteria, fungi, plant seeds, and tiny animals landing on the islands, all being borne by air or ocean currents. The South American finches obviously did not arrive until much later; but by then, with lots of large plant species well established, what these birds encountered was an environment rich with possibility, for there were *no birds of their kind* on the islands. Imagine that you are a graphic artist, working in a big city with lots of other graphic artists, most of whom *specialize* in this or that (magazines, Internet websites, etc.). Now you and a few other graphic artists move to a new city in which there are few graphic artists, but a good number of *possibilities* for graphic arts work. You would thus be able to specialize fairly easily—to fill a *niche* or working role in this new environment—because many of these niches would not yet have been taken. Just so did Darwin's finches rush in to fill previously unoccupied niches on the Galapagos Islands—this plant seed, this insect, this mating environment.

Such a situation is ripe with possibilities for change (meaning speciation) because, while niches are in flux, there is a good deal of shaping of species to environment. But more of this occurred on the Galapagos, because this niche-filling was taking place on 25 separate *islands*. The birds can fly from one island to another, but the islands nevertheless represent a geographic barrier to bird interbreeding, and you know what follows from this: allopatric speciation.

The Galapagos finches exemplify something known as **adaptive radiation**: The rapid emergence of many species from a single species that has been introduced to a new environment. The finches radiated out to fill new niches on the islands, with populations adapting to the environments over time. In summary, two conditions that are conducive to speciation are specialization (of food source or environment) and migration to a new environment, particularly when there are no closely similar species in that environment.

Is Speciation Smooth or Jerky?

Charles Darwin knew that in looking at large-scale evolutionary changes, such as the change from sea-dwelling to land-dwelling life-forms, what is striking is the lack of "transitional forms" in the fossil record. Not many creatures have been found that seem to lie in between the sea-dwelling and land-dwelling forms. Darwin thought that this was simply a problem of not having collected enough fossils in the right places; that in time, the transitional forms would show themselves.

Scientists have found some transitional-form fossils since Darwin's time, but not many. Indeed, the fossil record seems to speak against the mode of evolution as Darwin had envisioned it. He imagined evolution as a series of infinitesimally small changes accumulating in a slow, steady way in populations of a species until in time one species could be seen to have diverged into two.

What the fossil record exhibits again and again, however, is not evidence of slow, steady change in a species, but rather indications of enormous lengths of time in which so-called **stasis** occurs—in which species stay exactly the same or undergo some minute modifications. After this, there appears to be an *abrupt* change to a new species.

In 1972, two young scientists proposed that it was time to take the fossil record at face value. Why isn't it plausible, they said, that species experience stasis for long periods of time, after which speciation then takes place in relatively brief bursts? This is the **theory of punctuated equilibria** proposed by Stephen Jay Gould and Niles Eldredge. (Periods of stasis are the equilibrium that species generally live in, with these then being "punctuated" by the speciation events.)

So, how rapid is speciation under this view? It can take place in thousands of years rather than the millions during which a species is in stasis. If that is the case, it would explain why there are so few transitional forms in the fossil record: They aren't around for long enough to leave many fossils (**see Figure 18.8**).

There has been a great debate about the theory of punctuated equilibria ever since it was proposed. Some scientists have supported it, others have opposed it. Its detractors (sometimes called "gradualists") have noted that only the "hard parts" of species (bone and so forth) are preserved in the fossil record, while much evolutionary change must concern "soft parts." Other critics allege that the theory proposes nothing new, but merely puts terminology to ideas that already existed (in population genetics) about how rapidly species can evolve.

Some recent experimental evidence is consistent, however, with the idea of punctuated equilibria. Relying on the ability of bacteria to reproduce rapidly, scientists looked at 10,000 generations of them over a period of 4 years—freezing samples at various points in time—and found that these cells tended to stay the same size for long periods of time and then undergo rapid growth spurts. Other researchers have asked how much *genetic* change is required to bring about speciation and found that, in comparing two species of flower at least, the answer is very little. They found that changes in as few as eight genetic locations may account for the differences seen between two species of monkey flower that differ in color and ac-

cordingly attract different pollinators. If so few changes are required for speciation, then it could proceed more rapidly than if many changes were necessary.

One possibility is that speciation is both jerky (punctuated) and smooth (gradual), depending on the species and the characteristics being examined. Changes in "morphology" or physical structure are based on genetic changes, as you have seen. It could be that genetic changes are gradual while the resulting morphological changes come about in spurts.

18.4 The Categorization of Earth's Living Things

This chapter began by noting the importance of the species concept—of being able to say that this organism is fundamentally separate from that. To have a species concept means, of course, that there has to be some means of *naming* separate species. You have probably noticed that most of the species considered in this chapter have been referred to by their scientific names; you didn't just look at a fruit fly, but rather *Rhagoletis pomonella*. To the average person, such names may be regarded as evidence that scientists are awfully exacting—or just plain fussy. Why the two names? Why the Latin? Can't we just say fruit fly?

To take these questions one by one, scientists can't just say "fruit fly" for the same reason a person can't say "the guy in the white shirt" while trying to identify a player at a tennis tournament. There are lots of guys in white shirts at tennis tournaments and there

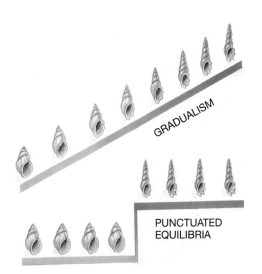

Figure 18.8
Is Speciation Smooth or Jerky?
According to the gradualism model, speciation occurs through gradual change over long periods of time. Contrast this with the punctuated equilibria model, which proposes that long periods of stasis may periodically be "punctuated" by rapid bursts of speciation.

GRADUALISM

PUNCTUATED EQUILIBRIA

are lots of different kinds of flies in the world. It is important to be able to say *which* fly we are talking about. The importance of this can be seen with *Rhagoletis pomonella* itself, which is a major pest in the apple industry. Tell some apple growers that you have spotted flies on their apples and you may get a shrug of the shoulders; tell them you have spotted *R. pomonella* and you may get a very different reaction.

The Latin that is used stems from the fact that this naming convention was standardized by the Swedish scientist Carl von Linné in the eighteenth century, at a time when Latin was still used in the Western world in scientific naming. (In fact, Linné is better known by the Latinized form of his name—Carolus Linnaeus.)

Linnaeus recognized the confusion that can result from having several common names for the same creature, and thus devoted himself to giving specific names to some 4,200 species of animals and 7,700 species of plants—all that were known to

exist in his time. Many of the names he conferred are still in use today.

The fact that there are *two* parts to scientific names—a **binomial nomenclature**—points to the central question of *groupings* of organisms on Earth. Consider the domestic cat, which has the scientific name *Felis silvestris*. The first part of its name is its genus (plural *genera*), which designates a group of closely related, but still separate species. It turns out that, worldwide, there are eight other species of small cat within the *Felis* genus, such as the small cat *Felis nigripes* ("black-footed cat") found in Southern Africa.

Taxonomic Classification and the Degree of Relatedness

The practical importance of the genus classification is that, if we know that two organisms are part of the same genus, then to know something about one of them is to know a good deal about the other. But what is the basis for placing organisms in such a category? Modern science classifies organisms largely in accordance with how closely they are *related*. In this context, "related" has the same meaning as it does when the subject is extended families. If people have the same mother, they are more closely related than if they shared only a common *grandmother*, which in turn makes them more closely related than if they had only a common great-great grandmother, and so on.

In the same way, species can be thought of as being related. Domestic dogs (*Canis familiaris*) are very closely related to gray wolves (*Canis lupus*), with all dogs being descended from these wolves. Indeed, by some reckonings dogs and gray wolves are the same species (*Canis lupus*), because they will interbreed in nature. Such differences as exist between them are the product of a mere 10,000 years or so of the human domestication of dogs. Scientists believe, then, that 15,000 years ago there was but *one* species (the gray wolf) that gave rise to both today's wolf and today's dog. A close relation indeed.

Domestic dogs are also related, however, to domestic *cats*; if we look far back enough in time, we can find a single group of animals that gave rise to both the dog and cat lines. This does not mean going back 10,000 years, however; it means going back perhaps

Figure 18.9
Classifying Living Things
The classification of the house cat, *Felis sylvestris*, based on the Linnaean system.

LINNAEAN SYSTEM OF CLASSIFICATION
KINGDOM (Animalia)
PHYLUM (Chordata)
CLASS (Mammalia)
ORDER (Carnivora)
FAMILY (Felidae)
GENUS (Felis)
SPECIES (sylvestris)

60 million years. Thus, there obviously is a big difference in how closely related dogs and wolves are, as opposed to dogs and cats. Establishing such *degrees* of relatedness is what the whole system of scientific classification is about. There is a field of biology, called **systematics**, that is concerned with the diversity and relatedness of organisms; part of what systematists do is try to establish the truth about who is more closely related to whom. They thus study the evolutionary history of groups of organisms.

Setting aside for the moment the difficulty in *determining* what such evolutionary histories are, there obviously is a tremendous cataloguing job here, given the number of living and extinct species mentioned earlier. Given this diversity, a grouping or **taxonomic system** is employed in order to classify every species of living thing on earth. There are eight basic categories in use in the modern taxonomic system: **species**, **genus**, **family**, **order**, **class**, **phylum**, **kingdom**, and **domain**. The organisms in each category make up a **taxon**.

A Taxonomic Example: The Common House Cat

If you look at **Figure 18.9**, you can see how this taxonomic system works in connection with the domestic cat. As noted, this cat is only one species in a genus (*Felis*) that has eight other living species in it. The genus then is a small part of a family (Felidae) that has 14 other genera in it (panthers, snow leopards, and others). The family is then part of an order (Carnivora) that includes not only big and small "cats," but other carnivores, such as bears and dogs. On up the taxa we go, with each one being more inclusive than the one beneath it until we get to the highest category in this figure, the kingdom Animalia, which includes all animals. (Later you'll look at the "supercategory" above kingdom, called domain; but for simplicity's sake, kingdom will be the highest-level taxon considered for now.)

18.5 Constructing Evolutionary Histories: Classical Taxonomy and Cladistics

So how do systematists go about putting organisms in these various groups? What evidence do they use in constructing their evo-

lutionary histories? The answer is radiometric dating, the fossil record, DNA sequence comparisons—all the things reviewed in Chapter 16 that are used to chart the history of life on Earth.

If you look at **Figure 18.10**, you can see the outcome of some of this work, an evolutionary "tree" for one group of organisms, in this case one of the major groups of mammalian carnivores. You can see that the tree is "rooted," about 60 million years ago, with an ancestral carnivore whose lineage then split two ways: to dogs on the one hand and everything else on the other. One interesting facet of this evolutionary history is how closely related bears are to the aquatic carnivores. All such histories are hypotheses about evolutionary relationships, with each such hypothesis known as a **phylogeny**.

Classical Taxonomy Looks for Similarities

The question is: How did systematists use the kinds of evidence noted in this chapter to construct this phylogeny? How do they interpret evidence to make conclusions about who is descended from whom and, in rough terms, at what time? There is a system for interpreting evolutionary evidence, handed down from the time of Darwin, that is sometimes referred to as *classical taxonomy*. In thinking about the organisms in this family tree, classical systematists would look first

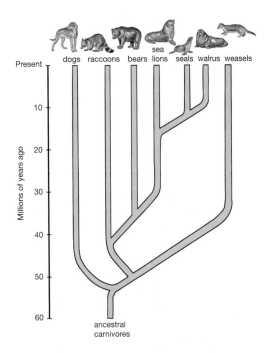

Figure 18.10
A Family Tree for a Selected Group of Mammalian Carnivores

and foremost at the physical form or "morphology" of the animals in question, as preserved in the fossil record, and compare it to the morphology of modern animals. Skull shape, "dentition" (teeth patterns), limb structure, and much more would be considered. They would also look at *where*

a HOMOLOGY: Common structures in different organisms that result from common ancestry

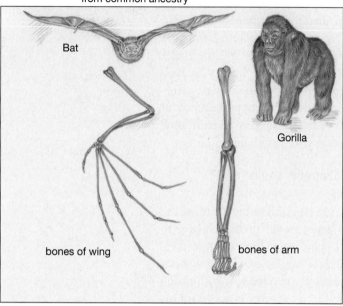

b ANALOGY: Characters of similar function and superficial structure that have *not* arisen from common ancestry

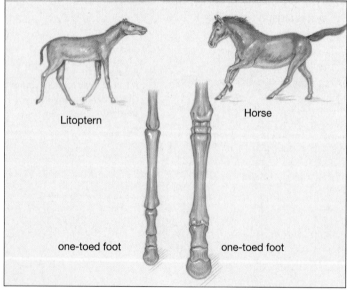

Figure 18.11
Related, as Opposed to Similar

a Bats and gorillas both have the same bone composition in their forelimbs—one upper bone, joined to two intermediate bones, joined to five digits—because they share a common ancestor. Of course these forelimbs are used today for very different functions.

b Horses and the extinct litopterns both have one-toed feet that serve the same function, but the one-toed condition was derived independently in the two creatures—in separate lines of descent.

the ancient forms existed, compared to modern forms. More recently, they have used molecular techniques, such as examining DNA or protein sequences in different living species to determine the relatedness among them. In essence, classical systematists would be looking at how many similarities one group has with another and, on that basis, trying to judge relatedness among them. The word *judge* is used advisedly here, because at the end of the day in the classical system, subjective judgments usually need to be made about who is more closely related to whom. It is a matter of weighing one piece of evidence against another, but who's to say which piece of evidence is more important?

Obscuring the Trail: Convergent Evolution

One of the things that makes classical interpretation of such evidence problematic is that similar features may arise *independently* in several evolutionary lines. A bedrock of systematic classification is the existence of **homologies** in organisms, which can be defined as common structures in different organisms, resulting from a shared ancestry. In Chapter 16 (Figure 16.17), a strong homology was noted in forelimb structure that exists in organisms as different as a gorilla, a bat, and a whale. If you look at **Figure 18.11**, you can see the gorilla and the bat again.

Now, however, consider an extinct group, the litopterns, that once lived in what is now Argentina. Like the modern-day horse, they roamed on open grasslands. Over the course of evolutionary time they developed legs that were extraordinarily similar to the horse's, right down to having an undivided hoof. In fact, litopterns are only very distantly related to the horse, but the leg similarities were enough to fool at least one nineteenth-century expert who thought he had found in litopterns the ancestor to the modern horse.

What is exhibited in the legs of the litoptern and horse is an **analogy**: A feature in different organisms that is the same in function and superficial appearance. Note the difference between it and a *homology*, which is the same structure in different organisms with a common ancestor. When nature has shaped two separate evolutionary lines in analogous ways, what has occurred is **convergent evolution**. (The litoptern and horse lines converged in the design of their legs.) But analogous features have nothing to do with common descent; they merely show that the same kinds of environmental pressures lead to the same kinds of designs. The problem for systematics is that analogy can be confused with homology; analogies can make us believe that organisms share common ancestry when in fact they do not. How common is analogy? To take one perhaps extreme example, eyes are thought to have evolved 38 separate times in animals.

Another System for Interpreting the Evidence: Cladistics

The confusion of analogy, and the subjectivity inherent in classical taxonomy, helped bring about the formation of another system for establishing relatedness. Developed in the 1950s by the German biologist Willi Henning, it is called **cladistics** (from the Greek *klados*, meaning "branch"). In practice, cladistics has become the core of most of the phylogenetic work going on today.

In **Figure 18.12** you can see a **cladogram**, which is an evolutionary tree constructed within the cladistic system. Note that there is a very simple branched line and that no time-scale is attached to it, as was the case with the evolutionary tree of carnivores. Cladistics concerns itself first with lines of descent—with the *order* of branching events. Once this order has been established, efforts can be made to fix events in time; but that is a secondary concern. First and foremost, this cladogram is a proposed answer to the question: Among the animals lizard, deer, lion, and seal, which two groups of animals have the most recent common ancestor? Then, which other two have the *next* most recent common ancestor, and so on. By extension, the question is: Who is more closely related to whom? Note that these questions are being asked about only four of the five animals listed. You'll get to the role of the hagfish in a moment.

Cladistics Employs Shared Ancestral and Derived Characters

The starting point for cladistic analysis is the difference between so-called ancestral characters and derived characters. If you look at any group of species (any taxon), its **ancestral characters** are those that existed in an ancestor common to them all. There are about 50,000 species of animals on Earth

who possess a dorsal vertebral column (better known as a backbone). The ancestor to all fishes, reptiles, and mammals had such a feature, which makes the vertebral column an ancestral character for all these groups. On the other hand, only *some* vertebrates then went on to become animals that have four limbs (and are thus tetrapods). The tetrapod feature is thus a **derived character** of the vertebrate state; it is a character *unique* to taxa descended from a common ancestor.

In Figure 18.12, you can see a continuation of this process of ever-more selective grouping in that only some mammals are carnivorous feeders, and only some members of this taxon are *aquatic* mammalian carnivores. The presumption in cladistics is that shared derived characters are evidence of common ancestry and that the *more* derived characters any two organisms share, the more recently they will have shared a common ancestor, compared to other organisms under study. At its most fundamental level, cladistics becomes a matter of counting shared derived characters. Seals and lions share *three* derived characters (tetrapod structure, mammary glands, and carnivorous feeding), while the lizard has only one derived character that it shares with all the other animals (tetrapod structure). Cladistics infers from this that the lion and the seal have a more recent common ancestor than do any other two groups in the cladogram—that their branching from this common line of descent came later than any other branching in the line. And the hagfish?

Figure 18.12
A Simple Cladogram
Among the lizard, deer, lion and seal, which two are most closely related? The tentative answer displayed in this cladogram is the lion and the seal. The basis for this assertion is the number of unique or "derived" characters these two creatures share compared to any other two organisms under consideration. (They share three: tetrapod structure, mammary glands, and carnivorous feeding.) Such a system pays no attention to superficial similarities among organisms. Seals swim in the water, like fish, but they share far more derived characters with lions than they do with fish—enough to persuade scientists that seals have *returned* to the water in relatively recent times, following a period in which their ancestors lived on land.

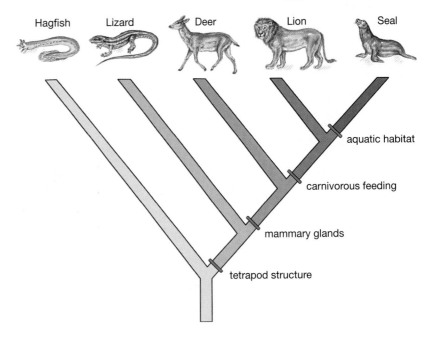

Hagfish   Lizard   Deer   Lion   Seal

aquatic habitat

carnivorous feeding

mammary glands

tetrapod structure

It is a starting-point organism for the mammalian carnivores, known as an *outgroup*. If we are to say that the animals under consideration have derived characters, they must be derived *compared* to some other organism—in this case, the hagfish. The examples here employ visible, physical traits, but in practice modern cladistics relies heavily on comparisons among molecular sequences, such as the number of shared, derived base pairs in lengths of DNA or RNA.

Note that this counting has done away with one of the problems of classical taxonomy: It has removed an element of subjectivity. There is a firm rule for inferring relatedness—the number of shared derived characters—which means that judgment need not play a part in this analysis.

Should Anything But Relatedness Matter in Classification?

Strict cladistic analysis can produce results that run headlong into common-sense notions of what is "different" and "the same." Which is more closely related: lizards and crocodiles on the one hand, or birds and crocodiles on the other? It may surprise you to learn that, in terms of evolution, it is birds and crocodiles. Birds split off from a dinosaur-crocodile line within the last 200 million years or so, while the split between the lizard and crocodile lines came far earlier. *Classical* taxonomy puts crocodiles, lizards, dinosaurs, and turtles into a single class (Reptilia), while putting birds in a different class (Aves; **see Figure 18.13**). Cladistic analysis rejects this classification, on grounds that crocodiles and birds have a more recent common ancestor than do crocodiles and lizards.

This gets us to an important point. Classical analysis holds that, in classifying creatures, something counts *besides* evolutionary relationships—in this case, the special qualities of birds (flight, warm-bloodedness)—while cladistics says the only criterion for placement is phylogeny. At root, then, cladistics is not about classification; it is about establishing lines of descent and only classifies species in accordance with the lines that emerge. Classical analysis is concerned with both evolutionary relationships and with classification.

On to the History of Life

The millions of species that exist on Earth today took billions of years to evolve. During that time, life went from being exclusively microscopic and single-celled to being the staggeringly diverse entity it is today. How did this happen over time? And how did we go from no life to life? And what is the order of appearance for life's living things? These are the subjects of Chapter 19.

a Classical view of relationships among tetrapods

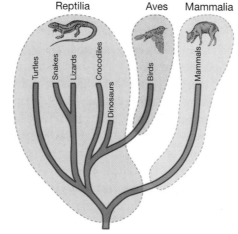

b Cladistic view of relationships among tetrapods

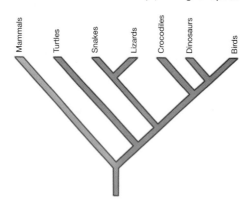

Figure 18.13
Are Birds Reptiles?

a Classical taxonomy employs three classes, Aves, Mammalia, and Reptilia, for birds, mammals, and reptiles, respectively. Thus crocodiles and birds are in separate classes—a recognition on the part of classical taxonomists of the unique features of birds, such as feathers.

b In cladistic analysis, by contrast, birds are grouped very closely with crocodiles and dinosaurs, in recognition of the recent shared ancestry of the three groups. The only consideration in this system of classification is who is more closely related to whom.

Chapter Review

Summary

18.1 What Is a Species?

- The most accepted definition of a species is derived from the biological species concept: "Species are groups of actually or potentially interbreeding natural populations that are reproductively isolated from other such groups." This definition does not apply to asexually reproducing organisms, such as bacteria. Additionally, hybrids—the offspring of two different species—can occur with some frequency, especially among plants.

18.2 How Do New Species Arise?

- Speciation can occur through branching evolution (cladogenesis) or nonbranching evolution (anagenesis). Branching evolution occurs when a single species splits and gives rise to one or more daughter species, with each species reproductively isolated from the others. Nonbranching evolution occurs when a single lineage evolves into a new species, presumed to be reproductively isolated from its ancestral form.

- Speciation comes about when populations of the same species cease to interbreed.

- Branching evolution may occur through allopatric speciation, which always involves a geographic separation of populations of a species, or through sympatric speciation, which does not involve geographic separation.

- A central factor in speciation is the development of reproductive isolating mechanisms, defined as any factor that, in nature, prevents interbreeding between individuals of the same species or of closely related species.

- One form of reproductive isolating mechanism is the geographic separation that takes place in allopatric speciation. For allopatric speciation to occur, however, populations must also develop intrinsic reproductive isolating mechanisms, defined as changes in their form or behavior that prevent them from interbreeding. Likewise, sympatric speciation always entails the development of intrinsic reproductive isolating mechanisms. At least six such mechanisms exist: ecological, temporal, behavioral, mechanical, gametic, and hybrid inviability or infertility.

18.3 When Is Speciation Likely to Occur?

- Speciation can occur rapidly with adaptive radiation into new niches, as with the migration to the Galapagos or Hawaiian islands.

- There is considerable debate among evolutionary biologists regarding the pace of evolution. The gradualist model holds that evolution proceeds at a steady rate, while the punctuated equilibria model holds that organisms undergo long periods of almost no change (stasis) followed by relatively brief episodes of rapid speciation.

18.4 The Categorization of Earth's Living Things

- The most important criterion for classifying species is their degree of relatedness. A binomial nomenclature is used in referring to any given species, with the first of its names identifying its genus and the second identifying the species within this genus (i.e., *Felis silvestris* for the common house cat). Genus and species fit into a larger framework of *taxonomy*, meaning the classification of species. The most commonly used groupings in this taxonomy are species, genus, family, order, class, phylum, and kingdom. The highest-level category, domain, is sometimes also used.

18.5 Constructing Evolutionary Histories: Classical Taxonomy and Cladistics

- Homologous structures are common structures in different organisms that result from a shared ancestry. Homologous structures provide strong evidence regarding relatedness among species. One problem with their use is that they can be confused with analogous structures: similar features that developed independently in separate lines of organisms (as with the legs of modern horses and extinct litopterns).

Key Terms

adaptive radiation 356

allopatric speciation 350

anagenesis 349

analogy 360

ancestral character 361

binomial nomenclature 358

biological species concept 347

cladistics 361

cladogenesis 349

cladogram 361

class 359

convergent evolution 360

derived character 361

domain 359

extrinsic isolating mechanism 350

family 359

genus 359

homology 360

intrinsic isolating mechanism 350

kingdom 359

order 359

phylogeny 359

phylum 359

polyploidy 352

reproductive isolating mechanisms 350

speciation 349

species 359

stasis 357

sympatric speciation 354

systematics 359

taxon 359

taxonomic system 359

theory of punctuated equilibria 357

• Classical taxonomy uses shared, derived characters to classify organisms; but it also uses other criteria, such as the possession of unique physical characters (as with feathers on birds). The cladistic system of taxonomy is concerned only with degrees of relatedness; it classifies organisms strictly on the basis of shared, derived characters.

Understanding the Basics

Multiple-Choice Questions

1. In the biological species concept, the factor that defines a species is
   a. geographical separation
   b. reproductive isolation
   c. hybridization
   d. cladogenesis
   e. its behavior

2. In allopatric speciation (select all that apply)
   a. Allele frequency changes are shared between populations.
   b. Populations are geographically isolated.
   c. Gene flow between populations is stopped by physical barriers.
   d. Geographically isolated populations accumulate genetic changes over time that eventually block successful interbreeding.
   e. Populations are directly adjacent.

3. Factors that can be problems in determining the true relationships among organisms may include (select all that apply):
   a. homologous traits
   b. analogous traits
   c. convergent evolution
   d. poor fossil record
   e. shared derived traits

4. Possible intrinsic reproductive isolating mechanisms include (select all that apply)
   a. mating at different times of year or times of day
   b. geographical isolation
   c. incompatibility between eggs and sperm that prevent fertilization
   d. singing a different mating song than another species
   e. producing fertile hybrids

5. Match the reproductive isolating mechanism with how it works to reduce successful interbreeding:
   a. temporal isolation   populations live in different environments
   b. ecological isolation   populations are separated by physical barriers
   c. geographical isolation   mating occurs at different times of day or year
   d. gametic isolation   egg and sperm do not recognize one another
   e. hybrid sterility   progeny of a cross are unable to reproduce

6. In ocean-dwelling creatures like sea urchins or corals, sperm and eggs are released into the water by males and females and fertilization occurs externally. If several closely related species of coral live in the same location, what reproductive isolating mechanisms are likely to be effective at preventing interbreeding? (Select all that apply.)
   a. behavioral isolation
   b. gametic isolation
   c. temporal isolation
   d. ecological isolation
   e. geographical isolation

7. Order the following categories into the appropriate biological hierarchy, with the most specific category first.
   a. family   d. genus
   b. species   e. class
   c. phylum

8. Adaptive radiations (select all that apply)
   a. are most likely to occur on mainlands
   b. are examples of rapid speciation that occur when a set of empty habitats are encountered
   c. concern the evolution of heat loss in vertebrates experiencing stressful climates
   d. are exemplified by the Galapagos finches
   e. often occur on islands

Brief Review

1. What is convergent evolution, and why is it a problem for phylogenetic analyses?

2. Contrast the gradual and punctuated equilibria models of long-term evolution. What type of evidence has been used to support the punctuated equilibria theory?

3. What is the difference between anagenesis and cladogenesis?

4. Describe one major difference between allopatric and sympatric speciation. What is one thing that they have in common?

5. State the biological species concept.

6. Why is the evolution of intrinsic reproductive isolation mechanisms required for two groups to be called separate species? Why isn't simple geographical isolation sufficient?

Applying Your Knowledge

1. Do you think that speciation and/or extinction are likely to be more frequent in species that are highly specialized in the resources they use than in more generalized species? Why or why not?

2. Populations of some kinds of organisms may be isolated by a barrier as small as a roadway, while other organisms require a much larger physical barrier to block gene flow. What features of organisms might determine how readily they become geographically isolated? What impact might these features have on how often or how rapidly speciation is likely to occur in these groups?

3. In an adaptive radiation, one species may colonize a previously empty environment, such as an island, and diversify evolutionarily into a set of closely related species that occupy a very wide range of habitats. For example, in the Darwin's finches on the Galapagos Islands, one species fills the role of woodpecker by using a stick to tap on trees and dig out insects. Do you think that such a species would have been likely to evolve in an area that already had woodpeckers? Why or why not?

4. Living things are classified into a set of hierarchical categories. Why is this more useful to biologists than simply giving everything a single part name (or a number) and eliminating some of the categories?

Group Activity

Form a group of four students. Imagine that you are the first scientists to explore a large island. In the course of your investigation, you discover that two large populations of flying birds coexist on the island. The two populations seem similar in general appearance, but you can detect small but consistent differences in plumage. Describe some of the possible mechanisms that might have led to the speciation of these two populations. If they are not different species, what explanation would your group give for the difference in plumage?

19

A Slow Unfolding
The History of Life on Earth

The history of life written in layers of rock.
(page 370)

So far apart but once so near.
(page 373)

Part of our family tree.
(page 386)

Within 5 minutes of its

birth, a newborn wildebeest on the plains of central Africa can be off and running with its herd (**see Figure 19.1**). By contrast, human infants probably will not be able to take a step by themselves until they are about a year old. Female lions reach sexual maturity at about 3 years, killer whales at about 7, but human males and females do not come to sexual maturity until they are past 10. Even in comparison to our fellow mammals, humans mature at a notably slow rate—a trait we share to some degree with our most closely related fellow primates, the chimpanzees.

At first glance, it might seem puzzling that later primate evolution took this turn. From what we have learned of natural selection, wouldn't it make sense that survival would *decrease* for offspring that developed more slowly? What is the fitness payoff in having young that are essentially helpless for several years and then immature for many more? With respect to human beings, here is one possible answer. As a species, we have survived not so much because of our physical capabilities, but because of our wits. Our success is the product of our remarkable capacity to *learn*. But learning takes time and it is arguably best undertaken by minds that remain for a long time in a state of what might be called flexible immaturity. (Think of how easily young children acquire a second language in comparison with adults.) Among our human ancestors, it may be that those who survived tended to be those who could learn the most, and that those who could learn the most were those who took the longest to mature.

Under this view, we carry our evolutionary heritage with us in the form of delayed

Figure 19.1
No Time to Lose
A newborn wildebeest struggles to its feet immediately after its birth on the plains of Africa. Wildebeest predators, such as hyenas, often concentrate their efforts on the newborn.

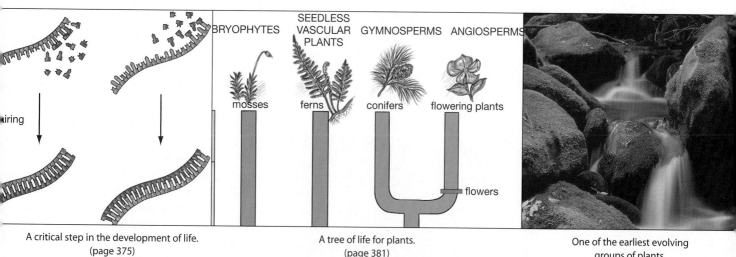

A critical step in the development of life.
(page 375)

A tree of life for plants.
(page 381)

One of the earliest evolving groups of plants.
(page 380)

maturity. We did not leave this heritage behind in the African savanna, where human beings first evolved; we can see it every time we look at a 2-year-old. In the same way, we can see evolutionary vestiges in us from a much earlier time as well. Note the way we walk, our *left arm* swinging forward as our *right leg* does the same. Almost all four-limbed animals walk this way. One view is that we inherited this from the elongated, four-finned fish we evolved from. They came onto the muddy land by slithering in an S-pattern, the left-front fin going back as the left-rear fin went forward, then the reverse (**see Figure 19.2**). Evolution may also be with us, some scientists speculate, in the lower-back pain that afflicts so many people in middle age and beyond. Our ancestors began walking upright probably no more than 5 million years ago; as this notion has it, we're still getting used to the change.

Note the sweep of evolution in all this. Maturation is with us from one period, walking style from another. If we look at the genetic code—*this* DNA sequence specifying *that* amino acid—we have within us a heritage that stretches back to the beginning of life on Earth, because this code is shared by nearly every living thing. What you will do in this chapter is trace some of these linkages and divergences. You'll look at how evolution has proceeded in its largest scale; how life began, and how it diversified into the major forms of life we see today.

19.1 The Geologic Timescale: Life Marks Earth's Ages

We can begin this inquiry by getting a sense of life's time line, which you can see in **Figure 19.3**. This is the geological timescale, which divides earthly history into broad **eras**, shorter **periods**, and shorter-still **epochs**. The scale begins with the formation of the Earth about 4.6 billion years ago and runs to "historic time," meaning time in the last 10,000 years.

Though this is a geologic timescale, you saw back in Chapter 16 that, in the days before radiometric dating, scientists measured the *relative* ages of geologic layers by looking at the fossils they contained. Looking at a certain layer and seeing fossils of a type of bony fish, a scientist knew he or she was looking at a Devonian layer, for example. Fossils, then, *defined* geologic layers (or "strata"), which is another way of saying that life defined the various ages of the Earth (**see Figure 19.4** on page 370). Within this framework, what marks the boundary between one geologic era or period and the next is some sort of *transition* in living things. One part of this transition is not surprising: New life-forms come with a new era or period. What may be surprising, however, is what often lays the groundwork for these new forms: death on a grand scale in "major extinction events." Six of them are noted in Figure 19.3, with all of them defining the end of an era or period. (The relatively recent Pleistocene Extinction, which brought about the demise of such large mammals such as the wooly mammoth, was not a major extinction event. One of the reasons it is notable, however, is that human beings played a part in it, through hunting.) The most famous extinction event, the **Cretaceous Extinction**, occurred at the boundary between the Cretaceous and Tertiary periods. This was the extinction, aided by the impact of a giant asteroid, that brought about the end of the dinosaurs, along with many other life-forms. The greatest extinction event of all, however, occurred at the end of the Paleozoic era. This was the **Permian Extinction**, in which as many of *96 percent* of all species on Earth were wiped out (see "Physical Forces That Have Shaped Evolution" on page 372).

Out of such extinctions, however, come new forms of life. With old forms gone, relatively rapid evolution can take place, as species fill depopulated niches (see Chapter 18, pp. 356–357). Such adaptive radiation can be seen with the evolution of our own lineage, the mammals. Prior to the extinction of the dinosaurs, we mammals existed in a few species, generally rat-sized insect hunters. With the death of our fearsome enemies in the Cretaceous Extinction, we were able to evolve into the many forms that exist today.

Figure 19.2
Our Evolutionary Heritage
The gait exhibited by almost all four-limbed animals, including human beings, stems from the reversing S-pattern that our evolutionary ancestors used in getting around when they were making the transition from aquatic to terrestrial life. Here a salamander employs this same motion in getting from one place to another.

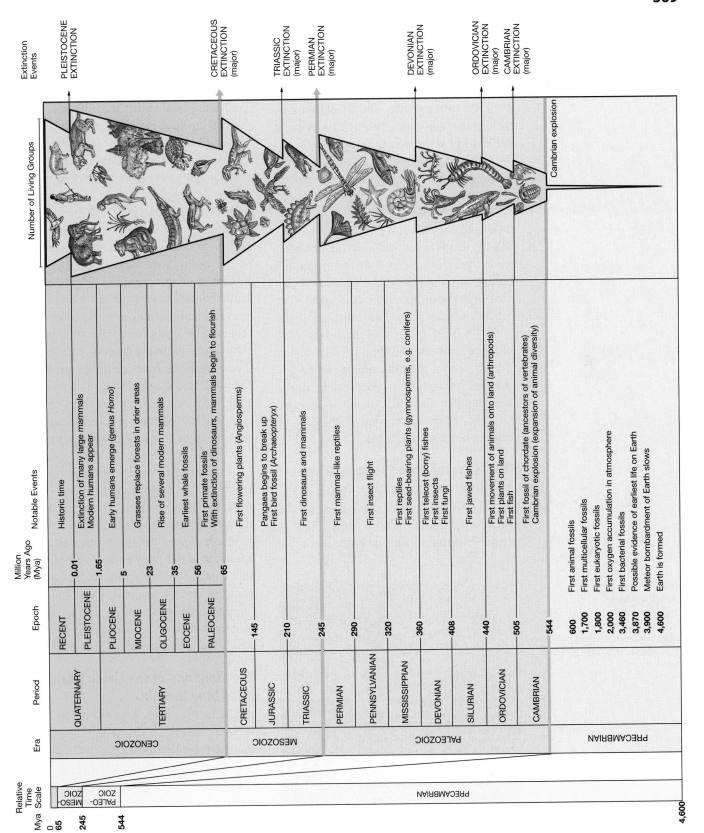

Figure 19.3
A Timescale for Earth and Its Living Things

Transition Does Not Always Mean Death: The Cambrian Explosion

Looking again at the eras in Figure 19.3, there is one exception to the transition-means-extinction rule. The transition between the Precambrian and Paleozoic eras marks not a major extinction, but a seeming explosion in the diversity of animal forms. The fossil record tells a tale of evolution proceeding at a dizzying pace for a short period that began about 544 million years ago (Mya), though this idea recently has been challenged, as you'll see. This "Cambrian Explosion" stands in marked contrast, however, to what preceded it. Look at the enormous amount of time that the Precambrian Era takes up in the timescale. In it, life evolved at such a slow pace that we scarcely have words for it. ("Glacial" greatly overstates things.) For more than *half* the time life has existed on Earth, it consisted of nothing but single-celled microscopic forms. Conversely, all the birds, reptiles, fish, plants, and mammals that exist today came about in the last 14 percent of evolutionary time.

What Is "Notable" in Evolution Hinges on Values

One final word about the timescale and the path you will follow in this chapter. Picking out the "notable events" in evolution is in-evitably an exercise in making value judgments. The notable events covered in this chapter will lead you along a line that begins with microscopic sea creatures and ends with human beings. The value judgment that guides this path is that students have a great interest in knowing about their own evolution, and that this lineage arguably presents the broadest sweep of evolutionary development. Such a course of study may leave the impression, however, that all of evolution amounts to a march toward the development of human beings, who then get to occupy the highest branch of an evolutionary tree.

In reality, we occupy one ordinary branch among a multitude of branches. Under notable events in the Cenozoic era, the table could just as easily have listed milestones in the evolution of fish or birds, or the emergence of social insects such as bees. *These* creatures have continued evolving in the modern era, along with mammals, but in lines separate from us. Focusing on a line that leads to humans is like focusing on a railroad route that leads from, say, Baltimore to Denver, while ignoring the multitude of lines that intersect with it or that are separate from it. We regard the Baltimore-Denver route as special because we are *interested* in it, but that does not mean that it is fundamentally *different* from any other line. Some special qualities have evolved in human beings (notably our large brains), but the same could be said for almost any species. In phylogenetic terms, we simply lie at the tip of one evolutionary branch, while bees lie at another, and cactus plants lie at still another.

The Kingdoms of the Living World Fit into Three Domains

In tracing the history of life, you'll be looking to some extent at all the major life-forms that have evolved on Earth. Thus it will be helpful to have at the outset a sense of what these major life-forms are. This book employs a system of three **domains** of life, which then have within them the "kingdoms" noted in Chapter 18. The three domains are **Archaea**, **Bacteria**, and **Eukarya**.

Archaea is a domain populated entirely by microscopic, single-celled organisms, many of whom live in "extreme" environments, such as boiling-hot vents on the ocean floor. Bacteria is likewise a domain of single-celled microscopic creatures, but these are the bacteria that

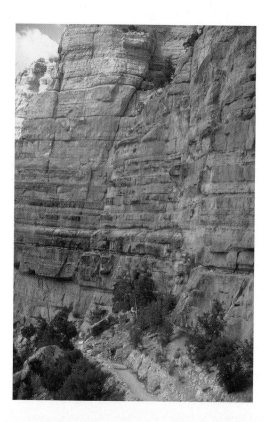

Figure 19.4
Revealing Layers
The history of life can be traced though fossilized life forms found in layers of sediment, seen here in America's Grand Canyon.

are more familiar to us. The third domain, Eukarya, then includes all the organisms that are most familiar to us—plants, animals, and fungi—along with another group that is less familiar, the protists. (The protist kingdom has within it lots of single-celled water dwellers, but also includes some large organisms, such as giant sea kelp.)

19.2 Tracing the History of Life on Earth: How Did Life Begin?

Taking a historical approach to tracing life on Earth, the first question that arises is one of the toughest: How did life begin? Darwin himself thought about this and imagined that life began in what he called a "warm little pond" in the early Earth. As it turns out, however, the early Earth had no warm little ponds; what it had was an environment more akin to hell on Earth.

You have seen that Earth was formed about 4.6 billion years ago. This took place in a process of "accretion," in which ever-larger particles clumped together: cosmic dust to gravel, gravel to larger balls, larger balls to objects the size of tiny planets. One consequence of such development was that the accretion now called Earth was periodically being slammed into by *other* large accretions—meteorites and comets, though some of them may have been as large as Mars. If we could look, then, at the Earth in its first 600 million years or so, we would see an enormously hot planet whose early seas were periodically vaporized to a depth of 100 meters or so because of the violent impact of the meteorites.

In What Kind of Environment Did Life Begin?

The meteorite bombardment seems to have slowed by about 3.9 billion years ago. With this change, we are left with a large question about what Earth was like. Scientists believe that the sun shone then some 20–30 percent less brightly than it does now. But the Earth could have remained hot or been very cold then, depending on how much heat-trapping carbon dioxide existed in Earth's atmosphere at that point. If it was cold, the Earth's seas may have been frozen to a depth of a few hundred meters.

These two views lead to different ideas about where life arose. Was it in very hot seas or in frigid seawater beneath a thick layer of ice? Most researchers in the field today seem to favor the hot-sea theory, with a critical piece of evidence being that the very earliest organisms seem to have lived in high-temperature environments.

What Was the Source of the Raw Materials for Life?

Whatever environment gave birth to life, it had to have in it the chemical raw materials that could be used in forming life's critical molecules. You saw in Chapter 3 that life's informational molecules (DNA and RNA) are composed of certain kinds of building blocks (nucleotides), while its proteins are composed of other kinds of building blocks (amino acids). The raw materials for such building blocks could have been the gasses methane and ammonia. These substances certainly would have been spewed out (or "outgassed") by Earth's early volcanoes; the question is whether they would have been quickly broken down in the ultraviolet light, coming from the sun, that bathed the early Earth's atmosphere.

Some scientists believe that building blocks such as amino acids could have arrived ready-made, delivered by the meteorites and comets that smashed into the young Earth. It's clear, from meteorites that have landed on Earth in recent decades, that such objects *can* carry organic molecules (**see Figure 19.5**). Research has cast doubt, however, on whether such accidental "seeding" from outer space could have brought a sufficient *quantity* of organic materials to get life going.

Life May Have Begun in Boiling Water

A third possibility for the supply of organic materials in the early Earth is the methane and hydrogen sulfide that gush out even today from deep-sea vents on the floors of the world's oceans. This fact can then be tied

Figure 19.5
Space Travelers
Fragments of the Murchison meteorite, which landed on the township of Murchison in Australia in 1969. The meteorite contained water and organic molecules such as amino acids. Some scientists have hypothesized that such organic materials could have seeded life on Earth.

Physical Forces That Have Shaped Evolution: Climate, Extraterrestrial Objects, and Continental Drift

Climate Change

The living world is very much affected by the physical forces operating on the Earth. You can see this in connection with the mass extinctions we noted in the main text. What has caused these extinctions? Evidence links at least four of those listed in Figure 19.3 to episodes of global climate change, specifically global cooling. (The four are the Ordovician, Devonian, Permian, and Cretaceous extinctions.) All kinds of things happen with global cooling: Glaciers develop, wiping out everything in their path and tying up huge amounts of water in ice. This reduces sea levels, thereby killing many organisms that live on the continental shelves. The oceans themselves get colder, particularly at greater depths, which is its own kind of killing event. As may be apparent, the factor of loss of habitat is intimately tied to the issue of global climate change. Global cooling is not the only factor involved in mass extinctions, but it is thought to be the most important factor.

Extraterrestrial Objects

In the public mind, the Cretaceous Extinction that finished off the dinosaurs had a single, nearly instantaneous cause: The explosive impact of an enormous asteroid, 10 kilometers (6.5 miles) in diameter, that struck Mexico's Yucatan peninsula 65 million years ago. There is no doubt that this catastrophic event happened; what is in doubt is that it was the sole or even predominant cause of the Cretaceous Extinction. For one thing, fossil evidence indicates that many Cretaceous life-forms, including the dinosaurs, were in severe decline *before* the asteroid struck. Secondly, as just noted, we have evidence that global cooling occurred in Cretaceous times. If there is a consensus view on this issue, it is that the asteroid probably was responsible for the nearly instantaneous demise of some species, but that in other cases it was merely the final blow in extinctions that already were underway.

At the very least, the Cretaceous Extinction shows that evolution can be affected by sudden, catastrophic events that have nothing to do with Earth's normal physical processes. A fascinating point is raised by this: Fitness, in

Figure 1
Continental Drift
During the history of the Earth, the tectonic plates that make up the surface layer of Earth have moved with respect to one another, causing the continents to spread apart and slam together.

CONTINENTAL DRIFT

Paleozoic era
(250 million years ago)

PANGAEA

Mid-Mesozoic era
(150 million years ago)

LAURASIA

GONDWANA

Late Mesozoic era
(70 million years ago)

North America
Eurasia
Africa
South America
India
Australia
Antarctica

Present

North America
Eurasia
Africa
South America
Australia
Antarctica

the evolutionary sense, can change in an eyeblink. The qualities that made the dinosaurs so dominant in the Triassic Period were of no use once the asteroid struck. The game of survival had changed. Now *other* qualities came to the fore in terms of survival. Among these may have been small size, which mammals had, perhaps to their great benefit.

Continental Drift

As noted in Chapter 18, the primary engine driving speciation is the geographic separation of populations. Geographic separation was described as taking place in scales that were both small (a stream changing course) and large (a glacier's movement). It turns out that geographic sepa-

ration can take place on a much larger scale yet, in that whole *continents* have come together and separated again during Earth's history.

If you look at **Figure 1**, you can see the land masses of the world as they have existed at four separate times, 250 million years ago, 150 Mya, 70 Mya, and the present. In the first of these, nearly all the Earth's continental mass was part of a single supercontinent, Pangaea, which stretched from the South to North Pole. By the Early Cretaceous, Pangaea was separating along the lines you see in the figure. This in turn gave way to our current division which, like its predecessors, is a configuration in flux.

These are only four "snapshots" in a process of **continental drift** that has produced lots of other alignments. How does such movement come about? In essence, the Earth's crust and part of its upper mantle are divided into a series of plates that move laterally over the globe. Hold two blocks together and slide them past each other; then move them away from each other, then *into* each other, and you get the idea of the kinds of movements that plates can have relative to one another.

The movement of the plates can be thought of in terms of a conveyor belt. Halfway between North America and Europe there is a long underwater mountain chain, the Atlantic Ridge. This ridge has in its middle a deep valley that forms the border between the North American and Eurasian Plates. Hot material in the Earth's interior is constantly being moved up into the valley from either plate, after which, as new crust, it moves eastward on one side of the ridge but *westward* on the other. What we have, then, is a spreading of the ocean floor in opposite directions. *Continents* sit atop this spreading material, and they are being moved in opposite directions—Europe to the east, North America to the west—as if they were objects on conveyor belts (**see Figure 2**).

(continued)

a Paleozoic era Pangaea (250 million years ago)

b Today

Figure 2
Similar fossils are found on the east coast of South America and the west coast of Africa. How can this be?

a 250 million years ago, the two continents were contiguous and formed part of the giant continent Pangaea.

b Animals that were alive then are now found on opposite sides of the ocean because the ocean floor has been spreading westward and eastward along the central Atlantic Ridge. The ocean floor is spreading because hot material from the interior of Earth is constantly being moved up into the central Atlantic Ridge, even today.

Physical Forces That Have Shaped Evolution: Climate, Extraterrestrial Objects, and Continental Drift (continued)

The Impact of Continental Drift on Evolution

Continental drift has had profound effects on evolution. You've already noted the role global cooling has played in mass extinctions. So what causes global cooling? One of the factors is the development of glaciers, which as it turns out can grow only when continents are located at or near the Earth's poles. This *was* the case 300 million years ago, then it was *not* the case 200 million years ago, and now it is again. By moving landmasses on and off the poles, continental drift has helped change the climate. As noted, glaciers lock up water, causing sea levels to drop, killing life on the continental shelves.

Apart from these factors, there is the more obvious separating and mixing of *populations* through these land movements. For example, the breakup of the supercontinent Pangaea eventually meant the creation of island continents such as Australia and Antarctica. Until the coming of humans, Australia had scarcely any placental mammals—mammals whose embryos are nourished by a placenta (see Chapter 27). Since prehistoric times, however, Australia has had an abundance of marsupial (or "pouched") mammals, such as kangaroos. We can read evolutionary history in this distribution. Marsupials developed in Pangaea before its breakup, in what is now North America and Western Europe, and spread to Australia by way of what is now Antarctica. Meanwhile, placental mammals were spreading too; but few were able to reach Australia before it broke away as an island continent. Placental mammals seem to outcompete marsupials wherever the two varieties meet. This could not happen in Australia, however, because continental drift had *isolated* its marsupial populations, leaving them free to evolve into the diversity of species we see today, such as kangaroos, koalas, and wombats.

together intriguingly with the earlier observation about Earth's earliest organisms being heat-loving. The evidence for this is not fossils, but instead the genetic makeup of creatures that live today at the vents—microscopic archaea and bacteria, some of which thrive at temperatures of 105° Celsius or 247° Fahrenheit, which is above the temperature at which water boils (**see Figure 19.6**). This idea has a good deal of support among origin-of-life researchers today: That life began in the "pre-biotic soup" of hot-water systems—in the deep-sea vents or the kind of hot-spring pools that exist in Yellowstone Park today (**see Figure 19.7**). Though life is fragile in some respects, it may be that it had its beginnings in boiling, sulphurous water—quite a different thing from Darwin's warm little pond.

Wherever it began, life cannot have been created in a single step from basic compounds such as methane and hydrogen sulfide. There is a huge gap between these materials and organisms as complex as the archaea. If organic molecules did not come from outer space, early Earth had to be a kind of natural chemical lab in which ever more complex organic molecules were produced from simpler ones. A major portion of origin-of-life research involves creating, in the laboratory, simulations of early-Earth environments and then watching to see if life's building blocks will come together in them. Scientists now have scenarios that can account for the creation of most, but not all, of these compounds.

The RNA World

Any living thing must have at least one critical feature: the ability to reproduce itself. Today, a key player in reproduction is DNA. In Chapter 14 (pp. 276–277), our "genome," or inven-

Figure 19.6
Hot Habitat
Material pours forth from a hot-water vent on the floor of the Atlantic Ocean. The fluid being emitted is mineral-rich enough to support bacteria and archaea that in turn form the basis for a local food web. A deep-sea crab can be seen at lower center.

tory of DNA, is considered as a kind of cookbook containing recipes (DNA sequences) that lead to products (proteins) that undertake a wide variety of tasks. Yet, as noted there, the DNA cookbook is powerless to do anything by itself. When it comes time for DNA to be copied, the double helix cannot unwind itself or pair its bases with messenger RNA or edit the resulting mRNA "transcript." Instead, all these tasks are carried out by the *proteins* known as enzymes. Hence there is a chicken-and-egg dilemma: Enzymes can't be produced without DNA's information, but DNA can't copy itself without the activity of enzymes. So, how did life get going if neither thing can function without the other?

In the late 1960s, researchers hit on a solution that has come to have a good deal of support. In the earliest life-forms, a *single* molecule performed both the DNA and enzyme roles. This molecule was RNA (ribonucleic acid), which is portrayed back in Chapter 14 as a kind of genetic middleman, ferrying DNA's information to the sites where proteins are put together. A critical piece of evidence about RNA's role in early life came in 1983, when researchers discovered the existence of enzymes that are composed of RNA instead of protein. Called **ribozymes**, these are multitalented performers. They can encode information *and* act as enzymes (by, for example, facilitating bonds that bind RNA units together).

This leads researchers to the presumed existence of the "RNA world"—a living world that consisted solely of self-replicating RNA molecules. The central question facing origin-of-life research today is this: How do we get to the RNA world from the chemical raw materials? RNA itself is so complex that many researchers believe it had to have taken over the role of a more simple precursor.

The Step at Which Life Begins

The critical step in life's development comes in imagining RNA or its precursor beginning to carry out replication of itself. *This* is the step at which life can be said to have begun. Given a molecule that can make copies of itself, a *line* of such molecules can come into existence; given the *mistakes* that are bound to take place in such copying, molecules can be produced that differ slightly from one another in being, say, more resistant to the elements. Molecules that have such an advantage will leave more daughter molecules, which might just as well be called offspring. What happens, in short, is self-replicating molecules evolving through natural selection.

Lots of scenarios have been proposed for how this replication got going. A central assumption in most of them is that RNA or its precursor took part in the base pairing that takes place today (**see Figure 19.8**). If a *strand* of RNA exists, a *complementary* strand can easily

Figure 19.7
Only Seemingly Inhospitable
The hot-water Grand Prismatic Pool in Yellowstone National Park, shown here in an aerial view, gets its blue color from several species of heat-tolerant cyanobacteria. The colors at the edge of the pool come from mineral deposits. The "road" at the top of the pool is a walkway for visitors.

Figure 19.8
How Did Self-Replication Get Started?
A ribozyme-like precursor of RNA may have been the first molecule to replicate itself. Free nucleotide bases pair with their complementary bases to form a complementary strand. If the two strands then separate, the process can repeat itself and many RNA strands can be formed. This process occurs today with the help of enzymes, but the process may have occurred millions of years ago with only the RNA itself to catalyze the reactions.

be produced through base pairing. This is a reaction that takes place now, catalyzed by modern-day enzymes. Imagine in the early Earth, however, a ribozyme precursor that could carry out a similar sort of complementary pairing. With its catalytic powers, such a strand could bring about binding of a complementary chain to itself, then separation of the two strands, then a continuation of the process. The challenge for scientists today is to figure out what substances could have carried this replication out, and under what conditions.

"Directed Evolution" of Ribozymes

In trying to figure out what the earliest replicating molecules were, scientists today are employing "directed evolution." They grow huge numbers of ribozyme chains in a test tube, many with different sequences, and select those sequences that are better at carrying out life's tasks, such as bringing together RNA nucleotides. Then these selected strands are allowed to replicate. Fascinatingly, the goal of some of these scientists is to create ribozymes that can replicate by themselves—that is, that can replicate without the researchers adding off-the-shelf enzymes at the proper times. So what? This can be stated in a less technical way. The goal of these scientists is to *create life* in the laboratory; to produce a ribozyme that is able to reproduce itself over generations without human intervention. What is the difference between this and life as we have defined it on the young

Figure 19.9
The Tree of Life
The universal ancestor at the base of the tree gave rise to two domains of organisms whose cells lack nuclei—Bacteria and Archaea. The third domain of life is then the Eukarya, whose earliest kingdom was the protists. From these protists all plants, animals, and fungi evolved.

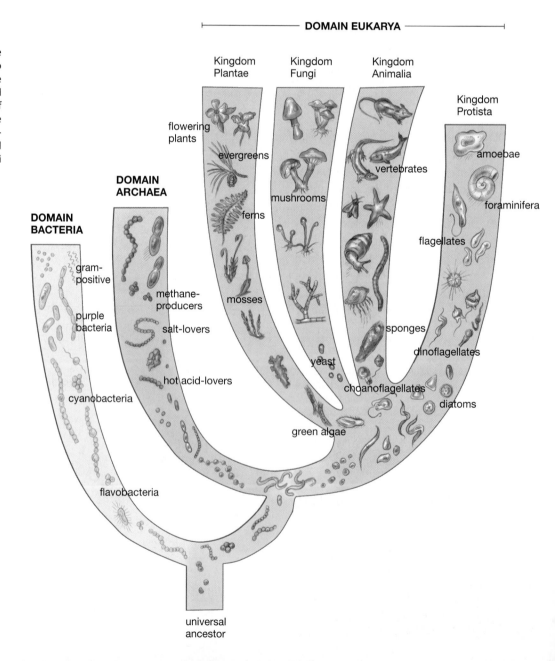

Earth? If scientists are successful at this, humans will have created life not in the Frankenstein-like form that has so often been imagined, but in the form of simple molecules like those that are the ancestors to us all.

Adding Life's Elaborations to Replication

There is a lot more to living things than replication, of course. Even the most primitive living things today are encased in the protective linings called plasma membranes; they have a sophisticated means of extracting energy from their environments; they can get rid of their wastes, and so forth. It may be that an elaboration such as an outer membrane developed along with the replication. But it is also possible that life first took the form of "naked RNA" or its precursor. Whatever the sequence of events, once life's elaborations are joined to replication, we move from simple molecules to the cellular ancestors of today's organisms.

19.3   The Tree of Life

Once life was established, how did it evolve? If you look at **Figure 19.9**, you can see the biggest picture of all. At the bottom you see a "universal ancestor," which is the organism—a successor to one of our replicating molecules—that gave rise to all current life. The evolutionary line that leads from the universal ancestor then branches out to give us, on the one hand, the domain Bacteria and, on the other, a line that leads to both the domain Archaea and the domain Eukarya. Archaea were long thought of as simply a different kind of bacteria, but modern research has shown these organisms to be distinct from bacteria in fundamental ways. Nevertheless, there are great similarities: Bacteria and archaea are essentially single-celled, and no archaeic or bacterial cell has a nucleus. Conversely, *all* the organisms in the domain Eukarya have nucleated cells, and are thus "eukaryotes."

As noted earlier, you'll be learning about life as it exists within the "kingdoms" of domain Eukarya. Looking at the chart, you can see that the earliest kingdom in Eukarya is Protista—mostly one-celled, water-dwelling creatures—and it is from Protista that the other three eukaryotic kingdoms

evolve. An unknown species within Protista gave rise to all the fungi that exists on Earth today; another grouping of protists called the *choanoflagellates* probably gave rise to all of the animal kingdom; and the protists we call green algae gave rise to all of today's plants.

19.4   A Long First Period: The Precambrian

The Slow Pace of Change in the Precambrian

With this picture in mind, let's start tracing life from its cellular beginnings. As noted, living things started small and stayed that way for a long, long time. Recall that the Earth was formed about 4.6 billion or 4,600 million years ago (Mya). The earliest undisputed evidence of life we have so far—bacterial microfossils from Australia—dates back 3,460 million years. There is some tentative physical evidence of life that dates back more than 3,800 million years, but let's stay with the sureness of the bacterial fossils and say that this is the *latest* evidence we have of life's appearance on Earth. After this, more than 1,600 million years pass before we have evidence of *any other* form of life. The new form that does arise then is single-celled algae; after that another 100 million years pass before we get the first *multi*-celled organisms, which are algae again. Evolution took its time in the Precambrian. The years rolled by in the millions, the land was barren, and the oceans were populated mostly by creatures too small to see with the naked eye.

Notable Precambrian Events

The Initiation of Photosynthesis

None of this is to say, however, that "nothing happened" in the Precambrian era. Those of you who have read Chapter 8 know how critical photosynthesis is for life on Earth. This capability first came about in the Precambrian era—and fairly early in the era at that. It began in bacteria at about the time the first bacterial fossils left their imprint, which is to say about 3,500 Mya. This event was a turning point. Had this capacity not been developed, evolution would have been severely limited, for the simple reason that there would have been so little to eat on Earth. The earliest organisms subsisted mostly on organic material

What Is It?
See p. 378.

in their surroundings, but the supply of this material was limited. In photosynthesis, the sun's energy is used to *produce* organic material—in our own era, the leaves and grasses and grains on which all animal life depends. Large organisms require more energy and, through photosynthesis, a massive quantity of energy-rich food was made available.

Oxygen in the Atmosphere: A Dramatic Change on Earth

One particular kind of photosynthesis, again beginning in the Precambrian, had a second dramatic impact on evolution. The earliest undisputed fossils are called *cyanobacteria*. It turns out that these were the first creatures to produce *oxygen* as a by-product of photosynthesis (**see Figure 19.10**). For us the word *oxygen* seems nearly synonymous with the word *life*, but until about 2,000 Mya there was *no*

oxygen in the atmosphere. When it finally did arrive, through the work of the cyanobacteria, its effect was anything but life-giving. For most organisms it was a deadly gas, producing what has been termed an "oxygen holocaust" and establishing a firm rule for life on Earth: Adapt to oxygen, stay away from it, or die.

Adapting to Oxygen: Bacteria and Eukaryotes Strike Up a Relationship

It is in the creatures that adapted to oxygen that we find one of the most interesting stories in evolution. Those of you who went through Chapter 7, on cellular energy harvesting, will recall that most of the energy that eukaryotes get from food is extracted in special structures within their cells, called *mitochondria*. Though mitochondria are tiny organelles within eukaryotic cells, they have characteristics of free-living *cells*, specifically free-living bacterial cells. As it turns out, that is probably what they once were. Ancient bacteria that could metabolize oxygen took up residence in early eukaryotic cells and eventually struck up a mutually beneficial relationship with them (**see Figure 19.11**). The bacteria benefited from the eukaryotic cellular machinery, and the eukaryotes got to survive in an oxygen-rich world. In a second bacterial invasion, algae and their plant descendants got not only mitochondria, but the chloroplasts in which photosynthesis is carried out. (See "The Stranger Within: Lynn Margulis and Endosymbiosis" in Chapter 4, page 81.)

One More Effect of Oxygen: The Blockage of Ultraviolet Light

The oxygen revolution had one more momentous consequence: Molecules of oxygen came together to form the gas called *ozone*, which rose through the atmosphere to form the ozone layer, giving Earth, for the first time ever, protection against the ultraviolet radiation that comes from the sun. Marine creatures had been shielded from this radiation by the ocean itself; but prior to the formation of the ozone layer, there was no chance for life to develop on *land*.

Figure 19.10
Home to the Oldest Fossils
Layered sediments called stromatolites contain the fossilized remains of cyanobacteria that flourished more than 3 billion years ago in what is now western Australia.

a
eukaryotic cell

bacterial cell

b
eukaryotic cell

mitochondrion

c

Figure 19.11
Adapting to Oxygen
The "powerhouses" of our cells—the organelles called mitochondria—have several characteristics of free-standing bacteria.

a There is general agreement among scientists that our mitochondria originally were bacteria that invaded eukaryotic cells many millions of years ago. These bacteria benefited from the eukaryote's cellular machinery, and the cell benefited from the bacteria's ability to metabolize oxygen.

b Over time, the bacteria came to be integrated into the host cells, replicating along with them. Two symbiotic organisms had now become a single organism.

c Today, the mitochondria in your cells turn the energy from food into a form that allows you to run, think, and turn the pages of this book. These artificially colored mitochondria functioned in a liver cell of a mouse.

19.5 The Cambrian Explosion: A Real Milestone or the Appearance of One?

As the Precambrian was coming to a close, life consisted of bacteria, archaea, and many kinds of protists (the earliest eukaryotes), all of which lived in the ocean. By about 600 Mya, however, we get the first evidence of *animals* in the seas. A formal definition of animals is given in Chapter 20. Suffice to say now that animals have the characteristics you would expect: They all are multicelled, and they get their nutrition not from performing photosynthesis (like algae) but from consuming organic material or other organisms.

There currently is a debate about whether the earliest animal fossils represent a line of creatures that died out, or are instead related to today's animals. Whatever the case, these early animals can be seen as a kind of early stirring; a first blip on the screen bringing news of something big to come. To judge by the fossil record, about 70 million years after the emergence of these animals, we get it: A tidal wave of new animal forms, the like of which has never been seen before or since in evolution. This is the Cambrian Explosion, which began about 544 Mya.

Recalling the taxonomic system you went over in Chapter 18, you may remember that just below the "domain" and "kingdom" categories there is the **phylum**, which can be defined informally as a group of organisms that shares the same body plan. There are perhaps 30 phyla in the animal kingdom today, such as Porifera (sponges), Arthropoda (insects, crabs) and Chordata (ourselves, salamanders). With one exception, the fossil record indicates, every single one of these came into being in the Cambrian Explosion. This seeming riot of sea-floor evolution was actually more extensive than even this implies, in that a good number of phyla that *don't* exist today—phyla that have become extinct—also seem to have first appeared in the Cambrian. Some of these creatures are so bizarre that it's surprising that Hollywood hasn't mined the Cambrian archives for new ideas on monsters (**see Figure 19.12**). As if all this weren't enough, the Cambrian Explosion appears to have taken place in a very short period of time relative to evolution's normal pace; perhaps as little as 5 million years.

But did it? The Cambrian Explosion is so extreme that one of the primary challenges to evolutionary biology has been to explain why such a thing came about—or why it came about only once. Note the series of events: Only a few animal forms before the Cambrian, then the explosion, then almost no new animal forms since. In the 1990s, several lines of evidence suggested something else: That animal forms actually began to diverge well back in the *Pre*cambrian. Under this view, all the Cambrian Explosion amounted to was an explosion of forms big and hard enough to leave *fossils*. The forms themselves appeared earlier and more gradually, this idea goes, but were too small and fragile to leave imprints behind.

If the Cambrian Explosion did in fact occur, what could have sparked it? Lots of ideas have been proposed to explain it. One of the best-received is that the explosion was triggered by the rise in atmospheric oxygen. To get bigger creatures, you need more oxygen, and levels of atmospheric oxygen may have reached a critical threshold about this time. Once larger forms *could* appear, adaptive radiation took place on a grand scale. So many niches were available that there was an explosion of forms. Once all the basic niches were taken, however, this frenzy of new forms not only came to a stop, it was pruned back (in the sense that some Cambrian phyla became extinct).

19.6 The Movement onto the Land: Plants First

The teeming seas of the early Cambrian period stand in sharp contrast to what existed on land at the time, which was no life at all. Earth was simply barren—no greenery, no birds, no insects. When life did come to the land, the first intrepid travelers were plants, whose earliest fossilized remains on land date from some 460 Mya. Theirs was a gradual transition that began with marine algae in the ocean, then continued with freshwater algae, then freshwater algae that came to exist in *shallow* water, living partially above the waterline. When the transition to full-time living on land came, it was to damp environments.

Figure 19.12
Product of the Explosion in Animal Forms
A model of an *Opabina regalis*, one of the now-extinct animals that arose in the flurry of evolutionary activity known as the Cambrian Explosion. *Opabina* had five eyes and a flexible proboscis beneath them.

Adaptations of Plants to the Land

Such a change required a lot of adaptation. Aquatic algae did not have to deal with water loss or the crushing effects of gravity, but land plants did. One of the plants' responses to the water problem was to develop a waxy outer covering, called a *cuticle*, that could retain moisture; while an initial response to gravity was to stay low. Some of the most primitive land plants are mosses that often hug the ground like so much green carpet (**see Figure 19.13**).

Then there was the problem of reproduction. When green algae reproduce sexually, their gametes (eggs and sperm) float off from them as individual cells, after which they are brought together by ocean currents or the cells' own movement. Once this happens, the resulting embryos develop completely on their own. An embryo that matured this way on *land*, however, would dry out and perish. Plants adapted to the land by developing a protection mechanism for both embryos and gametes. We can see this clearly in some modern-day descendants of the early plants. Sperm develop in pollen grains, which have an outer coat, while eggs develop

Figure 19.13
Lying Low
Moss covering rocks along a stream in Tennessee's Great Smoky Mountain National Park. Mosses are members of the earliest-evolving group of plants, the bryophytes.

inside the parent plant and are fertilized there, producing an embryo that likewise begins development in the plant. The maturation of embryos *within* a parent is a key characteristic that separates plants from algae.

Another Plant Innovation: A Vascular System

All plants share this reproductive characteristic, but plants then go on to diverge from each other in some very basic ways. If you look at **Figure 19.14**, you can see how the major divisions of plants evolved. The most primitive land plants, the **bryophytes** (represented by today's mosses), had no vascular structure—meaning a system of tubes that transports water and nutrients. Plants without such a system have very limited structural possibilities. They can grow out, but they lack ability to grow *up* very far, against gravity. By contrast, the ancestors of today's *ferns* developed a vascular system; with this, over the ensuing 100 million years, a variety of **seedless vascular plants** evolved, including huge seedless trees such as club mosses, some of them 40 meters (130 feet) tall.

Plants with Seeds: The Gymnosperms and Angiosperms

The Gymnosperms

Even as the seedless vascular plants were reaching their apex, a revolution in plants was well underway as some of them developed *seeds*. With the first seed plants, called **gymnosperms**, offspring no longer had to develop within a delicate, exposed plant on the forest floor; instead they developed within a tough seed. In modern gymnosperms this is best exemplified by the common pinecone, which contains many seeds. Gymnosperms also represent the final liberation of plants from their ocean past. Mosses and ferns had sperm that could make the journey to eggs only through *water*. Not much water was needed; a thin layer would do, as with the last of a morning's dew on a fern leaf. With seed plants, however, the water requirement is done away with entirely. Sperm move to the egg inside pollen grains, which can be carried by the *wind*. This has obvious ramifications for the dispersal of such plants. Several types of gymnosperms developed beginning about 350 Mya. Easily the most widespread of these today are the pine and fir trees, which cover huge stretches of the Northern Hemisphere.

The Last Plant Revolution So Far: The Angiosperms

About the time the dinosaurs were beginning their slide to extinction, the first flowering on Earth occurs, with the development of the flowering plants known as the **angiosperms**. Evolving about 140 Mya, the angiosperms eventually succeed the gymnosperms as the most dominant plants on Earth. Today there are about 700 gymnosperm species, but some 260,000 angiosperm species, with more being identified all the time. Angiosperms are not just more numerous than gymnosperms, they are vastly more diverse as well. They include not only magnolias and roses, but oak trees and cactus, wheat and rice, lima beans and sunflowers.

A Relationship between Plants and Animals

Angiosperms mark the beginnings of a remarkable relationship between plants and animals. Angiosperm pollen, which contains sperm cells, can be dispersed not only by the wind, but by bees, birds, bats—anything that brushes against the pollen-laden anthers of these plants. This is of such reproductive value to angiosperms that they evolved all manner of brightly colored petals and nectar to help ensure this contact. Pollen taken by an animal carrier to a *second* plant sets in motion the fertilization of its female gametes. In another innovation, angiosperms house their fertilized eggs in *fruit*, which is simply a mature ovary that encases both a seed and the embryo inside it. Thus plants began producing another form of food, which is obviously attractive to animals; but there is a benefit for plants as well. The fruit that, say, a bird eats is digested, but the tougher seeds within it are not. They are deposited by the bird and land on the ground in the middle of a patch of fertilizer commonly known as bird poop. Thus do we get seed dispersal through animal carriers.

With the rise of angiosperms, then, plants and animals enter into a mutually dependent relationship—one that has done much to direct the evolution of many species in both kingdoms.

19.7 Animals Follow Plants onto the Land

The movement of plants onto the land made it possible for animals to follow. With plants came food and shelter from the sun's rays. The first animals moved to land about 20 million years after the first plants (back before there were even gymnosperms). Recalling the division of the animal kingdom into various phyla, it was the phylum of arthropods that first moved to land; about 440 Mya, a creature similar to a modern centipede laid down the oldest set of terrestrial animal tracks that have been found so far. It makes sense that arthropods were the first animals to come onto land, because the hallmark of all of them is a tough external skeleton (an "exoskeleton"), which can prevent water loss and guard against the sun's rays.

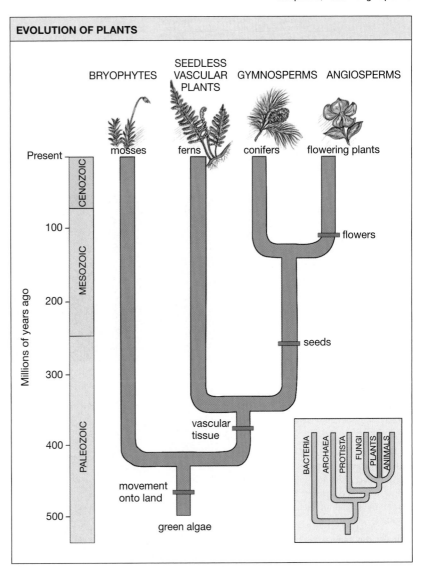

Figure 19.14
How Plants Evolved
The tree of life for bryophytes, seedless vascular plants, gymnosperms, and angiosperms.

What Is It?
See p. 385.

Insects onto Land

Other kinds of arthropods, the insects, came on land some 50 million years after the first arthropod immigrants. The insect pioneers were wingless, perhaps being something like modern silverfish. Insect flight would not develop for tens of millions of years more, but when it did insects had the skies to themselves for 100 million years. The more primitive flying insects are represented today by dragonflies, whose wings cannot be folded back on their body (**see Figure 19.15**). The compactness that came with foldable wings gave later insects the advantage of being able to get into small spaces. (Think of a common housefly.) In both winged and wingless forms, the insects took to the land with a vengeance. About 1.4 million species of living things have been identified on Earth to date, of which *half* are insects.

Vertebrates Move onto Land

Primitive Fish First

At about the time the first insects were coming out of the sea, the first vertebrates slithered onto land in the form of primitive "lobed-finned" fishes. The lobe-*fins* these creatures possessed are the precursors to the four *limbs* that are found in all tetrapod vertebrates, including ourselves (**see Figure 19.16**). It is a line of such tetrapods—the amphibians, the reptiles, and the mammals—that you'll follow for the rest of your walk through evolution. Here is the order of emergence among these forms. The lobe-finned fish give rise to amphibians, and early amphibians give rise to reptiles. Early reptiles then diverge into several lines, one of which leads to the dinosaurs and another of which leads to mammals. Dinosaurs eventually die out, of course, but before they do, a lineage that will live on—the birds—diverges from them (**see Figure 19.17**). Keep in mind the train-line metaphor employed earlier; in following these creatures, you are tracing only a couple of lines from among a huge number that exist.

Amphibians and Reptiles

In **Figure 19.18**, you can see an artist's rendering of *Ichthyostega*, the earliest known amphibian, which has a tail much like that of its lobe-finned ancestors. *Amphibian* literally means "double life," which is appropriate because in these creatures we can see the pull of both the watery world from which they came and the land onto which they moved. Modern-day frogs can serve as an example here. Many species of frog spend their adult lives as air-breathing land dwellers, but must return to the water to reproduce. Female frogs deposit eggs directly into the water, and the males then deposit sperm on top of them. The young that survive from this clutch go on to live as swimming tadpoles, complete with gills and a tail. After a time, however, both these features disappear, even as lungs start to develop, along with ears for hearing and legs for hopping. The tadpole has become a frog—a double life indeed.

Figure 19.15
The First to Take Flight
Dragonflies, such as this dew-covered individual, are descendants of the insects that first developed the ability to fly. Flying insects that evolved later could fold their wings back, as with today's housefly.

Figure 19.16
Vertebrates onto Land

a Our ray-finned fish ancestors lacked bones in their fins.

b The lobe-finned fish that evolved from ray-finned fish had small bones in their fins, that enabled them to pull out of the water and support their weight on tidal mudflats and sandbars.

c Finally, the early amphibians lost fin rays altogether, and had only bones in their limbs.

a RAY-FINNED FISH   **b** LOBE-FINNED FISH   **c** AMPHIBIAN

A Critical Reptilian Innovation: The Amniotic Egg

In the transition from amphibian to *reptile*, there is a severing of the amphibian ties to the water through a new kind of protection for the unborn. Amphibian eggs require a watery environment; lacking any sort of outer shell, they dry out if taken out of the water, thus killing the embryo inside. The early reptiles solved this problem with the remarkable adaptation of the **amniotic egg**: an egg that has not only a hard outer casing, but an inner "padding" in the form of egg "whites" and a series of membranes around the growing embryo. These membranes help supply nutrients and get rid of waste for the embryonic reptile. With this hardy egg, the tie to the water had been broken; reptiles could move *inland*.

Amphibians were the sole terrestrial vertebrates for about 30 million years, beginning about 350 Mya. We think of these creatures now in terms of small frogs and salamanders, but the amphibians who roamed the Earth in the Pennsylvanian period were often the size of modern-day pigs or crocodiles. The carnivores among them probably fed on small early reptiles, but reptiles ultimately became the hunters rather than the hunted. Thriving everywhere from water to dry highlands, they grew in numbers and diversity.

One line of reptiles evolved, about 220 Mya, into the most fearsome creatures ever to walk the Earth—the dinosaurs (**see Figure 19.19** on page 384). For 155 million years after their appearance, no other line of creatures dared challenge them for food or habitat. Not all dinosaurs were huge and many did not eat meat. Yet even a leaf-eater such as *Apatosaurus* must have commanded a good deal of respect, because it measured more than 26 meters or 85 feet from head to tail and weighed more than 30 tons. By comparison, a modern African elephant may weigh about 7 tons.

From Reptiles to Mammals

Scurrying about in the underbrush when the dinosaurs reigned supreme were several species of insect-eating animals, none bigger than a rat. These animals fed their young on milk derived from special female mammary glands, however, and they probably possessed fur coats over their skins—an important feature for animals restricted to feeding at night, when the dinosaurs were less active. These were the mammals, the first of whom appeared at about the same time as the first dinosaurs. In an example of being "born at the wrong time," these small mammals then lived in the shadow of the reptilian behemoths for more than 100 million years. Toward the end of the Cretaceous period, they began to radiate out into more environments and consequently evolved into more forms.

Figure 19.17
How Vertebrates Evolved
A tree of life for lobe-finned fish, amphibians, reptiles, mammals, and birds.

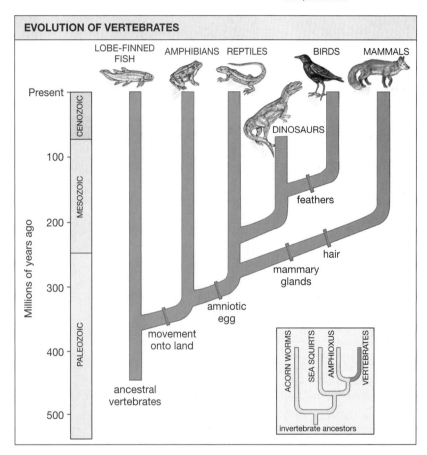

Figure 19.18
Transitional Animal
An artist's conception of *Ichthyostega*, the earliest known amphibian, seen at upper left climbing on the log and at center. Sometimes thought of as a "four-legged fish," *Ichthyostega* spent most of its time in water and probably used its back legs mostly for paddling. All land creatures with backbones, including human beings, ultimately trace their ancestry to tetrapod fishes such as *Ichthyostega*.

Figure 19.19
How Did *T. rex* Walk?
Previously, scientists thought *Tyrannosaurus rex* walked in the upright posture seen on the left. More recent research has persuaded scientists, however, that the "tyrant reptile" walked with the plane of its body parallel to the ground, as seen in the skeleton on the right. When standing still it may have sometimes raised itself into the upright position to look around.

b

a

Figure 19.20
Primate Characteristics
These modern-day lemurs in Madagascar exhibit several characteristics common to most primates: Opposable digits that enable grasping, front-facing eyes that allow binocular vision, and a tree-dwelling existence.

Figure 19.21
Several Types of Primates

a An animal called a loris from India (*Loris tardigradus*), is representative of an early lineage of primates, called prosimians, that also includes lemurs.

b A wooly spider monkey (*Brachyteles arachnoides*) from the Atlantic rainforest in Brazil. Note the tail, which is capable of grasping objects. Only New World monkeys have such tails.

c A lowland silverback gorilla (*Gorilla gorilla*), native to Africa.

At that point fate intervened: The asteroid that helped bring the reign of the dinosaurs to a crashing close 65 Mya also brought the mammals out of hiding. They radiated into niches far and wide, becoming the dominant form of large land animal, a status they retain to this day. Some of their members even returned to the sea, as with whales and seals. An obvious question here is: Why did the mammals survive the Cretaceous Extinction while the dinosaurs died out? Alas, we have no clear answers. Pure size may have had something to do with it; small creatures seem to survive extinction events better than large ones.

The Evolution of Mammals

So, how did mammals evolve from reptiles in the first place? As usual, the transition was a gradual one. Mammal-like reptiles had feet placed more nearly under their bodies (like mammals) as opposed to splayed out at an-

a

b

c

gles (like reptiles), and they may have developed some body hair. The transition also entailed something you've seen before: more protection for the unborn (recall that reptiles developed the amniotic egg). With the most advanced mammals, the young began developing entirely *inside* the mother before birth, sustained by the placental system in the womb.

The Primate Mammals

About five million years after the end of the dinosaurs, a species of placental mammals gave rise to the order of mammals called *primates*. If you look at **Figure 19.20**, you can see, in the lemur of Madagascar, a modern-day descendant of these earliest primates. Three things characteristic of most primates are apparent in this picture: large, front-facing eyes that allow for the binocular vision (which enhances depth perception); limbs that have an opposable first digit, like our thumb (which makes grasping possible); and a tree-dwelling existence. There are about 230 species of primates living today (**see Figure 19.21**), which is a small portion of the 4,600 species of mammals. It's a tiny number of species indeed compared to, say, the 60,000 species of mollusks, or the 750,000 species of insects. Yet because of one primate species, *Homo sapiens*, the primates have had an effect on living things all out of proportion to their numbers. **Figure 19.22** sets forth a probable primate family tree.

19.8 The Evolution of Human Beings

Figure 19.23 (page 386) could be thought of as a kind of "close-up" of the last fork in the primate tree pictured in Figure 19.21. Starting from a common primate ancestor, one branch of this fork gave rise to chimpanzees, while the other is the line that leads to human beings. The human line then branches out in a fairly complex way, with one tip of it getting us to modern human beings, designated *Homo sapiens*. Drawn up as little as 3 years ago, such a tree would have looked very different from this one, and one drawn up three years from now will likely have more changes yet. Paleoanthropology—the study of humans,

based on fossils—is a field whose conclusions are still very much in flux.

All of the species pictured in Figure 19.23 are referred to as *hominids*, which is to say, members of the family of man-like creatures (*Hominidae*). You can see however, that our own genus, *Homo*, is but one of four represented. The genus *Ardipithecus*, at the bottom of the tree, is represented by only one species, *Ardipithecus ramidus*, whose 4.4 million-year-old fossils were found in Ethiopia in 1994. At the moment there is some controversy over whether *A. ramidus* deserves the "offshoot" status given in the figure or instead should lie directly between the common primate ancestor and the next genus up, *Australopithecus*. Whatever the case, it is a form that lies very close to the common primate ancestor. *Australopithecus* is the genus that indisputably gave rise to the human genus *Homo* and to the *Paranthropus* genus, now extinct along with all the members of the *Australopithecus* genus.

What Is It?
Answer: Moth wing scales (*Urania ripheus*) magnified 53 times.

**Figure 19.22
How Primates Evolved**

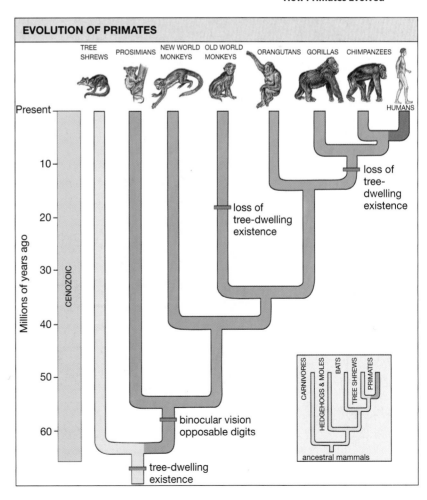

With the possible exception of *A. ramidus*, all the creatures in this tree walked upright on two legs—they were all "bipedal"—as opposed to our closest relatives, the chimpanzees and gorillas, who are "knuckle walkers." You can get some sense of how these various creatures differed by looking at the artist's renderings of several of the hominids in **Figure 19.24**. Apart from the obvious, what also changes with movement up the tree is brain size: The cranial volume of *Australopithecus afarensis* is about 450 cubic centimeters—about that of a chimpanzee—while that of modern *H. sapiens* is about 1,400 cubic centimeters. We can get a good idea of what evolution had produced with *A. afarensis* by noting that scientists have referred to it as a "bipedal chimpanzee." It may be obvious that *A. afarensis* is a pivotal species in our lineage. Above it, we get the split that leads to the larger *Paranthropus* genus on the one hand and the smaller *Homo* genus on the other. Easily the most famous *A. afarensis* is "Lucy," whose skeleton was found in Ethiopia in 1974 by Donald Johanson and

his colleagues. The remains of the 3-foot-7-inch Lucy have been dated to 3.18 Mya.

Human Beings Emerged in Africa

We have a good idea where it is that the hominid line originated: In Africa, which is where the whole human tree is rooted. In South Africa and in the Great Rift Valley that runs through much of eastern Africa, scientists have unearthed the most important fossils in human evolution. The part of Africa that includes Kenya, Tanzania and Ethiopia is assumed to be the cradle of the human species (**see Figure 19.25**).

The Descent from the Trees

As noted earlier, nearly all primates are tree dwelling (or "arboreal"). At some point, our hominid ancestors clearly abandoned the arboreal lifestyle of the typical primate, coming down to dwell on the dry savannas of East Africa. At one time it was thought that the early hominids were *forced* to leave the trees by a climate change that dried up our forest homes. The problem with this hypothesis, modern research has shown, is that there was no environmental change at the time the hominids were appearing; the vegetation of East Africa remained fairly stable during that period. What seems likely is that bipedalism predated our descent from the trees, and that our ancestors lived for a time in both worlds. Lucy was bipedal and yet had the long arms that would have made her a pretty good tree climber.

Homo ergaster: A Form More Like Ours

Change to physical forms that are more like ours comes at several steps in the hominid line, but the most dramatic change by far comes with the rise of *Homo ergaster*, whose best-preserved remains, in the form of an adolescent called "Turkana Boy," date from about 1.6 Mya. Experts estimate that, had he grown to maturity, Turkana Boy would have reached a height of 2 meters (6 feet). He had a much more modern face and limbs typical of those humans who dwell in arid climates. We were clearly down from the trees by this point.

The Neanderthals

You may have noticed that there are two species in the later *Homo genus*—*H. erectus* and *H. neanderthalensis*—that are evolution-

**Figure 19.23
How Humans Evolved**

a

b

c

Figure 19.24
Extinct Members of the Hominid Family

a A reconstruction of an *Australopithecus afarensis* couple.

b An artist's conception of *Homo erectus*.

c An artist's conception of a male Neanderthal.

ary dead ends. Of these, *H. neanderthalensis*, better known as "Neanderthal Man," is the more famous. It first appeared in Europe and Asia no more than 200,000 years ago and survived until about 30,000 years ago in Western Europe. Modern humans may have played a role in the Neanderthals' extinction by, for example, outcompeting them, but this is a matter of considerable debate. An even more intense debate existed for years about whether the Neanderthals were members of our evolutionary line. This question seems to have been settled in 1997, however, when scientists succeeded in sequencing DNA recovered from a Neanderthal fossil. This work showed that Neanderthals evolved separately from *Homo sapiens*. The likely scenario is what you see in the human family tree: *H. neanderthalensis* evolved from *H. heidelbergensis*, perhaps in Europe from a group of heidelbergians who migrated there from Africa. Meanwhile, our own species arose from *H. heidelbergensis* in Africa in a separate line.

Homo sapiens

The rise of fully modern humans is a relatively recent event, occurring only a little over 100,000 years ago. The earliest *H. sapiens* fossils that exist were found in South Africa and date from 117,000 years ago. From our African origins we then spread out across the world. Whatever we may have in common with other species, it is for certain that we have characteristics that make us unique: our unrivaled brain size, our superior tool-making ability, and our capacity for language among them. These qualities evolved over much time, as you've seen. As

this review of the history of life on Earth comes to a close, it's perhaps worth noting that one of the things evolution has wrought in us is a species capable not only of painting and writing and composing, but of comprehending the very thing that gave rise to us—evolution.

On to the Diversity of Life

In reviewing the history of life in this chapter, you necessarily looked, in some small way, at almost all of Earth's major life-forms—bacteria, animals, plants, and so forth. In the chapter coming up, you'll take a more detailed look at each of these life-forms. This is a tour that will make plain the astounding diversity that exists in the living world.

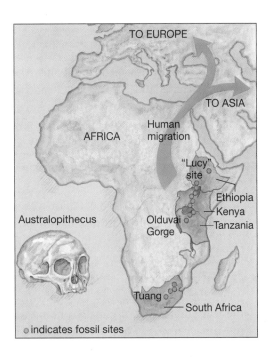

Figure 19.25
Africa Is the Cradle of the Humans
The oldest human fossils have been found in East and South Africa. Most scientists believe that humans originated in these regions, then migrated north and east into Europe and Asia, and eventually into the Americas.

Chapter Review

Key Terms

Summary

19.1 The Geologic Timescale: Life Marks Earth's Ages

- Geological time is divided into eras, shorter periods, and shorter-yet epochs, which are defined by the kinds of life-forms that have existed in them.

- What is notable in evolution hinges on values. The evolutionary process that led to human beings is similar in many ways to evolution as it occurred in all other lineages.

- Every living thing on Earth belongs to one of three domains of life: Archaea, Bacteria, or Eukarya. Both bacteria and the archaea are single-celled, microscopic life-forms; the domain Eukarya is composed of four kingdoms: plants, animals, fungi, and protists.

19.2 Tracing the History of Life on Earth: How Did Life Begin?

- Life on Earth began between 3.5 and 4 billion years ago with the origin of self-replicating molecules. There currently is a debate regarding the conditions and sequences of events that led to the formation of these molecules. One hypothesis that has a good deal of support among researchers is that life began in a "pre-biotic soup" of hot-water systems—in the deep-sea vents or the kind of hot-spring pools that exist in Yellowstone Park today.

19.3 The Tree of Life

- Life today can be traced to a universal ancestor that gave rise to the domains Archaea, Bacteria, and Eukarya. The earliest kingdom in Eukarya was the protist kingdom, whose organisms gave rise to the animal, plant, and fungi kingdoms.

19.4 A Long First Period: The Precambrian

- The oldest undisputed fossils we have are of bacteria dating from 3.4 billion years ago. During roughly half the time life has existed on Earth, it consisted of nothing but bacteria and archaea.

- Oxygen came to exist in the Earth's atmosphere through the activity of photosynthesizing organisms, originally photosynthesizing bacteria. Photosynthesis meant

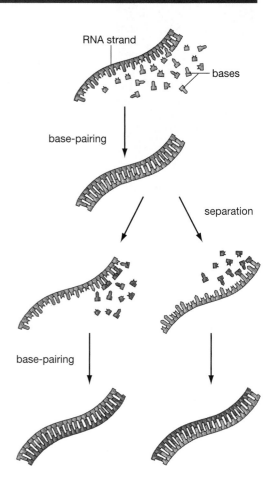

an abundance of food on the Earth and a concentration of atmospheric oxygen high enough that it set a strict condition for all life: Adapt to oxygen, stay separate from it, or perish.

19.5 The Cambrian Explosion: A Real Milestone or the Appearance of One?

- The fossil record indicates a tremendous, rapid expansion in the number of animal forms in a "Cambrian Explosion" that began about 544 Mya, but this notion is now being challenged.

19.6 The Movement onto the Land: Plants First

- Plants, which evolved from green algae, made a gradual transition to land about 460 Mya, followed by animals. Major transitions in plant life came with the development of seed plants—first the gymnosperms (represented by today's conifers) and then the angiosperms (flowers, food crops, many tree varieties).

a RAY-FINNED FISH   b LOBE-FINNED FISH   c AMPHIBIAN

19.7 Animals Follow Plants onto the Land

- Four-limbed vertebrates (tetrapods) moved onto land about 400 Mya in the form of lobe-finned fishes, which gave rise to amphibians (represented by today's frogs). Reptiles later branched off from amphibians, and mammals branched off from reptiles.

- About 60 Mya, there arose an order of mammals called the *primates*, characterized by large, front-facing eyes, limbs with an opposable first digit (thumbs in today's human beings) and a tree-dwelling existence.

19.8 The Evolution of Human Beings

- About 5 Mya, a common ancestor gave rise to both chimpanzees and to the evolutionary line that led to human beings. Africa is the birthplace of the human species, *Homo sapiens*, which emerged in modern form a little over 100,000 years ago.

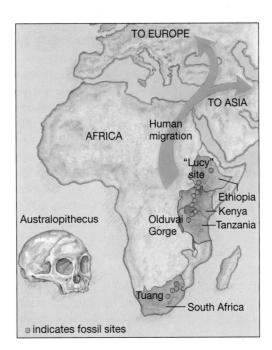

Understanding the Basics

Multiple-Choice Questions

1. The following organisms or events appeared at different times in the evolution of life. Put them into the correct temporal order, starting with the first to appear:
   a. first insects
   b. first dinosaurs
   c. first bacterial fossils
   d. first animal fossils
   e. first seed-bearing plants
   f. first flowering plants
   g. free oxygen accumulation in the atmosphere
   h. first land plants
   i. first birds

2. Periodic major extinctions (select all that apply)
   a. can eliminate the vast majority of species present at that time on Earth.
   b. are usually caused by meteor impact
   c. open up new evolutionary opportunities
   d. can lead to major evolutionary transitions in the dominant forms of life on Earth
   e. have occurred only twice in Earth's history

3. The first life on Earth may have been (select all that apply)
   a. eukaryotic cells
   b. ribozyme-like cells
   c. self-replicating DNA
   d. formed on land
   e. formed at extreme temperatures

4. Which of the following characteristics of animals and/or plants are adaptations to life on land? (select all that apply)
   a. multicellularity
   b. pollen and seeds
   c. amniotic egg
   d. muscles
   e. internal fertilization

5. Which of the following animals are tetrapods?
   a. birds
   b. frogs
   c. fish
   d. pigs
   e. insects

6. True or False: Dinosaurs were often hunted by humans.

7. Photosynthesis was a huge innovation in the history of life because (select all that apply)
   a. Free oxygen permitted more rapid movement by anaerobic bacteria.
   b. Metabolism using oxygen provides more energy.
   c. Photosynthesis resulted in the production of more food than had been previously available.

   d. It resulted in blockage of UV light, increasing habitability of the land for terrestrial organisms.
   e. Metabolism using oxygen uses more energy.

8. Mitochondria and chloroplasts (select all that apply)
   a. are thought to have arisen as symbiotes
   b. are very similar to bacteria
   c. are very similar to algae
   d. evolved in eukaryotes
   e. evolved in Archaea

9. Which of the following organisms have an amniotic egg? (select all that apply)
   a. grasshopper
   b. lizard
   c. salamander
   d. fish
   e. bird
   f. elephant
   g. turtle

10. The first lineage of terrestrial vertebrates was _____. The first lineage of vertebrates that was fully terrestrial was _____.
    a. lobe-finned fish, reptiles
    b. salamanders, frogs
    c. amphibians, reptiles
    d. amphibians, birds
    e. frogs, mammals

Brief Review

1. What are two differences between domain Bacteria and domain Eukarya?

2. What is the amniotic egg, and why was it a breakthrough for terrestrial animals? In which vertebrate class did the amniote egg evolve?

3. What are two characteristics of primates that are not found in other mammals? Why are they important for the evolution of hominids?

4. Which two groups are more closely related, humans and chimps or humans and orangutans?

5. Were Neanderthals the direct ancestors of humans?

6. When did life begin, and how much time passed between the origin of the first bacterial cells and the colonization of land? What was the world like during that time?

7. Describe three characteristics that evolved in mammals that had not been present in earlier organisms.

Applying Your Knowledge

1. Mitochondria and chloroplasts within eukaryotic cells evolved from bacteria that were ingested and took up residence in early eukaryotic cells. How did this partnership benefit each party?

2. What was the Cambrian Explosion, and why was it notable in the history of life? What could have caused it, and why hasn't a similar explosion happened again?

3. How did the evolution of angiosperms initiate an evolutionary interaction be- tween animals and plants? Discuss three important effects of plants on animals or animals on plants. Do you think that the Earth would have as many animal species if angiosperms had not evolved?

4. Describe two problems experienced by plants that colonized land. What types of "solutions" evolved in plants for each of these problems? What problems did ani- mals experience in colonizing land, and what types of solutions evolved in animal lineages?

Group Activity

The picture of human origins and evolution is constantly changing. At the library or on the Internet, look up some recent articles on human evolution. Summarize the findings that extend or even contradict what you have learned in this book, and compare your re- sults with those of your classmates. What types of evidence do scientists use to study human evolution? What are the current "big questions" in the field of human evolution?

20

Pond Dwellers, Log Eaters, and Fast Movers
The Diversity of Life

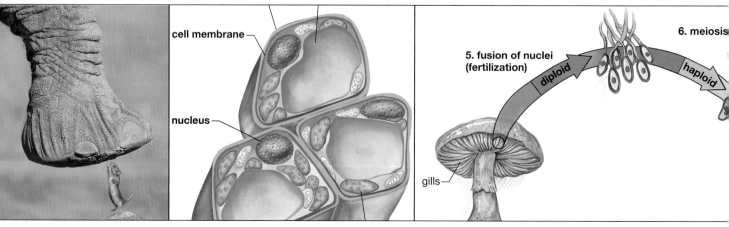

Creatures large and small.
(page 394)

An important type of cell.
(page 412)

Life cycle of a fungus, part 2.
(page 407)

The smallest living things

discovered to date are a type of bacteria that measure about two-hundredths of a micrometer in diameter—that is, two-hundredths of one-millionth of a meter in diameter. When we turn to the *biggest* living things, elephants or whales might come to mind, but it turns out that animals aren't even in the running for this title. At a maximum length of 27 meters (or about 89 feet), the blue whale is the largest animal that has ever existed, but it's small compared to California's coastal redwood trees, which might reach a height of 100 meters or about 330 feet. This is large indeed, but consider that a giant sea kelp once was found that was 274 meters in length, meaning it was about 900 feet long. It may be, though, that in terms of size, both kelp and redwoods have to take a back seat to a life-form normally thought of as quite small. In 1992, researchers reported finding a fungus growing in Washington State that runs underground through an area of 1,500 *acres* and that arguably is a single organism.

Now, how about the *deepest*-living things? The current record-holders are bacteria that were found in the early 1990s—in a combination of oil drilling and scientific ex-ploration—living almost 3 kilometers beneath the Earth's surface, which is about 1.9 miles down. Food and water are so scarce at such depths that these bacterial cells may live in a kind of suspended animation, dividing into "daughter" cells perhaps once a year, or even once a *century*. Turning to the highest living things, 12 species of bacteria have been found to be living in the Himalayas 8,300 meters or 27,000 feet above sea level, which is just a couple thousand feet lower than the peak of Mount Everest. A species of chickweed survives in the Himalayas at over 20,000 feet.

All kinds of life-forms—clams, tube worms, archaea—live kilometers beneath the surface of the oceans, near the "hot-water vents" that spew out water and minerals from the Earth's interior onto the ocean floor. It is the microbes called archaea of these deep-sea vents that set the pace for life in a hot environment. An average, midday temperature in the Sahara Desert would be so *cold* it would end the reproduction of the archaean *Pyrolobus fumarii*, which reproduces within the hot-water vent walls at temperatures of up to 113° Celsius, meaning that this creature lives in boiling water.

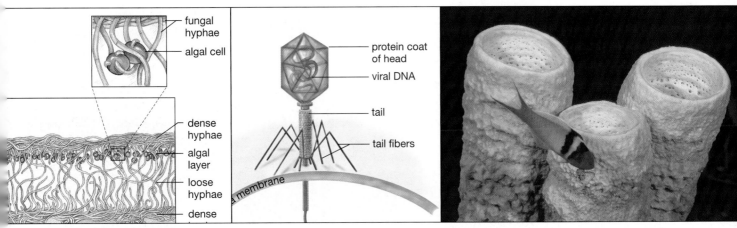

A mutually beneficial arrangement.
(page 409)

Invader of living things.
(page 396)

An unfamiliar animal—and not the fish.
(page 409)

Finally, what about the oldest living things? Humans occasionally live past the age of 120, but this is an eyeblink compared to the lives of some plants. There is a Sequoia tree in California that was growing 3,500 years ago, when the pharaohs ruled Egypt. But even it is a youngster compared to the oldest living thing that is clearly a single organism, a bristlecone pine tree (*Pinus longaeva*) living in the dry slopes of Eastern California that has been dated at 4,900 years.

Looking at life this way—in its biggest or deepest or oldest extremes—is one way to get at the incredible diversity of it. But it does not begin to do justice to life's variety, because life is diverse in so many diverse *ways*. Honeybees do a "dance" to let their hivemates know the location of a food supply; pufferfish, when threatened, swallow water very fast and inflate themselves into a spiky round ball that scares away predators. Pacific salmon live years in the ocean only to make one arduous, upstream journey to spawn in the fresh waters they were born in, after

which they die. Look closely at almost any creature and you are likely to find something only slightly less dramatic (**see Figure 20.1**).

The purpose of this chapter is to provide some sense of how this incredible diversity is divided up—to introduce you to life's large-scale categories. It was noted in Chapter 19 that all living things can be placed into one of three "domains"—Archaea and Bacteria, whose members are microscopic; and Eukarya, whose members range from the microscopic to the gigantic. Domain Eukarya is in turn is divided into four "kingdoms," which are plants, animals, fungi, and another called protists. (See "The Unsolvable Taxonomy Problem, facing page, for a discussion of problems inherent in any such categorization.) In this chapter, you will take a brief look at each of these categories of living things, with a detailed review of plants and animals then coming in succeeding chapters. To begin your tour of the living world, however, let us start with a category of reproducing entities that most biologists would say lie

Figure 20.1
An Amazing Diversity in the Living World

a A mouse beneath the foot of an elephant.

b Aquatic snails, moving across kelp.

c A rain forest fungus, growing from a fallen log.

d A tiny, aquatic animal, called a water-flea, swimming near a form of green algae called *Volvox*.

a

b

c

d

The Unsolvable Taxonomy Problem

This book uses a classification system that divides life first into three domains—Bacteria, Archaea, and Eukarya—and then into four kingdoms within the eukaryotic domain. But there are other ways of classifying all the creatures of the living world.

If you look at **Figure 1**, you can see both the three-domain structure (a) and another that employs six kingdoms (b). Note that this second system has no domains at all; instead, its highest taxa are the kingdoms Eubacteria, Archaebacteria, Protista, Plantae, Fungi, and Animalia, with Eubacteria corresponding to the domain Bacteria and Archaebacteria corresponding to the domain Archaea. It may be apparent that the Archaea and Bacteria have a lesser status in this second system. Rather than comprising two of life's three domains, they each comprise only a kingdom in a system that has six. Beyond this, the Archaea are thought of as bacteria in this latter system, whereas in the three-domain system they are a separate kind of organism.

Which system is "right?" Arguments can be made on both sides. No votes are taken on this issue, and there are highly respected biologists in both camps. Some specialists feel that the three-domain system best defines life's most fundamental distinctions, while other experts believe that the six kingdoms do a better job.

Another, more long-standing problem in taxonomy has to do with the catch-all kingdom Protista, whose members, mostly single-celled, are defined by what they are *not*—not plants, not animals, not fungi—which is a curious kind of definition. We would prefer to define a grouping by what its members *are*, which is to say what characteristics its members share, but the Protista are so diverse that no one can come up with such a definition. By some definitions, green algae should be in the kingdom Plantae, but in this text all algae are included in Protista. Here there clearly is no "right" answer. The fundamental problem is that Earth's creatures are so diverse that it's impossible for all of them to fit neatly into categories devised by human beings. We can have sensible taxonomic systems, but we can't have perfect ones.

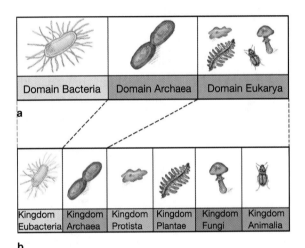

Figure 1
Different Ways of Categorizing Life

a The three-domain system.

b The six-kingdom system. Both systems take all living things into account, but group them differently.

---

just outside of life, though they have a great effect on living things.

20.1 Viruses: Making a Living by Hijacking Cells

If a living thing is too small to be seen and can cause disease, most people would put it in the vague category known as "germs." But there is a fundamental distinction to be made between *two* kinds of infectious organisms. On the one hand there are bacteria, which may be very small but nevertheless are *cells*, complete with a protein-producing apparatus, a mechanism for extracting energy from the environment, a means of getting rid of waste—in short, complete with everything it takes to be a self-contained living thing. On the other hand there are viruses, which by themselves possess none of these features. Indeed, viruses can be likened to a thief who arrives at a factory he intends to rob possessing only two things: the tools to get inside and some software that will make the factory turn out items that he can use.

The factory is a living cell, and the software that the virus brings is its DNA or RNA, which it may inject into the "host" cell. Once inside a cell, most viruses first destroy the host's ability to use its own genetic machinery; then the viruses get to work expressing *their* genes with this machinery, meaning they turn out *viral* proteins that will facilitate the production of multiple copies of the virus. Some viruses accomplish this feat by splicing their genetic material into the host's DNA. Ultimately, hundreds or thousands of clones of the original invading virus are produced, which often go on to **lyse** or burst forth from the infected cell, killing it in the process. Once outside, random chemical motion brings these viral particles into contact with yet more cells, and the whole process—the so-called **lytic cycle**—starts over again.

If you look at **Figure 20.2**, you can see an illustration and micrograph of one particular virus, known as T4, which preys on bacteria. (Viruses come in many shapes; it just so happens that this one looks like an invading spacecraft.) T4 attaches to its host via the tail fibers on its six spikes; its DNA is contained in its "head," which is sheathed in a protein coat. Its tail can contract, acting like a hypodermic needle to inject its DNA into the host cell. Once inside, this viral DNA uses the cell's own machinery (its enzymes) to turn out T4 virus clones in a multistep process. The viral DNA is replicated, even as new viral protein coats are being put together.

Then these components are assembled to create the finished product: 100 or so clones of the virus (**see Figure 20.3**). Finally, a lytic enzyme is produced that causes the host cell to lyse, thus releasing these clones to the environment outside the cell. From the time T4 enters the cell to the time its clones burst forth, perhaps 20 minutes will have elapsed.

The life cycles of some viruses can take a course that doesn't involve lysis; in the so-called **lysogenic cycle**, the viral DNA integrates with the host DNA and then stays there, replicating each time the host cell divides. The virus ends this arrangement, however, if its host cell is threatened by, say, starvation; then the virus "bails out," removing itself from the host genome and entering into the lytic cycle, its clones eventually bursting out of the cell.

Given how formidable viruses are, it is interesting to contemplate how little there is to them. They are so small that thousands of them can typically fit within a bacterial cell. They are so genetically simple that T4 has a DNA complement that is only about 40,000 base pairs in length, compared to the *E. coli* bacterium's genome of 4 million base pairs, and the human genome of 3 billion base pairs. Some viruses amount to nothing *but* their genome and the protein coat that surrounds it. There are even virus-like entities, called **viroids**, that lack a protein coat; they are simply small strands of infectious RNA. There are more complex viruses, to be sure, but in no case is a virus complex enough to replicate by itself. Because they can carry on scarcely any of life's basic functions unaided, viruses aren't even classified as living things by most scientists.

The Trouble Viruses Cause

But what trouble from something that is not even alive! Viruses cause smallpox, chickenpox, measles, mumps, rabies, polio, herpes, rubella, and some forms of cancer and hepatitis, to say nothing of common colds and flus. Then of course there is AIDS. From this, and the review of the techniques of viruses, it may seem as though all of them would be unstoppable; yet we know that people get over colds and flus, that herpes doesn't kill us, and that polio has largely been conquered in the developed world. What puts the brakes on these cellular invaders? In human beings

Figure 20.2
One Variety of Virus

a The T4 virus looks like a spacecraft that has landed on the surface of a bacterium.

b A viral invader lands. An artificially colored T4 virus injects its DNA into an *E. coli* bacterium (in blue).

a

protein coat
of head

viral DNA

tail

tail fibers

plasma membrane

viral DNA being
injected into
bacterium

b

it is our immune system and the vaccines we have developed to prime this system's responses.

For the Defense: The Immune System and Vaccines

The human immune system provides two main lines of defense against viruses. One is the production of antibodies, which are proteins capable of recognizing virus particles

LIFE CYCLE OF A VIRUS

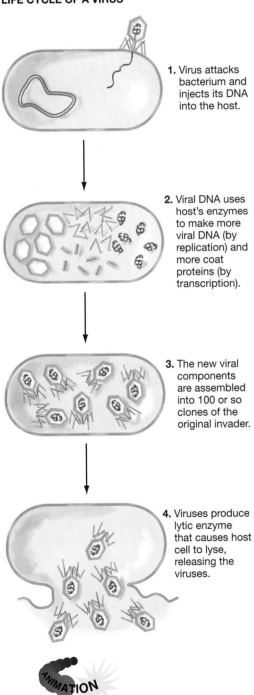

1. Virus attacks bacterium and injects its DNA into the host.

2. Viral DNA uses host's enzymes to make more viral DNA (by replication) and more coat proteins (by transcription).

3. The new viral components are assembled into 100 or so clones of the original invader.

4. Viruses produce lytic enzyme that causes host cell to lyse, releasing the viruses.

Figure 20.3
How Viruses Multiply
A lytic virus life cycle.

and binding to them, thus blocking their life cycle. The other is our so-called T-cell complex, which is capable of recognizing *cells* that have been infected with viruses and then destroying these cells. Critically, both these arms of the immune system have a kind of memory. The immune system's *initial* response to an infection by a given virus may be relatively slow, but once the virus is cleared from the body, the immune system retains clones of the antibodies and T cells it developed to fight the infection. This means that the *next* time this virus invades, the body can mount a rapid response, clearing the virus before harmful effects set in. This is why almost nobody gets chickenpox twice. The problem with this system is that some viruses do permanent damage on their first invasion (as with polio destroying the nerve cells that allow us to walk), while other viruses can remain hidden from the immune system, only to emerge later (as with herpes).

The theory behind vaccinations is that it is possible to elicit an initial immune response to a given virus without giving the disease itself to the person being inoculated (**see Figure 20.4**). How? By, for example, injecting *killed* virus particles—whose shape alone elicits the immune response—or *attenuated* viruses, meaning viruses that haven't been killed but that have been weakened so that they no longer can cause disease. The Salk polio vaccine is an example of killed virus, and the Sabin polio vaccine an example of attenuated virus.

How Did Viruses Originate?

All the organisms you looked at in Chapter 19 fit into a grand evolutionary framework: Archaea and bacteria were the first to appear on Earth, followed by the protists, who gave rise to animals, plants, and fungi. So where do the viruses fit in? No one knows. There are several hypotheses regarding the evolution of viruses, but all of them are weakened by a critical hole in the evidence, which is that there are no viral fossils. Viruses have no "hard parts" to make impressions, which severely hampers our ability to know when any virus first appeared. The most widely accepted theory is that viruses represent "renegade" DNA or RNA. We know that genomes often contain stretches of DNA that can move from place to place within the genome. At a minimum these "mobile elements"

Figure 20.4
Protection for the Future
A four-year-old receives a vaccination.

always encode an enzyme that can snip them out of one location in a genome and patch them into another. Imagine, then, such an element breaking away from its genome entirely, and taking with it (or acquiring) the genes necessary to fabricate a protein coat. In this way, viruses may have come into being. Now let's move on to the first domain of life.

20.2 Domain Bacteria: Masters of Every Environment

Domain Bacteria is made up entirely of the microbes known as *bacteria*—single-celled organisms that, like viruses, are "germs" as people commonly understand that term. It is true that bacteria are responsible for lots of diseases: tuberculosis, syphilis, gonorrhea, cholera, tetanus, and leprosy, not to mention all manner of food poisonings and blood-borne infections. The bacterially caused bubonic plague, still with us today, wiped out a *third* of Europe's population in just 4 years in the fourteenth century. Yet in the larger picture, only a very small proportion of bacteria cause human disease. The majority either don't affect us at all or are actually beneficial—not only to us but to all of the living world. Indeed, it is hard to imagine how life on Earth could continue without bacteria.

Intimate Strangers: Humans and Bacteria

To look first at the human relationship with bacteria, from the time we travel down our

mother's birth canal we are bathed in these organisms, both outside and inside our bodies. Outside, we are covered in bacteria from head to toe, with heavy bacterial concentrations in the armpits and scalp. Inside, bacteria exist in the entire digestive tract, from mouth to anus. In the mouth alone, there are a presumed 600 *species* of bacteria, with the number of individual bacteria in a person's mouth probably exceeding the number of people who have ever existed. Further down, about half the contents of our colon are bacteria as are perhaps a quarter of our feces by weight. It may be discomforting to think about having been colonized in this way, but consider that some of the intestinal bacteria are producing sugars and vitamins that our bodies can use. Moreover, by making a living in various digestive environments, these bacteria are effectively occupying almost all the available bacterial niches, thus leaving few places for harmful bacteria to get started.

Bacterial Roles in Nature

In the larger living world, bacteria fill lots of roles. The element nitrogen is a component of all living things, being found in all the primary amino acids that make up proteins. But animals get their nitrogen only by eating plants, or by eating *animals* that have eaten plants. And how do plants get nitrogen? To a great extent from bacteria, which alone among organisms can take the *gas* nitrogen from the atmosphere, metabolize it, and produce nitrogen-containing compounds that can be taken up by plants.

Beyond this, it is not by magic that dead trees, the bones of dead animals, or for that matter orange peels, are broken down and then recycled back into the Earth. The Earth's refuse is broken down by *organisms*, and some of the most important of these "decomposers" are bacteria. A spectacular meeting of waste and bacteria can be found in the sewage treatment plants of most cities, where a kind of production line of different kinds of bacteria is employed in treating solid and liquid waste (**see Figure 20.5**).

These are but a few examples of how bacteria are part of the fabric of life. We occasionally can't live with these tiny organisms but we certainly can't live without them, which is a good thing because we have no choice. Bacteria exist in great numbers al-

Figure 20.5
Bacterial Purification
Waste water coming to this treatment plant goes first to the airless storage tanks in the background. There, a type of bacteria that thrive in an oxygen-less environment break the waste into gases and a humus-like material. The product then is transported to the ponds in the foreground where another type of bacteria break down the solid material in it, yielding carbon dioxide. The bacteria and solids are then filtered out of the water before it is discharged into a river or the sea.

most anywhere there is food and water, and they can withstand all kinds of hostile assaults on their environments. As long as life exists on Earth, bacteria will be among the living.

Common Features of Bacteria

If you look at **Figure 20.6**, you can see the three most common shapes that bacteria take—spherical, rod-shaped, and spiral-shaped—and you can see a comparison of some bacterial sizes with those of the viruses you just looked at and with the protists you'll be looking at soon. Spherical bacteria are known as **cocci** (singular, *coccus*), which you may have heard of in connection with the streptococcus bacterium that causes strep throat. Rod-shaped bacteria are called **bacilli** (singular, *bacillus*). Spiral-shaped bacteria include the **spirochetes**, an example of which is the *Treponema pallidum* bacterium, which causes syphilis (see Figure 20.6).

Nobody knows how many species of bacteria come in these shapes and others. Using a DNA base-pairing technique, two Norwegian researchers looked at a single gram (about 1/28 ounce) of forest soil from their country and estimated that there were between 4,000 and 5,000 bacterial species living in it, a number that doubled when a gram of shallow marine sediment was tested. Given such abundance, it is useless to even speculate about how many bacterial species there may be in total.

So what characteristics do bacteria share? Here's a list.

- *No cell nucleus.* All bacteria are prokaryotes, meaning organisms whose cells lack a nucleus. As noted in Chapter 4, the nucleus is an internal compartment of a cell, bound by a membrane, that contains the cell's primary complement of DNA. Organisms whose cells have a nucleus are called *eukaryotes*, a category that includes all the organisms on Earth *except* bacteria and archaea.

- *No other organelles and no cytoskeleton.* A nucleus is an example of an **organelle**, or membrane-bound compartment within a cell. You may remember from Chapter 4 that eukaryotic cells also have such organelles as mitochondria, which act as energy transfer centers, and lysosomes, which act as disposal and recycling centers. Bacteria do not have such organelles.

- *Single chromosome.* The DNA of bacteria exists in a single, circular chromosome, which makes bacteria **haploid** organisms. By contrast, the chromosomes of eukaryotes generally exist in *pairs* for at least part of the eukaryote's life cycle, making eukaryotes **diploid** organisms during this time.

- *Asexual reproduction.* Bacteria reproduce by a simple cell-splitting or **binary fission**, with each pair of "daughter" cells being an exact replica of the parental cell.

cocci (spherical)

spirochetes (spiral-shaped)

bacilli (rod-shaped)

bacteria *(E. coli)* 2 μm long

protist *(Paramecium)* 75 μm long

viral clones (T4) 0.2 μm long

Figure 20.6
Types of Bacteria

a Both spherical cocci bacteria and rod-shaped bacilli bacteria can be seen in this micrograph, which shows *Streptococcus* bacteria, artificially colored orange, and *E. coli* bacteria, colored pink. These bacteria can cause infections, but they are also inhabitants of healthy human digestive and respiratory systems.

b Several of the spirochete bacterium *Treponema pallidum*, which causes syphilis, shown attaching to cell membranes in the testes, 22 hours after infection.

c Size comparison of E. coli bacteria on the left, with T4 virus clones on the right and a Paramecium protist in the middle.

Modes of Nutrition: How Organisms Get What They Need to Survive

All living things need energy and nutrients. The question is, how do they get them? What is their "nutritional mode," as biologists would put it?

The most fundamental distinction in nutritional mode separates groups known as autotrophs from those called heterotrophs. **Autotroph** means "self-feeding." Autotrophs can manufacture their own food, defined as some form of organic (carbon-containing) molecule that can be broken down to yield energy. All organisms that are not autotrophs are **heterotrophs**, meaning "other feeders" that cannot manufacture their own food, but must get it from elsewhere, as animals do.

The idea of heterotrophs is not strange to us—it's what we are, after all—but the notion of autotrophs may be a little more exotic. Those who have gone through Chapter 8 on photosynthesis will recall, however, that almost all *plants* are autotrophs in that they manufacture their own food, initially a sugar. To do this, they need a carbon source, because carbon is the "backbone" of the organic molecules we know as food, and they need an *energy* source that can drive the complex process of photosynthesis. The carbon source for plants turns out to be carbon dioxide (CO_2), obtained from the atmosphere, while the energy source is the rays of the sun (the "photo" of photosynthesis). Thus do we see exemplified a *two-part* requirement for nutrition that holds for all organisms: on the one hand a source of carbon and on the other an energy source.

Within this framework there are four nutritional modes, which you can review in **Table 1**. The two most important of these modes are **photoautotrophy**, which is the nutritional mode of plants, and **chemoheterotrophy**, in which organic materials (better known as food) act as both carbon supplier *and* energy source. This is another way of saying that the cereal you ate this morning supplied you, as a chemoheterotroph, with both carbon and high-energy electrons.

There is also **chemoautotrophy**, in which organisms—some bacteria and archaea—get their carbon from

- *Single-cell existence.* Bacteria exist in the main as single-celled organisms, though they can form colonies under some circumstances. Each bacterial cell, in other words, is a self-contained living thing.

Fighting Disease-Causing Bacteria with Antibiotics

In general, the medicines used to kill disease-causing bacteria are **antibiotics**, meaning a substance produced by one organism that is toxic to another. There are antiviral and antifungal medicines that can properly be called antibiotics, but in the main when we speak of antibiotics, we are talking about medicines that kill bacteria but that do not work against viruses. Two things are worth observing about antibiotics. First, scientists do not design them from scratch; instead, they are derived from compounds produced by living things. This makes sense because organisms such as fungi have been involved in a war of survival with bacteria for hundreds of millions of years, and they have accordingly evolved antibacterial defenses. The best-known antibiotic in the world, penicillin, is the product of a mold, which is a type of fungus. Second, any antibiotic that humans use must exploit the *differences* between human and bacterial cells; otherwise the antibiotic would kill human cells along with the bacterial cells. With respect to penicillin, the critical difference is that bacterial cells have a cell wall, while human cells do not. Penicillin blocks a bacterial enzyme that helps construct this wall; the result is that, after penicillin has been administered, bacterial cells burst as they are being constructed, during cell division.

Modes of Nutrition: Bacteria Do It All

By now, it may be apparent that bacteria are primitive organisms in many ways; they multiply by simple division, they have no organelles, they exist mostly as unspecialized individuals. Yet if the *diversity* of a grouping of organisms is thought of as an advanced characteristic, then bacteria are very advanced indeed compared to most of the other types of organisms you will be looking carbon dioxide, like plants, but power the production of food from this carbon by oxidizing (pulling electrons from) such inorganic materials as hydrogen sulfide and ferrous ions. A small number of bacteria and archaea practice **photoheterotrophy**, in which the sun supplies the energy but the carbon comes from surrounding organic material.

Nutritional Mode	Carbon Source	Energy Source	Practiced by
AUTOTROPHY			
Photoautotrophy	Carbon dioxide (CO_2)	The sun's rays	Almost all plants, some bacteria, and many protists
Chemoautotrophy	Carbon dioxide	Inorganic compounds such as hydrogen sulfide and ferrous iron	Some bacteria and archaea
HETEROTROPHY			
Photoheterotrophy	Organic material	The sun's rays	A few bacteria and archaea
Chemoheterotrophy	Organic material	Organic material	Almost all animals, all fungi, most bacteria, many protists, and a few plants

at. It is in the means by which they get their nutrition that the breadth of bacteria is most apparent; different kinds of bacteria utilize all the modes of nutrition known to living things, whereas humans, for example, utilize only one. You can read about these methods in "Modes of Nutrition" above. Now let's proceed to the second of life's three domains, Archaea.

20.3 Domain Archaea: From Marginal Player to Center Stage

As little as 10 years ago, any biology textbook you picked up would likely have referred to the archaea as archae*bacteria*, and most would have mentioned the archaea as a kind of marginal life-form, existing in only a few extreme habitats, such as hot springs or salty lakes. Today, the archaea are recognized by many experts as constituting their own domain in life—standing alongside bacteria and eukarya—and they have belatedly been recognized as existing in abundance in lots of

environments. The importance of the archaea stretches beyond this, however, in that, of all the organisms living today, archaea may have been the first to exist. Whether this means that archaea are ancestral to all other existing life-forms remains to be seen— many researchers have their doubts—but it is certain that archaea lie near the trunk of life's family tree.

Why did it take so long to recognize archaea as ancient, widespread, and in a class by themselves? With respect to their uniqueness, one problem was that archaea are superficially similar to bacteria: They are microscopic, basically single-celled, and their cells lack a nucleus (making them prokaryotes, along with the bacteria). In addition, it is hard to learn much about archaea because it is so difficult to "culture" them, meaning to grow populations of them in a laboratory. The lab environment is so different from the extreme environments inhabited by many archaea that cultures of them are quickly taken over by common bacteria.

Figure 20.7
Carl Woese's Tree of Life
Woese has argued that the archaea are sufficiently distinct from both bacteria and eukaryotes to warrant the status of a third branch of life, the domain Archaea.

Figure 20.8
Looking for Life in Extreme Environments
Researchers from the Laboratory of the Molecular Biology of Extremophiles, located in France, on a research outing in New Zealand prospecting for microscopic extremophiles. Their research group studies archaea and bacteria that live at temperatures close to 100° Celsius or 212° Fahrenheit—the temperature at which water boils.

Beyond these things, however, there was a long-term resistance among scientists to seeing archaea as anything but an odd type of bacteria. Indeed, a single individual, Carl Woese of the University of Illinois, had to wage a lonely battle lasting years before his claims about the uniqueness of the archaea were generally accepted (**see Figure 20.7**). In 1996 there came a triumphant confirmation of his ideas about the archaea with the sequencing of the entire genome of an archaean species, *Methanococcus jannaschii*. This work revealed that an amazing 56 percent of *M. jannaschii*'s genes were completely unknown to science—they were unlike anything seen in either bacteria or in eukaryotes. With respect to the remaining 44 percent of the genome, some *M. jannaschii* genes worked like those of eukaryotes, while others worked like those of bacteria.

All this speaks to the uniqueness and evolutionary place of the archaea, but what about the widespread distribution of them? In the 1980s, scientists developed a technique that allowed them to do a kind of census taking of archaea out in nature. This method revealed that archaeans live in extreme environments, such as the Yellowstone hot springs, in more numbers than had been thought. But the real surprise came when more expansive environments were tested; archaea turn out to account for almost a third of the microscopic organisms living in the surface waters off Antarctica, for example, and huge populations of them are believed to inhabit deep-ocean waters.

Prospecting for "Extremophiles"
The diverse species of archaea that do inhabit extreme environments have in recent years become the target of a kind of new-age prospecting, with "miners" being scientists from chemical and biotechnology firms. The archaea they are looking for are **extremophiles** ("extreme-lovers"), flourishing in conditions that would kill most organisms, such as high heat, high pressure, high salt, or extreme pH (**see Figure 20.8**).

To live in tough environments, extremophiles, which include both archaea and bacteria, must produce *enzymes* that function in these environments. And, through modern biotechnology processes, enzymes can be isolated from an organism and then turned out in quantity in factories. Enzymes from the hardy extremophiles can then be put to use in *man-made* extreme environments, such as the inside of a washing machine or the confines of a laboratory container. Lest you think these are fanciful examples, note that one U.S. company has already put a clothing detergent on the market that contains a cleaning additive that is simply an enzyme from a heat-loving extremophile.

20.4 Domain Eukarya: Protists, Plants, Fungi, and Animals
You have thus far looked at two of life's domains, Bacteria and Archaea, which leaves one more to review—Eukarya, meaning the eukaryotes of the world: protists, plants, fungi, and animals. As you've seen, all eukaryotes have cells with nuclei, and almost all have several other organelles. Beyond this, however,

there is very little that is common to every eukaryote. Unlike bacteria or archaea, most eukaryotes are capable of sexual reproduction, but asexual reproduction is carried on to some extent in every eukaryan kingdom.

Domain Eukarya is special because of its incredible diversity of form and capabilities. Bacteria and archaea may be diverse at the molecular level, but as whole organisms they are all basically single-celled microscopic individuals. The eukaryotic domain has some single-celled microbes in it too, but it also includes birds and orchids and whales and mushrooms.

As already noted, the eukaryotic domain is divided into four individual kingdoms: Protista, Plantae, Fungi, and Animalia. If you look at **Table 20.1**, you can see some of the major features of each kingdom. For the rest of the chapter, you'll be looking at these kingdoms individually.

20.5 Kingdom Protista: An Undefinable Collection

As you progress through the eukaryotic kingdoms, you will see that we have fairly clear definitions of what a plant or an animal or a fungus is. Not so with the protists, which are a kind of grab-bag category of organisms that must be defined in terms of what they are *not*: They are eukaryotic organisms that do not have all the defining characteristics of plants or animals or fungi. About 100,000 species of protists are known to exist, most of them single-celled, and all of them living in environments that are at least moist, if not fully aquatic—oceans, lakes, a damp forest floor. Many protists are parasites, as with the *Giardia* intestinal parasite much feared among campers.

Some protists perform photosynthesis (like plants), while others ingest food in order to live (like animals); still others are capable of moving back and forth between the two nutritional modes, depending on the availability of sunlight and nutrients. Meanwhile, other protists reproduce more like fungi. In short, there are plant-like, animal-like, and fungus-like protists, which squares with what you learned in Chapter 19 about evolution of the eukaryotes: Plants evolved from one kind of protist (green algae), animals from another (probably a grouping called *choanoflagellates*), and fungi from still another (though scientists are not sure

Table 20.1
Features of Kingdoms in the Eukaryotic Domain

Kingdom	Reproduction Method	Means of Nutrition	Multicellular?
Protista	Usually asexual, but many protists have both sexual and asexual reproductive phases	Both photoautotrophy (photosynthesis) and chemoheterotrophy (by ingestion of organic material) are common	Usually not, though both colonial multicellularity and true multicellularity exist
Fungi	Spores that are produced sexually or asexually	Always chemoheterotrophy by means of external release of enzymes that break down organic material, followed by absorption of this material	Usually, though unicellular forms also exist
Plantae	Usually sexual, though vegetative reproduction (for example, by cuttings) allows asexual reproduction	Almost always photoautotrophy	Always
Animalia	Generally sexual, but can be asexual	Always chemoheterotropy, almost always by ingestion	Always

which). You can see several kinds of protists in **Figure 20.9**.

In an example of how protists can bend our ideas about categories of living things, **Figure 20.10** shows *Dictyostelium discoideum*, a protist that has a fascinating multiple personality. At one point in its life cycle, this creature exists in the form of individual microscopic cells that live in soil, mostly by ingesting bacteria. When their food runs low, however, these cells begin aggregating, producing a migrating "slug" that has anterior and posterior ends and that moves toward light. Eventually this aggregation forms a stalk-like reproductive structure that releases spores that will develop into new individual cells.

Means of Mobility in Protists

Many single-celled protists get to sunlight or nutrients by means of a flagellum (plural *flagella*), which is a long, whip-like extension of the cell that propels the cell by its movement. Other protists have cilia—hair-like cellular extensions that exist in profusion on the cell surface. Beating in rhythm, they move the cell toward food or away from noxious substances. Still other protists, the amoebas, move by extending a

pseudopodium or "false foot," which is simply a portion of the cell cytoplasm that is cast out and then anchored, after which the rest of the cytoplasm streams into the extension. *Dictysoltelium discoideum* is one of these amoeboid protists. More fearsome is the amoeba *Entamoeba histolytica*, which can contaminate drinking water, causing amoebic dysentery.

True Multicellularity: A Division of Labor

Some Protists Are Very Large: The Giant Kelp

Lest you think of all protists as microscopic, however, consider the giant kelp of California's coastal waters. These protists, which are one type of brown algae, are capable of *growing* 60 meters a year—almost 200 feet. You also saw at the start of the chapter that giant kelp have been found that can lay claim to being the world's longest organisms.

For an organism to grow to the size of kelp, obviously it must be composed of many cells, rather than just one. Bacterial cells are capable of coming together in several kinds of aggregations; yet it is only the eukaryotes, such as protist algae, that have what is known as true multicellularity. This means a state in which multiple cells that are aggregated have different capabilities, resulting in a *division of labor* among cells. Your pancreas cells perform functions quite different from those of your brain cells. By contrast, none of the bacteria in bacterial aggregations are capable of performing functions different from any *other* bacteria in a body. Such bacterial aggregation is thus an example of mere **colonial multicellularity**. The general view among evolutionary biologists is that colonial protists were the bridge between unicellular and true multicellular life.

Single-Celled Algae: The Importance of Phytoplankton

Algae, which are protists, exist in single-celled as well as multicelled forms. The importance of the single-celled variety to life in general is great, because they form a large proportion of Earth's **phytoplankton**, meaning microscopic photosynthesizing organisms that drift in the upper layers of the oceans or freshwater bodies. Phytoplankton

Figure 20.9
Protists Small and Large

a A microscopic protist called *Vorticella*.

b Towering examples of giant kelp off the California coast, with fish swimming nearby.

a

b

perform much of Earth's photosynthesis and occupy a key position in water-based food chains. Growing in great abundance, they are in turn eaten by **zooplankton**, which are the small animals and other heterotrophs that drift or move weakly in the upper layers of bodies of water. Phytoplankton and zooplankton (together called just plain **plankton**) are eaten by larger animals, which are eaten by still larger animals. Blue and humpback whales feed on plankton directly, taking in great mouthfuls of ocean water and filtering out the plankton that rush in. Having looked at one kingdom in the eukaryotic domain, let's now move on to another.

20.6 Kingdom Fungi: Life as a Web of Slender Threads

If mushrooms are considered as an example, fungi may seem at first simply different kinds of plants. After all, they are **sessile**, or fixed in location, just as plants are, and they most often live in the soil. Yet there is a very basic distinction between plants and fungi: Almost all plants are photoautotrophic—they make their own food through photosynthesis—while all fungi are heterotrophs, meaning they must consume existing organic material in order to live.

When you think about what fungi are consuming, the term *nuisance* (or worse) may come to mind. Fungi cause mold, mildew, dry rot, ringworm, vaginal yeast infections, athlete's foot, and diaper rash for starters. The devastating Irish Potato Famine of the 1840s was caused by a fungus, *Phytophthora infestans*, that ruined one of Ireland's staple food sources. This agricultural blight is still with us today, as are the fungi commonly known as corn smut and wheat rust. To state the obvious, many fungi are parasites, feeding on living hosts but not killing them. Fungi can also pose mortal threats to people, such as AIDS patients, whose immune systems have been weakened.

The Important Roles of Fungi in Nature

Their invader role notwithstanding, fungi also are indispensable to humans and other living things. The mold that gives us penicillin is a fungus, as is the yeast we use to make bread rise and beer ferment. More generally, fungi join bacteria as the major "decomposers" of the living world, breaking down organic material such as garbage or fallen logs and turning it into inorganic compounds that are recycled into the soil. Without this activity Earth would long ago have turned into a massive garbage heap of dead organic material. Beyond this, fungi are involved in a critical cooperation with plants. Some 80 percent of seed plants have fungi associated with their roots, something you'll learn more about shortly. In sum, we could say of fungi what we did of the bacteria: We could do without some of them, but we could not live without others. Now let's look at some of the qualities that are common to most members of this kingdom, which contains about 100,000 named species (**see Figure 20.11** on page 406).

- *Fungi largely consist of slender, barely visible tubes, called* **hyphae**, *that collectively make up a web called a* **mycelium**. If you look at **Figure 20.12** (page 406), you can see the nature of this structure. You may have thought of a mushroom as the main part of a fungus, but in reality it is merely a reproductive structure that sprouts up from the larger web below.

Figure 20.10
Multiple Identities

a The protist *Dictyostelium discoideum*, a cellular slime mold, exists for part of its life as a collection of individual cells that consume bacteria. Reacting to a reduced food supply, these cells start aggregating as shown here.

b The aggregation that forms takes on the appearance of a slug and can move across the forest floor.

c This organized collection of cells then transforms itself into a tower-shaped reproductive structure whose cap releases spores that will be dispersed to start life elsewhere as new individual cells. *Dictyostelium* is of great interest to biologists because it provides insight into how cells work together to form a multicellular organism.

c

a

b

• *The cells that make up hyphae tend to be very porous relative to one another, a quality that allows for rapid fungal growth.* Individual hyphae cells usually are separated from one another by structural dividers called *septa*, but these septa have microscopic openings that allow for a fairly free flow of cytoplasm between one fungal cell and the next. When combined with the mycelium structure, this quality brings about one of the most notable characteristics of fungi: They *grow toward* their food supply, and they can do so very rapidly. Because of the porous quality of the hyphae cells, a fungus can move a great many cellular resources, such as proteins, right to the point of growth. Did it ever seem to you that a mushroom sprouted in your yard overnight? It probably did.

• *Fungi get their nutrition by dissolving their food externally and then absorbing it.* A fungus spore lands on a log, begins to grow, and then starts releasing enzymes that break down the wood; this digestion takes place *outside* the fungus, after which the fungus absorbs the resulting nutrients. Note how this differs from animals which ingest their food by some means—engulfing it or biting off pieces of it—but then digest it *internally*. Some fungi attack prey almost as an animal would; their hyphae take the form of lasso-like loops that can be constricted around worms that pass through them (**see Figure 20.13**). Even here, however, the nutritional mode is to invade the trapped animal with special hyphae, dis-

**Figure 20.11
Forms of Fungi**

a Cup fungus in a Costa Rican rain forest.

b Bridal veil fungus.

c Shelf fungus on a decaying log.

a

b

c

spore

hyphae

**Figure 20.12
Structure of a Fungus**
Fungi are composed of tiny slender tubes called *hyphae*. The hyphae form an elaborate network belowground, called a *mycelium*. The same hyphae also form a reproductive structure aboveground, called a *mushroom*.

charge the dissolving enzymes, and then absorb the digested nutrients.

- *Fungi are almost always multicellular.* The exception to this is the yeasts, which are unicellular and thus do not form mycelia.

A Phase of Life Unique to Fungi: Dikaryotic Cells

Nearly all the cells in the human body, with the exception of sperm and egg cells, are "diploid," meaning that they have *paired* sets of chromo-somes within them. Conversely, many protists live all or most of their lives in a haploid condition—they have but a *single* set of chromosomes per cell. This diploid/haploid dichotomy runs throughout nature, and it exists in fungi as well. Fungi are "rule breakers" though, because in addition to having haploid and diploid phases in their life cycles, most of them go through a phase that lies *in between* these two states: a **dikaryotic** ("two nuclei") phase in which each cell of a growing mycelium has *two* haploid nuclei within. Think of it this way. In human reproduction, when sperm and egg come together, the fusion of their cytoplasms is followed immediately by the fusion of their *nuclei* (the two sets of haploid chromosomes). In the mushroom, conversely, there is a pause between cytoplasm fusion and chromosome fusion—the dikaryotic phase, in which single cells have a common cytoplasm but *two* nuclei. This phase may be long-lived; it can go on for years, in fact, and ends only when the most visible manifestation of the mycelium, the mushroom, sprouts aboveground.

If you look at **Figure 20.14**, you can see how this comes about. This is the life cycle of

Figure 20.13
Trapped by a Fungus
A tiny worm, called a nematode, is trapped within the hyphae of a lasso fungus, and will soon be consumed by it.

Figure 20.14
Life Cycle of a Fungus

1. Two haploid mushroom spores from different organisms are blown through the air and land on ground suitable for growth.

2. They begin to germinate, which is to say, to grow haploid mycelia underground in their respective locations by simple cell division. These two fungi are of the of the same species, but they happen to have cells that are of complementary "mating types," meaning they are capable of fusing. (The two types are signified by the blue and red colors of their cell nuclei.)

3. Upon coming in contact with one another, hyphae of their respective mycelia then fuse, but note that it is only the cytoplasm of their cells that fuses, while the nuclei remain separate. Thus begins the dikaryotic phase of the life cycle—single cells with two nuclei. All the daughter cells of this original fusion event have this structure. Also note that the two organisms have now become one.

4. The mushroom is a reproductive structure, initially made up entirely of dikaryotic cells.

5. The accordion-like gills on the underside contain cells whose *nuclei* will fuse, producing cells that are not dikaryotic, but diploid.

6. This diploid phase is short-lived, however; meiosis occurs very quickly, yielding haploid cells that sprout out from the gills; these are spores that will drop off the bottom of the mushroom in tremendous numbers to be carried away by the wind and land on the ground, there to germinate and start the cycle all over.

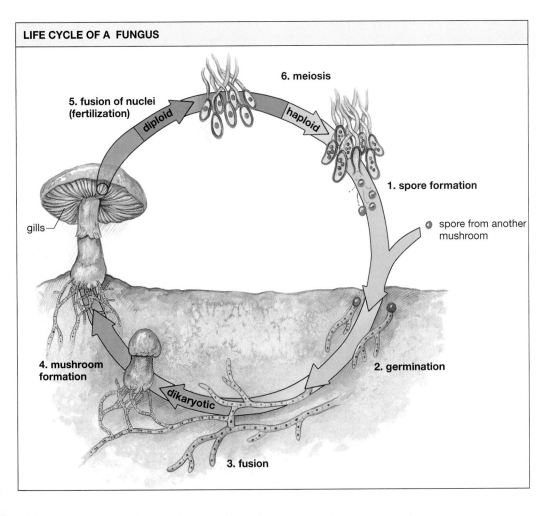

LIFE CYCLE OF A FUNGUS

6. meiosis

5. fusion of nuclei (fertilization)

diploid

haploid

1. spore formation

spore from another mushroom

gills

2. germination

4. mushroom formation

dikaryotic

3. fusion

a mushroom, which is representative of one of four major types of fungi. If you looked at all four types, you'd see different reproductive structures and life cycles, but the mushroom cycle will give you a general idea of how fungi reproduce.

Fungal Associations: Lichens and Mycorrhizae

Lichens

Everyone has seen the thin, sometimes colorful coverings called *lichens* that seemingly can grow almost anywhere—on rocks as well as on trees; in Antarctica as well as in a lush forest (**see Figure 20.15**). As it turns out, lichens are not a single organism. They are always two organisms living as one: a fungus and then either an alga or a bacterium that is nestled within the fungus. If you look at **Figure 20.16**, you can see that this association is structured as a kind of sandwich: An upper layer of densely packed fungal hyphae on top; then a zone of less dense hyphae that includes a layer of algal or cyanobacterial cells; then another layer of densely packed hyphae on the bottom that sprout extensions down into the material the lichen is growing on (such as a rock). The example here is an association between a fungus and an alga, which is the case in about 90 percent of lichens. In such a relationship the fungal hyphae either wrap tightly around the algal cells or actually extend into them.

A lichen is mostly a fungus, then, but it is a fungus that could not grow without its algal partner. Why? Because the alga carries out photosynthesis and then proceeds to supply the fungus with some of the nutrients photosynthesis provides. It is generally assumed that what fungi do for algae in this relationship is keep them from drying out. It may be, however, that they're doing nothing at all for the algae, in which case the fungus should be considered an algal parasite, rather than a partner.

Mycorrhizae

As noted earlier, about 80 percent of seed plants form an association with fungi. Here, both partners clearly are benefiting in a linkage that involves plant roots and fungal hyphae. What the fungi get from this is food, in the form of carbohydrates that come from the photosynthesizing plant; what the plants get is minerals, growth hormones, and water, all absorbed by the fungal hyphae. So important is this relationship to plants that some species of trees, such as pine and oak, cannot grow without fungal partners. Other plants will have stunted growth without a fungal partner.

Such a relationship is technically an "infection" of plant roots by fungal hyphae, in that the hyphae generally grow into the plant roots. In a few instances the hyphae wrap around the root but do not penetrate it. In either case, the root/hyphae combinations are known as **mycorrhizae**. The plant benefits from this relationship as the branched network of hyphae spreads out, thereby greatly increasing the volume of soil that the plant draws its nutrition from.

Figure 20.15
Flourishing in Different Environments
Lichens growing on **(a)** a boulder in the State of Washington's Olympic National Park, **(b)** a birch tree in Alaska's Kenai Peninsula, and **(c)** a rock in Alaska's Arctic National Wildlife Refuge.

a

b

c

20.7 Kingdom Animalia: Multicelled Movers

Animals are so familiar to us that you might think we wouldn't need to define them. Yet consider that, along with fish and monkeys and snails, the animal kingdom includes

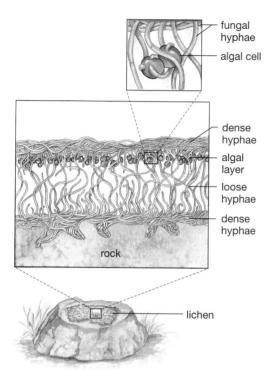

Figure 20.16
The Structure of a Lichen
A lichen is not a single organism. It is composed of a fungus and an alga (or sometimes a bacterium) living in a mutually beneficial arrangement. The fungal hyphae form a dense layer on the top and the bottom and a loose layer in the middle, within which the alga is nestled. The fungus provides a moist, protective environment for the alga, and the alga, which performs photosynthesis, provides food for the fungus.

water dwellers called *rotifers* that are as microscopic as most protists; marine creatures called *hydras* that can look more like plants than animals; and sponges, which have none of the usual organs that we associate with animals (a mouth, a gut; **see Figure 20.17**). This is not to say, however, that our common-sense notions of animals are meaningless, as we can see when we think about the defining characteristics of animals.

- *Movement.* Unlike plants or fungi, which are **sessile** or fixed in one place, animals generally are mobile. Lots of exceptions exist, however, such as sponges and the ocean-dwelling filter feeders called *bryozoans*.

- *Bilateral symmetry.* Animals usually are *different* front-to-back and top-to-bottom, but *symmetrical* side-to-side. Put another way, your head is different than your feet, and your chest is different than your back, but your left *side* is very similar to your right. Such **bilateral symmetry** is an understandable outcome of being mobile, given the way our ocean-dwelling ancestors encountered the world. They moved into new environments headfirst, and often had different environments beneath them as opposed to above them (ocean floor beneath, just plain ocean above). Meanwhile, each of their sides moved essentially through the *same* environment. There are, however, many animals that have **radial symmetry**, meaning a symmetry in which body parts are distributed evenly around a central axis.

What Is It?
See p. 410.

Figure 20.17
Unfamiliar, but Animals Nevertheless

a One large and two small freshwater rotifers dine on some algae. The large rotifer's head is at left; its stomach is green from ingested algae.

b A green hydra with tentacles.

c A warm-water tube sponge.

a

b

c

What Is It?
Answer: Life's diversity. A star-nosed mole peers through a lacy curtain of green moss. Twenty-two tentacles surrounding its nose are studded with nerve endings.

What Is It?
See p. 412.

- *Embryonic development and multicellularity.* All animals develop from embryos that go through a "blastula" stage, meaning a ball of cells that becomes differentiated into several kinds of tissues (see Chapter 26). Given this, there are no single-celled animals.
- *Heterotrophic, ingestive nutrition.* Almost all animals must consume existing organic matter in order to live. With some exceptions, they digest food internally.
- *Sexual reproduction.* Reproduction among animals generally is sexual reproduction, usually meaning the fusion of haploid gametes from a male and a female of a species to produce a diploid fertilized egg that then grows into a mature organism. Nevertheless, there are many exceptions. Some species of flatworms reproduce by **fission**, simply dividing themselves, with each of the resulting parts then becoming a whole flatworm. Cut off an arm and a small part of the central disk of a sea star, and a whole new sea star will be reproduced through **regeneration** from that body part. The relatively rare "drones," or males of honeybee colonies are haploid individuals, produced by **parthenogenesis**, meaning they develop from an egg, produced by the queen bee, that has never been fertilized.
- *A nervous system.* In multicellular organisms, mobility goes hand in hand with having a nervous system. Mobile organisms must react to their environment in more complex ways than do sessile creatures, and they must do so very quickly. There are mobile bacteria and protists, of course, but these are almost all *single-celled* organisms whose movements can be controlled through basic cell machinery. In larger organisms, a specialized set of cells is required to get messages quickly from one *part* of the organism to another—a nervous system, in short. Nervous systems arguably are the crowning glory of the animal kingdom, because they make possible the fantastic array of capabilities seen in animals, from night navigation in bats to thinking in human beings.

Dividing the Kingdom: The Animal Phyla

You saw in Chapter 19 that the animal kingdom is divided into about 30 major groupings, called *phyla*, with each **phylum** representing a group of animals whose members share a basic body plan. Surveying the animal phyla can be a humbling experience from a human perspective, because several varieties of worms, for example, are different enough from one another that they each get their own phylum, while the animals most familiar to us—fish, reptiles, birds, mammals—are all similar enough that they are lumped into a single phylum. This phylum is Chordata, named for the nerve cord that runs along the back or "dorsal" side of every chordate at some stage in its development. If you look at **Table 20.2**, you can see a selected set of animal phyla.

The life cycles of animals exhibit a tremendous diversity, with no one cycle being representative of the kingdom as a whole. Earlier in the book, you looked in great detail at the life cycle of one animal, a member of the Phylum *Chordata*, the Class *Mammalia*, the Order *Primates*, the Family *Hominidae*, and the Genus *Homo*. The particular species you looked at was *Homo sapiens* (see Chapters 9 and 10). You'll look at the physical makeup and functioning of animals in Chapters 23 through 27.

20.8 Kingdom Plantae: The Foundation for Much of Life

In your review of the living world so far, you've seen that human beings have reason to feel ambivalent about bacteria, fungi, and even protists in that there is much to appreciate in these organisms but a good deal to fear as well. When it comes to plants, however, it's hard to find much to fear—but easy to find much to appreciate. For starters, we humans are utterly dependent upon plants for food. Try naming something you eat that isn't a plant or that didn't come from creatures that do eat plants, and you will end up with a very short list. Then there is the atmospheric oxygen that plants produce as a by-product of photosynthesis. As you'll be seeing in Chapter 21, these two critical contributions actually are just the start of how plants affect the world we live in.

What Are the Characteristics of Plants?

We all have a good intuitive sense of what plants are, though there are lots of plants that don't adhere to our preconceptions.

Plants make their own food, but there are plants that do this while also being carnivorous, such as the Venus flytraps. Then there are a few plants that don't perform photosynthesis at all—for example, mistletoe—that are plant parasites. Allowing for such exceptions, plants generally share the following characteristics.

Plants make their own food, but there are plants that do this while also being carnivorous, such as the Venus flytraps. Then there are a few plants that don't perform photosynthesis at all—for example, mistletoe—that are plant parasites. Allowing for such exceptions, plants generally share the following characteristics.

Plants make their own food, but there are plants that do this while also being carnivorous, such as the Venus flytraps. Then there are a few plants that don't perform photosynthesis at all—for example, mistletoe—that are plant parasites. Allowing for such exceptions, plants generally share the following characteristics.

- *Plants are fixed in place, photoautotrophic, multicelled, and mostly land-dwelling.* We know that plants are fixed in one spot (sessile), and that they make their own food through photosynthesis (they are photoautotrophic). That they are multicelled follows from the fact that they develop from embryos, which are multicelled by definition. Recall from Chapter 19 that plants are descended from green algae, which live in water. As a group, then, plants made a transition to the land from water, and land is where most of them are found today—though some have made the transition *back* to water, as with water lilies.

**Table 20.2
Animal Phyla**

Phylum	Members Include	Live In	Common Characteristics
Porifera	Sponges	Usually ocean water, occasionally fresh water	Sessile, no organs, all cells function independently to some extent. Radial symmetry. System of pores, through which water flows, serves to trap microscopic organisms.
Cnidaria	Jellyfish, hydras, sea anemones, corals	Usually ocean water, sometimes fresh water	Radially symmetrical. Have mouth, gut, nervous system; stinging cells for predation and defense; usually tentacles. Two-stage life cycle: sessile polyps and swimming medusas. Coral reefs are aggregations of generations of coral skeletons.
Mollusca	Clams, mussels, snails, slugs, octopuses, squids	Ocean, fresh water, moist land environments	Bilaterally symmetrical. Three-part body, including a head, a central visceral mass containing internal organs, and single foot (divided into tentacles in octopus and squid). Most have shells, though some do not.
Annelida	Earthworms, leeches, tube worms	Ocean, fresh water, moist soil	All annelids are segmented worms – the "rings" on earthworms mark their segments.
Arthropoda	Spiders, mites, crabs, shrimp, centipedes, insects	Many aquatic and land environments	Segmented bodies with jointed, bilateral appendages (e.g., fly wings, lobster claws). External skeleton made of chitin. Of the estimated 1.25 million named animal species on Earth, about 1 million are arthropods.
Echinodermata	Sea stars, sea urchins, sand dollars	Ocean, almost always ocean floor	Radial symmetry, internal skeleton, hollow tube feet that function in water regulation and in typically slow movement.
Chordata	Lampreys, fish, amphibians, birds, reptiles, mammals	Many aquatic and land environments	Nervous system develops from longitudinal, dorsal nerve cord. All have gill slits at some point in development. Most have a segmented vertebral column, making them vertebrates. For the past 350 million years, Earth's dominant land animals have been a succession of vertebrates (amphibians, then reptiles, then mammals).

What Is It?
Answer: Life's diversity. The world's largest flower, *Rafflesia keithii*. This specimen in Borneo is almost a meter (about 3 feet) across.

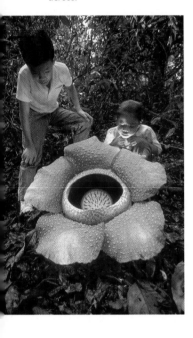

- *Plant cells have cell walls, a specialized set of organelles, and contain a high proportion of water.* Both plant and animal cells have a plasma membrane around their periphery, but plant cells have something else as well: A **cell wall**, which surrounds the plasma membrane and is composed in large part of complex compounds called **cellulose** and **lignin** that help give cell walls their strength (**see Figure 20.18**). All cell walls help plants regulate their intake of water—they help them keep what they have, but not take in too much. This regulation is even more important to plants than animals; a typical animal cell may be 70 to 85 percent water, but for plant cells the water proportion is likely to be 90 to 98 percent. Plant cells have a set of organelles, called *plastids*, that animal cells lack; these plastids include the chloroplasts that are the actual sites of photosynthesis. (See Chapter 4 for details of plant cell structure and Chapter 8 for photosynthesis.)
- *Successive generations of plants go through what is known as an alternation of generations.* This can best be explained by a comparison with human reproduction. Looked at one way, there is a haploid "phase of life" for human beings in that the gametes (eggs and sperm) that human

beings produce are haploid cells, meaning they have but a single set of chromosomes. When these haploid cells fuse, in the moment of conception, what's produced is a *diploid* fertilized egg, meaning an egg with

ALTERNATION OF GENERATIONS

**Figure 20.19
The Alternation of Generations in Plants, Compared to the Human Life Cycle**

Almost all cells in human beings are diploid or 2n, meaning they have paired sets of chromosomes in them. The exception to this is human gametes (eggs and sperm), produced through meiosis, which are haploid or 1n, meaning they have a single set of chromosomes. In the moment of conception, a haploid sperm fuses with a haploid egg to produce a diploid zygote that grows into a complete human being through mitosis. In plants, conversely, diploid mature plants—the multicellular sporophyte fern in the figure—go through meiosis and produce individual spores that, without fusing with any other cells, develop into a separate generation of the plant. This is the multicellular gametophyte shown. This gametophyte then produces its own gametes, which again are eggs and sperm. Sperm from one plant fertilizes an egg from another, and the result is a diploid zygote that develops into the mature sporophyte generation. The alternation between the sporophyte and gametophyte forms is called the *alternation of generations*.

**Figure 20.18
Characteristics of Plants**

All cells have an outer membrane, but plant cells have a wall external to this membrane. The compounds cellulose and lignin, which help make up the cell wall, impart strength to it. Plant cells have a higher proportion of water in them than do animal cells, with much of this water located in an organelle called a *central vacuole*. The sites of photosynthesis in plants are the organelles called *chloroplasts*.

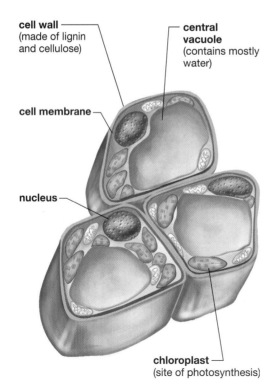

cell wall
(made of lignin and cellulose)

central vacuole
(contains mostly water)

cell membrane

nucleus

chloroplast
(site of photosynthesis)

a *paired* set of chromosomes. This egg gives rise to more diploid cells through cell division, eventually resulting in a whole new human being.

A typical *plant* in its diploid phase will, like human beings, produce a specialized set of haploid reproductive cells. Instead of these being the gametes that humans produce, however, what's produced are **spores**: Single reproductive cells that, without fusing with anything, have the ability to grow into a new generation of plant strictly through cell division. There is thus a separate phase of life that in some types of plants is self-sustaining. This generation of plant eventually does produce gametes, which then fuse and create a diploid zygote, which in turn develops into a diploid plant like the one we started with (**see Figure 20.19**). Because there is a single species that alternates back and forth between two different forms, the plant life cycle manifests what is known as the **alternation of generations**.

On to a More Detailed Look at Plants

Plants are so fundamental to our everyday lives that it is worth taking a more detailed look at how they are structured and how they function—and at what variety there is in the plant world. These are the subjects of the next two chapters.

Chapter Review

Summary

- All living things on Earth can be classified as falling into one of three domains of life, Bacteria and Archaea, whose members are microscopic; or Eukarya, whose members range from the microscopic to the gigantic. Domain Eukarya is in turn divided into four kingdoms: plants, animals, fungi, and protists.

20.1 Viruses: Making a Living by Hijacking Cells

- There is controversy over whether viruses are living things. To carry out almost any of life's basic functions, viruses must use the molecules and structures of the cells they invade. Some viruses employ a lytic life cycle, while others employ both a lytic and lysogenic life cycle. In the lytic life cycle, clones of the invading virus burst forth from an infected cell and then go on to infect more cells. In the lysogenic life cycle, the virus may for a time integrate itself in the DNA of the host cell and then replicate with it, leaving the cell through lysis only when threatened.

20.2 Domain Bacteria: Masters of Every Environment

- Though they can cause human disease, bacteria are indispensable to all life on Earth. They are largely responsible for moving the nutrient nitrogen into the living world and, along with fungi, are the major decomposers of the living world. Bacteria are essentially a single-celled lifeform and reproduce through simple cell division.

20.3 Domain Archaea: From Marginal Player to Center Stage

- Archaea were once thought to be a form of bacteria but have recently been shown to differ greatly from bacteria in their genetic makeup and metabolic functioning. Some archaea thrive in extreme environments, such as hot springs. The chemical products they produce are now being exploited by human beings.

20.4 Domain Eukarya: Protists, Plants, Fungi, and Animals

- Eukaryotes exhibit an incredible diversity of form and function; most reproduce sexually, though members of all eukaryotic kingdoms can reproduce asexually as well.

20.5 Kingdom Protista: An Undefinable Collection

- Kingdom Protista, part of the eukaryotic domain, is made up of an assortment of organisms so diverse that no one definition

Key Terms

alternation of generations 413

antibiotic 400

autotroph 400

bacillus 399

bilateral symmetry 409

binary fission 399

cell wall 412

cellulose 412

chemoautotrophy 400

chemoheterotrophy 400

coccus 399

colonial multicellularity 404

dikaryotic 407

diploid 399

extremophile 402

fission 410

haploid 399

heterotroph 400

hypha 405

lignin 412

lyse 396

lysogenic cycle 396

lytic cycle 396

mobile 409

mycelium 405

mycorrhiza 408

can be applied to all protists. Many protists are single-celled aquatic creatures, but large ocean-dwelling algae are protists, as are many parasites and soil-dwelling microbes. The aquatic protists known as *phytoplankton* are important in photosynthesis and as the base of aquatic food chains.

20.6 Kingdom Fungi: Life as a Web of Slender Threads

- Fungi consist largely of slender tubes, called *hyphae*, that grow toward their food supply, releasing enzymes that digest the food externally, after which the fungus absorbs the nutrients. Along with bacteria, fungi are nature's primary decomposing organisms. Most are multicellular. Some fungi live in a mutually beneficial relationship with plants, the plants supplying the fungi with food produced in photosynthesis, the fungi using their hyphae to absorb water and nutrients for the plants. The combinations of plant roots and fungal hyphae are called *mycorrhizae*.

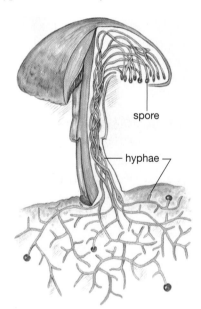

spore

hyphae

20.7 Kingdom Animalia: Multicelled Movers

- Animals are notable for their movement, their bilateral symmetry, their sexual reproduction, and their highly evolved nervous systems. All are multicellular.

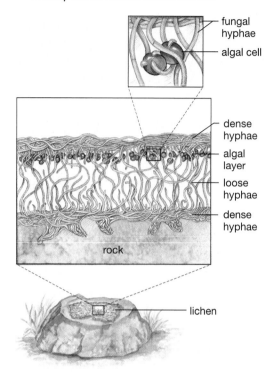

fungal hyphae

algal cell

dense hyphae

algal layer

loose hyphae

dense hyphae

rock

lichen

20.8 Kingdom Plantae: The Foundation for Much of Life

- Plants are the foundation for much of life on Earth, because they are responsible for much of the living world's food production and oxygen generation. All plants are multicelled; most are fixed in one spot and carry out photosynthesis. Plants reproduce through an alternation of generations.

Understanding the Basics

Multiple-Choice Questions

1. Characteristics shared by all bacteria include (select all that apply)
   a. rod shape
   b. no nucleus
   c. circular chromosome
   d. sexual reproduction
   e. one organelle

2. What do all eukaryotes have in common? (select all that apply)
   a. multicellularity
   b. cell organelles
   c. nucleus
   d. sexual reproduction
   e. circular chromosomes

3. Important positive roles of fungi in life on Earth include (select all that apply)
   a. decomposition
   b. photosynthesis
   c. assist in water uptake in plants
   d. used in baking and brewing
   e. join with algae to make lichen

4. Match the organisms on the left with the descriptions on the right:
   a. hyphae — symbiotic infection of plant roots
   b. mycorrhizae — filamentous strands of porous material
   c. cell walls — ancestors of animals, plants, and fungi
   d. archaea — found only in plants
   e. protists — often occupy extreme environments

5. Defining characteristics of plants include (select all that apply)
   a. bilateral symmetry
   b. hyphae
   c. cell walls
   d. nervous system
   e. heterotrophic, internal digestion

6. True or false: Starfish, worms, birds, and fish are all in different phyla.

7. Which of the following are important characteristics of plants?
   a. cell wall
   b. lignin for support
   c. circular chromosomes
   d. hyphae
   e. the primary sources of food on Earth

8. Differences and similarities between viruses and bacteria include (select all that apply)
   a. Viruses are not cells, and bacteria are cells.
   b. Bacteria are living cells, while viruses arguably are not.
   c. Viruses and bacteria can both self-replicate.
   d. Viruses and bacteria can both be controlled by antibiotics.
   e. Viruses and bacteria are both always pathogenic.

9. True or false: Most bacteria are harmful.

10. True or false: Meiosis occurs in plants and animals, but not in fungi.

Brief Review

1. What is an extremophile?

2. How do vaccines work to prevent viral diseases?

3. Why do most biologists classify viruses as nonliving?

4. Discuss two reasons why phytoplankton are so important to life on Earth.

5. Describe two common associations that fungi make with other organisms. What does each party in each of these associations do for the other, and what do they gain themselves?

6. How does nutrition in plants, animals, and fungi differ?

7. Given their structure, why are raw plants crunchy?

Applying Your Knowledge

1. What would life on Earth be like if there were no decomposers? Would evolution of life have followed the same course? What might have been different?

2. Discuss the statement, "If success is having the ability to eat anything or live anywhere, then bacteria are the most successful organisms on Earth."

3. If new viruses can form as "renegade" DNA that transposes out of eukaryotic genomes, might some viruses be more closely related to various eukaryotes than they are to other viruses?

4. What is the main difference between colonial multicellularity and true multicellularity? Why is this important for the evolution of eukaryotes?

Group Activity

1. Some vaccines are made from killed viruses, while others are made from "attenuated" or weakened viruses. At the library or on the Internet, look up articles or other information on the history of the two types of polio vaccine, then divide into groups and summarize the pros and cons of each approach. Note the reasons some parents give for refusing to have their children vaccinated at all. Have group members find other examples of vaccines made in each of the two ways and discuss the reasons for the choice.

2. At the library or on the Internet, read about the discovery of hydrothermal vents, and describe how the life-forms found there changed our views of evolution and our understanding of the diversity of life.

A Bounty That Feeds Us All: Plants

The spectacular *Cephaelis elata* (common name:"Hotlips").

21

Introduction to Plants

Plant Structure, Signaling, Defenses, and Responses to the Environment

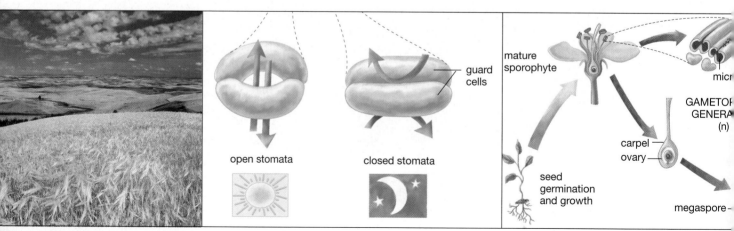

Amber waves of grain.
(page 421)

Guard duty.
(page 429)

Part of a life cycle.
(page 432)

About 500 years ago, Europeans first sailed to the Western Hemisphere and soon after began colonizing it. Now here's a question to ponder about this event: Why did things work out this way instead of, say, the Aztecs sailing to Europe and colonizing *it*? One obvious answer is that Europe was much more advanced technologically than was the New World: There were differences in ships, navigational equipment, weapons. This, however, just pushes the question back a step, prompting us to ask why this technological gap should have existed.

There are lots of possible answers to this question, but let's note one hypothesis that has been put forth by the UCLA physiologist Jared Diamond. It is that Europe developed faster in part because it got lucky in what might be called the plant sweepstakes.

21.1   The Importance of Plants

Human beings first developed agriculture—the systematic planting and care of food plants—about 10,000 years ago in what is known as the Fertile Crescent, meaning a crescent-shaped swath of territory stretching from modern-day Turkey through modern-day Iran. Wherever agriculture has developed, it has had a tremendous effect on human culture, turning societies of hunters and gatherers into societies in which there is enough food that people can *specialize* in terms of vocation. This is critical because a culture that has metalworkers and accountants and soldiers advances technologically at a rapid rate compared to one that doesn't.

Now, as it turns out, farming not only developed first in the Fertile Crescent, but it spread relatively rapidly from it, as you can see in **Figure 21.1** (page 420). Within 3,000 years, agriculture that was based mostly on innovations made in the Fertile Crescent was being practiced throughout much of Europe. In the New World, by contrast, agriculture developed later and spread much more slowly. Why this difference?

First, pure genetic luck was involved. One of the "founder" crops of the Fertile Crescent was a form of wheat that, like almost all food crops, existed originally as a wild plant. As it happens, there are very few genetic differences between the wild variety of this wheat, which yielded little food, and the domesticated variety, which yielded a good deal

The force of gravity.
(page 439)

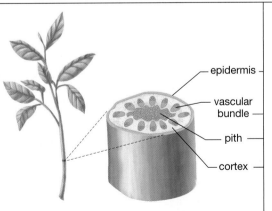

A stem for support, storage, and transport.
(page 431)

epidermis

vascular bundle

pith

cortex

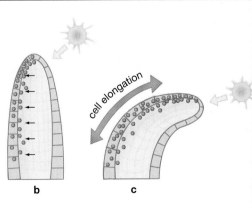

cell elongation

b

c

The power of light.
(page 438)

419

more. By contrast, there was a large genetic difference between the wild variety of the main New World food crop, teosinte, and its domesticated descendant, maize (or corn). Take the members of two societies, one located in the Fertile Crescent and another in the New World, and let each group do the same thing with their respective wild plants, which is simply pick the grains from the most food-laden individual plants each year. As a result of this unconscious human activity, the *seeds* from these selected plants would then *germinate*, or sprout from the ground, in such places as the garbage dumps around human settlements. Both situations result in the rise of nutrition-laden strains of crops by means of this unintended breeding. In the Fertile Crescent, however, these strains appear *quickly*—within a few hundred years—because of the short genetic distance between wild and domestic varieties. In the New World, the transition takes much longer because many more generations of hearty maize plants had to be "selected for" (picked) and "planted" (the excess discarded) to get the domesticated maize.

This explains why the Fertile Crescent was first with agriculture, but why the rapid *spread* of agriculture from it? Pure geographic luck, says Diamond. Europe has a largely east-west orientation, while the New World has a predominately north-south orientation. Plants can spread east to west without going through much climate change, meaning the same domestic strains can spread rapidly. The situation is different for the spread of plants either north or south, because they must adapt to changes in latitude (and temperature) as they go in either direction, which means a slower spreading.

In short, as this hypothesis has it, Europe won the technological race in part because it had a lucky head start in the agricultural race. This is a hypothesis, not established fact; but if it's true, think of its meaning. We need hardly note the tremendous effect that European colonization had on human affairs from the 1500s through the present. It may be that the fates of both colonizers and those who were colonized were determined in significant part by the ticket they held in the plant sweepstakes.

Other Roles of Plants

Whatever their role may have been in history, plants are vital to almost all life at a much more basic level. They are the most important photosynthesizers on Earth, and it is photosynthesis that makes Earth a planet of living *surplus*—of biomass that is here for the harvesting by other life-forms. A single mature maple tree can produce over 100 kilograms (or 220 pounds) of leaves a year. As noted before, if you try to name something you eat that isn't a plant or that doesn't itself depend on plants for food, you'll come up with a very short list. There also is the oxygen that plants produce as a by-product of photosynthesis and, for humans, the lumber and paper that trees provide.

Beyond this, plants are a kind of anchoring environmental force. Their roots prevent soil erosion, while their leaves absorb such pollutants as sulfur dioxide and ozone (**see Figure 21.2**). One of the main greenhouse gasses thought to be warming Earth is carbon dioxide, and plants absorb carbon dioxide in order to perform photosynthesis. Thus, there is not only a *production* side to atmospheric carbon dioxide—in significant part the pollution human beings produce—there is also a *consumption* side to it, in the amount plants absorb. One proposed way to fight greenhouse warming, therefore, is to plant trees. The maple tree mentioned earlier will absorb about 450 kilograms (about 1,000 pounds) of carbon dioxide from the atmosphere over the course of a single summer.

Figure 21.1
The Origins of Agriculture in the Old World
Farming originated almost 10,000 years ago in the "Fertile Crescent" east of the Mediterranean Sea. It then spread rapidly on a mostly east-west axis, helping to spur the rapid cultural evolution for which Europe is known. The lines indicate how far food production had spread by the dates shown. (Redrawn from Jared Diamond, "Location, Location, Location: The First Farmers," *Science*, November 14, 1997, p. 1243.)

2500 years ago

British Isles

3800 y.a.

Germany

France

5000 y.a.

Spain

Italy

Black Sea

Caspian Sea

Greece

6000 y.a.

7000 y.a.

Turkey 8000 y.a.

Mediterranean Sea

Iran

Egypt

21.2 Four Main Categories of Plants: Bryophytes, Seedless Vascular Plants, Gymnosperms, and Angiosperms

Botanists traditionally have partitioned the plant kingdom into a dozen "divisions," but for our purposes, it is convenient to separate the members of the plant kingdom into a mere four types. These are the *bryophytes*, which in-clude mosses; the *seedless vascular plants*, which include ferns; the *gymnosperms*, which include coniferous ("cone bearing") trees; and the *angiosperms*, a vast division of flowering plants—by far the most dominant on Earth today—that includes not only flowers such as orchids, but also oak trees, rice, and cactus (**see Figure 21.3**). After a brief review of the bryophytes, seedless vascular plants, and gym-nosperms, most of this chapter and all of the next is devoted to the angiosperms.

Figure 21.2
Uses Galore
Plants are used by human be-ings in innumerable ways.

a Tapping a rubber tree in Sri Lanka.

b Wheat in a field.

c A cabin made of wood.

a

b

c

a

b

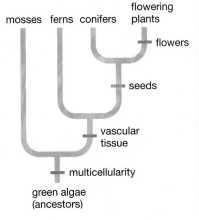

mosses ferns conifers flowering plants

flowers

seeds

vascular tissue

multicellularity

green algae (ancestors)

c

d

Figure 21.3
Four Main Varieties of Plants
All four types are multicellular, but only the most primitive of them, **(a)** the bryophytes (moss in the figure), lack a fluid-transporting vascular system. **(b)** Ferns, representing the seedless vascular plants, have a vascular system but do not have seeds. **(c)** Gymnosperms (conifers in the picture) do utilize seeds. **(d)** Flowering plants (blossoming pear trees in the picture) produce seeds and are responsible for a number of other plant innovations, among them fruit.

Bryophytes: Amphibians of the Plant World

Look in low-lying, wet terrain and you are likely to see a carpet-like covering of moss; in countries such as Ireland, whole fields of peat moss grow and are harvested to be used as fuel for heating and cooking. Mosses are the most familiar example of a primitive type of plant that falls under the informal classification of **bryophyte**, which can be defined as a type of plant lacking a true vascular system (**see Figure 21.4**). What's a vascular system? A network of tubes within the plant that serves to transport fluids. You can get a good idea of what the bryophytes are like by looking at one type of them, the mosses.

Mosses are representatives of some of the earliest plants that made the transition from water to land and, as such, they can be conceptualized as plants that made only a partial break with aquatic living. In making this transition, they had to deal with something their fully aquatic relatives didn't, which is the effect of gravity. Lacking a vascular system, they could not transport water and other substances very *far* against the force of gravity; thus they must lie low, hugging the surface to which they are attached while spreading out horizontally to maximize their exposure to sunshine. Nevertheless, the mosses are hearty competitors in some tough environments, such as the arctic tundra and the cracks in sidewalks. In keeping with their aquatic origins, they have sperm that can get to eggs only by swimming through water. Not much water is necessary; a thin film left over from the morning dew will suffice. But some water must be present. Bryophytes have no roots at all, but instead use single-celled extensions called *rhizoids*—which absorb water and minerals—to anchor themselves to their underlying material, which may be soil or rock or wood. For more on the minerals plants absorb, see "What Is Plant Food?" page 424.

Seedless Vascular Plants: Ferns and Their Relatives

What plants are the vascular plants? The answer is simple: all plants except the bryophytes. To put this another way, only the bryophytes lack the network of fluid-conducting tubes that make up a *vascular system*, meaning a kind of plumbing system for plants. One part of this system distributes water—upward, from root through leaf—and one part distributes the food produced in photosynthesis, along with hormones and other compounds.

Seedless vascular plants means just what it sounds like: Plants that have a vascular system, but that do not produce seeds as part of reproduction. Easily the most familiar representatives of the seedless vascular plants are ferns, with their often beautifully shaped leaves (**see**

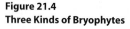

Figure 21.4
Three Kinds of Bryophytes

(a) Mosses, **(b)** liverworts, and **(c)** hornworts. These plants have no vascular tissue so tend to be small. The sperm of bryophytes must travel through water to get to eggs. For this reason bryophytes generally are found in moist environments.

a

b

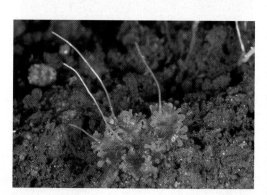

c

Figure 21.5). These plants are a step further in the direction of separation from an aquatic environment. The vascular system allows the plants to grow *up* as well as *out*—something you'd expect would develop in organisms that are competing for sunlight—and it allows for roots that extend into the ground, where they serve their absorptive function. Despite this evolutionary innovation, the sperm of the seedless vascular plants, like that of the bryophytes, needs to move through water to fertilize eggs.

The First Seed Plants: The Gymnosperms

There are two kinds of seed plants, the gymnosperms and the angiosperms, but putting things this way makes it sound as though these plants are on a kind of equal footing with the bryophytes and seedless vascular plants. In reality, the **gymnosperms**, which are nonflowering seed-bearing plants, took over from the seedless vascular plants about the time the dinosaurs came to dominance, only to find themselves replaced as the dominant plants when the angiosperms began flourishing, about 80 million years ago. Today

there are only about 700 gymnosperm species; but the gymnosperm's presence is considerable, especially in the northern latitudes, where they exist as vast bands of coniferous trees, including pine, Douglas fir, and white spruce. Conifers such as these provide most of the world's lumber (**see Figure 21.6**).

Reproduction through Pollen and Seeds

When the gymnosperms are grouped together with the flowering plants as *seed* plants, it's apparent that they have been very good competitors, compared to the bryophytes and seedless vascular plants. Why should this be so? Part of the answer lies in how the male gametes (the sperm) of the seed plants get to the female gametes (the egg): Residing in tiny structures called pollen grains, sperm are carried through the air or on animals, rather than being limited to *swimming* to female gametes, as is the case with the seedless plants. Given this, the seed plants can propagate over great distances, which gives them an obvious competitive advantage over the seedless plants.

Another advantage is the seeds themselves, which can be thought of as tiny packages of

Figure 21.6
Gymnosperms

a This spruce tree is a member of the grouping of gymnosperms known as conifers, which account for about three-quarters of all gymnosperm species. The conifers also include redwood, pine, juniper, and cypress trees.

b There are other types of gymnosperms as well, however. Pictured are the leaves and fruit of the maidenhair tree, *Ginkgo biloba*, which are used today in herbal preparations, including some purported to improve memory.

a

b

a

b

c

Figure 21.5
Three Kinds of Seedless Vascular Plants

(a) Fall-colored ferns in New Hampshire, **(b)** horsetails, and **(c)** club mosses. Because these plants have vascular tissue, they are able to grow taller than most bryophytes. Like bryophytes, however, they do not produce seeds.

424

What Is Plant Food?

As noted several times in this book, plants are able to make their own food through photosynthesis if they have just four things: water, sunlight, carbon dioxide, and a few nutrients. Water, sunlight, and probably even carbon dioxide are no mystery to you by now, but what exactly are these nutrients you've been reading about?

A **nutrient** is simply a chemical element that is used by living things to sustain life. Recall from Chapter 2 that "elements" are the most basic building blocks of the chemical world. Silver is an element, as is uranium, but these are not nutrients because living things do not need them to live. In the broadest sense, carbon, oxygen, and hydrogen are nutrients. And it turns out that almost 96 percent of the weight of the average plant is accounted for by these three elements, which come to plants primarily from water and air.

Plants need at least another 13 elements to live, however, and not all of these are supplied by water and air. When we think of what is commonly called plant food, we are generally referring to those nutrients that plants can *use up* in their soil and that thus must be replenished

to assure continued growth and reproduction. Of these, the most important three are nitrogen, phosphorus, and potassium, whose chemical symbols are N, P, and K. When you look at a package of an average plant fertilizer, you will see three numbers in sequence on it (for example, 10-20-10). What these refer to are the percentage of the fertilizer's weight accounted for by these nutrients. The growth of plants usually is enhanced by a fertilizer that has equal ratios of the three elements, but if increased *blooming* is your goal, an increased phosphorus ratio generally is recommended, as with the 10-20-10 example.

It is not just houseplants or lawns that benefit from fertilizer, but farm crops as well. Indeed, the planting of a crop on a parcel of land generally requires a large investment in fertilizer because a crop such as wheat removes a great deal of N, P, and K from the soil in a single season. Historically, such fertilizers used on farms were **organic**, meaning they came from decayed living things, such as the fish that Native Americans famously taught the Pilgrims to use when planting corn. Nowadays, however, commercial fertilizers generally are **inorganic**, meaning

food and protection. If you look at **Figure 21.7**, you can see one example of a gymnosperm seed, this one from a pine tree. **Figure 21.8** then shows you how this seed fits into the life cycle of the pine, which is a representative gymnosperm. Following the sequence through, once the sperm has reached the egg and fertilized it, an embryo begins to develop from it, in the same way that a human egg starts developing into an embryo once human sperm and egg have come together. In the pine tree example, however, this embryo is developing inside a seed: a tough coat that has packed inside it a

food supply for the growing embryo—a kind of sack lunch of stored carbohydrates, proteins and fats. In sum, a **seed** is a plant structure that includes a plant embryo, its food supply, and a tough, protective casing.

The two major groups of seed plants—the gymnosperms and the angiosperms—turn out to differ in the kind of seeds they have. Angiosperm seeds come wrapped in a layer of tissue that gymnosperm seeds do not have. This layer is called fruit, and you'll be reading more about it shortly. By comparison, gymnosperm seeds are naked. Indeed, gymnosperms get their name from the Greek words *gymnos*, meaning "naked," and *sperma*, meaning "seed."

21.3 The Flowering Plants: Nature's Grand Win-Win Invention

The plants you'll be looking at from here on out are the flowering, seed-bearing plants, the **angiosperms**. The 260,000 known species of angiosperms can be defined by the way they reproduce and the way they develop as embryos, and you'll be looking at these characteristics shortly (**see Figure 21.9**). But for now,

Figure 21.7
A Gymnosperm Seed
Seeds are tiny packages of food and protection. They come in many shapes and sizes, but they all contain an embryo, some food, and a protective seed coat.

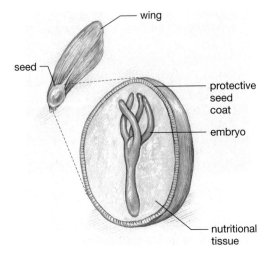

they are mixtures of pure elements within binding materials, produced by chemical processes.

Nature is, of course, perfectly capable of producing a green bounty in such places as forests and marshes without the aid of any human-made fertilizer. Decaying plant and animal matter puts nutrients back into the soil, and bacteria are able to fix nitrogen, meaning to absorb it from the air and transform it into a form that plants can take up. Looked at one way, however, houseplants and farm crops are *isolated* plants whose participation in this web of life is very limited and who thus require a helping hand from humans in order to remain robust.

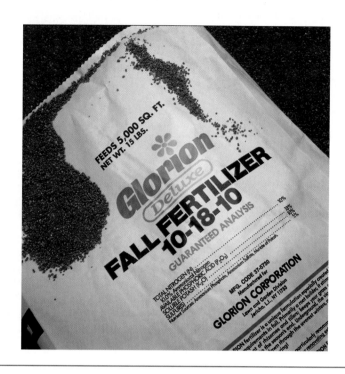

Figure 1
Needed Nutrients
Plants make their own food from sunlight, water, carbon dioxide, and a few nutrients, some of which come from the soil. The most important of these nutrients are nitrogen (N), phosphorus (P), and potassium (K). Because garden plants and houseplants are isolated from natural ecosystems, they must often get these nutrients in the form of fertilizer.

Figure 21.8
The Life Cycle of a Gymnosperm
Pine trees have two kinds of cones. The small male cones produce pollen (the same yellow dust that causes many people with allergies to sneeze in the spring), and the larger female cones produce eggs. When pollen is carried by the wind onto the female cone, the sperm within the pollen fertilizes the egg within the cone. An embryo then begins to develop inside a seed, which falls to the ground. Once conditions are suitable, the seed germinates and a whole new pine tree begins to grow.

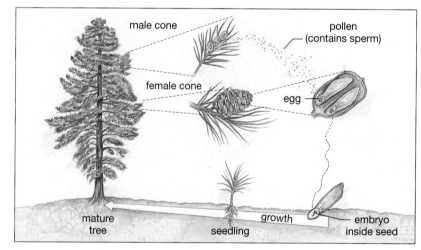

Figure 21.9
Angiosperm Variety

a Calla lilies on the California coast.

b Cholla cactus in Arizona.

c Corn in a field.

a

b

c

let's look at the *outcomes* of these characteristics—at why this vast grouping of plants is so special and so important to us.

a

b

c

Pollination Is an Important Plant-Animal Interaction

For an angiosperm such as a new maple tree to come into existence, a two-part process must take place. First, sperm—developing inside a pollen grain—must get to egg and fertilize it, which produces the embryo, encased in a seed coat. Second, this seed must land on a patch of earth somewhere and begin to germinate.

Now, how is it that sperm gets to egg so that it can be fertilized in the first place? With gymnosperms such as pine trees, pollen grains are carried by *wind* to the female reproductive structure, residing in the familiar pinecone. By contrast, the angiosperms developed a vast array of attractants that induce *animals* to carry pollen from one plant to another—to pollinate them. Not all angiosperms are pollinated in this way; some are pollinated by wind, and a few aquatic species rely on water currents. But most are pollinated with the help of animals, be they insects, birds, mammals, or even snails (**see Figure 21.10**).

The most important attractant that flowering plants use to encourage their animal couriers is nectar, which is essentially sugar-water. But angiosperms don't stop with nectar as an attractant. So important is pollination to them that they have developed a host of what might be called pollination marketing strategies. How about a *fragrance* that, say, bees are sensitive to? How about *colors* that they're attracted to? Insect-pollinated

a This is what we see (normal sunlight).

b This is what the bee sees (ultraviolet light).

c This is what the bee "thinks."

Figure 21.11
Food Lies This Way
Animals may be attracted to a particular flower by its fragrance, color, or pattern of visual elements. Insects are guided into the nutrients in the center of the flower by color patterns called nectar guides, visible to an insect—because it perceives ultraviolet light—but not to us. **(a)** What a flower looks like to a human being. **(b)** What a flower looks like to an insect that can perceive ultraviolet light. The circular pattern is the nectar guide and the light-colored flecks on the petals are pollen grains.

flowers generally have both sweet fragrances and coloration patterns, called nectar guides, that are nature's equivalent of homing signals and landing lights. They serve to get an animal to the flower in the first place and then into the right *position* for feeding and pollination (**see Figure 21.11**).

What's the value of all this for the flowering plants? The wind-borne pollination that gymnosperms practice can be thought of as a kind of scatter-shot approach: If the wind *happens* to blow a pollen grain onto the female reproductive structure, then pollination occurs. Think how much more directed things are with animal pollination; what exists, essentially, is door-to-door service. This is one of the reasons that angiosperms are by far the dominant plants on Earth.

Seed Endosperm: More Animal Food from Angiosperms

The second big step in angiosperm reproduction is for the product of this fertilization, the embryo (inside its seed), to begin germinating from the ground. Gymnosperm embryos store food reserves in structures within the seed called *cotyledons*. Angiosperm embryos store food this way too, but also store food in an additional form, called **endosperm**, that often surrounds the embryo (**see Figure 21.12**). It is this endosperm tissue that supplies much of the food for human beings throughout the world. Rice and wheat grains consist in large part of the endosperm meant to sustain the plant embryo. The white "meat" and liquid "milk" of the coconut are also forms of endosperm.

Fruit: An Inducement for Seed Dispersal

Some flowering plants then go on to add something else: tissue, whose function is to be attractive to animals as a means of getting

the plant seeds *dispersed*. This is fruit, as that term is commonly understood—the flesh of an apricot or cherry, for example. As such, it represents one more piece of bounty that plants provide to animals. Yet plants get something out of this relationship as well. Imagine a bear who consumes a wild berry, which consists not only of the fruit flesh, but of the seeds it surrounds. The fruit is digested by the bear as food, but the *seeds* are tough; they are passed through its system intact to be deposited, with bear feces as fertilizer, at a location that may be very remote from the place where the bear *ate* the berry. *Voilà!* Seed dispersal at what may be a promising new location for a berry plant (**see Figure 21.13**).

Given pollination and the varieties of food that angiosperms produce, it's easy to see that, with the rise of the flowering plants, animals and plants entered into a much more interdependent relationship than had existed before, one that is still evolving today. Taking a step further back, if we ask which plants are the most important in shaping our world, the answer is the angiosperms.

21.4 The Structure of Flowering Plants

The Basic Division: Roots and Shoots

Now let's look at the ways in which the flowering plants are structured. When approaching this subject, it's common to think in terms of different cell and tissue types, and you'll be looking at plants that way in Chapter 22. For now, however, let's look at the larger-scale structures of these plants,

CORN

— seed coat
— endosperm
— embryo

Figure 21.12
Food for the Seedling, Food for Humans
Endosperm is a food reserve in seed. It feeds the embryo before it can make its own food. It is this same endosperm that nourishes humans throughout the world in the form of corn, wheat, rice, and even coconut.

Figure 21.13
Seed Carriers
Angiosperms have taken advantage of the mobility of animals not only to transfer pollen, but also to disperse seeds.

a Some seeds are wrapped in tasty fruit that is consumed by animals, such as bears.

b Other seeds come wrapped in burrs and spines that stick to the fur of animals and are carried away.

a

b

starting with a simple, two-part division that rhymes: roots and shoots. Plants live in two worlds, air and soil, with their root system belowground and their shoot system above (**see Figure 21.14**).

The function of the root system is straightforward: Grow to water and minerals and absorb them from the soil and then begin transporting them up through the rest of the plant. In addition, most roots are anchoring devices for plants, and some also act as storage sites for food reserves.

The function of the shoot system is more complex. Photosynthesis takes place in this system, primarily in leaves, which must thus be positioned to absorb sunlight. Plants are in *competition* with one another for sunlight, which is a primary reason so many plants are tall. The food that is produced in photosynthesis must be distributed throughout the plant. It is not just food production that is centered in the shoot system, however; it is plant reproduction as well. It is within flowers and their derivatives that we find all the components of reproduction you've been reading about—seeds, embryos, pollen, and so forth. Now let's look in more detail at the components of the roots and shoots systems.

Roots: Absorbing the Vital Water

If you look at **Figure 21.15**, you can see pictures of the two basic types of plant root systems, one a **taproot system** consisting of a large central root and a number of smaller lateral roots, and the second a **fibrous root system** consisting of a number of roots that are all about the same size. **Figure 21.16** then shows you the taproot of a young sweetcorn plant, with a common addition to root systems. These are **root hairs**—the threadlike extensions of main roots that greatly increase the absorptive surface area of roots. Each root hair actually is an elongation of a single outer cell of the main root. Between main roots and root hairs, the root structure of a given plant can be extensive. One famous analysis, conducted in the 1930s on a rye plant, concluded that the main and lateral roots alone, if laid end to end, would have totaled some 622 kilometers, which is about 386 miles. Meanwhile, the root hairs collectively were 10,620 kilometers long, meaning the plant had almost 6,600 miles of root hairs! This from a plant whose *shoot* stood 8 centimeters (about 3 inches) off the ground.

Why do plants lavish such resources on their roots? For one thing, water doesn't come to roots nearly as much as roots must come to water; roots must continually *grow*

Figure 21.14
A Life-Form in Two Parts
Flowering plants live in two worlds, with their shoots in the air and their roots in the soil. Although flowering plants differ enormously in size and shape, they generally possess the external features shown.

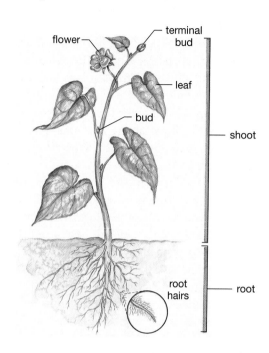

Figure 21.15
Two Root Strategies

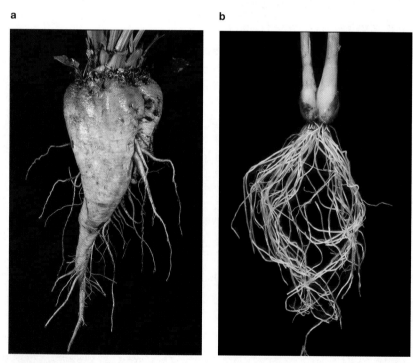

a The taproot system of this sugar beet shows a large central root and a number of smaller lateral roots.

b The fibrous root system of this onion shows a number of roots that are all about the same size.

to new water supplies. Why such a great need for water? To perform photosynthesis, plants must have a constant supply of carbon dioxide, which enters the plants through microscopic pores, called *stomata*, that exist mostly on the underside of plant leaves. Open stomata, however, don't just let carbon dioxide in; they let the plant's water *out*, as water vapor. To accommodate this fact while still keeping vital tissues moist, plants continually pass water through themselves, from roots, up through the stem, into the leaves (at which point it exits as water vapor). Through this process, known as **transpiration**, more than 90 percent of the water that enters a plant evaporates into the atmosphere. Because the roots of a single tall maple tree can absorb about 220 liters (or nearly 60 gallons) of water per *hour* on a hot summer day, the scale of transpiration obviously is immense. Thus do we see the importance of root development.

The food storage aspect of roots has a meaning for human beings. Sweet potatoes and carrots, which are roots, can represent stored food for people as well as for plants.

Shoots: Leaves, Stems, and Flowers

Leaves: Sites of Food Production

The primary business of leaves is to absorb the sunlight that drives photosynthesis, which is why most leaves are thin and flat. This leaf shape maximizes the surface area that can be devoted to absorbing sunlight while minimizing the number of cells in leaves that are irrelevant to photosynthesis. Beyond this, through their tiny stomata, leaves serve as the plant's primary entry and exit points for gasses. As noted, the most important gas that's entering is carbon dioxide, which is one of the starting ingredients for photosynthesis. What's exiting is the by-product of photosynthesis, oxygen, and the water vapor.

The broad, flat leaves that are so common in nature have in essence a two-part structure—a **blade** (which we usually think of as the leaf itself) and a **petiole**, more commonly referred to as the leaf stem. If you look at the idealized cross section of this leaf in **Figure 21.17b**, you can see that the blade can be likened to a kind of cellular sandwich, with layers of **cuticle**, or waxy outer covering, on the outside, and a layer of epidermal cells just inside them. In the leaf's interior there are **vascular bundles**, which bear some relation to animal veins in that they are part of a transport system. Then there are several layers of mesophyll cells. It is these cells that are the sites of most photosynthesis.

Figure 21.17
Site of Photosynthesis

a Leaves tend to be broad and flat to increase the surface area exposed to sunlight.

b Photosynthesis occurs primarily within the mesophyll cells in the interior of the leaf. The vascular bundles carry water to the leaves and carry the product of photosynthesis, sugar, to other parts of the plant. It is the pores called *stomata*, mostly on the underside of leaves, that allow for the passage of gases in and out of the leaf

c Guard cells of the stomata control the opening and closing of the stomata. When sunlight shines on the leaf during the day, the guard cells engorge with water that makes them bow apart, thus opening the stomata. When sunlight is reduced, water flows out of the guard cells, causing the stomata to close.

Figure 21.16
Root Hairs
Root hairs enormously increase the surface area of a root. The greater the surface area, the more fluid absorption can occur. Shown here are root hairs on the taproot of a sweet-corn plant.

What Is It?
See p. 432.

Figure 21.18
Leaves Come in Many Sizes, Shapes, and Colors
Simple leaves have just one blade. In compound leaves, the blade is divided into little leaflets. Some leaves are so modified that they are not even recognizable as leaves—for example, the spines of a cactus.

Now note, on the underside of the blade, the openings called **stomata** (singular, *stoma*). These are the pores, noted earlier, that let water vapor out and carbon dioxide in—but for most plants only during the day. When the sun goes down photosynthesis can no longer be performed, and it is not cost-effective for a plant to lose water without gaining carbon dioxide. The stomata thus close up until photosynthesis begins again the following day. If you look at Figure 21.17c, you can see how this opening and closing of stomata is achieved: Two "guard cells," juxtaposed against one another like sides of a coin purse, are arranged around the stomata. When sunlight strikes the leaf, these cells engorge with water, which makes them bow apart. Then, in the absence of light, water flows out and the door closes again. A given square centimeter of a plant leaf may contain from 1,000 to 100,000 stomata.

Figure 21.18 gives you some idea of the varied forms of leaves. In addition to the so-called simple leaf we've been looking at, there are compound leaves in which individual blades are divided into a series of "leaflets." Pine needles and cactus spines are likewise actually leaves. The cactus, in particular, has reduced the amount of leaf surface it has as a means of conserving water. So how does it perform enough photosynthesis to get along? It carries out most of its photosynthesis in its stem.

Stems: Structure and Storage

We all generally understand what the stem of a plant is, though it is less generally appreciated that the trunk of a tree is simply one kind of plant stem. The main functions of stems are to give structure to the plant as a whole and to act as storage sites for food reserves. In addition, water, minerals, hormones, and food are constantly shuttling through (and to) the stem, with water on its way up from the roots and food on its way down from the leaves to the rest of the plant.

If you look at **Figure 21.19**, you can see a cross section of one type of plant stem, showing the vascular bundles or veins you first saw in the leaves; and the outer, or *epidermal, tissue* just as the leaves had. You can also see so-called *ground tissue* of two types: an outer cortex and an inner pith, both of which can play a part in food storage and wound repair and provide structural strength to the plant.

Flowers: Many Parts in Service of Reproduction

Flowers are the reproductive structures of plants. A single flower generally has both male and female reproductive structures on it, which might make you think that a given plant would fertilize *itself*. This is indeed the case with some plants, such as the pea plants reviewed in Chapter 11. However, because the evolutionary benefit of sexual reproduction is to get the genetic diversity that comes with *mixing* genetic material from different organisms, natural selection has worked against self-fertilization by endowing many flowering plants with mechanisms to reduce the incidence of it. For example, the male pollen of a given plant might be genetically incompatible with the plant's female reproductive structures. Some of the illustrations you'll be seeing show a plant fertilizing itself, but this is done only for visual simplicity.

compound leaves

simple leaves

leaves modified as spines

leaves modified as tendrils

If you look at **Figure 21.20**, you can see the components of a typical flower. Taking things from the bottom, there is a modified stem, called a pedicel, which widens into a base called a receptacle, from which the flowers emerge. Flowers themselves can be thought of as consisting of four parts: sepals, petals, stamens, and a carpel. The **sepals** are the leaflike structures that protect the flower before it opens. (Drying out is a problem, as are hungry animals.) The function of the colorful **petals** is to announce "food here" to pollinating animals.

The heart of the flowers' reproductive structures are the stamens and the carpel. If you look Figure 21.20, you can see that the **stamens** consist of a long, slender **filament** topped by an **anther**. These anthers contain cells that ultimately will yield pollen grains and the sperm inside them. At maturity, these pollen grains will consist of three cells surrounded by an outer coat.

It is these pollen grains that will be released and then carried—perhaps by a pollinating bee or bird—to the carpel of another plant (or perhaps of the same plant). As Figure 21.20 shows, a **carpel** is a composite structure, composed of three main parts: The **stigma**, which is the tip end of the carpel, on which pollen grains are deposited; the **style**, a slender tube that raises the stigma to such a prominent height that it can easily catch the pollen; and the **ovary**, the area in which fertilization of the female egg and then early development of the plant embryo take place.

21.5 How Flowering Plants Function

Having looked at the structure, or anatomy, of the angiosperms, let's now go over the activities of some of these components. Think in terms of systems here: the reproductive system, the transport system, the hormonal system, the communication system, and the defense system. In addition, you'll look briefly at the nature of plant growth. Because you've just reviewed flower anatomy, let's start with reproduction.

Reproduction in Angiosperms

Sexual Reproduction

The function of a pollen grain is the fertilization of an egg. There are cells in the anthers

of the original flower that undergo the type of cell division reviewed in Chapter 10, meiosis, thereby producing **microspores** that are haploid—that contain only a single set of chromosomes. As you can see in **Figure 21.21** (page 432), each pollen grain develops from one of the microspores inside the anther. In the type of plant pictured in the figure, the grain will consist, by the time it leaves the anther, of a tough outer coat and three cells: one **tube cell** and two **sperm cells**. Though the pollen grain never gets much more complex than this, it is not just a component of the original plant; it is its alternate generation on the male side. (For more on the alternation of generations in plants, see page 412 in Chapter 20.)

You can see in Figure 21.21 what the activities of the pollen grain's cells are in the context of the angiosperm life cycle. When

Figure 21.19
The Stem and Its Parts
Stems provide support to the rest of the plant, act as storage sites, and conduct fluids. Vascular plants have a "plumbing" system that transports food, water, minerals, and hormones. The vascular bundles in the figure are groups of tubes, running in parallel, that serve this transport function.

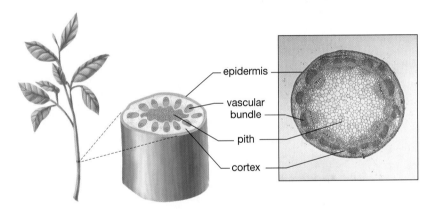

Figure 21.20
Introduction to the Flower
Flowers are composed of four main parts: the sepals, petals, stamens, and a carpel. The sepals protect the young bud until it is ready to bloom. The petals attract pollinators. The stamen are the reproductive structures that produce pollen (which contains sperm cells). The carpel is the female reproductive structure that contains the egg within the ovary.

the pollen grain leaves the anther, it is bound for the stigma of another plant (or its own plant). But for a grain to merely *land* on a stigma doesn't mean that anything has been fertilized. The tube cell in the grain must then begin to *germinate* on the stigma, forming a **pollen tube** that grows down through the style. Once this has taken place, one of the sperm cells in the grain travels through the tube, gets to the female egg, and only *then* is there fertilization. (The other sperm moves down along with the first, but then spurs the growth of the food reserves for the growing embryo, about which you'll learn more in Chapter 22.)

And how does the egg arise that is fertilized by the the sperm? Together with its supporting cells, the egg is the plant's alternate generation on the female side. Inside the plant's ovary, there is a cell that also goes through meiosis, thus producing a **megaspore** (*mega*, because it's bigger than the male microspore). It in turn gives rise to several kinds of cells, one of which is the egg that the sperm from the pollen grain will fertilize.

Watch closely now, because once this happens—once sperm (from pollen grain) has fertilized egg (from megaspore)—it is the beginning of the kind of flowering plant we started with. We have alternated back to this generation of plant. Many a step remains before arriving at something that *looks* like the original plant: The fertilized egg (now called a zygote) must first have a seed develop around it, and that seed must be released and then land on a suitable patch of earth, there to germinate and grow to full flowering plant. But the step at which this new generation appears is the point at which sperm fertilizes egg.

To put some terms to these generations, note that both the megaspore and pollen grain are haploid spores. Because *spores* are what the original, flowering plant produced, this generation is the **sporophyte generation**. Meanwhile, the pollen and megaspore generation produced *gametes*—sperm or egg—and are thus the **gametophyte generation**. When the sperm and egg came together, they fused their haploid sets of chromosomes and thus produced a sporophyte embryo with a doubled, or diploid, set of chromosomes. A detailed account of angiosperm reproduction can be found in Chapter 22, beginning on page 464.

Asexual Reproduction in Angiosperms: Vegetative Reproduction

We are used to the human mode of reproduction, which—pending the arrival of cloned human beings—is always *sexual* reproduction: At some point, sperm must fertilize egg to produce a new embryo. However, plants, including angiosperms, have another option, **vegetative reproduction**, which is *asexual* reproduction. This is familiar to us in the form of "cuttings" that can be taken from one houseplant to start another. A growing cutting represents a new plant that is an exact genetic replica of the original plant; no alternate generation or fertilization is required. One sporophyte plant has been grown from another.

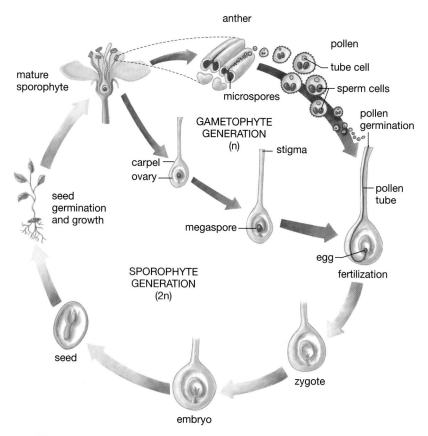

Figure 21.21
The Angiosperm Life Cycle
The mature sporophyte flower produces many male microspores within the anther, and a female megaspore within the ovary. The male microspores then develop into pollen grains that contain the male gamete, sperm. Meanwhile the female megaspore produces the female gamete, the egg. Pollen grains, with their sperm inside, move to the stigma of a plant. There the tube cell of the pollen grain sprouts a pollen tube that grows down toward the egg. The sperm cells then move down through the pollen tube, and one of them fertilizes the egg. This results in a zygote that develops into an embryo protected inside a seed. This seed leaves the parent plant and sprouts or "germinates" in the earth, growing eventually into the type of sporophyte plant that began the cycle.

Human beings are motivated to put such vegetative reproduction to use when nature produces a particularly *valuable* plant, generally through mutation. Navel oranges, for example, seem to have arisen through mutation on a conventional orange tree that was grown in Brazil in the nineteenth century. If pollen from that tree had helped fertilize an egg from another tree, we may well have *lost* the advantage this unique tree provided—fat, navel oranges—through the shuffling of genetic material that comes with sexual reproduction. The trick was to use cuttings (or "grafts") from the original tree to produce one *identical* generation of orange tree after another. Vegetative reproduction is also common in nature. The roots of aspen trees, for example, produce a form of shoot known as a sucker which, if physically separated from the parent plant, will grow into a new aspen tree. Oftentimes, then, a stand of aspen trees amounts to a group of clones of one another (**Figure 21.22**).

Plant Plumbing: The Transport System

In looking at leaves and stems earlier, you saw that they have something called *vascular bundles* running through them. This term refers to collections of tubes through which fluid materials move from one part of the plant to another. This transport system bears obvious similarities with the circulation systems of animals, but there are differences as well. Many animals employ blood as a transport medium, and animals such as ourselves have a transport *pumping* device, called a heart. Plants do not have blood, and they have no pumping system. (In fact, plants have no "moving parts" at all.) Yet think of the transport job plants have to do. Water that is transpired from the top of a redwood tree may have made a journey of more than 100 meters straight up—more length than a football field! What kind of power is at work here?

There are two essential components to the plant transport system. First there is **xylem**, which can be defined as the tissue through which *water* and dissolved minerals flow in vascular plants. (*Tissue* means a group of cells that perform a common function.) Second, there is **phloem**, which is the tissue through which the *food* produced in photosynthesis—mostly sucrose—is con-

ducted, along with some hormones and other compounds. As noted before, water is making a directional journey from root through leaf, but the plant must be able to transport food and hormones everywhere within itself.

If you look at **Figure 21.23** on page 434, you can see an idealized view of a plant's transport system as it would appear within a stem. You'll also see that the vascular bundles noted earlier are composed of bundles of linked xylem and phloem tubes that run in parallel.

Xylem is composed of two types of fluid-conducting cells, *vessel elements* and *tracheids*, which have different shapes as you can see. Upon reaching maturity, these cells do the rest of the plant the favor of dying. The cells' content is cleared out, leaving strands of empty cells stacked one on top of another—leaving tubes, in other words. The walls of these cells remain, however, and indeed are reinforced with the materials cellulose and lignin, which provide the *load-bearing* capacity that allows something as massive as a redwood tree to be so tall. (See "Keeping Cut Flowers Fresh," page 435.)

Phloem is also composed of two types of cells, but the arrangement here is very different. One type of phloem cell—a *sieve element*, which does the actual nutrient conducting—doesn't undergo cell death upon maturity, but it does lose its cell nucleus, meaning nearly all its DNA. Those of you

**Figure 21.22
Reproduction without Sex**
Stands of aspen trees—the yellow trees in the picture—often are clones of one another, reproducing without the fusion of eggs and sperm from different individuals.

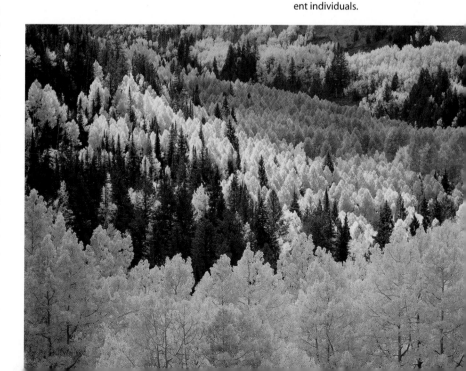

who went through the genetics unit may wonder how any cell could function without its DNA information center, but there is an answer. Each sieve element cell has associated with it one or more *companion cells* that retain their DNA and that seem to take care of all the housekeeping needs of their sister sieve elements. So, why do sieve elements lose their nuclei? Apparently to make room for the rapid flow of nutrients through them. In Chapter 22 you'll look at the means by which phloem and xylem conduct their respective materials.

Communication: Hormones Affect Many Aspects of Plant Functioning

We're so used to associating hormones with *animal* functioning that it may come as a surprise to learn that plants also have hormones. Taken as a whole, plant hormones do many of the same things that animal hormones do. Their most important roles are to regulate growth and development and to integrate the functioning of the various plant parts. In less abstract terms, hormones make buds grow and leaves fall and fruit ripen. This goes hand in hand with helping plants respond to their environments—to heat and cold, munching goats, and voracious insects.

Though most people have an intuitive sense of what hormones are, a definition might be helpful. **Hormones** are chemical messengers; they are substances made in one part of an organism that influence the functioning of

another part of the organism. In animals, hormones are generally synthesized in well-defined organs, called *glands*, whose main function is to produce hormones (for example, the thyroid or adrenal gland). In plants, however, hormone production is a more diffuse process, taking place not in glands, but in collections of cells that carry out a range of functions.

Five major plant hormones have been discovered to date. Three of these actually are *classes* of hormones, and two are individual hormones (**see Table 21.1** on page 437 for a list of all these hormones). You can read about the hormone ethylene in "Ripening Fruit Is a Gas," on page 436. Let's go over just one other plant hormone, a member of the family of hormones known as *auxins*, to give you some idea of what plant hormones do.

An Auxin Gives the "A" Shape to Trees

Why do Christmas trees have their characteristic "A" shape? (**See Figure 21.24**.) Thanks in large part to the effects of an auxin known as IAA. Unlike animals, plants do not grow globally, over their whole surface, but instead confine most of their vertical growth to special regions, called *apical meristems*, that exist at the *tips* of the roots and shoots. **Figure 21.25** shows you how this works out in connection with the shoots; the shoot apical meristem gives rise to a series of growth *modules*: more stem, one or more leaves, and a lateral bud that forms in tandem with each leaf. These lateral buds have meristem tissue

Figure 21.23
Fluid-Transport Structure of Plants

Fluids move through plants in sets of tubes called *vascular bundles*. Two types of tissue, running in parallel, can be found in each vascular bundle: xylem, the tissue through which water and dissolved minerals flow; and phloem, the tissue through which the food produced in photosynthesis flows. Xylem is composed of two kinds of fluid-conducting cells, vessel elements and tracheids. Phloem is composed of sieve elements and their companion cells.

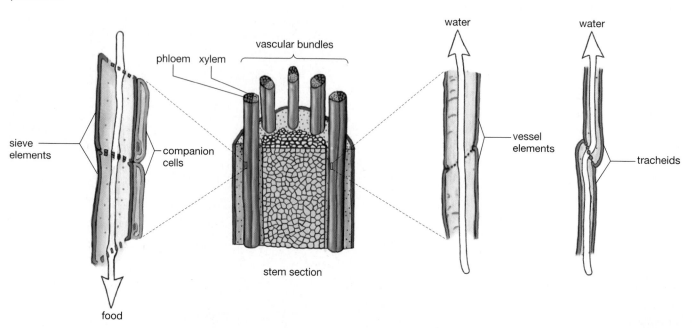

Keeping Cut Flowers Fresh

As already noted, xylem is the plant tissue through which water moves *up,* from roots through leaves. The flowers we put in vases in our homes have lost their roots, of course, but they haven't lost their xylem, which continues to function long after the flower has been picked. Given this, flowers can last a long time indoors; but we can maximize their stay if we follow a few simple rules.

First, realize that the liquid in the xylem is under negative pressure—its natural tendency is to move up *into* the stem, not to flow out of it. As such, if the stems are cut when they are out of water, *air* gets sucked up into the cut ends, creating air bubbles (called *embolisms*) that can then get trapped in the xylem and keep water from rising up through it. When this happens, flowers can wilt, even when their stems are submerged in clean water. Recutting the stem under water (or under a steady stream from the faucet) can remove this blockage. Better yet, cut the stems under water the first time.

Beyond this, acidic sugar-water, such as can be found in citrus-flavored soft drinks, will prolong the life of some flowers by keeping bacterial growth down; changing water frequently is a good idea, particularly when it starts to look gummy or discolored. Keep your arrangement out of direct light or heat, and remove dead and dying flowers in the arrangement, because the hormone ethylene is given off by dying flowers and in many cases will hasten the demise of the healthy ones in the bunch. (For another role of ethylene, see "Ripening Fruit Is a Gas," page 436.)

---

in them and can grow into new branches, but they normally don't *when they lie close to the apical meristem.* Why? The apical meristem is producing IAA, which works in tandem with ethylene to suppress the growth of these buds. But, the farther away from the apical meristem, the *smaller* the concentration of IAA that is available to the lateral buds. The result is a greater budding of branches at the *base* of a tree, and a tapering of this growth going from the base up through the tree's apex. This phenomenon, called **apical dominance**, is most pronounced in certain conifers, such as the Douglas firs often used for Christmas trees. Meanwhile, trees that don't have such a strict IAA gradient don't get the strict "A" shape. This maintenance of apical dominance is just one of the many things that IAA does. To give you some idea of the complexity of its function, it may serve to *suppress* the emergence of lateral buds, as noted, but it works to *promote* the cellular elongation that is important in plant growth.

Plant Growth: Indeterminate and at the Tips

Having looked a little at how plant growth is affected by hormones, we can note some basic characteristics of plant growth. To visualize how it works, imagine a boy of about 8

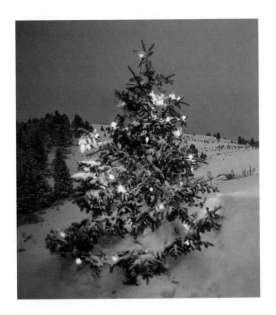

Figure 21.24
Why the "A" Shape?
The effects of apical dominance are clearly visible in a Christmas tree. The apical meristem tissue at the tip of the tree produces a hormone called IAA that inhibits the growth of lateral branches. Because the concentration of the inhibitor IAA is highest at the top near the apical meristem, and lowest at the bottom, the branches are very short near the top, but longer toward the bottom.

Figure 21.25
Modular Growth
Vertical growth in plants occurs in "modules." Each module includes a new portion of stem, at least one leaf, and a lateral bud that can give rise to a branch or a flower. Here alternating green and black segments indicate modules and we go from one module in week one to four modules in week four.

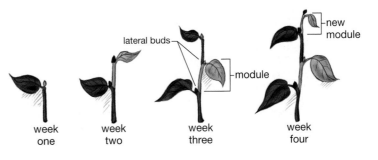

lateral buds

new module

module

week one

week two

week three

week four

Ripening Fruit Is a Gas

That banana you had? The one you put in that plastic bag, thinking you'd make it part of your lunch? It already had a few black spots on it by the time it went in the bag, but then you forgot about it for lunch and, by the time you looked at it again, it was more black than yellow and something you were no longer interested in eating. The question is: Why did it go downhill so *fast*?

The answer is that those little black spots on the banana result from concentrations of a gas called *ethylene*—a plant hormone that can ripen lots of fruits. The bag you put the banana in became a kind of gas chamber, with the ethylene concentrations inside being very high. Moreover, ethylene stimulates its own production; the more ethylene there was, the more that got produced. Given this contagious quality to ethylene, you can see why it is true that one bad apple can spoil the bunch.

Ethylene doesn't work on all fruits; grapes, strawberries, and cherries, for example, are unaffected by it. But apples, avocados, and tomatoes join bananas in being very sensitive to it. If you wish to ripen one of these latter fruits fast, putting it in a plastic bag will give you an edible fruit faster than putting it on the window sill.

But what does it mean for a fruit to be ripe? Let's match up our intuitive sense of this with the underlying physical changes. Compared to unripened fruits, ripe fruits tend to be sweeter (starches are converted to sugars); softer (a secretion of enzymes softens cell walls); and more fragrant (ripening releases compounds that our nose detects as sweet-smelling). Ethylene helps bring about all these changes, basically by upping the tempo at which a plant carries out its metabolic processes. The reason a green banana turns to yellow is that ethylene hastens the breakdown of the green chlorophyll in the banana, which allows us to see yellow pigments that had been there all along.

For years, commercial fruit growers have employed ethylene, or test-tube compounds that help release it, in order to *time* the ripening of their fruit. This is why we can pick green tomatoes at the source and then ripen them when they get closer to market. It's also possible to *suppress* ethylene production by storing fruit in a room that has high concentrations of carbon dioxide and low concentrations of oxygen. In this way, a fruit such as apples can be kept on hold for months before being brought to ripeness with a dose of this gas.

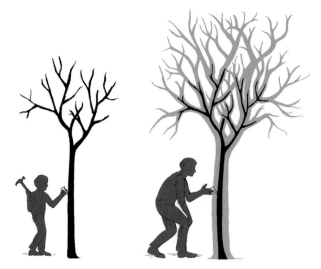

When Billy was eight...   When Billy was eighteen...

Figure 21.26
The Basic Characteristics of Plant Growth
If a young boy drives a long nail into a tree and returns 10 years later, two things will be apparent: The nail is at the same height it used to be despite the increased height of the tree, and it is almost completely buried in the tree. What does this say about tree growth? Trees do not elongate throughout their whole length, but only at their tips in "primary growth." Further, they widen laterally in "secondary growth."

years old who decides to drive a long nail into the trunk of a secluded tree. He doesn't drive the nail in completely, but lets it stick out a bit. After 10 years pass, the boy happens to walk by the same tree; upon seeing it, a couple of things occur to him (**see Figure 21.26**). First, he has grown but the nail is in the same place; he would have to bend down now to pull the nail out, whereas when he drove it in he was standing up straight. Second, he would have a hard time actually pulling the nail out now because not much of it is visible; the tree has grown around it. Third, if the nail's height is any indication, the trunk of the tree may not have moved up, but the *top* certainly has; it's much taller than he remembered it.

The lessons here are that, in their vertical growth, plants do not grow the way people do—throughout their entire length—but instead grow only at their tips (of both shoots *and* roots, though the boy couldn't see the

Table 21.1
Plant Hormones

Hormone	Major Functions	Where Found or Produced in Plant
Auxins	Suppression of lateral buds; elongation of stems; growth and abscission (falling off) of leaves; differentiation of xylem and phloem tissue	Root and shoot tips; young leaves
Cytokinins	Stimulate cell division; active in the development of plant tissues from undifferentiated cells	Roots
Gibberellins	Stem elongation, growth of fruit, promotion of seed germination	Seeds, apical meristem tissue, young leaves
Ethylene	Ripening of fruit, retardation of lateral bud growth, promotion of leaf abscission	Fruits, leaves, stems
Abscisic Acid	Induces closing of leaf pores (stomata) in drought; promotes dormancy in seeds; counteracts growth hormones	Young fruit; leaves

roots). In addition, as the half-buried nail attests, this tree has carried out the *lateral* growth that botanists refer to as *secondary growth*, which you'll look at next chapter.

Beyond these things, it turns out that a plant's growth is indeterminate: In most plants it can go on indefinitely at the tips of roots and shoots. By contrast, animal growth is generally determinate: It comes to an end at a certain point in development. Imagine if, say, the tips of your fingers just kept growing long after your trunk and arms had reached their full extension.

Defense and Cooperation

Plants may be immobile and toothless, but they are not defenseless. Indeed, they have developed a formidable defense arsenal (**see Figure 21.27**). One variety of plant defenses, the structural defenses, are familiar to us because they're so visible—think, for example, of the spines on a cactus or the thorns on a rose. A second variety of defenses, the chemical defenses, are not so visible but are potent nevertheless. The chomping a beetle does on a potato plant induces the plant to produce substances (called *protease inhibitors*) that are believed to give the insect an ache in its gut that it won't forget. Other plants, when attacked by a fungus, release antifungal compounds that spread not only

at the site of infection, but throughout the plant.

Beyond these self-defense mechanisms, plants can warn *other* plants of dangers in their midst. Airborne signaling compounds can be produced from the leaf surface of a plant under attack, waft to a neighboring plant, and prompt it to start producing compounds that are noxious or harmful to the would-be plant-eater. Some plants can even call in animal allies. Corn plants that are attacked by army worms release a substance

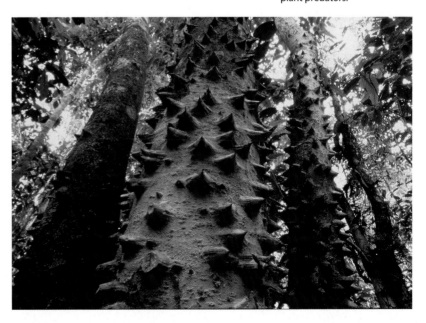

Figure 21.27
Silent and Fixed in One Spot, but Not Defenseless
The spiny bark of this Costa Rican tree, a member of a genus *Zanthoxylum*, protects it from plant predators.

Figure 21.28
Working Arrangement
Plants need nitrogen to grow. Some get it through a cooperative relationship with bacteria that are nitrogen-fixing, meaning they can convert atmospheric nitrogen into a form living things can use. Shown are the root nodules (the small spheres) on the roots of a soybean plant. Bacteria make a home in the nodules, where they fix nitrogen that is then used by the plant.

that attracts parasitic wasps, which then devour the worms.

Plants also enter into cooperative relationships with other organisms. In Chapter 20 you saw that most plant roots are linked to the thin tubes, called *hyphae*, that extend from fungi. What the fungi get from this is nutrition from the photosynthesizing plant; what the plants get is a greatly expanded water and mineral absorption network. Other species, notably the legumes, such as soybeans, have developed partnerships with nitrogen-fixing bacteria, which is to say bacteria that are able to take nitrogen out of the atmosphere and put it into a form that living things can use. The bacterium infects the plant through a root hair, directs the construction of its own lodgings (called a *root nodule*), and then begins converting atmospheric nitrogen to ammonia. This process must take place in the absence of oxygen, which the plant sees to by eliminating the available oxygen in the nodule. Lots of crops require applications of human-manufactured nitrogen, but legumes such as beans and alfalfa can do fine on their own (**see Figure 21.28**).

21.6 Responding to External Signals

Plants may lack a nervous system, but this doesn't mean that they can't respond to their environment. They must do so, actually, be-

cause their life depends on it. They sense the march of the seasons by measuring available light and the surrounding temperature; they use internal clocks to synchronize their development and physical functioning; they respond to light, gravity, touch, heat, cold, and pH, among other things. Let's conclude the chapter by reviewing the plant's interactions with light, gravity, touch, and the seasons as examples of these environmental responses.

Responding to Light: Phototropism

Many plants will bend toward a source of light—meaning even a candle—with the value of this perhaps being obvious: Plants produce their own food, and that production depends on light. Thus a plant must avoid having its sunlight *blocked* by another plant or some other physical object in its surroundings. As it turns out, it is IAA, produced in the shoot-tips of growing plants, that controls this **phototropism**, meaning a curvature of shoots in response to light. When light strikes one side of the shoot, it causes IAA to migrate to the other side, where the IAA acts to promote the *elongation* of cells on this side. The effect is to make the shoot curve toward the light (**see Figure 21.29**).

Responding to Gravity: Gravitropism

When a seed first starts to sprout, or "germinate," its roots and shoots must go in very specific directions: The root must go *down*, toward the water and minerals the plant needs, and the shoot must go *up* toward the sunlight. Imagine a seed that has simply fallen on the ground; it would not do for either root or shoot to grow *horizontally*; each must grow vertically. If you look at **Figure 21.30**, you can see dramatic evidence of a plant's ability to sense which way is up and which way down. The impatiens in Figure 21.30a was placed on its side and, within 16 hours, the shoot began curving upward, as you can see. Figure 21.30b shows the *root* of a germinating sweet-corn seed, likewise oriented horizontally (by researchers), but quickly beginning to bend downward. What's being exemplified here is **gravitropism**, meaning a bending of a plant's root or shoot in response to gravity. Figure 21.30c again shows the root of a

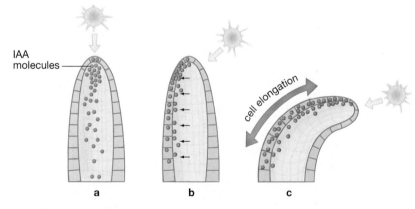

IAA molecules

cell elongation

a   b   c

ANIMATION

Figure 21.29
Plants Respond to Light

a When sunlight is overhead, the IAA molecules produced by the apical meristem are distributed evenly in the shoot.

b Once the sunlight shines on the shoot at an angle, the IAA molecules move to the far side and induce the elongation of cells on that side.

c Cell elongation results in the bending of the shoot toward the light.

sweet-corn plant, only this one has had its very tip, or *root cap*, snipped off. Here there is no bending in accordance with gravity, but rather a continuation of straight, horizontal growth. Conclusion? Cells or substances in the root cap are essential for root gravitropism.

But how do roots "know" which way is down, and how do they orchestrate a course correction when they need one? Logically, there must be at least two elements at work here: first, a gravitational *sensing* mechanism and then a means of *responding* to gravitational cues, such that corrective bending points the root toward the center of Earth. For years, the leading candidate for the sensing was a group of starch-storing organelles within plant cells called *amyloplasts*. Like a bubble mechanism in a carpenter's level, amyloplasts can be seen to move up or down within the cell in response to gravity, and researchers know that they are abundant in certain cells of the root cap. So far so good, but some genetically engineered plants, specially designed to *lack* amyloplasts, turned out to exhibit gravitropism anyway. As it stands, therefore, we don't know what the gravity-sensing mechanism is in plants. With respect to the response mechanism, what is at work is something similar to what is seen in phototropism: relatively greater cell elongation along one *side* of a root or shoot, such that the bending takes place.

Responding to Contact: Thigmotropism

We've all seen plants that manage to climb upward by encircling the stem of another plant. Whole stems can undertake such encircling, but it is often a thin, modified leaflet called a **tendril** that does so (**see Figure 21.31**). But how is a tendril able to wrap around another object? Once again through differential growth—more rapid growth on one side of a tendril than on the other. Contact with the object is perceived by outer, or epidermal, cells in the tendril. This sets into motion the differential growth, which probably is controlled by the hormones IAA and ethylene. This process is called **thigmotropism**, meaning growth of a plant in response to touch. Plants with this capability can piggyback

on other plants in order to get more access to sunlight.

Responding to the Passage of the Seasons

Dormancy in Winter

The profusion of brightly colored leaves in the fall is one of the great seasonal markers in temperate climates such as exist in the American Midwest and most of the East Coast. As green leaves turn red, gold, and purple, we know that autumn has arrived (**see Figure 21.32**, page 440). Trees that exhibit a coordinated, seasonal loss of leaves are called **deciduous** trees. But why should these broad-leafed trees lose their leaves, while the evergreen pines and firs keep the modified leaves known as needles?

First and foremost, there is relatively little water available in winter (most of it being frozen in the ground), and flat-leafed trees transpire more than do evergreens. In addition, because of the way wind flows over flat leaves, they lose more *heat* than do pine needles. Thus, the deciduous trees have evolved

a

b

c

Figure 21.30
Plants Respond to Gravity
a Potted impatiens demonstrate the effects of gravitropism. The plant on the right was laid on its side 16 hours before the photo was taken, by which time the shoot had curved, beginning to grow upward.

b Similarly, the root of a germinating sweet-corn seed begins to curve downward only hours after being placed on its side. How do these plants sense which way is up and which way is down?

c One hint is that if the very tip, or "root cap" of the root is removed, the root no longer bends downward. Researchers are still trying to determine what it is in the root cap that is responding to gravity.

Figure 21.31
Moving up by Curling around
This pea plant is able to grow higher through means of its tendrils curving around the structure in the photo. This is an example of thigmotropism in action.

a strategy that is a matter of straight economics. Such winter photosynthesis as they might perform would not be worth the water loss that would result. The strategy thus becomes to lose the leaves, grow new ones next spring, and in the meantime perform no photosynthesis but exist instead completely on food reserves. This state, in which growth is suspended and there is a low level of metabolic activity, is called **dormancy**. Evergreens exhibit dormancy too, in low temperatures, but deciduous trees are locked into it until the coming of spring.

Limitation of Loss in the Deciduous Strategy

The deciduous strategy, of first losing leaves and then growing them back, may at first seem very wasteful, akin to a company destroying much of its machinery each year, only to rebuild it months later. But consider how economical trees are about this. They *reclaim* the nutrients they have stored in their leaves. Proteins are broken down by enzymes and shipped, along with the nutrients, to storage cells in the stem and roots. By the time a leaf is ready for **abscission**, or detachment from the tree, it is little more than an empty shell of cell walls.

Before this happens, however, the change in coloration takes place. It may surprise you to learn there is less *adding* of new colors here than an *unmasking* of colors that existed in the leaves all along. With the approach of fall, green chlorophyll, which is the main pigment in photosynthesis, begins to break down in the leaf; with it gone, the yellow and orange colors provided by the carotenoid pigments become visible. New colors are added as well; blue and red come with the synthesis of pigments called *anthocyanins*. (These pigments are largely responsible for the stunning red color of many maples in fall.) The combination of carotenoids and anthocyanins gives us the final result, which is multihued leaves.

Photoperiodism

So, how does a deciduous tree sense when it's time to begin preparing for cold weather? In general it is an interaction between two factors. First there is cold itself, which brings about changes in metabolism. In addition, there is a phenomenon known as **photoperiodism**, which is the ability of a plant to respond to intervals of day and night. This is another example of the plant's ability to respond to its environment, but it should not be confused with the "phototropism" or *bending* toward the light that was discussed earlier. In photoperiodism, plants are making seasonal adaptations based on the length of the nights they are experiencing. Most plants in the Earth's temperate regions and many in the subtropics exhibit photoperiodism, which can affect such processes as flowering as well as the onset of dormancy.

To get a feel for how photoperiodism works, consider ragweed, a plant that is the bane of many hay-fever sufferers. After sprouting in the spring and reaching a certain level of maturity, ragweed flowers only when it is in darkness for 10 hours *or more* per day. When would this threshold be crossed? As the nights begin to get longer. Ragweed thus joins a group of plants, called long-night plants, whose flowering comes only with an *increased* amount of darkness. As you can imagine, most of these are like ragweed in that they flower only in late summer or early fall. Their counterparts are the so-called short-night plants, which flower only when the nights get short enough—that is to say, when nighttime hours are *decreasing* as a proportion of the day. Predictably, these plants tend to flower in early to midsummer.

Figure 21.32
Seasonal Marker
Brightly colored fall leaves in Michigan's Ottawa National Forest. Plants have a number of mechanisms that allow them to respond to the passage of the seasons.

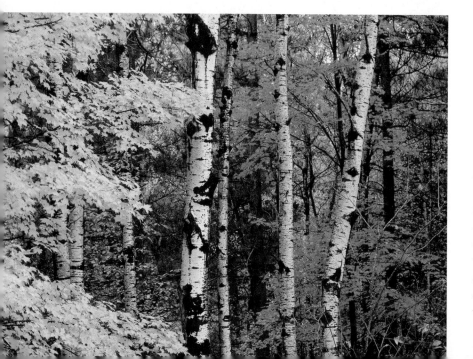

The mechanism by which photoperiodism works is still something of a mystery. Long-night plants, as their name implies, need a certain uninterrupted period of darkness to be able to flower. Shine a light of a certain wavelength on many of these plants at night, even for a couple of minutes, during the time of year in which we would expect them to flower, and the flowering will be suppressed. Botanists have long known of a pigment called *phytochrome* that provides the plant with information about the relative amounts of daylight or darkness, yet its role is clearly not controlling in the process. For us, the important point to note is that plants have a way of sensing the seasons and reacting to them.

On to a More Detailed Picture of Plants

This chapter has served to introduce you to the basic features of plants—their varieties, structures, and the most important systems that operate in them. There is more to say, however, about many of the topics that haven't been gone over (such as how trees grow *out* as well as up). Chapter 22 goes into more detail on plant structure and functioning, including the function of reproduction.

Chapter Review

Summary

21.1 The Importance of Plants

- Plants are vital to almost all life on Earth. The photosynthesis they carry out indirectly feeds many other life-forms. The oxygen they produce as a by-product of photosynthesis is vital to many organisms. The lumber and paper that trees provide is important to human beings. Plants act as an anchoring environmental force by preventing soil erosion and absorbing pollutants.

21.2 Four Main Categories of Plants

- There are four principal groups of plants: bryophytes, seedless vascular plants, gymnosperms, and angiosperms. Bryophytes include mosses; seedless vascular plants include ferns; gymnosperms include coniferous trees, such as pine; and angiosperms include a wide array of plants, such as orchids, oak trees, rice, and cactus.

- Bryophytes are representative of the earliest plants that made the transition from water to land. They lack a true fluid transport or vascular system and thus tend to be low-lying. Bryophyte sperm can get to eggs only by swimming through water.

- Seedless vascular plants have a vascular system but do not produce seeds in reproduction. Their sperm must move through water to fertilize eggs.

- Gymnosperms are seed-bearing plants that do not produce flowers. There are only about 700 gymnosperm species, but their presence is considerable, particularly in northern latitudes where gymnosperm trees, such as pine and spruce, often dominate landscapes. The sperm of gymnosperms is encased in pollen grains, which are carried to female reproductive structures by the wind. Gymnosperms (and angiosperms) produce seeds in carrying out reproduction. Seeds are structures that include a plant embryo, its food supply, and a tough, protective casing.

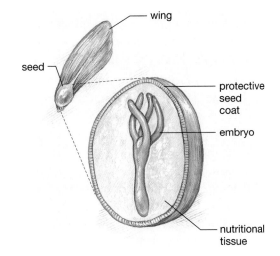

Key Terms

21.3 The Flowering Plants: Nature's Grand Win-Win Invention

- Angiosperms, or flowering plants, are easily the most dominant group of plants on Earth, with some 260,000 species having been identified to date.

- Sperm-containing angiosperm pollen grains often are transferred from one plant to another by animals such as bees and birds. Such animal-assisted pollination is unique to angiosperms. To induce animals to carry out this pollination, angiosperms produce nectar and have developed attractive colorations and fragrances.

CORN

seed coat

endosperm

embryo

- Angiosperm seeds contain tissue called *endosperm* that functions as food for the growing embryo. Endosperm supplies much of the food that human beings eat. Rice and wheat grains consist largely of endosperm. Angiosperm seeds are unique in the plant world in possessing a layer of tissue called *fruit*. Fruit that is edible functions in plant seed dispersal, because animals will eat and digest the fruit but then excrete the tough seeds inside.

21.4 The Structure of Flowering Plants

- Plants live in two worlds, above ground and below it. As such, their anatomy can be conceptualized as consisting of the aboveground shoots and the belowground roots.

- Roots absorb water and nutrients, anchor the plant, and often act as nutrient storage sites.

- Leaves serve as the primary sites of photosynthesis in most plants. Leaves have a profusion of tiny pores, called *stomata*, that open and close in response to the presence or absence of light. In this way, the stomata control the flow of carbon dioxide into the plant and the flow of oxygen and water vapor out of the plant.

- Stems give structure to plants and act as storage sites for food reserves.

- Flowers are the reproductive structures of plants, with most flowers containing both male and female reproductive parts. The male reproductive structure, called a *stamen*, consists of a slender filament topped by an anther. The anther's chambers contain the cells that will develop into sperm-containing pollen grains. The female reproductive structure, the *carpel*, is composed of a stigma, on which pollen grains are deposited; a tube called a style, and a structure called an ovary, where fertilization of the female egg and early development of the resulting embryo take place.

21.5 How Flowering Plants Function

- Pollen grains develop from microspores inside the plant's anthers. At maturity, each pollen grain consists of two sperm cells and one tube cell. Pollen grains from one plant generally land on the stigma of a second plant and then germinate, developing a pollen tube that grows down through the second plant's style. Sperm cells travel through the pollen tube, with one of the sperm cells reaching the egg in the ovary of the second plant and fertilizing it. Prior to this, the egg has developed from a megaspore within the ovary. Following fertilization, the egg develops into an embryo, surrounded by a seed, that can grow into a new generation of plant.

- Plants can reproduce asexually, through such means as grafting. This is known as *vegetative reproduction*.

- Fluid transport in plants is handled through two kinds of tissue: *xylem*, through which water and dissolved minerals flow; and *phloem*, through which the food the plant produces flows, along with hormones and other compounds. Xylem is composed of two types of fluid-conducting cells—vessel elements and tracheids—while phloem is composed of cells called sieve elements and their related companion cells.

- Plant hormones regulate plant growth and development and integrate the functioning of various plant structures. Many fruits ripen under the influence of the plant hormone ethylene, while the hormone IAA is important in controlling plant growth.

- Plants do not grow vertically throughout their length, but instead grow almost entirely at the tips of both their roots and shoots. Some plants, such as trees, thicken through lateral or "secondary" growth. The growth of most plants is indeterminate, meaning it can go on indefinitely.

- Plants have formidable defenses, both structural (such as cactus spines) and chemical (such as antifungal compounds). Plants enter into cooperative relationships with other organisms. Most plant roots are linked to fungal extensions called *hyphae*. This relationship brings added water and nutrients to the plant (from the hyphae) and food to the fungi (from the photosynthesis the plant performs).

21.6 Responding to External Signals

- Plants will bend toward a source of light through the process of *phototropism*, meaning a curvature of shoots in response to light. Plants are able to sense their orientation with respect to the Earth and direct the growth of their roots and shoots accordingly—roots into the Earth, shoots toward the sky. This is called *gravitropism*.

Some plants can climb upward on other objects by making contact with them and then encircling them in growth. This is *thigmotropism*, meaning growth of a plant in response to touch. Differential growth on one side of the stem makes possible phototropism, gravitropism and thigmotropism.

- In temperate climates, deciduous trees exhibit a coordinated, seasonal loss of leaves and enter into a state of dormancy, existing on stored nutrient reserves in colder months.

- Plants can sense the passage of seasons and time their reproductive activities accordingly. One mechanism that assists in this process is *photoperiodism*, which is the ability of a plant to respond to intervals of day and night. Some plants that exhibit photoperiodism are long-night plants, meaning those whose flowering comes only with an increased amount of darkness—in late summer or early fall. Others are short-night plants, meaning those that flower only with a decreased amount of darkness—in early to midsummer.

Understanding the Basics

Multiple-Choice Questions

1. Bryophytes such as mosses and liverworts differ from vascular plants in that they
   a. produce xylem and phloem
   b. have rhizoids, rather than true roots
   c. are flowering plants
   d. conduct photosynthesis
   e. have a seed coat

2. Some seedless vascular plants, such as relatives of horsetails, grew up to 60 feet tall during the "Coal Age" in the Paleozoic era. Which evolutionary innovation allowed them to tower above their ground-hugging predecessors, the bryophytes?
   a. They had stomata.
   b. They were no longer dependent on water for fertilization.
   c. They developed vascular tissue, constituting an efficient plumbing system.
   d. They could grow in the shade.
   e. They lacked true roots.

3. Gymnosperms and angiosperms did away with the dependence on water for fertilization. The structures that enabled them to do so are
   a. stomata   d. guard cells
   b. seedcoat   e. pollen
   c. embryos

4. A number of structural and reproductive innovations contributed to the runaway success of flowering plants in the late Cretaceous period. These traits, lacking in the land plants that preceded the angiosperms, include
   a. presence of xylem and phloem
   b. seed as a unit of dispersal
   c. protection of the seed inside ovary walls, to give rise to the fruit
   d. the evolution of pollen, as a means for ferrying the male gamete to the female gametes
   e. all of the above

5. Stomata are a double-edged sword for plants, because while they are necessary, they also present hazards. Plants need stomata
   a. to draw in water through roots
   b. to acquire atmospheric carbon dioxide for photosynthesis
   c. to dry out plant tissues through transpirational loss of water
   d. to gain warmth from the sun
   e. to draw in water from the air

6. Vascular plants absorb water and minerals
   a. from the air, using stomata
   b. from the soil, using rhizoids
   c. from the soil, using root hairs
   d. from the air, using parenchyma cells
   e. from the air, using leaves

7. The sucrose produced by photosynthesis is transported from the leaf to other parts of the plant through
   a. sieve tubes
   b. vessel elements
   c. tracheids
   d. rhizoids
   e. guard cells

8. Which of the following statements is true?
   a. Spores are diploid.
   b. Gametes are single-celled, haploid structures.
   c. Spores participate in fertilization directly, fusing to give rise to a diploid zygote.
   d. Gametes are diploid.
   e. Spores are the plant's genes.

9. Phototropism and gravitropism lead to bending of shoots or roots toward or away from the stimulus (light or gravity, respectively). Such bending is brought about by
   a. increased growth over the general surface of the root or shoot
   b. differential growth on opposite sides of the root or shoot axis
   c. decreased growth at the root or shoot apex, and increased growth at the base
   d. muscular contractions, leading to a curvature in the root or shoot
   e. all of the above

10. Photoperiodism allows some plants to respond to _____ by timing such things as _____ in accordance with the amount of _____ they are exposed to.
   a. predators, dormancy, moisture
   b. competition, growth, moisture
   c. the seasons, growth, toxins
   d. other plants, flowering, ethylene
   e. the seasons, flowering, darkness

Brief Review

1. List the distinguishing features of the main groups of plants: bryophytes, seedless vascular plants, gymnosperms, and angiosperms.

2. What functions do roots perform? What are the main types of roots produced by flowering plants?

3. List the main parts of a flower, and the functions performed by each part.

4. Compare the life span of some perennial plants and that of people; contrast the way people grow, and how plants grow and develop over their lifetimes.

5. Give examples of mutually beneficial relationships between plants and bacteria, and plants and fungi.

6. Distinguish between the gametophyte and sporophyte of land plants.

Applying Your Knowledge

1. Consider the colonization of land by plants. What do you suppose were the biggest problems these plant pioneers faced in the terrestrial environment compared to their erstwhile aquatic life? Would amphibians, the first vertebrates to invade the land, also have faced some or all of these problems?

2. Although bryophytes led the land invasion, they were—and continue to be—limited in their distribution range by the challenges of the terrestrial lifestyle. What adaptive processes did vascular plants evolve that progressively allowed them to succeed in drier environments?

3. As flowering plants emerged in the late Cretaceous, their evolutionary fate was influenced by their interactions with the animal world. Describe how, and why, plants have recruited the assistance of animals in furthering their own reproductive success.

4. Gardeners often "pinch off" the terminal shoot apex to stimulate bushiness in young plants. Explain the physiological basis for this common horticultural practice.

5. Why do broad-leaved plants lose their leaves in autumn in the colder parts of the world, and how do they "know" when it is time for them to shed leaves?

Group Activity

1. Think back on all of your waking hours yesterday, from the moment you got up to the time you fell asleep at the end of the day. Make a list of all the plant-based products you encountered, consumed, used, or were affected as you went through the day. Organize them by the type of industry—large-scale or small-scale agriculture, food industry, forestry, for example—that put that plant product within your reach. Also include a category for non-industry-based plant products (some pussy willows you gathered in the woods, for example). Compare your list with those prepared by your classmates.

2. Since the dawn of agriculture, people have moved plants about the globe. The discovery of the Americas, in particular, bestowed an enormous botanical wealth on the rest of the world. Research the origins of common crop plants, and produce a list of plants native to the New World that are now in cultivation worldwide. You might want to start with visions of your Thanksgiving table. Each member of the group should then pick one crop plant and describe its effect on the history or cuisine of a nation.

22

Common Themes in the Green World

Plant Tissues, Growth, Transport, and Reproduction

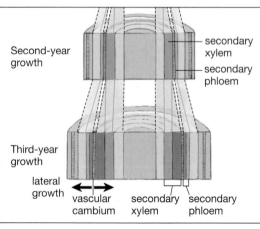

How woody plants grow wider.
(page 457)

Food carriers.
(page 462)

Plant pipes.
(page 453)

Growing out.
(page 457)

Insects that eat plants are

not much of a surprise, but plants that eat insects? About 500 species of flowering plants are carnivorous, with at least one variety, the tropical pitcher plant *Nepenthes*, able to capture animals up to the size of a small bird. But the most famous plant predator of them all undoubtedly is the Venus flytrap (*Dionaea muscipula*), which is native to the wetlands of North and South Carolina. So how does this plant, which has no muscles and is fixed in one spot, end up consuming insects?

Each trap is a single leaf folded in two, with interlocking "fingers" of the two halves able to snap together in less than half a second, imprisoning any small wanderer. But what keeps the leaf halves from wasting their energy by snapping together in pursuit of wind-borne blades of grass or other objects? The chemical reaction that puts the trap in motion is set off by a series of trigger hairs on the inside of each of the leaf's two sections. The trick is that *two* trigger hairs must be tripped in order for the halves of the leaf to snap shut; *insects* wandering around inside the leaf are likely to trip two hairs, but blades of grass are not. And how does the plant slam its "jaws" shut without muscles? When the trigger hairs are tripped, they set off a chemical reaction that engorges cells at the base of the leaf with water. This rapid movement causes the leaf halves to come together. Enzymes are then released that begin to digest the struggling insect.

It may surprise you to learn that Venus flytraps make their own food through photosynthesis, just like any other plant. So why do they need these side orders of insects? For nutritional supplements; like many other plant carnivores, Venus flytraps grow naturally in mineral-poor soil, and the insects they catch are a good source of nitrogen and phosphate.

The world of plants is filled with species as unique in their own way as the Venus flytrap is. Queen Victoria water lilies (*Victoria regia*) have circular leaves that can be nearly 2 meters (about 6 feet) in diameter and that can serve as a floating platform for a person if pressed into service (**see Figure 22.1** on page 448). The seeds of orchids are so tiny that they are disseminated in nature by ants, while the seeds of the double coconut come wrapped in a fruit that might be 0.6 meter (about 2 feet) across.

Food and water take a trip.
(page 464)

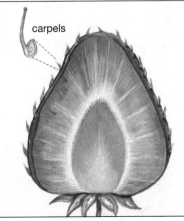

Where are the strawberry seeds?
(page 470)

Pass the flapjacks.
(page 462)

Figure 22.1
Floating Platforms
A woman plays a violin while standing atop a Victoria water lily at the Missouri Botanical Garden in St. Louis at the turn of the century. The lilies are native to South America. *Yrupe*, the term for the lilies among some South American Indians, can be translated as "big water tray."

Yet there is a unity to plant life, particularly in the case of the flowering plants, or *angiosperms*, that you'll be looking at in this chapter. Their common features and characteristics range from their cell types all the way to aspects of their reproduction. The goal in this chapter is to cover some of these commonalities in four areas: cell and tissue types, growth, plant plumbing (meaning the plant fluid transport system), and reproduction.

You will be looking at flowering plants from the cell types to the larger systems. First, however, let's revisit the look at the whole plant that was first presented in Chapter 21, so that you can "find your place" when reading about a plant's component parts. If you look at **Figure 22.2**, you can see a diagram of a typical plant, with its two-system division—roots and shoots—and the various parts of the plant that lie within these two systems. The function of each of these parts will become clear to you as you go through the details on them.

22.1 Two Ways of Categorizing Flowering Plants

The Life Spans of Angiosperms: Annuals, Biennials, and Perennials

The constituent parts of plants are often organized one way in a given kind of plant, but another way in a different variety. So what are these varieties? Here are a couple of ways of *categorizing* plants, the significance of which will become apparent as you go along.

One important question that might be asked about any plant is: How long does it take for it to go through its entire life cycle—from being a seed germinating in the ground, through growth, flowering, seed dispersal, and then death? Plants that go through this cycle in 1 year or less are known as **annuals**. Some ex-

Figure 22.2
Anatomy of a Flowering Plant (an Angiosperm)

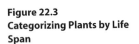

Figure 22.3
Categorizing Plants by Life Span

a Annuals such as this tomato plant live for only one year.

b Biennials such as this carrot plant have a two-year life span.

c Perennials such as this white oak tree live for many years, some for many hundreds of years. Plants with long life spans tend to be woody and larger than plants with short life spans.

a

b

c

 amples of these are food crops, such as toma-toes and the commercial grains. Plants that go through the cycle in about two 2 years are **biennials**, which include carrots and cabbage. (Have you ever seen the flowers of a carrot plant? Probably not; we *pick* the taproots called carrots after 1 year, meaning their shoots don't get a chance to flower.) Plants that live for many years are known as **perennials**, a category that includes trees, woody shrubs such as roses, and many grasses. **Figure 22.3** shows some representative annuals, biennials, and perennials.

A Basic Difference among Flowering Plants: Monocotyledons and Dicotyledons

A second distinction among the flowering plants has to do with their anatomy, meaning the arrangement of the structures that make them up. In this respect, there are two broad classes of flowering plants: monocotyledons and dicotyledons, which are almost always referred to as monocots and dicots. A cotyledon is an embryonic leaf, present in the seed, as you can see in **Table 22.1** on page 450. **Monocotyledons** are plants that have one embryonic leaf; **dicotyledons** are plants that have two.

This distinction in the embryos then makes for major differences in the mature plants: Their roots are different, their leaves are different, their transport tubes, or vascular bundles, are arranged differently (**Figure 22.4**). More than 75 percent of all flowering plants are the broad-leafed dicots. But given the 260,000 known species of flowering plants, this still leaves more than 50,000 species of narrow-leafed monocots, which include most of the important food crops. Table 22.1 shows you some of the differences between these two kinds of plants.

With these distinctions in mind, you're now ready to start looking at the basic elements that go into making up flowering plants.

22.2 There Are Three Fundamental Types of Plant Cells

You saw in Chapter 4 that all living things are made up of cells, and plants are no exception to this rule. As it happens, there are three fundamental types of plant cells that, alone or in combination with each other, go on to make up most of the plant's **tissues**, meaning *groups* of cells that carry out a common function.

Parenchyma Cells

First, there are **parenchyma cells**, which have thin cell walls and are easily the most abundant type of cell in the plant, being found almost everywhere in it. Accordingly, these cells have a lot of different functions, forming the flesh of many fruits, the outer surface of most young plants, parts of leaves that are active in photosynthesis, and much of the "ground" tissue you'll be hearing about shortly. Further, parenchyma are generally alive at maturity, a quality that might hardly seem worth mentioning except that some plant cells are dead in their mature, functioning state.

Sclerenchyma Cells

This dead-at-maturity condition characterizes a second type of plant cell, **sclerenchyma cells**, which have thick cell walls that help the plant return to its original shape when it has been deformed by some force, such as the push of wind or animals. Sclerenchyma cells are found in parts of the plant that are mature—that will not be growing further, and that need the support that sclerenchyma can provide. These cells have **cellulose** running through their walls, a compound that can be likened to the steel bars in reinforced concrete. When cellulose is joined by another tough compound called *lignin*, the result is

MediaLab Link
Monocotyledons, dicotyledons, see p. 478.

**Figure 22.4
Categorizing Plants by Physical Features**

a Corn is an example of a monocot.

b Geraniums are an example of a dicot. See Table 22.1 for a list of the anatomical differences between monocots and dicots.

a

b

cells that can bear the crushing weight of massive trees (**see Figure 22.5**).

Collenchyma Cells

The third type of cell, **collenchyma cells**, can be thought of as support cells, like sclerenchyma cells, but with the function of stretching or elongating in parts of the plant that are growing, such as young leaves and stems. Collenchyma cells combine the properties of parenchyma and sclerenchyma cells, performing some of the functions of each of these cell types. They provide mechanical strength in areas undergoing active growth, and can also participate to some extent in photosynthesis and wound repair.

Parenchyma as Starting-State Cells

Both collenchyma and sclerenchyma are derived from parenchyma cells; that is, some parenchyma cells differentiate *into* collenchyma or sclerenchyma cells. Parenchyma are thus a kind of starting-state cell, but their capabilities are greater than this. They can be transformed "backward," in a sense, into the type of embryonic cells that give rise to the whole plant. You may have wondered in Chapter 21 how we can get a new plant by taking a cutting from an existing plant and placing it in the ground. The answer is that when a plant is cut, parenchyma cells lying close to the cut are transformed into *growth* cells of a type you'll look at shortly, and these can sprout a new root. This obviously gives plants a flexibility not possessed by animals such as ourselves. If one of our fingers is cut off, we do not grow a new finger, to say nothing of a whole new person.

22.3 The Plant Body and Its Tissue Types

It was noted earlier that tissues are groups of cells that carry out a common function. So what kinds of tissues are there in angiosperms? The short answer is dermal, ground, vascular, and meristematic, with each type of tissue generally being composed of one or more of the cell types you just looked at. Dermal tissue can be thought of as the plant's outer covering, vascular tissue as its transport or "plumbing" tissue," meristematic tissue as its growth tissue, and ground tissue as almost everything else in the plant.

First: A Distinction between Primary and Secondary Growth Tissue

Before you proceed further with this topic, first note a basic distinction in plant tissue that has to do with the way plants grow. Some plants are capable only of what is called **primary growth**, meaning growth at the tips

---

**Table 22.1
A Cotyledon Is an Embryonic Leaf**

Lots of differences result from whether a plant is a monocot (has one embryonic leaf) or a dicot (has two embryonic leaves).

	MONOCOTS	DICOTS
Embryonic leaves	one cotyledon	two cotyledons
Mature leaves	narrow leaves, parallel veins	branching veins, broad leaves
Roots	fibrous root system	tap root system
Vascular bundles	scattered throughout stem	arranged in ring in stem
Type of growth	only primary growth	may have secondary woody growth
Flower parts	multiples of three	multiples of four or five
Examples	orchids, wheat, rice, bananas	oak and maple trees, cacti, sunflowers

of their roots and shoots that principally increases their *length*. In contrast, some plants also exhibit something known as **secondary growth**, which can be thought of as the *lateral* growth, or thickening, that occurs in **woody plants**. Secondary growth exists in, for example, trees, but not in orchids or strawberry plants. Non-woody plants such as an orchid are known as **herbaceous plants**, meaning those that never develop wood (or bark) and that thus contain only primary tissue. It is this *primary* tissue that you will look at first, with a review of secondary tissue to come shortly.

If you look at **Figure 22.6**, you can see the location of all four kinds of primary tissue in a dicot. Looking at the stem in cross section, you can see that ground tissue is well named, because visually it forms a kind of background against which you can see the tubes of the vascular tissue. Dermal tissue then is at the periphery of the plant, forming a "skin" layer around the ground tissue. The placement of the meristematic tissue is discussed shortly. First, let's look in a little more detail at dermal, ground, and vascular tissue.

Dermal Tissue Is the Plant's Interface with the Outside World

Plants can't move, they can't bite, and, as you saw in Chapter 21, they must carefully control their water supply. All this makes their **dermal tissue**, or *epidermis*, very important to them. This outer coat is generally only one layer of cells thick, but it serves to protect the plant and control its interaction with the outside world (**see Figure 22.7**, page 452). A plant's first line of defense against outside predators is a waxy coating called the **cuticle** that covers the epidermis of the plant's aboveground parts (its shoot). The cuticle is a kind of waterproofing, serving to keep water in, and an invader-proofing, helping to keep infecting bacteria and fungi out.

Plants need to exchange gasses with their environments, however. They take in carbon dioxide and emit oxygen and water vapor through the microscopic pores called stomata, found mostly on the underside of their leaves. (You went over the functioning of stomata in Chapter 21, starting on page 430.) Stomata also represent a kind of weak point for plants, however, because they provide a passageway for microbial invaders and a conduit for excessive loss of moisture.

Dermal tissue also forms several kinds of extensions that protrude from the plant's surface. These are "hair" cells, also known as trichomes. You learned about one variety of them in Chapter 21: root hairs, meaning the thread-like outgrowths of individual epidermal root cells. As we noted there (page 428), these wispy outgrowths greatly increase the

Figure 22.5
Plant Cell Walls Are Composed of Composite Materials, as Is Reinforced Concrete
The cellulose fibers of plant-cell walls are very strong in tension, as are the steel bars in reinforced concrete. Lignin is similar to the matrix of concrete in that it binds the fibers together and resists compression. Both plant-cell walls and reinforced concrete are important in providing support for structures that are both tall and heavy.

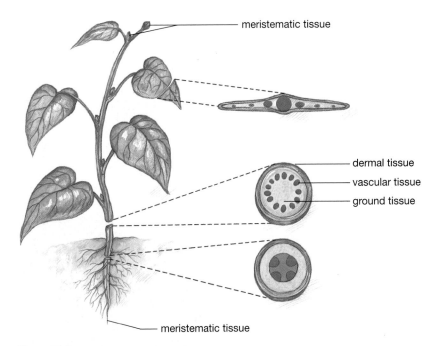

Figure 22.6
Four Types of Tissue in Primary Plant Growth
The dermal tissue (bright green) is found on the outside of the plant and is the interface between the plant and its environment. The ground tissue (pale green) is found throughout the plant and gives the plant its shape, stores food, and is active in photosynthesis. The vascular tissue (bright red) transports water, nutrients, and sugar throughout the plant. The meristematic tissue (blue) occurs at the tips of shoots and roots and at leaf bases, and gives rise to the entire plant, through the production of cells that develop into the other three tissue types.

plant's ability to absorb water and minerals. Other types of trichomes serve not to absorb substances, but to secrete them; it is trichomes that secrete the toxic chemicals that some plants use to ward off plant-eating animals.

Ground Tissue Forms the Bulk of the Primary Plant

Most of the primary plant is ground tissue, which can play a role in photosynthesis, storage, and structure (**see Figure 22.8**). Large parts of ground tissue may be simple parenchyma, but other parts are combinations of cell types.

Vascular Tissue Forms the Plant's Transport System

As noted in Chapter 21, the plant's vascular system has two main functions. First, it must transport water and minerals; second, it must transport the food made in photosynthesis. In line with this, there are two main components to the plant vascular system. First there is xylem, which is the tissue through which water and dissolved minerals flow. Second, there is phloem, which is the tissue that conducts the food produced in photosynthesis,

along with some hormones and other compounds (**see Figure 22.9**). The movement of water through xylem is directional—from root up through stem and then out through leaves as water vapor. Meanwhile, the food made in photosynthesis must travel through phloem to every part of the plant, though, as a practical matter, in temperate climates the net flow is downward in summer (from food-producing leaves) but upward in early spring (from root and stem storage sites.)

Xylem and phloem are arranged in **vascular bundles**, which is to say collections of xylem and phloem tubes that run in parallel in the stem—xylem toward the inside of the stem, phloem toward the outside. The bundles are arranged in different ways within a stem depending on whether the plant is a monocot or a dicot; the dicot bundles are configured in a circle, the monocot in an irregular pattern. This is why dicot ground tissue often is conceptualized as existing in the two regions, which you can see in **Figure 22.10**: a **cortex** that is outside a ring of the vascular bundles and a **pith** that's within it, with the pith cells being specialized for storage.

Figure 22.7
Where Plant Meets Environment
Dermal tissue serves as the interface between a plant and the environment around it. The waxy cuticle and the single layer of cells in the epidermis work together to protect the plant and to control interactions with the outside world.

a Specialized epidermal cells called guard cells regulate the opening of the plant pores called stomata, thus controlling gas exchange between the plant and its environment. Here, guard cells have opened two stomata. In most plants, the majority of stomata are found on the leaves.

b Some epidermal cells have hairlike projections, called trichomes, that serve various functions. Two kinds of trichomes are visible in this rose plant. The larger of the two varieties, with the bulbous tips, help secrete chemicals that guard against plant-eating predators.

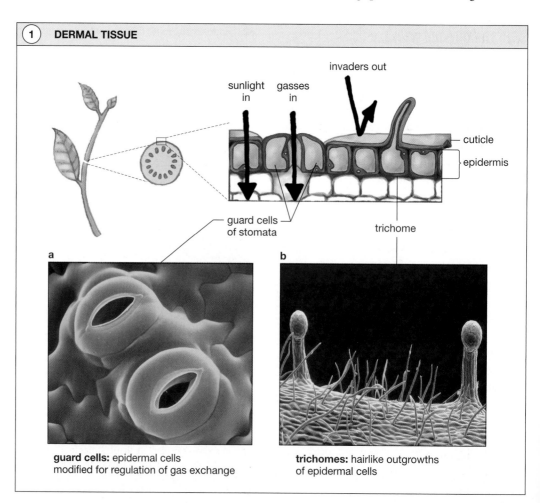

1 DERMAL TISSUE

sunlight in
gasses in
invaders out
cuticle
epidermis
guard cells of stomata
trichome

a guard cells: epidermal cells modified for regulation of gas exchange

b trichomes: hairlike outgrowths of epidermal cells

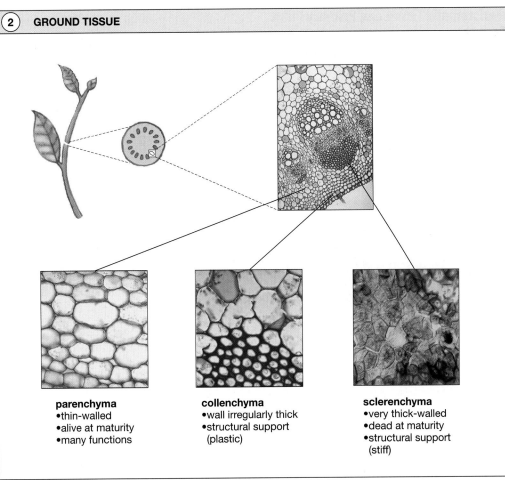

② GROUND TISSUE

parenchyma
- thin-walled
- alive at maturity
- many functions

collenchyma
- wall irregularly thick
- structural support (plastic)

sclerenchyma
- very thick-walled
- dead at maturity
- structural support (stiff)

Figure 22.8
Ground Tissue
Most of a plant is made of ground tissue, which is composed of parenchyma, collenchyma, and sclerenchyma cells. The examples of these cells shown here come from several different plants. In a cross section of a stem, thin-walled parenchyma can be found in most of the interior. Collenchyma—at the bottom of the picture, surrounded by parenchyma—are support cells that can elongate in growing parts of the plant. Very thick-walled sclerenchyma, such as these from a pear, can provide strength to a plant.

③ VASCULAR TISSUE

vascular bundle

xylem

phloem

tracheid

vessel element

sieve element

companion cell

Figure 22.9
Materials Movers
Vascular tissue transports fluids throughout the plant. Xylem is composed of cells called tracheids and vessel elements that transport water and dissolved minerals. Phloem is composed of sieve elements (and their companion cells), which transport the food produced during photosynthesis. These cell types are stacked upon one another to form long tubes.

Meristematic Tissue and Primary Plant Growth

So you've seen three types of primary plant tissues. The question now is: How do these arise? How is it that a plant gets its vertical growth? The short answer is that the fourth type of plant tissue, meristematic tissue, gives rise to all the other tissue types. In considering the role of meristematic tissue, recall from Chapter 21 that, in their vertical growth, plants do not grow the way people do—throughout their entire length—but instead grow only at the tips of their roots and shoots. In addition, a plant's growth is *indeterminate*: In most plants it can go on indefinitely at the tips of roots and shoots, as opposed to animal growth, which generally comes to an end at a certain point in development.

Figure 22.10
Two Types of Flowering Plants

These cross sections show the organization of primary tissues in a monocot and a dicot.

a Monocots (in this case an onion) have vascular bundles arranged in an irregular pattern within ground tissue.

b Dicots (in this case a buttercup) have vascular bundles arranged in a ring that separates the ground tissue into pith on the inside of the bundles and cortex on the outside.

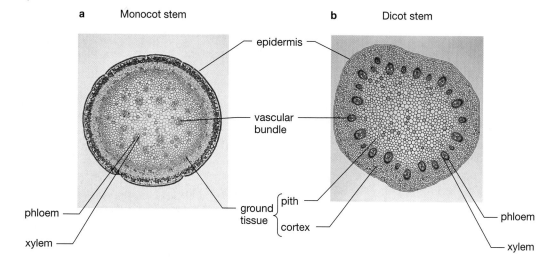

a Monocot stem   b Dicot stem

epidermis
vascular bundle
phloem
xylem
ground tissue — pith, cortex
phloem
xylem

Figure 22.11
Meristem Tissue

Growth originates from meristem tissue. The shoot and root elongate at the apical meristems while new branches originate at the lateral buds. Meristem tissue gives rise to more dermal, ground, and vascular tissue, as well as more of itself.

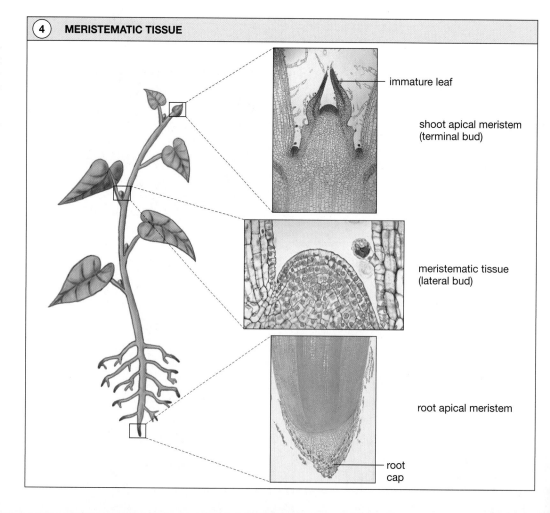

4 MERISTEMATIC TISSUE

immature leaf

shoot apical meristem (terminal bud)

meristematic tissue (lateral bud)

root apical meristem

root cap

22.4 How a Plant Grows: Apical Meristems Give Rise to the Entire Plant

Putting indeterminate growth together with the concept of growth at the tips, it follows that there must be plant cells at root and shoot tips that remain perpetually young, or, more accurately, perpetually embryonic. That is, these cells are able to keep dividing, thereby giving rise to cells that then differentiate into all the primary cell and tissue types you've been reading about. Indeed, everything in plants (both their primary and secondary tissues) develops ultimately from these cells, which collectively form the **apical meristems** of plants.

If you look at **Figure 22.11**, you can see the apical meristem locations in a typical plant. Note that in this plant with a taproot, each lateral root tip has its own apical meristematic tissue that is capable of giving rise to yet more roots. In the shoot, plants confine their growth to many shoot apices at the tip of each stem. Why? The better to compete for precious sunlight. It is the **shoot apical meristem** that gives rise to all the cells that allow this vertical growth. Note, however, that there is a second location for meristematic *tissue* in the shoot: the area nestled between leaf and stem. It's here that we find lateral buds (sometimes called *axillary buds*). Any **bud** is an undeveloped shoot, composed mostly of meristematic tissue. A **lateral bud** is meristematic tissue that may give rise to a branch or a flower (which obviously ends up growing laterally from the stem). Lateral buds also serve, however, as an *insurance policy* in that they can switch roles. Should the plant's apical meristem become damaged, one of the lateral buds steps in to assume the role of the shoot apex, thus allowing the plant to maintain its vertical growth. As long as the original apical meristem is intact, however, it generally will produce hormones that *suppress* the growth of the lateral buds near it, leaving them dormant. The shoot apex is itself sometimes referred to as a **terminal bud**, particularly when *it* is dormant, as with trees in winter.

We can think of plant growth in terms of growth modules, each consisting of an internode (the stem between leaves), a node (the area where a leaf attaches to stem), and one or more leaves (**see Figure 22.12**).

A Closer Look at Root and Shoot Apical Meristems

Shoot Meristems

Now let's look a little more closely at these important collections of cells from which all else comes, the shoot and root apical meristems. The right-hand illustrations in Figure 22.11 show you what shoot apical meristem, lateral bud, and root apical meristem look like up close.

The shoot apical meristem, as you can see, is a dome-shaped collection of cells that has, overlapping its top, two budding leaves that serve a protective function. Notice also the lateral bud meristems at the base of these leaves. How did the apical meristem produce them? It left them behind, in a sense, as it continued its upward growth.

Root Meristems

Because roots are pushing their way into the ground, root apical meristems are located not at the very tip of the root, but just *behind*, or "sub-apically" from, the tip. The root apical meristems are adjacent to a collection of cells called a **root cap** that both protects the meristematic cells and secretes a lubricant that helps ease the way for the root in its progress. Toward the middle of the apical meristem, there is a collection of cells called the quiescent center. These slowly dividing cells amount to another kind of insurance policy for the plant. In instances in which the meristematic cells have become damaged, quiescent center cells will start dividing more rapidly, producing new meristematic cells. They are, in short, a reserve that goes into action in times of trouble. They are well suited for this because, in their dormancy, they are more resistant than apical meristem cells are to injury from such influences as toxic chemicals.

What Is It?
See p. 458.

MediaLab Link
Roots, stems, flowers, leaves, see p. 478.

**Figure 22.12
Vertical Growth in Plants Occurs in Modules**
Each module (indicated in the figure by alternating green and black segments) consists of an internode, a node, and one or more leaves. Here we move from one module in week one to four modules in week four.

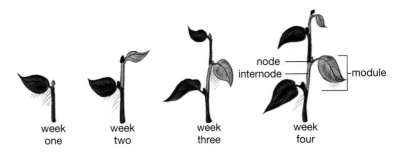

week one   week two   week three   week four

node
internode —module

How the Primary Tissue Types Develop

You've looked so far at three of the plant's primary tissues—the ground, dermal, and vascular tissues. And you know that the fourth type of primary tissue, meristematic tissue, gives rise to all these. So how do plants get from meristem to the others? Not in a single step, it turns out—not directly from apical meristem cells to, say, fully functioning vascular tissue. Instead, there is a transition involving three main changes: *More* cells are produced through cell division; cells *elongate* through cell growth; and cells *differentiate* into the three kinds of tissue. These three changes are overlapping—cells are elongating even as they are differentiating—but they do come in roughly this order. You can see this by looking at **Figure 22.13**. In a gradual transition with no sharp boundaries, the apical meristem gives way to a **zone of cell division**, then there is a **zone of elongation**, and finally a **zone of differentiation**, which is where the ground, dermal, and vascular tissues fully take shape. Note that root hairs don't appear until the zone of differentiation. Though the three kinds of tissues *finish* taking shape in the zone of differentiation, they actually *begin* this process just above the apical meristem (in the zone of cell division). It's accurate to think of the tissues here as being precursors to the three tissue types.

If you look at **Figure 22.14**, you can see the practical effect of a zone of elongation: It is where most of the primary plant growth occurs. There is some growth with the addition of more cells down near the apical meristem, but most of the vertical expansion of the plant takes place because of cell elongation. Such elongation has the effect of extending the roots into the soil and the stem into the air.

Intercalary Meristems Keep Grasses Growing

There is one important variation on primary growth through apical meristems. What about plants such as grasses that are constantly losing their tops through such means as trampling or grazing? Grasses get their growth from tissue known as **intercalary meristems**, which are found at the base of *each node*. There thus exists a series of vertical growth tissues that are intercalated or interspersed between regions of nondividing cells as we go up the plant. One practical effect of this is that when we mow the top half of a blade of grass, it grows back. A second is that, because these plants are growing at *many* points along their length (rather than

**Figure 22.13
Tissue Development from Apical Meristems**
Apical meristems give rise to the primary tissues. In development, cells first are primarily engaged in division, just above the apical meristem, then elongation, and finally differentiation. The end result is three fully formed kinds of tissue.

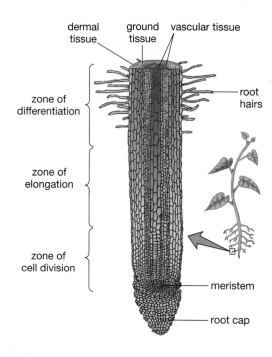

**Figure 22.14
Growth by Cell Elongation**
Much of the elongation of the shoot and root of a young plant results from cell elongation rather than from cell division. Thus it is cells just above the root tip, in the zone of elongation, that provide most of the plant's growth, as seen in the figure. The very tip of the root (where cell division occurs) has elongated very little, and the cells in the upper root have completed most of their elongation. This same pattern holds true for the shoot.

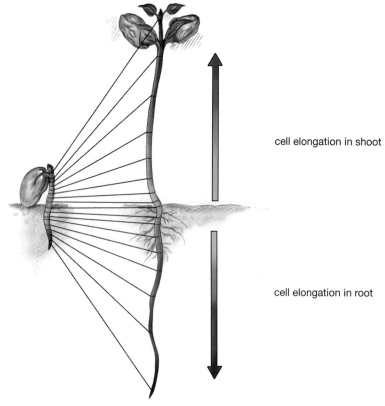

just their apex), they can manifest very *rapid* growth. We can easily see nodes in the giant grass known as bamboo, which exhibits intercalary growth (**see Figure 22.15**).

22.5 Secondary Growth Comes from a Thickening of Two Types of Tissues

So far, you've been looking at a plant's vertical or primary growth. In the herbaceous plants that's all the growth there is. But you've seen that some plants—the woody plants—grow *out* as well as up and down; they have primary *and* secondary growth, in other words (**see Figure 22.16**). So how does this secondary growth come about? Once again, through cells that are meristematic. In essence, secondary growth yields three new kinds of tissue: a different type of xylem and phloem—secondary xylem and secondary phloem—and several layers of outer tissue that collectively go under the familiar name of *bark*.

At the start, it's worth noting that secondary growth in plants is related to two of the plant characteristics that we reviewed at the start of the chapter. First, secondary growth almost always takes place in perennials, rather than annuals (though there are plenty of perennials that exhibit only primary growth, as with grasses.) Second, with very few exceptions, the monocots described earlier have only primary growth. Meanwhile, dicots are more often woody, though many are herbaceous.

Secondary Growth through the Vascular Cambium: Secondary Xylem and Phloem

You saw earlier that the plant's vascular tissues are arranged into bundles of pipe-like vessels, with the food-carrying phloem pipes toward the outside of the stem and the water-carrying xylem pipes toward the inside. In a cross section of a dicot plant stem, these pipes form concentric rings. If you look at the plant another way—through its entire *length*—you can see that these rings take the shape of cylinders that are nested, as if these layers of tissue were a series of open-ended cans, one inside the next (**see Figure 22.17**).

Now notice that woody plants develop a thin layer of tissue *between* the primary xylem and primary phloem cylinders. This is

the **vascular cambium**, one of the additional types of meristematic tissue that brings about secondary growth.

The vascular cambium has one group of cells in it, called *ray initials*, that produce exactly what their name indicates: *rays* of parenchyma cells that carry water

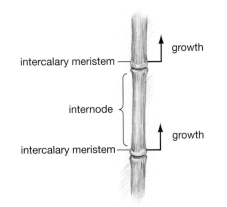

Figure 22.15
Intercalary Meristems Keep Grasses Growing
One reason that bamboo grows so rapidly is because growth occurs at several points along its length at once. The intercalary meristems responsible for this growth are located at the base of each internode.

Figure 22.16
Growing Up versus Growing Out
Herbaceous plants grow at their tips, mostly vertically, through primary growth. Woody plants, however, grow not only vertically but also laterally, through what is called secondary growth. Dicots can be either woody or herbaceous, but almost all monocots are herbaceous.

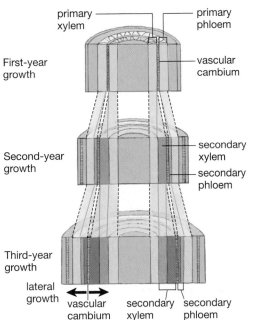

Figure 22.17
Vascular Cambium and Secondary Growth
Vascular cambium is meristematic tissue that produces secondary xylem tissue to its interior and secondary phloem tissue to its exterior. Over the years, the thickening of a tree comes about primarily because of the growth of secondary xylem.

What Is It?
Answer: A European ash twig, nature's spring doll.

and dissolved compounds through the *width* of the plant. (If you've ever looked at the stump of a tree, you've probably noticed fine lines that seem to radiate out, from near the center to the periphery of the trunk. These are the tree's transport rays, made up of parenchyma cells that form streaks of living cells in an expanse of dead xylem. You can see an example of them in Figure 22.20.)

More important for our purposes, the vascular cambium is something like a cell factory that is continually pushing out two different kinds of cells to either side of itself. The result is two new cylinders of tissue: **secondary xylem**

Figure 22.18
How Vascular Cambium Thickens the Plant
The cells of the vascular cambium divide, producing one cell that differentiates into either xylem or phloem and another cell that remains vascular cambium. This process continues over and over again, causing the stem of the plant to thicken.

Figure 22.19
Annual Rings Tell Time and More

a The thick secondary xylem (the wood) of a tree contains many concentric rings, with the dark line of each ring representing the end of one year's growth. The xylem rays transport water and nutrients laterally through the trunk.

b This close-up of the annual rings from an actual tree shows how growth can vary with environmental conditions. The earlier growth rings on the right are thicker than the rings on the left. The tree grew less in each of the later years because it was being affected by acid rain.

c This micrograph of a pine tree shows the structural basis for annual rings. On the right—toward the inside of the trunk—are smaller cells that grew in late summer and early fall, when water was less plentiful. The very narrow cells at right-center create the dark line of an annual ring. Immediately to the left of these densely packed cells are the larger cells that grew the following spring, when water was again plentiful. These larger cells are visible as the lighter wood in the trunk.

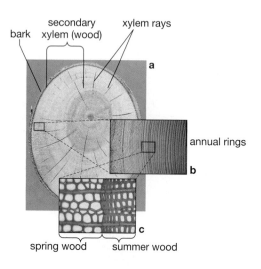

to the inside of the vascular cambium, and **secondary phloem** to the outside. Secondary xylem and phloem initially do just what primary xylem and phloem do: transport, through the *length* of the plant, water through xylem and food through phloem. **Figure 22.18** shows you how the vascular cambium pulls off its feat of two-way xylem/phloem production. Note in the figure that the cells of the vascular cambium divide, producing one cell that differentiates into either xylem or phloem and another cell that remains vascular cambium. This process continues over and over again, causing the stem of the plant to thicken.

Secondary Xylem Is Responsible for Most of a Plant's Widening

It is the secondary xylem that is responsible for most of the widening of trees and other woody plants. Indeed, another term for secondary xylem is **wood**, a material that comprises about 90 percent of an average tree. After an initial period of carrying water, the secondary xylem cells will cease doing this, but are still valuable to the tree because of their strength.

Almost everyone is familiar with the **annual rings** of trees in temperate climates such as ours. But what is each visible ring? **Figure 22.19** tells the tale. Each ring represents the abrupt change between the secondary xylem of late summer and the secondary xylem of the following spring. In late summer, the

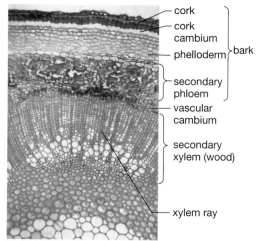

Figure 22.20
Secondary Growth Tissues
The secondary xylem, the wood with which we are all familiar, is responsible for water transport and the structural support of the plant. Xylem rays are lateral conduits for water. Bark is composed of four layers of tissue: secondary phloem, phelloderm, cork cambium, and cork.

xylem cells are smaller and have thicker walls because they're getting less water. Come next spring, with its plentiful water, and the cells are much larger (and hence not as dark). Thus we can look just to the outside of the "line" of any given ring and say "spring began here." To learn more about what we can tell by looking at a cross section of a felled tree, see "A Tree's History Can Be Seen in Its Wood," on page 460.

Secondary Growth through the Cork Cambium: The Plant's Periphery

With the secondary xylem cells pushing out from the interior of the tree or shrub, the plant's original outer covering (its epidermis) can't last. It splits apart, eventually to be shed altogether, usually within a year. Before it goes, however, the plant develops a new kind of outer covering, one that has the ability to renew itself constantly as the plant continues to grow. This quality may sound familiar by now, and indeed **cork cambium** represents the final type of meristematic tissue you'll be encountering—one that gives rise to the plant's outer tissues (**see Figure 22.20**).

The cork cambium itself is a product of other tissue in the stem—in a mature tree, the secondary phloem tissue you just looked at. Cork cambium is once again secondary meristemic tissue, which is like a cell factory that is continually pushing out different kinds of cells to either side of itself. The main product of this activity goes to the *outside* of the cambium in the form of **cork** cells. Once these cells are produced, they go on to infuse their cell walls with

a waxy substance that acts as nature's own waterproofing and invader-proofing. Cork cells are, however, born to die, in a sense, in that their genetic blueprint brings about a so-called programmed cell death. It is in this dead, mature state that they serve their protective function. In a single growing season, the cork cambium can produce layer upon layer of cork cells, the result being a well-protected tree. To its interior, the cork cambium can produce, in a few species, a layer of parenchyma cells called phelloderm.

What Is Bark?

So, there is the cork cambium and its two products (cork and phelloderm). Then *inside* this three-part structure, there is the secondary phloem. Put together, all four parts constitute a region of a woody plant that goes by a familiar name: **bark**. Put another way, bark is everything outside the vascular cambium. Figure 22.20 shows all the layers of tissues in secondary growth.

If a tree is "ringed" by having a strip removed around its entire circumference—as hungry animals might do in winter—it might live without its cork and the cork cambium interior to it and the phelloderm inside that. If this cutting continues into the younger, food-conducting secondary phloem, however, that would be the end of the tree, because it could no longer move its energy stores to where they're needed. Indeed, this kind of cutting describes the "girdling" that is a common way to kill unwanted trees (**see Figure 22.21**).

To conclude this section on plant growth, **Figure 22.22** shows you how the entire plant grows, with the starting point for both secondary and primary growth being apical meristem tissue.

Figure 22.21
Death from Lack of Nutrients
This tree has been girdled by someone who has the intention to kill it. The bark was removed all the way down to the secondary phloem, the tissue layer that transports food from one part of the plant to another.

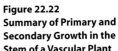

Figure 22.22
Summary of Primary and Secondary Growth in the Stem of a Vascular Plant
The "precursor" tissues that arise from the apical meristem are those that take shape in the zones of cell division, elongation, and differentiation noted earlier in the chapter.

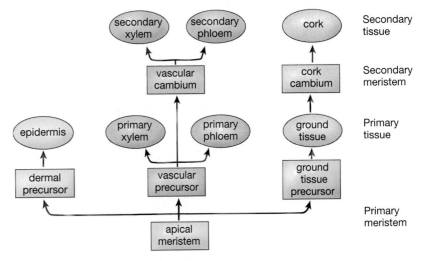

A Tree's History Can Be Seen in Its Wood

Everyone knows about looking at the cross section of a felled tree trunk and counting its "rings" to gauge the age of the tree. But what other stories does this kind of tree wood have to tell?

First, are the rings wider in some years than others? To the extent that the tree grew more in a given year, it will have a wider ring. The wider the ring, the more likely it is that optimal conditions for growth existed during that year: plentiful rainfall, enough sun, a lack of pests. A series of unusually narrow rings probably means just the opposite (**see Figure 1**).

Next, are the rings perfectly concentric, or are they *eccentric* at some point in the tree's growth? Is a given set of circles wide on, say, the left as you look at the trunk, but narrower on the right? If so, it may be because some force—for example, pressure from another tree—was bending the tree trunk toward the left at that point in its history. Because trees prosper by being as tall as they can, they will resist being pushed over. They do so by growing extra wood on one *side* of themselves. The wider areas in a given set of circles represent this extra wood, known as **reaction wood**.

Look at the center of any large log, and the wood you see there is almost always darker than the surrounding wood. Why? The inner wood—which like all wood is secondary xylem—has long since closed down with respect to its water-transport function. The cells that make up the water columns were always dead in their mature, water-transporting state; but eventually each column was severed altogether, meaning it quit conducting water. The problem is that an *empty* xylem column is a *dangerous* column as far as the tree is concerned, because a fungus could grow right up through it. Therefore, trees have developed a plugging mechanism: Surrounding living cells push material into the columns, sealing them off. Over time, these cells produce oils, lignin, and microbe-fighting substances and then die themselves. The result is a center of a tree that is completely dead—and dark in col-oration—but strong in its load-bearing strength and resistant to bacteria and fungi. This is the tree's **heartwood**. Its counterpart is the tree's **sapwood**—the lighter, outer wood whose xylem tubes are still conducting xylem sap.

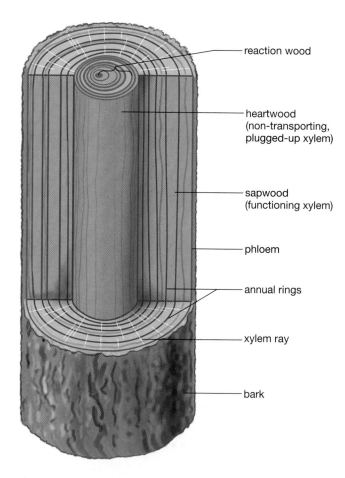

Figure 1
The Anatomy of Wood

reaction wood

heartwood
(non-transporting,
plugged-up xylem)

sapwood
(functioning xylem)

phloem

annual rings

xylem ray

bark

22.6 How the Plant's Vascular System Functions

You have had ample opportunity already to look at the plant's vascular or "plumbing" system as it relates to plant growth, but now it's time to take a closer look at how the vascular system itself works, starting with the water-conducting xylem and then continuing to the food-conducting phloem.

How the Xylem Conducts Water

Remember the statement at the top of the chapter that a good many plant cells are dead in their mature, functioning state? You saw this once already with cork cells; here is another variety of these cells, which form most of the xylem. Two types of cells make up the water-conducting portions of both the primary and secondary xylem. These are tracheids and **vessel elements**. Upon reaching maturity, these cells die, and their contents are then cleared out, leaving a strand of empty, thick-walled cells stacked one on top of another—leaving tubes, in other words (**Figure 22.23**). The walls of these xylem cells remain, however; indeed, *these* are some of the sclerenchyma cells noted earlier that have the strong cellulose reinforcement in their cell walls. The upshot is not only a transport system but a *load-bearing* system, which is why something as massive as a tree can grow so tall. As noted earlier, this load-bearing function remains in a given group of xylem cells long after they have ceased to act as a conduit for the water and dissolved nutrients, sometimes known as **xylem sap**.

A tube metaphor actually is more accurate for the vessel elements than for the slender, tapered tracheids, because the vessel elements not only lose their internal contents, but may lose part or all of the cell-wall barriers between adjoining cells as well. The result is a great capacity to transport water relative to the tracheids, whose cell walls remain intact. Any two tracheid cells will have matched pairs of perforations that line up with one another, allowing water to move from one tracheid to another.

The Value of Having Tracheids *and* Vessel Elements

You may remember from Chapter 21 that, through evolutionary time, flowering plants took over as the dominant variety of plant from the gymnosperms (which include coniferous trees). One reason this happened was the development of vessel elements; almost all angiosperms have vessel elements *and* tracheids, while almost all woody gymnosperms possess only tracheids. As a result, angiosperms can simply transport more water than can gymnosperms. You may wonder, though: If vessel elements are so much more efficient than tracheids at transporting water, why has evolution preserved tracheids in angiosperms? As it turns out, compared with vessel elements, tracheids are less susceptible to trapping air bubbles, or "embolisms," that can arise in flowing water. Given this, tracheids may be another kind of insurance policy for plants as they face various environmental changes.

Suction Power Moves Water through Xylem

In some trees, water coming from the roots goes through a vertical journey of up to 100 meters, or about 300 feet, before it gets to the topmost leaves of the tree. What kind of force can move water such a long way? Plants spend no energy at all in this process. Instead, it is **transpiration** itself—the loss of water from a plant, mostly through the leaves—that moves water through xylem. As water vapor evaporates into the air, it creates a region of low pressure, and this pulls a continuous column of water upward (**see Figure 22.24**). Ultimately, the energy source for this

Figure 22.23
Water Carriers
The fluid-conducting cells of the xylem are tracheids and vessel elements. Both kinds of cells stack up to form tubes.

Figure 22.24
Suction Moves Water through Xylem
When water evaporates from leaves, a low pressure is created that pulls more water up to fill the void. Water is so cohesive that it moves up in a continuous column. The sun's energy, rather than the plant's energy, fuels this process.

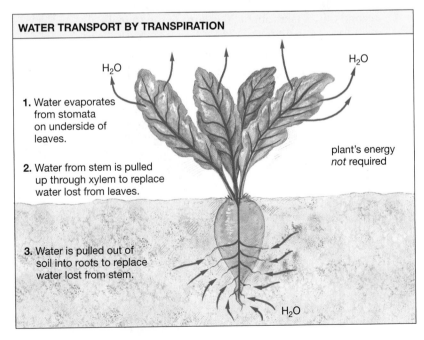

WATER TRANSPORT BY TRANSPIRATION

H_2O

H_2O

1. Water evaporates from stomata on underside of leaves.

2. Water from stem is pulled up through xylem to replace water lost from leaves.

plant's energy *not* required

3. Water is pulled out of soil into roots to replace water lost from stem.

H_2O

The Syrup for the Pancakes Comes from Xylem

In early March, when winter is winding down in New England and Eastern Canada, farmers who own sugar maple trees are kept busy in a game of drilling and watching. What they are drilling is a series of two-to-three-inch holes, about three of them per maple tree. What they are watching is the weather, because they need a series of freezing nights followed by warmer days to get what they want. Such nights prompt the maple trees to convert nutrients they have stored into sugar; the warmer days then bring about a flow of this sugar, upward in sap through the tree's xylem. This material can then be "tapped" through spigots inserted into the drilled holes. The clear white sap that drips out—as much as 150 liters (about 40 gallons) of it per tree per year—is the starting material for pure maple syrup. Vermont is the chief maple-syrup-producing state in the United States, with an output in 1996 of just over 2 million liters (about 550,000 gallons). It is Canada, however, that is the world's largest maple syrup producer.

The "maple syrup" that most Americans buy in stores actually has very little maple syrup in it (perhaps 2–3 percent, the bulk being made up of corn syrup and syrup from regular sugar). This is understandable, because pure maple syrup is expensive, costing about $25–$35 per gallon. One reason for this expense is that harvesting the syrup is a labor-intensive business; each tap must be put in by hand, just as it was hundreds of years ago (**see Figure 1**). Beyond this, it takes about 50 gallons of sap to make a single gallon of maple syrup. The sap must be boiled down, until its sugar concentration reaches 66.5 percent.

Too many taps can kill a tree, but experienced syrup harvesters know how to tap the same tree year after year—potentially for hundreds of years—without harming it. The 150 liters of sap a tree can yield up sounds like a lot, but it may account for only 10 percent of a tree's total supply of sugar.

In the text, you've read much about the plant's food supply flowing through *phloem*, not xylem. But in sugar maples and some other deciduous trees, the food produced in summer's photosynthesis is stored as starch in xylem cells, to be broken down into sugar and released to the growing upper portions of the tree in early spring.

Figure 1
Source of Maple Syrup
Maple trees in Vermont being tapped for their xylem sap, which will become the key ingredient in maple syrup.

Figure 22.25
Food Carriers
The micrograph shows a cross section through some sieve elements (looking at them from above), surrounded by a number of companion cells. A plant's phloem sap moves from one sieve element to another through the perforations visible in the cells.

movement is the sun, whose warm rays power the evaporation of water at the leaf surface.

Food the Plant Makes Is Conducted through Phloem

Now let's turn from water-and-nutrient transport to food transport. What is the nature of the food that flows through phloem? It's mostly sucrose—better known as table sugar—that is dissolved in water. Hormones, amino acids, and other compounds move through phloem as well, but sucrose and water are the main ingredients of this material, which is known as **phloem sap**.

The phloem cells that this sap moves through are called **sieve elements**. Unlike xylem cells, sieve elements are alive at maturity, only in a curiously altered state: Each one has lost its nucleus, which you may remember is the DNA-containing information center of a cell. Sieve elements continue to function as living cells, however, because each one has associated with it one or more **companion cells** that retain their DNA and that seem to take care of all the housekeeping needs of their sister sieve elements (**see Figure 22.25**). The association between the two kinds of phloem cells is so intimate that when sieve elements die out, as they often do at the end of a growing season, their companion cells die as well. The reason sieve elements lose their nuclei seems to be to make room for the rapid flow of nutrients through them. The reason they are called *sieve* elements is that the ends of adjoining cells are studded with small holes, much like a sieve. Stacked on top of the other, these cells form the plant's food pipes, the **sieve tubes**.

Sugar-Pumping and Water Pressure Move Nutrients through the Phloem

The motive power for the movement of food through phloem is different from that of water through xylem; with nutrient transportation, the plant does not get a free ride on transport (via transpiration). Instead, to load the sucrose it makes into the phloem, the plant must expend its own energy (**see Figure 22.26**). After this, a principal player in moving fluid through the phloem is a phenomenon you've encountered before—osmosis.

Osmosis and the Movement of Phloem

You can review the process of osmosis in Chapter 5 (pp. 98–102), but here's a brief summary. Any time there is a membrane, such as the lining of a cell, that is immersed in water and that is semipermeable—that some molecules can pass through but others can't—this creates the conditions for osmosis. There is a net movement of water across the membrane to the *side* in which there is a greater concentration of solutes, or substances that can be dissolved in liquid (**see Figure 22.27**). In the case of plants, the solute is sucrose, which moves to the outside of the sieve elements and is then *pumped* into them through special membrane channels. (The

pumping is what requires energy from the plant.) This results in a higher concentration of sucrose inside the cells than outside, and there is a net movement of *water* into the cells because of osmosis. The effect is the same as water coming into a garden hose from a spigot: The water moves through its path of least resistance—through the hose and out the nozzle. Just so is this solution of water and sucrose moved through the sieve tubes under the power of this pressure flow

Figure 22.26
How Food Moves through the Plant
Sucrose is produced in the leaves (source) by photosynthesis, after which the plant expends its own energy to load the sucrose into the phloem. Osmosis is then a key factor in transporting the sucrose through the plant for use and storage (sink).

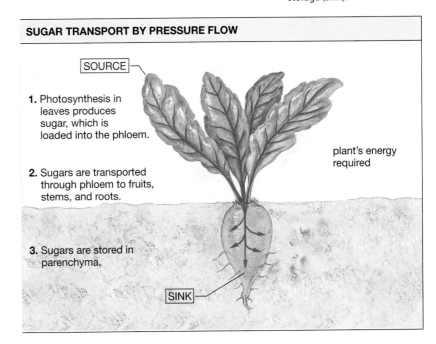

SUGAR TRANSPORT BY PRESSURE FLOW

SOURCE

1. Photosynthesis in leaves produces sugar, which is loaded into the phloem.

plant's energy required

2. Sugars are transported through phloem to fruits, stems, and roots.

3. Sugars are stored in parenchyma.

SINK

a

salt

water

semipermeable membrane

b

OSMOSIS

Figure 22.27
Osmosis in Action

a An aqueous solution divided by a semipermeable membrane has a solute—in this case salt—poured into its right chamber.

b As a result, though water continues to flow in both directions through the membrane, there is a net movement of water toward the side with the greater concentration of solutes in it.

(**see Figure 22.28**). The name given to this hypothesis about phloem sap movement is the **pressure-flow model**, first developed in the 1920s by German plant researcher Ernst Münch.

The loading site of this system is called a **source** and the unloading site is called a **sink**, but these sites could also be called the "producer" and the "consumer," or perhaps "producer" and "bank." Sources tend to be sugar-producing leaves, while sink sites tend to be storage tissue such as roots, along with growing fruits and stems. At the sink, sucrose moves passively or is actively pumped from sieve element cells to sink cells (the roots, for example). This triggers osmosis once again, only this time the net water flow is *out* of the sieve tube, because more solute is now outside the tube than is inside. This flow creates a pull that, like the push at the source, helps move the phloem sap along.

Water Flows from Xylem to Phloem and Back Again

You saw earlier that phloem and xylem tubes are stacked together in bundles. These transport systems are not sealed off from each other, however, but instead are closely interconnected. Remember the phloem sources, where water flowed into the sieve tube? Well, where did that water come from? From the adjacent xylem. And in the phloem sinks, when water flows out of the sieve tubes, most of it flows into adjacent xylem, where it joins water coming from the roots and moves on up with it (see Figure 22.28a). With this, let's take our leave of transport in plants in order to cover another critical topic, reproduction and development.

22.7 Sexual Reproduction in Flowering Plants

As almost everyone knows, the human reproductive cells called eggs and sperm fuse—in a moment often referred to as *conception*—thereby creating a single cell that starts dividing, eventually producing a new human being. Imagine, however, that instead of eggs and sperm, human beings produced *different* kinds of reproductive cells; ones that—on their own, without coming together—grew into a separate form of human life that looked very different from the humans we're used to. Then imagine that *this* is the generation that produced the eggs and sperm (the gametes) that give rise to human beings. Beyond this, imagine that the human eggs and sperm were brought together not by sexual intercourse between humans themselves, but by the wind, or by other animals. In a nutshell, this describes some of the major differences between plant

XYLEM   PHLOEM

SOURCE

sugar

water

leaf cell

1. active transport of sugar into phloem (requires ATP)

2. water follows by osmosis

3. pressure gradient moves fluid down phloem

SINK

sugar

water

root cell

4. active or passive transport of sugar into root cell (may require ATP)

5. water follows by osmosis

vessel elements   sieve elements

Figure 22.28
Food Transport through Phloem and Phloem-Xylem Linkage
In food transport, on the right, (1) sugar produced within leaves is actively pumped by the plant into the sieve elements of the phloem. (2) Water then "follows" the sugar in, under the power of osmosis. This creates (3) a pressure gradient that moves the sugar-water compound through the phloem. At the sink, (4) sugar moves either passively or through pumping, this time out of phloem cells, with (5) water following, again through osmosis. Because water flows into the phloem at the leaves and out of the phloem at the root, they are regarded as the "source" and the "sink" of the phloem, respectively. The phloem and xylem systems are related. Water returns to the xylem from the phloem (at the bottom of the figure) and is transported back to the leaves. Meanwhile, it is water transported to the phloem from the xylem (at the top left of the figure) that provides the water that flows through the phloem.

and animal reproduction. Because plant reproduction has an added step—an intervening generation that produces gametes—it is a bit more complicated than animal reproduction, but fascinating to look at precisely because it differs from what we're used to.

Flowers Are the Reproductive Parts of Plants

To begin, consider a brief reminder regarding the reproductive parts of plants, their flowers, which you can see in **Figure 22.29**. You may recall from Chapter 21 that the heart of the flowers' reproductive structures are the stamens and the carpel. A stamen consists of a long, slender filament and then are topped by an anther, whose chambers contain cells that give rise to the male pollen. The **carpel**, meanwhile, is a composite structure, composed of three parts: the stigma, which is the tip end of the carpel, on which pollen grains are deposited; the style, a slender tube that raises the stigma to such a prominent height that it can easily catch the pollen; and the ovary, the area in which fertilization of the female egg and then development of the next generation of plant will take place.

Flowering Plants Reproduce through an Alternation of Generations

Critical to an understanding of how a new generation of maple trees or roses arises from an existing generation is the alternate generation described earlier. Maple trees or roses don't produce gametes (eggs and sperm) that go on to fuse and then give rise to a new generation that looks like the old one. Instead, they produce **spores**—single cells that, without fusing with anything, grow into their own generation of plant. It is *this* generation of plant that produces the gametes that will once again result in a maple tree or rose bush.

Our trouble in grasping this is that the generation growing from spores is a *tiny* generation compared to the parent generation. In addition, it is a generation that spends almost its entire life in (or on) the parent generation of plant. In short, from a human perspective this separate generation looks like some small component of the original generation.

The back-and-forth between types of generations in plants gives us a term that was reviewed in Chapter 20—the **alternation of generations** (see page 412). Because the generation that is familiar to us—the maple tree or rose bush—produces spores, it is known as the sporophyte generation. These spores then give rise to the gametes (eggs and sperm) that, upon fusing, will develop into maple trees and rose bushes once again. Hence, this is the gametophyte generation. For a view of how sporophyte and gametophyte generations differ not just in angiosperms but in all types of plants, see **Figure 22.30** (page 466). In the more primitive plants, such as mosses, it is the gametophyte generation that is larger and more dominant, but things reverse with movement up the evolutionary line. By the time we reach angiosperms, the sporophyte generation of, for example, a tree might be 50 meters (about 165 feet) tall, while its gametophyte generation is microscopic.

Look now at **Figure 22.31** on page 466 to get an overview of angiosperm reproduction. As you proceed through the story, you'll be looking at various pieces of this figure in greater detail.

The Sporophyte and Gametophyte Generations Are Genetically Different

Apart from being different in size and appearance, the sporophyte and gametophyte generations are genetically very different. Once again, there is a generational back-and-forth. Let's look at this on the male side,

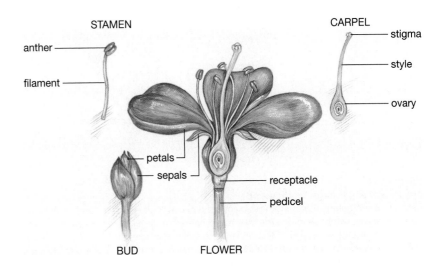

Figure 22.29
Anatomy of a Flower, the Reproductive Structure of an Angiosperm

466

SPOROPHYTE DOMINANCE (2n)

green algae   mosses   ferns   gymnosperms   angiosperms

GAMETOPHYTE DOMINANCE (n)

Figure 22.30
Generational Dominance Varies with the Type of Plant
In more primitive plants, such as mosses, the gametophyte generation is larger and more dominant. In the most recently evolved plants, the angiosperms, the sporophyte generation may be relatively small (a tomato plant) or massive (a tree), but the gametophyte generation generally is microscopic and almost wholly dependent on the sporophyte generation. (Though not plants, green algae are included for comparative purposes.)

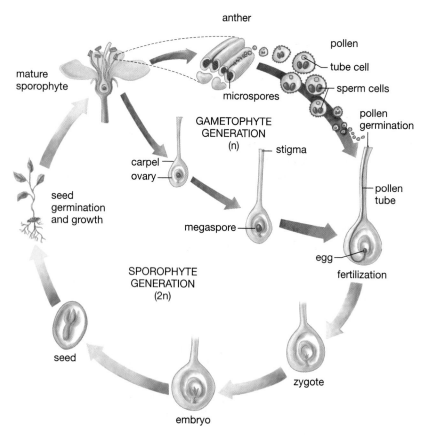

Figure 22.31
Overview of Angiosperm Reproduction
The mature sporophyte flower produces many male microspores within the anther and a female megaspore within the ovary. The male microspores then develop into pollen grains that contain the male gamete, sperm. Meanwhile the female megaspore produces the female gamete, the egg. Pollen grains, with their sperm inside, move to the stigma of a plant. Moving through a pollen tube, the sperm reaches the egg and fertilizes it. This results in a zygote that develops into an embryo protected inside a seed. This seed leaves the parent plant and sprouts or "germinates" in the earth, growing eventually into the type of sporophyte plant that began the cycle.

starting with the sporophyte of a hypothetical flowering plant (**see Figure 22.32**). Cells within the anthers, called **microspore mother cells**, give rise to spores by undergoing *meiosis*. Students who have been through the genetics section will recall that the normal state for the cells of most complex organisms is diploid, meaning that cells have chromosomes that come in matched *pairs*—a state sometimes referred to as 2n. This is the state of all the cells in the sporophyte generation. In meiosis, the diploid microspore mother cells will divide twice, in the process reducing their number of chromosomes by half, so that the microspores that result have only a single set of chromosomes each (making them haploid or n). It is these microspores that will develop into pollen grains. Keeping an eye on the forest as well as the trees, it's worth noting that the line between one generation and the next is crossed after the microspore mother cell has undergone meiosis. Once the microspores have been produced, the result is the gametophyte generation.

Chromosome reduction takes place in the production of the female gametophyte as well. Here, meiosis, as well as development of the female gametophyte, take place entirely inside the ovary that lies at the base of the plant's carpel. In this case, a single diploid cell—called a **megaspore mother cell**—undergoes meiosis, yielding female megaspores. (Why is the female spore "mega" compared to the male's "micro"? Because the female spore is bigger.)

As you can see in Figure 22.32, both the male and female gametophyte generations are multicelled structures (though just barely), with these cells arising by mitosis, or common cell replication, from the original single haploid spores. Because mitosis is a process that makes exact copies of cells, this means that *every cell* in the gametophyte generation is haploid. In sum, the gametophyte generation is haploid (n) while the sporophyte generation is diploid (2n).

Development of the Male and Female Gametophyte Generation

The Male Gametophyte

So, for our purposes, the starting point for plant reproduction is the mature sporophyte plant that gives rise to both male and female gametophyte generations. Figure 22.32 also shows you *how* the male and female gametophyte generation develop in angiosperms. On the male side, things are pretty simple. Meiosis in the microspore mother cells yields four haploid microspores (the start of the gametophyte generation). Each spore develops into a pollen grain that initially consists of

Figure 22.32
Development of the Male and Female Gametophyte Generation

The male gametophyte generation takes shape inside the chambers of the pollen sacs, where the diploid microspore mother cells will undergo meiosis, each one of them thereby giving rise to four haploid microspores. It is these microspores that represent the new, haploid generation of the plant on the male side. The single cell in each microspore goes through cell division, thereby producing two cells, a tube cell and a generative cell. Before or during this time, a protective coat develops around the microspore. The combination of the cells and protective coat is the pollen grain. At some point, the generative cell in the grain divides into two sperm cells. In the figure pictured, this happens before the pollen grain leaves the anther. With this cell division a mature male gametophyte has developed, consisting of an outer coat plus three cells—two sperm cells and a tube cell. The pollen grain is then released from the anthers to make its way to a stigma.

For the female gametophyte generation, development begins inside a structure of the parent sporophyte plant called an ovule, located deep within the carpel of the plant. From a central mass of cells in the ovule, a single diploid megaspore mother cell undergoes meiosis, producing four haploid megaspores; this marks the beginning of the female gametophyte generation. Three of the megaspores then die, leaving a single megaspore. The megaspore undergoes mitosis, eventually producing six cells with a single nucleus each and one cell with two nuclei. These seven cells form the embryo sac, which is the mature female gametophyte. Note that one of the seven cells in the embryo sac is the egg that will undergo fertilization by one of the sperm in the pollen grain.

three things: an outer coat, a tube cell, and a generative cell. This working pollen grain can be thought of as an *immature* male gametophyte plant. At some point, the generative cell divides once, resulting in two *sperm* cells. (In the type of plant you're looking at, this division comes before the pollen grain is released from the sporophyte plant, but in other species the division comes later.) Now having three cells, the pollen grain is released from the sporophyte plant and makes its way, by wind or bird or insect, to the stigma of another plant. This is what **pollination** is. You'll get to its story in a second.

The Female Gametophyte

On the female side, note in Figure 22.32 that there develops from the ovary wall a structure called an *ovule*—a part of the sporophyte plant—that will enclose the female gametophyte generation. Note that the ovule consists of the protruding stalk, a central core of cells, and couple of layers of tissue called **integuments** that wrap around the central cells, almost coming together except for a tiny opening they leave, called a **micropyle**. From the ovule's central mass of diploid cells, a single cell, the megaspore mother cell, undergoes meiosis, thereby producing four megaspores. This action represents the crossover into the gametophyte generation. Then three of the

megaspores will die, leaving the one that remains as an immature female gametophyte plant.

The remaining megaspore now undergoes three rounds of mitosis, but of a curious sort: There is the doubling of the cell's DNA (in the nucleus), but no subsequent cell division, as is usually the case. The result is a single cell that has eight haploid nuclei. This cell eventually does undergo cell division, but once again there is a twist. Six of the cells that result will each have a single nucleus, but one will have two nuclei. You can see, in Figure 22.32, the most common arrangement of these seven cells. Collectively, they form the **embryo sac**, which can be thought of as the mature female gametophyte plant. At one end is the cell of greatest importance, the egg, flanked by a couple of cells on either side. On the opposite end there are three other cells; and in the middle, a very large central cell, which is the cell that has two nuclei.

Fertilization of Two Sorts: A New Zygote and Food for It

The stage is now set for male to come into contact with female. The pollen grain, with its two sperm cells and one tube cell, now lands on the stigma of a plant with an egg and embryo sac matured down in the ovule (**see Figure 22.33**). The tube cell then spouts

Figure 22.33
Double Fertilization in Angiosperms
When the male pollen (shown greatly enlarged) lands on the stigma, the tube cell within it sprouts a pollen tube, through which the two sperm pass in moving to the ovule. One sperm cell fertilizes the egg, forming a zygote—the new sporophyte plant. The other sperm cell fuses with the two nuclei of the central cell, forming the endosperm that will later nourish the embryonic plant.

pollen grain — tube cell
— sperm cells
stigma

style

pollen tube
sperm cells

micropyle
ovary
ovule with female gametophyte
egg (n)

fusion of one sperm cell with nuclei of central cell to form endosperm (3n)
fusion of one sperm cell with egg to form zygote (2n)

Pollination   **Pollen tube growth**   **Double fertilization**

something: a pollen tube, which tunnels its way down through the style, eventually gaining access to the female reproductive cells within the ovule. The two sperm cells travel down this tube to achieve fertilization. (Why *two* sperm cells? Stay tuned.) In sum, in its most mature state, the male gametophyte generation consists of three cells, one of which grows a tube and two of which are sperm cells—quite a different thing from the sporophyte generation with its billions of cells.

The pollen tube grows through the ovule's micropyle and ejects its two sperm cells into the embryo sac. One of these sperm fuses with the egg, yielding the zygote that is the beginning cell in the new generation of plant—another sporophyte generation like the tree or rose bush we started with. This is no different in principle than what takes place with *human* life, where egg and sperm fuse, producing a single-celled zygote that develops into a whole human being. And, as with human egg and sperm, remember that the plant egg and sperm were both haploid, containing a single set of chromosomes each. Their fusion produces a diploid zygote—one with a *paired* set of chromosomes—that will divide over and over, resulting in a diploid sporophyte plant.

And what of the other sperm? Here is one of the beauties of flowering plants. This sperm enters the central cell with its two nuclei, producing a cell with *three* nuclei (making it triploid or 3n.) This too is a fertilization, only what it sets in motion is the development of *food* for the embryo—endosperm, which is tissue rich in nutrients that can be used by the embryo that will now develop from the zygote. Thus two fertilizations have taken place. This is known as **double fertilization** in angiosperms—a fusion of gametes and of nutritive cells at the same time. Double fertilization is an innovation that belongs almost solely to the flowering plants. As it happens, it has been a boon to all kinds of *animals*, including people. It is endosperm, or the tissue it develops into (the cotyledon) that supplies much of the food we human beings consume. The rice and wheat grains we're familiar with consist in large part of endosperm meant to sustain the plant embryo (**see Figure 22.34**).

22.8 Embryo, Seed, and Fruit: The Developing Plant

The Development of a Seed

Egg fertilization produces a single-celled zygote. The question is, how does the plant develop from this zygote into an embryo-containing *seed* that falls to the ground, eventually giving rise to a growing sporophyte plant? Many steps are involved here; what follows are only the highlights. First, recall that the female gametophyte, the embryo sac, was nearly surrounded by ovule tissue called *integuments*. These will now develop into the **seed coat** that will protect the embryonic plant until it germinates in the ground. What once was an ovule with a gametophyte inside is now a seed: a reproductive structure of a plant that includes an outer seed coat, an embryo, and nutritional tissue for the embryo (the endosperm). This is a reproductive structure that bears similarities to one employed by some animals, the egg (**see Figure 22.35**).

Figure 22.34
Endosperm Is Food for the Embryo
In monocots such as corn, endosperm persists in the mature seed, providing nutrition for the embryo inside. The cotyledon in the figure is the corn plant's embryonic leaf.

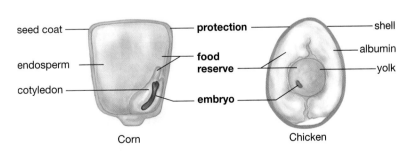

Figure 22.35
A Kernel of Corn and a Chicken Egg Are Similar in Some Ways
Both are home to an embryo, both contain plenty of food for development, and both offer protection for the embryo in dry, land environments.

Figure 22.36
Fruits Come in Many Forms
A fruit, the mature ovary of a flowering plant, surrounds a seed or seeds. Thus **(a)** the flesh of an apricot fits this definition of fruit, but so does **(b)** the pod of a pea, which has several seeds inside. The structure we commonly think of as **(c)** the strawberry fruit actually is the strawberry plant receptacle, with each receptacle having many fruits on its surface. What are commonly thought of as strawberry "seeds" are tiny strawberry plant carpels, each complete with its own fruit and seed. Fruits are important not only because they protect the seeds but also because they attract animals that disperse the seeds.

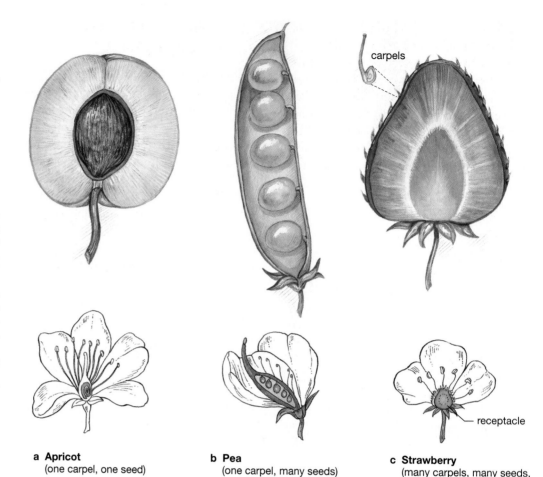

a Apricot
(one carpel, one seed)

b Pea
(one carpel, many seeds)

c Strawberry
(many carpels, many seeds, one receptacle)

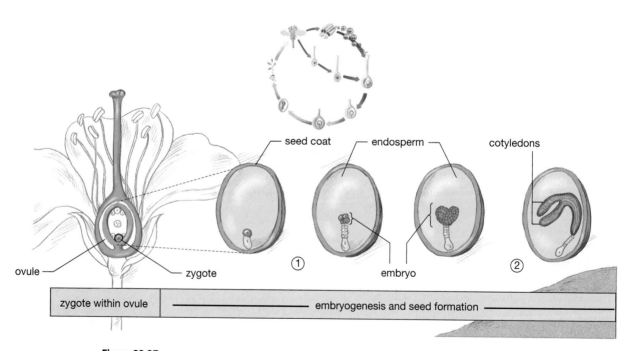

Figure 22.37
Development of a Plant Embryo, Germination of a Plant Seed
Egg and sperm have come together in this dicot to produce a single-celled zygote, which divides repeatedly in developing as the plant embryo. The zygote is surrounded by the plant ovule, which develops into a seed.

1. The zygote divides into two cells, one of which is the embryo (green in the figure). The other cell (yellow in the figure) develops into a structure that will push the embryo into the endosperm.

2. The embryo starts to develop its cotyledons or embryonic leaves. Most of the endosperm eventually is taken up by these cotyledons. (Because this plant is a dicot, it has two cotyledons.)

The Development of Fruit

Something more will now develop *outside* the seed coat. Recall that the ovule was surrounded by the ovary of the flower. The ovary wall now develops into **fruit**. Strictly speaking, fruit is the mature ovary of any flowering plant. Given this, *all* flowering plants have an ovary, but it's clear that most of the 260,000 species of flowering plants are not fruits as we commonly understand that term. It follows that the botanical definition of fruit differs from the common definition; under the botanical definition, fruits can take on lots of forms. The pod of a pea plant is a fruit, for example, as is the outer covering of a kernel of corn (though here fruit has fused with the corn seed coat, leaving scarcely anything separate that can be recognized as fruit). On the other hand, fruits can take exactly the form we're used to, as with the flesh of an apricot or cherry. To add another twist, where is the fruit in a strawberry? The answer, shown in **Figure 22.36**, may surprise you.

Fruits Serve in Protection and in Seed Dispersal

What is common to many fruits is that they serve a protective function for the underlying seed, and many act as *attractants* for seed dispersal. As noted in Chapter 21 (page 427), the wild berry that a bear eats consists not only of the fruit flesh, but of the seeds it surrounds. The fruit is digested by the bear as food, but the *seeds* pass through the bear's digestive tract. Thus has this plant engineered seed dispersal at what may be a promising new location for a berry plant. Plants with such attractants *time* their fruit ripening, so that the fruit is not attractive before the seed is ready for dispersal.

Development of the Embryo and Germination of the Seed

While fruit and seed coat are maturing, the embryo inside them is also developing. **Figure 22.37** shows you the outline of the development of a plant embryo, this one a dicot plant. Recall that the process starts with a single cell, the zygote, encased within

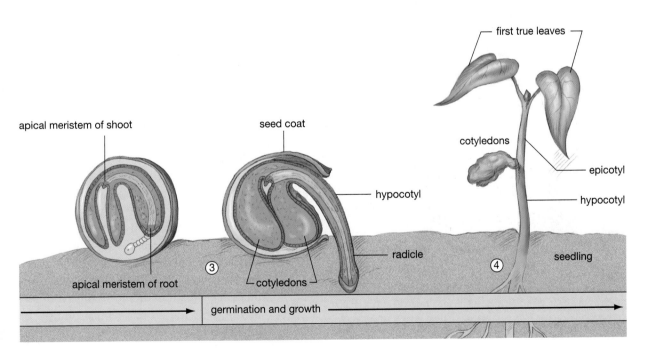

3. Having made its way to the ground, the seed is ready to "germinate" or start sprouting. The seed coat splits, allowing the emergence of the hypocotyl (tissue below the cotyledons), which includes the plant's radicle, or early root structure.

4. Root and shoot have both sprouted. The plant's cotyledons, having completed their function, will wither away. The plant sprouts its first true leaves and begins the process of photosynthesis. The reproductive cycle is ready to begin again.

the structure that will become the seed, the ovule. The zygote divides, yielding two cells, one of which becomes the embryo. The other cell gives rise to a paddle-shaped structure that pushes the growing embryo into the endosperm.

Eventually, the embryonic tissue develops two **cotyledons**, or embryonic leaves. When the embryo is still tightly encased in the seed coat, the cotyledons can take on an important role you've seen before, which is to provide nutrients to the growing embryo. You may say: But isn't this the function of the endosperm? In most monocots (for example, the grains) the endosperm retains this function, but in most dicots endosperm tissue is *taken up* by the cotyledons as they develop. The foods we know as beans are comprised of embryos with large cotyledons (**see Figure 22.38**).

The seed that encases the embryo eventually is expelled from the sporophyte parent plant, and it makes its way into a suitable patch of earth, there to germinate, or to sprout both roots and shoots. First to emerge from the seed is the root structure or radicle, which is attached to hypocotyl, which can be thought of as all the plant tissue below the cotyledons. Then there emerges the **epicotyl**, meaning the tissue above the cotyledons. This epicotyl gives rise to the first true leaves of the plant. The seed has sprouted into a new sporophyte plant.

Seed Dormancy Can Be Used to a Plant's Advantage

One of the survival strategies of plants is the use of dormancy, or a prolonged low metabolic level in seeds. Seeds can *postpone* germinating until they have favorable conditions around them, such as the proper temperature or amount of light. What generally sets seed germination in motion is an uptake of water from the surrounding environment, but what *allows* the seed to take up this water is what tells the tale in dormancy. Some seeds simply need to dry out, once they've shed the fruit from around them; others require a scraping or abrasion before they will take up water; and still others require the action of enzymes from an animal's gut, or even charring by fire.

Once the plant has germinated and matured a little, you could take a cross section of its stem and find concentric circles of the tissue types mentioned near the start of this chapter: dermal, vascular, and ground. Meanwhile, meristematic tissue could be found at the tips of the root and shoot, allowing the plant to grow further. With this, we have come full cycle: Another generation of plants is maturing.

On to Animals

In Chapters 21 and 22, you have looked at a group of living things, the plants, whose members are silent, fixed in one spot, and have no moving parts. Now it's time to shift gears. There are animals that are silent and animals that are fixed in place, but you'll also find with animals nature's only sound-makers and its supreme travelers. The next five chapters are concerned with the functioning of animals in general, but they focus on one animal that in some ways is very familiar to us—the human animal.

Figure 22.38
Cotyledons Serve a Nutritive Function
In a dicot such as this bean plant, the endosperm food reserves are taken up by the embryonic leaves or cotyledons of the embryo, which then serve as the mature seed's food reserves.

Chapter Review ▬▬▬▬▬

Summary

22.1 Two Ways of Categorizing Flowering Plants

- Plants can be categorized by how long it takes them to go through a cycle from germination to death. Those that go through this cycle in a year or less are annuals; those that go through in about two years are biennials; those that live for many years are perennials.

- A cotyledon is an embryonic leaf, present in the seed. Angiosperms are classified according to how many cotyledons they have—one in the case of monocotyledons, two in the case of dicotyledons. This distinction in embryos results in major differences in the anatomy and functioning of monocot and dicot plants.

22.2 There are Three Fundamental Types of Plant Cells

- There are three fundamental types of cells in plants that, alone or in combination, make up most of the plant's tissues, meaning groups of cells that carry out a common function. These three cell types are parenchyma, sclerenchyma, and collenchyma.

22.3 The Plant Body and Its Tissue Types

- Some plants are capable only of primary growth, meaning growth at the tips of their roots and shoots that primarily increases their length. Plants that exhibit only primary growth are herbaceous plants, composed solely of primary tissue. Other plants exhibit both vertical growth and lateral, or secondary growth. These are the woody plants, composed of primary and secondary tissue.

- There are four tissue types in the primary plant body: dermal, vascular, meristematic, and ground. Dermal tissue can be thought of as the plant's outer covering, vascular tissue as its "plumbing," meristematic tissue as its growth tissue, and ground tissue as almost everything else in the plant.

22.4 How a Plant Grows: Apical Meristems Give Rise to the Entire Plant

- The entire plant develops from meristematic cells in regions called *apical meristems*. Meristematic cells remain perpetually embryonic, able to give rise to cells that differentiate into all the plant's tissue types.

- Shoot apical meristems give rise to the entire shoot of the plant. In addition to providing for vertical growth, shoot apical meristems produce meristematic tissue called *lateral buds* at the base of leaves that can give rise to a branch or flower. Root apical meristems are located just behind a collection of cells at the very tip of the root, called the *root cap*.

- The plant's tissue types develop in stages from meristematic cells. This development takes place in a series of regions adjacent to the apical meristem. In a gradual transition, the apical meristem gives way to a zone of cell division, followed by a zone of elongation (in which developing cells lengthen) followed by a zone of differentiation (in which cells fully differentiate into different tissue types).

22.5 Secondary Growth Comes from a Thickening of Two Types of Tissues

- Secondary growth in plants takes place through the division of cells in two varieties of meristematic tissue: the vascular cambium, which gives rise to secondary phloem and xylem; and cork cambium, which gives rise to the outer tissues of woody plants. Secondary xylem, also known as wood, is responsible for most of a tree's widening.

- Looking at a tree from the secondary phloem outward to the tree's periphery, four tissues constitute the tree's bark: secondary phloem, phelloderm, cork cambium, and cork. The cork cells are dead in their mature state and provide layers of protection for the tree.

22.6 How the Plant's Vascular System Functions

- Two types of cells make up the water-conducting portions of xylem tissue. These

Key Terms

are tracheids and vessel elements, both of which are dead in their mature, working state. Vessel elements, which exist almost solely in angiosperms, conduct more water than tracheids. Their existence in angiosperms is one of the reasons for the angiosperms' dominance in the plant world.

- Water movement through xylem is driven by transpiration, meaning the loss of water from a plant, mostly through the leaves. As water evaporates into the air, it pulls a continuous column of water upward through the plant.

- Sucrose is the main product that flows through phloem. The fluid-conducting cells in phloem are called *sieve elements*, which lack cell nuclei in maturity. Each sieve element has associated with it one or more companion cells, which retain their nuclei and which seem to take care of the housekeeping needs of their related sieve elements.

- Plants must expend their own energy to load the sucrose they produce into the phloem. Dissolved sucrose then moves through the plant through the power of osmosis and the fluid pressure that results from it.

22.7 Sexual Reproduction in Flowering Plants

- All plants, including angiosperms, reproduce through an alternation of generations. A sporophyte generation (the familiar tree or flower) produces haploid spores that develop into their own generation of plant, the gametophyte generation. The male gametophyte is the pollen grain, consisting in maturity of an outer coat, two sperm cells, and one tube cell. The female gametophyte consists at maturity of an embryo sac composed of seven cells, one of which is the egg. The female gameto-

phyte is housed inside a structure of the parent sporophyte plant called an *ovule*.

- Fertilization of the egg by sperm requires that a pollen grain land on the stigma of a plant. The tube cell of the pollen grain then germinates, sprouting a pollen tube that grows down through the sporophyte plant's stigma and style, eventually reaching the female reproductive cells inside the ovule. The sperm cells inside the pollen grain travel down through the pollen tube and one of the sperm cells fertilizes the egg in the ovule, producing a zygote. With this, the new sporophyte generation of plant has come into being.

- The second sperm cell in the pollen grain enters the central cell in the embryo sac, setting in motion the development of food for the embryo, endosperm tissue. This second fertilization completes the process of double fertilization—a fusion of gametes and of cells producing nutritive tissue.

22.8 Embryo, Seed, and Fruit: The Developing Plant

- With fertilization, the ovule integuments that surrounded the embryo sac begin to develop into the seed coat that will surround the growing sporophyte embryo. The ovary that surrounded the ovule then starts to develop into a layer of tissue that will surround the seed—fruit, which is defined as the mature ovary of a flowering plant. Under this definition the pod of a pea plant is fruit, as is the flesh of an apricot.

- The seed with its fruit covering, which encases the embryo, eventually is expelled from the sporophyte parent plant and then germinates in the earth, sprouting both roots and shoots.

Understanding the Basics

Multiple-Choice Questions

1. Dermal tissue _____, while vascular tissue _____.
   a. protects the plant, conducts fluids
   b. provides strength, facilitates growth
   c. facilitates reproduction, protects the plant
   d. facilitates growth, provides strength
   e. protects the plant, facilitates reproduction

2. Biennials, like cauliflower, radishes, and beets,
   a. go through their entire life cycle in one season
   b. go through their entire life cycle in about two years
   c. grow and bloom each season, for many years
   d. have the shortest life span, among annuals, biennials, and perennials
   e. go through their entire life cycle twice each year

3. Sclerenchyma cells
   a. lack cellulose
   b. function in metabolic processes such as photosynthesis
   c. are generally dead at maturity
   d. are commonly found in rapidly growing parts of the plant
   e. are found exclusively in roots

4. Secondary growth
   a. leads to increased height of the plant
   b. is more widespread in herbaceous plants than in wood shrubs and trees
   c. leads to an increase in the girth of the plant
   d. is brought about through cell divisions in the cuticle
   e. is growth that occurs in a plant's second year

5. Which of the following statements about the ascent of water in tall trees is true?
   a. Plants must expend metabolic energy to power the ascent of water into treetops.
   b. The evaporation of water from the leaf surface is required to lift water from the roots to the canopy of a tall tree.
   c. The ascent of water in the xylem is fastest when the stomata are closed.
   d. The formation of embolisms, or air bubbles, in xylem elements increases the speed of ascent of water.
   e. All of the above are true.

6. Phloem sieve tubes develop a positive pressure because
   a. Osmotic uptake of water occurs when sugars are loaded into the sieve elements.
   b. Sieve tubes lack a nucleus.
   c. Sieve tubes are dead at maturity.
   d. Sucrose is a bulky molecule that exerts a pressure against the sieve tube walls.
   e. All of the above

7. Endosperm is formed when
   a. A sperm nucleus from the pollen fuses with one haploid nucleus from the embryo sac.
   b. Two nuclei from the embryo sac fuse with each other.
   c. A sperm nucleus from the pollen fuses with the egg nucleus in the embyro sac.
   d. An egg nucleus from the pollen fuses with a diploid nucleus in the embryo sac.
   e. A sperm nucleus from the pollen fuses with two haploid nuclei from the embryo sac.

8. Which of the following statements about spores is true?
   a. Spores are diploid structures that give rise to gametes.
   b. Spores are multicellular, haploid structures that give rise to gametes.
   c. Spores are part of the sporophyte generation.
   d. Spores give rise to multicellular haploid gametes.
   e. Spores give rise to single-celled diploid gametes.

9. Which of these is a gametophyte plant? (More than one answer may be correct.)
   a. petal
   b. stigma
   c. embryo sac
   d. pollen grain
   e. sepal

10. The difference between a seed and a fruit is that

    a. Fruit is composed of triploid tissue; seeds are diploid.

    b. Seeds are always hard; fruits are always soft.

    c. Seeds develop from ovules while fruit tissue is derived from ovary walls.

    d. Annuals produce seeds while perennials produce fruits.

    e. Fruits are annuals while seeds are perennial.

Brief Review

1. Compare monocots and dicots with respect to their anatomy, morphology, and seed structure.

2. What are the three main cell types in a plant, and what distinct functions are performed by each?

3. What is a meristem? Describe the organization of the meristem, and the zonation in cell activity, in a root tip.

4. Compare the structure and function of xylem- and phloem-conducting elements.

5. How are growth rings in a tree trunk created? Why is it that you can tell the age of a tree by counting the number of growth rings?

6. Describe the process of double fertilization in plants.

Applying Your Knowledge

1. If you were to regularly shear a lilac bush down to a stub, you would probably kill it. Yet, mowing after mowing, the grass in your lawn continues to grow. Is there a fundamental difference in the location of growing points in the lilac and the grass? Explain.

2. Many grocery stores try to improve on Mother Nature by dipping fruits and vegetables in natural and synthetic waxes to prolong their shelf life. What is the role of the waxy cuticle in plants? Would you expect desert plants to have a thicker or thinner cuticle than plants that occur in moist areas?

3. If you were to pull out the "strings" on the ridges of a celery stick, you'd be peeling off strands of almost pure collenchyma cells. Why is this cell type especially prevalent in large but tender (nonwoody) parts of the plant, such as the leaf stalks of rhubarb and celery?

4. Beavers often kill young trees by girdling them, that is, removing a strip of bark from the circumference of a tree trunk. Why does this bark removal kill a tree? Which part of the tree would die first, the root or the shoot?

5. Vessels are wider than tracheids, and therefore more efficient as water conduits. Even though flowering plants have evolved vessels, they also produce a good number of tracheids, producing even greater numbers of them late in the season in temperate areas. Can you think of a reason why this might happen?

Group Activity

1. With a group of your peers, assemble a collection of different types of seeds, fruits, and vegetables, with each member of the group contributing some. (You could do this as a thought exercise as well, with each person contributing names or drawings of various plant foods.) For each item in your collection, determine whether the portion that we derive calories from belongs in one or more of these categories: endosperm, cotyledon, entire ovule (seed), ovary wall (flesh of fruit), flower parts or buds, storage root, storage stem, leaves or storage leaves.

2. In doing activity 1, you might come up against this question: What are vegetables? *Vegetables* is not a technical term, and therefore lacks a scientific (botanical) definition. Arrange your collection from activity 1 into two groups: foods that would be popularly called vegetables and foods most people would describe as fruits. Do you notice any common features at all within each category? (Hint: Think sweetness.) Now identify from each group those foods that are fruits in the botanically accurate sense.

MediaLab

Why Do We Need Plants Anyway?
The Importance of Plant Diversity

It is speculated that a meteorite struck the Earth near the Gulf of Mexico about 65 million years ago. After that event, there is no fossil record of any animal species over 65 pounds surviving. Most animals were probably not killed by rocks or tidal waves, but died when Earth's plant life could no longer support their needs. A significant meteor impact would send millions of tons of dust and debris into the atmosphere, blocking the sun's rays from reaching Earth's surface, significantly reducing photosynthesis, and ulti-mately leaving little food to support animal life. This event illustrates the essential role of plants in creating an environment that supports animal life. We simply couldn't survive without them. To learn more, the *CD Activity* will introduce you to the basics of plant structure. In the *Web Investigation*, you will see how plants have adapted these basic structures to specific habitats and conditions. Finally, in the *Communicate Your Results* section, you will synthesize and apply your knowledge to the plant life around you.

This *MediaLab* can be found in Chapter 22 on your CD-ROM and Companion Website (http://www.prenhall.com/krogh).

CD ACTIVITY

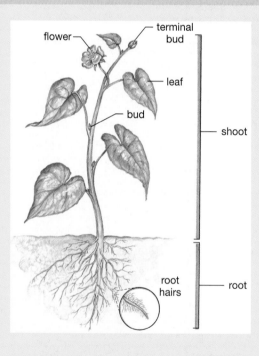

Plants, because of their ability to trap solar energy in a chemical form, are a food source for all other living organisms. They are pivotal members of every ecosystem and have adapted their basic body parts to suit these various environmental conditions. This *CD Activity* reviews the four principal organs of plants:

1. *Roots—anchor the plant; absorb and transport water and minerals.*

2. *Stems—support the plant aboveground and transport nutrients.*

3. *Leaves—gather sunlight for photosynthesis while minimizing water loss.*

4. *Flowers—provide the site of sexual reproduction.*

WEB INVESTIGATION

Investigation 1

Estimated time for completion = 15 minutes

There are two major groups of flowering plants, the monocots and the dicots, whose lineages diverged early in their evolutionary history. To contrast how these two groups of plants are constructed, select the Keywords **DICOT** and **MONOCOT** on your CD or Website for this *MediaLab*. Then compare the architecture and anatomy of the tomato plant, a dicot, and the rice plant, a monocot. How many differences can you observe in these two plants? Examine the roots, stems, leaves, and flowers.

Investigation 2

Estimated time for completion = 5 minutes

Flowering plants appeared on Earth about 150 million years ago. Since then, they have moved into an extraordinary number of habitats. Over generations, plant species have altered their form and function to become better adapted to living in varying habitats. Select the Keyword **SILVERSWORD** on your CD or Website for this *MediaLab* to view a striking example of adaptation by a plant species found in Hawaii. How do the different forms of these plants suit their environment?

Investigation 3

Estimated time for completion = 5 minutes

The structures of flowering plants are the culmination of millions of years of plant evolution. Features such as vascular tissue and enclosed seeds have appeared in various plant groups at various times in the evolutionary history of plants. Select the Keyword **FOSSIL PLANTS** on your CD or Website for this *MediaLab* to determine when these plant features first appeared in the fossil record.

COMMUNICATE YOUR RESULTS

Now that you've gained some understanding of plant structures and functions by doing the preceding activities, consider these questions.

Exercise 1

Estimated time for completion = 15 minutes

Use your observations from the *Web Investigation 1* to construct a chart illustrating the differences between the architecture and anatomy of dicot and monocot plants. You could start with something like Table 22.1, but you should be able to greatly expand on the differences. To do this exercise most efficiently, you may want to systematically work your way from flowers to roots, or any other order you would prefer.

Exercise 2

Estimated time for completion = 20 minutes

Using the adaptations you observed of the silversword species to environments in the Hawaiian Islands, answer this question: How would you construct a plant to be well adapted to the following environment?

- *Poor-quality soil, frequently disturbed by rock slides*
- *Growing season about 8 months long*
- *Annual rainfall of about 18 inches, most of it during the first third of the growing season*
- *About 8 hours of sunlight at the start of the growing season; about 16 hours by the last half*

After deciding how to construct this plant so that it will be adapted to its environment, draw your hypothetical plant. Then, list how its adaptations help the plant cope with its environmental hardships.

Exercise 3

Estimated time for completion = 10 minutes

Using the information you obtained from *Web Investigation 3*, construct a graph illustrating when the various plant features first appeared in the fossil record. Your chart should extend from the time when the first land plants appeared to the present day.

What Makes the Organism Tick: Animal Anatomy and Physiology

A human fetus during the fourteenth week of gestation.

23

Introduction to Animal Anatomy and Physiology

The Integumentary, Skeletal, and Muscular Systems

Muscles help us move.
(page 498)

This helps to support us.
(page 487)

How does a muscle contract?
(page 499)

Almost all human beings

are constantly in contact with other human beings, of course. As a result, most of us take for granted such basic human capabilities as speaking, walking, reasoning, or listening. Anyone who has ever tried to devise a *machine* that can carry out these human activities would tell you, however, that human capabilities should be looked at with more wonder than indifference. At tremendous cost of time and money, IBM devised a computer capable of being a world-class chess player, but how about devising a computer that can play chess *and* recognize faces *and* speak in fluid sentences? How about devising a computer, in other words, that had a few of the capabilities of an average 10-year-old? There are differing schools of thought about what limits there might be to machine or "artificial" intelligence. One view, however, is that computers will never be able to think as humans do, in significant part because of what computers lack: a body. Think of what the body provides in the way of learning. First, wonderful sensory inputs—such as seeing and hearing—that *allow* us to learn in diverse ways. Second, a set of needs—for food, warmth, and so forth—that *drive* us to learn from the moment of our birth.

As remarkable as any one human capability may be, the truly wondrous thing is the body as a whole. Our bodies have a system for acquiring energy (a mouth, a digestive tract) a system for eliminating waste (a bladder, etc.), a system for movement (muscles), a system for moving materials around internally (blood vessels), and lots more. It is these systems that you will be reading about in the next five chapters. The idea is to give you some sense of the workings of the human machine.

23.1 The Sciences of Anatomy and Physiology

Zoology is the study of animals, and one of the ways of thinking about animals is in terms of the makeup and functioning of their bodies. Anatomy (literally "to cut up") is the study of the body's internal and external structure and the physical relationships between the body's constituent parts. Physiology is the study of how these various structures work. Anatomy tells us, for example, where the heart is, how it is structured and how it is situated in relation to other organs and tissues. Physiology describes to us the *mechanisms* of the heart's functioning.

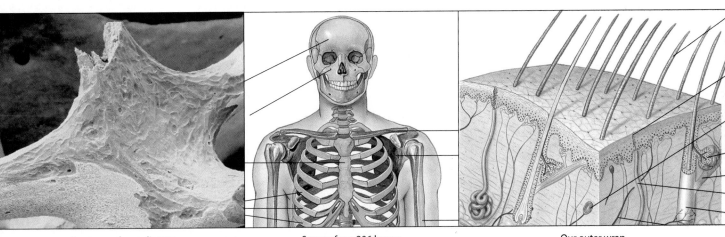

A common bone disease.
(page 496)

Some of our 206 bones.
(page 497)

Our outer wrap.
(page 493)

Animals vary tremendously in size and complexity; there are sponges, corals, various types of worms, and then vertebrates such as sharks, birds, and humans. Because human beings are the most-studied and best-understood animal in the animal kingdom, you'll primarily be looking at humans in this discussion of animal anatomy and physiology.

23.2 What Are the General Characteristics of Humans?

There are obvious differences between a human being and, say, a bird; but there are also some similarities. Indeed, human beings have a good number of characteristics in common with many other animals. Going from the general to the specific, three of these shared characteristics are an internal body cavity, an internal skeleton, and internal temperature regulation.

Internal Body Cavity

An internal body cavity, or **coelom**, not open to the external environment, is characteristic of vertebrates and most invertebrates. Humans and other mammals have two primary body cavities. First, there is a dorsal ("back") cavity that is further divided into cranial and spinal cavities, into which our brain and

spinal cord fit (**see Figure 23.1**). Second, there is a ventral ("front") cavity that includes a thoracic or "chest" portion that contains the lungs and heart, and a lower abdominopelvic cavity that contains such organs as the stomach, liver, pancreas, and intestines. Between the thoracic and abdominopelvic cavities lies a muscular sheet called the *diaphragm*. Body cavities allow for flexibility; with them, internal organs can expand and contract without disrupting the activities of neighboring organs.

Internal Skeleton

Humans have an internal skeleton that supports the body and allows for continuous body growth up to maturity. We humans are described as **vertebrates**, because our internal skeleton contains a backbone, or vertebral column, made up of individual bones called *vertebrae*. The vertebrae protect the spinal cord by forming the spinal cavity, while the skull protects the brain by forming the cranial cavity.

Internal Temperature Regulation

Most fish, amphibians, and reptiles are **ectothermic**, meaning that their primary source of body heat is external to them (*ecto*

**Figure 23.1
Body Cavities**

a A side or "lateral" view of the two main cavities in the human body—the dorsal and ventral cavities—and the smaller cavities within them. The dorsal body cavity is bounded by the bones of the skull and vertebral column and includes both the cranial and spinal cavities. The ventral cavity includes a thoracic ("chest") cavity and an abdominopelvic cavity, which is separated from the thoracic cavity by a muscular diaphragm.

b A front or "ventral" view of the body's cavities.

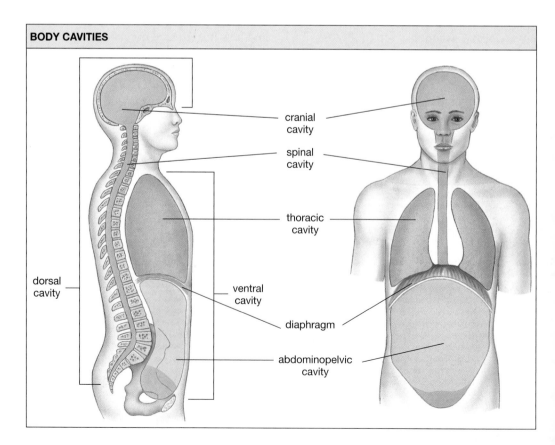

BODY CAVITIES

cranial cavity

spinal cavity

thoracic cavity

dorsal cavity

ventral cavity

diaphragm

abdominopelvic cavity

means "outside," and *therm* means "heat"). These animals are unable to regulate their body temperature and so exist at the same temperature as the air or water surrounding them. Thus a lizard may warm itself for the day's activities by lying in the sun. The term *cold-blooded*, which is often applied to ectotherms, is misleading because ectotherms may have warm body temperatures—if they are in warm environments. Likewise, they can maintain a *stable* internal temperature, but only if their environment does.

In contrast, humans, other mammals, birds, and some fish and insects warm their bodies primarily through the activities of their own metabolism. Because this warming is internal, these animals are referred to as **endothermic** (*endo*, "inside"). There are both costs and benefits to endothermy. One benefit is that endotherms can be fully functional in lots of environments. Imagine that, like lizards, we had to warm ourselves in the sun before we could be fully active. The price we pay for this functionality, however, is energy usage. Pound per pound, in a resting state a human being burns up many more calories than would, say, a crocodile. And because endotherms burn up proportionately more calories, they must obtain proportionately more food.

23.3 Animal Architecture and Organization

One distinguishing characteristic of animals is that they are multicellular; put another way, there is no such thing as a single-celled animal. With multicellularity comes size, and with size comes complexity. As animals became larger through evolutionary time, they also became more structurally complex. In a sense they had to because, as you learned in Chapter 4 ("The Size of Cells," page 73), as anything grows in size, its volume increases more than its surface area. With large size, then, not all the cells of an organism can have access to the outside world. Thus, new *systems* had to evolve that would allow cells far from the surface to get needed nutrients and gasses and to expel their own wastes safely. In addition, most large organisms need muscles in order to move and some type of skeleton for general support and for muscle attachment.

This division of labor resulted in the formation of several types of specialized **tissues**, meaning groups of cells that have a common function. In more complex animals, tissues are in turn arranged into various types of **organs**, meaning complexes of several kinds of tissues that perform a special bodily function. (The heart is an organ that has muscle tissue within it, but it also has another type of tissue called epithelial tissue.) Organs then go on to form **organ systems** that make up the whole organism. Contractions of the heart push blood into a network of blood vessels. The heart, blood, and blood vessels form the cardiovascular *system*, which is one of eleven systems the human body has.

Tissues, organs, and organ systems are three levels of organization within animals. You may remember from Chapter 1 that the levels of organization in living things do not *start* with tissues. Tissues are made of cells, cells have within them tiny working structures called *organelles*, and organelles are in turn made of the building blocks called *molecules*. At the bottom of this chain is the fundamental unit of matter, the atom. In this chapter, the focus is on the levels of organization that run from tissues through organ systems. **Figure 23.2** shows the various levels of structural organization in the human body, using the cardiovascular system as an example.

23.4 The Animal Body Has Four Basic Tissue Types

So, what kinds of tissues are there in animals? You'll be reading about a lot of different tissue types in this chapter, but it turns out there are only four *fundamental* tissue types: epithelial, connective, muscle, and nervous. All the other tissue types are subsets of these fundamental types.

Figure 23.2
Levels of Organization in the Human Body
Atoms form molecules, which in turn combine to form organelles, which combine to form cells, such as heart muscle cells. A group of cells that performs a common function is a tissue; here cells have formed heart muscle tissue (whose common function is heart muscle contraction). Two or more tissues combine to form an organ, such as the heart. The heart is one component of the cardiovascular system; other parts of the system are blood and blood vessels. All the organ systems combine to create an organism, in this case a human being.

atom
(hydrogen)

molecule
(water)

organelle
(mitochondrion)

cell
(muscle)

tissue
(muscle)

organ
(heart)

organ
system
(cardiovascular)

Epithelial Tissue

Epithelial tissue covers all the exposed body surfaces of animals. The surface of your skin is epithelia, for example. But epithelia also lines your internal passageways—such as your digestive tract—and forms **glands**, which are organs that secrete one or more substances.

Epithelial tissue may consist of a single cell layer (a simple epithelium) or be made up of more than two layers (a stratified epithelium). In their shape, epithelial cells may be squamous, meaning flat; cuboidal, meaning square-shaped; or columnar, meaning rectangular. **Figure 23.3** shows some of the epithelial tissues in the human body. Because the exterior surface of an epithelial cell may be exposed, its *interior* surface needs to be firmly anchored to deeper body tissues. Epithelial cells are attached to what is called a **basement membrane**, which is made up of a dense network of protein fibers and filaments. This network also keeps large molecules in the underlying connective tissue from reaching the epithelium.

Epithelial tissues have several functions in humans and other animals. All materials that enter or leave the body must contact epithelial cells. Why? Because these cells form the internal and external *surfaces* of animal organs. In skin, epithelial cells can serve as an effective barrier to the outside world, in large part because they fit together tightly, with portions of their cell membranes interlocking. This is one of the body's key means of keeping invaders and harmful substances out. In addition, in land vertebrates such as ourselves, epithelial cells that lie on our body's surface produce the waterproofing protein keratin.

Epithelial glands in animals as diverse as flatworms, earthworms, and snails produce mucus secretions that aid body movement and, in humans, ease the movement of foods through the digestive tract.

Glands made up of epithelial tissue come in two main types: **Exocrine glands**, which secrete their materials through tubes or "ducts" onto an epithelial surface; and ductless **endocrine glands**, which release their materials directly from cells into surrounding tissues or into the bloodstream. The perspiration on your skin comes through a duct; thus your sweat glands are exocrine glands. Meanwhile endocrine glands secrete many **hormones**—molecules produced by cells in one location that have an effect on cells at other locations.

Connective Tissue

The second type of animal tissue, connective tissue, is never exposed to the external environment. Connective tissues function in supporting and protecting the body and its organs, in transporting materials within the body, in storing energy, and in defending the body. Unlike epithelial tissues, which consist almost entirely of cells, connective-tissue cells are separated from each other by extracellular material they produce. Indeed, a prime function of most connective tissue cells is to produce such materials, which are composed of protein fibers and a "ground substance" that varies in consistency from a fluid to a solid. See **Figure 23.4** for the different types of connective tissues in the human body.

Loose Connective and Adipose Tissue

Not surprisingly, connective tissues are classified according to their surrounding extracellular material. Loose connective tissue is composed of a fluid-like ground substance and fibers of a protein called collagen, which provide strength and flexibility. This tissue is

**Figure 23.3
Some Epithelial Tissues in the Human Body**

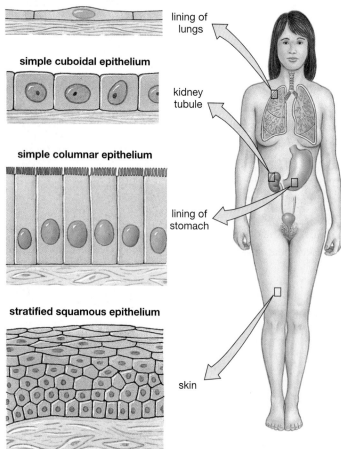

simple squamous epithelium

simple cuboidal epithelium

simple columnar epithelium

stratified squamous epithelium

lining of lungs

kidney tubule

lining of stomach

skin

the packing material of the body. It fills spaces between organs, provides cushioning, and supports the epithelia. **Adipose tissue**, or fat, is a type of loose connective tissue that contains large numbers of fat cells. It too provides cushioning, but it also insulates the body and stores energy.

Fibrous and Supporting Connective Tissue

Containing less ground substance and packed with collagen fibers, fibrous connective tissue is tough and strong. Such tissues form tendons that attach skeletal muscle to bones, ligaments that connect one bone to another, and capsules that surround organs and enclose joint cavities.

Supporting connective tissues, the most rigid connective tissues, make up bone and cartilage. The extracellular material of these tissues contains a dense ground substance and closely packed fibers. In humans, **cartilage** makes up the outer portions of the nose and ear, where it provides support and flexibility. It is also found between the bones of the spine and at the ends of limb bones, where it protects the bones by reducing fric-

tion and acting as a shock absorber. Then there is **bone**, which is a supporting connective tissue whose ground substance contains mineral deposits primarily made of calcium (thus making it calcified). These minerals, along with flexible collagen fibers, make bone both strong and resistant to shattering.

Fluid Connective Tissue

It may be difficult to think of blood as a tissue at all, but it is one more type of connective tissue—a fluid connective tissue. Blood is composed of a distinctive population of cells (blood cells) suspended in a *watery* ground substance (called *plasma*) in which tissue proteins and other materials are dissolved. In humans, blood contains red blood cells, which carry oxygen; white blood cells, which are important to the immune system; and platelets, which are cell fragments that aid blood clotting.

Some fluid is constantly pushed out of the smallest blood vessels and into the surrounding body tissues, where it is known as **interstitial fluid**. At a certain point this fluid enters special vessels, called **lymphatics**, that return it to the

**Figure 23.4
Connective Tissues in the
Human Body**

488

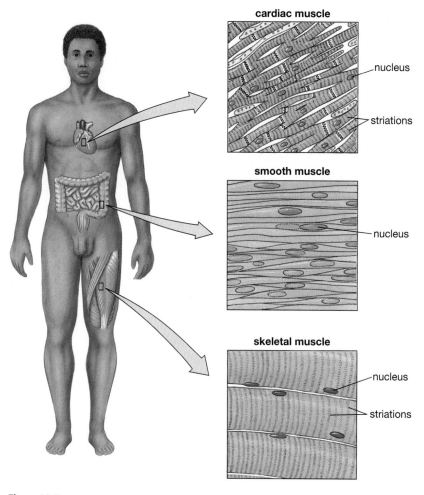

Figure 23.5
Muscle Tissue in the Human Body
The three types of muscle tissue are cardiac, smooth, and skeletal.

blood circulation. While in the lymphatics, the fluid constitutes a second kind of fluid connective tissue, **lymph**. Before being returned to the blood, lymph is checked for signs of injury or infection by cells of the immune system.

Muscle Tissue

Muscle tissue is specialized in its ability to shorten, or contract. It does so through the interaction between two contractile protein elements or "fibrils," actin and myosin. Muscle in which the fibrils are regularly arranged in stripe-like patterns is called striated muscle, while muscle lacking such an arrangement is called nonstriated muscle. Striated muscles cells form **skeletal** and **cardiac muscle tissues**, and nonstriated muscle cells form **smooth muscle tissue**.

Most of the cells in the living world contain a single nucleus, and this is the case with smooth muscle tissue and the striated cardiac muscle tissue. Skeletal muscle tissue, conversely, is made up of long cells, each of which typically contains hundreds of nuclei. In humans and other vertebrates, skeletal muscle is always associated with bones and is responsible for movement of the body—it is the muscle that makes up the biceps, for example. Meanwhile, smooth muscle is responsible for contractions in blood vessels, air passageways in the lungs, and hollow organs

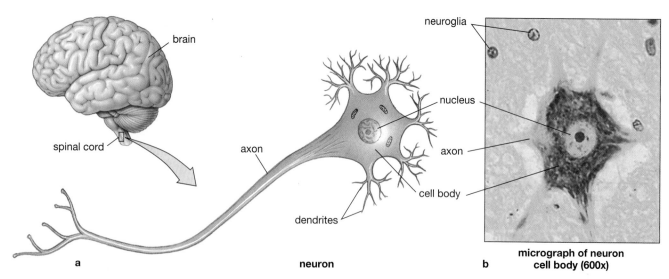

Figure 23.6
Nervous Tissue in the Human Body

a The functional unit of nervous tissue is the neuron, which contains a single axon that carries signals from the cell body; and dendrites, which carry signals to the cell body. Most neurons are surrounded by an array of supporting neuroglia cells. This neuron happens to be located in the spinal cord.

b A micrograph of a neuron and associated neuroglia.

such as the uterus. As might be apparent from this list of tasks, smooth muscle is not under voluntary control. Cardiac muscle tissue is well named, because it is found only in the heart; its function is to pump blood throughout the body. In **Figure 23.5**, you can see illustrations of these three muscle tissue types in humans. You'll get more detail on how muscles work later in this chapter.

Nervous Tissue

Nervous tissue is specialized for the rapid conduction of electrical impulses. It is made up of two basic types of cells; **neurons** and some associated helper cells called **neuroglia**, which you can see in **Figure 23.6**. Neurons are the functional units of nervous tissue, while neuroglia physically support the neurons, aid

their nourishment and insulation, and defend nervous tissue from infection. A typical neuron has a cell body, which contains the nucleus; a single long extension called an *axon*, along which information generally is transmitted *from* the neuron; and many extensions called **dendrites**, along which information generally is transmitted *to* the neuron. In humans, axons may reach a meter in length.

23.5 A Summary of the Organ Systems of the Human Body

As noted earlier, combinations of the four tissue types just reviewed go on to form various organs in the body, and these organs then work in concert to form the body's organ systems. **Table 23.1** summarizes the 11 organ systems of the human body, while

Table 23.1
Organ Systems in the Human Body

Organ System	Composed of	Main Functions
Integumentary	Skin, hair, nails, associated exocrine glands	Protection, water and temperature insulation, fat storage
Skeletal	Bone and cartilage	Support, mineral storage, red blood cell production
Muscular	Skeletal muscles	Movement, posture, balance, maintenance of body temperature
Nervous	Brain, spinal cord, sense organs and interconnecting nerves	Sensing, communicating, activating
Endocrine	Endocrine glands, hormone-producing cells	Maintenance of internal state, regulation of development, maintenance of sex characteristics
Lymphatic	Lymphoid organs (e.g., thymus gland) and lymphatic vessels	Protection against invading organisms
Cardiovascular	Heart, blood, blood vessels	Transport of nutrients, gases, and hormones; waste removal
Respiratory	Lungs and air passageways	Delivery of oxygen to the blood, removal of carbon dioxide from the blood
Digestive	Mouth, stomach, small and large intestine, digestive (exocrine) glands	Digestion of food, elimination of waste
Urinary	Kidneys, ureters, urethra	Elimination of liquid wastes; regulation of water balance
Reproductive	Testes, penis, associated glands in men; ovaries, uterine tube, uterus, vagina and associated glands in women.	Reproduction

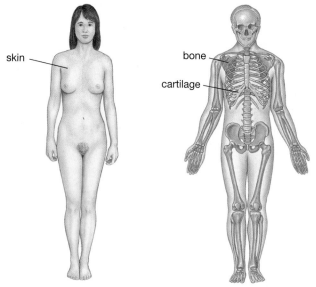

a   The Integumentary System

b   The Skeletal System

skin

bone

cartilage

c   The Muscular System

d   The Nervous System

skeletal muscles

brain

spinal cord

nerves

e   The Endocrine System

f   The Lymphatic System

pituitary gland
thyroid gland
parathyroid glands
thymus
pancreas
adrenal glands
gonad

thymus
lymph nodes
spleen
lymphatic vessels

Figure 23.7
The Organ Systems of the Human Body

Figure 23.7 gives you a schematic view of the systems. What follows is a brief look at each of these systems, grouped according to the functions they perform. The balance of this chapter is then devoted to a more detailed look at the three systems in the first group—the integumentary, skeletal, and muscular systems. In Chapters 24 and 25, you'll be taking more detailed looks at the systems in group 2 and group 3.

Organ Systems 1: Body Support and Movement: The Integumentary, Skeletal, and Muscular Systems

- The **integumentary system**, or integument (Figure 23.7a), is made up of the skin and associated structures, such as glands, hair, and nails. This system protects the body from the external environment and assists in the regulation of body temperature.
- The **skeletal system** (Figure 23.7b) forms an internal supporting framework for the body. This is a complex framework; humans have about 206 bones in their body, along with a large number of cartilages. This system also includes the connective tissues and ligaments that connect the bones at various joints. The individual bones store minerals, and some bones, such as those of the skull and ribs, also protect delicate tissues and organs.
- The **muscular system** (Figure 23.7c) includes all the skeletal muscles of the body that are under voluntary control—about 700 of them. It is the contractions of skeletal muscle that produce movement of the body. Muscles also maintain posture and balance, support soft tissues, and help maintain body temperature, in that our human bodies are largely warmed by the heat given off from muscle contraction. The smooth and cardiac muscle tissues noted earlier are part of different systems; cardiac muscle, for example, is part of the cardiovascular system.

Organ Systems 2: Coordination, Regulation, and Defense: The Nervous, Endocrine, and Lymphatic Systems

- The nervous *tissue* reviewed earlier is made up of neurons and their helper neuroglial cells. This tissue then goes on to make up the **nervous system** (23.7d), which includes the brain, spinal cord, sense organs such as the eye and ear, and all the nerves that interconnect these organs and link the nervous system with other systems.

- The **endocrine system** (Figure 23.7e) bears comparison with the nervous system, in that both systems are in the business of sending signals throughout the body. The endocrine system, however, works through the substances we call hormones, many of which are produced in specialized glands, such as the thyroid or adrenal glands. Hormones generally travel through the bloodstream and often function to maintain stability in the body—by conserving water or maintaining temperature, for example.

- The **lymphatic system** (Figure 23.7f) has a great overlap with the immune functions of the body. The system consists of a widespread network of lymphatic vessels that collect extracellular fluid and deliver it, as the lymph just mentioned, to the blood vessels of the cardiovascular system. While the fluid is in the lymphatic system, it is checked for signs of infection. The lymphatic system also contains lymphoid organs, which produce or support specialized immune cells that monitor and destroy disease-causing invaders. These organs include the small lymph nodes and the larger thymus gland, tonsils, and spleen.

Organ Systems 3: Transport and Exchange with the Environment: The Cardiovascular, Respiratory, Digestive, and Urinary Systems

- The **cardiovascular system** (Figure 23.7g) consists of the heart, blood, and blood vessels. It also includes the bone marrow, where red blood cells are formed. The cardiovascular system transports nutrients, dissolved gases, and hormones to tissues throughout the body. It also carries waste products from the body's tissues to sites of filtration, such as the kidneys.

- The **respiratory system** (Figure 23.7h) includes the lungs and the passageways that carry air to them. These passageways begin at the nasal cavities, and continue

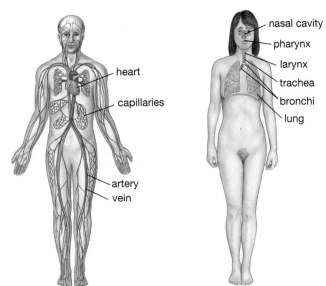

g   The Cardiovascular System   **h   The Respiratory System**

heart
capillaries
artery
vein

nasal cavity
pharynx
larynx
trachea
bronchi
lung

salivary gland
pharynx
esophagus
liver
stomach
gallbladder
small intestine
large intestine
anus

kidney
ureter
urinary bladder
urethra

i   The Digestive System   **j   The Urinary System**

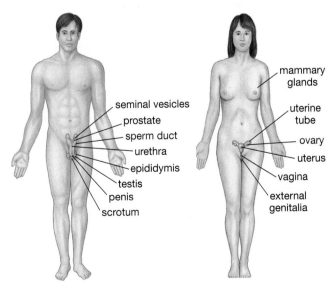

seminal vesicles
prostate
sperm duct
urethra
epididymis
testis
penis
scrotum

mammary glands
uterine tube
ovary
uterus
vagina
external genitalia

k   The Male Reproductive System   **l   The Female Reproductive System**

through the pharynx (throat), larynx (voicebox), and trachea (windpipe). Through this system, oxygen comes into the body and carbon dioxide is expelled from it.

- The central feature of the **digestive system** (Figure 23.7i) is the digestive tract, a long tube that begins at the mouth and ends at the anus. Along the way, there is also the stomach, the small intestine, and the large intestine. Associated with the tract through its length are a number of accessory glands, including the salivary glands, the liver, and the pancreas. These glands, along with the stomach and the small intestine, secrete digestive enzymes that help break down food so that it can be absorbed and used by the body.

- The major function of the **urinary system** (Figure 23.7j) is the elimination of waste products from the blood through the formation of urine. The system is made up of the kidneys, where urine production occurs; tubes called the ureters, which carry urine to another part of the system, the urinary bladder; and the urethra, which conducts urine from the bladder to the exterior of the body.

- Humans have two **reproductive systems**—one for males and another for females (Figure 23.7k–l). The two are linked, of course, in that together they produce offspring. Both systems are discussed in detail in Chapter 27.

23.6 The Integumentary System: Skin and More

Now let's look in greater detail at the first three organ systems, grouped under "body support and movement," starting with the integumentary system. The primary component of this system is the organ we call skin. Just as it may have been difficult to think of blood as a tissue, it may also seem strange to think of the skin as an organ, because it is not the kind of small, easily definable structure we associate with the term. However, skin actually constitutes the largest and most visible organ of the human body. Together with its related structures (hair, nails, and a variety of exocrine glands) it makes up the integumentary system.

Skin does more than just cover the body. It also protects underlying tissues and organs from infection or puncture; it controls the evaporation of body fluids and regulates heat loss; and it stores fat and synthesizes vitamin D. Meanwhile the exocrine glands that are accessories to the skin excrete materials such as salt, water, oils, and milk; and specialized nerve endings in the skin detect touch, pressure, pain, and temperature.

The Structure of Skin

The **skin** is an organ in two parts: A thin, outer epithelial covering, the **epidermis**, and a thicker underlying **dermis**, composed mostly of connective tissue. Beneath the dermis—and not part of the skin, proper—is the subcutaneous layer, or hypodermis, made up of loose connective tissue that attaches the skin to deeper structures, such as muscles or bones. A cross section of human skin can be seen in **Figure 23.8**. One way to look at the skin is as a complex organ composed of all four basic tissue types described earlier: epithelial, connective, nerve, and muscle. You'll learn shortly how all four tissue types work together in skin.

The Outermost Layer of Skin, the Epidermis

The outermost of our skin layers, the epidermis, itself consists of several different cell layers that form a stratified squamous epithelium. (Remember a stratified epithelium is one with more than one layer, and squamous epithelial cells are *flat* cells.) Because the outermost skin layer, the epidermis, is continually being rubbed or scraped by objects, its cells are worn away as a result. This means that epidermal cells must be constantly replenished. As cells divide in the epidermis, they are pushed outward, toward the surface, to replace cells that have worn away. As these new epidermal cells are pushed farther outward by even newer cells, they increase their production of **keratin**, the water-resistant protein mentioned earlier. The uppermost, or surface, layer of the epidermis is made up of flattened, dead epithelial cells that are tightly joined and filled with large amounts of keratin—an effective barrier for water control and invader protection.

It takes about 14 days for a cell to move outward from the deepest to the outermost

What Is It?
See p. 493.

layer of the epidermis. The dead, surface cells usually remain in their exposed position for an additional two weeks before they are shed or washed away.

A Skin Layer beneath the Epidermis, the Dermis

The skin's deeper layer, the **dermis**, is filled with accessory structures derived from the epidermis, such as hair follicles and sweat glands. The uppermost region of the dermis consists of loose connective tissue that supports and nourishes the epidermis in several ways. For example, it contains the capillaries and nerves that supply the surface of the skin. Nerve fibers control blood flow, adjust gland secretion rates, and monitor the sensory receptors of touch, pain, pressure, and temperature that are scattered throughout the dermis.

Integumentary Structures Associated with Skin

Hair

Hair projects above the surface of the skin almost everywhere except over the sides and soles of the feet, the palms of the hands, the sides of the fingers and toes, the lips, and portions of the external genital organs.

Hairs originate in tiny complex organs called **hair follicles**. The epithelium at the base of a follicle surrounds the **hair papilla**, a peg-like structure of connective tissue containing capillaries and nerves. Ribbons of smooth muscle (arrector pili muscles) extend from the upper dermis to the hair follicle. When stimulated, the arrector pili can pull on the follicles and elevate the hairs. Contraction of the arrector pili may be caused by emotional states, such as fear or rage, or as a response to cold, producing the characteristic "goose bumps" associated with shivering. As you might guess, the arrector pili muscles are *involuntary* muscles, which is why we can neither "give ourselves" goose bumps nor stop them from forming.

Glands

The skin contains two types of duct-containing or exocrine glands, sebaceous glands and sweat glands (see Figure 23.8). **Sebaceous glands** produce a waxy, oily secretion called **sebum**, which secreted into hair follicles and directly onto the surface of the skin, lubricates the hair shaft and inhibits bacterial growth in the surrounding area.

The sweat glands of the skin come in two varieties: apocrine sweat glands and merocrine (or eccrine) sweat glands. Apocrine sweat glands are found in the armpits, around the nipples, and in the groin. Like sebaceous glands, apocrine sweat glands also

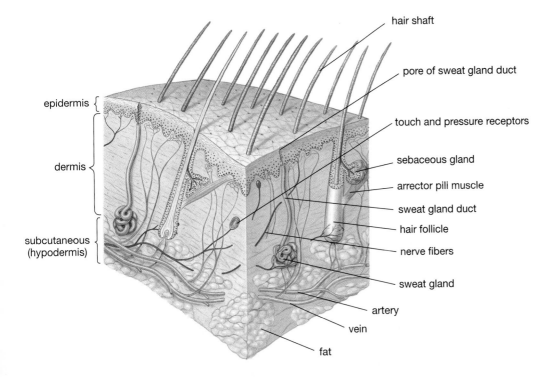

Figure 23.8
Our Outer Wrap: Components of the Integumentary System

What Is It?
See p. 495.

empty into the hair follicles. At puberty, these coiled tubular glands begin discharging a thick, cloudy secretion that becomes odorous when broken down by bacteria. In other mammals this odor is an important form of communication; for us, whatever function body odor may have is usually masked by deodorants. The milk-secreting mammary glands of the breast are modified apocrine sweat glands.

Merocrine (eccrine) sweat glands are far more numerous and widely distributed than apocrine glands. Merocrine sweat glands produce the clear secretion called *sweat*, also known as perspiration. The primary function of this secretion is to cool the surface of the skin and to reduce body temperature. The perspiration produced by merocrine glands is more than 99 percent water, but it contains a mixture of other compounds, including salts—which give sweat its salty taste—and waste products such as urea.

Nails

Nails form over the tips of the fingers and toes, where they protect the exposed tips. The body of the nail covers the nail bed, but nail growth occurs from the nail root, an epithelial fold not visible from the surface. The deepest portion of the nail root lies very close to the bone of the fingertip. A portion of the outer epidermal layer of the fold extends over the exposed nail nearest the root, forming the cuticle. Underlying blood vessels give the nail its pink color.

23.7 Body Support and the Skeleton

Now let's turn from skin to bone and cartilage. The human skeletal system (composed of both bone and cartilage) performs more functions than just the obvious one of support. For example, the bones serve as storage sites for valuable minerals such as calcium and phosphate. In addition, some bones store lipids within their internal cavities and produce red blood cells and other blood elements. Most of the human skeletal system is bone, but cartilage, which does not contain blood vessels, exists in certain places—at the end of our noses and ribs and in our outer ears, for example. Some vertebrates, such as sharks, have a skeleton that is made entirely of cartilage.

Structure of Bone

Bone is largely composed of osseous tissue (bone tissue), which is one of the two types of supporting connective tissue noted earlier. But each bone of the skeleton is considered to be an organ because it contains not only connective tissue but other tissue types as well. Like other connective tissues, osseous tissue is made up not only of specialized cells but also of extracellular fibers and a ground substance. Bone tissue contains large deposits of hard calcium compounds that give the tissue its characteristic solid, stony feel. The remaining bone mass is mainly collagen fibers, with bone cells and other cells providing only around 2 percent of the mass of a bone.

Large-Scale Features of Bone

The typical features of a long bone such as the humerus of a limb are shown in **Figure 23.9**. A long bone has a central shaft, or diaphysis, and expanded ends, or epiphyses (singular, epiphysis). The diaphysis surrounds a central cavity, filled with **marrow**, a loose connective tissue that comes in two forms: red marrow, which is blood-forming tissue; and yellow marrow, which is made up of fat. The ends (epiphyses) of adjacent bones form joints with each other and are covered by slippery cartilages called *articular cartilages*. There are two types of bone tissue. Compact bone is relatively solid, whereas **spongy bone** is porous and less dense. Both types are present in long bones, with compact bone forming the diaphysis, and spongy bone filling the epiphyses (see Figure 23.9a).

Small-Scale Features of Bone

Dropping down to the microscopic level and looking at bone *cells*, three different types are involved in the growth and maintenance of bone. **Osteoblasts** are immature bone cells that are responsible for the production of new bone. They secrete the organic material that becomes the bone matrix. Once surrounded by this material, they reduce their production of it. This marks their transition to osteocytes, the second type of bone cell. Osteocytes are mature bone cells; they *maintain* the structure and density of normal bone by continually recycling the calcium compounds around themselves. Then there is the third type of bone cell, the osteoclasts, which could be thought of as the demolition team of bone tissue. They release enzymes into the bone matrix that break it down, thus liberating the stored minerals. This process, essential in bone growth, also helps regulate calcium and phosphate concentrations in body fluids. Osteocytes are the most numerous bone cells in a person whose bone growth is complete, but bone is a very dynamic tissue. At any given moment, osteoclasts are removing bone and osteoblasts are adding to it. Beginning in middle age, the body may start removing more bone than it adds, a condition that particularly affects women. You can read more about this in "Doing Something about Osteoporosis While You Are Young," on page 496.

Within both compact and spongy bone, osteocytes exist in microscopic cavities called lacunae (which means "little lakes"). Within compact bone, the basic functional unit is the **osteon** or Haversian system (Figure 23.9c). The essence of bone growth is that,

within each osteon, osteoblasts produce layer after layer of concentric cylinders of bone— much like layers of insulation wrapped around a pipe. These layers surround what could be thought of as the hole in the pipe, which is a central or Haversian canal. These canals parallel the long axis of the bone and have running through them a kind of support system for the bone—blood vessels and nerves. Unlike compact bone, spongy bone has no osteons. It consists of an open network of interconnecting, calcified rods or plates. These open areas of spongy bone provide space for red marrow.

The Human Skeleton

Taking a step back from the microscopic world, you can see in **Figure 23.10** (page 497) that there are a lot of bones in the skeletal system—206 of them in all. (Only the major bones are visible in the figure.) There are two main divisions to the skeletal system: The **axial skeleton**, whose 80 bones include the

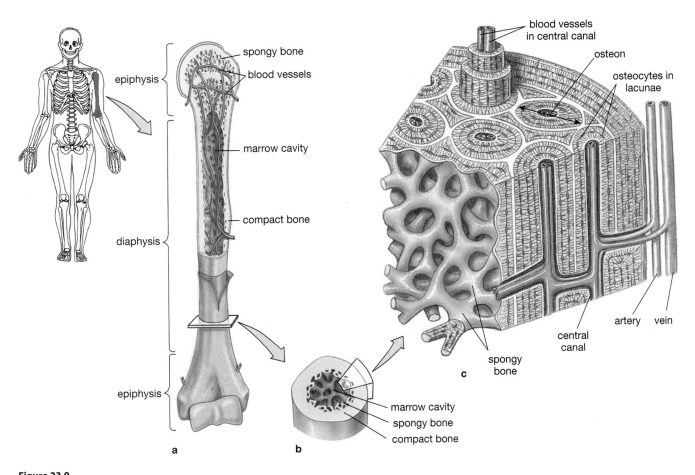

Figure 23.9
Large- and Small-Scale Features of a Typical Bone
The building units of compact bone, the osteons, run parallel to the long axis of the bone. Spongy bone does not contain osteons, but instead is composed of an open network of calcified rods or plates.

Doing Something about Osteoporosis While You Are Young

Osteoporosis is a thinning of bones that leaves its victims susceptible to bone fractures—breaks or cracks in bones, especially bones of the hips, spine, and wrist (**see Figure 1**). It is generally thought of as a condition that affects women, and indeed 80 percent of the Americans who suffer from it are female. It is further thought of as a condition related to age, and this too is true in a sense, in that it is menopause—with its related hormonal changes—that generally brings about the onset of osteoporosis.

Most of the attention paid to osteoporosis has to do with prevention of it *in middle age*, which is to say prevention therapies that are undertaken beginning with menopause. Activities that generally keep us healthy—exercise, proper diet, not smoking—can play a part in this therapy. But the most well-known means of treating osteoporosis is one that involves replacing the hormones whose levels drop off with menopause. In hormone replacement therapy, women generally take pills or use skin patches that contain estrogen and a synthetic form of the hormone progesterone.

There is no doubt that hormone replacement therapy significantly reduces the risk of developing osteoporosis, and it seems to have a wealth of other benefits as well, most notably a reduction in the risk of heart disease. Against this, hormone replacement therapy appears to *raise* the risk of developing breast cancer. Given these complexities, there is no blanket recommendation about hormone replacement therapy. Instead, decisions need to be made on the basis of individual risk factors. (Is there a history of heart disease or breast cancer in the family, for example?)

There is another side to the prevention of osteoporosis, however—one that can involve *young* women, rather than middle-aged women. The main text in this chapter notes that bone is a dynamic tissue; at any given moment, the bone cells called *osteoclasts* are removing bone while the cells called *osteoblasts* are adding to it. It is the *balance* in this activity that changes through life. In a 15-year-old, there is much more bone being added than removed; in a 30-year-old these two activities are roughly balanced out; in a post-menopausal woman there is much more bone being removed than being added.

Human females attain almost all their growth in height by late adolescence, but they are still gaining in bone density up until about the age of 30. The critical thing is that the entire period until age 30 represents a *window of opportunity* to develop bone mass. And the more bone mass a woman has by 30, the less risk she will have of osteoporosis after menopause. Once the window of opportunity has closed, however, there is no possibility of raising *peak* bone density, meaning the highest bone density a woman will ever have. Whatever a woman achieves in this regard, she will achieve by age 30.

Most bone is deposited by late adolescence. But medical researcher Robert Recker and his colleagues at Creighton University found in the early 1990s that a group of women they tracked over several years added almost 13 percent to their bone mass during their 20s. Looking for lifestyle factors that might make a difference in adding bone, the researchers found three: physical exercise, intake of calcium and protein, and use of oral contraceptives. None of the women studied were heavy exercisers or had particularly high calcium intake. Thus, the research suggested that simple, modest increases in exercise and calcium during a woman's 20s might yield benefits decades later, when menopause begins and bones begin to thin.

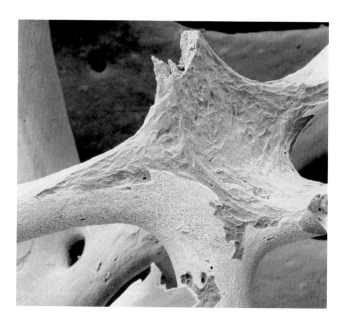

Figure 1
Thinning Bone
Pictured is the fractured thighbone (femur) of a patient with osteoporosis. Note that the lighter areas of the bone have a grainy, pock-marked quality. The dark spots in these areas are places that have lost bone mineral. (×85)

skull, the vertebral column, and the rib cage that attaches to it; and the **appendicular skeleton**, whose 126 bones include those of our paired appendages—the arms and the legs, along with the pelvic and pectoral "girdles" to which they are attached.

Joints

Joints, or articulations, exist wherever two bones meet. Each joint represents a compromise between the need for strength and the need for mobility. When movement is not required, joints can be very strong. For ex-

ample, joints such as the immovable sutures in the skull hold its separate elements together as if they were a single bone. Conversely, the ball-and-socket joint at the shoulder permits a wide range of arm movement that is limited more by the surrounding muscles than by joint structure. The shoulder joint itself is relatively weak, however, and as a result shoulder injuries are rather common.

Highly movable joints are typically found at the ends of long bones, such as those of the arms and legs. You can see a view of such

Figure 23.10
Solid Framework
A view of the human skeleton, which has two divisions, axial and appendicular. The major bones of the axial division are listed on the left, and the major bones of the appendicular division on the right.

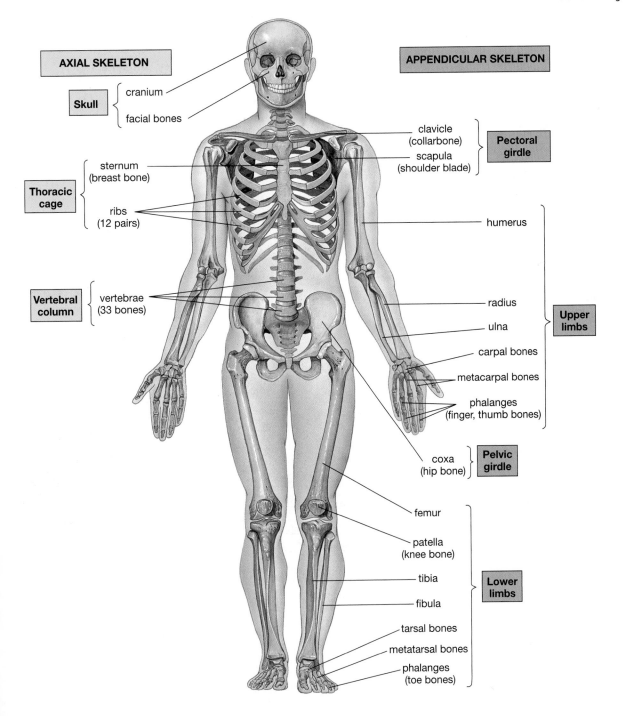

AXIAL SKELETON

Skull { cranium
facial bones }

Thoracic cage { sternum (breast bone)
ribs (12 pairs) }

Vertebral column { vertebrae (33 bones) }

APPENDICULAR SKELETON

clavicle (collarbone)
scapula (shoulder blade) } **Pectoral girdle**

humerus

radius
ulna
carpal bones
metacarpal bones
phalanges (finger, thumb bones) } **Upper limbs**

coxa (hip bone) } **Pelvic girdle**

femur
patella (knee bone)
tibia
fibula
tarsal bones
metatarsal bones
phalanges (toe bones) } **Lower limbs**

a joint, the human knee joint, in **Figure 23.11**. Under normal conditions the bony surfaces of such a joint do not contact one another, for they are covered with special articular cartilages. Such a *hinge joint* is surrounded by a fibrous articular capsule, and the inner surfaces of the joint cavity are lined with a fluid-producing membrane that provides additional lubrication. The knee joint has additional padding in the form of cartilage that lies between the opposing articular surfaces. Such shock-absorbing cartilage pads are called menisci. Also present are fat pads, which protect the articular cartilages

and act as packing material. When the bones move, the fat pads fill in the spaces created as the joint cavity changes shape. Additionally, ligaments that join one bone to another may be found on the outside or inside of the articular capsule.

23.8   Muscles and Movement

Bones may provide a scaffolding for the body, but how is it that this scaffolding is capable of movement? The answer, of course is muscles, specifically skeletal muscles, which are numerous and large enough that they account for about 40 percent of our body weight. (When we talk about the "meat" in, say, cattle, we are mostly talking about skeletal muscle.) Like bones, skeletal muscles are organs in that they contain a variety of tissues—connective and neural tissues as well as skeletal muscle tissue. At each end of a muscle, the fibers of the outer layer come together to form **tendons** that attach skeletal muscles to bones. You can see a typical skeletal muscle in **Figure 23.12**.

Makeup of Muscle

Each muscle has within it a number of oval-shaped bundles called fascicles. Inside each fascicle boundary there is a collection of muscle cells, but the term *cell* here may be a little misleading, because any one of these cells can be as long as the muscle itself—a gigantic length relative to the strictly micro-

patella

fat pad

joint cavity

articular capsule

ligament

menisci

articular cartilage

ligament

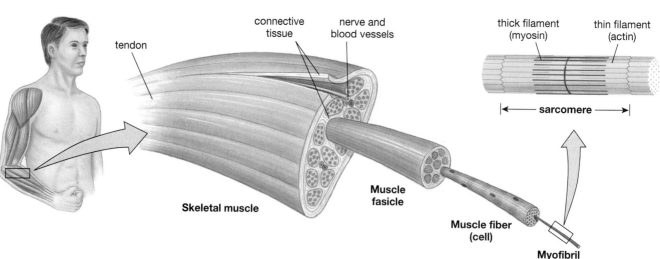

connective tissue

nerve and blood vessels

thick filament (myosin)

thin filament (actin)

sarcomere

tendon

Skeletal muscle

Muscle fasicle

Muscle fiber (cell)

Myofibril

Figure 23.12
Structure of Skeletal Muscle
Elongated muscle cells or "fibers" are grouped into oval-shaped bundles called fascicles. Each fiber is composed of many thin myofibrils and each myofibril has within it a collection of alternating thick and thin protein strands called *filaments*. The thin filaments are made of the protein actin, the thick filaments of the protein myosin. Each myofibril is divided up into a series of repeating functional units called *sarcomeres*.

scopic dimensions of most cells. Because skeletal muscle cells are elongated, they are referred to as **fibers**. A look inside one of these fibers reveals that it is composed of more strands—hundreds of long, thin structures called **myofibrils** that run the length of the cell (see Figure 23.12.) Each myofibril, in turn, has a large collection of strands inside it; these are assemblies of two kinds of protein strands that alternate with one another: thin filaments made of the protein *actin* and thick filaments made of the protein *myosin*. Along their length, these fibers are then divided into a set of repeating units, called **sarcomeres**, that are the functional units of the muscle.

How Does a Muscle Contract?

The structure of the sarcomere provides a key to understanding how a muscle fiber contracts. Thick filaments lie in the *center* of the sarcomere. Thin filaments are attached to either *end* of the sarcomere and extend toward the center, where they overlap with the thick filaments (**Figure 23.13**). In contraction, the thin filaments *slide toward the center* of the sarcomere, causing the unit to shorten—which is to say, causing the muscle to contract.

But what enables the thin filaments to slide? The thick myosin filaments have numerous club-shaped "heads" extending from them which are capable of alternately binding

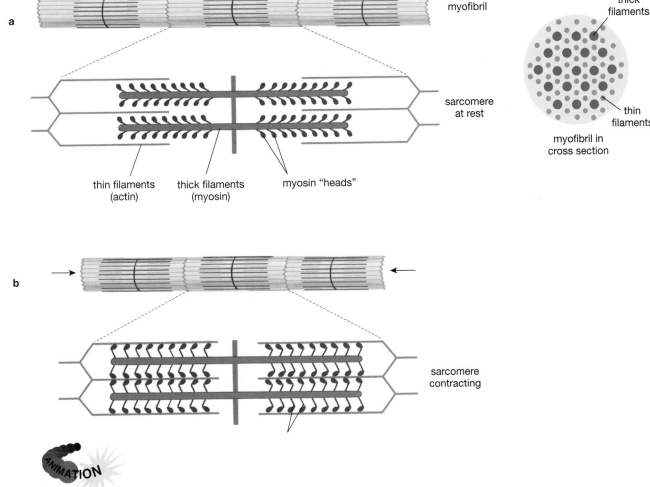

Figure 23.13
How a Skeletal Muscle Contracts

a The myofibril contains several sarcomeres, one of which is shown at rest. The thin filaments within the sarcomere, made of actin, overlap with adjacent thick filaments, made of myosin. The thick filaments have club-shaped outgrowths, called *heads*. The myofibril is also shown in cross section, at a location where thick and thin filaments overlap.

b Muscle contraction is a matter of the thin actin filaments sliding together within a sarcomere. This comes about when the myosin heads attach to the actin filaments and pull them toward the center of the sarcomere. The myosin heads pivot at their base and are capable of alternately binding with, or detaching from, the thin filaments.

Figure 23.14
Major Superficial Muscles of the Body

a Anterior view.

b Posterior view.

with or detaching from the adjacent thin actin filaments, creating so-called **cross-bridges**. The myosin heads pivot at their base, as if they were on a hinge. Here's the order of their activity: Attach to the actin filament, pull it toward the center of the sarcomere, detach from the actin, reattach to it and pull again. This whole process is powered by ATP.

Major Muscles of the Body

You can see an overview of the muscular system in **Figure 23.14**. The general appearance of the skeletal muscles provides clues to their primary function. Muscles involved with locomotion and posture work across joints, producing skeletal movement. Those that support soft tissue form slings or sheets between relatively stable bony elements.

On to the Nervous, Endocrine, and Immune Systems

Having looked at the basic characteristics of the body and at three of its organ systems, we're now ready to move on to three other systems, which function in communication, regulation, and defense against invaders. These are the nervous, endocrine, and immune systems of the body.

Chapter Review

Summary

23.1 The Sciences of Anatomy and Physiology

- Anatomy is the study of the body's internal and external structure and the physical relationships between the body's constituent parts. Physiology is the study of how the body's various structures work.

23.2 What Are the General Characteristics of Humans?

- Human beings share a number of characteristics with many other animals. Three of these are an internal body cavity, called a *coelom*; an internal skeleton; and internal temperature regulation.

- Like all mammals, human beings are endothermic, meaning that human bodies are warmed primarily by their own metabolism. Conversely, most fish, amphibians, and reptiles are ectothermic, meaning that their primary source of body heat is external to them.

23.3 Animal Architecture and Organization

- Animal bodies have several levels of organization. Tissues are groups of cells that have a common function. In some animals, tissues are arranged into organs: complexes of several kinds of tissues that perform a special bodily function. Organs in turn form organ systems, such as the cardiovascular system. The human body has 11 organ systems.

23.4 The Animal Body Has Four Basic Tissue Types

- There are four fundamental tissue types in the human body: epithelial, connective, muscle, and nervous.

- Epithelial tissue covers all the exposed body surfaces of animals; it also forms glands, which are organs that secrete one or more substances.

- Glands come in two main types: exocrine glands, which secrete materials through ducts; and endocrine glands, which release their materials directly from cells into surrounding tissues or the bloodstream.

- Connective tissues function in support and protection of the body and its organs, in the transport of materials within the body, in energy storage, and in defense of the body. A prime function of most connective tissue cells is to produce extracellular material composed of protein fibers and a ground substance that varies in consistency from a fluid to a solid.

- Connective tissues are classified according to their surrounding extracellular material. There is loose connective tissue, which provides strength and flexibility as a packing material of the body. Fibrous connective tissues form tendons and ligaments, along with capsules, which surround

Key Terms

adipose tissue 487

appendicular skeleton 497

axial skeleton 495

basement membrane 486

bone 487

cardiac muscle tissue 488

cardiovascular system 491

cartilage 487

coelom 484

cross-bridge 501

dendrite 489

dermis 492

digestive system 492

ectothermic 484

endocrine gland 486

endocrine system 491

endothermic 485

epidermis 492

exocrine gland 486

fiber 499

gland 486

hair follicle 493

hair papilla 493

hormone 486

integumentary system 490

interstitial fluid 487

organs and enclose joint cavities. Supporting connective tissues make up bone and cartilage. Fluid connective tissue includes blood—blood cells suspended in a watery ground substance called *plasma*—and lymph, which functions primarily in bodily defense.

- Muscle tissue is specialized in that its protein elements or fibrils have the ability to contract. There are three primary types of muscle tissue: skeletal muscle, cardiac muscle, and smooth muscle. Skeletal muscle is always attached to bone and is under voluntary control; it is the muscle that is responsible for movement of the body. Smooth muscle is responsible for contractions in blood vessels, air passages, and hollow organs such as the uterus; it is not under voluntary control. Cardiac muscle exists only in the heart and is responsible for its beating.

- Nervous tissue is specialized for the rapid conduction of electrical impulses. It is made up of two basic types of cells: neurons, which are the functional units of nervous tissue; and neuroglia, which are support cells for neurons.

23.5 A Summary of the Organ Systems of the Human Body

- The integumentary, skeletal, and muscular systems function in body support and movement. The integumentary system is made up of skin and such associated structures as glands, hair, and nails. The skeletal system forms an internal supporting framework for the body. The muscular system includes all the skeletal muscles of the body that are under voluntary control.

- The nervous, endocrine, and lymphatic organ systems function in coordination, regulation, and defense of the body. The nervous system includes the brain, spinal cord, sense organs such as the eye and ear, and all the nerves that interconnect these organs and link the nervous system with other systems. The endocrine system functions in bodily regulation, working through hormones. The lymphatic system has a great overlap with the immune functions of the body; it consists of a network of lymphatic vessels that collect extracellular fluid and deliver it, as lymph, to the blood vessels. While in the lymphatic vessels, lymph is checked for signs of infection. The lymphatic system also produces cells that function in defense of the body. Lymphatic system organs include the lymph nodes, thymus gland, tonsils, and spleen.

- The cardiovascular, respiratory, digestive, and urinary organ systems function in transport and in exchange with the environment. The cardiovascular system consists of the heart, blood, blood vessels, and the bone marrow—where red blood cells are formed. The respiratory system includes the lungs and the passageways that carry air to them. Through this system, oxygen comes into the body and carbon dioxide is expelled from it. The central feature of the digestive system is the digestive tract, a tube that begins at the mouth and ends at the anus. Other digestive system organs include the stomach, the small and large intestines, and such accessory glands as the salivary glands, the liver, and the pancreas. The urinary system functions to eliminate waste products from the blood through the formation of urine. It is made up of the kidneys, the ureters that carry urine to the urinary bladder, and the urethra, which conducts urine from the bladder to the exterior of the body.

- Humans have two reproductive systems—one for females and one for males.

23.6 The Integumentary System: Skin and More

- The primary component of the integumentary system is the organ called *skin*. Integumentary structures associated with skin are hair, nails, and a variety of exocrine glands.

- In addition to covering the body, skin protects underlying tissues and organs, controls the evaporation of body fluids, regulates heat loss, stores fat, and synthesizes vitamin D. Exocrine glands associated with skin excrete such materials as water, oils, and milk. Specialized nerve endings in the skin detect touch, pressure, pain, and temperature.

- Skin is an organ in two parts: a thin outer epithelial covering, the epidermis; and a thicker underlying dermis, composed mostly of connective tissue. Epidermal skin cells must constantly be replenished. The surface layer of the epidermis is composed of tightly interlocked cells that produce a water-resistant protein called *keratin*. The dermis is filled with accessory structures, such as hair follicles and sweat glands, and contains capillaries and nerves that support the surface of the skin.

- Hairs originate in tiny organs called *hair follicles*. The skin contains two types of exocrine glands: sebaceous glands, that produce an oily secretion; and sweat glands. Sweat glands come in two varieties: apocrine glands—found in the armpits, around the nipples, and in the groin—which produce a thick, cloudy secretion that empties into hair follicles; and the more numerous merocrine glands, which produce perspiration.

- Nails form over the tips of the fingers and toes, where they protect the exposed tips.

23.7 Body Support and the Skeleton

- The human skeletal system is composed of bone and cartilage. It functions in support, as a storage site for minerals, and as a production site for red blood cells and other blood elements.

- Each bone of the skeleton is considered to be an organ because it contains not only connective tissue—mostly the supporting connective tissue called *osseous* tissue—but other tissues as well. Like other connective tissues, osseous tissue is made not only of specialized cells, but of extracellular fibers and a ground substance.

- The typical bone features a long central shaft, called a *diaphysis*, and expanded ends, called *epiphyses*. The diaphysis surrounds a central cavity filled with a loose connective tissue called *marrow* that comes in two forms: red marrow, which is blood-forming; and yellow marrow, which is made up of fat.

- There are two types of bone tissue: compact bone, which is relatively solid; and spongy bone, which is porous and less dense. Compact bone forms the diaphysis of long bones; spongy bone fills in the epiphyses.

- Three different types of cells are involved in bone growth and maintenance. Osteoblasts are responsible for the production of new bone; osteocytes are mature bone cells that maintain the structure and density of normal bone; and osteoclasts release enzymes into the bone matrix that breaks it down.

- In compact bone, the basic functional unit is the osteon. Osteoblasts produce layers of concentric cylinders of bone, which are wrapped around a central canal of the osteon. The central canal, which parallels the long axis of the bone, has blood vessels and nerves running through it.

- The human skeleton has 206 bones in all, arranged in two main divisions: an axial skeleton, whose 80 bones include the skull, the vertebral column, and the rib cage; and the appendicular skeleton, whose 126 bones include those of the arms and legs and the pelvic and pectoral girdles to which they are attached.

- Joints or articulations exist wherever two bones meet. Normally, the bony surfaces of highly movable joints do not meet directly, but are instead covered with special cartilages and pads that provide shock absorption and lubrication.

23.8 Muscles and Movement

- A given skeletal muscle cell can be as long as the muscle itself; because of their elongation, these cells are called *fibers*. Muscle fibers are composed of thin structures called *myofibrils* that are in turn composed of assemblies of two kinds of protein strands that alternate with one another: thin filaments made of the protein actin and thick filaments made of the protein myosin. These filaments are divided into a set of repeating units called *sarcomeres*.

- In contraction, thin muscle filaments slide toward the center of the sarcomeres, causing the unit to shorten. The thick myosin filaments have numerous club-shaped heads that can bind with and detach from the adjacent thin actin filaments. The myosin filaments thus bring about contraction by attaching to the actin filaments, pulling them toward the center of the sarcomeres, detaching, and then pulling again.

Understanding the Basics

Multiple-Choice Questions

1. Which of the following organs is not found in the abdominopelvic cavity?
   a. stomach
   b. liver
   c. lungs
   d. pancreas
   e. intestines

2. The four basic types of tissue in the human body are
   a. epithelia, connective, muscle, nervous
   b. epithelia, blood, muscle, nervous
   c. skin, bone, muscle, nervous
   d. skin, connective, heart, nervous
   e. epithelia, organ, connective, muscle

3. Endocrine glands differ from exocrine glands in that
   a. they have ducts
   b. they are ductless
   c. they produce only perspiration
   d. they consist of epithelia and muscle tissue
   e. they are found only in the liver

4. Large muscle fibers that are multinucleated, striated, and voluntary are found in
   a. cardiac muscle tissue
   b. skeletal muscle tissue
   c. smooth muscle tissue
   d. rough muscle tissue
   e. all four types of muscle tissue

5. The skeletal system includes 206 bones and several
   a. tissues
   b. muscles
   c. nerves
   d. organs
   e. cartilages

6. The lungs and the passageways that carry air make up the
   a. cardiovascular system
   b. respiratory system
   c. lymphatic system
   d. urinary system
   e. connective system

7. The following structures are all associated with the integumentary system except
   a. nails
   b. sweat glands
   c. hair
   d. dermis
   e. salivary glands

8. The two types of bone tissue are
   a. spongy bone and cartilage
   b. dense bone and compact bone
   c. compact bone and spongy bone
   d. compact bone and periosteum
   e. cartilage and ligaments

9. The cells that maintain bone structure are
   a. osteoblasts
   b. osteoclasts
   c. osteocytes
   d. osteons
   e. osteolytes

10. The appendicular skeleton consists of the bones of the
    a. pectoral and pelvic girdles
    b. skull, thorax, and vertebral column
    c. arms, legs, hands, and feet
    d. limbs, pectoral girdle, and pelvic girdle
    e. arms, legs, and neck

11. A skeletal muscle contains
    a. connective tissues
    b. blood vessels and nerves
    c. skeletal muscle tissue
    d. all of the above
    e. none of the above

Brief Review

1. Beginning with the molecular level, list in correct sequence the levels of organization of the human body from the simplest level to the most complex.

2. What fluid connective tissues are found in the human body?

3. What is keratin?

4. List three functions of the skeletal system.

5. List three primary functions of the muscular system.

6. Which organ system is composed of ductless glands whose secretions coordinate cellular activities of other organ systems?

7. Why does the epidermis continually shed its cells?

8. A sample of bone shows concentric layers surrounding a central canal. Is this sample from the shaft (diaphysis) or the end (epiphysis) of a long bone?

9. Why does skeletal muscle appear striated when viewed under a microscope?

Applying Your Knowledge

1. Why is the same stratified epithelial organization found in the mucous membranes of the pharynx, esophagus, anus, and vagina?

2. Lack of vitamin C in the diet interferes with the body's ability to produce collagen. What effect might this have on connective tissue?

3. The repair capability of cartilage is low compared with that of bone. What accounts for this difference?

Group Activity

Divide into groups of three or four students. Vertebrates have evolved many modes of locomotion as a result of differences in skeletal and muscle systems. After examining a drawing of the forelimbs of a cat, bat, gorilla, and whale (Figure 16.12, page 000) answer the following questions:

1. List all the differences and similarities that you see in the diagram.

2. What are the advantages and disadvantages of each difference that your group observed?

3. Knowing the mode of locomotion for each organism (fly, walk, swim), describe which factors of the muscle and skeletal systems provide for an animal to be fast or slow or affect what the animal eats?

24

Control and Defense

The Nervous, Endocrine, and Immune Systems

Budding killers.
(page 532)

A knee-jerk reaction.
(page 514)

The nature of a nerve.
(page 509)

One of the central qualities

that separates living beings from inanimate objects is that living beings have the ability to respond to their external environment and to maintain a relatively stable internal environment. To be alive means to monitor what's around us, to pursue what we need, and to adapt to change. We vertebrate animals are able to do these things largely because of the complex and marvelous workings of three bodily systems. One is the nervous system, with its ability to respond almost instantaneously to external signals or internal desires; another is the endocrine or hormonal system, which allows us to maintain a physical steady state over time; a third is the immune system, which allows us to deal with the microscopic invaders that mount their attacks on us every day. These three systems are the subject of this chapter.

24.1 Overview of the Nervous System

The basic function of the nervous system is to monitor an animal's internal and external environment, integrate the sensory information the animal receives in this process, and coordinate the voluntary and involuntary re-

sponses of many systems throughout the body. All of these functions are performed by **neurons**—the key functional cells of the nervous system—with major assistance from **neuroglia** cells, which surround, support, and protect neurons.

The major subdivisions of vertebrate nervous systems are shown in **Figure 24.1** (page 508). The two major divisions are, first, the **central nervous system (CNS)**, consisting of the brain and spinal cord, which are responsible for integrating and coordinating sensory data and issuing action or *motor* commands. The brain is also the seat of higher functions, such as intelligence, memory, and emotion. All communication between the CNS and the rest of the body occurs via the nervous system's second major subdivision, the **peripheral nervous system (PNS)**, which includes all of the neural tissue outside the CNS.

The peripheral nervous system then has an **afferent division**, which brings sensory information to the CNS. It also has an **efferent division** that carries motor commands to muscles and glands, which collectively are known as the body's **effectors**. Within the peripheral system's efferent division, there are two subsystems: The **somatic nervous system**, which provides

A nerve impulse is transmitted.
(page 511)

Central to the immune system.
(page 530)

Fluids on the move.
(page 527)

voluntary control over skeletal muscles; and the autonomic nervous system, which provides involuntary regulation of smooth muscle, cardiac muscle, and glands. The autonomic nervous system is further subdivided into the **sympathetic division** (which generally has stimulatory effects) and **parasympathetic division** (which generally has relaxing effects).

Lest all this seem too confusing, let's think of it in terms of a real-world occurrence: a bell going off for a high-school class on a day when you had to give a report in front of the class. Sensory receptors in your ears (cells that are part of the peripheral nervous system) perceived the sound of the bell and transmit-

ted a message through afferent auditory nerves to your brain (part of the central nervous system), which processed the information and transmitted motor commands through the peripheral nervous system's efferent division to two places. One was to the effectors we call *skeletal muscles* (in your legs and arms), commanding them to move so you could get to class; the other was to the effectors called *adrenal glands*, prompting them to produce adrenaline (because this was a public performance). It was the somatic nervous system that issued the command to move, but it was the autonomic nervous system that issued the command for adrenaline release.

Cells of the Nervous System

All the functions of the nervous system involve neurons communicating—either with one another, or with cells of other types, such as muscle or hormonal cells. The helper neuroglial cells regulate the environment around the neurons, provide a supporting framework for neural tissue, and act as defenders of neural tissue. Although neuroglia are much smaller than neurons, they outnumber neurons by about 10 to 1.

On the basis of the functions they perform, we can distinguish three types of neurons: sensory neurons, motor neurons, and interneurons.

Sensory neurons do just what their name implies: They sense conditions both inside and outside the body and convey information about these conditions to neurons inside the CNS. Sensory information from the external environment includes touch, temperature, and pressure sensations, as well as the more complex senses of sight, smell, hearing, and taste. Internal sensory information concerns such things as the positions of skeletal muscles, the activities of the digestive, respiratory, and urinary systems, and sensations of deep pressure and pain.

Motor neurons carry instructions from the CNS to other tissues, organs, or organ systems. The targets of information for motor neurons are the effectors noted earlier, such as skeletal muscles; they are called effectors because they can effect *change*, as with skeletal muscle contracting after it receives a neural signal.

The third class of neurons, **interneurons** (or association neurons), are located entirely

Figure 24.1
Divisions of the Nervous System
Information about the body and its environment comes to the nervous system from its sensory receptors (for example, cells in the ears), which are part of the peripheral nervous system. This information then goes through the afferent division of the peripheral nervous system to the brain and spinal cord, which constitute the central nervous system. The brain and spinal cord process information and issue motor commands through the peripheral nervous system's efferent division. These commands go to "effectors" such as skeletal muscles and glands. The peripheral nervous system's efferent division has two systems within it: The somatic nervous system, which provides voluntary control over skeletal muscles; and the autonomic system, which provides involuntary regulation of smooth muscle, cardiac muscle, and glands. The autonomic system is further divided into sympathetic ("fight-or-flight") and parasympathetic ("rest-and-digest") divisions.

within the brain and spinal cord. As their name implies, they interconnect other neurons. They are responsible for the analysis of sensory inputs and the coordination of motor outputs.

Anatomy of a Neuron

Figure 24.2 shows a typical motor neuron. Its cell body contains all the basic elements of cells in general—a nucleus, organelles such as mitochondria, and so forth. Extensions projecting from the cell body include a variable number of **dendrites**, which receive signals coming *to* the cell body; and a single, large **axon** that carries signals *from* the cell body. Each neuron has an outer plasma membrane, just like any other cell; what sets the neuron's membrane apart is the kind of sensitivity it has to chemical, mechanical, or electrical stimulation. Such stimulation often leads to the generation of electrical signals— meaning nerve impulses—that are conducted along the axon. At the tips of an axon or its branches, there may be structures called **synaptic terminals**, which are part of a **synapse**—a site where a neuron communicates with another cell. Neurons can be extraordinarily *long* cells; the axon that allows the movement of your little toe starts in a cell body that lies partway up your spinal cord.

The Nature of Neuroglia

Neuroglia, which are found in both the CNS and PNS, have no information processing ability of their own. In the CNS, one type of neuroglia is protective in that it can ingest invaders or foreign molecules. A second group of neuroglia covers tiny blood vessels within the brain, forming what is known as the **blood-brain barrier**, which keeps some substances from passing from general circulation into the brain. Other neuroglia wrap their cell membranes around the axons of neurons in the brain and spinal cord. Such a covering is called **myelin**, and an axon covered in this way is said to be myelinated.

Axons that are myelinated carry nerve impulses faster than those that are not. Myelin is fat- or lipid-rich, and areas of the brain and spinal cord containing myelinated axons are glossy white. Thus do we get the term the **white matter** of the CNS, which contains mostly axons, whereas areas dominated by neuron cell *bodies* are **gray matter**. If you look at **Figure 24.3**, you can see what a

myelinated axon looks like. In the PNS, the most important neuroglia are called *Schwann cells*, which wrap around the axons of neurons, making them myelinated.

Nerves

A **nerve** is a bundle of axons in the PNS that transmit information to or from the CNS. Any given bundle may stem from a collection of sensory neurons, a collection of motor neurons, or a mixture of the two types of neurons. Like muscles, nerves are covered by and contain three layers of connective tissue and contain blood vessels that supply nutrients and oxygen. In a dissection, nerves are whitish string-like structures, easily visible to the naked eye.

Figure 24.2
The Neuron
Structure of a motor neuron, in this case innervating a skeletal muscle.

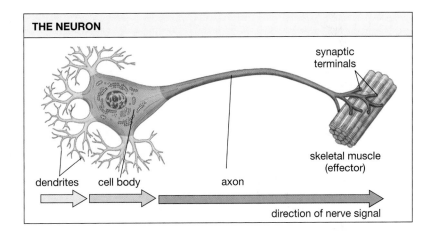

THE NEURON

dendrites   cell body   axon

synaptic terminals

skeletal muscle (effector)

direction of nerve signal

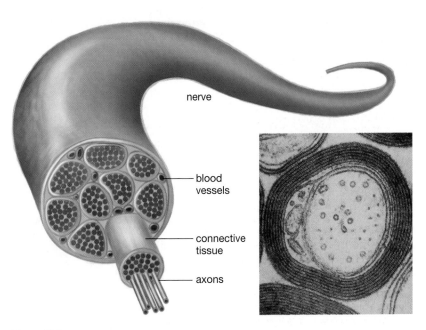

nerve

blood vessels

connective tissue

axons

Figure 24.3
Looking at an Axon
The micrograph at right shows a cross section of a vertebrate axon that is wrapped in multiple layers of myelin covering.

24.2 How Does Nervous System Communication Work?

Nerve impulses travel between a given neuron and other neurons, as well as between neurons and effectors, such as muscle cells or gland cells (as Figure 24.2 indicates). But nerve signals must also travel *within* a given neuron—along the length of its axon. This communication occurs through changes in an electrical property of the neuron's outer lining or plasma membrane.

Nerve Signal Transmission within a Neuron

Those who read Chapter 5 will recall that the plasma membrane of any cell regulates the passage of substances in and out of the cell. Because the plasma membrane is not permeable to charged particles (such as ions), the membrane separates charged particles that lie inside the membrane from those that lie outside. There are, however, protein channels in the membrane that can open or close like gates, allowing for the controlled flow of charged particles across the membrane. A resting cell has a greater positive charge outside itself than inside; this is so because there is an excess of positively charged compounds (ions) on the outside and an excess of negatively charged compounds (ions and protein molecules) on the inside. Because oppositely charged particles attract, this difference in overall charge across the two sides of the membrane amounts to a form of stored energy that can be used in transmitting nerve signals; it is potential energy that, fittingly enough, is called a membrane potential (**see Figure 24.4**).

In response to stimulation a neuron receives from other neurons, its permeability to charged particles can change. Recall that the cell membrane has protein "gates" that can open or close. Upon stimulation, positively charged sodium ions (Na^+), which lie in abundance outside the cell, are let inside in greater numbers through gates that open up. After a very brief time, the influx of Na^+ is great enough that the inside of the cell is more positive than the outside. This very activity then causes the Na^+ gates to shut and a group of gates for another positively charged ion, potassium (K^+), to open up more fully. However K^+ initially exist in greater numbers *inside* the plasma membrane than outside. With the opening of their gates, they rush out of the cell and the positive charge that existed on the inside of the cell begins to decline. Eventually, enough K^+ have rushed out that the cell returns to its resting state, with a net positive charge outside the membrane and a net negative charge inside. This entire process takes place in a few thousandths of a second (milliseconds).

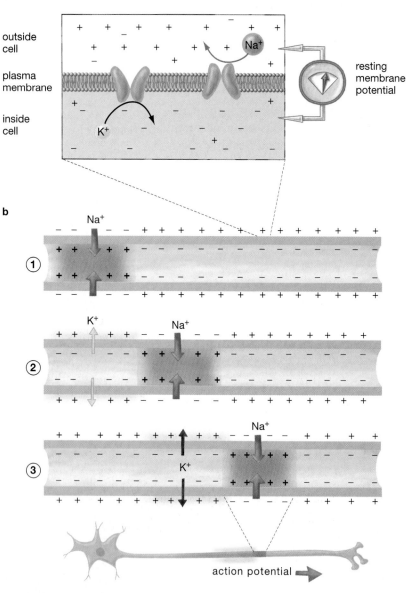

a

b

Figure 24.4
Nerve Signal Transmission within a Neuron

a Electrical energy is stored across the cell membranes of resting neurons because there is an excess of positively charged compounds on the outside of the membrane and an excess of negatively charged compounds on the inside. This stored energy is called a *membrane potential*. Protein channels (shown in green) that can allow the movement of electrically charged ions across the membrane remain closed in a resting cell, thus maintaining the membrane potential.

b 1. Nerve signal transmission begins when, upon stimulation, some protein channels open up, allowing a movement of positively charged sodium ions (Na^+) into the cell. For a brief time, the inside of the cell becomes positively charged. **2.** The Na^+ gates close and the gates for positively charged potassium ions (K^+) open up, allowing a movement of K^+ out of the cell. With this, there is once again a net positive charge outside the membrane. The influx of Na^+ at one point in the cell membrane then triggers the same sequence of events in an adjacent portion of the membrane—note the influx of Na^+ next to the outflow of K^+. **3.** The nerve signal continues to be propagated one way along the axon by means of this action potential.

Transmission along the Whole Axon: The Action Potential

You have just looked at the alteration of membrane potential at *one spot* along an axon. But how does a nerve signal get transmitted down the *length* of the axon? The key to such impulse transmission is that an influx of sodium ions at an initial location on the membrane triggers reactions that cause *neighboring* portions of the membrane to permit sodium ions to rush into the cell in their location (see Figure 24.4). What occurs, in other words, is a chain reaction that moves down the entire length of the axon membrane, again in milliseconds. This **action potential** is what a nerve signal amounts to within an axon. Action potentials have been compared to what happens with a lighted fuse: The heat of the spark causes the neighboring section of the fuse to catch fire, thus moving the spark along.

We expect a fuse spark to move in one direction, of course, and the same is true of an action potential. In the case of a fuse, this one-way movement comes about because the burnt part of the fuse cannot be burnt further. In the case of nerve signals, the portion of an axon that has potassium (K^+) ions flowing out of it cannot simultaneously have its sodium (Na^+) channels opened up. In effect, there is a short *refractory* period in which a given portion of an axon is incapable of transmitting a nerve signal. Thus the signal always moves away from the original point of stimulation.

All action potentials are of equal strength and duration. An all-or-none principle applies to neurons: A given stimulus either triggers a full-blown action potential or none at all.

Communication between Cells: The Synapse

Having traveled the length of an axon, an action potential then reaches the synaptic terminals (the branched tip of the axon). How does the signal then get to the *next* neuron (or muscle or gland cell)? This question was once very intriguing to biologists, because it was clear that a neuron does not touch the downstream cell. The axon of one neuron comes *close* to touching the dendrites or cell body of another neuron (within about 20 millionths of a meter), but there is inevitably a small intervening space, or gap, between them. There are thus three entities involved in cell-to-cell transmission: the "sending" cell, the "receiving" cell, and the gap between them. The location where all three come together is the area in which cell-to-cell communication takes place, and it is called a **synapse**.

Structure of a Synapse

Every synapse has three basic parts: (1) the **synaptic cleft**, which is the space between the two cells; (2) the **presynaptic neuron**, which is the transmitting neuron; and (3) the **postsynaptic neuron** (or other effector cell type), which receives the action potential. Upon arrival of the action potential at a synaptic terminal, the presynaptic neuron releases a chemical, called a **neurotransmitter**, which is stored in small sacs or vesicles in the synaptic terminal. When released into the synaptic cleft, the neurotransmitter travels over to the receiving cell and binds to receptors in that cell's membrane. This binding stimulates the opening of sodium channels, thereby allowing positively charged sodium ions to cross the *post*synaptic cell membrane, setting in motion the generation of an action potential in that cell. (**See Figure 24.5** for an example involving the neurotransmitter acetylcholine.) So, why doesn't neurotransmitter *keep* stimulating the postsynaptic cell? One means of control is that the postsynaptic cell releases into the synaptic cleft enzymes that inactivate the neurotransmitter.

**Figure 24.5
Structure and Function of a Synapse**
As a nerve impulse reaches the synapse, molecules of a neurotransmitter (such as acetylcholine) are released into the synaptic cleft from synaptic terminals of the presynaptic neuron. The means of this release is that small membrane-bound vesicles—each filled with thousands of neurotransmitter molecules—fuse with the outer membrane of the presynaptic neuron, and in so doing release their contents to the synaptic cleft. The neurotransmitter molecules then move across the cleft and stimulate receptors in the receiving cell's membrane, thus initiating a nerve impulse within it.

presynaptic neuron

synapse

postsynaptic neuron

synaptic cleft

synaptic terminal

arrival of nerve impulse

initiation of new impulse

neurotransmitter receptors

vesicles containing neurotransmitter molecules (such as acetylcholine)

The Importance of Neurotransmitters

There are many different neurotransmitters. Some, like acetylcholine, almost always have stimulatory effects on postsynaptic cells. Others, like dopamine in the CNS, are usually inhibitory. The importance of neurotransmitters is immense. As just one example of their power, you may have heard of a widely prescribed class of drugs called SSRIs (selective serotonin reuptake inhibitors) that are aimed at fighting depression in human beings. (The drugs Prozac and Paxil are in this class.) Serotonin is a neurotransmitter found in the brain. Presynaptic cells have means of taking serotonin back into themselves once they have released it into the cleft. SSRIs work by reducing this "reuptake." The result—an increased amount of serotonin in the synaptic cleft—is often a reduction of depression.

24.3 The Spinal Cord

Now that you've looked at how nerve impulses are transmitted at the cellular level, it's a good time to take a step back and look at the nervous system in its larger dimensions. Let's start by examining the basic structure and function of the spinal cord.

The spinal cord serves two major functions. First, it can act as a communication center on its own, receiving input from sensory neurons and directing motor activities in response, with no input from the brain. These are our so-called reflexes. Second, it is the major pathway for sensory impulses going to and from the brain.

The spinal cord extends from the base of the brain to an area just below the lowest rib, measuring approximately 45 cm (18 inches) in length and about 14 mm or (0.5 inches) in width (**see Figure 24.6a**). The entire spinal cord consists of 31 segments, each connected to a pair of **spinal nerves**—one nerve for the left side of the body, one for the right. The spinal nerves can be grouped into several classes according to the region of the body they serve, as Figure 24.6 illustrates.

The most striking feature of a cross section of the spinal cord is a rough H- or butterfly-shaped area around the narrow central canal. This H is the gray matter of the spinal cord, composed mostly of the cell bodies of neurons and their supporting neuroglial cells. Surrounding the spinal cord's gray matter is its white matter. White matter, as you've seen, is largely neural *axons*, and this is the case with the spinal cord. The **central canal** of the spinal cord

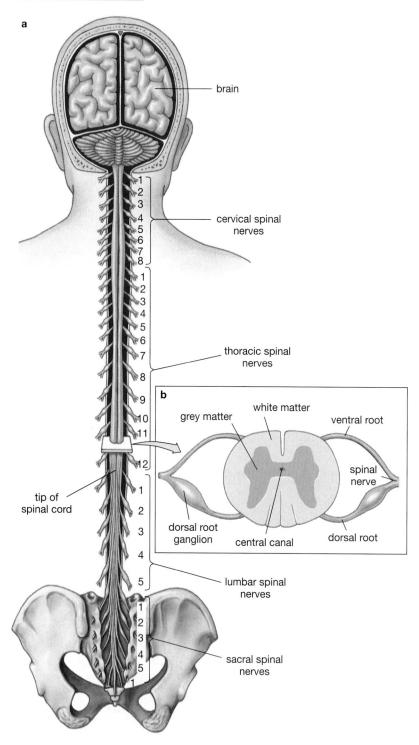

Figure 24.6
Structure of the Spinal Cord

a The numbers to the right of the spinal cord identify 31 pairs of spinal nerves, which can be grouped according to the body part they control. The eight cervical nerves control the head, neck, diaphragm and arms; the 12 thoracic nerves control the chest and abdominal muscles; the five lumbar nerves control the legs; and the five sacral nerves control the bladder, bowels, sexual organs, and feet.

b Cross section through the lower thoracic region of the spinal cord, showing the arrangement of gray and white matter, the central canal, the ventral roots, the dorsal roots, and the dorsal root ganglia.

that you see in Figure 24.6b is a space filled with **cerebrospinal fluid**, which circulates not only in the spine, but in the brain, supplying nutrients, hormones, and white blood cells and providing protection against jarring motion.

The Spinal Cord and the Processing of Information

Sensory input is, of course, coming *to* the spinal cord from various sensory receptors in the body, while motor commands must go *from* the spinal cord to various effectors. The motor neurons that are sending the motor commands lie in the gray matter of the spinal cord, as do the interneurons discussed earlier. The axons of these motor neurons leave the spinal cord through the ventral roots you can see in Figure 24.6b.

Now consider sensory neurons, which are responsible for transmitting the information that goes to the spinal cord. Their cell bodies lie *outside* the spinal cord in the **dorsal root ganglia** you can see in Figure 24.6b. (A **ganglion** is any collection of nerve cell bodies in the PNS.) The axons of these sensory neurons may extend all the way to distant sensory receptors in the body—say, to a touch receptor located in your arm. The receptor itself may be a dendrite that feeds into the axon, or it may be a specialized cell that communicates with the axon across a synapse. Either way, once stimulated, the sensory neuron axon transmits a message to the sensory neuron cell body in a dorsal root ganglion. This message then goes to the spinal cord, through the dorsal root you can see in Figure 24.6b. In sum, motor commands leave the spinal cord through a ventral root, while sensory information comes to the spinal cord through a dorsal root. Inside the spinal cord, sensory and motor neurons may link up directly (through synapses) or they may link up with interneurons. Note that both dorsal and ventral roots come together, like fibers being joined in a single cable, to form a given spinal nerve.

Quick, Unconscious Action: Reflexes

The sensory/motor division of labor is clearly reflected in simple body **reflexes**, which can be defined as automatic nervous system responses, triggered by specific stim-uli, that help us avoid danger or preserve a stable physical state. If you accidentally touch a hot stove, you automatically pull back your hand. No conscious thought of "Oh, I've touched a hot stove; I'd better pull my hand back," is necessary; the reaction is automatic. The neural wiring of a single reflex is called a reflex arc. A reflex arc begins at a sensory receptor, runs through the spinal cord, and ends at an effector, such as a muscle or gland. In the simplest reflex arc, a sensory neuron links directly to a motor neuron. Because only one synapse is involved in such a situation, simple reflexes control the most rapid motor responses of the nervous system. **Figure 24.7** on page 514 shows one of the best-known examples, the knee jerk, or patellar reflex, in which a properly placed sharp rap on the knee produces a noticeable kick. Note that this exemplifies what you saw earlier: A sensory receptor is stimulated, thus prompting a signal to move through the axon of a sensory neuron and then into the spinal cord. There the sensory neuron is linked via a synapse to a motor neuron, which issues a command that is carried out by an effector—in this case a skeletal muscle.

Many other reflexes may have at least one interneuron placed between the sensory receptor and the motor effector. Due to the larger number of synapses, these reflexes have a longer delay between a stimulus and response. However, they can produce far more complicated responses because the interneurons can control several different muscle groups simultaneously.

24.4 The Autonomic Nervous System

Effectors, such as muscles, respond to motor commands that come through the "outgoing" or efferent division of the PNS. But such commands can be either voluntary or involuntary. As noted earlier, there is a division of the PNS over which we have conscious control; it is called the **somatic nervous system**, and it is responsible for the voluntary control of skeletal muscle. The division of the PNS over which we do *not* have conscious control is called the **autonomic nervous system**, and it is very important; imagine having to consciously control your breathing. This system controls the involuntary regulation of smooth muscle, cardiac muscle, and glands.

What Is It?
See p. 514.

The Sympathetic and Parasympathetic Divisions of the Autonomic Nervous System

The autonomic nervous system exercises its control through two "divisions," the sympathetic and parasympathetic. The **sympathetic division** is often called the *fight-or-flight* system, because it usually stimulates tissue metabolism, increases alertness, and generally prepares the body to deal with emergencies. The **parasympathetic division** is often regarded as the *rest-and-digest* system, because it conserves energy and promotes sedentary activities such as digestion. In **Figure 24.8**, you can see the effects these two divisions have on various bodily functions. Although some organs are innervated by one division or the other, most vital organs receive dual innervation, meaning instructions from both autonomic divisions. Where dual innervation exists, the two divisions often have opposing effects, keeping the body's stability mechanisms working in a dynamic balance.

24.5 The Human Brain

The adult human brain is far larger and more complex than the spinal cord. It contains almost 98 percent of the neural tissue in the body. An average adult brain weighs 1.4 kg (a little over three pounds), but brain size varies considerably between individuals. The brains of males are generally about 10 percent larger than those of females (because of the males' greater average body size), but there is no correlation between brain size and intelligence.

The brain consists of gray and white matter. **Gray matter** is made up of nerve cell bodies that exist as individual working groups called *nuclei*, and as a layer at the surface of the brain called the **cerebral cortex**. **White matter**, made up of bundles of axons, forms "tracts" that surround the nuclei and connect them with the cerebral cortex. Like the spinal cord and its spinal nerves, the brain can communicate directly with other body tissues and organs by means of cranial nerves that extend from the brain.

Six Major Regions of the Brain

There are six major regions in the adult brain: (1) cerebrum, (2) thalamus and hypothalamus, (3) midbrain, (4) pons, (5) cerebellum, and (6) medulla oblongata. **Figure 24.9** (page 516) shows the locations of these brain regions and their general functions. Now let's take a look at each of them.

Figure 24.7
Steps in a Reflex Arc

1. Stimulus (tapping) arrives and a receptor is activated.

2. The signal from the receptor reaches a sensory neuron cell body, in the dorsal root ganglion near the spinal cord; the signal moves to a sensory neuron/motor neuron synapse in the spinal cord.

3. Information processing of the event takes place, prompting a signal to be sent through the motor neuron.

4. The motor neuron signal stimulates the effector (the quadriceps muscle) to contract. Note that CNS processing for this reaction was handled entirely in the spinal cord; the brain was not involved.

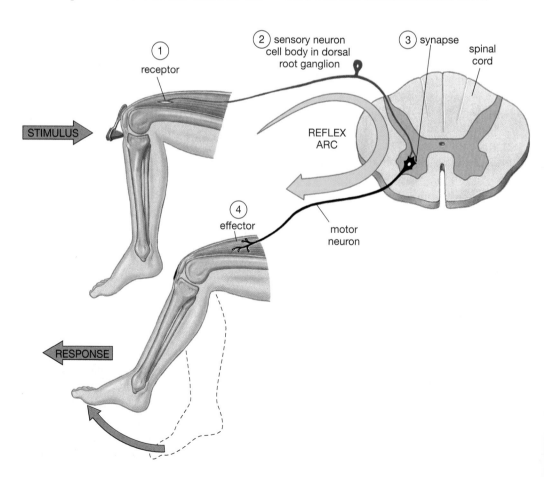

- The **cerebrum** is divided into large left and right **cerebral hemispheres**, the outer portion of which is the cerebral cortex. Conscious thought processes, sensations, intellectual functions, memory storage and retrieval, and complex motor patterns originate in the cerebral hemispheres.

- The **cerebellum** adjusts ongoing movements on the basis of sensory information and stored memories of previous movements. Its activities help us adjust postural muscles, maintain balance and equilibrium, and reproduce learned movements such as swinging a golf club and playing the piano.

In looking at the brain, if you take away the cerebrum and the cerebellum, what remains is the brain stem, whose structures are shown in Figure 24.9b. The brain stem contains important processing centers and relay stations for information passing to or from the cerebrum or cerebellum, and also controls vital functions such as breathing, and activities of the heart and digestive system. Here are its constituent parts:

- The **thalamus** contains relay and processing centers for incoming sensory information. Lying below it is the **hypothalamus**, which contains cell bodies involved with emotions, autonomic function, and hormone production. A narrow stalk connects the hypothalamus to the pituitary gland, which is part of the endocrine system. The hypothalamus is the primary link between the nervous and endocrine systems.

Figure 24.8
Involuntary Control of Bodily Functions
The autonomic nervous system has two divisions, the sympathetic and the parasympathetic, which exercise automatic control over the body's organs, generally in opposing ways. The parasympathetic generally has inhibitory or relaxing effects, while the sympathetic has stimulatory effects. Axons of the parasympathetic division emerge not only from the spinal cord, but from the brain as well.

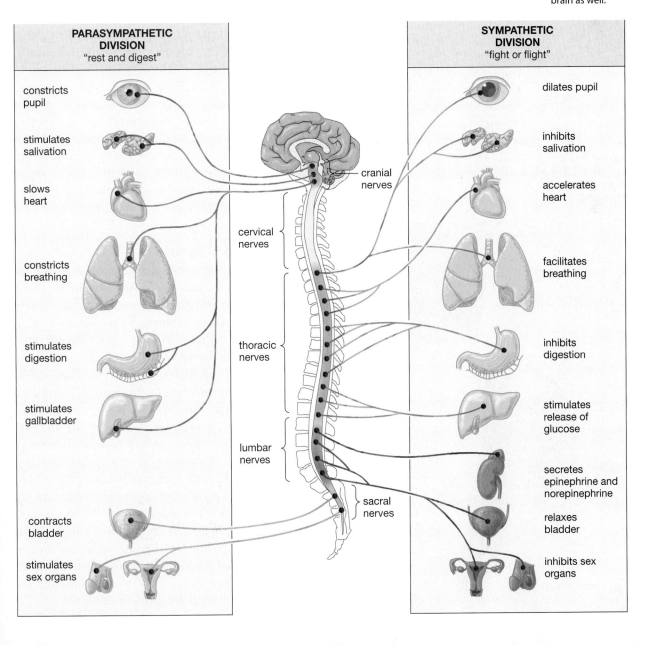

- The **midbrain** processes seeing and hearing information and generates involuntary motor responses related to muscle tone and posture. It also contains centers involved with maintaining consciousness.

- The term **pons** refers to a bridge, and the pons serves this function, connecting, with relay centers in it and axons that run through it, several other structures of the brain. The pons also contains cells involved with the involuntary control of breathing.

- One structure the pons is connected to is the **medulla oblongata**, the segment of the brain that is attached to the spinal cord. The medulla oblongata relays sensory information to the thalamus and brain-stem centers; it also contains major centers concerned with the regulation of autonomic functions, such as heart rate, blood pressure, respiration, and digestive activities.

The brain and spinal cord contain internal cavities filled with the cerebrospinal fluid noted in connection with the spinal cord. In the brain, this fluid provides cushioning for delicate neural structures and support—the brain essentially floats in the cerebrospinal fluid. Though the brain weighs about 1.4 kilograms (about 3 pounds) in air, it weighs only about 50 grams (about 2 ounces) when supported by the cerebrospinal fluid. As noted earlier, the cerebrospinal fluid also transports nutrients, chemical messengers, and waste products.

24.6 The Endocrine System: Hormones and How They Work

Let's now shift the focus from the nervous system to the endocrine system. Though the two systems are separate, they work closely together, as you'll see; and they are alike in function in that both are concerned with coordination and control in the body.

Endocrine system cells secrete chemicals called **hormones** that travel mostly through the circulatory system and affect the activities of *other* cells. Hormonal production and secretion take place to a significant extent within specialized organs called **endocrine glands**. The major endocrine glands are shown in **Figure 24.10**. Some of these glands, such as the pituitary gland, secrete nothing but hormones; others, such as the pancreas, have many other functions in addition to hormone secretion. Not all hormone-producing structures are glands. For example, the heart, kidneys, stomach, liver, small intestine, placenta, and adipose (fatty) tissue also secrete hormones.

Types of Hormones

Most hormones are composed of amino acids, the building blocks of proteins. These amino-acid-based hormones range in size from single amino acid molecules to long chains that form proteins. **Steroid hormones**, on the other hand, are lipid-based hormones, built from cholesterol molecules. They are released by the reproductive organs and the outer layers (or cortex) of the adrenal glands, which sit atop the kidneys. Another group of lipid-based hormones is called *prostaglandins*. Produced from fatty acids by most tissues of the body, these hormones coordinate cellular activities and affect body processes, such as blood clotting, that take place in extracellular fluids. Unlike most hormones, the prostaglandins do not travel throughout the body; thus they affect only nearby cells. For this reason, they are sometimes called *local hormones*.

Figure 24.9
The Brain

a Lateral view of the left surface of the brain, showing the cerebrum and cerebellum.

b The structures of the brain stem, and the endocrine system's pituitary gland. The medulla oblongata connects the brain with the spinal cord.

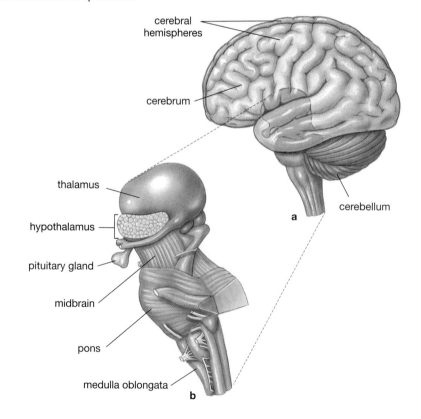

cerebral hemispheres

cerebrum

cerebellum

a

thalamus

hypothalamus

pituitary gland

midbrain

pons

medulla oblongata

b

Actions of Hormones

You've noted that hormones are substances that, having been released by cells in one location, then travel short or long distances through the body (generally through the circulatory system) to affect the activities of cells in *other* locations. Thus, each hormone is said to have specific **target cells**. Most target cells have receptor molecules protruding from their outer membranes; it is these receptors to which specific amino-acid-based hormones attach, as you can see in **Figure 24.11** on page 518. (The steroid-based hormones work by passing through a cell's plasma membrane, after which they bind with a receptor protein inside the cell.) Because a given *type* of target cell may occur in multiple locations throughout the body, a single hormone can simultaneously affect the activities of multiple tissues and organs.

As noted in Chapter 3, proteins have a great diversity of functions in living things, and it is this diversity that enables living things to be so complex. Structural proteins determine the general shape and internal structure of a cell, and the special proteins called *enzymes* speed up cellular activities, in effect enabling the cell's chemical processes to go forward. Hormones alter the workings of a target cell by changing the identities, activities, or quantities of its important enzymes and structural proteins. As an example, the hormone testosterone—found in differing amounts in males and females—stimulates the production of enzymes and proteins in skeletal muscle fibers, thus increasing muscle size and strength.

24.7 How Is Hormone Secretion Controlled?

Hormones are major players in a general process through which the body maintains a stable state. Like people in a large city, the cells, tissues, and organ systems of the body exist together in a shared environment. And just as city dwellers would suffer if the air in their environment became highly polluted or too hot or cold, so cells would be harmed if the *fluid* environment in which they exist became too hot or cold or acidic. The body flourishes in a stable environment, and it has developed an elaborate set of mechanisms to maintain stability. **Homeostasis**, derived from the words *homeo* ("unchanging") and *stasis* ("standing"), refers to an organism's tendency to maintain a relatively stable internal environment in the midst of a constantly changing *external* environment. And it turns out that a single process is almost entirely responsible for maintaining homeostasis in the body.

Negative Feedback

Most people are aware of how a home heating system works. Falling temperature causes a thermostat to turn on a furnace. To look at this another way, there was a **stimulus** (cold air) that brought about a response (furnace operation). The *product* of this response is hot air. When enough of this hot air circulated to raise the temperature, the thermostat sensed this and shut the furnace down. Now let's think of your body, which has—as an example—a certain amount of calcium circulating through it in the bloodstream. When levels of calcium in the blood fall too low (stimulus), some glands in your body, the parathyroid glands, sense this and secrete a hormone that causes your bones to release calcium that they have stored up

**Figure 24.10
The Major Hormone-
Secreting Glands of the Body**
The hypothalamus is part of the nervous system, but releases hormones and plays a central role in endocrine regulation.

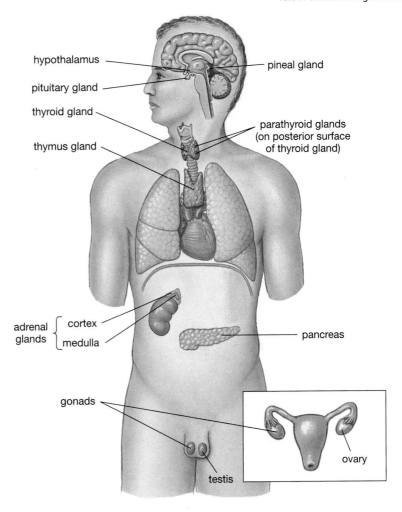

hypothalamus

pituitary gland

thyroid gland

thymus gland

pineal gland

parathyroid glands
(on posterior surface
of thyroid gland)

adrenal glands — cortex, medulla

pancreas

gonads

testis

ovary

(response; see **Figure 24.12**). The product of *this* stimulus-response chain is released calcium, and when enough of it is circulating, the parathyroid glands sense this and stop releasing their calcium-liberating hormone. Thus, with both the thermostat and the parathyroid glands, their responses to a stimulus bring about a decrease in their activity. In other words, their responses feed back on their activity in a negative way—they reduce it. This is **negative feedback**, which can be defined as a process in which the elements that bring about a response have their activity reduced by that response. The calcium example is only one of the countless negative feedback loops that exist in the body.

Positive Feedback

Far less frequently, the body employs **positive feedback**, in which the initial response *magnifies* the stimulus that gave rise to it. For example, the primary stimulus that begins labor and delivery in childbirth is the stretching of the womb, or uterus, by the growing fetus. This stretching stimulates a control center in the mother's brain, and the result is the release of a hormone that brings about muscle contractions in the wall of the uterus. These contractions begin moving the fetus toward the birth canal. The product of the stimulus is thus movement, but this product causes *more* uterine stretching, which brings about additional stimulation of the control center. This kind of cycle, a posi-

tive feedback loop, can be broken only by some external force or process—in this instance, the delivery of the newborn infant. Is it obvious why, in physiology, negative feedback is so much more common than positive feedback? Positive feedback brings about instability, which rarely benefits organisms; negative feedback supports homeostasis, which almost always benefits organisms.

A Hierarchy of Hormonal Control: The Hypothalamus

Though negative feedback is the ultimate control mechanism in the release of hormones, it's accurate to view a good part of the endocrine system as a hierarchy that has a structure in the *brain*, called the *hypothalamus*, sitting on top. In this system, the hypothalamus releases hormones that control a "master" gland, called the *pituitary gland*; the pituitary then releases hormones that control *several other* endocrine glands. These glands then release hormones that go to work on target cells, be they in the kidneys, the muscles, or elsewhere.

The hypothalamus gets input from several sources, which is why it's accurate to view negative feedback as the real controller of hormonal secretion. Sensory nerves feed into the hypothalamus, and it senses bloodstream levels of a number of substances in the body. In response to these inputs, the hypothalamus releases hormones that then control much of the endocrine system's functioning. **Figure 24.13** (page 520) shows you the systems governed by the hypothalamus. Note all that the hypothalamus does. First, it acts as an endocrine organ itself, releasing hormones you'll look at shortly called *ADH* and *oxytocin*. Second, it exercises control, via the nervous system, over the endocrine cells of the inner portion (or medulla) of the adrenal glands, which sit on top of the kidneys. Upon stimulation from the hypothalamus, the adrenal medulla releases into the bloodstream the hormones norepinephrine and epinephrine, the latter of which is better known as *adrenaline*. These are the fight-or-flight hormones that increase heart activity, blood pressure, and free up energy reserves for immediate use. Beyond these activities, the hypothalamus secretes regulatory hormones. These are the special hormones that regulate the endocrine secretions in the pituitary gland—specifically the anterior pituitary gland. There are two classes of regulatory hor-

Figure 24.11
Hormones and Their Target Cells

For a hormone to affect a target cell, that cell must have receptors that can bind the hormone. The binding of hormone to receptor then initiates a change in the target cell's activity. The figure shows an amino-acid-based hormone that affects skeletal muscle tissue. The hormone would not affect, for example, nervous tissue, because only muscle tissue has the appropriate receptors to respond to this hormone.

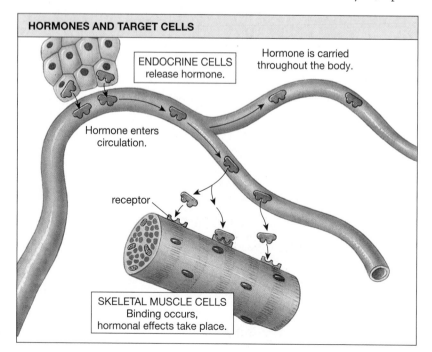

HORMONES AND TARGET CELLS

ENDOCRINE CELLS release hormone.

Hormone is carried throughout the body.

Hormone enters circulation.

receptor

SKELETAL MUSCLE CELLS
Binding occurs, hormonal effects take place.

mones that come from the hypothalamus: releasing hormones (RH), which *stimulate* the production of pituitary hormones; and inhibiting hormones (IH), which *prevent* the production and secretion of pituitary hormones.

The Pituitary Gland Link

The pituitary gland itself is a small, oval structure that is suspended from the brain's hypothalamus by a slender stalk (see Figure 24.9). It is divided into two distinct regions, the anterior and posterior regions, which differ greatly in function.

The pituitary secretes eight different hormones. Two of these are secreted by the posterior pituitary, which receives its hormones directly from the hypothalamus and then stores these hormones and later releases them. The other six are released by the anterior pituitary, and four of these then regulate the production of hormones by other endocrine glands. In sum, the anterior pituitary synthesizes its *own* hormones, and releases them under control of the hypothalamus, while the posterior pituitary is simply a storage-and-release site for hormones synthesized by the hypothalamus. Figure 24.13 shows the hormones released by the pituitary gland and the target organs for them.

The Anterior Pituitary Gland

The endocrine cells of the **anterior pituitary** are surrounded by a network of the tiny blood vessels called *capillaries*. Move up to the hypothalamus, and it too has a capillary bed in its lower portion. Running between these two capillary beds are a small number of blood vessels. Here's how the hypothalamus-pituitary system works. Nerve cells in the hypothalamus synthesize hormones—in vanishingly small amounts—that travel down axons and are released into the hypothalamic capillaries; from there they travel down the linking blood vessels to the anterior pituitary capillaries, and then move into the pituitary gland itself. In the pituitary these hormones—sometimes called *releasing factors*—stimulate pituitary cells to produce hormones. Various *cell types* in the pituitary, each synthesizing a different hormone, are responsible for the number of hormones that this master gland produces. These hormones then circulate out into the bloodstream and head toward their target cells. Here are the six hormones that the anterior pituitary releases:

1. **Thyroid-stimulating hormone (TSH)**, which targets the thyroid gland and triggers the release of thyroid hormones.
2. **Adrenocorticotropic hormone (ACTH)**, which stimulates the release of steroid hormones by the outer cells of the adrenal gland (the adrenal cortex). In particular, ACTH stimulates the cells producing hormones called glucocorticoids.
3. **Follicle-stimulating hormone (FSH)** promotes egg development in women and stimulates the secretion of **estrogens**, steroid hormones produced by cells of the ovary. In men, FSH production supports sperm production and testosterone secretion in the testes.
4. **Luteinizing hormone (LH)** induces ovulation (egg release) in women and promotes the ovarian secretion of estrogens

Figure 24.12
Calcium levels in the bloodstream are regulated in part by the parathyroid glands. When these glands sense that calcium levels are too low (stimulus in the first drawing), they secrete a hormone that causes the bones to release calcium. In the second drawing, the stimulus is high calcium levels, once again sensed by the parathyroid glands. These increased calcium levels cause the parathyroid glands to halt their production of the calcium-liberating hormone.

NEGATIVE FEEDBACK

STIMULUS: low calcium levels in blood

Hormone released

blood Ca++
low   high

bone   parathyroid

RESPONSE: release of calcium into blood

STIMULUS: high calcium levels in blood

Hormone stops being released

blood Ca++
low   high

RESPONSE: release of calcium is shut down

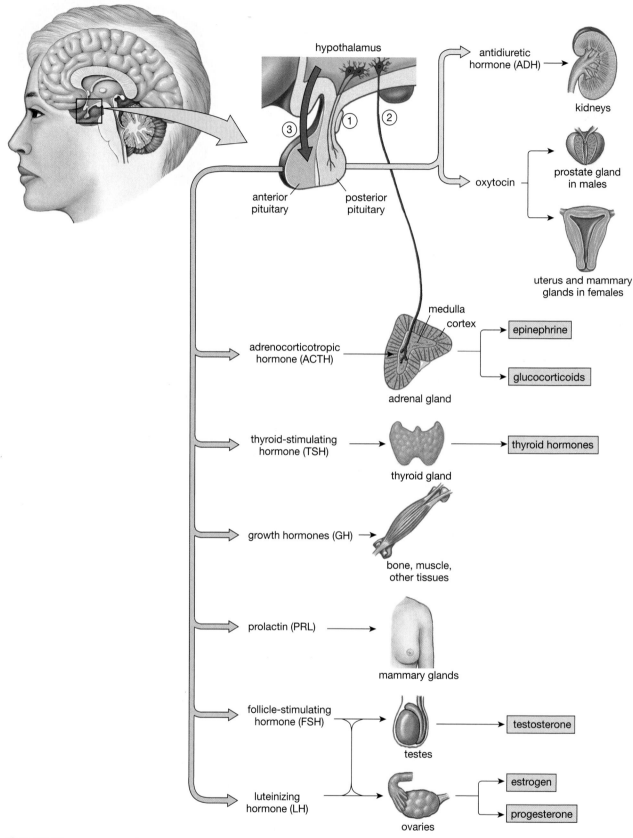

Figure 24.13
Pivotal Player in Hormone Secretion
The structure of the brain called the hypothalamus directs much of the body's endocrine activity.

1. It secretes antidiuretic hormone (ADH) and oxytocin, which are released by the posterior pituitary gland.

2. It exercises direct control over the release of epinephrine and norepinephrine from the medulla of the adrenal glands.

3. It releases regulatory hormones that either stimulate or prevent the production of several hormones by the anterior pituitary gland. The functions of the body's various hormones are summarized in Table 24.1.

and the **progestins** (such as proges-terone) that prepares the body for possi-ble pregnancy. FSH and LH are called gonadotropins, because they regulate the activities of the male and female gonads or sex organs. In males, LH stimulates testosterone production.

5. **Prolactin**, which means "before milk," stimulates the development of the mam-mary glands and their production of milk in the female. Prolactin has no known effects in the human male.

6. **Growth hormone**: Human growth hor-mone (hGH), or somatotropin, stimulates overall body growth through cell growth in a two-step linkage. Human growth hor-mone stimulates the production in the liver of another hormone, somatomedin, which directly spurs on cell growth. Human growth hormone has its greatest effects on muscular and skeletal develop-ment, especially in children.

The Posterior Pituitary Gland

The hormones the **posterior pituitary** stores and releases are produced by two different groups of nerve cells within the hypothala-mus. One group manufactures antidiuretic hormone (ADH), while the other produces oxytocin. These secretions are transported within nerve fibers that extend to the poste-rior pituitary.

Antidiuretic hormone (ADH) is released when there is a rise in the concentration of solutes in the blood or a fall in blood volume or pressure. ADH acts to decrease the amount of water lost at the kidneys. It would most definitely be released if we had to make a long, hot walk with no water.

In women, **oxytocin** (a name that means "quick childbirth") stimulates smooth muscle cells in the uterus and spe-cial cells of the mammary glands. The stim-ulation of uterine muscles by oxytocin helps maintain and complete normal labor and childbirth. After the infant is born, oxytocin triggers the release of milk from the mother's breasts. In the male, oxytocin stimulates the smooth muscle contraction in the walls of the prostate gland. This ac-tion may be important prior to ejaculation. **Table 24.1** (page 522) is a summary of the body's most important hormones and their effects.

24.8 The Immune System: Defending the Body from Invaders

It's time to turn now to the third organ sys-tem reviewed in this chapter, the immune system. It may go without saying that the human body is constantly under attack by microbes. There are viruses, bacteria, fungi, and other single-celled life forms, such as the *Giardia* that is so well-known for contami-nating the water in outdoor areas. The human body has multiple means of defend-ing itself against such invaders. (The military metaphor is irresistible when talking about the immune system because the actions it's involved in are so much like war.)

Two Basic Types of Defense: Nonspecific and Specific

The body's defensive strategies fall into two basic types. First, there are general nonspecific defenses that do not discrimi-nate between one invader and the next. These provide the body with a defensive ca-pability known as *nonspecific resistance*. The body's second main strategy, its specific de-fenses, provides protection against *particular* invaders. A specific defense may protect against infection by one type of bacteria, but ignore other bacteria and viruses. Specific defenses are dependent upon the immune system's ability to develop a kind of memory regarding invaders it has faced. What these defenses produce is a state of long-lasting protection known as specific resistance, which is better known as **immunity**.

Before going into details, here are a couple of terms that you'll see repeatedly in this re-view of the immune system. One is the word **antigen**, which is any foreign substance that elicits an immune system response. A bacter-ial cell is, of course, a foreign substance; but it is likely to be certain *proteins* on the surface of the bacterial cell that set off the human im-mune response. In general, antigens are the molecules that are part of a foreign body—generally proteins or polypeptide chains. Thus, a single invader could "present" more than one antigen to the immune system.

Also note that the immune system is re-liant on several types of cells and proteins to get its work done. Among cells, several vari-eties of **white blood cell** are critical to immune function. You'll be hearing the terms *T cell, B*

Table 24.1
Hormones of the Endocrine System: Their Sources and Effects

Gland/Hormone	Effects
Hypothalamus	
Releasing hormones	Stimulate hormone production in anterior pituitary
Inhibiting hormones	Prevent hormone production in anterior pituitary
Anterior Pituitary	
Thyroid-stimulating hormone (TSH)	Triggers release of thyroid hormones
Adrenocorticotropic hormone (ACTH)	Stimulates adrenal cortex cells to secrete glucocorticoids
Follicle-stimulating hormone (FSH)	*Female:* promotes egg development; stimulates ovaries to produce estrogen *Male:* promotes sperm production
Luteinizing hormone (LH)	*Female:* produces ovulation (egg release); stimulates ovaries to produce estrogen and progesterone *Male:* stimulates testes to produce androgens (e.g., testosterone)
Prolactin (PRL)	Stimulates mammary gland development and production of milk
Growth hormone (GH)	Stimulates cell growth and division
Posterior Pituitary	
Antidiuretic hormone (ADH)	Reduces water loss at kidneys
Oxytocin	Stimulates contraction of smooth muscles of uterus and release of milk in females; stimulates prostate gland contraction in males
Thyroid	
Thyroxine	Stimulates general rate of body metabolism
Calcitonin	Reduces calcium ion levels in blood
Parathyroid	
Parathyroid hormone (PTH)	Increases calcium ion levels in the blood
Thymus	
Thymosins	Stimulate development of white blood cells (lymphocytes) in early life
Adrenal Cortex	
Glucocorticoids	Stimulate glucose synthesis and storage (as glycogen) by liver
Mineralocorticoids	Cause the kidneys to retain sodium ions and water and excrete potassium ions
Androgens	Produced in both sexes, but functions uncertain
Adrenal Medulla	
Epinephrine (E)	Also known as *adrenaline*; stimulates use of glucose and glycogen and release of lipids by adipose tissue; increases heart rate and blood pressure
Norepinephrine (NE)	Also known as *noradrenaline*; effects similar to epinephrine
Pancreas	
Insulin	Decreases glucose levels in blood
Glucagon	Increases glucose levels in blood
Testes	
Testosterone	Promotes production of sperm and development of male sex characteristics
Ovaries	
Estrogens	Support egg development, growth of uterine lining, and development of female sex characteristics
Progesterones	Prepare uterus for arrival of developing embryo and support of further embryonic development
Pineal Gland	
Melatonin	Delays sexual maturation; establishes day/night cycle

cell, neutrophil, and macrophage, among others. All these are different varieties of white blood cell. **Table 24.2** presents some of the cells and proteins involved in the body's defense.

24.9 Nonspecific Defenses of the Immune System

In dividing the immune system into its nonspecific and specific defense functions, it makes sense to start with the nonspecific variety, because it is the body's first line of defense. Here's a list of the body's nonspecific defenses:

- Physical barriers
- Phagocytic cells
- Immunological surveillance
- Interferons
- Complement
- Inflammation
- Fever

Nonspecific defenses generally deny entrance to, or limit the spread of, microorganisms or the harmful substances—called toxins—that they can secrete.

Physical Barriers

To cause trouble, a disease-causing organism—known as a **pathogen**—must enter the body tissues; and that means crossing an *epithelium*, or tissue that covers an exposed body surface. The epithelial covering of the skin,

described in Chapter 23, has multiple layers, a waterproof keratin coating, and a network of tight seams that lock adjacent cells together. These specializations create a very effective barrier that protects underlying tissues.

The exterior surface of the body is also protected by hairs and glandular secretions that provide some protection against mechanical scraping (or "abrasion"), especially on the scalp. Secretions from sebaceous and sweat glands flush the surface of the skin, washing away microorganisms and chemical agents. Some secretions also contain bacteria-killing chemicals and destructive enzymes, called *lysozymes*, which can be found in tears and saliva.

The epithelial tissue lining the digestive, respiratory, urinary, and reproductive tracts is more delicate; but it is equally well defended. Mucus bathes most surfaces of the digestive tract, and the stomach contains a powerful acid that can destroy many potential pathogens. Mucus moves across the lining of the respiratory tract, urine flushes the urinary passageways, and glandular secretions do the same for the reproductive tract.

Phagocytes

As noted earlier, white blood cells play an important role in the immune system. Some of these white blood cells are **phagocytes**, meaning cells that can ingest other cells, parts

Table 24.2
Cells and Proteins Active in the Human Immune System

White Blood Cells	Other Cells Involved in Defense	Proteins Involved in Defense
• Phagocytes (cells that can ingest other cells) —Microphages ("small eaters") Neutrophils Eosinophils —Macrophages ("large eaters"), which can become antigen-presenting cells (APCs) • Lymphocytes —Natural killer cells —B Cells, which specialize into: Plasma cells Memory B cells —T Cells, which specialize into: Helper T (CD4) cells Cytotoxic (killer) T cells Suppressor T cells	Mast cells (connective tissue cells that release histamine and heparin)	• Lysozymes (destructive enzymes found in bodily fluids) • Complement proteins • Antibodies (receptors on surface of B cells; later released by plasma cells derived from B cells) • Interferons (released by immune cells and infected cells; interfere with viruses; stimulate other parts of the immune system) • Interleukins —Interleukin-1 (released by antigen-presenting cells; helps activate helper T cells) —Interleukin-2 (released by helper T cells; spurs activation of B cells and cytotoxic T cells)

of cells, or other particles. In body tissues, phagocytes remove not only parts of the body's own cells that have become cellular debris but also pathogens, by means of phagocytosis or "cell eating." There are two general classes of phagocytic cells. **Microphages**, or "small eaters," are the white blood cells called *neutrophils* and *eosinophils*. Though normally found in the circulating blood, these phagocytic cells can leave the bloodstream and enter tissues subjected to injury or infection. **Macrophages**, or "large eaters," are large cells derived from white blood cells called *monocytes*. Almost every tissue in the body shelters resident or visiting macrophages.

Immunological Surveillance

The immune system generally ignores normal cells in the body's tissues, but attacks and destroys abnormal cells. The constant monitoring of normal tissues is called immunological surveillance. This surveillance primarily involves a type of cell known as a **natural killer (NK) cell**, a special variety of lymphocyte, which is a variety of white blood cell. NK cells are sensitive to the presence of antigens on the body's own cells. These antigens in effect say "abnormal cell membrane here." When NK cells encounter these antigens on a cancer cell or a cell infected with viruses, they secrete proteins that kill the abnormal cell by destroying its cell membrane. So, why do we get cancer? Some cancer cells avoid detection by the system; they can multiply and spread without interference by NK cells.

Interferons

Interferons are small proteins released by activated lymphocytes and macrophages, and by tissue cells infected with viruses. These proteins cause normal, uninfected cells to produce antiviral compounds that interfere with viral replication. They also stimulate macrophages and NK cells into action.

Complement

Complement refers to a group of proteins in the blood that "complements," or supplements, the action of *antibodies*, a group of immune system proteins that will be reviewed shortly. Complement activation enhances phagocytosis, destroys the membranes of invading cells, and promotes the defensive reaction known as *inflammation*.

Inflammation

Inflammation is a localized tissue response to injury. It produces the well-known sensations of swelling, redness, heat, and pain. Inflammation can be produced by any stimulus that kills cells or damages loose connective tissue. When stimulated by local tissue damage, special connective tissue cells called **mast cells** release—into the interstitial fluid—special substances called *histamine* and *heparin* that initiate the process of inflammation. **Figure 24.14** follows the different actions that occur during inflammation of the skin. These actions serve to slow the spread of pathogens at sites away from the injury, to prevent the entry of additional pathogens, and to set into motion a wide range of defenses that can overcome the pathogens and make permanent tissue repairs. The repair process is called *regeneration*. Note that one of the later actions in the inflammation process is "activation of specific defenses." You'll learn more about these defenses shortly; for now, just note that the immune system almost always works together as an integrated whole. Though inflammation is a nonspecific defense, a skin puncture that causes inflammation will activate specific defenses as well.

Fever

A **fever** is a continued body temperature greater than 37.2°C (99°F). The hypothalamus contains groups of nerve cells that regulate body temperature and act as the body's thermostat. Macrophages, pathogens, and bacterial toxins, among other things, can reset the thermostat and cause a rise in body temperature. Within limits, a fever may be beneficial, because high body temperatures speed up the activities of the immune system. However, high fevers (over 40°C, or 104°F) can disrupt or damage many different organ systems.

24.10 Specific Defenses of the Immune System

Interferons and natural killer cells do not discriminate between one invader and the next. However, **immunity** or *specific resistance* is the resistance to injuries and diseases caused by *specific* foreign chemical compounds and pathogens. In the inflammation response, some parts of the immune system took actions that they would have employed against any invader. But while a skin puncture may be a general type of injury, *specific* viruses, or bacteria, or fungi would take advantage of it. Thus, the body's response to such an injury involves not only nonspecific immunity, but specific immunity. Specific immunity comes in two varieties, innate and acquired. Innate immunity is inherited; it is present at birth and has no relation to previous exposure to the antigen involved. Such inborn immunity

is what protects us from diseases that may infect farm animals or pets. For example, humans are not subject to the same diseases as goldfish, and vice versa. In contrast, an acquired immunity is an acquired resistance to infection. It may develop in either of two ways, passively or actively, as you'll now see. The relationships between innate immunity and the different forms of acquired immunity are diagrammed in **Figure 24.15**.

**Figure 24.14
The Inflammation Process**

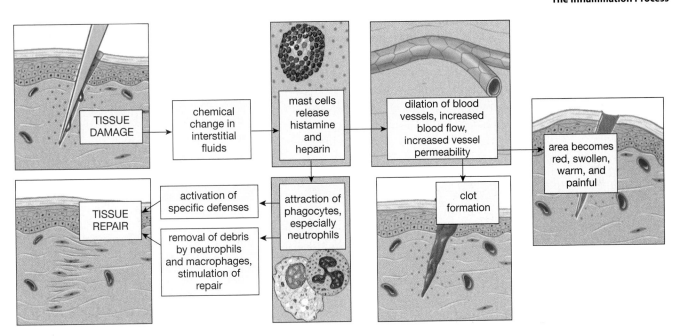

**Figure 24.15
Types of Immunity: Acquired and Innate**

Passively acquired immunity is acquired by the administration of disease-fighting substances produced by another individual. This occurs naturally when substances produced by the mother provide protection against infections that could attack a fetus. Such substances may also be *administered* to fight or prevent infection. For example, antivenoms are used to treat bites or stings suffered by a person, while immunoglobulin (antibody) shots are given to increase resistance to possible infection. Importantly, these means of immunity provide only short- or medium-term resistance to harm—lasting as much as several years—because disease-fighting compounds produced *outside* our own bodies are eventually degraded in our system. Truly long-lasting immunity is something that our bodies develop with their own resources.

Actively acquired immunity is developed as a result of natural exposure to an antigen in the environment, or from deliberate exposure to an antigen. It is the means by which we get long-lasting protection against invaders. Such immunity normally begins to develop after birth, and it is continually up-dated and broadened as individuals encounter pathogens or other antigens that they have not been exposed to before. One form of actively acquired immunity is produced artificially by the immunization shots given to children and adults to protect them against specific diseases, such as smallpox. Let's now begin to look at how actively acquired immunity works.

24.11 Antibody-Mediated and Cell-Mediated Immunity

A key to understanding the body's actively acquired immunity lies in what takes place with vaccination against a disease like polio. In vaccination, a person is not being treated with a long-lasting substance that is toxic to viruses or bacteria. Rather, what is being injected or swallowed is an antigen—at least part of the disease-causing organism itself. So why is it that, having been inoculated in this way, a person is unlikely to contract polio? The answer is that as far as the body's immune system is concerned, the person *did* contract polio when he or she was inoculated. The immune system was presented with polio antigens, and it then proceeded to mount an attack on this invader. What "getting polio" means, then, is not having the poliovirus come into our bodies; it means being *harmed* by the virus before the immune system can clear it from our system. This is what can happen when the full-blown virus invades our system by natural means before we are inoculated against it.

The critical thing is that once we are inoculated with a vaccine, the poliovirus is nearly powerless to harm us. Not powerless to come into our bodies, mind you; powerless to harm us because the body is now *ready*. As a product of the original "attack" (the vaccination), the immune system is put into a permanent heightened state of readiness with respect to the poliovirus. What vaccines do, then, is elicit readiness. But how does the immune system stay ready for specific invaders? For that matter, how does it work to defeat them when they come? That is the subject of this section on immune system coverage.

There are two major components to the body's specific resistance system. The first is called **antibody-mediated immunity**. The second is called **cell-mediated immunity**.

**Figure 24.16
Development of T Cells and B Cells**
Key players in the body's specific defense system are T cells and B cells, both of which start out as a type of white blood cell, called a *lymphocyte*, in bone marrow. Some of these lymphocytes migrate to the thymus gland, where they differentiate into T cells (thymus cells). Others that remain in the bone marrow develop into B cells. Most T and B cells then migrate to lymphoid tissues, such as lymph nodes and the spleen, where they will serve their immune function.

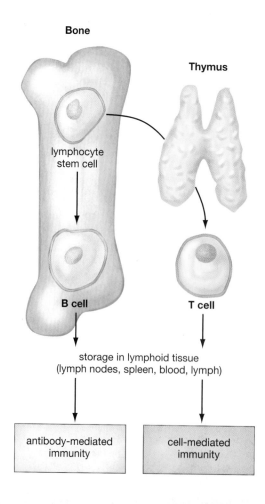

Bone

Thymus

lymphocyte stem cell

B cell

T cell

storage in lymphoid tissue
(lymph nodes, spleen, blood, lymph)

antibody-mediated immunity

cell-mediated immunity

Both systems begin in the same place, in a sense, in that the white blood cell called the *lymphocyte* is the basis for cells that are critical to both systems. All white blood cells originate in the bone marrow, and lymphocytes are no exception. Some lymphocytes, however, migrate to the body's thymus gland (located just above the heart), which is where these lymphocytes differentiate into the main cells of *cell-mediated* immunity, the **T-lymphocyte cells** or just **T cells** (with the T standing for "thymus" cell).

The Body's Lymphatic System

Meanwhile, lymphocytes that remain in the bone marrow can develop into the central cells of *antibody*-mediated immunity, the **B-lymphocyte cells** or **B cells**. Mature T and B cells are most often found in the body's lymphatic system organs—the spleen and the lymph nodes, for example—where they are prepared to latch onto invaders that might come into the system. The development of T and B cells is tracked in **Figure 24.16**. The body's **lymphatic system** is laid out for you in **Figure 24.17**. Think of it as a system of ves-

sels that picks up fluids from outside the body's cells (interstitial fluid) and eventually channels this fluid into the body's circulatory system. While *in* the lymphatic system these fluids pass through lymphoid organs, which are packed with immune system cells. This is why, when you get sick, your "lymph glands" are swollen—they are filled with an abnormally large number of cells that are busy fighting off the infection.

From this point forward, the antibody-mediated and cell-mediated immune operations are treated separately. The two systems work together to a degree, but for now let's look at them in isolation. After separate tours of the two systems, you'll take a step back and see how the immune system functions as a whole. Let's start now with antibody-mediated immunity.

24.12 Antibody-Mediated Immunity in Detail

As you might imagine, a key player in antibody-mediated immunity is the **antibody**, which can be defined as a circulating immune

MediaLab Link
Antibodies, T cells, B cells, see p. 538.

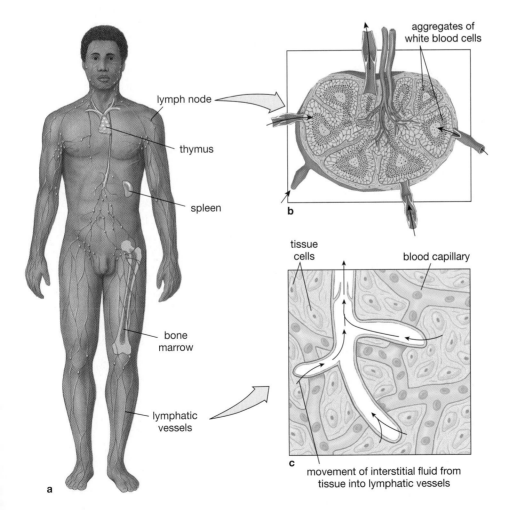

Figure 24.17
The Human Lymphatic System

a Fluid moving through lymphatic vessels passes through lymphatic organs, such as lymph nodes or the spleen, where the fluid is inspected by an array of immune system cells.

b Detail of fluid moving through a lymph node.

c Detail of interstitial fluid moving into lymphatic vessels.

system protein that binds to a particular antigen. From this definition, it may be apparent that antibodies and antigens are closely linked. (So closely linked that one is named for the other; the word *antigen* is short for "antibody generating.") To a first approximation, a given antibody will bind to a specific antigen and to no other antigen. In their binding capacity, you will see antibodies in two guises: first as *receptors* on the surface of B cells, extending in a generalized Y-shape from the cells' outer membranes; second as *free-standing* molecules produced by B cells, moving through the bloodstream. In their receptor identity, antibodies are also known as antigen receptors.

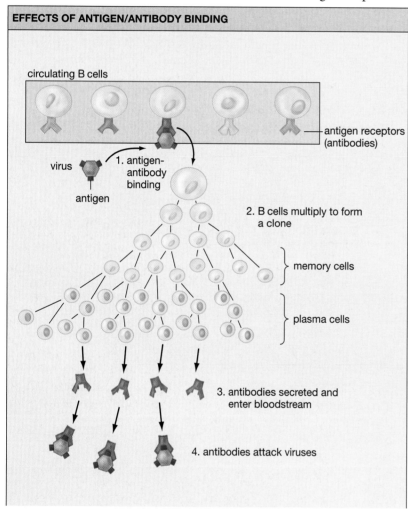

EFFECTS OF ANTIGEN/ANTIBODY BINDING

circulating B cells

antigen receptors
(antibodies)

virus

1. antigen-
antibody
binding

antigen

2. B cells multiply to form
a clone

memory cells

plasma cells

3. antibodies secreted and
enter bloodstream

4. antibodies attack viruses

Figure 24.18
B Cells and Antibody-Mediated Immunity

1. The antigen-receptors (antibodies) of a particular B cell allow it to bind with a virus displaying a specific antigen.

2. This action spurs the B cell to produce a huge number of clones of itself by dividing rapidly. These B cells differentiate into two types: memory cells and plasma cells. Memory cells remain in the body long after the infection has passed, ready to quickly produce more B cells of the same type, should the same invader come again.

3. Plasma cells secrete free-standing antibodies, which enter the bloodstream.

4. Antibodies produced by the plasma cells attack the invader.

The Fantastic Diversity of Antibodies

It's common knowledge that there is no such thing as the common cold; rather, there are countless *varieties* of common cold, each caused by a different virus. Then there are all the other potential invaders, such as bacteria, fungi, and so forth. How can the immune system cope with such a variety of foes? One of its key strengths is that it produces millions of different kinds of B cells, each having a type of antigen receptor on it that is specific to only one antigen. Look at one B cell and it will have antigen receptors that latch onto an antigen only on *this* virus; look at a *different* B cell and it has a *different* set of receptors. This mind-boggling complexity is made possible by DNA arranging that takes place back in the bone marrow, when the precursors of B cells are being formed. The result is millions of varieties of B cells, each expressing its own specific antigen receptors (antibodies).

A good many of these "soldiers" will never see battle for the simple reason that the body is never invaded by a pathogen displaying their type of antigen. Some B-cell types are called on to fight, however; and when they do, they have the marvelous capacity not only to make more soldiers like themselves, but to create a permanent set of sentries for a given invader, as you'll now see.

The Cloning of Selected Cells in Antibody-Mediated Immunity

Lying in readiness in the lymph nodes or circulating through the bloodstream, a B cell encounters, say, a virus, one that presents an antigen that can be bound by this B cell's antigen receptors (meaning antibodies). If you look at **Figure 24.18**, you can see what happens next. This very binding causes the B cell to start dividing very rapidly. Each new B cell of this type gives rise to more identical B cells, creating a selected "clone," or a huge number of identical cells of this type. Such numbers are necessary because the *virus* will be multiplying rapidly as well.

Fighters and Sentries: The Differentiation of B Cells

After a time, these B cells then differentiate into two types of cells. One variety is called a **plasma cell**, which exists in the lymph nodes and is specialized to produce anti-bodies. These antibodies are not the cell-surface receptors you've been looking at, however. Instead, they are free-standing antibodies that leave the plasma cells by the millions—they are *exported* from the B cells—after which they move through the bloodstream and fight the invader directly. The other type of B cell is called the **memory B cell**, and it is the body's permanent sentry. Long after the initial war with this invader is over, this specific variety of memory B cell remains in the body. Should this invader come in a *second* time, these memory cells will be ready to produce more plasma cells to mount a very quick defense. This is why almost no one gets chicken pox twice; it is why being inoculated with antigens from the poliovirus protects us permanently from the damage caused by this crippler.

If you look at **Figure 24.19**, you can see the typical time-course of exposure to antigen and production of antibodies. What is known as the "primary immune response" has reached a substantial level seven days out from first exposure. Look at the second exposure, however, to see what a difference this first exposure has made. Now there is a much greater response in a much shorter time.

Action of the Antibodies

But what is it that the circulating antibodies do to combat the invader? First, because they are binding to the invader's antigens, the antibodies can sometimes prevent the invader from attaching to anything *else*, which stops its spread. They can also cause an "agglutination" of antibodies and antigens in an antigen-antibody "complex" that renders the invaders inactive. Finally, the antibody binding can trigger the activities of another immune system component, the complement proteins you looked at earlier. Complement can bind to the antibody at the same time the antibody is bound to the antigen. Critically, complement binds to an antibody *only* when the antibody is bound to an antigen. This dual binding sets into motion the production of a cascade of complement proteins, the final members of which can poke holes in the cell walls of bacteria.

24.13 Cell-Mediated Immunity in Detail

It may go without saying at this point that antibody-mediated immunity is quite spectacular, but it also turns out to be limited in that it works strictly on foreign organisms and molecules. The problem with this is that many viruses (as an example) are successful in invading the body's *own cells*, which then become part of the problem. To deal with this threat, the body has developed another line of acquired immunity, the cell-mediated immunity mentioned earlier.

You may recall that the central player in cell-mediated immunity is the *T*-lymphocyte or *T* cell. T cells come in several varieties, three of which are **cytoxic T cells** (or **killer T cells**), **helper T cells**, and **suppressor T cells**. Let's now go through the central actions of the cell-mediated immunity system, taking as a starting point a body that has been infected with a virus.

Macrophages Bearing Invaders: Antigen-Presenting Cells

A given body cell that has been infected with a common virus puts out one of the interferon proteins, noted earlier in the discussion of nonspecific immunity. This protein acts as a chemical signal saying, in effect, "infected cell here," and the result is an attack by the natural killer and macrophage cells noted earlier. The macrophages actually serve two functions at this point. First, they ingest infected cells by engulfing them; second, they then display *fragments* of proteins from the invading virus on their own surface. You won't be surprised to learn that these fragments will now serve as antigens. When macrophages play this role, they are sometimes referred to as **antigen-presenting cells** or **APCs**.

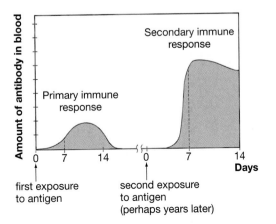

first exposure
to antigen

second exposure
to antigen
(perhaps years later)

Figure 24.19
Prepared for an Invasion
The memory cells produced by the body during a first attack by an invader allow it to mount a faster, more vigorous defense should the same invader attack a second time.

① Macrophage ingests virus

② Macrophage, now an APC, presents viral fragments as antigens

Macrophage

Antigen-presenting cell

③ Helper T cell binds to APC

④ This binding activates helper T cell, stimulating production of other immune cells

Helper T cell

Antigen-presenting cell

APC

cytotoxic T cell

suppressor T cell

cell-mediated immunity

B cell

antibody-mediated immunity

Figure 24.20
T Cells and Cell-Mediated Immunity

1. A cell eating macrophage ingests a virus.

2. The virus is displayed as an antigen within one of the macrophage's receptors, making the macrophage an antigen-presenting cell, or APC.

3. A helper T cell binds to the antigen-APC binding site.

4. This binding produces activated T cells that put several other processes in motion: the production of clones of helper and cytotoxic T cells specific to the invader, the production of suppressor T cells, and the production of B-cell clones.

Figure 24.21
An Overview of Cell-Mediated and Antibody-Mediated Defenses and Their Link with Nonspecific Defense

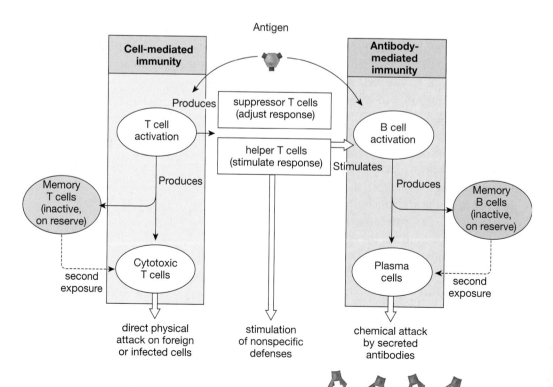

If you look at **Figure 24.20**, you can see how the next interaction plays out. The helper T cell now locks onto the APC; but as usual with the immune system, there is great specificity in this. The APC is displaying the viral protein fragment within one of its own membrane proteins, and the helper T cell cannot bind with the APC unless this dual condition is met: A viral antigen must be bound up with what we can call a self-protein of the APC. This makes an antigen/self-binding site that is unique in two ways: Only this specific invader would prompt the site's creation, and only one variety of the body's tens of millions of varieties of helper T cells would bind with it. In short, we once again have a tremendously specific immunity, this time mediated by cells. After binding with the helper T cell, the APC also puts out a protein, interleukin-1, that helps activate helper T cells.

Activated Helper T Cells Are Central to the Immune Response

The "activated" helper T cells now secrete a protein, interleukin-2, that puts several other processes into motion. First, this secretion spurs production of clones of helper and cytotoxic T cells. These clones include not only active helper and cytotoxic T cells but also *memory* cells, for future invasions. Second, the interleukin-2 also stimulates production of suppressor T cells, which multiply more slowly than do cytotoxic T cells. Ultimately the suppressor cells will inhibit the production of cytotoxic T cells, once the immediate infection battle has passed. Third, the interleukin-2 helps activate B-cell immunity (by stimulating clone development).

By now you may be saying: So where's the invader-fighting in all this? The answer is in the clone of cytotoxic T cells, which are called "killer" T cells for a good reason. Using their own special receptors, they latch onto infected cells in the body and puncture their outer membranes, causing them to "lyse" or burst. Note that cytotoxic cells are not binding with the APCs, as the helper cells did. These killers are binding with the body's own infected cells. From this, it may be apparent why this is called *cell-mediated* immunity. In the end, it is a housecleaning involving not the antibodies described before, but several sets of cells.

Taking a step back from cell- and antibody-mediated immunity, in **Figure 24.21** you can see how the two systems work together, along with nonspecific defenses.

24.14 Allergies and Autoimmune Disorders

Clearly, a critical factor in cell-mediated immunity is the body's ability to distinguish between "self" (its own uninfected cells) and "non-self" (the infected cells). Otherwise, the immune system would attack the body's own healthy cells. Despite safeguards built into the system, the immune system does attack either harmless substances or the body itself with disturbing frequency.

T cells learn to distinguish self from non-self during the embryonic or newborn phases of life. Developing during this period in the thymus, some T cells may never "learn" to distinguish correctly between self and non-self. Conversely, it may be that an invading virus is able to incorporate some of the body's own proteins into itself, meaning that "self" would now be read as "non-self." For whatever reason, the body does sometimes attack itself—with devastating consequences. If you look at **Figure 24.22**, you can see the results of the disease known as *rheumatoid arthritis*, which may be a result of helper T cells attacking the wrong targets. There is also multiple sclerosis, in which the immune system starts attacking the axon-covering myelin you read about when looking at the nervous system.

Figure 24.22
When the Body Attacks Itself
Rheumatoid arthritis has so disfigured this person's hands that simply signing a greeting card has become a difficult act. Rheumatoid arthritis is an autoimmune condition.

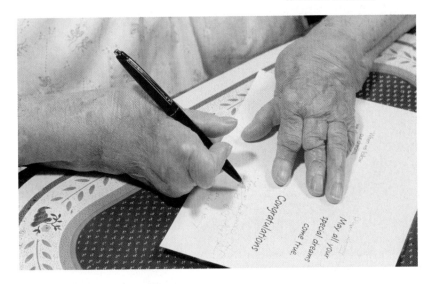

Beyond this there are **allergies**, in which a foreign substance (termed an **allergen**) causes the body to start releasing the histamine and heparin noted earlier in connection with inflammation. Allergies can run from life-threatening (asthma) to merely annoying (sneezing and sniffling). Allergens need not be living things, but almost always are at least derived from living things, because it is protein molecules that are serving as the antigens setting off the immune attack. Thus pollen, dust mites, foods, and fur are common allergens. Because allergens commonly elicit histamine release, it makes sense that a common treatment for allergies is the class of substances called *antihistamines*.

24.15 AIDS: Attacking the Defenders

The disease called acquired immune deficiency syndrome, or **AIDS**, is devastating precisely because it attacks the very immune system that normally protects the body. Within the immune system, it is the helper T cells noted earlier that are decimated by the human immunodeficiency virus, or **HIV**, which is the cause of AIDS.

HIV works by attaching itself to a helper T cell, injecting its genetic material into it, and then hijacking the T cell's own protein-making machinery. The cell then begins turning out *HIV* proteins; eventually, complete virus particles are assembled within the T cell. These then leave the cell, destroying it in the process. They then latch onto other helper T cells, infecting them. If you look at **Figure**

MediaLab Link
AIDS, HIV, see p. 538.

Figure 24.23
The Invaders Emerge
Artificially colored AIDS virus particles bursting forth from a stricken helper T cell. The cell will be killed by this budding of virus particles from it.

24.23, you can see a picture of AIDS virus particles emerging from an infected T cell.

You saw earlier that helper T cells are critical not only in getting cell-mediated (T-cell) immunity going, but in stimulating the antibody-mediated system as well (B-cell immunity). As a result, when the helper T cell "count" drops too low, the AIDS victim falls prey to a host of afflictions that the immune system normally would make quick work of. A fungal infection that most people wouldn't even realize they had can become life-threatening to an AIDS sufferer. Many cells have so-called cell identity markers on their surface—proteins that serve to identify what type of cell they are. Helper T cells have what are known as CD4 glycoproteins on their surface, and this is the nomenclature that you will hear employed in connection with AIDS. A critical question becomes: What is the patient's CD4 count? From your review of the immune system, you probably can see why this question would be so important.

New Hope for AIDS Victims: Protease Inhibitors

In March 1996, the U.S. Food and Drug Administration approved a new class of drugs aimed at treating AIDS. Called protease inhibitors, the drugs work by blocking the action of an HIV enzyme called *protease*, which the virus needs to carry out its replication. Used in "cocktail" combinations with other drugs, the protease inhibitors have thus far had a dramatic effect in the United States. The number of AIDS-related deaths in the United States dropped from about 21,000 in the first half of 1996 to about 12,000 in the first half of 1997. AIDS victims treated with the new drug combinations show both increases in their CD4 counts and dramatic decreases in the numbers of HIV particles in their systems. The new drugs are not a cure, and it remains to be seen what their long-term effects will be; but thus far they have been good news indeed in connection with this scourge.

On to Transport and Exchange

Having looked at the amazing capabilities of the nervous, endocrine, and immune systems, next you'll look at four other systems that are just as important in maintaining a healthy body. Chapter 25 is concerned with blood, breathing, digestion, and elimination.

Chapter Review

Summary

24.1 Overview of the Nervous System

- The key functional cells of the nervous system are neurons, which are supported by neuroglial cells.

- The two major divisions of the human nervous system are the central nervous system (CNS), consisting of the brain and spinal cord; and the peripheral nervous system (PNS), which includes all the neural tissue outside the CNS. The PNS has an afferent division, which brings sensory information to the CNS; and an efferent division, which carries motor (action) commands to muscles and glands—the body's effectors. Within the PNS efferent division are two subsystems: the somatic nervous system, which provides voluntary control over skeletal muscles; and the autonomic nervous system, which provides involuntary regulation of smooth muscle, cardiac muscle, and glands. The autonomic system is further divided into the sympathetic division, which generally has stimulatory effects; and the parasympathetic division, which generally has relaxing effects.

- There are three types of neurons. Sensory neurons sense conditions inside and outside the body and convey information to neurons inside the CNS. Motor neurons carry instructions from the CNS to other tissues, organs, or organ systems. Interneurons, located entirely within the CNS, interconnect other neurons.

- Neurons have extensions called *dendrites* that receive signals coming to the neuron cell body, and a single large extension, called an *axon*, that carries signals from the cell body. At the tips of an axon there may be a structure called a *synaptic terminal* that is part of a synapse—a site where a neuron communicates with another cell.

- Neuroglia have no information processing ability, but serve to protect neurons and to facilitate communication by them.

- A nerve is a bundle of axons in the PNS that transmits information to or from the CNS.

24.2 How Does Nervous System Communication Work?

- An electrical charge, called a *membrane potential*, exists across the plasma membrane of neurons. It represents a form of stored energy that is used in transmitting nerve signals. The basis of nerve signal transmission is that the plasma membrane's permeability to charged substances can change upon stimulation.

- Transmission along the entire axon of a neuron takes place through an action potential, in which a change in membrane permeability at one point along the axon changes the permeability at an adjacent point. The action potential moves in one direction along the axon, away from the initial point of stimulation.

- A nerve signal moves from one neuron to another (or to a muscle or gland cell) across a synapse, a site of nerve cell communication that includes a "sending" presynaptic neuron, a "receiving" postsynaptic neuron or effector cell, and a synaptic cleft, which is the space between the two cells. Communication takes place via a chemical, called a *neurotransmitter*, released into the synaptic cleft by the presynaptic neuron.

24.3 The Spinal Cord

- The spinal cord can receive input from sensory neurons and it can direct motor activities in response; it is also the major pathway for impulses going to and from the brain.

- Spinal cord motor neurons, which are sending motor commands to various effectors in the PNS, have cell bodies that lie within the gray matter of the spinal cord. The axons of the motor neurons leave the spinal cord through ventral roots. Sensory neurons, which transmit information to the spinal cord, have their cell bodies outside the spinal cord, in the dorsal root ganglia.

- Sensory neurons are stimulated via sensory receptors, which may be dendrites that feed into an axon of a sensory neuron, or may be specialized cells that communicate with the axon across a synapse.

Key Terms

action potential 511
actively acquired immunity 526
adrenocorticotropic hormone (ACTH) 519
afferent division 507
AIDS 532
allergen 532
allergy 532
anterior pituitary 519
antibody 527
antibody-mediated immunity 526
antidiuretic hormone (ADH) 521
antigen 521
antigen-presenting cell (APC) 529
autonomic nervous system 513
axon 509
blood-brain barrier 509
B-lymphocyte (B cell) 527
cell-mediated immunity 526
central nervous system (CNS) 507
cerebellum 515
cerebral cortex 514
cerebral hemisphere 515
cerebrospinal fluid 513
cerebrum 515
complement 524
cytotoxic T cells 529
dendrite 509
dorsal root ganglion 513
effectors 507
efferent division 507
endocrine gland 516
estrogen 519
fever 524
follicle-stimulating hormone (FSH) 519
ganglion 513
gray matter 509
growth hormone 521

Upon stimulation, a message goes from sensory receptor, through the axon of the sensory neuron, through a dorsal root ganglion and the dorsal root, and finally to a site within the spinal cord where the sensory neuron links either with a motor neuron or with an interneuron.

- Reflexes are automatic nervous system responses, triggered by specific stimuli, that help us avoid danger or preserve homeostasis.

- The neural wiring of a single reflex, called a *reflex arc*, begins with a sensory receptor, runs through the spinal cord, and ends at an effector—such as a muscle or gland.

24.4   The Autonomic Nervous System

- The division of the PNS over which we do not have conscious control, the autonomic nervous system, has sympathetic and parasympathetic divisions. The sympathetic division is often called the *fight-or-flight* system, because it generally activates bodily functions. The parasympathetic division is often called the *rest-and-digest* system, because it conserves energy and promotes sedentary activities. Most bodily organs receive input from both systems.

24.5   The Human Brain

- The brain contains almost 98 percent of the human body's neural tissue. It consists of gray matter, made up of nerve cell bodies and existing primarily as a layer at the surface of the brain called the *cerebral cortex*; and of white matter, made up of bundles of nerve cell axons.

- There are six major regions in the adult brain: the cerebrum, thalamus and hypothalamus, midbrain, pons, cerebellum, and medulla oblongata. The cerebrum is divided into right and left cerebral hemispheres and is the seat of conscious thought processes, intellectual functions, and memory storage and retrieval.

- The brain stem is all the structures of the brain except the cerebrum and cerebellum. The four primary structures of the brain stem are active in controlling involuntary bodily activities, in relaying information, and in processing sensory information.

24.6   The Endocrine System: Hormones and How They Work

- Endocrine system cells secrete chemicals called *hormones* that travel mostly through the circulatory system and that affect the activities of other cells. Hormonal production and secretion takes place to a significant extent within endocrine glands.

- Most hormones are amino-acid-based hormones; but there are also steroid hormones, built from cholesterol molecules. Some hormones, called *local hormones*, do not travel throughout the body, but affect only nearby cells.

- Each hormone has specific target cells upon which it works. Amino acid hormones link to their target cells via receptors that protrude from the target cells' outer membranes. This binding produces changes in the target cell's production of proteins. Steroid hormones pass through a cell's plasma membrane, after which they bind with a receptor protein inside the cell.

24.7   How Is Hormone Secretion Controlled?

- *Homeostasis* refers to the body's tendency to maintain a relatively stable internal environment in the midst of a constantly changing external environment.

- Homeostasis is maintained primarily through negative feedback, which is a process in which the elements that bring about a response have their activity reduced by that response.

- Infrequently, the body employs positive feedback, in which the initial response magnifies the stimulus that gave rise to it. Positive feedback promotes instability, while negative feedback promotes stability.

- The endocrine system's release of hormones is fundamentally controlled by negative feedback; but a hierarchy of control exists for much of the system, in which the brain structure called the *hypothalamus* (1) acts as an endocrine organ; producing its own hormones, which are released by the posterior pituitary gland (2) exercises control, via the nervous system, over the hormonal activity of the medulla of the adrenal glands, and (3) releases regulatory hormones that control secretion of hormones by the anterior pituitary gland. The hypothalamus is the central link between the body's nervous and endocrine systems. The pituitary is regarded as the master gland of the endocrine system.

24.8 The Immune System: Defending the Body from Invaders

- The body's immune system protects it against invading organisms and toxic substances.

- The body has two primary defensive strategies: nonspecific defenses that do not discriminate between one invader and the next, and specific defenses that provide protection against particular invaders. Specific defenses can produce immunity, defined as a state of long-lasting protection against particular invaders.

- An antigen is any foreign substance—often a protein from an invading organism—that elicits an immune system response.

- Several varieties of white blood cell are important to immune function; these include T cells, B cells, and macrophages.

24.9 Nonspecific Defenses of the Immune System

- The body's nonspecific defenses include physical barriers, phagocytic ("cell-eating") cells, immunological surveillance, interferon, complement, inflammation, and fever.

24.10 Specific Defenses of the Immune System

- Specific immunity can be either innate or acquired. Innate immunity is present at birth and has no relation to exposure to an antigen. Acquired immunity can develop either actively or passively. Passively acquired immunity is achieved through the administration of disease-fighting substances produced by another individual. Actively acquired immunity develops as a result of natural or deliberate exposure to an antigen. One form of actively acquired immunity is produced artificially by immunization shots.

24.11 Antibody-Mediated and Cell-Mediated Immunity

- There are two major components to actively acquired immunity: antibody-mediated immunity and cell-mediated immunity.

- White blood cells called *lymphocytes* are fundamental to both types of resistance, because lymphocytes develop into both T cells (which are central to cell-mediated immunity) and B cells (which are central to antibody-mediated immunity).

- Mature T and B cells are most often found in organs of the body's lymphatic system, which is a system of vessels that captures interstitial fluids, subjects these fluids to scrutiny by immune system cells, and returns the fluids to the circulatory system.

24.12 Antibody-Mediated Immunity in Detail

- An antibody is a circulating immune system protein that binds to a particular antigen. Antibodies exist as (1) receptors on the surface of B cells that bind to specific antigens, and (2) as free-standing invader-fighting proteins that are secreted by B cells and move through the bloodstream.

- Antibody-mediated immunity begins with the binding of an antigen to a B-cell receptor (antibody). This binding produces a cascade of effects, including the production of a B-cell clone—a huge number of identical copies of the B cell that bound to the antigen.

- The B cells produced differentiate into plasma cells, which produce infection-fighting antibodies, and memory B cells, which remain in the body long after the initial infection from the specific invader has passed, providing permanent immunity against it.

24.13 Cell-Mediated Immunity in Detail

- Cell-mediated immunity provides protection in instances in which the body's own cells are harmful after having become infected by an invader.

- The central player in cell-mediated immunity is the T cell, which comes in three varieties: cytotoxic (or killer) T cells, helper T cells, and suppressor T cells.

• An infected cell secretes an interferon protein that elicits an attack by natural killer and macrophage cells. Macrophages activated in this way ingest infected cells and display fragments of proteins from the invader, bound with macrophage receptors, on the macrophage's surface. Macrophages that play this role are known as *antigen-presenting cells*, or APCs.

• Helper T cells, binding with the combined antigen/APC binding site, activate the T cells, bringing about a cascade of activities that includes the production of helper, cytotoxic, and suppressor T cells, and the stimulation of B-cell clone development. The T-cell activity produces memory T cells for future invasions by the specific pathogen. Cytotoxic T cells bind with infected cells and puncture their outer membranes.

24.14 Allergies and Autoimmune Disorders

• The immune system attacks the body's own healthy tissues with disturbing frequency, producing such conditions as rheumatoid arthritis and multiple sclerosis.

• Allergies often are an immune system disorder in which a foreign substance, termed an *allergen*, causes the body to release infection-fighting histamine and heparin.

24.15 AIDS: Attacking the Defenders

• AIDS is devastating because it attacks the very immune system that normally protects the body.

• The cause of AIDS, the human immunodeficiency virus or HIV, attaches itself to helper T cells, making copies of itself inside and destroying the helper T cells in the process. Because helper T cells are central to specific immunity, the result can be a great vulnerability to common pathogens.

Understanding the Basics

Multiple-Choice Questions

1. The peripheral nervous system (PNS) is made up of
   a. the brain and spinal cord
   b. the brain and cranial nerves
   c. all of the nervous tissue outside of the CNS
   d. all of the nervous tissue inside the CNS
   e. all of the above

2. Which of the following is *not* a response of the sympathetic division of the ANS?
   a. dilation of the airway passages
   b. dilation of the pupils
   c. increased heart rate
   d. increased digestive activity
   e. All of the above are responses to the sympathetic division of the ANS.

3. Conscious thought processes and all intellectual functions originate in the
   a. cerebrum
   b. cerebellum
   c. thalamus
   d. medulla oblongata
   e. hypothalamus

4. The regulation of autonomic function, such as heart rate and blood pressure, occurs in the
   a. cerebrum   d. pons
   b. cerebellum   e. medulla oblongata
   c. thalamus

5. The ventricles in the brain are filled with
   a. blood
   b. air
   c. cerebrospinal fluid
   d. nervous tissue
   e. lymphatic fluid

6. Cells of different tissues and organs respond to the same hormone if they have the same kind of _____ molecules.
   a. receptor   d. protein
   b. target   e. all of the above
   c. DNA

7. Steroid hormones are produced from
   a. amino acids
   b. fatty acids
   c. proteins
   d. cholesterol molecules
   e. plasma cells

8. What is the ultimate source of control in the endocrine system?
   a. the hypothalamus   d. positive feedback
   b. genes   e. negative feedback
   c. the pituitary gland

9. Inflammation
   a. aids in the temporary repair at an injury site
   b. slows the spread of pathogens
   c. produces swelling and redness
   d. produces heat and pain
   e. all of the above

10. B cells can be activated only by
    a. pathogens
    b. exposure to a specific antigen
    c. fever
    d. cells infected with viruses
    e. bacteria

Brief Review

1. Three functional groups of neurons are found in the nervous system. What role does each fill?

2. What advantage does dual innervation of a bodily organ provide?

3. List the important functions of cerebrospinal fluid.

4. What is the primary difference in the way the nervous and endocrine systems communicate with their target cells?

5. How are the body's specific defenses different from its nonspecific defenses?

6. The medulla oblongata is one of the smallest regions of the brain, yet damage there can cause death, whereas similar damage in the cerebrum might go unnoticed. Why?

7. What is the difference between negative feedback and positive feedback?

8. Why is the pituitary gland referred to as the "master gland" of the human body?

9. How would a lack of helper T cells affect the antibody-related immune response?

10. What types of cells would be affected by a decrease in the monocyte-forming cells in the bone marrow?

Applying Your Knowledge

1. A police officer has just stopped Bill on suspicion of driving while intoxicated. Bill is first asked to walk the yellow line on the road and then asked to place the tip of his index finger on the tip of his nose. How would these activities indicate Bill's level of sobriety? What part of the brain is being tested by these activities?

2. Julie is pregnant and is not receiving any prenatal care. She has a poor diet consisting mostly of fast food. She drinks no milk, preferring soft drinks instead. How will this situation affect the activity of Julie's parathyroid glands?

3. What kind of immunity protects the developing fetus and how is it produced?

Group Activity

Many viruses, bacteria, fungi, and invertebrate animals have developed sophisticated strategies for evading and resisting the immune response. Can you guess what some of these strategies might be? Form a group of four students, and try to design the structures and behaviors that might have emerged in these invaders through natural selection. Assign two members of your group to design a structural solution and a behavioral solution that an invader might use to avoid being recognized by the immune system. The other two members of the group should design one structure and one behavior that might resist immune attack even after recognition. Next, bring the entire group together and discuss the strengths and drawbacks of your proposed solution. Your final product should be a "designer organism" that has multiple ways of evading immune recognition and attack. Be sure to specify your organism's size, life cycle, structure, and food source.

MediaLab

How Does Your Body Fight the Flu?
Understanding the Immune System

In 1972, a child named David Vetter was born, and then spent his 12 short years encased in a plastic bubble. The bubble protected him from the millions of microbes we all battle daily. David had a disease called SCID (severe combined immune deficiency, found in 1 in 100,000 people born each year), that affects the body's ability to produce a functional immune system. This *CD Activity* will help you explore how a normal immune system functions to spare us a similar fate. In the *Web Investigation*, you'll gather information concerning the actions of antibodies and their success in fighting two of these biological battles, against influenza and HIV. In the *Communicate Your Results* section, you will discuss and analyze how well our bodies may be fighting these battles.

This *MediaLab* can be found in Chapter 24 on your CD-ROM and Companion Website (http://www.prenhall.com/krogh).

CD ACTIVITY

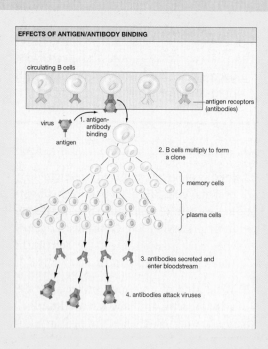

EFFECTS OF ANTIGEN/ANTIBODY BINDING

circulating B cells

antigen receptors (antibodies)

virus

1. antigen-antibody binding

antigen

2. B cells multiply to form a clone

memory cells

plasma cells

3. antibodies secreted and enter bloodstream

4. antibodies attack viruses

There are two main defensive mechanisms of our immune system, the antibody-mediated and cell-mediated immune responses. This *CD Activity* will illustrate the actions of the two types of white blood cells, B- and T-lymphocytes, which coordinate these responses.

Activity

1. *First, you'll learn how the B-lymphocytes produce antibodies, proteins that move through your body's fluids attacking any suspended particles that fit their binding site.*

2. *In contrast, you'll learn that the T-lymphocytes produce killer T cells, capable of destroying any cells that have already succumbed to infection.*

WEB INVESTIGATION

Investigation 1

Estimated time to complete = 15 minutes

In your everyday experiences, you know that getting the flu is at least an inconvenience, but you may not be aware that influenza can be deadly. Select the Keyword **PANDEMIC** on your CD or Website to read an article describing the 1918 influenza outbreak that killed over 20 million people. Then, click the Keyword **FLUNET** to visualize the development and ex- tent of flu outbreaks in the United States and around the world. Click on "epidemic activity" on the left side of the screen; you can select the geographic area you want data from, the specific type of virus (leave this at "Epidemiological Activity" for now), the format for presenting the data, and the time period for the data. Go back in time and try to find any patterns that indicate a worldwide flu outbreak like the pandemic of 1918.

Investigation 2

Estimated time to complete = 10 minutes

Select the Keyword **ANTIBODIES** on your CD or Website for this *MediaLab* to read an article describing the general structure and activities of antibodies. Pay particular attention to how the structure of antibodies helps them recognize a huge number of different antigens, and what functions are associated with the five different classes of antibodies.

Investigation 3

Estimated time to complete = 20 minutes

By this time you've probably heard at least a little about the HIV virus that causes AIDS. Select the Keyword **CDC** on your CD or Website for this *MediaLab* to access data for HIV and AIDS. To see a report of cumulative AIDS cases in the United States, click on "Cumulative Cases." How many have there been in the United States? How are these cases distributed through

different age groups, or races in the United States? Go to "Exposure Categories" for the data on how the virus has been spread. What is the most frequent mode of transmission in the United States? Is it the same for all age groups? Races? Now click on the Keyword **EUROPE** or **ASIA** to compare the U.S. data with data from other parts of the world.

COMMUNICATE YOUR RESULTS

Exercise 1

Estimated time to complete = 10 minutes

Going back through the years, can you find a pattern (from the Websites in *Web Investigation 1*) that indicates a worldwide flu season? Are there flu seasons within regions of the world? Is the outbreak of flu as widespread from year to year? Recently, which years have had the worst flu outbreak? Fast-forward now, about 10 years. You are working for a government agency that has been charged with creating policy to stem the spread of a new influenza strain, which has been rapidly spreading to pandemic proportions during the past four months. What is your plan?

Exercise 2

Estimated time to complete = 10 minutes

Use the description of antibody function from *Web Investigation 2* to compose a 100-word description of the immune system's response when presented with an antigen such as a

particular influenza strain the first time, or a second or third time. What types of antibodies are produced at each exposure? What antibodies regulate the robust response to the second exposure? Why does influenza repeatedly attack human populations, frustrating our immune systems?

Exercise 3

Estimated time to complete = 15 minutes

Construct bar graphs illustrating differences in the occurrence of HIV/AIDS in the United States with its occurrence in 10 other countries around the world. Use the bar graphs to answer the following questions: Are the differences you find as striking between two developed countries versus a developed country and an undeveloped country? How about between Western countries and countries in Asia? Where is the occurrence of HIV/AIDS the highest? How likely are you to meet someone with HIV/AIDS in your own age group in the United States versus one of the other countries?

25

Transport, Nutrition, and Exchange

Blood, Breath, Digestion, and Elimination

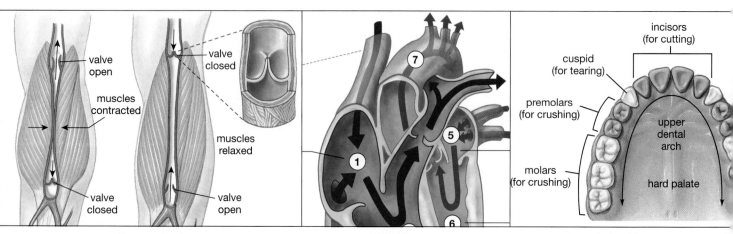

One way street.
(page 548)

A hearty pumper.
(page 545)

Open wide.
(page 554)

Many times in this text, you

have had occasion to think of life in terms of its building blocks, with one set of those blocks being the cells that help make up organisms. One problem with this metaphor is that cells are living things and thus, unlike real building blocks—bricks or cinder blocks—they have ongoing maintenance requirements. They have to obtain energy and get rid of wastes; in many cases they have to obtain oxygen. They have to have a source of nutrients such as nitrogen and phosphorus. When life consisted of nothing but single-celled creatures, every cell had access to the world outside itself, which meant that food could enter, waste could be expelled, and gases could be "exchanged" in a fairly simple manner. In a human being, conversely, many of our 10 trillion or so cells spend their entire existence wrapped inside layers of tissue which themselves are wrapped inside other layers of tissue.

During the course of evolution, this access problem, as it might be termed, was dealt with by the development of physiological *systems*: a special bodily apparatus for delivering nutrients, another for delivering oxygen, and another for removing waste, for example. The capabilities of these three systems—digesting food, taking in oxygen, and getting rid of waste—are subjects of this chapter. These systems are all involved in carrying out various exchanges of materials between a living thing and its surroundings. One additional system then links all three of these systems together. It is the cardiovascular system, which—through blood—picks up gases, nutrients, and wastes from one location in the body and brings them to others. The essential subject of this chapter is how the body's transport and exchange operations function as a coherent life-support system. If you look at **Figure 25.1** on page 542, you can see a representation of how these systems are linked. Let's start this review with the cardiovascular system.

25.1 The Cardiovascular System and Body Transport

Those of you who went through Chapter 7, on energy, know that the reason oxygen is necessary to organisms such as ourselves is that oxygen is in the energy transfer business. Large living things need large amounts of energy, and oxygen is a critical element in

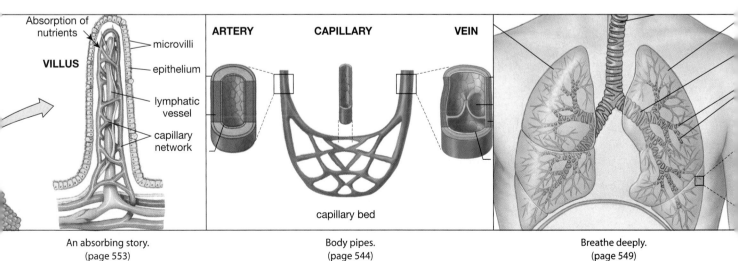

An absorbing story.
(page 553)

Body pipes.
(page 544)

Breathe deeply.
(page 549)

helping them obtain it. But what is the distribution mechanism for oxygen? After entering the body through the lungs, how is it delivered to every cell in the body?

The answer is the cardiovascular system, which can be compared to the cooling system of a car. A car uses antifreeze and water as its circulating fluids; the cardiovascular system has blood. A car may use a water pump to circulate its fluids; the cardiovascular system uses a pump called the heart. A car has an assortment of conducting pipes; the cardiovascular system does too, in the form of blood vessels.

Apart from oxygen, the cardiovascular system also transports other materials needed and excreted by the body's cells. These materials include nutrients, vitamins, carbon dioxide, hormones, waste products, and molecules and cells specialized for body defense. Let's start this tour of the cardiovascular system by looking at the material that runs through it—blood.

Blood Has Two Major Components: Formed Elements and Plasma

The volume of blood in the body varies from 4 to 5 liters (about 1 to 1.3 gallons) in an adult woman to 5 to 6 liters in an adult man. The temperature of the blood in the body is roughly 38°C (100.4°F), slightly higher than normal body temperature. Blood pH averages 7.4 and is therefore slightly alkaline or "basic," as opposed to acidic. Blood is also sticky, cohesive, and somewhat resistant to flowing freely. Such characteristics determine the **viscosity** of a fluid. Blood is about five times "thicker," or more viscous, than water.

Collect a sample of blood from a human being, prevent it from clotting, and spin it in a centrifuge and it will separate into two layers. The lower layer, accounting for some 45 percent of the total volume, consists of blood *cells* and cell fragments—material that collectively is known as the formed elements of the blood (**see Figure 25.2**). The yellowish, upper layer, which makes up 55 percent of the volume of blood, is called **plasma**, and it is mostly water. In the body, the formed elements are suspended in the plasma.

Formed Elements

Formed elements include **red blood cells (RBCs)**, which transport oxygen to, and carbon dioxide from, every part of the body; the far less numerous **white blood cells (WBCs)**, which are components of the immune system; and **platelets**—small fragments of cells that have pinched off from larger, complete cells. Platelets contain cellular material that includes enzymes and other substances, collectively called **clotting factors**, that are important to the blood-clotting process.

The formed elements (including platelets) are produced through a process called *hemopoiesis*, which means "to make blood." Almost all blood cells and platelets are formed within bone marrow.

Red Blood Cells Red blood cells (RBCs) are also known as **erythrocytes**—from *erythros*, which means "red." Red blood cells are red because they contain the iron-containing protein **hemoglobin**, which is the element in the red blood cell that oxygen binds to. Blood as a whole is deep red because RBCs constitute 99.9 percent of all the formed elements in it. Indeed, the *number* of RBCs in the blood of an average person staggers the imagination. One cubic millimeter of blood—an amount that might scarcely be visible—contains roughly 4.8 to 5.4 million RBCs. So numerous are these cells that they

Figure 25.1
The Transport and Exchange Systems of the Body
The respiratory system moves oxygen into and carbon dioxide out of the body. The digestive system puts the food we eat into a form that can be transported throughout the body. The urinary system filters bodily fluids, removing waste while conserving water and other materials. The cardiovascular system is central to all the others, in that it transports materials to and from them.

THE TRANSPORT AND EXCHANGE SYSTEMS

Respiratory system
(O_2 in, CO_2 out)

Digestive system
(nutrients in)

Urinary system
(wastes out)

Cardiovascular system
(transport)

make up roughly one-third of all the cells in the human body.

Numerous though they may be, RBCs are much different from most cells. A mature red blood cell primarily consists of a cell membrane surrounding a mass of hemoglobin. Viewed another way, an RBC has neither the nucleus nor many of the organelles that are standard equipment in almost all eukaryotic cells. Because of this difference, the RBC has a life span of about 120 days, which is relatively short. About 1 percent of the circulating erythrocytes are replaced each day, meaning that approximately 180 million new erythrocytes enter the circulation *each minute*.

White Blood Cells White blood cells are also known as **WBCs** or **leukocytes** (*leukos* means "white"). They are quite different from RBCs because they have nuclei, but lack hemoglobin. Those of you who read about the immune system in Chapter 24 were introduced there to the pivotal role of white blood cells in the body's defense system.

A typical cubic millimeter of blood contains 6,000 to 9,000 white blood cells—a tiny number relative to the red blood cells present in such a volume. Most of the WBCs in the body are, however, found in body tissues; those in the blood represent only a small fraction of the total population. For WBCs, the bloodstream is used primarily for transportation to sites of invasion or injury. WBCs also move under their own power by extending parts of their cytoplasm as a kind of cellular foot. This allows them to move along the walls of blood vessels and, when outside of the bloodstream, through surrounding tissues. When problems are detected, these cells leave the circulation and enter the affected area, getting out of the bloodstream by squeezing between the cells that form the blood-vessel wall. Once the WBCs reach the problem area, they engulf pathogens, cell debris, or other materials by **phagocytosis** or "cell eating." You went over the types of WBCs and their functions in Chapter 24 (see pages 523–524).

Platelets Platelets can be thought of as enzyme-bearing packets. They are not cells, but are fragments of cells that have broken away from a set of unusually large cells that exist in the bone marrow. The function of the enzymes in platelets is to aid in blood clot-

ting at sites of injury. Apart from releasing these enzymes, the platelets facilitate clotting by clumping together at the site of an injury, forming a temporary plug that slows bleeding while the process of clotting continues.

Blood's Other Major Component: Plasma

All the formed elements you've just read about are suspended in blood's other major component, plasma, which is 92 percent water. Proteins and a mixture of other materials are dissolved in this fluid as well. Most blood proteins are too large to cross through the walls of the tiny blood vessels called *capillaries*, and they thus remain trapped within the circulatory system. This large concentration of plasma proteins is one of the chief differences between plasma and the **interstitial fluid** found outside the circulatory system.

There are three primary classes of plasma proteins:

- **Albumins**, which constitute roughly 60 percent of the plasma proteins, are synthesized by the liver and function in the transport of both hormones and the components of fats known as fatty acids.

Figure 25.2
The Composition of Blood
Blood that is spun in a centrifuge separates into its two primary constituent parts, formed elements and plasma. Formed elements are composed of red blood cells, white blood cells, and the cell fragments called *platelets*. Blood plasma—the liquid in which the formed elements are suspended—is composed mostly of water, but also contains proteins and other materials.

BLOOD SAMPLE

FORMED ELEMENTS

PLASMA

55%

45%

Formed elements	
Red blood cells	99.9%
White blood cells Platelets	0.1%

Plasma	
Water	92%
Plasma proteins	7%
Other solutes	1%

Red blood cells

White blood cells

Platelets

What Is It?
See p. 546.

Figure 25.3
The Structure of Blood Vessels
Both arteries and veins have three layers to them: an inner endothelium, a layer of smooth muscle, and a layer of connective tissue. By contracting or relaxing, the smooth muscle can change the diameter of the vessels. Many veins contain valves that prevent a backflow of the blood that is moving through them toward the heart. Arteries and veins are linked by intervening capillary beds. A capillary generally consists of a single layer of endothelial cells.

• **Globulins**, which make up 35 percent of the dissolved proteins, come in two forms that have important functions. The immunoglobulins or **antibodies** that you saw so much of in Chapter 24 attack foreign proteins and pathogens (disease-causing organisms). Meanwhile, the **transport proteins**, also synthesized by the liver, carry small ions, hormones, or compounds that might otherwise be filtered out of the blood at the kidneys. Some of these transport proteins are important in carrying lipids that normally do not dissolve in water. These proteins are called **lipoproteins**; that name may sound familiar, because an important consideration in health is the relative abundance of the high-density lipoproteins (or **HDLs**) a person has circulating compared to low-density lipoproteins (or **LDLs**). For more on these proteins, see Chapter 3, page 58.

• **Fibrinogen**, the third type of plasma protein, functions in blood clotting. It, too, is made by the liver. Under certain conditions, fibrinogen molecules interact to form large, insoluble strands of fibrin, the basic framework molecule in a blood clot. In the absence of steps taken to prevent clotting in a blood sample, fibrinogen is converted to fibrin. The fluid left after clotting is known as **serum**.

The remaining 1 percent of the plasma consists of such compounds as electrolytes (ions), organic nutrients, organic wastes, and hormones. The two major ions dissolved in plasma are sodium (Na^+) and chloride (Cl^-), whose combined form (NaCl) is better known to us as table salt. Some of the other ions important for normal cellular activities include potassium (K^+), calcium (Ca^{2+}), and iodide (I^-). Nutrients include glucose, fatty acids, and amino acids. They are used by cells for production of the energy-currency molecule, ATP, as well as for growth and maintenance. Wastes dissolved in the plasma are carried to sites of excretion. For example, the **urea** produced by the liver's breakdown of amino acids is carried to the kidneys.

Blood Vessels

Having looked at what moves through the cardiovascular system, let's now look at the plumbing that the blood courses through, our network of blood vessels.

Blood pumped by the heart is carried within tubes of varying diameter to all parts of the body. Blood vessels carrying blood *away* from the heart are the **arteries**, while the vessels carrying blood back *to* the heart are the **veins**. The farther from the heart these vessels are, the smaller they tend to be. The smallest of all blood vessels, the **capillaries**, connect the arteries with the veins. It is through the thin walls of the capillaries that gases, nutrients, and wastes are exchanged between the blood and the body's tissues.

The walls of arteries and veins contain three distinct layers. The innermost layer is made up of a group of flat epithelial cells, an endothelium. The middle layer contains smooth muscle tissue. When these smooth muscles contract, the vessel decreases in diameter (an action known as **vasoconstriction**); when they relax, the diameter increases (an action termed **vasodilation**).

This mechanism plays a part in regulating our blood pressure; as diameter decreases, pressure increases. The outer layer of arteries and veins consists of a stabilizing sheath of connective tissue (**see Figure 25.3**). These multiple layers give arteries and veins considerable strength, and the muscular and elastic components permit controlled alterations in diameter.

Arteries and Veins

The largest arteries are extremely resilient, elastic vessels with diameters of up to 2.5 centimeters or about 1 inch. The smallest arteries, the **arterioles**, have an average diameter of about 30 micrometers (μm), which is about a thousandth of an inch. The vessel walls of veins are thinner than arteries, because veins do not have to withstand much

ARTERY　　　　CAPILLARY　　　　VEIN

connective tissue

muscle

endothelium

connective tissue
valve
muscle

endothelium

capillary bed

blood pressure. Most of this pressure has been spent by the time blood arrives at the veins from the capillaries. In the limbs, medium-sized veins contain valves that prevent a backflow of blood, thus ensuring a one-way flow, toward the heart.

Capillaries

Capillaries are the only blood vessels whose walls are thin enough to permit exchanges between the blood and the surrounding interstitial fluids. A typical capillary consists of a single layer of endothelial cells and has an average diameter of only 8 μm. Capillaries exist in interconnected networks called **capillary beds**, as you'll see shortly.

Systems of Circulation

Vertebrates such as ourselves employ the blood vessels you've just reviewed in what is called a **closed circulation** pattern of blood transport—one in which blood is delivered from one set of vessels to another. In contrast, many invertebrates have an **open circulation** that lacks capillaries. In such systems, arteries carry blood into open spaces, called *sinuses*. Blood from such sinuses flows around cells and is then returned to a heart or hearts through openings in veins. Segmented worms (annelids) have both open and closed circulation, while arthropods and most mollusks have an open circulation.

25.2 The Heart and the Circulation of Blood

What makes the blood flow through this system of closed circulation? The answer is the muscular pump we call the heart. A small organ, roughly the size of a clenched fist, the heart actually is shaped something like a Valentine heart in its major outline. It lies near the back of the chest wall, directly behind the bone known as the *sternum*, and sits at an angle with its blunt tip pointed downward and to the left side of the body.

As noted earlier, a central task of the circulatory system is to get oxygen to all the various bodily tissues. For blood to carry oxygen to tissues, the blood first has to *take up* the oxygen that is brought into the body through the lungs. So imposing is this task of oxygenating blood that the body has an entire circulation loop devoted to this function. Once oxygenated in this loop, blood then returns to the heart, where it is pumped out to the rest of our tissues in the body's second major circulation loop.

To put these things more formally, there are two general networks of blood vessels in the cardiovascular system. One of these is the **pulmonary circulation**, in which the blood flows between the heart and the lungs. Then there is the **systemic circulation**, in which blood is moved between the heart and the rest of the body. Now let's see how this two-network system works (**see Figure 25.4a**).

Figure 25.4
Two Circulatory Networks and How the Heart Supports Them

a Veins in the systemic circulation bring blood into the right side of the heart; it is pumped out via the pulmonary circulation to the lungs, where it is oxygenated, and then returns to the left side of the heart. It is then pumped out from the left side of the heart, through the systemic circulation, to the rest of the body.

b Deoxygenated blood enters (1) the right atrium of the heart through the superior and inferior venae cavae. The right atrium pumps the blood into (2) the right ventricle, which contracts, pumping the blood to the lungs, via (3) the pulmonary arteries. After being oxygenated in the lungs, the blood returns to heart through (4) the pulmonary veins and moves into (5) the left atrium, which pumps the blood into (6) the left ventricle. The powerful contractions of the left ventricle pump the blood out through (7) the aorta, after which it is distributed to the rest of the body.

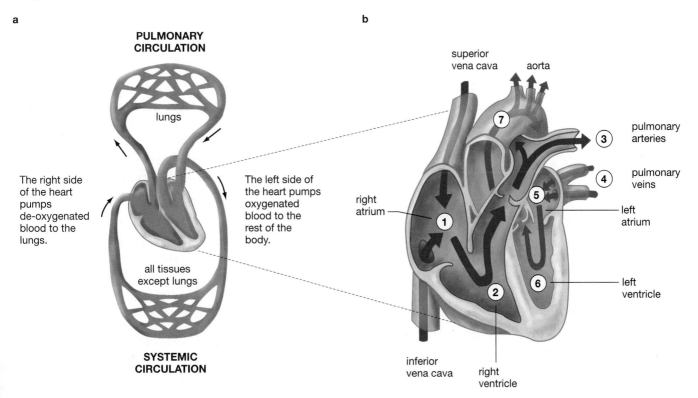

a

PULMONARY CIRCULATION

lungs

The right side of the heart pumps de-oxygenated blood to the lungs.

The left side of the heart pumps oxygenated blood to the rest of the body.

all tissues except lungs

SYSTEMIC CIRCULATION

b

superior vena cava

aorta

pulmonary arteries

pulmonary veins

right atrium

left atrium

left ventricle

inferior vena cava

right ventricle

The human heart contains four muscular chambers, two associated with the pulmonary circulation and two associated with the systemic circulation. These are the left atrium and right ventricle (associated with the pulmonary circulation) and the right atrium and left ventricle (associated with the systemic circulation). A look at the path of blood through these chambers will give you a clear idea of how blood moves through the body.

Following the Path of Circulation

Let's take as the starting point the blood entering the right atrium (on the *left* side of **Figure 25.4b**) from the veins called the *superior* and *inferior venae cavae*. These veins receive the blood returning to the heart from the upper and lower portions of the body, respectively. This is blood that is returning to the heart after having distributed oxygen and other blood-borne materials to the rest of the body. As such it is deoxygenated blood, actually dark red in color, but indicated by the blue color of the veins it flows through. After the right atrium receives this blood, it pumps it the short distance to the right ventricle, which contracts, pushing the blood into the pulmonary arteries. (These are arteries instead of veins, because blood moves *away* from the heart through them.) This blood is now bound for the lungs, where it will pick up oxygen. Having done so, it returns to the heart through the pulmonary veins—veins that in this case are carrying oxygen-*rich* blood, which is bright red. These veins both empty into the left atrium. The blood then is pumped into the left ventricle, which contracts and sends the blood coursing up into the enormous artery called the **aorta**. This vessel has branches stemming from it that

will carry the blood to all the tissues of the body except the lungs.

Taking a step back from this, you can see that the right side of the heart is pumping blood to the lungs, while the left side of the heart is pumping blood to all the body's other tissues, as shown in Figure 25.4a. All four chambers of the heart pump blood, but the right and left *atria* are essentially blood collection sites that pump blood only into their adjacent ventricles. The right *ventricle* is pumping blood to the lungs, so there is some greater contraction power needed there. But the *left* ventricle is pumping to all the rest of the body, and its contractions are by far the most powerful. When we "take our pulse," what we are feeling is the surge of blood produced by the contraction of the left ventricle.

Control of Blood Flow through Valves

Because both ventricles are contracting with such force, you may wonder why blood doesn't go back into their respective atria with each beat. The answer is that there are valves between the atria and ventricles that allow the blood to flow only one way—from atrium to ventricle. Occasionally, because of disease or genetic bad luck, people do have this kind of backflow into the atria. The fluid turbulence that results can be heard as a kind of gurgling sound, known as a "heart murmur" (which is not necessarily a sign of a serious heart problem). Valves also exist to prevent backflow between the ventricles and the arteries they pump blood into. The heart's set of valves is responsible for the familiar "lub-dub" sound of the heartbeat. The "lub" is the sound of the valves closing between the atria and ventricles, while the "dub" is the sound of the valves closing between the ventricles and the arteries.

25.3 The Heart's Own Blood Supply: What Is a Heart Attack?

The heart is in some ways a unique organ; but like any other organ, it too needs a supply of blood and the oxygen and nutrients that come with it. Indeed, the heart is a voracious user of these materials because it essentially is a set of muscles that never stop working. Blood does not come to the muscles of the heart through any of the chambers you've been reviewing. (Think about the two ventricles; one is sending blood to the lungs,

Figure 25.5
Critical Vessels
Damage to the coronary arteries brings about the heart attacks and heart conditions that are responsible for about half of all deaths in the United States. The coronary arteries begin with a right and left coronary artery that branch off from the aorta. These arteries then undergo branching themselves, thus supplying blood to the whole heart. Blockage of the left anterior descending coronary artery, shown in the drawing, figures in almost half of all heart attacks; this vessel supplies blood to heart muscle in the left ventricle.

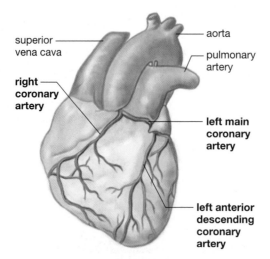

superior vena cava

right coronary artery

aorta

pulmonary artery

left main coronary artery

left anterior descending coronary artery

the other to the rest of the body, but not directly to the heart itself.) The way blood does come to the heart is through two arteries that branch off from the aorta just after it emerges from the left ventricle. Because these arteries encircle or "crown" the heart before they start branching, they are known as the **coronary arteries** (see Figure 25.5).

It is difficult to overstate the importance of the coronary arteries, because about half of all deaths in the United States today are caused by a blockage of one or more of them. Such blockages generally are initiated by a buildup of small masses of fatty material called *plaque* that accumulate between the smooth muscle and inner lining of an artery. As a result of the blockage, the blood supply can be cut off altogether to given collections of heart-muscle cells. Small or large groups of these "myocardial" cells can die when this happens; the technical term for this event is a myocardial infarction, but it is better known as a **heart attack**. Death can come suddenly from such an event in several ways. The heart muscles may no longer be able to contract with sufficient force to pump blood through the body; or a wild, irregular rhythm, called a *ventricular fibrillation*, can be set off.

The terrible consequences that can result from having a blockage of one small vessel within the body ought to make us think twice about high-fat diets, smoking, and lack of exercise, all of which are contributing factors to heart disease. It's worth noting that while heart attacks are generally events of middle or old age, the buildup of fatty plaques seems to proceed from childhood forward. In other words, it is never too early to adopt a healthy lifestyle.

25.4 Getting the Goods to and from the Cells: The Capillary Beds

Having looked at the players involved, you're now ready to see how the cardiovascular system actually carries out its central functions of bringing materials to tissues and taking materials away from them. Propelled by the heart through the aorta, blood flows through the tube-like arteries. The large-diameter arteries branch repeatedly, gradually decreasing in size until they become the smaller-diameter **arterioles**. Arterioles then branch into the delivery vehicles of the cardiovascular system,

the capillary beds. If you look at **Figure 25.6a**, you can see a representation of one of these beds. Note that at one end, the bed extends from the body's arterial system, but that at the other it is feeding into the body's system of *veins*—the venous system that is bringing blood *back* to the heart. Looking at the venous system as it begins with the capillaries, it initially consists of the tiny **venules**, which then merge into veins; smaller veins eventually merge into larger ones that finally come together into the largest veins of all, the two vena cavae that feed directly into the heart.

This explains "blood-out" (through the arteries) and "blood-back" (through the veins), but the real action in the circulatory system takes place in the capillaries that lie *between* the arteries and the veins. Capillaries are so small in diameter that red blood cells must squeeze through them in single file; and as you've seen, each capillary is composed of but a single layer of epithelial cells. Think now about a line of oxygenated red blood cells, suspended in plasma and moving through a capillary along with hormones, glucose, ions, and other materials needed by the cells *outside* the capillary.

Figure 25.6
Cardiovascular System Delivery Vehicle
Capillary beds are the sites of the exchange of materials between blood vessels and body's tissues.

a An idealized view of how blood flows from arteries to arterioles, through capillary beds (losing oxygen along the way) and then into the venules and veins that return it to the heart.

b Blood pressure moves materials out of the capillaries near the arterial end of capillary beds, while osmotic pressure moves materials into the capillaries near the venous end of the beds.

c Micrograph of red blood cells moving in single file through a capillary.

The oxygen, glucose, and all the rest move out through the thin wall of the capillary itself and into the **interstitial fluid**—the liquid in which both the capillary and the cells around it are immersed. It is from this fluid that these needed materials will make their way to the surrounding cells. The distance the materials must travel to the cells is not far, however; so extensive is the body's capillary network that no cell is farther than two or three cells away from a capillary. Oxygen leaves the red blood cells and diffuses across the capillary lining and into the interstitial fluid. (The red blood cells themselves are too big to pass out.) Glucose and small ions such as sodium likewise are distributed in this manner. Meanwhile, carbon dioxide and nitrogenous wastes are flowing *into* the capillaries, with this movement, like that of oxygen and other materials, driven by a concentration gradient. (For more on concentration gradients, see Chapter 5, page 98.)

Forces That Work on Exchange through Capillaries

Water and some of the small molecules in it tend to flow *out of* a capillary bed at its arterial end, but *into* a bed at its venous end. This is the result of two opposing forces at work at any given point along a capillary. Blood pressure, which results from the heart's contractions, tends to move materials out of the capillary. At the arterial end of a bed, this pressure is relatively strong—strong enough to bring about a net movement of water out of the capillary—but the narrowness of the capillaries means that this pressure is largely spent by the time blood approaches the venous end of a capillary bed. At that point another force, osmotic pressure, tends to move materials *into* the capillary: Proteins too big to leave the capillary result in an osmotic gradient that pulls on fluids (for more on osmosis, see Chapter 5, page 98). At the venous end of a capillary, then, the force of osmosis overcomes the force of blood pressure. Most of the water that leaves the capillary beds is returned to them in this way, though some remains as interstitial fluid to be taken up by the vessels of the body's lymphatic system.

Muscles and Valves Work to Return Blood to the Heart

Because blood pressure has dropped to such low levels by the time the blood gets to the venous system, you may wonder how it can get back to the heart. The answer is that our muscles are essentially doing double duty. In contracting, they squeeze the veins in a way that moves the venous blood along. Blood can move only *toward* the heart in this system, because the veins have a series of valves in them that block movement of blood away from the heart (**see Figure 25.7**).

25.5 The Respiratory System and the Exchange of Gases

One-celled organisms and other simple life-forms (such as flatworms) do not need to breathe in the way that we do. They rely instead on the process of diffusion in order to bring oxygen in and move carbon dioxide out of themselves. Larger organisms have evolved various specialized organs, within which oxygen and carbon dioxide are exchanged with their environment. In aquatic habitats, gills serve this purpose for several different groups of animals (segmented worms, arthropods, mollusks, and fishes). Land-dwelling vertebrates, meanwhile, have evolved various types of lungs for the exchange of gases.

The Functions of Respiration

People can live perhaps a couple of months without food, and generally a few days without water. But if they go 5 or 6 minutes without

Figure 25.7
One-Way Flow to the Heart
Skeletal muscle contraction is the primary force that drives blood back to the heart through the venous circulation. A system of valves guards against the backflow of this blood.

Valves allow blood to go forward...

...but not backward

valve open

muscles contracted

valve closed

valve closed

muscles relaxed

valve open

breathing, death is usually the result. This is because, as you've seen, oxygen is in the energy transfer business; without oxygen, our cells simply don't have enough energy to function.

Cells may need to obtain oxygen to live, but they *generate* carbon dioxide as a consequence of living and need to dispose of it. Thus do we see the two central functions of respiration: capturing oxygen and disposing of carbon dioxide. Additional functions of respiration include defending the body against invasion by microorganisms, producing sounds for speaking, assisting in the regulation of blood volume and pressure, and assisting in the control of body fluid pH.

Structure of the Respiratory System

The lungs are one part of a **respiratory system** that also includes the nose, nasal cavity, and sinuses; the pharynx (throat), the larynx (voice box), and the trachea (windpipe); and the conducting passageways, called *bronchi* and *bronchioles*, that lead to the lungs (**see Figure 25.8**). The end points of the bronchioles are a profusion of tiny air sacs called **alveoli**. Alveoli are like capillaries in a sense, because they are structures that lie at the ends of conducting passageways and are specialized for diffusion. Like the capillaries, the alveoli are thin enough to allow the diffusion of materials in and out of themselves.

Steps in Respiration: Breathing

So how does respiration work? First there is **pulmonary ventilation**, or breathing, meaning the physical movement of air into and out of the lungs. A single breath, or *respiratory cycle* consists of an inhalation, the movement of air into the lungs, and an exhalation, the movement of air out of the lungs. This movement

**Figure 25.8
Components of the
Respiratory System**

a In the magnification of the alveoli, the capillary network that surrounds all alveoli has been cut away, on the right, to reveal the structure of the alveolar sacs.

b Artificially colored micrograph of a section of a human lung showing alveoli at the left and a larger bronchiole at right.

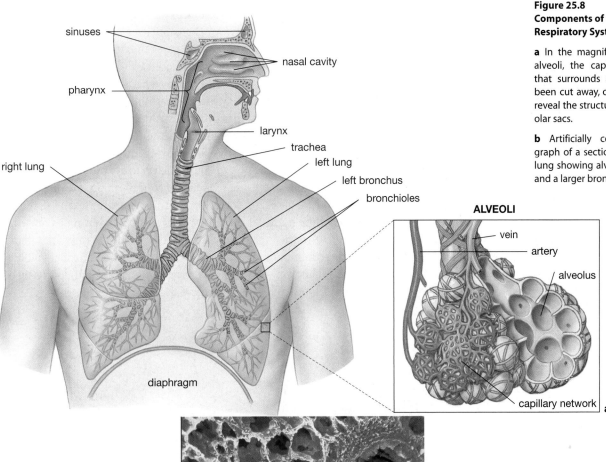

sinuses

nasal cavity

pharynx

larynx

trachea

right lung

left lung

left bronchus

bronchioles

diaphragm

ALVEOLI

vein

artery

alveolus

capillary network

a

b

of air is controlled by differences in air pressure between the lungs and the external environment. At the start of a breath, pressures inside and outside the lungs are identical and there is no movement of air. When the chest or "thoracic" cavity enlarges, the pressure inside the lungs drops, and air enters the respiratory passageways (**see Figure 25.9**). Enlargement of the thoracic cavity involves the contractions of the diaphragm, aided by muscles of the rib cage (the *external intercostal muscles*). In exhalation, the downward movements of the rib cage and upward movement of the diaphragm reduces the size of the lungs. Pressure inside the lungs now exceeds atmospheric pressure, and air moves out of the lungs.

During quiet breathing, as when you're sitting and reading, exhalation is a passive process. That is, when the respiratory muscles used during inhalation relax, the stretched elastic fibers of the lungs recoil and air is forced out. When you are breathing heavily, contractions of another set of rib cage muscles (the *internal intercostal muscles*) and the abdominal muscles assist in exhalation.

Steps in Respiration: Exchange of Gases

Pulmonary ventilation is just one step in respiration. Next there is the exchange of respiratory gases, which occurs in two general processes. External respiration is the diffusion of gases between the alveoli and the blood. Think of it; the oxygen you inhale is needed in all the internal tissues of the body. How is it going to get there? It diffuses through the thin layer of alveoli and into an adjacent capillary; inside the capillary, oxygen is bound to one of the red blood cells that are so numerous within the blood. Thus locked up in the blood, the oxygen is ready to move with it into the heart, which pumps the oxygenated blood out to the rest of the body. At some point, the oxygen is dropped off, to be used by a bodily tissue.

Meanwhile, at the same capillary-alveoli interface, carbon dioxide that has come *from* the body's tissues moves into the alveoli from the capillaries. From the alveoli, it is expelled into the environment outside the body (**see Figure 25.10**). Carbon dioxide is moving to the external environment in this process, while oxygen is moving from the external environment; hence, these are forms of external respiration.

But there is also internal respiration, meaning the two-way diffusion of gases between the blood that flows through the capillaries and the interstitial fluid that lies outside the capillaries. Because cells are immersed in the interstitial fluid, they also can use the two-way movement of gases between the capillaries and interstitial fluid to obtain the gases they need and to dispose of their waste gases.

In both internal and external respiration, oxygen and carbon dioxide diffuse independently of each other—the exchange of one is not linked to the exchange of the other. The gases are moving down their concentration gradients, which is to say moving from an area of their higher concentration to an area of their lower concentration. In the external respiration of carbon dioxide, for example, more carbon dioxide exists inside the capillaries than in the alveoli; thus the carbon dioxide is "following" its concentration gradient in diffusing out of the capillaries and into the alveoli.

What Is It?
See p. 551.

**Figure 25.9
How We Breathe**

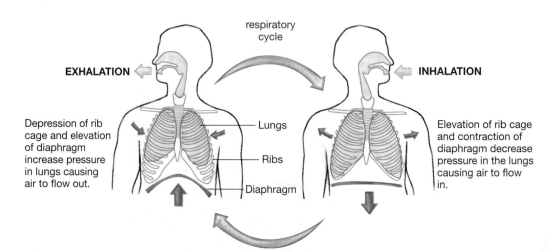

respiratory cycle

EXHALATION

Depression of rib cage and elevation of diaphragm increase pressure in lungs causing air to flow out.

Lungs

Ribs

Diaphragm

INHALATION

Elevation of rib cage and contraction of diaphragm decrease pressure in the lungs causing air to flow in.

Steps in Respiration: Oxygen and Carbon Dioxide Transport

So how does blood transport the two gases you've been looking at? In other words, how do these gases get loaded into and out of blood? Recall that blood is made up of both formed elements (cells and cell fragments) and plasma (a watery, protein-filled fluid). Almost all the oxygen that moves from the alveoli to the blood in the capillaries is bound to molecules of the protein hemoglobin (Hb) that exist in red blood cells.

In contrast to this simplicity, the carbon dioxide that is being transferred *to* the alveoli is transported by the blood in three different ways after being picked up from the interstitial fluid. As **Figure 25.11** shows, 93 percent of the carbon dioxide that comes into capillaries enters red blood cells, meaning that only 7 percent remains dissolved in the blood plasma. In the red blood cells, some of the carbon dioxide binds directly to hemoglobin, but most (70 percent) is rapidly converted within the red blood cells to bicarbonate ions (HCO_3^-). These newly formed ions then move out of the red blood cells and into the blood plasma. Hydrogen ions that are produced by this reaction are absorbed by hemoglobin molecules. Upon arrival at the capillaries in the lungs, this process is reversed: The bicarbonate and hydrogen ions combine and reform carbon dioxide, which diffuses into the alveoli; the hemoglobin-bound carbon dioxide is released, diffusing into the alveoli; and the carbon dioxide that had been dissolved in the plasma diffuses into the alveoli.

25.6 The Digestive System

Let's shift gears now, moving from respiration to digestion. The digestive system is in one sense very simple. Its central feature is a muscular tube that passes through the body from the mouth to the anus. The system consists of the tube, the **digestive tract**, and various accessory organs, including the salivary glands, gallbladder, liver, and pancreas. Through the food we eat, the digestive system supplies the energy and chemical building blocks for growth and maintenance.

The Digestive Process in Overview

The central function of the digestive system is to get the foods the body ingests into a form the body can use; the system must also collect and then rid the body of the waste that remains once the useful material from the food has been removed. You've seen already that nutrients are delivered to tissues throughout the body via the circulatory system. So the real question for the digestive system is: What does it need to do to food to get it into a form that can *move into* the

What Is It?
Answer: Artificially colored micrograph of normal human lung tissue showing alveoli and red blood cells. (×250)

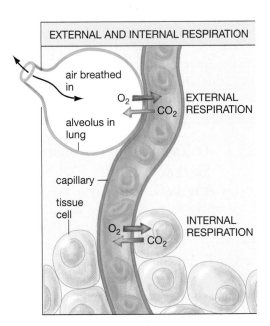

Figure 25.10
External and Internal Respiration
External respiration involves the movement of gases between capillaries and the alveoli—oxygen moving from the alveoli to the capillaries, and carbon dioxide moving to the alveoli from the capillaries. Internal respiration involves the exchange of gases between capillaries and the cells outside them.

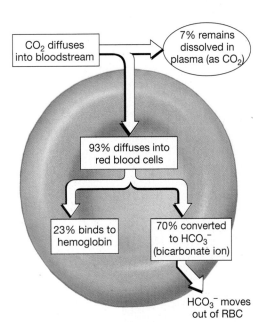

Figure 25.11
How Carbon Dioxide Is Transported in the Blood

circulatory system? For carbohydrates and proteins, the pathway is straightforward. Partway through the organs called the *small intestines*, carbohydrates and proteins have been broken down enough that they move out through the inner lining of the intestine and into adjacent capillaries. This network of capillaries then feeds into a common vein that goes to the liver. Most nutrients, then, are carried straight from the small intestine to the liver. In a controlled release, the liver then sends these nutrients out (through another vein) to the heart, which pumps them out to all the tissues of the body.

The digestion of fats takes a somewhat more complicated route, as you'll see. But the principle remains the same; fats must be broken down to such a point that they can leave the digestive tract for the rest of the body. For all foods, "breakdown" means just what it sounds like: a transition from large to small. The transition goes from big, visible bites of food to smaller bits, and eventually from big *molecules* of food to the building blocks that make them up. It is only these smaller molecules—simple sugars, amino

acids, and fatty acids—that are ready to leave the digestive tract for the rest of the body. Now take a look at the steps in digestion and the components of the digestive system, after which you'll follow the path of digestion. The functions of the digestive system involve a series of interconnected steps:

- *Ingestion.* The entry of food into the digestive tract through the mouth.
- *Mechanical processing.* The physical manipulation of solid foods, first by the tongue and the teeth and then by swirling and mixing motions of the digestive tract.
- *Digestion.* The chemical breakdown of food into small molecules that can be absorbed by the digestive lining.
- *Secretion.* The release of water, acids, enzymes, and buffers from the digestive tract lining and accessory organs, aiding digestion.
- *Absorption.* The movement of small organic molecules, electrolytes, vitamins, and water across the digestive inner lining—or epithelium—into capillaries or lymphatic vessels.
- *Excretion.* The elimination of waste products from the body. In the digestive tract, these waste products are compacted and discharged through defecation.

The lining of the digestive tract also plays a defensive role by protecting surrounding tissues against the corrosive effects of digestive acids and enzymes, and pathogens that are either swallowed with food or reside within the digestive tract.

25.7 Components of the Digestive System

Figure 25.12 shows the locations and functions of the accessory glands and different subdivisions of the digestive tract. The digestive tract begins with the oral cavity and continues through the pharynx, esophagus, stomach, small intestine, and large intestine before ending at the rectum and anus. (The small and large intestines are further divided into various regions that are described later.) Other terms used to describe the digestive tract are **alimentary canal** and **gastrointestinal tract**. Generally, the gastrointestinal tract, or **GI tract**, refers to the lower portion of the digestive tract—the stomach and intestines.

Figure 25.12
Components of the Digestive System

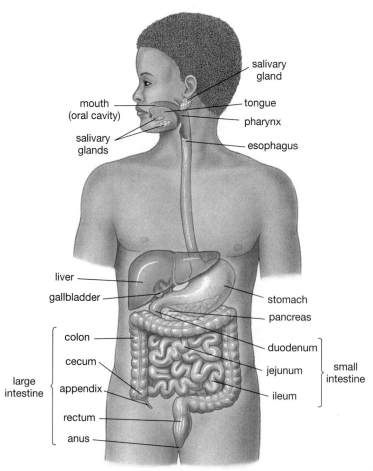

- salivary gland
- mouth (oral cavity)
- tongue
- pharynx
- salivary glands
- esophagus
- liver
- gallbladder
- stomach
- pancreas
- colon
- duodenum
- cecum
- jejunum } small intestine
- large intestine
- appendix
- ileum
- rectum
- anus

The Digestive Tract in Cross Section

Thinking again of the digestive tract as a muscular tube, let's look at a cross section of this tube, within the small intestine, to see how it is structured.

As you can see in **Figure 25.13**, the outermost layer of the tract is connective tissue known as the serosa, inside of which there is a layer called the muscularis externa, which is composed of two sets of smooth muscle. In a process called **peristalsis**, these muscles take turns contracting, which produces waves of contraction that push materials along the length of the digestive tract. Inside the muscularis externa there is the submucosa, a layer of loose connective tissue that contains large blood vessels and a network of nerve fibers, sensory neurons, and motor nerve cells. This nerve tissue helps control and coordinate the contractions of the smooth muscle layers, and also helps regulate the secretions of digestive glands. Then there is the epithelial layer that makes up the internal lining of the tract. It is called the mucosa, because it is constantly bathed in mucous secretions. Along most of the length of the digestive tract, the mucosa has a large number of folds that increase the surface area available for absorption and permit ex-

pansion like an accordion—helpful after a large meal. Additionally, in the small intestine, the mucosa forms fingerlike projections called **villi** (singular *villus*) that greatly increase the area for absorption. It is the capillary and lymph vessels that run through the villi, just outside the epithelium, that nutrients move into after having been sufficiently digested. The internal space of the digestive tract is called its **lumen**.

The Oral Cavity

Let's now look at the process of digestion from start to finish, beginning with the part that's most familiar to us, the mouth.

The mouth opens into the oral cavity, the part of the digestive tract that ingests, or receives, food. Food is, of course, tasted and broken up into smaller pieces in the oral cavity through the mechanism of chewing. This mechanical processing is joined by a lubrication, through mucus and salivary secretions, that also takes place in the oral cavity. The digestion of food—the breakdown by chemical means—starts here as well, as enzymes in the saliva start to break down carbohydrates.

The hard (front) and soft (rear) palates provide a roof for the oral cavity, while the tongue forms its floor. The dividing line

Figure 25.13
Structure of the Digestive Tract
The digestive tract is a tube formed of several layers of tissue, each performing a different function. Note that the absorption of nutrients takes place when food that has been sufficiently digested moves across the epithelium and is taken up by either a capillary or a lymphatic vessel inside a villus.

Absorption of nutrients

VILLUS

microvilli

epithelium

lymphatic vessel

capillary network

SMALL INTESTINE

Mucosa: highly folded lining of intestine where absorption occurs

Submucosa: connective tissue containing blood vessels and nerves

Muscularis externa: circular and longitudinal muscle layers used for peristalsis

Serosa: connective tissue

between the oral cavity and pharynx extends between the base of the tongue and the dangling **uvula** (the structure at the back of the throat that looks like a little punching bag). The muscular tongue is used for manipulating food during chewing, preparing it for swallowing, and sensing its temperature and taste. Most of the tongue lies within the oral cavity, but the base of the tongue extends into the pharynx.

Three pairs of salivary glands produce up to 1.5 liters, or almost half a gallon of saliva each day. Saliva is more than 99 percent water, but it also contains ions, pH neutralizers, waste products, antibodies, and the digestive enzymes already mentioned. Between meals, small quantities of saliva are continually released onto the surfaces of the oral cavity. These secretions clean the oral surfaces and, through the action of a bacteria-destroying enzyme, reduce the numbers of bacteria.

The Structure and Function of Teeth

The action of chewing breaks down tough connective animal tissues and plant fibers, and helps saturate the materials with salivary lubricants and enzymes. Adult teeth are shown in **Figure 25.14a**. There are four different types of teeth. Incisors, the blade-shaped teeth found at the front of the mouth, are useful for clipping or cutting. The pointed cuspids also called *canines* or *eyeteeth*, are used for tearing or slashing. Bicuspids (or premolars) and molars have flattened crowns with prominent ridges for crushing and grinding food.

During infant development, two sets of teeth begin to form. The first to appear are the deciduous teeth, also known as *baby teeth*. Beginning at the age of five or six, these teeth are gradually replaced by the adult permanent teeth. As we age, three additional teeth appear at the back of each side of the upper and lower jaws, extending the length of the tooth rows and bringing the permanent tooth count to 32. The third molars, or wisdom teeth, may not erupt before age 21 because of abnormal positioning or inadequate space in the jaw. To avoid problems later in life, wisdom teeth are often surgically removed.

A cross scan of a typical adult tooth is shown in Figure 25.14b. The bulk of each tooth consists of a mineralized material called dentin. It surrounds the connective tissue that makes up the central pulp cavity. The pulp cavity receives blood vessels and nerves through a narrow root canal at the tooth's base, or root, which sits within a bony socket. The neck of the tooth marks the boundary between the root and the crown. The dentin of the crown is covered by a layer of enamel—the hardest substance manufactured by living things.

The Pharynx and Esophagus

The **pharynx**, or throat, transmits solid food, liquids, and air. The pharyngeal muscles cooperate with muscles of the oral cavity and esophagus to initiate the process of swallowing. The muscular contractions during swallowing force the food along the esophagus and into the stomach.

The **esophagus** is a muscular tube that begins at the pharynx and ends at the stomach. It is approximately 2 centimeters or about 0.8 inch in diameter.

Figure 25.14
Adult Teeth

a Permanent teeth of an adult, showing the upper dental arch (roof) of the mouth.

b A typical adult tooth.

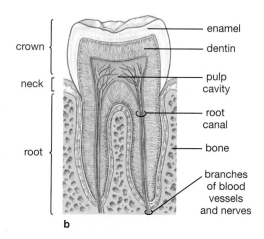

The Stomach

The stomach's main functions include (1) the temporary storage of ingested food, (2) the mixing of ingested foods, and (3) the breaking of chemical bonds (the digestion) of proteins through the action of acids and enzymes. What leaves the stomach is a mixture of food and gastric juices; this thick, soupy material is called **chyme**.

Figure 25.15 shows the muscular stomach. The **pyloric sphincter** muscle at one of its ends regulates the flow of chyme between the stomach and small intestine. The dimensions of the stomach are extremely variable. When empty, it resembles a tube with a narrow cavity. When full, it can expand to contain 1 to 1.5 liters or up to almost 0.5 gallon of material. (The fact that it can hold material in this way is the reason we don't have to eat all the time.) This degree of expansion is possible because the stomach wall contains thick layers of smooth muscle, and the mucosa of the relaxed stomach contains numerous folds, called *rugae*. As the stomach expands, the smooth muscle stretches and the folds gradually disappear.

The Stomach's Digestive Juices

The stomach is lined by an epithelium filled with mucous cells. The mucus they produce helps protect the stomach lining from the acids, enzymes, and abrasive materials it contains. Shallow depressions called gastric pits open onto the internal surface of the stomach. Secretions from underlying gastric glands pass through the openings of the gastric pits. Each day, these glands produce about 1,500 milliliters or 45 ounces of gastric juice, which contains hydrochloric acid (HCl) and pepsin, an enzyme that breaks down proteins. The hydrochloric acid lowers the pH of the gastric juice, kills microorganisms, and breaks down cell walls and connective tissues in food.

The Small Intestine

The small intestine is about 3 meters (almost 10 feet) long and ranges in diameter from 4 centimeters (1.6 inches) at the stomach to about 2.5 centimeters (1 inch) at the junction with the large intestine. Although longer than the large intestine, it is called the small intestine because its overall diameter is smaller. It is made up of three regions, as you can see in

Figure 25.16. The **duodenum** is the 25 centimeters (10 inches) of the small intestine closest to the stomach. It receives chyme from the stomach and secretions from the pancreas and liver. The jejunum, the next region, is about 1 meter (3 feet) in length. Most of the chemical digestion and nutrient absorption occurs in the jejunum. The third segment, the ileum, is the longest and averages 2 meters (6 feet) in length. The ileum ends at a sphincter muscle that controls the flow of chyme from the ileum into the large intestine. On average it takes about 5 hours for chyme to pass through the small intestine.

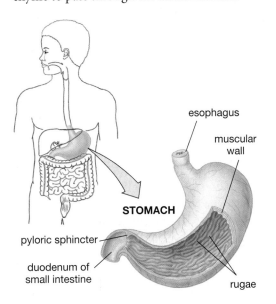

**Figure 25.15
External and Internal Views
of the Stomach**

esophagus

muscular wall

STOMACH

pyloric sphincter

duodenum of small intestine

rugae

**Figure 25.16
Structure of the Small
Intestine**

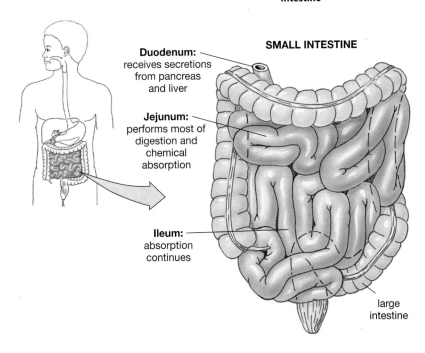

SMALL INTESTINE

Duodenum: receives secretions from pancreas and liver

Jejunum: performs most of digestion and chemical absorption

Ileum: absorption continues

large intestine

Most of the important digestive processes are completed in the small intestine, where the final products of digestion—simple sugars, fatty acids, and amino acids—move out of the digestive system, along with most of the system's water content. Approximately 80 percent of all absorption takes place in the small intestine, with the rest divided between the stomach and the large intestine.

The Pancreas

The pancreas, shown in **Figure 25.17**, lies behind the stomach, extending laterally from the duodenum. It is an elongated, pinkish-gray organ about 15 centimeters (six inches) long that is primarily an exocrine organ—one that secretes its materials through ducts. However, about 1 percent of its cells are the endocrine cells that secrete two substances directly from cells into the bloodstream. One of these substances, insulin, moves blood sugar out of the bloodstream and into cells. The other, glucagon, releases glucose from storage in skeletal muscles and the liver. The exocrine secretions of the pancreas are digestive enzymes and buffers that are carried to the duodenum by the **pancreatic duct**, which passes through the duodenal wall along with the common bile duct from the liver and gallbladder. Pancreatic exocrine enzymes are broadly classified according to their intended targets. **Lipases** work on lipids, **carbohydrases** digest sugars and starches, and **proteases** break proteins apart.

The Liver and Gallbladder

Setting aside the skin, the liver is the largest organ in the body, being about the size of a football and weighing about 1.5 kg (3.3 pounds). In addition to its digestive role, this large, firm, reddish-brown organ provides a wide range of other essential services vital to life.

The liver lies directly under the diaphragm. As Figure 25.17 shows, it is divided into two large lobes, the left and right. Lodged within a recess under the right lobe of the liver is the **gallbladder**, a muscular sac that stores and concentrates a digestive substance called *bile*. Produced by the liver, **bile** aids in the digestion of fats and can move into the small intestine through either of two routes. It always passes through passageways, called *hepatic ducts*, within the liver until it eventually leaves the liver through the common hepatic duct. The bile in this duct may either (1) flow into the small intestine's duodenum through the common bile duct, or (2) enter the cystic duct that leads to the gallbladder and only later flow into the small intestine (through the common bile duct). Why would it take one course or the other? A sphincter at the intestinal end of the common bile duct stays opens only at mealtimes. At other times, bile backs up through the cystic duct for storage within the expandable gallbladder before moving to the small intestine. Liver cells produce roughly 1 liter (about a quart) of bile each day.

Functions of the Liver

The liver plays a central role in regulating body metabolism. It can do this because, as you've seen, all the blood leaving the absorptive areas of the digestive tract flows through the liver before going into general circulation. This enables liver cells to extract absorbed nutrients or toxins from this blood before they reach the rest of the body, and also to monitor and adjust the circulating levels of nutrients. Because the liver is the first stop for much digested material, it is also the first stop for toxins, such as alcohol (which is one of the few substances to move directly from the stomach to the liver without going through the small intestines). It's not surprising, then, that people who have abused their bodies with too much alcohol can over time find themselves with damaged livers.

The liver also functions as the largest blood reservoir in the body. In addition to nutrient-laden blood arriving through the hepatic portal vein described earlier, the liver

**Figure 25.17
Accessory Glands in
Digestion: The Liver,
Pancreas, and Gallbladder**

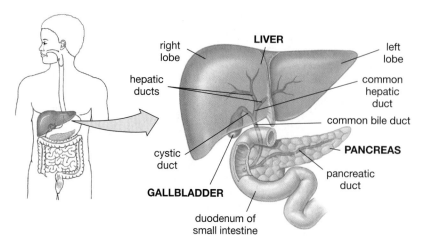

also receives about 25 percent of the cardiac output (through the hepatic artery). Phagocytic cells in the liver constantly remove aged or damaged red blood cells, debris, and pathogens from the body's circulation.

The Large Intestine

The large intestine begins at the end of the third section of the small intestine, the ileum, and ends at the anus. As you saw in Figure 25.12, the large intestine lies below the stomach and liver and almost completely frames the small intestine. If you look at **Figure 25.18**, you can see the structure of this organ in greater detail. The main function of the large intestine is to hold and compact the refuse produced by the rest of the digestive system (as solid waste, or feces). It also absorbs water and important vitamins produced by resident bacteria. The large intestine is called "large" only because it has a wider diameter than the small intestine, but it is only about 1.5 meters (5 feet) long—about a quarter the length of the small intestine. Often called the *large bowel*, it is divided into three major regions: (1) the cecum; (2) the colon; and (3) the rectum. Let's look at these in order.

Three Regions of the Large Intestine: Cecum, Colon, and Rectum

Material (chyme) arriving from the ileum of the small intestine first enters an expanded chamber, called the **cecum**. A muscular sphincter, the ileocecal valve, guards the connection between the ileum and the cecum. The cecum usually has the shape of a rounded sac, and the slender, hollow vermiform appendix attaches to it, below the ileocecal valve. The appendix walls contain lymphatic tissue and it is infection of this tissue that produces the symptoms of appendicitis. The longest portion of the large intestine is the **colon**. Visible bands of longitudinal muscle produce the series of pouches characteristic of the colon. The lower portion of the colon empties into the rectum.

The reabsorption of water is an important function of the large intestine. Although roughly 1,500 milliliters (45 ounces) of watery material arrives in the colon each day, some 1,350 milliliters of water move back out of it, leaving only about 150 milliliters (4.5 ounces) that is ejected with the feces. In addition to water, the large intestine absorbs

vitamins produced by bacteria that live within the colon.

Bacterial action breaks down protein fragments remaining in the feces into various compounds. Some of these are reabsorbed, processed by the liver into relatively nontoxic compounds, and eventually excreted at the kidneys. Others are responsible for the odor of feces or result in the generation of hydrogen sulfide (H_2S), a gas that produces a "rotten egg" odor.

Some carbohydrates are not digested and arrive in the colon intact, where they provide a nutrient source for bacteria. The metabolic activities of these bacteria are responsible for intestinal gas, or flatus. Beans often trigger gas because they contain a high concentration of indigestible polysaccharides for the bacteria to work on. Movement of chyme through the large intestine occurs very slowly, allowing hours for the reabsorption of water.

The **rectum** forms the end of the digestive tract. The rectal chamber is usually empty except when powerful peristaltic contractions force fecal materials out of the

**Figure 25.18
The Large Intestine**

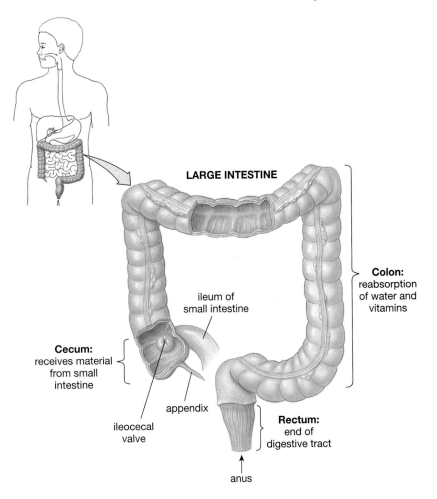

LARGE INTESTINE

ileum of
small intestine

Cecum:
receives material
from small
intestine

ileocecal
valve

appendix

anus

Colon:
reabsorption
of water and
vitamins

Rectum:
end of
digestive tract

lower part of the colon. Stretching of the rectal wall then triggers the defecation reflex. **Defecation**, the excretion of solid wastes, occurs through the rectal opening, the anus.

25.8 Different Digestive Processes for Different Foods and Nutrients

A typical meal contains a mixture of carbohydrates, proteins, lipids, water, electrolytes (ions), and vitamins. Water, electrolytes, and vitamins can be absorbed directly into circulation, but large organic molecules, such as starches and proteins, must be broken down through digestion before absorption can occur. Once absorbed, these molecules can be used by the body as building blocks for other molecules or to generate the body's "energy currency" molecule, ATP.

Organic molecules are usually complex chains of simpler molecules. A complex carbohydrate, such as a starch, is made of simple sugars. In a protein, meanwhile, the building blocks are amino acids, and in lipids they are fatty acids and glycerol molecules. Digestive enzymes split the bonds between the different building-block molecules in a process involving water called **hydrolysis** ("to split with water.") (For more on carbohydrates, lipids, and proteins, see Chapter 3, beginning on page 45.)

Carbohydrates

Carbohydrate digestion begins in the mouth through the action of salivary amylase. Amylase breaks down complex carbohydrates into smaller fragments, producing a mixture primarily composed of two- and three-sugar molecules. Salivary amylase travels with the meal and continues to digest the starches and glycogen in it for an hour or two before stomach acids render it inactive. In the duodenum, the remaining complex carbohydrates are broken down through the action of pancreatic amylase.

Further digestive activity occurs on the surfaces of the intestinal microvilli, where digestive enzymes reduce the two- and three-sugar molecules to simple single sugars. *Now* the carbohydrates exist in a form that is ready to pass into circulation. The simple sugars pass through the cells of the intestinal wall into the interstitial fluid of the villi.

They then enter the intestinal capillaries and are transported directly to the liver by the hepatic portal vein.

Proteins

Proteins are relatively large and complex molecules, and different regions of the digestive tract cooperate to take proteins apart. Stomach cells secrete acid and the protein-splitting enzyme pepsin. The strongly acidic environment of the stomach provides the proper pH for pepsin to begin the process of reducing the large protein molecules into smaller fragments.

After the chyme enters the duodenum and the pH has risen (become less acidic), pancreatic enzymes come into play. Working together, different protease enzymes from the pancreas break the protein fragments into a mixture of short amino acid chains and individual amino acids. Enzymes on the surfaces of the microvilli complete the process by breaking the short amino acid chains into single amino acids. In this form, they are absorbed by the intestinal cells and then diffuse into intestinal capillaries and are carried to the liver.

Lipids

The most common dietary lipids are **triglycerides**. A triglyceride molecule consists of three fatty acids attached to a single molecule of glycerol. Triglycerides and other dietary fats are relatively unaffected by conditions in the stomach and enter the duodenum in the form of large, lipid drops. Bile breaks the large drops apart in a process called **emulsification**, creating tiny droplets that now present enough surface area for pancreatic lipase to begin breaking the triglycerides apart. Although the resulting lipid fragments are absorbed by the intestinal cells, they do not then enter intestinal capillaries. Instead, they are attached to proteins that aid their transport. Too large to enter the capillaries, the lipid-protein packages enter small lymphatic vessels that underlie the intestinal villi (Figure 25.13). From these vessels, the lipids proceed along the larger lymphatic vessels, and finally enter the venous circulation.

Water and Vitamins

Each day 2 to 2.5 liters (or 2.1 to 2.65 quarts) of water enter the digestive tract in the form of food and drink. The salivary, gastric, intestinal, and accessory gland secretions provide another 6 to 7 liters (6.3 to 7.4 quarts). The water total, then, is about 9.5 liters or 2.5 gallons a day; of this, a remarkably small amount, about 150 milliliters or 4.5 ounces, goes out of the body with the fecal wastes, the rest being absorbed into the body from the digestive system.

Vitamins are organic compounds required in very small quantities for critical chemical reactions. There are two major groups of vitamins: water-soluble vitamins (those that dissolve in water; the C and the B vitamins) and fat-soluble vitamins (those that dissolve in fat; A, D, E, and K).

There is plenty of fat (or lipid) material in the body for the fat-soluble vitamins to dissolve into, such as the plasma membranes of cells, which are largely lipid. This means that our bodies can *store* substantial amounts of the fat-soluble vitamins. This can be very helpful, because we might be able to go for months at a time on stored reserves. It also means, however, that we can get too *much* of a fat-soluble vitamin such as A or D, with potentially harmful consequences. (Too much A can cause liver damage; too much D can cause calcium deposits.) There's little danger of this with the water-soluble vitamins, because they readily move between fluid stores and the bloodstream, with excess amounts excreted in the urine.

25.9 The Urinary System in Overview

Here our subject changes once again (for the last time in this chapter), as your tour of the body's systems moves from digestion to the elimination of liquid wastes. All organisms create waste materials simply by living; the question every living thing confronts is how to get rid of them. Single-celled creatures can simply dump their waste materials into their external environment through exocytosis. (See Chapter 5, page 105.) In all highly organized animals, excretory *organs* take over this task. If these organs are small and numerous, and scattered throughout the body, they are known as *nephridia*. Conversely, higher invertebrates and vertebrates have a single compact set of excretory organs that performs this function. These organs—the kidneys, the bladder, and others—are what you'll be looking at in this review of the body's urinary system.

Though gathering and discharging waste is an important function of the urinary system, it has other, equally important roles. For example, it regulates blood volume—how much blood we have within us—a function whose necessity becomes obvious when we think about consuming lots of liquids. Suppose the body simply incorporated all the water in these liquids into circulating blood? We would eventually swell and burst like a water balloon. Instead, the body *regulates* how much water it retains and how much it excretes, with the kidneys being a critical player in this process. The urinary system also controls ion concentration and maintains pH balance in the body. In short, it is important in preserving homeostasis. We almost never think of the urinary system in this regard, essentially because it works so well. But a person who suffers kidney failure and must rely on a kidney "dialysis" machine to filter wastes knows the value of the urinary system in a personal way.

In carrying out its activities, the urinary system is very much involved in the *conservation* of bodily resources. Indeed, the kidneys are as much in the conservation business as the urination business. They are conserving water, amino acids, ions, sugars—all kinds of things that the body needs. Consider that the kidneys will process about 180 liters (or almost 48 gallons) of fluid every day. Yet our urination is only about 1.8 liters, or about *half* a gallon daily. More than 47 gallons cycled back into the body each day! Conservation indeed. Now let's see how the kidneys and the larger urination system manage this feat.

Structure of the Urinary System

The organs of the urinary system are shown in **Figure 25.19** on page 560. The pivotal organs in the system are the two **kidneys** that produce **urine**. Urine leaving the kidneys travels along two tubes, the left and right **ureters**, to the **urinary bladder** for temporary storage. When urination occurs, contraction of the muscular bladder forces the urine through the **urethra** and out of the body.

25.10 How the Kidneys Function

The kidneys are located on either side of the vertebral column, at about the level of the eleventh and twelfth ribs. Not surprisingly, they are shaped like kidney beans. A typical kidney is about 10 centimeters (4 inches) long and about 3 centimeters (1 inch) thick.

In considering how the kidneys work, it's helpful to think first in terms of input and output. What comes into the kidney in the way of fluids? Blood from the circulatory system by way of the renal arteries. As you can see in Figure 25.19, these branch off from the aorta, one renal artery going to the left kidney, one to the right. (The word *renal* comes from *renes*, which means "kidneys"; you'll be seeing a lot of renal this and that.) Now, what comes *out* of the kidney? As you've seen, urine, which exits each kidney through its ureter. However, blood is also flowing out of the kidneys, through the renal *veins*. Thus, each kidney has one input (the renal artery) but *two* outputs (the ureter and the renal vein). This is a clue to what's happening inside. Blood is flowing into the kidney through the renal artery and out through the renal vein; but inside the kidney, some of the materials that come in with the arterial blood (the waste materials) are being channeled out of the blood vessels and into a *separate* system of vessels—the system whose output tubes are the ureters. Now let's see how this happens inside the kidney.

Seen in cross section in **Figure 25.20a**, the kidney can be divided into an outer **renal cortex**, an inner **renal medulla**, and a chamber called the **renal pelvis** whose numerous

branches collect the urine produced by the kidneys and channel it into the ureter.

The work of the kidney begins in its outer portion, or cortex. If you look at **Figure 25.20b**, you can see the basic working unit of the kidney, called a *nephron*. Each kidney has within it a million or so nephrons, each serviced by a single renal arteriole that has branched off from the entering renal artery. In the figure you can see that an arteriole branching off the renal artery enters a hollow chamber, called the **Bowman's capsule**, and then promptly expands into a knotted network of capillaries, called the **glomerulus**, that is surrounded by the capsule. *Emerging* from this network, there is again a single arteriole, but it once again branches into a network of capillaries that surrounds a convoluted nephron tubule that has both proximal ("near") and distal ("distant") portions. The **nephron**, then, is the functional unit of the kidneys, composed of a nephron tubule, its associated blood vessels, and the interstitial fluid in which both are immersed. At the "output" end of the nephron, the tubule is one of several that are emptying into a common **collecting duct**. Eventually, many of these ducts will themselves merge to feed into the renal pelvis, which feeds into the ureter.

Figure 25.21 (page 562) uncomplicates things a little by showing you what the nephron tubule structure looks like without the surrounding blood vessels. You can see blood input at the Bowman's capsule, then the proximal portion of the nephron tubule, known properly as the **proximal tubule**. Then there is the U-shaped loop of Henle, which ends with the **distal tubule** that feeds into the collecting duct. Note that the Bowman's capsule is in the renal cortex, but that the loop of Henle dips deep into the renal medulla. Now, what happens in this structure?

First Kidney Function: Filtration in Bowman's Capsule

Blood flows into the glomerulus capillaries under high pressure. As a result, some of the plasma component of the blood moves out of the capillaries and into the Bowman's capsule. Water, amino acids, ions, and glucose contained in the blood—along with a waste product called *urea*—are among the substances that will make this transition. Together, they constitute a fluid, called **filtrate**,

Figure 25.19
The Urinary System in Overview

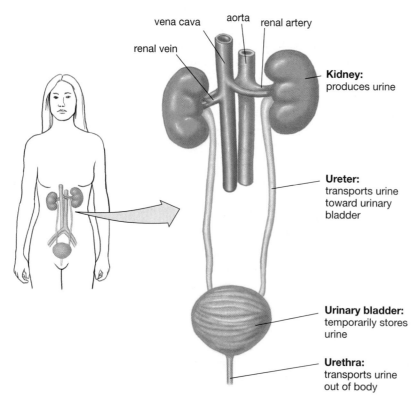

vena cava

aorta

renal artery

renal vein

Kidney:
produces urine

Ureter:
transports urine
toward urinary
bladder

Urinary bladder:
temporarily stores
urine

Urethra:
transports urine
out of body

that is funneled from the capsule into the proximal tubule. In short, a transfer has been made from the blood vessels to the kidney's nephron tubule. What is not part of this **filtration** process are materials that were too big to pass out of the glomerulus capillaries: red blood cells, proteins, and lipid molecules.

Second Kidney Function: Reabsorption from the Nephron Tubule

Remember how you saw earlier that the kidneys are in the conservation business to a significant extent? Well, this conservation begins immediately after the filtrate moves into the proximal tubule. Cells in the tubule actively pump sodium (Na^+) from the filtrate into the interstitial fluid surrounding the tubule. Chloride (Cl^-) and water follow the sodium, and all these substances move into the network of capillaries surrounding the tubule, along with amino acids and sugars, which also are pumped out of the tubule. What is happening, then, is **reabsorption** of valuable materials back into the bloodstream. Indeed, about two-thirds of the water—and nearly all the glucose and amino acids that initially became part of the filtrate—are now moving back to the circulatory system.

Third Kidney Function: Secretion into the Nephron Tubule

If you now jump ahead (past the loop of Henle) to the *distal* tubule, you can see a similar movement of water and salt out of the tubule and into the bloodstream. Notice, however, that substances are also moving into the tubule at this point: Hydrogen ions (H^+) and potassium ions (K^+) are among the compounds that will move from the capillaries into the distal tubule, in a process known as **secretion**. (Substances coming from the capillaries are being taken up by tubule cells, which are secreting the substances into the tubule.) Some poisons, and medicines such as penicillin will move into the tubules at this point, to be excreted in the urine.

Fourth Kidney Function: The Loop of Henle, the Collecting Duct, and the Conservation of Water

As noted earlier, at the proximal tubule about two-thirds of the water in the filtrate moved back to the circulatory system. What about the other third of the water? This is where the loop of Henle and the collecting duct come into play. The makeup of the loop allows for the selective movement of water and salt (NaCl) out along its length. In this process, the loop functions to maintain a concentration of salt in the interstitial fluid—a concentration that increases with depth in the medulla. This salt gradient exists in the fluid surrounding both the loop of Henle and the collecting duct. The result is a greater solute concentration outside the loop and collecting duct than inside them, particularly at the lower part of the loop. This means that water moves by osmosis out of the collecting duct, into the interstitial fluid, and from there back into the circulatory system. The arrows you see around the loop in Figure 25.21 show the movement of water and salt along its length.

When the filtrate reaches the collecting duct, it is called urine; with its loss of water in the collecting duct, the urine that flows through the duct may have as much as four

Figure 25.20
Structure of the Kidney and Its Working Unit

a A kidney in cross section.

b A nephron in the kidney, composed of a tubule and its associated blood vessels, which are immersed in interstitial fluid.

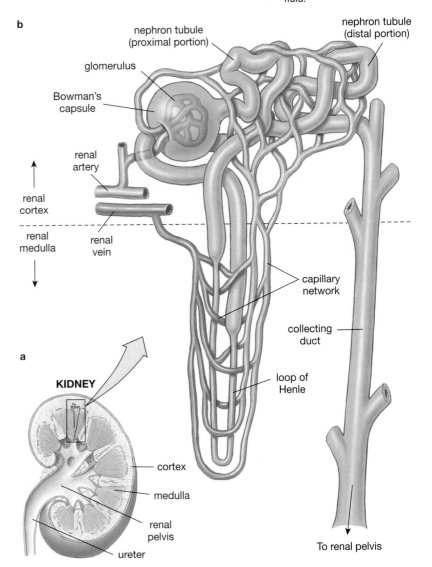

b

nephron tubule (proximal portion)

nephron tubule (distal portion)

glomerulus

Bowman's capsule

renal artery

renal cortex

renal medulla

renal vein

capillary network

collecting duct

loop of Henle

a

KIDNEY

cortex

medulla

renal pelvis

ureter

To renal pelvis

times the concentration of dissolved solvents as does blood plasma. In short, this whole system allows the body to rid itself of a large amount of waste with only a minimal loss of water and nutrients.

To recap, the kidneys function by undertaking four main processes:

- Filtration: Portions of the blood plasma move from the glomerulus capillaries into the nephron tubule by way of the Bowman's capsule.
- Reabsorption: Water and nutrients move from the tubule back to the blood vessels, while waste is retained within the tubule.
- Secretion: Hydrogen ions and various compounds (for example, toxins) move from the capillaries into the tubules.
- Concentration: The final solute concentration of the urine is increased through removal of water from the filtrate in the collecting duct, thus conserving water.

Hormonal Control of Water Retention

The kidneys don't just go along producing the same amount of urine in simple accordance with the amount of filtrate that flows into the Bowman's capsule. Rather, the kidneys are able to *control* how much water there is in urine. This means they control how much water we retain or urinate, with this regulation in turn being under hormonal control. This capacity can be critical for people who are, say, lost in the desert. In that situation, the body needs to *retain* as much water as possible. How does it do it? The heart has receptors in it that sense blood volume, while the brain's hypothalamus has receptors that sense the solute concentration in the blood. The concentration of solutes in our desert travelers' blood is likely to rise, while their blood volume will fall because they have been steadily losing water (in breathing,

Figure 25.21
Physiology of the Nephron

1. Blood flows into the glomerulus within the Bowman's capsule under pressure, driving some plasma components of the blood out of the glomerulus and into the capsule.

2. Conservation begins, as water, glucose, and other needed materials move from the proximal tubule back into the circulatory system.

3. Water and then salt move out of the tubule in the loop of Henle, producing an osmotic effect that moves more water out of the tubule, and into the interstitial fluid, at the collection duct.

4. Acids, toxins, and ions move from capillaries into the distal tubule, even as water and salt move from the tubule back into the bloodstream.

5. Water moves from the collecting duct into the interstitial fluid and from there back into the circulatory system, thus preserving the water. Urine that remains drains to the renal pelvis.

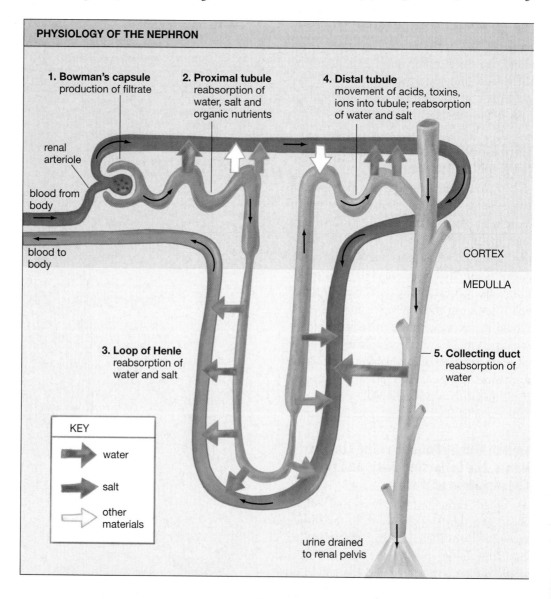

PHYSIOLOGY OF THE NEPHRON

1. Bowman's capsule
production of filtrate

2. Proximal tubule
reabsorption of water, salt and organic nutrients

4. Distal tubule
movement of acids, toxins, ions into tubule; reabsorption of water and salt

renal arteriole

blood from body

blood to body

3. Loop of Henle
reabsorption of water and salt

CORTEX

MEDULLA

5. Collecting duct
reabsorption of water

KEY

water

salt

other materials

urine drained to renal pelvis

perspiring, and so forth). When these things happen, the hypothalamus orders the release of **antidiuretic hormone (ADH)** from the pituitary gland. ADH works on the distal nephron tubule and the collecting duct, having the same effect on both: It increases their permeability to water. This means that more water moves out of the tubule and collecting duct and thus back *into* circulation. Looked at another way, it means that the urine becomes more concentrated because it has less water in it. Should our travelers be saved, however, and then not be able to resist a fourth glass of lemonade, the pituitary's production of ADH will be shut down. The tubule and collecting duct will *retain* more water, meaning that more of it will be urinated. Along these lines, if you've ever wondered why 12 ounces of beer makes a person have to urinate more than does 12 ounces of lemonade, it's because the alcohol in the beer suppresses the production of ADH.

25.11 Urine Transport, Storage, and Excretion

Having seen how the kidneys produce urine, let's now look at the steps by which urine is then excreted from the body. Recall that urine is transported to the urinary bladder by the pair of tubes called the **ureters**—one stemming from each kidney. The ureters are muscular tubes that extend for a distance of about 30 centimeters, or 12 inches. The wall of each ureter contains an inner expandable epithelium, a middle layer of smooth involuntary muscle, and an outer connective tissue covering. Starting at the kidney, about every 30 seconds a peristaltic wave of muscle contraction squeezes urine out of the renal pelvis, along the ureter, and into the urinary bladder. Occasionally, solids composed of calcium deposits, magnesium salts, or crystals of uric acid develop within the collecting tubules, collecting ducts, or ureters. Such kidney stones can be very painful; in addition to obstructing the flow of urine, they may also reduce or eliminate filtration in the affected kidney.

The Urinary Bladder

The **urinary bladder**, shown in **Figure 25.22**, is a hollow muscular organ that stores urine. Its dimensions vary with the volume of stored urine, but a full urinary bladder can contain up to 800 milliliters of urine, which is about 27 ounces. The area surrounding the urethral entrance, called the *neck* of the urinary bladder, contains the internal sphincter muscle that provides involuntary control over the discharge of urine from the bladder.

The Urethra

The urethra extends from the urinary bladder to the exterior of the body. Its length and functions differ in males and females. In the male (shown in Figure 25.22), the urethra extends to the tip of the penis and is about 18 to 20 centimeters (7 to 8 inches) in length. The initial portion of the male urethra is surrounded by the prostate gland. In addition to transporting urine, the male urethra carries sperm cells and the reproductive secretions from the prostate and other glands. In contrast, the female urethra is very short, extending 2.5 to 3.0 centimeters (about 1 inch), and transports only urine. In both sexes, the urethra contains a circular band of skeletal muscle that forms an external sphincter. Unlike the internal sphincter, the external sphincter is under *voluntary* control.

Urine Excretion

The urge to urinate usually appears when the bladder contains about 200 milliliters (about 7 ounces) of urine. We become aware of this need through nerve impulses sent to the brain from stretch receptors in the wall of the urinary bladder. The stimulation of these receptors also results in involuntary contractions of the urinary bladder (which is why we "have to go" to the bathroom). Such contractions increase the fluid pressure inside

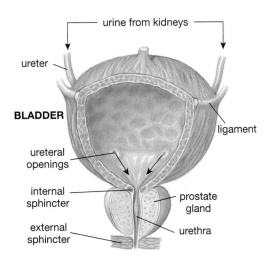

urine from kidneys

ureter

BLADDER

ligament

ureteral openings

internal sphincter

prostate gland

external sphincter

urethra

**Figure 25.22
The Urinary Bladder in a Male**

the bladder, but urine ejection cannot occur unless both the internal and external sphincters are relaxed. We control the time and place of urination by voluntarily relaxing the external sphincter. When this sphincter relaxes, so does the internal sphincter. If the external sphincter does not relax, the internal sphincter remains closed, and the bladder gradually relaxes. A further increase in bladder volume begins the cycle again, usually within an hour. Once the volume of the urinary bladder gets large enough, sufficient pressure is generated to force open the internal sphincter. This leads to an uncontrollable relaxation in the external sphincter, and urination occurs despite voluntary opposition. Normally, less than 10 milliliters (about a third of an ounce) of urine remain in the bladder after urination.

As noted earlier, an average of about 1.8 liters (half a gallon) of urine is excreted daily.

Normal urine is a clear, bacteria-free solution with a yellow color.

On to Development and Reproduction

Animals breathe and move and fight off invaders in ways that have been reviewed over the last three chapters. Each generation of animals gives rise to another, of course, through the workings of another specialized system, the reproductive system. Between the moment of conception and the moment of birth, an animal must also go through an amazing transformation—in the case of humans, from a single cell to a baby in full cry. How do single sets of male and female cells interact to form a new living thing? How does this new life then make the transition from one-celled existence to trillion-celled existence? These are the subjects of the next two chapters.

Chapter Review

Key Terms

alimentary canal 552

alveolus 549

antibody 544

antidiuretic hormone (ADH) 563

aorta 546

arteriole 544

artery 544

bile 556

Bowman's capsule 560

capillary 544

capillary bed 545

carbohydrase 556

cecum 557

chyme 555

closed circulation 545

clotting factor 542

collecting duct 560

colon 557

coronary artery 547

cortex 560

defecation 558

digestive tract 551

Summary

25.1 The Cardiovascular System and Body Transport

- The cardiovascular system transports materials to and from the body's cells. These materials include oxygen, carbon dioxide, nutrients, vitamins, hormones, waste products, and molecules and cells specialized for the body's defense. These materials are transported within blood.

- The two major components of blood are formed elements and blood plasma. Formed elements include red blood cells, white blood cells, and cell fragments called *platelets*. Blood plasma is the liquid medium in which the formed elements are suspended.

- Red blood cells constitute 99.9 percent of all the formed elements in blood and are responsible for transporting oxygen to the body's tissues and carbon dioxide from them. White blood cells come in many varieties and are active in defending the body from invaders. Platelets—cell frag-

ments broken off from larger cells—assist in blood clotting at injury sites.

- Plasma is 92 percent water, but it also contains a number of dissolved materials, including proteins. There are three primary classes of plasma proteins: albumins, which transport hormones and fatty acids; globulins, which aid the immune system and serve as transport proteins; and fibrinogen, which functions in blood clotting. Other plasma compounds include electrolytes, organic nutrients, organic wastes, and hormones.

- Blood vessels carrying blood away from the heart are arteries; blood vessels returning blood to the heart are veins. The smallest of the blood vessels, the capillaries, connect the arteries with the veins.

- Vertebrates such as humans utilize closed circulation—circulation in which blood is delivered from one set of vessels to another. This contrasts with the system of open circulation, used by many invertebrates, in which arteries carry blood into open spaces, called *sinuses*. The blood then flows around cells and is returned to a heart or hearts through openings in veins.

25.2 The Heart and the Circulation of Blood

- Blood is transported out to the various tissues of the body through the muscular contractions of the heart.

- There are two general networks of blood vessels in the cardiovascular system. One is the pulmonary circulation, in which blood flows between the heart and the lungs. The other is the systemic circulation, in which blood circulates between the heart and the rest of the body.

- The human heart contains four muscular chambers—two associated with pulmonary circulation, and two associated with systemic circulation. The left atrium and right ventricle are associated with pulmonary circulation, while the right atrium and left ventricle are associated with systemic circulation.

- Blood returning to the heart from the systemic circulation flows into the right atrium from veins called the *superior* and *inferior vena cava*. It then flows into the right ventricle, from which it is pumped into the pulmonary arteries, which carry it to the lungs. It returns to the left atrium through the pulmonary veins and then flows into the left ventricle, from which it is pumped to the rest of the body.

- A series of valves that open and close ensures that blood flows only one way through the heart.

25.3 The Heart's Own Blood Supply: What Is a Heart Attack?

- About half of all deaths in the United States today are caused by the blockage of one of the coronary arteries, which supply heart tissue with blood. Such blockages can cut off the blood supply to collections of heart-muscle cells, which can die if deprived of blood for long enough. The result is a heart attack.

- Heart cells that have died are replaced by scar tissue, which cannot take part in the heart's pumping function. High-fat diets, smoking, and lack of exercise all are contributing factors to heart disease.

25.4 Getting the Goods to and from the Cells: The Capillary Beds

- Arteries near the heart branch into smaller-diameter arterioles, which feed into the delivery vehicles of the cardiovascular system, the capillary beds. The capillary beds then feed back into the body's system of veins that returns blood to the heart. The venous system begins with venules that merge into larger veins.

- Materials carried by arteries that are needed by the body's tissues move directly out of the capillaries (through their thin walls) and into the interstitial fluid—the liquid in which both the capillary and the cells that surround it are immersed. Meanwhile, carbon dioxide and wastes from cells flow into capillaries from the interstitial fluid.

- The movement of materials out of the capillaries occurs primarily near the arterial end of the capillary beds and is driven by blood pressure; movement of materials into the capillaries occurs primarily near the venous end of the capillary beds and is driven by osmosis.

- Blood pressure is at low levels by the time blood has moved through the capillaries. It is returned to the heart by the contraction of muscles, which squeeze the veins in a way that moves the venous blood toward the heart. A system of valves in the veins blocks venous movement away from the heart.

25.5 The Respiratory System and the Exchange of Gases

- The respiratory system functions primarily to capture oxygen and to dispose of carbon dioxide. It also aids in immune functions, in producing sounds for speaking, in regulating blood volume and pressure, and in controlling body fluid pH.

- The respiratory system includes the lungs; the nose, nasal cavity, and sinuses; the pharynx (throat); the larynx (voice box); the trachea (windpipe); and the conducting passageways, called *bronchi* and *bronchioles*, that lead to the lungs. The end points of the bronchioles are the air sacs called *alveoli*, which are the sites of oxygen and carbon dioxide exchange in the lungs.

- Steps in respiration include, first, breathing or pulmonary ventilation, meaning the movement of air into and out of the lungs. The second step is the exchange of respiratory gases that occurs in two general processes: external respiration, or the diffusion of gases between the alveoli and the blood; and internal respiration, meaning the two-way diffusion of gases between capillary blood and interstitial fluid.

- In both internal and external respiration, gas exchange occurs through diffusion down concentration gradients; the gases are moving from areas of their higher concentration to areas of their lower concentration.

- Almost all the oxygen that moves from the alveoli to the blood in the pulmonary capillaries is bound to molecules of the protein hemoglobin, which exists within red blood cells. Carbon dioxide is transported within the blood in several different ways; some dissolves in blood plasma, some binds to hemoglobin in red blood cells, and some is converted to bicarbonate ions that move from red blood cells into the blood plasma.

25.6 The Digestive System

- The central function of the digestive system is to get the foods the body ingests into a form the body can use; the system must also collect and then rid the body of the waste that remains once the useful material from the food has been absorbed.

- The process of digestion is a process of the breakdown of foods—from large pieces of food to smaller pieces, and then from relatively large molecules of food to smaller molecules. It is these smaller molecules that leave the digestive system for transport through the rest of the body.

- The central feature of the digestive system is a muscular tube, the digestive tract, that passes through the body from the mouth to the anus. The functions of the digestive system involve a series of interconnected steps: ingestion, mechanical processing, digestion, secretion, absorption, and excretion.

25.7 Components of the Digestive System

- The mouth opens up into the oral cavity, which receives or ingests food that is then broken up into smaller pieces through the mechanical action of chewing.

- There are four different types of teeth: incisors (used for clipping and cutting), cuspids (tearing, slashing), bicuspids, and molars (crushing and grinding).

- The pharynx (throat) transmits solid food, liquids, and air. The esophagus is a muscular tube that begins at the pharynx and ends at the stomach.

- The stomach is a muscular digestive organ whose functions include the temporary storage of food, the mixing of ingested foods, and the breaking of chemical bonds of proteins through chemical action. Gastric glands lining the stomach produce quantities of gastric juice that aid in digestion and kill microorganisms. The stomach empties chyme—a mixture of food and gastric juices—into the small intestine.

- The small intestine is the part of the digestive tract that receives chyme from the stomach, and completes the digestion of most of the food in it. Absorption is the movement of digested food out of the digestive tract; about 80 percent of all absorption takes place in the small intestine, with the rest divided between the stomach and large intestine. The small intestine is aided in its digestive function by the digestive secretions it receives from the pancreas and the liver.

- The pancreas is an organ that lies behind the stomach and that secretes both blood-borne hormones (insulin and glucagon) and substances that move through the pancreatic duct (digestive enzymes and pH buffers).

- The liver is a large organ that lies directly under the diaphragm and plays a central role in regulating the body's metabolism. All the blood leaving the absorptive areas of the digestive tract flows through the liver before going into general circulation. Thus the liver is the first stop for much digested material and can monitor and adjust circulating levels of nutrients. The liver also functions as the largest reservoir of blood in the body and serves to detoxify harmful substances before they reach the rest of the body.

- Lodged under the liver, the gallbladder is a muscular sac that stores and concentrates a substance produced by the liver, bile, which aids in the digestion of fats. Bile can move into the small intestine either from the gallbladder or directly from the liver.

- The large intestine begins at the small intestine and ends at the anus. The main function of the large intestine is to hold and compact the refuse produced by the rest of the digestive system (as solid waste, or feces). It also absorbs water and important vitamins produced by resident bacteria. The final portion of the colon, the rectum, forms the end of the digestive tract. The rectal chamber is usually empty except when peristaltic contractions force fecal materials out of the lower part of the colon. Stretching of the rectal wall then triggers the defecation reflex.

25.8 Different Digestive Processes for Different Foods and Nutrients

- A typical meal contains a mixture of carbohydrates, proteins, lipids, water, electrolytes (ions), and vitamins. Water, electrolytes, and vitamins can be absorbed directly into circulation, but large organic molecules, such as starches and proteins, must be broken down through digestion before absorption can occur. Different types of nutrients are digested in different ways.

- Both carbohydrates and proteins are broken down into their component molecules (simple sugars and amino acids, respectively) and move through intestinal cells, diffusing into intestinal capillaries. They then move through the capillaries to the hepatic portal vein, which carries them to the liver. Conversely, because of their large size, lipids do not enter intestinal capillaries. Instead, they enter lymphatic vessels after becoming attached to proteins. From these vessels, lipids enter the larger lymphatic vessels, after which they enter venous circulation.

- Vitamins are organic compounds required in small amounts for chemical reactions in the body. There are two major groups of vitamins: water-soluble vitamins (those that dissolve in water; the C and the B vitamins) and fat-soluble vitamins (those that dissolve in fat; A, D, E, and K). Humans can store fat-soluble vitamins for significant periods of time, but we cannot store water-soluble vitamins. This means that we can consume an excess of fat-soluble vitamins, with potentially harmful consequences.

25.9 The Urinary System in Overview

- Vertebrates and higher invertebrates have a single, compact set of excretory organs—the kidneys. The kidneys filter bodily fluids, removing waste in the process, but conserving such bodily materials as amino acids, ions, sugars, and water. By regulating fluid retention, the kidneys serve to control the volume of blood in the body.

- The urinary system consists of two kidneys that produce urine; the left and right ureters that the urine travels through upon leaving the kidneys; the muscular urinary bladder, which temporarily stores urine; and the urethra, through which urine passes out of the body.

25.10 How the Kidneys Function

- The kidneys undertake four principal processes: (1) filtration, in which portions of the blood plasma move from the bloodstream to the kidney's nephron tubules; (2) reabsorption, in which water and nutrients move back into the blood vessels, while waste is retained within the tubules; (3) secretion, in which hydrogen ions, toxins, and other compounds move from the capillaries into the tubules; and (4) concentration, in which the final solute concentration of the urine is increased through the removal of water from the collecting duct, thus conserving water.

- Retention of water by the kidneys is regulated by the hypothalamus' control of the secretion of antidiuretic hormone (ADH) by the pituitary gland. An increased secretion of ADH means that more water will move out of the kidney's tubules and collecting ducts and back into circulation.

25.11 Urine Transport, Storage, and Excretion

- Frequent peristaltic waves of muscle contraction squeeze urine out of the renal pelvis, through the ureter and into the urinary bladder. The urethra extends from the urinary bladder to the exterior of the body.

- The area surrounding the urethral entrance contains the internal sphincter muscle that provides involuntary control over the discharge of urine. The urethra also contains a band of skeletal muscle under voluntary control, the external sphincter. We become aware of the need to urinate through nerve impulses sent to the brain from stretch receptors in the wall of the urinary bladder. Urine ejection cannot occur unless both the internal and external sphincters are relaxed; relaxation of the external sphincter prompts the internal sphincter to relax.

Understanding the Basics

Multiple-Choice Questions

1. The formed elements of blood include
   a. albumins, globulins, fibrinogen
   b. RBCs, WBCs, platelets
   c. plasma, fibrin, serum
   d. plasma, globulins, and lacteals
   e. RBCs, WBCs, PVCs

2. The amount of blood in the average person is in the range of _____ liters.
   a. 1 to 2   d. 8 to 10
   b. 2 to 4   e. 12 to 14
   c. 4 to 6

3. The two-way exchange of substances between blood and body cells occurs only through
   a. capillaries
   b. arteries
   c. veins
   d. arterioles
   e. valves

4. The oxygen-carrying protein within an erythrocyte is
   a. fibrin
   b. hemoglobin
   c. fibrinogen
   d. plasmid
   e. globulin

5. A heart attack can best be described as
   a. a sudden harmful contraction of the heart muscles
   b. an invasion of heart tissue by pathogens such as bacteria
   c. a death of cells in an area of the heart caused by a reduction of blood supply
   d. a failure of the nervous system to cause heart contractions
   e. a rapid, harmful expansion of blood volume supplied to the heart

6. The arteries of the pulmonary circulation differ from those of the systemic circulation in that they carry _____.
   a. oxygen, carbon dioxide, and nutrients
   b. oxygen and nutrients
   c. deoxygenated blood
   d. oxygenated blood
   e. platelets

7. The movement of air into and out of the lungs is
   a. external respiration
   b. internal respiration
   c. cellular respiration
   d. pulmonary circulation
   e. pulmonary ventilation

8. The part of the digestive tract that plays the primary role in digestion and absorption of nutrients is the
   a. large intestine
   b. cecum and colon
   c. stomach
   d. small intestine
   e. esophagus

9. An adult who has a complete set of permanent teeth has _____ teeth.
   a. 20   d. 36
   b. 26   e. 40
   c. 32

10. The urinary system includes all the following except
    a. kidneys
    b. gallbladder
    c. ureters
    d. urinary bladder
    e. urethra

Brief Review

1. What are the major types of functions performed by blood?

2. Why do capillaries permit the diffusion of materials, whereas arteries and veins do not?

3. What is external respiration?

4. Name and describe the four layers of the digestive tract, beginning with the innermost layer.

5. What is the primary function of the urinary system?

6. If you drank three glasses of fruit juice just because it tasted so good, how would you expect your body to react with respect to its production of antidiuretic hormone (ADH)?

7. In what form is most of the carbon dioxide transported to the lungs?

8. How is the small intestine adapted for the absorption of nutrients?

9. Why don't blood cells and large protein molecules in the plasma appear in the filtrate under normal circumstances?

Applying Your Knowledge

1. Which of the formed elements would you expect to see increase after donating a pint of blood?

2. Why is the wall of the left ventricle more muscular than the right ventricle?

3. Why do long-haul trailer truck drivers frequently experience kidney problems?

Group Activity

Divide into groups of four or six people. Have each member of each group place himself or herself into one of two classes: those who exercise regularly (30 minutes per day, at least four days per week) and those who do not. Now, take the resting heart rate of each member of the group for one minute and calculate the average rate for those who exercise and the average rate for those who don't. Then have all the members of the group jog in place or do some other activity for 90 seconds. Now calculate the average increase in heart rate for those who exercise regularly as opposed to the increase for those who don't. Then, try to answer the following questions.

1. Why should regular exercise make the heart beat more slowly? Why should it lessen the increase in rate that people experience when they do exercise?

2. Two of the biggest risk factors in heart disease are smoking and lack of exercise. How many in the group have neither of these risk factors? How many have one? How many have both?

26

An Amazingly Detailed Script
Animal Development

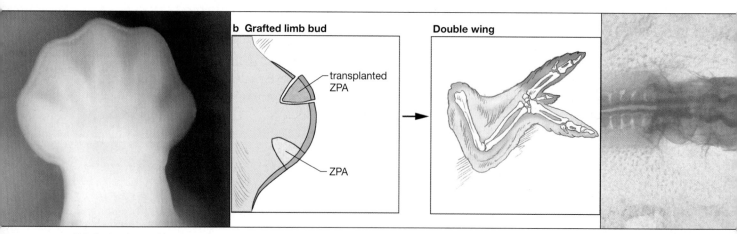

Fish, fowl, or human?
(page 579)

Sprouting two sets of digits.
(page 576)

What are somites?
(page 575)

Of all the marvels in the natural world, the most spectacular may be a complex process that goes under a simple name: development. The set of steps by which, for example, a human embryo goes from being a speck of microscopic cells to a newborn cradled in its mother's arms is so remarkable that even the old adage about "seeing is believing" is turned on its head. People who view pictures of such a work in progress can scarcely believe what they're seeing. It's not uncommon, months after the birth of a child, for parents to find themselves looking at the baby and then at each other in amazement, still wondering how such a thing is possible.

This process is all the more notable because it takes place through *self-assembly*. There is no outside hand here; no master artisan watching over each step and intervening where necessary. There is instead a starting set of instructions supplied by maternal and paternal DNA, and a starting set of proteins and other molecules that come in the maternal egg. From these elements, an embryo fashions itself, aided only by the nurturing environment of egg or womb.

How can such an event take place? As you'll see, in species as diverse as frogs, insects, and human beings a common, core set of steps is followed in development. In addition, it is becoming increasingly clear that nature has provided a fairly small number of "tools" that are employed in development—and that these, too, tend to be utilized across the different species.

26.1 General Processes in Development

Let's now begin by looking at the developmental progression that is common to much of the animal world, noting first *what* happens in development, and then going over *how* these things happen.

Two Cells Become One: Fertilization

In reviewing meiosis in Chapter 11, it was noted that in the formation of egg and sperm cells, what occurs is a *halving* of the cell's usual genetic complement. Each resulting egg or sperm contains 23 chromosomes, rather than the 23 *pairs* of chromosomes that exist in most cells in the body. Going forth with this reduced, or *haploid* set of chromosomes, egg and sperm then fuse in the moment generally referred to as **fertilization**,

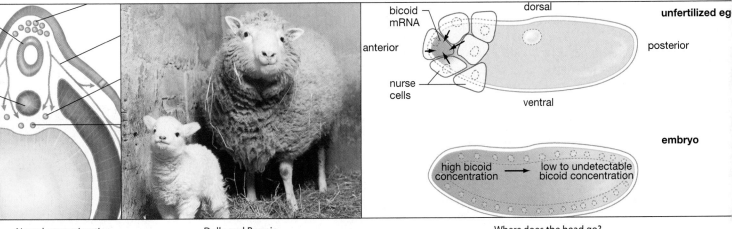

Neural crest migration.
(page 575)

Dolly and Bonnie.
(page 578)

Where does the head go?
(page 577)

more poetically known as conception. Because egg and sperm each bring a haploid set of chromosomes to this union, the fertilized egg contains a doubled, or *diploid*, chromosome set. The fertilized egg, also known as a **zygote**, is a cell, although a very special one: An entire human being will develop from it. Development at its simplest level is a process by which this single, original cell divides to become two cells, after which the two become four, the four eight, and so on. In the period after zygote formation, what develops is an **embryo**.

Three Phases of Embryonic Development

Embryonic development in animals can be divided into three phases.

- Cleavage: Cellular division results in a ball of cells with an interior cavity.
- Gastrulation: These cells rearrange themselves into three layers that will give rise to specific organs and tissues.
- Organogenesis: Organs begin to form.

First Phase: Cleavage

In **cleavage**, the zygote is repeatedly divided into smaller and smaller individual cells through cell division. The reason the cells get progressively smaller is that none of the usual cell *growth* occurs between these cell divisions.

The product of this process of cleavage is a **morula** (from the Latin for "mulberry"), a tightly packed ball of cells. This configuration soon changes, however, as the cells arrange themselves in one or more layers around a liquid-filled cavity. Several hundred to several thousand cells make up this structure, which is called a **blastula**; the cavity inside is called a **blastocoel** (see Figure 26.1).

In the original, egg, the distribution of yolk—which in development means both the yellow "yolk" and the surrounding "whites"—helps give the zygotes of many species a *polarity*, meaning that one end of the zygote is different from the other. Many zygotes have a **vegetal pole**, which is to say an end that contains a relatively greater proportion of yolk, and an **animal pole**, which has relatively less yolk and which lies closer to the cell's nucleus. As you'll see, this factor helps define the way the embryo develops.

Second Phase: Gastrulation

Anyone who has ever watched a play in rehearsal probably has witnessed a phenomenon that bears some comparison with the second phase in embryonic development—gastrulation. It's the moment in which the director says to the assembled cast, "Take your places." What happens then is a carefully directed movement of actors; some move only a little, whereas others go from one side of the stage to the other. Whatever the case, after all this movement is done, the actors—having assumed their places—are ready to take on their assigned roles.

Gastrulation is likewise a process of directed movement, in this case of groups of cells that move to particular places in the developing embryo. When this process is finished, the result is an embryo that is divided into three different *layers* of cells, each layer giving rise to particular organs or types of tissue (**see Table 26.1**).

The endoderm, mesoderm, and ectoderm are "germ" layers, meaning layers of tissue that *lead* to other kinds of structures. You can see in the table that these three layers of cells give rise to organs and tissues that have a rough kind of "inside/middle/outside"

**Figure 26.1
Cleavage of a Frog Egg**

a Repeated cell division of the zygote results in a tightly packed ball of cells called the *morula*. This configuration gives way to a blastula, in which the cells are arranged around a liquid-filled cavity, the blastocoel. The yolk at the vegetal end of the zygote impedes cleavage, so that fewer but larger cells result at the vegetal pole than at the animal pole of the morula and blastula. The figure at far right shows a cross section of the blastula, with its blastocoel visible.

b A micrograph of cleavage in a frog egg, when the egg is at the eight-cell stage.

a

CLEAVAGE

zygote   morula   blastula

blastocoel

animal pole

vegetal pole

cross-section of blastula

b

quality to them: Endoderm gives rise to interior tissues (the lungs), mesoderm to tissues exterior to these (skeletal systems), and ectoderm to tissues that are more exterior yet (nails, hair). This makes sense because, in gastrulation, endoderm takes its place as the innermost layer, mesoderm is the middle layer, and ectoderm is the outer layer.

One Example of Gastrulation: The Sea Urchin

Let's now look at the process of gastrulation in one organism, the sea urchin (**see Figure 26.2**). The sea urchin blastula amounts to a few thousand cells, arranged in a single layer around the liquid-filled blastocoel. Gastrulation begins at the blastula's vegetal pole. Some cells at the pole detach from the ring of blastula cells and move into the blastocoel. These are called the *primary mesenchyme cells*, and with their loss, the vegetal pole begins to pinch inward. This process, called *invagination*, eventually will carry the vegetal pole as much as halfway toward the blastula's animal pole. The structure that starts forming around this indentation is called the *archenteron*, or primitive gut. Cells growing at the tip of the archenteron begin to send out slender extensions of themselves called *filopodia*, which adhere to selected cells that lie "above" them on the rim of the blastocoel.

The contraction of these filopodia then effectively *pulls* the archenteron toward the animal pole, eventually moving the archenteron all the way to the pole itself.

The cells with the filopodia are called *secondary mesenchyme cells*; once they have gotten the archenteron to the top of the blastocoel, they move back down into the blastocoel cavity. There they will form the mesoderm, or middle layer, of the sea urchin—eventually its skeleton and muscles. Meanwhile, the mouth of the sea urchin will form in the place where archenteron moved up to meet the blastocoel lining. This mouth

Table 26.1
Three Layers of Cells That Are Assembled in Gastrulation and Some of the Structures They Give Rise to in Mammals

Cell Layer	Gives Rise to Organs or Types of Tissue
Endoderm	Lungs, liver, pancreas, lining of digestive tract, lining of bladder, and several glands
Mesoderm	Much of the body mass of higher animals, including most of the muscle and skeletal systems; blood, blood vessels, and bone marrow; the uterus and the sperm- and egg-producing gonads; the kidneys, and the cortex of the adrenal gland
Ectoderm	Entire nervous system; nails, hair, lenses of eyes, tooth enamel, several glands, the outer or epidermal layer of skin

GASTRULATION

Figure 26.2
Gastrulation in the Sea Urchin

a The sea urchin blastula amounts to a few thousand cells, arranged in a single layer around the liquid-filled blastocoel.

b Some cells at the blastula's vegetal pole move into the blastocoel, prompting the vegetal pole to pinch inward, moving toward the animal pole. The structure that starts forming around this indentation is the archenteron, or primitive gut.

c Mesenchyme cells, growing at the tip of the archenteron, send out slender extensions of themselves that adhere to cells on the rim of the blastocoel. These extensions contract, moving the archenteron to the animal pole.

d The mouth of the sea urchin forms where the archenteron contacted the blastocoel lining. A digestive tube then runs from this mouth to the rim of the vegetal pole, which develops into an anus. This apparatus is the inner layer or endoderm of the sea urchin. Mesenchyme cells move back down into the blastocoel, forming the middle layer or mesoderm of the sea urchin. The outer cells of the blastula form the sea urchin's outer layer or ectoderm.

will fuse with the archenteron, creating a tube that runs the length of the sea urchin, from mouth right down to the original vegetal pole; there, the rim of the invagination you saw earlier will develop into an anus. This entire feeding and digestive apparatus is the inner cell layer or endoderm of the sea urchin. Finally, the outer cells of the blastula form the outer cell layer or ectoderm of the organism, eventually becoming its skin and related organs.

Third Phase: Organogenesis

In vertebrates, an interplay of two of these three germ layers marks the first steps on the path to the development of organs—in this case, organs of the nervous system.

In talking about development in any organism, an important concept is *body planes*—imagined divisions that pass through the center of an animal, dividing it into two parts of roughly equal size. In **Figure 26.3a**, you can see that a frog embryo has been divided with two such planes. The first yields dorsal and ventral portions of the embryo; the second yields anterior and posterior ends. In the case of a frog, dorsal eventually will mean its back, ventral its belly; anterior will mean its head, and posterior its hindquarters.

Looking at Figures 26.3b and 26.3c, you can see what takes place toward the frog embryo's dorsal surface as gastrulation is nearing completion. First, a temporary rod-shaped support organ, the **notochord**, develops. One of its functions is to *induce*, or bring about, development in the ectodermal tissue that lies above it, toward the dorsal surface. (You'll hear more about such induction shortly.) This development begins with an elongation of the ectodermal cells to form a flattened surface called the *neural plate*,

which curls up and eventually folds all the way over on itself to form an enclosed, hollow **neural tube** that will lie below the dorsal surface. Developing from anterior to posterior, this structure eventually gives rise to both the brain and spinal cord.

Equally important are cells that break away from the periphery of the neural tube as it is folding over. These are the **neural crest** cells, which exist only in vertebrates and have two interesting properties: Groups of them migrate along specific routes to different locations throughout the embryo, passing through various kinds of tissues along the way; these groups then go on to develop into very different organs or tissues, depending on where they come to rest. For example, the medulla portion of the adrenal gland (which in humans sits atop the kidneys) is derived from neural crest cells. You can see the routes of these migrations and some of their outcomes in **Figure 26.4**. One consequence of a failure of migrating neural crest cells to come together is a gap that occasionally occurs between the nose and mouth in humans, which is known as a cleft lip.

As this ectodermal movement is taking place, mesodermal tissue on both sides of the notochord is developing into blocks called **somites** (see **Figure 26.5**) that will give rise to muscles as well as to the vertebrae that enclose the spinal cord.

A Theme in Development: From the General to the Specific

As complicated as all this may seem, it amounts to a brief look at important, but early periods of development in animals, with an emphasis on vertebrates. For the larger picture, it may be obvious that in the early stages of development, what takes place is a movement from the general to the spe-

**Figure 26.3
Body Planes in a Frog and Formation of the Frog Central Nervous System**

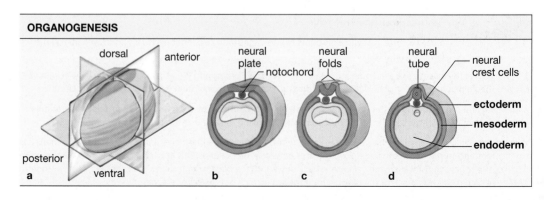

cific: from a small group of superficially similar cells, to three *layers* of cells, to tissues that proceed from these layers (notochord, somites) to actual organs formed from these tissues (with many organs being formed from several different kinds of cell layers).

26.2 What Factors Underlie Development?

Now that you've taken a brief look at what happens in early development, the next question is: *How* do these things happen? How do vegetal pole cells "know" they are vegetal pole cells? How do neural crest cells, starting from similar locations, then take different migratory routes through an embryo and end up serving as different kinds of cells?

The Process of Induction

You saw earlier that the notochord induces development of the ectodermal tissue that lies above it. Such **induction** is an important player in development in general. It can be defined as the capacity of some embryonic cells to direct the development of other embryonic cells. If you look at **Figure 26.6** on page 576, you can see how induction works in the well-studied example of chick wings. In essence, there is a relatively small group of cells that gives out positional information for the development of a portion of the wing. Researchers learned this by working with the "limb bud" that is transformed into the chick wing. They took cells from a posterior "zone of polarizing activity" (ZPA) in one limb bud and transplanted them to the *anterior* portion of a second limb bud. The result was mirror images of *two* sets of developing digits in wings, spreading out like two sets of human fingers developing from joined hands. In short, the cells in the ZPA induced the development of other cells, sending out a message that said, in effect, "begin digit development here."

The Interaction of Genes and Proteins

The induction process just described raises the question of what lies at the root of development. There are lots of cells in an embryo. How can some of them (such as the ZPA cells) control the development of others? It turns out that, at its most basic level, development is controlled by the interaction of

genes and proteins—the genetic regulation reviewed in Chapter 14. Biologists' understanding of how this interaction works in development has grown dramatically in the period from the early 1980s to the present. And the key "model" organism in this research

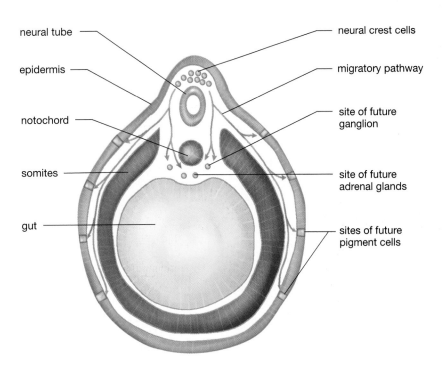

neural tube

epidermis

notochord

somites

gut

neural crest cells

migratory pathway

site of future ganglion

site of future adrenal glands

sites of future pigment cells

Figure 26.4
Neural Crest Migration in a Frog Embryo
This cross section through a frog embryo shows some of the pathways of neural crest cells as they migrate to various locations, prompting the development of different tissues and organs.

Figure 26.5
Somites
These regular, repeating blocks of mesodermal tissue, visible in this case on either side of the notochord of a chick embryo, will give rise to muscles and the vertebrae that enclose the spinal cord.

has been the tiny fruit fly, *Drosophila melanogaster*. So important has this work been that three developmental investigators who worked with *Drosophila*—Edward Lewis and Eric Wieschaus of the United States, and Christiane Nüsslein-Volhard of Germany—were awarded the Nobel Prize in Physiology or Medicine in 1995 for their discoveries (**see Figure 26.7**). Let's look at a little of what *Drosophila* research has uncovered.

Two Lessons in One Gene

There is a *Drosophila* gene, called *bicoid*, that codes for a protein that is also called bicoid.

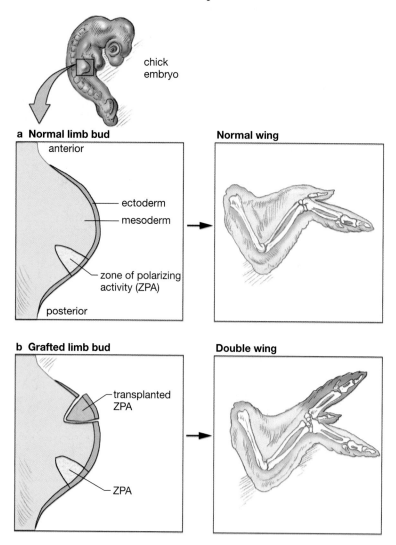

Figure 26.6
Cells Can Induce Development in Other Cells
Transplantation experiments with chick embryos revealed that a group of cells on the posterior portion of the chick's limb bud give out positional information for the development of digits in the chick wing.

a When cells from a zone of polarizing activity (ZPA) were left in their normal posterior location, the chick wing developed as usual.

b When limb bud ZPA cells were transplanted to the anterior portion of a second limb bud, the result was mirror images of two sets of developing wing digits, much like two sets of human fingers extending from a joined hand.

(It is a convention that the gene and protein have the same name; note, however, that the gene name is italicized while the protein is not.) This single gene offers a clear example of two principles that operate in development:

- A critical early step in development is the establishment of positional information in the embryo.
- Substances called *morphogens* are important in establishing this positional information.

Look at the unfertilized *Drosophila* egg pictured in **Figure 26.8** to see how these principles play out with respect to the bicoid protein. Recall first, the body planes described earlier. In Figure 26.8a, the top of the egg is its dorsal portion and the bottom its ventral portion, while the left side corresponds to its anterior portion and the right side to its posterior. Eventually, the anterior portion will be the head of the fly, the posterior its hindquarters, the dorsal portion its back, and the ventral portion its belly.

While the egg is still in the mother, a group of "nurse cells" surrounds its anterior portion and pump molecular products into it. These products include bicoid messenger RNAs (mRNAs)—genetic instructions copied from the *bicoid* gene that led to the bicoid protein. Looking at the figure, you can see that the *bicoid* mRNAs are initially found at the anterior end of the egg.

After the egg is fertilized, it begins dividing into many cells, of course; but these cells in the early *Drosophila* embryo can be thought of as part of an undivided egg. Once the *bicoid* mRNA molecules are translated into proteins, they diffuse through roughly 70 percent of the length of the egg, starting from anterior and stretching toward posterior. However, as you can see in Figure 26.8b, this bicoid protein exists in a *concentration gradient*, with the greatest concentration still being at the anterior end, a lesser concentration toward the center, and no detectable amounts found at the posterior end.

Positional Information

With this general background, you can now see how *bicoid* exemplifies one of the two principles of development noted earlier. What does the bicoid protein do? It imparts positional information to the embryo and is critical in setting in motion the development of structures in the anterior and center portions of the embryo. Recall that the far-anterior portion of the egg eventually becomes the head of the *Drosophila* fly. In experiments, however, scientists have taken bicoid-laden cytoplasm from the anterior portion of one egg, placed it at the *center* of an egg with no bicoid in it, and watched as the head and related structures developed at the center. An egg with no bicoid protein gives rise to a fly that never develops a head, though its posterior structures develop normally.

Morphogens

The *bicoid* gene is a **morphogen**: a diffusable substance whose local concentration affects the course of local development in an organism. Several other morphogens are operating in *Drosophila* in early development. Indeed, several positional "systems" are at work in the fly. What's important to note is that, with these systems, a three-dimensional *grid* is set up in which the destiny of various cells is determined by their position within this grid. How? Their exposure to different morphogens varies according to their position.

So, how do morphogens work? You may remember learning in Chapter 14 about proteins that feed back on DNA, controlling its transcription, which is to say turning genes on or off. These DNA-regulating proteins are called *transcription factors*. All of the *Drosophila* morphogens you've been reading about here are transcription factors. These morphogens, in other words, are proteins that, once brought into being by genes, then go on to help turn *other* genes on or off.

With this knowledge of morphogens under your belt, you can now understand the underlying basis for the chick wing development noted earlier. Recall that ZPA cells in the limb buds were able to induce the development of digits in the chick wing. These cells can play their controlling role because they produce a *morphogen* (called sonic hedgehog) that acts on surrounding cells, inducing their development.

A Hierarchy of Genetic Regulation

Once you begin to see the cascade of events that are controlled through genetic regulation, it's easier to understand how development moves from the general to the specific. Genes operate in a hierarchy in development. The first level of genes specifies broad positional patterns across the embryo (as with *bicoid*). Going down the hierarchy, this positioning gets finer-grained—now development is being specified within *sections* of the embryo. Finally, control is exercised over the development of specific body *structures*. Remember how development goes from cells, to layers of cells, to specific structures such as a mouth and a gut? The genetic cascade explains this pattern.

Figure 26.7
Advancing Human Knowledge of Development
Edward B. Lewis of the United States (left), Christiane Nüsslein-Volhard of Germany (center), and Eric F. Wieschaus of the United States (right), in Stockholm after receiving Nobel Prizes for research they did in developmental biology.

a

bicoid mRNA

anterior

nurse cells

dorsal

ventral

unfertilized egg

posterior

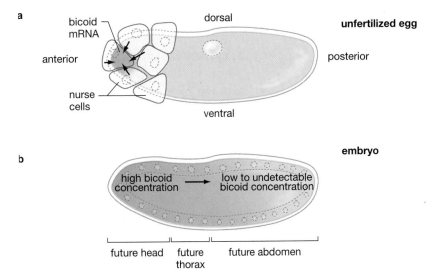

b

high bicoid concentration → low to undetectable bicoid concentration

embryo

future head   future thorax   future abdomen

Figure 26.8
Drosophila **Egg and Bicoid Gradient**
A fly begins to develop. **(a)** Nurse cells surround the anterior portion of an unfertilized *Drosophila* egg. The cells pump messenger RNA molecules for the bicoid protein into the egg. **(b)** The messenger RNAs diffuse through the now-fertilized egg, setting up a concentration gradient for the bicoid protein. This protein is a morphogen that imparts information regarding development of anterior and center-position structures in the fly. An egg without bicoid in it gives rise to a fly that never develops a head.

26.3 Cell States: Determined and Committed

Having reviewed a couple of examples of wings or heads being induced to grow where they normally wouldn't, you may have the impression that groups of cells are infinitely adaptable in what they can become; that given the right developmental factor (such as a morphogen), a cell could be induced to become anything at any point. This may not be the case, however.

By way of analogy, it might be helpful to think of humans and their acquisition of language. A newborn baby can learn to speak Chinese, English, or Dutch with equal ease. If the child is raised in an English-speaking household, English will be the lifelong language, barring some intervention; a switch could be made to Dutch or Chinese, and until about the time of puberty this can be done with relative ease. Once the child passes puberty, however, the window of adaptability slams shut; a new language can be acquired only though hard, slogging work. Cells are something like this in development. They are adaptable in early embryonic life, but then the window of adaptability begins to close as development proceeds.

The Shock of Dolly the Sheep

Until the late 1990s, the general view was that once the cell's window of adaptability was shut, it was *unalterably* shut; nothing could turn one kind of cell line into another. Liver cells were fated to give rise only to more liver cells, and bone cells to more bone cells.

Then, in 1997, came Dolly the sheep (**Figure 26.9**). You can read about how Dolly

Figure 26.9
Scientific Surprise
Dolly the sheep resulted from the fusion of an adult sheep udder cell and a sheep egg whose nucleus had been removed. Dolly is pictured here with her first lamb, Bonnie.

was cloned in Chapter 15 (page 289), but here is a summary. First a cell was removed from the udder of an adult sheep. Then an egg was taken from a second sheep and its nucleus was removed, which meant this egg lost its essential component of DNA. After this, the udder cell (which had DNA) and the egg (which didn't) were placed side by side. Then an electric current was applied to the egg. One effect of this was that the two cells fused into one. The stunning event, however, was that the mammary cell DNA, now housed inside the fused cell, then *reprogrammed* itself, so that this cell started to develop as a *zygote*. The result, 21 weeks later, was a perfectly normal lamb, Dolly, who has now given birth to a healthy lamb of her own.

The assumption of scientists had been that the DNA in an udder cell could bring about the development only of more udder cells, not of a zygote that could develop into a separate organism. How could the udder cell DNA be reprogrammed? The sheep egg may have had its nucleus removed, but it still had all its maternal *proteins*—egg proteins whose normal function is to bring about development of a zygote. These interacted with the udder cell DNA, reprogramming it so that it brought about zygote development. The proteins in an unfertilized egg represent a strictly maternal contribution to development—one that obviously is powerful. You saw an example of such a maternal contribution earlier with the *bicoid* mRNA molecules, pumped into the *Drosophila* egg by the surrounding nurse cells. These then went on to be important in the *Drosophila* fly's early development.

Determined and Committed Cells

The terms used to describe developmental cell states are *determined* and *committed*. A cell has become **determined** when it's possible to predict its fate. If scientists know that, left undisturbed, a given cell will become part of, say, a *Drosophila* antenna, then we know that this cell's fate has been determined. If at this point, however, it is still possible to *alter* this fate—by, for example, adding morphogens or other factors—then this cell has not yet undergone **commitment**. A cell is committed when it is no longer possible to alter its developmental fate. But what does commitment mean in the face of the Dolly results? Dolly was cloned from a cell that was committed—to being an udder cell.

Early embryonic cells are **totipotent**: They are capable of developing into any kind of cell in the organism. Indeed, in the very early stages of development in some species—up through about the eight-cell stage in mammals—each cell of the embryo is capable of developing into a complete organism. Then cells begin to specialize; they differentiate into specific kinds of cells. But in so doing, scientists thought, they lose the marvelous capability of being able to go in *any* direction in development. Dolly and other clones like her have now cast doubt on this idea. Only time will tell what lies in store in this exciting scientific arena.

26.4 Developmental Tools: Sculpting the Body

To this point, you've been considering development at the level of genes. Ultimately, however, the shaping of an organism takes place through the actions of complete cells. You may wonder how a hand, for example, can be shaped from a seemingly formless mass of cells. The answer lies in cell *capabilities* that allow for this sort of development. Three examples of these capabilities are

- Cell movement
- Cell adhesion
- Cell death

You saw examples earlier of **cell movement** and **cell adhesion** when reviewing the process of gastrulation in the sea urchin. Recall that cells growing at the tip of the primitive gut were able to send out slender extensions of themselves that pulled the gut toward the "top" of the blastocoel. These extensions were able to do this, however, only after *adhering* to selected cells that lay "above" them in the blastocoel. Several families of proteins, collectively called *cell adhesion molecules*, allow cells to carry out this trick of selectively gluing themselves onto other cells.

Cell death may not sound like a promising means of developing an organism, but consider that a sculptor may work not by *adding* material, but by chipping away at an initial mass of it. Thus it is in development, where cell death works to create spaces that define shapes. An example of this is the human hand which, but for the death of large numbers of cells between future fingers, would look more like the webbed foot of a duck. Cells that die in this fashion are not attacked by other cells, nor do they perish for lack of nutrients. Instead, they die through their own metabolic processes in what is sometimes called **programmed cell death** (see Figure 26.10).

26.5 Development through Life

It is convenient to think of animal development as something that happens before birth. But if this is the case, what are we to call the loss of "baby" teeth in childhood, the onset of puberty in the teenage years, or cessation of menstruation (menopause) among middle-aged women? Development never stops throughout the life span. In essence it is governed by the same factors that control the embryonic development you've looked at so far: the interaction of genes and proteins and the influence of environment upon them.

On to Human Reproduction

The steps of development that you have just looked at form a kind of middle phase to the process you'll be looking at next chapter, which is reproduction, specifically human reproduction. Coming *before* the development you've reviewed are the formation of the egg and sperm that come together to create a new human being. Coming *after* it are late-stage human embryonic growth and then birth itself. With this chapter, you know something about development; in Chapter 27 you'll see how development fits into the larger scheme of human reproduction.

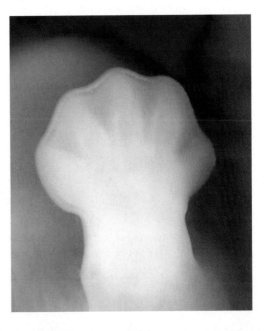

**Figure 26.10
Soon to Be Different**
The webbed hand of a human embryo, 40 days after conception. The programmed death of cells in the hand will bring about the emergence of five separate fingers.

Chapter Review ▮▮▮▮▮▮▮▮▮▮▮▮

Key Terms

Summary

26.1 General Processes in Development

- Development in animals begins with the creation of a zygote or fertilized egg. Once a zygote is formed, an embryo develops from it.

- There are three phases of embryonic development. The first is cleavage, which produces a ball of cells, called a blastula, with an interior cavity. The second is gastrulation, in which these cells rearrange themselves into three layers that give rise to specific organs and tissues. The third is organogenesis, in which organs begin to form.

- The first product of cleavage is a tightly packed ball of cells called a *morula*. The cells in the morula rearrange themselves into a structure composed of one or more layers of cells surrounding a liquid-filled cavity. This structure is the *blastula*; its interior cavity is called the *blastocoel*.

- Gastrulation produces an embryo composed of three layers of cells, the endoderm, the mesoderm, and the ectoderm. The endoderm gives rise to interior organs and tissues, such as the liver and bladder in mammals. The mesoderm gives rise to much of the body mass of higher animals, including most of the muscle and skeletal systems. The ectoderm gives rise to the entire nervous system of mammals, as well as various external tissues, such as hair and tooth enamel.

- In early organogenesis in vertebrates such as frogs, a rod-shaped support organ, called a *notochord*, develops near the dorsal surface and induces the development of ectodermal tissue that lies above it. A neural plate develops, folding over on itself to form a hollow neural tube that gives rise to both the brain and the spinal cord. Groups of neural crest cells break away from the periphery of the neural tube during its development and then migrate to different locations throughout the embryo. These groups go on to develop into different organs or tissues.

- The early stages of development represent a transition from the general to the specific. A small group of superficially similar cells gives rise to three different layers of cells, which give rise to tissues and then to specific organs.

26.2 What Factors Underlie Development?

- Induction is the capacity of some embryonic cells to direct the development of other embryonic cells.

- Development is fundamentally controlled by the interaction of genes and proteins. A key model organism in scientific understanding of this interaction is the fruit fly, *Drosophila melanogaster*.

- A morphogen is a diffusable substance whose local concentration affects the course of local development in an organism. The bicoid protein is a *Drosophila* morphogen that imparts positional information to the embryo, directly affect-

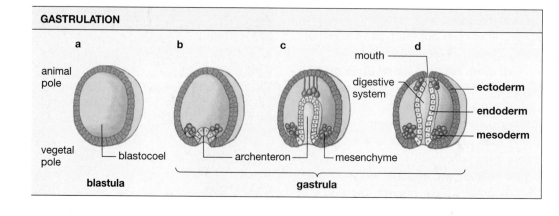

GASTRULATION

ing the development of its anterior and center portions. Morphogens generally are transcription factors (meaning proteins), brought into production by genes, that then affect the activity of other genes.

26.3 Cell States: Determined and Committed

- In nature, the developmental potential of cells decreases over time. In very early development, cells are totipotent, meaning they are capable of developing into any kind of cell in the organism. Cells progressively specialize, however, and in so doing may lose the ability to develop into different types of cells.

- A cell has become determined when, in nature, it is possible to predict its fate. A cell has undergone commitment when it is no longer possible to alter its developmental fate.

- Dolly, the cloned sheep born in 1997, has called into question the meaning of cell commitment. Dolly was cloned from a mature udder cell, which presumably had undergone commitment.

26.4 Developmental Tools: Sculpting the Body

- Three cell capabilities help shape the animal body in development. The first is cell movement. The second is cell adhesion, in which several families of proteins produced by cells have the capability to selectively adhere to other cells. The third is selective cell death, in which tissues take shape through the destruction of cells, often through programmed cell death.

26.5 Development through Life

- Development is sometimes thought of as ending at birth, but it continues through the life span of an organism. It is exemplified in human beings in such processes as puberty and menopause.

Understanding the Basics

Multiple-Choice Questions

1. The animal and vegetal poles of the egg are determined during the production of the
   a. sperm
   b. gastrula
   c. blastula
   d. egg
   e. nurse cell

2. All of the following tissues are mesodermal derivatives EXCEPT
   a. blood
   b. skin
   c. kidneys
   d. muscle
   e. bone marrow

3. Which of the following statements best describes the gastrula?
   a. hollow ball of cells
   b. newly fertilized zygote
   c. embryo with fully formed organs
   d. embryo with three distinct layers of cells
   e. embryo with a vegetal and animal pole

4. The process of gastrulation involves
   a. massive cell death
   b. little or no growth of the embryo
   c. cell reorganization and movement
   d. cell enlargement
   e. organogenesis

5. During gastrulation, secondary mesenchyme cells
   a. create the primitive gut
   b. induce the formation of somites
   c. cover the outside of the embryo eventually forming skin
   d. induce the formation of the nervous system
   e. move to the middle of the embryo, eventually forming skeleton and muscle

6. Organogenesis is accomplished, in part, by the interaction between various regions of the three germ layers. This interaction is referred to as
    a. reduction
    b. transformation
    c. translation
    d. induction
    e. fusion

7. The cells that move away from the developing notochord and eventually form many different organs and tissues are
    a. neural crest cells
    b. archenteron cells
    c. primary mesenchyme cells
    d. secondary mesenchyme cells
    e. endoderm cells

8. Substances, like the bicoid protein, that are essential in the establishment of positional information in the embryo are called
    a. nurse cells
    b. mRNAs
    c. morphogens
    d. receptor proteins
    e. enzymes

9. During the production of the unfertilized *Drosophila* egg, bicoid protein is produced by these cells
    a. sperm
    b. egg
    c. nurse cells
    d. endoderm cells
    e. mesoderm cells

10. Many maternal proteins, like bicoid, that help establish positional information function as
    a. RNA polymerase
    b. transcription factors
    c. ribosomes
    d. ATP
    e. tRNA

11. A cell that is capable of developing into any kind of cell is called
    a. committed
    b. developed
    c. specialized
    d. determined
    e. totipotent

Brief Review

1. Briefly describe the order of events that occur after fertilization to produce an embryo with three layers of cells.

2. Outline the steps involved in the production of the chick wing.

3. What are the roles of the mesenchyme cells during sea urchin gastrulation?

4. How do cell adhesion molecules influence early events during development?

5. What is a concentration gradient and how does it effect early development of the fruit fly, *Drosophila*?

6. How do cells become determined? Is this process reversible?

7. Describe the role cell death plays during the growth and development of an organism.

Applying Your Knowledge

1. Explain how the bicoid protein acts as a morphogen. What would happen to the developing *Drosophila* embryo if bicoid mRNA was introduced into the posterior region of the egg prior to fertilization?

2. The production of Dolly raised question of the ability of adult cells to be reprogrammed. On the other hand, it confirmed the notion that adult DNA can still direct the production of an entire organism. What crucial cell must be employed to allow both the adult cell and DNA to produce a new organism? Why?

3. Cell death is a common feature to the development of many structures in the growing embryo. This process of cell death, if unregulated, could cause severe problems. What types of problems might be observed if the cell death process is altered?

Group Activity

This activity analyzes the production of lenses during the development of the frog *Xenopus laevis*. The lens is derived from the ectoderm and is induced to form over the developing eye. This induction is quite specific, because we do not usually see lenses developing all over the outside of the frog. To examine lens induction, you and a group of your classmates performed a few experiments as follows: You took eye tissue of various ages and combined it with ectoderm (precursor of lens tissue) from embryos of various ages. When you used ectoderm from 12-hour-old embryos and combined it with eye tissue from 10-, 12-, 14-, 16-, and 18-hour-old embryos, lenses were induced only when the eye tissue was from 12- or 14-hour-old embryos. Conversely, when you used ectoderm from 8-, 10-, 12-, 14- and 16-hour-old embryos and combined this ecto-derm with eye tissue from 14-hour-old embryos, lenses were only induced when the ectoderm was from 12-hour-old embryos. As a group, analyze the data you obtained and then formulate answers to these questions.

1. Create a table to organize your data.

2. Which tissue is producing the morphogen?

3. When is the morphogen being produced?

4. When can the receiving tissue respond to the morphogen?

5. During what hour of development are the lenses produced?

6. Which tissue—the eye or the ectoderm—limits the window during which lenses can be produced?

27

How the Baby Came to Be
Human Reproduction

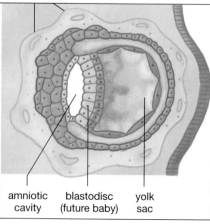

amniotic   blastodisc   yolk
cavity   (future baby)   sac

Implantation.
(page 599)

Many sperm come
in contact with the
accessory cells and
are capacitated.

Fertilization begins.
(page 596)

7 months

Lungs develop, eyelids open.
(page 601)

A complex network.
(page 600)

Memorable events mark

every human life, but for parents, one event that is sure to be indelibly etched into memory is the birth of a child. The mother's struggle in labor, the final "pushing" that brings the baby forth, the first real look at the infant and the infant's first look at the world; these are moments that, once experienced, are never forgotten.

The birth of a child is a beginning, of course, and yet it is also a culmination. Human beings must grow to a certain level of biological maturity before being capable of reproduction. After this, finding a partner may be difficult, and getting pregnant more difficult yet. Eventually, however, two very special "seeds"—*this* egg and *this* sperm, out of all the millions a man and a woman produce in a lifetime, fuse to create *this* child.

In this chapter you'll trace the biology of human reproduction, which might be defined as all the biological steps that must fall into place before a child can be born. The process of development you studied in Chapter 26 represents part of what you'll be looking at, but other elements are involved as well.

27.1 Overview of Human Reproduction and Development

Taking this story from the beginning means picking up a narrative that was begun in Chapter 10, when you went over the formation of the reproductive cells called **gametes**—eggs in females, sperm in males. In this chapter, the focus widens, taking into account the physical *setting* in which this formation takes place and the steps that bring egg and sperm together. Finally, you will look at human development from conception to birth. For information on some of the methods humans use to *prevent* conception, see "Methods of Contraception" on page 594.

The human male and female reproductive systems are functionally alike up to a certain point, in that they first develop their respective gametes and then deliver them to a location where they can fuse in the act known as fertilization or conception. The female reproductive system is called on to do all that the male system does—*and more*, however, in that, once fertilization takes place, the female system must then provide

An egg
is formed.
(page 588)

Ready for delivery.
(page 603)

Getting special treatment.
(page 602)

the environment for the embryo to develop, after which it must deliver the baby into the world. This takes place over a period of 38 weeks, on average.

Reproduction in Outline

The first part of reproduction, the formation and delivery of egg and sperm, is a straight-forward story in outline. All the eggs a woman possesses exist from birth in a pre-cursor form, called oocytes, that lie within two walnut-shaped structures called ovaries. These exist just right and left of center within the female pelvic area (**see Figure 27.1**). Oocytes sit in the ovaries within nurturing complexes of cells and fluids called **follicles**. On average, once every 28 days one follicle-oocyte complex is brought to a state of maturity such that the oocyte and some accessory cells that surround it rupture from their

Figure 27.1
Reproductive Anatomy of the Human Female
Eggs develop in the ovaries within nurturing complexes called *follicles*. Once every 28 days, on average, an egg is expelled from one ovary or the other and journeys through the uterine tube to the uterus. If it encounters sperm on the way, pregnancy will result when egg and sperm fuse.

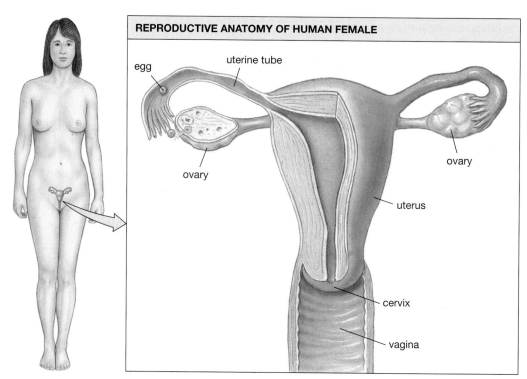

REPRODUCTIVE ANATOMY OF HUMAN FEMALE

egg

uterine tube

ovary

ovary

uterus

cervix

vagina

Figure 27.2
Reproductive Anatomy of the Human Male
Sperm are produced in the testes and transported to the epididymis, where they mature and are stored. With orgasm, the sperm are transported through the vas deferens and then the urethra, from which they are ejaculated. Along the way, they are joined by materials secreted by several accessory glands.

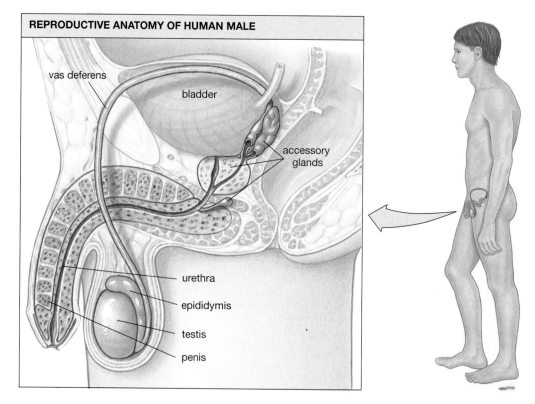

REPRODUCTIVE ANATOMY OF HUMAN MALE

vas deferens

bladder

accessory glands

urethra

epididymis

testis

penis

ovary and begin a slow journey to the uterus. This trip takes place inside the structure called the **uterine** (or **Fallopian**) **tube**. The release of the oocyte is called **ovulation**. If, during this trip, the oocyte encounters a male sperm, the two may fuse, in which case conception has occurred.

Sperm in men are produced, in unfinished form, in sets of tubules in the two male testes (**see Figure 27.2**). They then are transported to a structure called the epididymis (one for each testis) for further development. Then, with sexual excitation, they are transported in a loop: up and over the urinary bladder through two ducts, each called a **vas deferens**. The contents of these ducts then empty into a single duct, called the urethra, from which they are ejaculated into the female vagina. Along the way, materials secreted by several glands will join the sperm in a process that can be likened to tributaries feeding into a common stream. The resulting mixture of sperm and glandular materials is called **semen**.

Whereas the female releases an average of one oocyte per month, the male releases an average of 250 million sperm per ejaculation. Upon entering the vagina, sperm are able to propel themselves up into the uterus and from there into the uterine tubes. Of the hundreds of millions of sperm that begin the journey, a few dozen may encounter the female oocyte as it continues its journey from ovary to uterus (**Figure 27.3**). The first sperm to breach the coating that surrounds the oocyte fuses with it, while the rest are shut out through a kind of gate-slamming mechanism that operates almost instantaneously. When this one sperm has joined with an oocyte, conception has occurred. The now-fertilized egg still floats free in the uterine tube, where it begins to develop in the ways

described in Chapter 26: one cell becomes two, two become four, and so on. Four days after fertilization the developing embryo enters the uterus; a day or so more and it implants itself in the uterine wall.

Both the male and female reproductive systems are greatly influenced by the actions of hormones—chemical signaling molecules that often move through the bloodstream. Indeed, the reproductive organs are primary sites of production for some of the most important of these hormones, testosterone in men and estrogen and progesterone in women.

27.2 The Female Reproductive System

The Female Reproductive Cycle

Although hormones operate in both sexes, in females they bring about a reproductive *cycle* that repeats itself about once every 28 days (though it can range from 21 to 42 days). In males, conversely, there is no cycle, but instead a steady production of sperm and a destruction or removal of the sperm that are not ejaculated. Why a cycle for females and not for males? The answer lies partly in the two roles the female reproductive system must fulfill. It must form an egg *and* create an environment in the uterus in which a fertilized egg can develop. An important part of this uterine environment is essentially built up and then dismantled once every 28 days (except in those rare instances in which an oocyte is fertilized). The end result of the dismantling of the uterine environment is **menstruation**, which is a release of blood and the specialized uterine tissue it infuses, the **endometrium**. This is tissue that lines the interior of the uterus; it is tissue that *would have* housed an embryo, had one been implanted there.

a

b

Figure 27.3
Egg and Sperm
Artificially colored images of **(a)** a human egg and **(b)** a human sperm.

Why Does Menstruation Exist?

From this, it's reasonable to ask: Why this monthly cycle of buildup and destruction, rather than a one-time formation of a lasting embryonic environment? After all, menstruation has some obvious costs attached to it; apart from pain and discomfort, there are what might be called species-survival costs: a loss of fertile days and a loss of blood and other tissue. (On average women lose about 40 milliliters—or about 1.4 fluid ounces—of blood in a menstrual cycle, and about that much other tissue as well.) So, in the face of these costs, why has menstruation been preserved throughout human evolution? Scientists have no firm answers here, but we do have a couple of intriguing recent hypotheses. One is that menstruation serves to clear the uterus of harmful bacteria that can be transported in on sperm. (These bacteria exist on male and female genitalia and are able to attach themselves to ejaculated sperm.) The other hypothesis is that, costly as menstruation may be, in terms of expended energy it is not *as costly* as the alternative of maintaining an embryonic environment in the uterus.

How Does an Egg Develop?

As noted earlier, eggs develop from precursor cells called **oocytes**. If you look at **Figure 27.4**, you can see *where* oocytes develop: in the outer portion of each ovary. Each oocyte is surrounded by a sphere of accessory cells, called *follicular cells*, that eventually provide the oocyte with nutrients. The complex of oocyte and follicular cells is called an **ovarian follicle**; it is the basic unit of oocyte development.

Follicle Development and Ovulation

Every oocyte that matures into an egg must go through the developmental steps you see in Figure 27.4. This is a monthly process of selection, running from a large starting set of follicles, through fewer **primary follicles** that develop from them, to a greatly reduced number of **secondary follicles** that continue development, and finally to a single **tertiary follicle** that nurtures an oocyte through its time in the ovary. Responding midway through the menstrual cycle to hormonal influence, the tertiary follicle is squeezed to the extent that it ruptures, thus expelling the oocyte and its

Figure 27.4
Steps Leading to an Egg
An egg develops from precursor cells called *oocytes* that exist near the surface of the ovary. Oocytes join with nurturing complexes of cells to form ovarian follicles. On average, only one of these follicles matures—once every 28 days—through the stages of primary, secondary, and tertiary follicle. The oocyte contained in that follicle is expelled at the end of the process for a journey through the uterine tube. With its expulsion, the tertiary follicle is transformed into the corpus luteum, which for a time produces hormones that can aid in pregnancy.

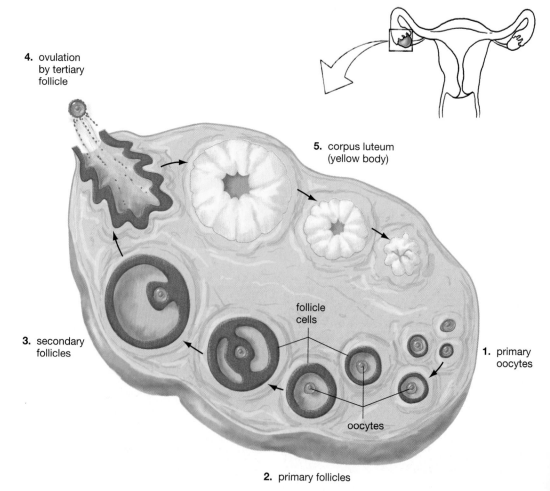

4. ovulation by tertiary follicle

5. corpus luteum (yellow body)

follicle cells

3. secondary follicles

1. primary oocytes

oocytes

2. primary follicles

surrounding accessory cells not only from the follicle but from the ovary as well. With this event, ovulation has occurred. It takes place in this way in *one* ovary each month, with the ovary being selected at random.

Movement through the Uterine Tube

Freed from its ovarian housing, but still surrounded by accessory cells, the oocyte—technically not yet an egg—is now swept into the uterine tube through the actions of the finger-like **fimbriae** lying at the tube's funnel-shaped end (**see Figure 27.5**). Responding to liquid currents and the whip-like cilia that beat toward the uterus, the oocyte makes a journey of about 10 centimeters (4 inches) through the uterine tube over the next four days. While within the tube, it may encounter sperm and become fertilized prior to entering the uterus.

Meanwhile the ruptured tertiary follicle develops into a body, called the **corpus luteum** ("yellow body"). This structure secretes hormones that help prepare the reproductive tract for pregnancy. If pregnancy does not occur, the corpus luteum begins degenerating after about 12 days. With the corpus luteum, you can see again the part that hormones play in the monthly female reproductive cycle. For details on this hormonal regulation, see "Hormones and the Female Reproductive Cycle" on page 590.

Changes through the Female Life Span

In between the creation of oocytes and the journey of one oocyte down the uterine tube lies a multistep process that could be characterized by the phrase: "Many are called, but few are chosen." Recall from Chapter 10 that all the oocytes a woman will ever possess are produced in her prior to birth; these oocytes in turn become associated with the nurturing follicle complexes. A 6-month-old female fetus has in its ovaries about 7 million follicles that are capable of becoming eggs. At birth, this number has been reduced to about 1 million, primarily because of the process of *atresia*, or the natural degeneration of follicles. By puberty, the number has declined to about 250,000. An unknown number of these follicles are "recruited" monthly for further growth through hormonal action, but only *one* of these will, each month, win a kind of competition that allows it to become the oocyte that bursts forth from the ovary. The only exception to this rule comes in the 1 percent or so of ovarian cycles that result in *multiple* ovulations, in which case fraternal twins, triplets, and other multiple births are possible. In sum, the oocyte that begins the journey down the uterine tube each month has been selected from more than 7 million initially created, and it is one of about 500 that will make this trip in a woman's lifetime (about 13 a year for 35 or 40 years).

Figure 27.5
Journey to Pregnancy
Surrounded by accessory cells, an oocyte is expelled from the ovary into the uterine tube, where it encounters sperm and is fertilized by one of them. Now a fertilized egg (or zygote), it begins the process of development and completes its journey with implantation in the uterine wall.

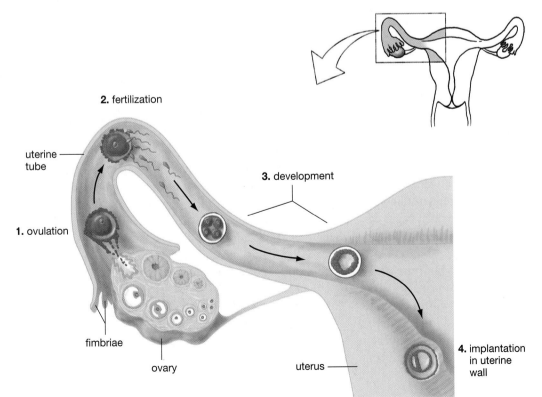

1. ovulation
2. fertilization
uterine tube
3. development
fimbriae
ovary
uterus
4. implantation in uterine wall

Hormones and the Female Reproductive Cycle

One of the more interesting aspects of human reproduction is that in females it is carried out through a *cycle*, which is to say a series of events that repeat themselves regularly—on average, once every 28 days. But how can the female reproductive system "keep time" like this? What is the clock that governs the cycle?

A real clock is actually a good analogy for what happens. If you look at an old-fashioned mechanical watch, what you'll see is a central spring that exerts a force through a series of gears. Given these parts, it takes 12 hours for the physical mechanism of the "small hand" on the watch to make its way through a complete cycle—to end up where it began. In women, the substances called *hormones* set in motion a group of physical processes that take about 28 days to complete. The last of these processes leaves the body right where it began, at the start of another cycle.

We actually should speak of female reproductive *cycles*, in the plural, because several are going on in reproduction, all of them interlinked. There is the ovarian cycle, which has to do with the development and release of oocytes; the menstrual or "uterine" cycle, which concerns the regular buildup and then breakdown of tissue in the inner lining of the uterus; and several hormonal cycles, which help govern both menstruation and egg creation.

If you look at **Figure 1**, you can see a time line for all these cycles and what occurs during them. To give you a feel for the processes at work, let's follow just the menstrual cycle, whose "first day," over on the far left, is defined as the first day of menstruation. What the graph tracks is the state of the tissue (the endometrium) lining the inside of the uterus. As you can see, beginning on day 1, this tissue goes through a breakdown that lasts three or four days, after which it begins a steady buildup that continues all the way to the end of the cycle—at which point it starts to break down again. Now look at what is going on in the ovarian cycle. This part of the graph is following the single "successful" oocyte (out of many that begin development) through its progression into the oocyte that bursts from the ovary on day 14.

The real lesson, however, is to think about the menstrual and ovarian cycles as they relate to *each other*: The endometrium is built up in the uterus as a means of creating an environment in which an egg could implant itself, should one be fertilized. If such an egg begins its journey on day 14 and then takes another 6 days or so to implant

itself in the uterus, you can see what stage the *endometrium* will be in at this point: near its maximum thickness (and at a highly productive stage of synthesizing a blood vessel system). Such an environment is just what an implanted embryo would need.

And what of the figures over on the right side of the ovarian cycle? They represent the corpus luteum, the "remains" of the follicle that the successful oocyte broke out of. It is much more than a spent vessel at this point, however, because hormones secreted by it will promote the growth of the endometrium. Without a pregnancy occurring, the corpus luteum carries out such secretion only for about a week. After that it begins to degenerate, and *this* means the degeneration of the endometrium, which washes out, along with some of the blood that supplied it, in the process called *menstruation*.

What hormones are secreted by the corpus luteum? Look now at the ovarian hormone cycle, and you'll see a line for "progestins"—a family of hormones whose most important member is progesterone. As you can see, the level of progestins takes a sharp turn upward just as the corpus luteum is formed—no surprise, because it is the corpus luteum that is secreting them.

There is a circularity to all this hormonal control, with the hormones involved influencing each other in different ways as the cycle goes on. Let's arbitrarily take as a starting point, however, first the development of a follicle and then the release of an egg from it.

One important hormone in the cycle is produced neither in the ovaries nor the uterus, but in the brain. A brain structure called the **hypothalamus** releases vanishingly small amounts of a hormone called **gonadatropin-releasing hormone**, whose function is spelled out in its name: it controls the release of *other* hormones, namely the gonadatropins.

Two gonadatropins are spurred into production in this way, **follicle-stimulating hormone (FSH)** and **luteinizing hormone (LH)**, both of them produced in a gland located beneath the hypothalamus, the **anterior pituitary gland**. You can see both FSH and LH levels charted in the gonadatropic hormone cycle in Figure 1.

After being secreted by the anterior pituitary, both FSH and LH have their primary early-cycle effects at the site of the ovarian follicles, FSH by promoting the development *of* the follicles, LH by stimulating the synthesis of a hormone *by* the follicles, **estrogen** (which actually is a family of hormones). The more the follicles develop, the

(continued)

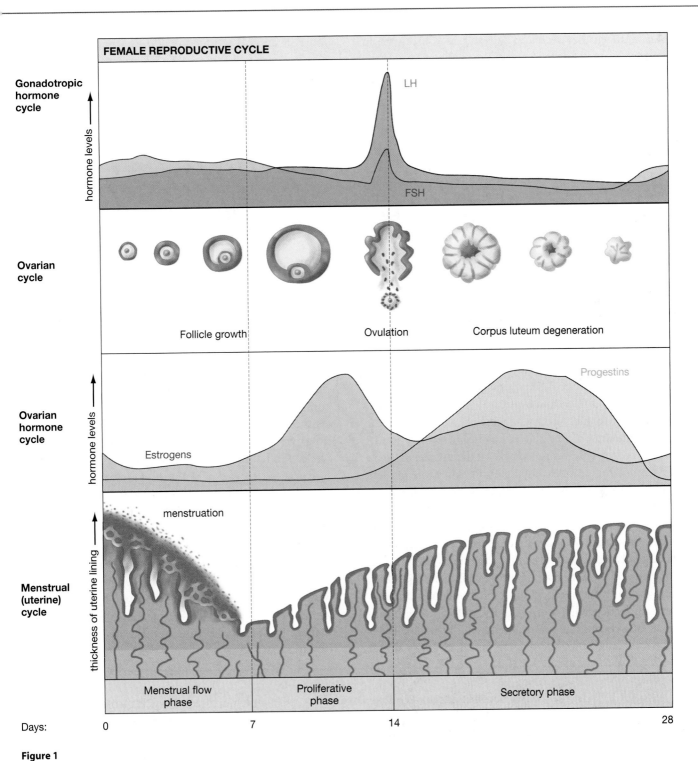

Figure 1
The Female Reproductive Cycles
Female reproduction is controlled by the interrelated secretions of a number of hormones.

Hormones and the Female Reproductive Cycle (continued)

more estrogen they secrete. The result, as you can see in Figure 1, is that estrogen levels slowly rise through the first week or so of the cycle. One thing estrogen is doing is promoting the growth of the endometrium in the uterus; another, however, is controlling the release of both FSH and LH through *feedback* mechanisms, which is to say mechanisms in which the level of one thing controls the level of another.

Estrogen feedback operates in two ways at two different phases of the cycle. Early on, the relatively low (but increasing) levels of estrogen act to *decrease* FSH—not enough to stop development of the follicles, but enough to slightly reduce FSH levels over the first 10 days or so of the cycle, as you can see in Figure 1. In the second week of the cycle, however, estrogen output is greatly increased thanks to the development of the tertiary follicle whose oocyte will burst forth from the ovary. At this level, estrogen acts to *increase* not only FSH production but LH production as well. The result, as seen in Figure 1, is the "spike" of LH production just before day 14 of the cycle. This hormonal surge is what causes the oocyte to break out from both follicle and ovary.

Now the former tertiary follicle has a new identity, the corpus luteum. In this role it has begun production of progesterone, which as noted acts on the uterine lining to develop it for embryo implantation. Then, in tandem with estrogen, progesterone has another effect. It acts on the hypothalamus and pituitary to *lower* FSH and LH levels. But it turns out that the corpus luteum needs LH in order to remain healthy. In the absence of pregnancy, falling LH levels spell the end for the corpus luteum. This in turn causes a decline in progesterone levels, which brings about the degeneration of the endometrium, essentially by depriving it of a blood supply. Now, in the absence of a corpus luteum or developed follicles, the levels of estrogen and progesterone are *low*. Such levels stimu-late FSH and LH production in the anterior pituitary, and this is where you came in: The cycle begins again, as LH and FSH stimulate the growth of another set of follicles.

So what keeps this cycle from continuing when pregnancy does occur? The implantation of an embryo in the endometrium causes release of yet another hormone

---

How can the female reproductive system "keep time"? What is the clock that governs the cycle?

(called human chorionic gonadotropin), which acts like LH in that it keeps the corpus luteum robust and generating progesterone and estrogen, which in turn maintains the endometrium.

With this knowledge under your belt, you're in a position to see what it is that **birth control pills** do. They are compounds that contain both synthetic estrogen and a synthetic progestin. Women generally take them beginning five days after the start of menstruation for three weeks (and during the fourth week take a chemically functionless "spacer" or nothing). Think about what is required at the beginning of a cycle to prompt the development of follicles: *low* levels of estrogen. (Recall that this low level stimulates the hypothalamus and pituitary to secrete follicle-stimulating hormone.) The pill provides a *higher* level of estrogen than exists naturally, thus suppressing the release of FSH. Second, the progestin contained in the pill suppresses LH release by the pituitary, which prevents the *ovulation* of any follicle that might manage to develop, because the LH surge is required for ovulation.

MediaLab Link
Female reproductive cycle, see p. 608.

The Consequences of Follicle Loss

Given that a woman begins puberty with about 250,000 follicles, you might think that her supply would never run out. This is not the case, because the process of atresia continues throughout the lifetime, not only for "recruited" follicles, but for other follicles as well. By the time a woman reaches her early fifties perhaps a thousand follicles remain; this scarcity is the primary factor that brings about **menopause**, or the cessation of the monthly ovarian cycle. Because follicles not only nurture oocytes, but are primary sites of estrogen synthesis as well, the loss of follicular activity helps cause the physical problems associated with menopause: the loss of muscle tone, the thinning of the vaginal walls, and a loss of bone deposition, which has been associated with **osteoporosis**, or a decrease in bone mass. One aid in combating these problems is estrogen-replacement therapy, which is the use of synthetic estrogen by women who are past menopause. Now let's look at the male reproductive system.

27.3 The Male Reproductive System

Figure 27.6 provides another view of the male reproductive system. Here you see the large-scale structures of the system. There is the **testis**, in which sperm begin develop-ment; the **epididymis**, in which sperm mature and are stored; and the **vas deferens**, which sperm move through in the process of ejaculation. (All three of these structures exist in pairs, though the second member of each pair is not visible in the figure.) Then there is the single **urethra**, which each vas

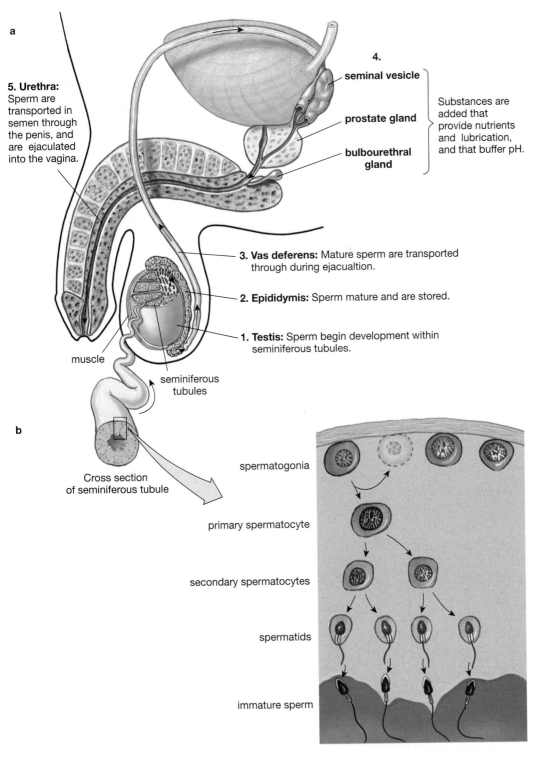

a

5. Urethra: Sperm are transported in semen through the penis, and are ejaculated into the vagina.

4.

seminal vesicle

prostate gland

bulbourethral gland

Substances are added that provide nutrients and lubrication, and that buffer pH.

3. Vas deferens: Mature sperm are transported through during ejacualtion.

2. Epididymis: Sperm mature and are stored.

1. Testis: Sperm begin development within seminiferous tubules.

muscle

seminiferous tubules

b

Cross section of seminiferous tubule

spermatogonia

primary spermatocyte

secondary spermatocytes

spermatids

immature sperm

Figure 27.6
Development and Delivery of Sperm

a After the sperm develop in the testes and mature in the epididymis, orgasm prompts them to be transported through the vas deferens and then the urethra. Secretions from accessory glands (the seminal vesicles, the prostate gland, and the bulbourethral glands) join with the sperm to form semen before ejaculation.

b How sperm mature. Sperm development begins within seminiferous tubules, located in the testes. Sperm start out as spermatogonia at the periphery of the tubules and end up as immature sperm in the interior cavity of the tubules. From there, they are transported to the epididymis for further development and storage.

Methods of Contraception

Selected methods of contraception are presented here. The rate of effectiveness for each method refers to effectiveness with correct use during every instance of sexual intercourse. In practice, effectiveness for each method tends to be lower, because use of the method may be incorrect or inconsistent. See "Hormones and the Female Reproductive Cycle" on page 590 for information on how birth control pills work. The type of birth control pill most commonly used has an effectiveness of 99.9 percent.

METHODS OF CONTRACEPTION

CONDOM

Made of latex, plastic, or animal tissue, condoms are put on the erect penis prior to intercourse and catch sperm after it is ejaculated. Condoms are the only method of contraception that helps protect both partners from some sexually transmitted diseases.

Effectiveness: 97 percent.

DIAPHRAGM

A soft, circular rubber barrier inserted into the woman's vagina before sex. Covering the cervix, it prevents sperm from reaching the egg. Along with the related cervical caps, diaphragms offer females some protection against some sexually transmitted diseases, such as gonorrhea and chlamydia.

Effectiveness: about 95 percent, when used with a spermicide (a chemical compound that immobilizes sperm).

INTRAUTERINE DEVICE (IUD)

A small device made in some instances of plastic and copper that is inserted into a woman's uterus by a physician. These IUD's can be left in place for several years. Other IUD's are made of plastic and contain a supply of a progestin hormone that is continuously released. These IUD's must be replaced once a year. Both types of IUD's help prevent fertilization of egg by sperm, in a process that may involve altering the movement of each type of gamete. Both types of IUD's also work to prevent implantation in the uterus of any embryo that might be produced.

Effectiveness: copper IUD, 99.4 percent; hormone-releasing IUD, 98.5 percent.

TUBAL LIGATION

A surgical procedure that generally results in the sterilization of a woman, meaning permanent infertility. In it, the uterine tubes are first cut and then tied, thus blocking eggs from moving to a position in which they could be fertilized by sperm. Does not affect feminine characteristics, hormonal production, or sexual performance. Very rarely, tubes will reconnect themselves, which accounts for the (low) failure rate of the procedure. It is possible to reverse the operation in some instances.

Effectiveness: 99.6 percent.

VASECTOMY

A surgical procedure that can result in the sterilization of a man, though vasectomies are sometimes reversible. In the procedure, the vas deferens are cut and tied, thus blocking the movement of sperm during ejaculation. Does not affect male hormonal production, physical capacity for erection, or ability to ejaculate. (Sperm make up a small portion of the semen that is ejaculated.) In rare instances, vas deferens will grow back together again, which accounts for the (low) failure rate of the procedure.

Effectiveness: 99.9 percent.

deferens empties into and through which sperm continue their journey. (Eventually the urethra is channeled through the penis.) There are also several accessory glands whose contributions are added to sperm as it proceeds: the **seminal vesicles**, the **prostate gland**, and the **bulbourethral glands**. (The seminal vesicles and bulbourethral glands also exist in pairs.)

Structure of the Testes

The testes are surrounded by a muscle whose contraction regulates the temperature of the sperm inside. Sperm development requires temperatures somewhat cooler than those found in the rest of the body; this is why the testes are *external* to the body's torso. Sperm cannot develop in temperatures that are too cold, however. Thus, when a male steps into, say, a cold pond, the muscle surrounding the testes will contract, drawing the testes up closer to the torso—and warming the sperm.

Figure 27.6a shows that the testes are divided into a small series of lobes, and that within each lobe there are a number of highly convoluted tubes, the **seminiferous tubules**. These are the sites of initial sperm development. If you now look at a single seminiferous tubule in cross section, in Figure 27.6b, you'll see how this development proceeds: from the outside of each tubule—where sperm precursors are in their *least* developed state—toward the interior cavity of each tubule, where sperm are in a *more* developed state. Eventually, the immature sperm are released into the interior cavity, and are transported up it, moving then to the epididymis for further processing and storage. Thus, there is an assembly-line quality to sperm production, starting with initial development at the periphery of the tubules, continuing in the interior of the tubules, and ending with storage in the epididymis. Older sperm that are not ejaculated are destroyed (by other cells) or eliminated in the urine.

Male and Female Gamete Production Compared

The production of sperm bears an interesting comparison with the development of eggs. The cells that give rise to sperm are called **spermatocytes**. In going back to the precursors of *these* cells, however, we see a significant difference between males and females. The cells that give rise to primary spermatocytes are called **spermatogonia**; they exist at the periphery of the seminiferous tubules and, through normal mitotic division, they actually lead to *two* kinds of cells. One is the primary spermatocyte, which goes on to develop into a sperm; the other, however, is another spermatagonium. Thus, spermatogonia are "sperm factories," if you will, that give rise not only to sperm, but to more sperm *factories* as well.

This, then, is the reason males *keep* producing sperm throughout their lives while, as we've noted, females have a fixed quantity of oocytes. Although male sex hormone production lessens with age, it is possible for male reproductive capability to remain intact until death. In contrast, as you've seen, female reproductive capability ends with menopause, generally when a woman is in her fifties, primarily because of a lack of oocytes.

If you look at Figure 27.6b, you can see how the male sperm progression works in the testes: from spermatagonium, through spermatocyte to spermatid, and finally to immature sperm.

Further Development of Sperm

As you can see in Figure 27.6a, the seminiferous tubules eventually feed into the epididymis, a single convoluted tubule (one for each testis) that is about 7 meters (about 23 feet) long. This structure serves as the site of storage and final sperm development and as a kind of materials recycling site for damaged sperm. By the time sperm arrive at the bottom portion of the epididymis, they are fully matured and "motile," or capable of movement.

Time Sequence, Selection, and Sperm Competition

How long does this whole process of sperm development take? About 2.5 months from start to finish. Perhaps 250 million sperm are produced in this assembly line each day, which you may recall is about the number that are ejaculated with each orgasm. (This latter number should be taken as a very rough average, however. It varies somewhat from person to person and greatly in accordance with how much time has elapsed since the last ejaculation.)

Why Are So Many Sperm Produced?

As noted, the watchword for oocyte selection could be "many are called but few are chosen." How much more extreme is the selection that goes on with sperm. The number that are expelled with *each ejaculation* vastly exceeds the number of oocytes a woman ever has stored in her ovaries. On average, however, only a single sperm per ejaculation will be able to fertilize an egg—assuming an egg is present in the uterine tube at the right time.

Why is there such an enormous difference between the number of sperm produced and the number that will fertilize eggs? As with the reason for menstruation, we can't say for sure, but we do have an interesting recent hypothesis put forward by two British scientists. It is that most sperm don't function to fertilize eggs; they function to block or destroy the sperm of *other males* (thus helping ensure the reproductive success of the male they came from). Part of the evidence supporting this idea is that perhaps half of all mature sperm are so ill-shapen that they seem incapable of making the long journey to the uterine tube, and are perhaps better suited to serve as "blockers" of other sperm.

From Vas Deferens to Ejaculation

Once they have matured, human sperm are stored in both the epididymis and the vas deferens. With sufficient sexual stimulation, ejaculation takes place through the contractions of muscles that surround the penis at its base. This contraction pushes the sperm through the urethra and out the urethral opening at the tip of the penis—at a speed of about 10 miles an hour.

Supporting Glands

As noted earlier, several other kinds of materials join with sperm before ejaculation to form the substance known as **semen**. Indeed, given the number of sperm ejaculated, it may be surprising to learn that sperm don't account for much of the volume of semen; some 95 percent of the ejaculated material comes from the accessory glands already mentioned: two seminal vesicles, the prostate gland, and two bulbourethral glands. It's helpful to think of sperm as getting "outfitted" for their travels by these glands just as they are exiting from the body; the glands supply both nutrients (which sperm will use to power their movement) and alkaline and lubricating substances (which neutralize the acidic environment of the vagina and facilitate transportation).

One of these supporting glands may sound familiar. The **prostate gland** surrounds the urethra near the bladder. Later in life, a significant percentage of men develop either enlarged prostate glands or, worse, the growth of cancerous cells in the prostate, which is to say prostate cancer. Given that the prostate encircles the urethra—and that the urethra transmits not only semen but urine as well—you can see why a warning sign of these conditions is reduced urine flow.

**Figure 27.7
Anatomy of a Spermatozoon and Steps in Fertilization**

a The "head" of each sperm includes not only its complement of chromosomes, but an acrosome, which contains enzymes that, when released externally, break down the layers of cells and other materials surrounding the egg. This release comes when the acrosome is "capacitated" by its contact with these surrounding cells. Behind the sperm head are the mitochondria that supply the energy for the motion of the tail that propels the sperm.

b The process by which sperm fertilize an egg.

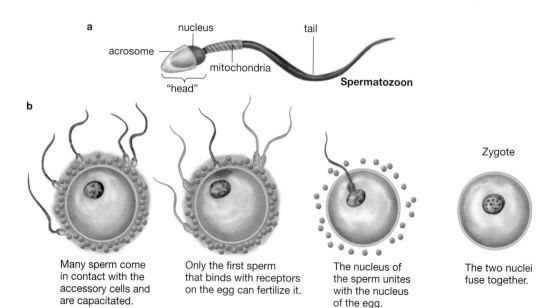

a

acrosome
nucleus
"head"
mitochondria
tail
Spermatozoon

b

Many sperm come in contact with the accessory cells and are capacitated.

Only the first sperm that binds with receptors on the egg can fertilize it.

The nucleus of the sperm unites with the nucleus of the egg.

Zygote

The two nuclei fuse together.

27.4 The Union of Sperm and Egg

To this point, you have reviewed the process by which male sperm and female egg are first created and then *positioned* so that they can come together in the process called fertilization. Now let's look at this moment of fusion in somewhat greater detail.

When the oocyte is expelled from the ovaries, there still is no solitary oocyte in transit, but rather an oocyte surrounded by a group of accessory cells and connective substances.

How Sperm Get to the Egg

Now consider the sperm (each one of which is known as a *spermatozoon*). As you can see in **Figure 27.7a**, each spermatozoon has a "head" that contains not only the nucleus and its vital complement of chromosomes, but an outer compartment, called an **acrosome**, which contains enzymes. These enzymes are released *externally*, so that they break down materials lying in their path, meaning the layers of accessory cells that surround the oocyte. To release these enzymes, the sperm need to be *capacitated*, which means they need to have a protective coating that covers the acrosome stripped away. This is accomplished by the action of substances secreted by the outer layers of cells surrounding the oocyte.

Many sperm are capacitated, but only one of them completes the next step, which is to latch onto the oocyte (**Figure 27.8**). This happens through an extension of the sperm cell that binds with receptors on the surface of the oocyte (Figure 27.7b). If you've ever wondered what separates one species of animal from a closely related, but different species, this is an important control point. Only sperm extensions and egg receptors that are chemically matched can bind with one another—and if there is no binding, there will be no offspring. Once this binding has occurred, the plasma membranes of sperm and oocyte fuse; the sperm is inside the oocyte, and sperm and egg nuclei combine. Fertilization is complete.

How Latecomers Are Kept Out

But what of the other dozens of sperm surrounding the egg? The fusion of the plasma membranes of egg and sperm slams a kind of gate shut faster than the blink of an eye.

Membrane channels in the egg open up, allowing sodium ions to rush in; this changes the electrical potential of the egg membrane, effectively shutting out any other sperm. This process, called the **fast block to polyspermy**, is a kind of first line of defense against *multiple* sperm entering the egg. It's critical that a block like this exists, because if several sperm did enter the egg, there would be no development of a zygote. (How could there be, since an egg would have, say, one and a *half* complements of genetic material within it?) The sexual activity that is so important to human reproduction also carries with it the risk of sexually transmitted disease, which you can read about on page 598.

27.5 Human Development Prior to Birth

Having seen how sperm and egg get together, you have reached a point in human reproduction that should be familiar: development, meaning the process by which a fertilized egg is transformed, step by step, into a functioning organism. As you'll see, processes you looked at in Chapter 26 in connection with frogs and flies have counterparts in human development. What may be surprising is how *early* in human development many of the critical events of development occur. Here is a list of some of the key developmental processes discussed in Chapter 26.

- Formation of the blastula, or hollowed-out body of cells (called a *blastocyst* in humans and other mammals)
- Gastrulation, in which the three primary layers of cells are formed
- Formation of the neural tube (which gives rise to the brain and nervous system)
- Organogenesis, in which the organs of the body start to take shape

MediaLab Link
Eggs, sperm, fertilization, see p. 608.

Figure 27.8
Near the Moment of Conception
A human sperm and egg at the moment the sperm binds the egg. Very soon after this, the sperm will enter the egg and its nucleus will fuse with the egg nucleus, producing a human zygote.

Sexually Transmitted Disease

Given the necessity of sex for the continuation of human life, it may seem surprising that so much risk can be associated with it, but such is the case. The root of the problem is that microorganisms that literally couldn't stand the light of day—that is, that couldn't exist for long in an exterior environment—are quite well adapted to life inside human tissue. More important, these organisms can be transmitted from one human being to another during the act of sex.

The risk of sexually transmitted diseases (STDs) varies in accordance with lots of things: the disease itself, the type of sexual activity a person engages in, and the precautions that are taken to avoid trouble. But one rule that holds true for all STDs and all behaviors is that as the number of partners goes up, so does risk. Here are the most common STDs, their prevalence in the United States, and the methods of treatment for them.

Disease	Caused by	Possible Consequences	Estimated Cases in United States	Treatment
Chlamydia	Chlamydia trachomatis organism—has features of both virus and bacteria	Spread to uterine tubes in women, causing pelvic inflammatory disease (PID), which can lead to sterility	3 million cases per year	Curable with timely use of antibiotics
Gonorrhea	Gonococcus bacterium	Arthritis, pelvic inflammatory disease	650,000 per year	Curable with timely use of antibiotics, though antibiotic-resistant strains now exist
Syphilis	Spirochete bacterium	Heart, nervous system, bone damage if allowed to progress	35,000 cases of primary and secondary infection per year	Curable with timely use of antibiotics, though antibiotic-resistant strains now exist
Genital herpes	Virus	Recurrent skin lesions, eye damage, pregnancy complications	1 million per year	Occurrence and severity of outbreaks can be lessened with treatment, but no cure exists
Genital warts	Human papilloma virus	Cervical infection in women, association with increased risk of cervical and penile cancer	5.5 million per year	Several treatments exist for removing warts, but virus may persist
AIDS	Human immunodeficiency virus (HIV)	Death, dementia, injury from a variety of opportunistic infections	57,000 new AIDS cases in 1996	No cure; treatments to reduce symptoms exist

Now, keeping in mind that a human pregnancy lasts an average of 38 weeks, here is the time line for these major developmental events. Blastocyst formation begins 5 days after conception; gastrulation 16 days after conception; neural tube formation begins in the third week after conception; and organogenesis begins in the fourth week. By the end of 12 weeks, immature versions of all the major organ systems have formed. To place some of these events in the context of what the mother experiences, organogenesis is occurring about the time the first menstrual period is missed, and formation of the immature organ systems is being completed at about the same 12-week point at which morning sickness generally begins.

The embryo's development over time can be contrasted with its *growth* over time. By the end of the third week—after blastocyst formation and gastrulation—a human embryo is perhaps 2 millimeters (or less than a tenth of an inch) in diameter. Skip to the twelfth week after conception, when major portions of organogenesis are concluding, and the fetus may still be only about 13 centimeters, or a little over 5 inches in length.

The message here is that development is concentrated early in pregnancy, while the simple growth of what *has* developed comes later. So pronounced are the changes during development that biologists use different terms for the growing organism over time. The original cell—the fertilized egg—is a **zygote**; about the time the blastocyst forms, it is referred to it as an **embryo**; once the major features of development have concluded, it is referred to as a **fetus**. The typical 9-month pregnancy is divided into periods of roughly three months each, called **trimesters**; the end of the first trimester is also the end of the embryonic period.

Early Development

You saw earlier that the zygote/embryo moves from ovary to uterus during the first four days after fertilization. It arrives in the uterus on day four as the mulberry-like *morula*, or tightly packed ball of cells noted in Chapter 26 (see Figure 26.2, page 573). Over the next two days the morula becomes the blastocyst, which implants itself in the endometrial lining. Actually, this "implantation" is more like an invasion. The blastocyst carves out a cavity for itself by releasing enzymes that effectively digest endometrial cells (**see Figure 27.9**). Implantation normally occurs in the dorsal wall of the uterus. Occasionally, however, a blastocyst attaches itself inside the uterine tube, or sometimes to the **cervix**, which is the lower part of the uterus that opens onto the vagina. This is an **ectopic pregnancy**, which the embryo usually does not survive.

Two Parts to the Blastocyst

Figure 27.9 shows you the arrangement of cells in the blastocyst. On one side of the blastocyst cavity is a group of cells called the **inner cell mass**; these are the cells that develop into the baby. The periphery of the blastocyst, however, is formed by cells called the **trophoblast**. These are the cells that etch their way into the uterine wall; with implan-

tation, they begin extending farther into the endometrium, establishing indirect links with the maternal blood supply. In time, these cells grow into the fetal portion of a

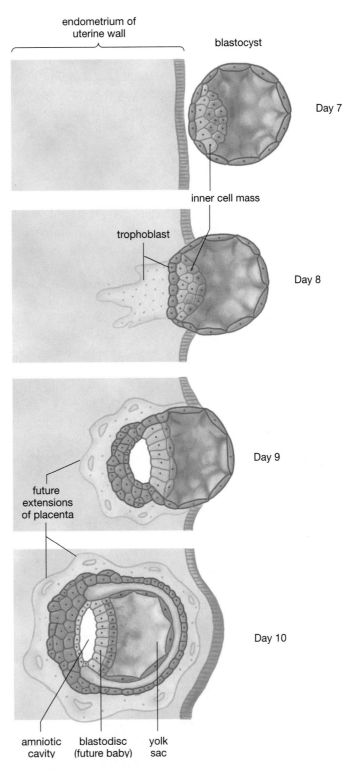

Figure 27.9
Implantation of the Embryo in the Uterus
Seven days after conception, the embryo, now in the blastocyst stage of development, arrives at the uterine wall. It is composed of both an inner cell mass, which develops into the baby, and a group of cells called a *trophoblast*, which carve out a cavity for the embryo in the uterine wall and then begin establishing physical links with the maternal tissue. The outgrowth of this linkage is the placenta.

What Is It?
See p. 602.

well-known structure in pregnancy, the **placenta**. This structure is a complex network of maternal and embryonic blood vessels and membranes. It allows nutrients and oxygen to diffuse *to* the embryo (from the mother), while allowing carbon dioxide and waste to diffuse *from* the embryo (to the maternal blood system). See **Figure 27.10** for a look at a mature placenta and uterine environment as they would exist at about two months. The tissue that links the fetus with the placenta is the **umbilical cord**, which houses fetal blood vessels that move blood between the fetus and the placenta.

The Origins of Twins

Having learned a little about fetal development, you're now in a position to understand something about the formation of twins. There are two points in early pregnancy at which identical twins develop. The first is after the initial cell division of the zygote—when the zygote becomes two cells, in other words—with each of these now-separated cells developing into a human being. The second is in the blastocyst stage when the inner cell mass seems capable of dividing into *two* masses, each able to give rise to a person. In contrast, for fraternal twins (or triplets, and so on) there must first be multiple ovulations in the mother, followed by multiple fertilizations from separate sperm of the father, and then multiple implantations of the resulting embryos in the uterus. From this, you can see the difference between fraternal and identical twins: **Fraternal twins** are like any other pair of siblings, except that they happen to develop at the *same time*. Meanwhile, **identical twins** are, strictly speaking, a single organism at one point in their development and as such have exactly the same genetic makeup.

Development through the Trimesters

If you look at the series of pictures and diagrams that make up **Figure 27.11**, you can see how an embryo and then fetus develops at selected points in the 38 weeks of pregnancy. Here's an overview of what happens through time.

First Trimester

As noted, blastocyst, gastrula, and neural tube formation all take place within the first month of conception; but much more happens during this period as well. By the 25th day, a primitive heart has formed in the human embryo and has begun pumping blood; the "somites" or repeating segments you read about in Chapter 26 have begun to take shape for muscular and skeletal formation.

Figure 27.10
Anatomy of the Placenta and Uterine Environment

a EMBRYO DEVELOPMENT:

3 weeks	4 weeks	5 weeks	6 weeks	7 weeks	8 weeks
2–3 mm	4 mm	8 mm	13 mm	18 mm	30 mm
Gastrulation and neural tube formation	Heart beats. Arm buds, tail, and gill grooves form. Somites form.	Eye development begins. Leg buds form. Brain enlarges.	Fingers and external ears form. Tail and gills disappearing.	Toes form. Bones begin to harden. Eyelids form.	Arms bend at elbows. Genitals begin to develop.

b FETUS DEVELOPMENT:

3 months	5 months	7 months	9 months
87 mm	190 mm	270 mm	350 mm
Well-defined neck appears. Genital formation complete. Sucking reflex appears.	All major organs formed. Blood cells form. Head and body hair appear. Movements felt by mother.	Lungs and lung circulation develop. Eyelids open. Fat deposited under skin. May be viable if born.	Fetus usually viable if born. Body hair lost, head hair well developed. Most senses are well developed.

Figure 27.11
Development through the Trimesters
(a) Development of the human embryo (3 weeks to 8 weeks). (b) Development of the human fetus (3 months to 9 months). Photos of a human embryo at 6 weeks (left), 11 weeks (middle), and 5 months (right).

Second Trimester

The fetus lengthens considerably during this period, but most of its weight gain takes place in the third trimester. Midway through a pregnancy, a fetus has achieved about half of its birth-*length* (which is about 50 centimeters or 20 inches on average), but only about 15 percent of its birth-*weight* (which averages 3.5 kilograms or 7.5 pounds).

The heartbeat of a fetus can be heard through a stethoscope during the second trimester, and the mother can feel the baby begin kicking by the end of the fourth month. Kicking is only one of several movements the fetus is capable of at this point; others include facial movements and the sucking reflex. By the end of the trimester, about half of all fetuses will be startled by loud sounds. A fine, white downy hair, called the *lanugo*, will cover the body of the fetus by the end of the fifth month, probably as an aid to getting an oily, protective coating to adhere to the skin.

Third Trimester

Along with the terrific growth that takes place during the third trimester, fetuses are running through a series of activities that can be thought of as practice for life outside the uterus (or womb). The fetus is surrounded by a protective and nutritive **amniotic fluid**, which fills its lungs. It therefore doesn't breathe, but it periodically exercises its breathing muscles during this period by moving amniotic fluid in and out of them.

Premature Babies and Lung Function The end of the second trimester and the start of the third also mark the critical period for a baby being able to survive *outside* the uterus. Clinically, any baby born before 37 weeks of gestation is *premature*; but many babies born between 35 and 37 weeks after conception require little special care. The *more* premature a baby is, however, the greater its risk of having difficulties with physical and cognitive development, at least for the first few years of life. In very premature babies, survival itself is called into question (**see Figure 27.12**). In general, babies cannot survive without having spent at least 22 weeks inside the uterus.

A critical factor in whether a premature baby can live outside the uterus is lung function. Working against the survival of very early "preemies" is the fact that, when they exhale, the air sacs in their lungs tend to close (or "collapse") and then stick together. Why doesn't this happen with the lungs of a full-term 38-week fetus? Because, beginning in the 28th week after conception, fetuses begin synthesizing a lung *surfactant*—a soap-like fatty substance that acts as a lubricant to keep lung tissue from sticking together. One of the great (and strangely unheralded) medical advances of the 1980s was the development of a synthetic surfactant by John Clements and his colleagues at the University of California, San Francisco. Infant deaths due to so-called respiratory distress syndrome (RDS) were cut roughly in *half* by this advance, thus saving the lives of thousands of premature babies.

27.6 The Birth of the Baby

As the fetus grows, the uterus becomes an increasingly cramped place, and the movements the fetus undertakes become much more restricted. In normal circumstances, late in the pregnancy its head will lodge near the base of the mother's spine (**see Figure 27.13**). In this "upside-down" position, the fetus is ready to make its entrance into the world. What triggers this journey? The immediate cause is **labor**: the regular contractions of the uterine muscles that sweep over the fetus from its legs to its head. The pressure this generates opens or "dilates" the cervix (ultimately to approximately 10 centimeters or about 4 inches). This creates an

Figure 27.12
Tough Going Early on
A pediatric nurse attends to a premature baby in an incubator.

opening large enough for the baby to pass through. These things must occur in stages, however. The first phase of the uterine contractions is given over to cervical dilation; then comes the expulsion of the baby and finally expulsion of the placenta (also called the *afterbirth*, for obvious reasons).

These events take place in response to a complex interplay of hormones. Interestingly, research on sheep and monkeys indicates that, in them at least, this hormonal cascade begins not with the mother but with the fetus. In sheep, it is fetal stress hormones (the "fight-or-flight" hormones) that trigger the release of the maternal hormones that in turn bring on labor.

On to Ecology

In this chapter and the four that preceded it, you followed the workings of the cells and tissues that allow animals to function in their marvelously diverse ways. But animals—and for that matter all the other varieties of living things—are not self-contained units, but rather are constantly *interacting* with each other. In addition, all the creatures of the living world are influenced by the physical forces that surround them—climate and soil characteristics and ocean temperature, to name a few. In the next two chapters, you'll be looking at life at the level of this integrated web. What you will be studying is ecology.

a

b

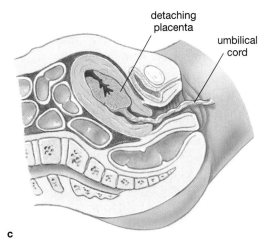

c

Figure 27.13
Birth of the Baby
With uterine contractions, the cervix dilates, allowing room for the baby to pass through the birth canal and to the outside world. The placenta and its associated fluids and tissues are expelled shortly afterward, as the "afterbirth."

Chapter Review

Key Terms

Summary

27.1 Overview of Human Reproduction and Development

- Both the human male and female reproductive systems must produce their respective gametes (sperm and eggs) and then deliver them to a place where they can fuse, in the moment of fertilization.

- The precursors to eggs, called *oocytes*, exist in the ovaries, which lie just left and right of center in the female pelvic area. Oocytes mature within nurturing complexes of cells and fluids called *follicles*. On average, once every 28 days, one follicle-oocyte complex is brought to a state of maturity. The egg and some of its accessory cells then rupture from the follicle and ovary and begin a journey through the uterine tube, to the uterus. The release of the oocyte is called *ovulation*. Conception can occur if the oocyte encounters sperm on the way to the uterus.

- Sperm are produced in the male testes and then pass through two types of ducts, the vas deferens and the urethra, after which the sperm are ejaculated from the penis into the female vagina. Materials secreted by three separate sets of glands will join with the sperm before ejaculation. The resulting mixture of sperm and glandular materials is called *semen*. The first of the sperm to reach the oocyte moving through the uterine tube may fertilize it. Development of the fertilized egg, called a *zygote*, then ensues.

27.2 The Female Reproductive System

- A portion of the female reproductive system is built up and then dismantled in a reproductive cycle that takes approximately 28 days to complete. The end-result of the dismantling is menstruation, which is a release of blood and the specialized uterine tissue it infuses, the endometrium. The endometrium, which lines the uterus, is tissue that an embryo implants itself in when pregnancy occurs.

- The basic unit of oocyte development is the ovarian follicle, a complex that includes the oocyte that develops into an egg and some surrounding accessory cells and materials.

- On average, a single follicle each month will develop into a tertiary follicle that ruptures, expelling the egg and some accessory cells from the ovary; this complex of cells then begins a journey through the uterine tube to the uterus. The ruptured tertiary follicle develops into the corpus luteum, which secretes hormones that prepare the reproductive tract for pregnancy.

- All the follicles a woman will ever possess are produced in her prior to birth. This number steadily decreases throughout her lifetime, primarily through the process of follicle degeneration called *atresia*. The scarcity of follicles a woman has in middle age is the primary factor that brings about menopause, the cessation of the monthly ovarian cycle.

27.3 The Male Reproductive System

- Sperm begin development in the testes and continue development and are stored in the adjacent epididymis. Sperm move from the epididymis, through the vas deferens and then the urethra in ejaculation. The accessory glands that contribute material to the sperm before ejaculation are the two seminal vesicles, the prostate gland, and the two bulbourethral glands.

- Within the testes, sperm development takes place in the seminiferous tubules. Development begins at the periphery of a given tubule, with sperm becoming more developed toward the interior of the tubule. Eventually, developing sperm are transported from the interior of a tubule to the epididymis for further development and storage.

- Sperm development in the testes begins with spermatogonia, which develop into spermatocytes, which give rise to spermatids and then to immature sperm. Spermatogonia give rise not only to spermatocytes, but to more spermatogonia. This self-generation of sperm precursors is the reason men keep producing sperm throughout their lifetimes. About 250 million sperm are produced each day—about the number of sperm ejaculated with each orgasm.

- The materials secreted by the seminal vesicles, the prostate gland, and the bulbourethral glands make up 95 percent of semen. These materials provide nutrients, pH neutralizers, and lubricating fluids for the sperm.

27.4 The Union of Sperm and Egg

- Sperm release enzymes externally that break down layers of accessory cells that surround the oocyte. To release these enzymes, the sperm must be capacitated, meaning they need to have a protective coating that surrounds their "head" stripped away. This happens through action of substances secreted by cells surrounding the oocyte.

- Extensions of the sperm cells bind with receptors on the surface of an oocyte, after which the nuclei of the two cells will fuse. This is the moment of conception. Fusion brings about an almost instantaneous change in the electrical potential of the egg membrane that effectively shuts out other sperm. This process, called the *fast block to polyspermy*, is a defense against multiple sperm entering the egg.

27.5 Human Development Prior to Birth

- Human development proceeds through the stages of animal development noted in Chapter 26: formation of the blastocyst, gastrulation, formation of the neural tube, and organogenesis or the formation of organs. Many of the most important stages in development are completed by the end of 12 weeks of pregnancy. The developing organism is termed a *zygote* at conception, an *embryo* when the blastocyst is formed, and a *fetus* at the end of three months of pregnancy.

- The blastocyst implants itself in the endometrial lining of the uterus during the first week of pregnancy.

- The blastocyst initially consists of an inner cell mass, which develops into the baby, and a group of cells called a *trophoblast*. The trophoblast is active in uterine implantation and the production of cells that extend into the endometrium, establishing links with the maternal blood supply. These trophoblast cells grow into the fetal portion of the placenta, a complex network of maternal and embryonic blood vessels and membranes. The placenta allows for the movement of nutrients, wastes, and gases between the developing embryo and the mother.

27.6 The Birth of the Baby

- The immediate cause of birth is labor, meaning the regular contractions of uterine muscles that sweep over the fetus from its legs to its head.

- The pressure that results from these contractions opens or dilates the cervix, thus creating an opening large enough for the baby to fit through.

Understanding the Basics

Multiple-Choice Questions

1. The precursors of the eggs produced in the ovary are referred to as
   a. gametes
   b. sperm
   c. oocytes
   d. small eggs
   e. oviducts

2. The structure that transports sperm up and over the urinary bladder to be delivered during sexual excitation is the
   a. epididymis
   b. oviduct
   c. penis
   d. testes
   e. vas deferens

3. The tertiary follicle ruptures midway through the menstrual cycle expelling
   a. one oocyte along with some accessory cells
   b. two oocytes along with some accessory cells
   c. one mature egg, ready to be fertilized
   d. one mature egg with some accessory cells
   e. the fourth follicle

4. Once the oocyte and its accessory cells have been expelled from the ovary, the fingerlike fimbriae act to
   a. move the incoming sperm toward the oocyte
   b. protect the oocyte from premature fertilization
   c. move the oocyte along the uterine tube
   d. force the oocyte into the uterus
   e. implant the newly fertilized egg in the uterus

5. The natural degeneration of follicles during a woman's lifetime is referred to as
   a. menopause   d. atresia
   b. ovulation   e. apoptosis
   c. puberty

6. Sperm development begins in the
   a. bone marrow
   b. epididymis
   c. seminiferous tubules
   d. vas deferens
   e. seminal fluid

Brief Review

1. Describe the hormonal changes that regulate the female reproductive cycle.

2. Discuss how the testes act to protect the developing sperm.

3. Menstruation represents a costly biological process. Why do scientists believe it occurs so frequently?

4. Compare and contrast the tertiary follicle and the corpus luteum.

7. These sperm factories not only give rise to sperm but they also give rise to more sperm factories called
   a. spermatocytes
   b. spermatogonia
   c. oocytes
   d. vas deferens
   e. prostate glands

8. The portion of the sperm that contains enzymes that are used to degrade materials found on the outside of the egg is the
   a. acrosome
   b. nucleus
   c. tail
   d. mitochondria
   e. chromosomes

9. The portion of the blastocyst called the trophoblast eventually forms the
   a. head of the baby
   b. legs of the baby
   c. entire baby
   d. nervous system of the baby
   e. placenta

10. Birth of identical twins results from unusual events during
    a. egg production
    b. sperm production
    c. early cell divisions after fertilization
    d. first trimester
    e. second trimester

5. What is the fast block to polyspermy? Why is it necessary?

6. Outline the events in human development that occur prior to implantation.

7. Define the term *premature* with respect to human development. What complications may arise during a premature birth?

8. How do birth control pills work?

Applying Your Knowledge

1. When a couple is having trouble conceiving a child, the first thing fertility specialists generally will look at is the fertility of the prospective father. What do you think a specialist would be looking for?

2. The fast block to polyspermy prevents more than one sperm from fertilizing the egg by quickly changing the ionic environment of the egg after the first sperm has successfully fertilized the egg. There is also a slow block to polyspermy. This block involves the production of a hard outer coating around the newly formed zygote. Why is this necessary?

3. Birth control pills manipulate hormone levels in the female to prevent ovulation. Could alterations in hormone levels be used to manipulate sperm production in males? Why? Why not?

4. Compared to molecules such as O_2 and CO_2, viruses are rather large and more complex. In order for a virus to enter a cell or cross a blood vessel it must be transported across. This transportation process is specific to certain cells. The virus that causes AIDs, HIV, can be transported only into a particular type of white blood cell called the helper T cell. Based on this information do you think that an HIV-positive mother will pass this virus to her child via the placenta?

Group Activity

Walk around your classroom and introduce yourself to three members of the opposite sex. Write the names of these individuals on a note card. Each name on your card represents a random sexual contact. Once everyone has made their introductions, your instructor will "expose" a certain percentage (about 1 out of 100) of the class to particular disease. These exposed students will then stand and read the names on their cards. The students whose names were read should now stand and read the names on their cards. This process will be repeated until all the students that were exposed to the infectious disease are standing. Then form small groups of three or four and formulate answers to the following questions.

1. Compare the results of this simulation to the spread of a sexually transmitted disease in a larger population.

2. Has this exercise altered your perception of sexually transmitted diseases? How?

MediaLab

Are Test Tube Babies the Solution? Understanding Reproductive Problems

Where do babies come from? This question is usually followed by parents squirming through the old-standby talk about birds and bees. Imagine, though, the difficulty some modern parents will have in describing their use of assisted reproductive technology, mechanical methods that remove the sex from sexual reproduction, like *in vitro* fertilization, egg and sperm donors, even sperm injection. These techniques seem to be the innocent product of a late twentieth-century desire to use scientific advances to make childbirth possible for couples who could never conceive naturally. But is there a more sinister reason for the increase in reproductive technology? Are we becoming less fertile as a species? To learn more, the *CD Activity* will help you review normal female and male reproduction, so that in the following *Web Investigation* and *Communicate Your Results*, you can evaluate how these processes might be sensitive to our modern lifestyle.

This *MediaLab* can be found in Chapter 27 on your CD-ROM and Companion Website (http://www.prenhall.com/krogh).

CD ACTIVITY

Human sexual reproduction involves multiple, highly coordinated steps. In addition to the events of meiosis to create haploid gametes, fertilization and embryonic development occur internally, so humans need accessory structures to direct sperm and egg fusion as well as structures to allow the embryo to access oxygen and food and get rid of wastes. This *CD Activity* leads you through the reproductive systems of males and females and also through the early internal development of a human embryo.

Activity

1. *First, you will learn about the formation and maturation of sperm, and the accessory glands needed to activate sperm for their journey to fertilize the egg.*

2. *Next, you will review the female menstrual cycle: the coordinated release of an oocyte and preparation for the female body to receive and nourish a developing embryo.*

3. *Finally, you will learn about the voyage of the fertilized egg into the uterus and the creation of the membranes used to access nutrients from the mother: the placenta, amnion, and umbilical cord.*

Failure in any one of these steps can result in an inability to conceive or to carry the fetus to term; a problem for increasing numbers of couples dealing with infertility. Could factors from our environment be responsible for this problem? The following *Web Investigation* and *Communicate Your Results* sections are meant to help you evaluate that question.

a
nucleus
tail
acrosome
mitochondria
"head"
Spermatozoon

b
Zygote

Many sperm come in contact with the accessory cells and are capacitated.

Only the first sperm that binds with receptors on the egg can fertilize it.

The nucleus of the sperm unites with the nucleus of the egg.

The two nuclei fuse together.

WEB INVESTIGATION

Investigation 1

Estimated time for completion = 5 minutes

The environmental movement in the past 30 years has awakened public concern over the effects of pollution on animal health and reproduction. Now, many are questioning whether the same human-made chemicals— including pesticides, plastics, and detergents— that interfere with normal hormone function in laboratory animals are causing a decline in human sperm counts. Select the Keywords **SPERM CONTROVERSY 1** or **2** on your CD or Website for this *MediaLab* to read two contradictory articles on declining sperm counts.

Investigation 2

Estimated time for completion = 5 minutes

What have fertility specialists concluded about the major causes of infertility in most couples? Select the Keyword **INFERTILITY** on your CD or Website for this *MediaLab* to view a chart of the major causes of infertility. Then select the Keyword **ART** to view a government site posting background information on couples seeking help from Assisted Reproductive Technology (ART).

Investigation 3

Estimated time for completion = 10 minutes

Many pregnant women fear birth defects caused by environmental substances (teratogens that interfere with normal development). Although only about 5 percent of birth defects are attributed to exposure to environmental substances, these environmental agents should be clearly recognized and avoided.

However, information on the identity of teratogens is often inaccurate, alarming, and not supported by scientific studies. Select the Keyword **PREGNANCY** on your CD or Website for this *MediaLab,* and list the substances that are suggested to be teratogens and are thus to be avoided during pregnancy. Then, select the Keyword **TERATOGENS** to view a list of known teratogens.

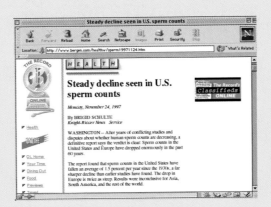

COMMUNICATE YOUR RESULTS

Now that you have been introduced to several sources of suspicion that the environment is adversely affecting human reproduction, use the following questions to evaluate the validity of these claims.

Exercise 1

Estimated time for completion = 15 minutes

Although members of the scientific community disagree over whether sperm counts are declining worldwide, they generally agree about an increase in testicular cancer and congenital birth defects in male reproductive systems. In this exercise, form a group and propose a hypothesis to explain these observations. Your group should search the Internet to find supporting studies, and use the Message Board at the Website for your textbook to post the Internet addresses.

Exercise 2

Estimated time for completion = 15 minutes

Test-tube babies, sperm injections, frozen eggs and sperm are all commonly covered in news stories; but we often hear more about technology than about patients. Using information from the Web sites in *Web Investigation 2,* write a 250-word portrait of the average woman (or man) who is using ART. Include answers to some of these questions: What is the most common procedure? How many ART patients are not pregnant after one try or cycle? How many couples experience multiple births after one cycle? What is the average age of females receiving this treatment?

Exercise 3

Estimated time for completion = 5 minutes

Hopefully it can be said that for every inaccurate Internet site there is a well-informed, scientifically credible site. Compare the list of chemicals to avoid during pregnancy that you made for *Web Investigation 1* with the information from the site listing human teratogens. In 50 words, express your conclusions about the first article.

The Living World as a Whole: Ecology

The bald eagle, an ecological success story. Once an endangered species, it now thrives.

28

An Interactive Living World
Populations and Communities in Ecology

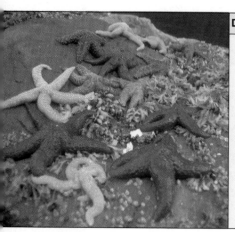

What is a keystone species?
(page 625)

DEFINITIONS OF BIODIVERSITY

High biodiversity

a Species diversity

many different species

Biodiversity.
(page 626)

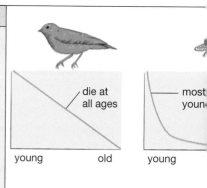

die at all ages

most youn

young   old   young

Why do most flies die young?
(page 621)

612

For those who already have

been through some earlier units of this book, a word that is a favorite in the humanities by now may have come to mind in connection with biology. It is "*microcosm*," which comes from the Greek *mikros* (tiny) and *kosmos* (universe): a tiny universe, which is also to say a world within a world.

What are biology's microcosms? Well, for one, in all living things, a *system* exists whereby DNA contains information that is then used for the production of proteins. Once put together, these proteins may make up some structure (your hair, for example) or get busy on some specific task (such as breaking down sugar). Critically, one of these tasks is turning on or off the very DNA that called the proteins into being in the first place. Thus it is a *self-regulating* system; it exists within a whole living being, but it has some of the characteristics of a tiny world of its own—a microcosm.

Now take a step up in scale and consider the parts of a human body. Drink some lemonade and your body will start producing insulin, which will help store away the blood sugar that is now coursing through your veins. *Stop* eating for awhile and your body will produce a hormone called *glucagon* that helps bring these stored food supplies *out* of reserve. This self-regulating system is thus a kind of world of its own—one that works in concert with, and yet that is in some ways separate from, the circulatory system, the nervous system, and on and on.

By moving up another level in scale, you arrive at a whole organism—a person in the example just given, but it could be a bat or a tree. You might be tempted to think that this is the end of the line for microcosms; that here you have reached the *macro*cosm, containing all microcosms. Think again, however, because the life of, say, a tree is profoundly influenced by *other* trees around it as well as by other kinds of organisms (insects, fungi). Under this view, a living thing is a kind of microcosm that exists within the larger world of living *things*.

Even after taking this larger world into account, there is still one more jump to make in scale. The whole of the living world exists within the context of the *nonliving* world and universe: the Earth's climate and gases, its sea currents and volcanoes, as well as the energy from the sun. Life doesn't exist apart from these things, but instead is very much conditioned

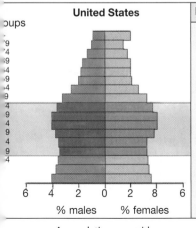

A population pyramid.
(page 622)

That's my branch.
(page 629)

Ecologist up in the air.
(page 614)

613

by them, though, as you'll see, some of these things are also conditioned by life.

28.1 The Study of Ecology

In your walk through biology so far, you have made it through the molecular world (DNA, for example) and through the world of organ systems (circulation, for example) and to some extent through whole organisms. Now it is time to take the last steps up; in this chapter and the one that follows, you will consider the largest realm of all in biology: the interactions that living things have with each other and with the environment around them. In short, you will study **ecology**.

Ecology Is Not Environmentalism

The "eco" in ecology comes from the Greek word *oikos*, meaning "house," and this is a pretty good summation of what you'll be studying—the house of the living world. In common speech, however, the word *ecology* has become synonymous with the conservation of natural resources or with the "environmental movement" in all its political dimensions. Under this view, you might assume that the job of an ecologist is to help preserve the natural world. In fact, however, the job of an ecologist is to *describe* the natural world in its largest scale; to say what is, rather than what should be. Are ecologists

environmentalists? Most are, but this shouldn't be surprising. We would no more expect ecologists to be indifferent to the environment than we would expect art historians to be indifferent to paintings. Moreover, there is a branch of biology, conservation biology, that is directly concerned with preserving natural resources. But the business of ecology per se is to tell us about the large-scale interactions that go on in the natural world. This work has an important use in that it provides the *information base* that society can use to make decisions about the environment. Does species diversity make for healthier natural communities? How many gray whales are there in the world? How important are prairie dogs to the prairie? To answer these questions, ecologists end up working in some places you might expect— in tents near the Arctic Circle or at treetop level in tropical rainforests (**Figure 28.1**). They also work in some locations you might not expect, however, such as in offices, hunched over computers, seeing how mathematical models of the natural world work out.

Path of Study

In this chapter and the next one, you'll follow a bottom-to-top approach by moving from smaller ecological units to larger ones. What are the scales of life that concern ecology? The smallest is that of the physical functioning, or **physiology**, of given organisms. At this level, ecologists are examining individual living things, but at all other scales they are looking at *groups* of organisms or species. For the ecologist, life is conceptualized as being organized into:

> Populations
> Communities
> Ecosystems
> The Biosphere

From Individuals to a Population

The smallest level of group organization is a **population**, defined in Chapter 21 as all the members of a single species that live together in a specified geographic region. The North American bullfrog (*Rana catesbeiana*) is a *species* that can be found from southern Canada to central Florida, but in a given pond it's possible that only a single *popula-*

Figure 28.1
An Ecologist on the Job
Ecologists spend a good deal of time going out into nature to study the conditions that exist in various ecosystems—however inaccessible those ecosystems may be. Here a researcher is studying tree branch growth in the rain forest canopy in Costa Rica.

tion of this species would be found. Of course it's possible to define our population to be all the bullfrogs in *two* ponds, or in two states, or two countries—whatever geographic region you wish.

From Populations to a Community

Going up the scale, if you take the populations of *all* species living in a single region, you have a **community**. Often the term *community* is used more restrictively, to mean populations in a given area that potentially interact with one another.

From a Community to an Ecosystem

If you add to the community all the *non*living elements that interact with it—rainfall, chemical nutrients, soil—you have an **ecosystem**. Ecosystems can be of various sizes; you'll be looking at a range that goes from fairly small (a single field, for example) up through the enormous ecosystems called **biomes** that may take up half a continent.

From Ecosystems to the Biosphere

The largest scale of life is the **biosphere**, which can be thought of as the interactive collection of all the Earth's ecosystems; given what you've reviewed about ecosystems, this means all life on Earth and all the nonliving elements that interact with life. Sometimes, however, the term *biosphere* is used purely in a territorial sense—to mean that portion of the Earth that supports life. If you look at **Figure 28.2**, you can see a graphic representation of ecology's scales of life. We'll now begin to look at ecology through these levels of biological organization, starting small, with the population.

28.2 Populations: Size and Dynamics

So what is it ecologists want to know about a population? Well, they need to know how to count it, how it is distributed over its geographical area, and how its size *changes* over time—what its population dynamics are, to use the term employed by ecologists.

Estimating the Size of a Population

The reason ecologists want to count the members of a given population is straightforward: Without such a count, there is no

Biosphere

Ecosystem

Community

Population

Organism

Figure 28.2
The Scales of Life That Concern Ecology
In this chapter, you will consider how organisms of one species are associated in populations, and how populations of different species are associated in communities. In Chapter 29, you will look at ecosystems, which include not only living community members but the non-living factors that interact with them (such as the rainfall in the ecosystem above). Chapter 29 also reviews the large-scale ecosystems called biomes and the biosphere, which is the interactive collection of all the Earth's ecosystems.

way to answer a question such as how much territory a group of cheetahs must have in order to flourish, or how fast a population of finches recovers from a drought. (How would you know about the latter unless you could compare the population after the drought with the population before it?)

With large, immobile species such as trees, taking a census can sometimes be easy; just mark off an area and count. Things become more difficult when the area under consideration is so large that not all species in it can be counted, but rather must be estimated on the basis of a population counted in a smaller representative area. Estimating becomes more difficult yet when the individuals are numerous and mobile, as with birds. Ecologists employ various means to estimate such populations; they count animal droppings within a defined area or survey bird populations as they migrate, for example.

Growth and Decline of Populations over Time

How is it that populations *change* size? As you begin to think about this question, it's worth going over a more general concept, which is the way a population of *anything* might increase in number. If you look at, say, the number of cars coming off the end of a production line, there might be 1,000 on Monday and then 1,000 on Tuesday, and so forth. The important thing to note is that the number of cars produced on Tuesday is not related to the number produced on Monday. The increase in the number of cars is thus an **arithmetical increase**: Over each interval of time (a day in this case), an unvarying number of new units (cars) is added to the population.

Now contrast this with what happens to living things. In most cases, each new unit (each living thing) is capable of playing a part in giving rise to *more units*, which certainly is not the case with cars. Population increases for living things are thus *proportional to* the number of organisms that already exist; this is what is known as an **exponential increase**. To put this another way, exponential growth is in operation when the number that *will* exist is a function of the number that *does* exist.

Table 28.1 gives an example of the difference between the two kinds of growth over a period of weeks, using cars for arithmetic growth and the tiny water flea *Daphnia* for exponential growth. Let's start out at the end of day 1 with the 1,000 cars produced that day and with 1,000 water fleas existing in an optimal laboratory environment.

So, *Daphnia* is a relatively slow starter, but its population quickly overwhelms that of the car population—not surprising, because the *Daphnia* population doubles every 3 days. If you look at **Figure 28.3**, you can see what these two kinds of growth look like when plotted on a graph.

Population Growth in the Real World

As noted, the *Daphnia* were growing in an "optimal" laboratory environment—plenty of food, habitat kept clean, no predators.

**Figure 28.3
Arithmetic and Exponential Growth**
When the same number of objects is produced in a given interval of time—in this case cars from a factory each day—arithmetical growth is at work. By contrast, the populations of most organisms, such as the water flea *Daphnia*, can exhibit exponential growth, at least for a time, meaning the population grows in proportion to its own size.

exponential growth
of water-flea
population

arithmetic growth
of car production

	Day 1	Day 4	Day 7	Day 10	Day 13	Day 16	Day 19	Day 22	Day 25
Table 28.1 Arithmetic Growth versus Exponential Growth									
Arithmetic Growth: Cars	1,000	4,000	7,000	10,000	13,000	16,000	19,000	22,000	25,000
Exponential Growth: *Daphnia*	1,000	2,000	4,000	8,000	16,000	32,000	64,000	128,000	256,000

Thanks to human intervention, this was a kind of paradise for water fleas, in other words. But could any population ever grow like this in the real world? In all cases the answer is "not forever," and in most cases the answer is "not for long," but some population growths can be dramatic while they last.

In 1859, Thomas Austin released 24 wild European rabbits (*Oryctolagus cuniculus*) onto his estate in southern Australia, near Melbourne, in order to provide more game for sport hunting. His was not the first attempt to introduce rabbits to Australia, but it was by far the most successful—or perhaps the most fateful. This alien species spread north and west in Australia like wildfire; within 16 years it had expanded its range through the entire latitude of the continent, a distance of almost 1,800 kilometers or 1,100 miles. And this invasion was not a matter of a rabbit here and a rabbit there; the European rabbits became a scourge that only recently has shown signs of being brought under control, through use of a virus (**see Figure 28.4**). Here is Eric Rolls' description in *They All Ran Wild* of the land east of Peterborough in South Australia in 1887:

> . . . for over a hundred miles most trees and plants had been killed. Out-alpa head-station was a thousand square miles of rabbits. . . . Rabbits crawled like possums in branches several feet off the ground. They had burrowed under the beautiful granite homestead; they had taken possession of the cellar; they were in the turkeys' coop; they ran in and out of the men's huts apparently without fear; they took no notice of the dogs nor the dogs of them.

There are lots of instances of such seemingly uncontrolled growth. Charles Darwin devoted some thought to the issue as a theoretical matter. He wrote that "There is no exception to the rule that every organic being naturally increases at so high a rate, that, if not destroyed, the Earth would soon be covered by the progeny of a single pair." To put a little finer point on this, a single pair of carrion flies (which are a little larger than a housefly) could in theory give rise to enough flies in a single year to cover the state of New Jersey to a depth of 1 meter (3 feet).

The Shape of Growth Curves

If you look at **Figure 28.5a**, you can see what growth of this type, **exponential growth**, looks like when plotted on a graph. This is called the **J-shaped growth form** because it

Figure 28.4
Rabbit Control in Australia
Rabbits run free on one side of a "dingo fence" in Australia. Note the contrast between the barren land where the rabbits feed and the grassland, to the left, from which the rabbits have been kept out.

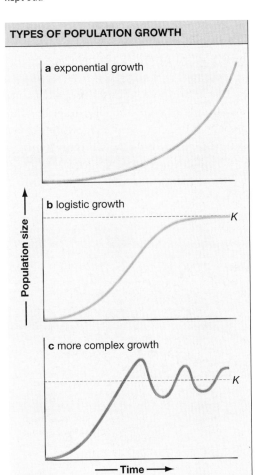

Figure 28.5
Three Models of Growth for Natural Populations

MediaLab Link
Overpopulation, see p. 642.

can be so extreme that when plotted it looks like the letter J. It may look familiar, because it is the growth form you just saw with *Daphnia*. As with *Daphnia*, there is relatively slow growth at first, then faster growth, then faster yet. Growth occurs, in other words, in which the *rate* of increase keeps accelerating. In the real world, the curve eventually drops off, often sharply, for reasons you'll soon go over.

Now look at **Figure 28.5b** to see a second model of growth for natural populations, the **S-shaped growth form**, which tracks **logistic growth**. This *starts* like exponential growth, but eventually the rate of growth slows and finally ceases altogether, stabilizing at a certain level, denoted here as *K* (which you'll learn the significance of shortly).

What has intervened to account for the difference between the dizzying increase of the J-shaped form and the moderate increase of the S-shaped form? All the forces of the environment that act to limit population growth, which in ecology are known as **environmental resistance**. Organisms will run out of food or have their sunlight blocked; temperatures will plunge or rain will cease; greater numbers of predators will discover the population; the wastes produced by the organisms will begin to be toxic to the population.

There are populations whose dynamics over time look something like either the J- or S-shaped curve, but in the real world, population size is likely to vary in more complex patterns. In **Figure 28.5c**, you can see one example of this.

Calculating Exponential Growth in a Population

You've seen that ecologists have ways to count natural populations; but is it possible for them to *predict* the size of populations, given certain pieces of starting information? The answer is yes, as you'll now see. Let's look first at a simplified means of calculating exponential growth, meaning the J-shaped curve. For now, assume the population is isolated—that is, that no individuals are moving in or out of it.

What first needs to be done is to pick a time period to evaluate population change. This period may be 40 minutes (for bacterial populations) or 50 years (for elephants), but some unit of time needs to be chosen for the analysis. The next step is to learn the difference between the birth rate and the death rate in the population over the time period you've picked. This, of course, raises the question: What's a birth rate? It is the number of individuals born in a given time period expressed as a proportion of the whole population. Thus, if a population of 1,000 had 100 births in a year, its birth rate would be 0.10 or 10 percent per year.

Figuring the death rate is similar; if a population had 80 deaths per year, its death rate would be 0.08 or 8 percent per year.

The *difference* between the birth and death rates, of course, is 2 percent. This number is the population's growth rate in an optimal setting, denoted as *r*, and it is very important. With it in hand, you can predict what the population's size will be in the future. Growing at 2 percent per year, the population of 1,000 will increase to 1,020 in the first year, and to about 1,480 in 20 years.

Growth rate is critical because a small change in it can mean a big change in population over time. Take a population of another species, once again with 1,000 individuals in total, but this time with an *r* of 6 percent, rather than 2 percent. In the first year this population would grow by 60 individuals, giving it 1,060 members as opposed to the first population's 1,020. Go 20 years *out*, however, and this population has about 3,200 individuals in it—it's now more than twice the size of the first population even though both were the same at the start. Meanwhile, if a population's *r* is less than zero—that is, if the number of deaths ex-

Figure 28.6
A Population's Potential for Growth: The Intrinsic Rate of Increase (*r*)
If the birth rate exceeds the death rate, then the population size will increase over time (*r* = 0.02, *r* = 0.06 in the example). If the birth rate equals the death rate, then the population size will stay constant (*r* = 0). Finally, if the birth rate is lower than the death rate, the population size will decrease over time (*r* = −0.05). Note what dramatic effects a small change in *r* can have over time, in this case from a population whose starting size is 500 individuals.

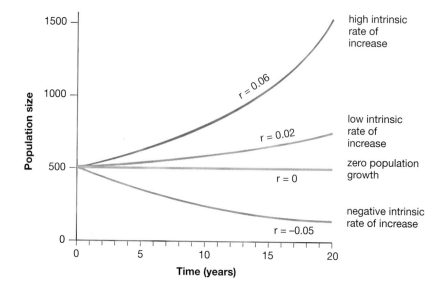

ceeds the number of births—the population is shrinking. If r is zero on the nose, meaning that births exactly match deaths in a given period, the population is at **zero population growth** (see Figure 28.6).

The r figure goes by another name, which is a population's **intrinsic rate of increase**: The rate at which a population would grow if there were no external limits on its growth. You could think of it as a population's potential for growth. Importantly, each species has a characteristic rate of increase. It won't surprise you to learn that this rate is higher for *Daphnia* than it is for whales; you can see some examples of an important factor in rate of increase, the time it takes to produce a new generation, in **Figure 28.7**.

Logistical Growth of Populations: Reality Makes an Appearance

You've now figured exponential or J-shaped growth, but you've seen that such growth never occurs for long in nature. Environmental resistance begins to assert itself and, in some instances, there will be a flattening out of the population increase and in time perhaps a complex pattern with the curve moving above and below the line denoted earlier as K. But what is K? What is this point around which the population hovers? It is the maximum population density of a given species that a defined geographical area can sustain over time. The term for this is an area's **carrying capacity**.

A Real-Life Example of Carrying Capacity

There is an island in the San Francisco Bay called Angel Island, just a short boat ride from the Golden Gate side of the Bay, that had a small population of deer introduced to it by humans early in the twentieth century. The problem was that the island lacked any natural *predators* for the deer, with the result that the deer population proliferated way beyond what the island's vegetation could support. Eventually, any hiker who frequented the island could see that the deer were emaciated, so many visitors did what you might expect: They fed the deer with what they brought along in their backpacks. Surviving through this artificial means, however, *more* deer came to live on the island and stripped it further of its vegetation, reducing its carrying capacity further. The main lesson here

is that there are limits to how many members of a given species an expanse of territory can support. Note also, however, that carrying capacity is not fixed; in the case of Angel Island, it was lowered (through the actions of human beings) while in other instances it might be raised for given periods (perhaps through abundant rainfall). The environments that many species live in tend to be stable over time, however, meaning that their carrying capacities will tend to be stable as well.

The K line can generally be established for a given population by estimating its numbers over an extended period of time. For many species, their population may exceed or "overshoot" K at times, but this will be offset by periods during which it is under K. Once K is known, it is valuable in predicting the change in a population's size, because it often serves as a *factor* that can be used in mathematical equations to calculate the limits to population growth. The higher a population is relative to K, the more severely its growth will be limited.

28.3 *r*-Selected and *K*-Selected Species

It's well known that for human beings, about 9 months will elapse between conception and birth; but it's less well known that for an elephant this same process takes almost 2 years. For any given human or elephant female, then, births are relatively few and far between. And once the offspring come, the

**Figure 28.7
How Long between Generations?**
An important factor in the population dynamics of any species is its "generation time," meaning the time that elapses between the birth of one generation and the birth of the next. Shown on the graph are the minimum generation times for a few selected species, along with a measurement of the length of each organism at the time it reproduces. The larger a species is, the longer its generation time tends to be. (Redrawn from J. T. Bonner, *Size and Cycle: An Essay on the Structure of Biology* [Princeton University Press, 1965.])

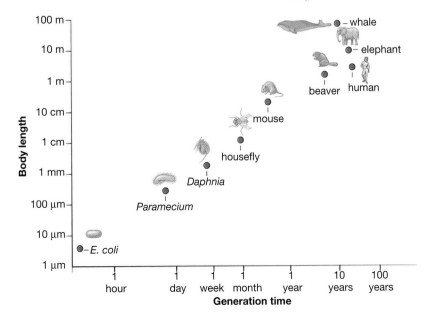

into type II, because they die off at a nearly constant rate through their life span.

28.4 Thinking about Human Populations

We'll look at the final elements in population dynamics in connection with the human population. The concepts involved could be applied to any population of living things, but here the focus is on human beings, so that you can consider population principles along with the real-world issue of human population growth.

Table 28.2
A Life Table for the United States

Taking a hypothetical group of 100,000 persons born in the United States in 1996, the table shows the proportion likely to still be living at the ages indicated and the average remaining lifetime for persons at each age. Note that life expectancy does not drop 10 years for each 10 years lived. The average person surviving to age 70 can expect to live until age 84, whereas the average person at age 10 could expect to live only to about age 77.

At Age	Number Still Living	Average Remaining Lifetime in Years
10	99,022	66.9
20	98,519	57.2
30	97,487	47.7
40	95,881	38.4
50	92,946	29.5
60	86,630	21.2
70	73,056	14.1
80	49,276	8.4
85+	33,629	6.1

Source: National Center for Health Statistics

Survivorship Curves Are Constructed from Life Tables

The survivorship curves you just looked at are constructed by developing what are known as **life tables** for the species in question. To create these, scientists divide the species' life span into suitable units of time—days for fruit flies, years for human beings—and see how likely it is that an average species member will be alive after a given number of days or years. The technique for doing this was not invented by biologists who wanted to know about the survival of flies, but by a nineteenth-century British actuary—a person trained to calculate probabilities—who knew the information would be useful for life insurance companies. How likely is it that a person will be alive at 10, or 20, or 80 years of age? If you look at **Table 28.2**, you can see the answer to this for a hypothetical group of 100,000 persons born in the United States in 1996.

Population Pyramids: What Proportion of a Population Is Young?

You've seen that population growth in the natural world can be exponential (for a time) because a population of living things grows in proportion to its own size. However, all members of a population do not count equally in calculating this growth. If scientists want to peer into the future of a given population, what they want to know is: What proportion of the population is *past* the age of reproduction as opposed to the proportion that is, or will be, of reproductive age? If you look at **Figure 28.10**, you can see the an-

Figure 28.10
Age Structure in Populations: Kenya and the United States, 1995
Each bar on the graphs represents a five-year age grouping of the population. Note the greater proportion of the Kenyan population that is of "pre-reproductive" age—an indication of greater population increases in the future than is the case with the United States. (Data from Population Reference Bureau.)

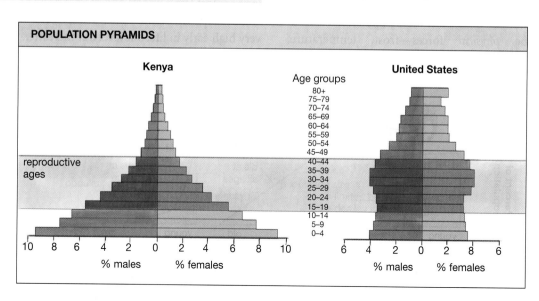

swer to this question, expressed as a "population pyramid" for two countries. Each bar on the graph represents a 5-year age grouping, or "cohort," of the population (those who are 0–4 years old, 5–9, and so on), with the length of each bar representing the relative size of the cohort. The reproductive age-range for humans generally is considered to be between 15 and 45. If you look at the graph for Kenya, you can see that it has many more individuals of reproductive and pre-reproductive age than of post-reproductive age. What does the future hold for Kenya? In all probability, a huge growth in the population. In contrast, the age-structure diagram for the United States shows that a much greater proportion of its population is at or beyond reproductive age.

Immigration and Population Change: The United States

Thinking globally, there is but one human population, of course. But to continue thinking along lines of population by country, two factors that haven't been mentioned come into play, which are the movement of individuals into a population (**immigration**) and the movement of individuals out of a population (**emigration**). Immigration is expected to have a great impact on the population of the United States, as you can see in **Figure 28.11**. According to projections released in 1997 by the National Research Council, if the United States allowed no net legal or illegal immigration, its 1995 population of 263 million would grow to about 307 million by the year 2050. At current net levels of immigration, however (about 800,000 people per year) the U.S. population is projected to grow to 387 million people by 2050—80 million more people than if there were no immigration, and an increase of 47 percent over the 1995 population. Note also the trend lines. With no immigration, the U.S. population actually would be declining after 2040, while with immigration the trend will be upward for many more years.

The Increase in World Population

Large as the U.S. growth figure is, it pales in comparison with the growth projected for many other parts of the world. Between 1990 and 1995, the annual growth rate for the more developed countries such as the United States was 0.4 percent, but for the less developed countries it was 1.88 percent. In parts of Africa, meanwhile, it approached 3.0 percent (though this high rate is now being drastically reduced in some African countries because of AIDS). Remember how, as noted earlier, small differences in the rate of growth stand to have tremendous impact *over time*? In 1950, developed countries accounted for 33 percent of the world's population; by 2025 they are expected to account for 15 percent of the total. By 2025, Bangladesh is likely to have a total population of 196 million people—this in a country that is about the size of Iowa, whose population in 1994 was about 2.8 million people (**see Figure 28.12**).

Figure 28.11
Immigration Can Have a Big Impact on Population Size
Current projections are that without immigration, the United States population would level out and begin to decrease by the middle of the next century. With its influx of immigrants, the U.S. population is expected to continue growing into the foreseeable future. (Data from *The New Americans*, National Research Council, 1997, Table 3.3.)

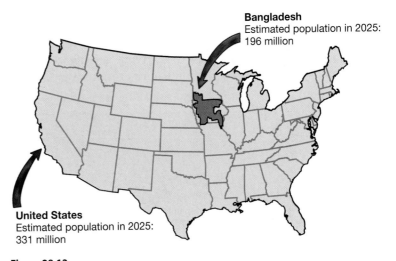

Bangladesh
Estimated population in 2025: 196 million

United States
Estimated population in 2025: 331 million

Figure 28.12
Small Area, Big Population
By 2025, Bangladesh, which is about the size of Iowa, may have 196 million people, compared to 331 million in all of the United States.

Human Population Growth over Time

If you look at **Figure 28.13**, you can see how human population has grown over time. For centuries our numbers grew relatively little, but then began an upward climb about 1700. This increase was nothing, however, compared to the rise that occurred beginning about 1950. Improved sanitation, better medical care, and increases in the food supply came together to produce the rate of growth you see. The Earth's human population didn't pass the 1 billion mark until 1804; it then took 123 years to double to 2 billion (in 1927), then 48 years to double to 4 billion (in 1974), and has now exceeded 6 billion. The curve you see that resulted from this growth may look familiar, because it is a more extreme version of the J-shaped curve, though greatly elongated on its left side. Were we looking at any species other than our own and saw this kind of growth curve, we would predict that this kind of increase could not go on for long; that environmental resistance would assert itself and that the population might even crash in one catastrophic sense or another.

The Earth's population is projected to grow from its present 6.2 billion to between 8.5 and 11 billion by the end of the twenty-first century. So, can the Earth support a population of 8 or 10 or 11 billion people? Do current famines and disease outbreaks represent environmental resistance to such population sizes? Do we see on the horizon any other signs of environmental resistance? These questions are a matter of spirited debate among ecologists, economists, and demographers. As may be apparent, this is a discussion about *carrying capacity*—in this case the Earth's carrying capacity for human beings. Some scholars think we have come nowhere near this capacity, while others think we have effectively exceeded it already.

Current Population Impacts on Quality of Life and the Environment

It may be that the notion of planetary carrying capacity is beside the point; that what we ought to be focusing on is the impact human population increase is having right now on the quality of human life and on the quality of Earth's environment. The relationship of population growth to the quality of human life is complex; suffice it to say here that there is great disagreement as to whether growth per se works against human well-being.

With respect to the environment, some scientists would argue that there is no greater single environmental threat than the continued growth of the human population. The basis for this argument is that population affects so many environmental issues: the use of natural resources, the amount of waste that is pumped into the environment daily, the reduction of species habitat, the decimation of species through hunting and fishing. Look at almost any environmental problem and you are likely to find human population growth playing a part in it. Other experts would argue that it is not population per se, but rather the use of resources *per person* that is of most pressing concern. Consider that the United States used less water in 1995 than it did in 1980, even though the U.S. population grew by 16 percent during the period. The use of resources per-person is much higher in the more-developed countries than in the less-developed countries. Given this, a concern for environmentalists is the strain that economic development will put on the Earth's environment irrespective of human population growth.

Figure 28.13
Human Population Increase through the Centuries
How many people can the Earth sustain?

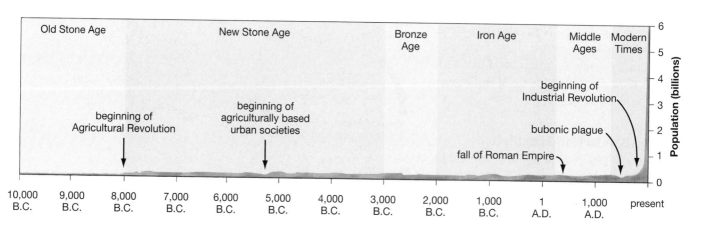

28.5 Communities: Looking at the Interactions of Many Populations

In this tour of ecology, you've so far been looking at species in isolation—at one population or another as it increases or decreases in size. In nature, of course, species live together in a rich mix of combinations. How rich? To take the high end of diversity, which is the tropical rainforest, Smithsonian biologist Terry Erwin found an estimated 1,700 species of beetle in a single tree in Panama. Not 1,700 beetles, mind you; 1,700 *species* of beetle.

In beginning to think about the interactions of these diverse populations, a helpful concept is that of the community. An **ecological community** generally is defined as all the populations that inhabit a given area; all the plants, animals, fungi, bacteria, protists—every living thing, in other words. The term also is used, however, to mean a collection of populations in a given area that potentially *interact* with each other.

So what is it ecologists would like to know about any community? For starters, they'd like to know what at least some of the species are and what their relative numbers are. They'd also like to know about the importance to the community of some of its individual members.

Large Numbers of a Few Species: Ecological Dominants

There's a tremendous variability in what mix of species would be found in different communities, but most communities tend to be dominated by only a few species. Forests tend to be populated by certain kinds of trees, stretches of prairie by certain kinds of grasses. The few species that are abundant in a given community are called **ecological dominants** (see Figure 28.14). These generally are plants but they can be other lifeforms, as is the case with the animals known as corals that are the ecological dominants in coral reefs.

Importance beyond Numbers: Keystone Species

Ecologists have long recognized that there are also species that may not be numerous in a given area, but whose *absence* would bring about significant change in the community. These organisms are called **keystone species** (named for the pressure-bearing "keystone" in an arch). The concept of the keystone species was introduced in the 1960s by marine ecologist Robert Paine, who, with his students, went to a shallow-water zone of the Pacific Ocean in Washington State and for six years regularly removed all sea stars in the genus *Pisaster* from a small area. Such a sea star may sound harmless enough, but *Pisaster* was, in fact, the **top predator** in the area. It preyed on other species, but no species preyed upon it (**see Figure 28.15**). The impact of the *Pisaster* removal was big: Before the change there had been 15 species in the area; after it there were 8. One species of mussel, freed from its former predator's control, took over much of the attachment space

Figure 28.14
Ecological Dominants

a A forest dominated by spruce trees in Bavaria, Germany.

b Dry grassland dominated by Mitchell grass in Queensland, Australia.

a   b

Figure 28.15
Keystone Species
When the predatory sea star *Pisaster ochraceus* was removed from a small area of rocky shore along the Pacific coast, the species composition of the community changed drastically, more so than if one of the other species had been removed. Pictured are several *Pisaster* sea stars on the Pacific northwest coast.

in the area, crowding out other animals, such as barnacles.

Over time, the keystone species concept has undergone some modification; where once scientists thought of keystones as always sitting at the top of food chains, they now recognize that organisms in other positions can take on a keystone role: beavers building dams, for example, or even lichens that are critical in getting communities going in the desert. Beyond this, it turns out that there are communities without keystones; remove any one species from such a community and its role will be taken over by another species.

Variety in Communities: What Is Biodiversity?

Apart from ecological dominants and keystones, a third element that ecologists pay attention to in any community is the *range* of species it has in it. This touches on the more general concept of **biodiversity**, which can be defined as variety among living things. In everyday speech, what biodiversity means is a diversity *of species* in a given area. This is an important measure of diversity, but it is only one among several. You can see why from an imaginary experiment that has been noted by ecologist Paul Ehrlich.

Suppose that you could get a few members of every species on Earth, but that you restricted each species to a single population housed somewhere (in a single zoo or an aquarium or botanical garden). Species *diversity* would not drop at all in such a scenario—you'd still have some of each kind of creature, after all—but the Earth quickly would become barren because what's needed is a rich *distribution* of species in populations across the planet. This distribution of populations is the second measure of biodiversity.

Figure 28.16
Three Types of Biodiversity

DEFINITIONS OF BIODIVERSITY

High biodiversity   Low biodiversity

a Species diversity

many different species   few species

b Geographic diversity

broad distribution of species   narrow distribution of species

c Genetic diversity

high genetic diversity within population   low genetic diversity within population

The third measure exists *within* populations or species. It is genetic diversity, meaning a diversity of "alleles" or variants of genes in a population. Without such diversity, populations are vulnerable to disease, their members may die young or suffer from a variety of inherited mental and physical afflictions. (See "Lessons from the Cocker Spaniel" in Chapter 17.) In summary, biodiversity means species diversity, distribution diversity, and genetic diversity (**see Figure 28.16**).

28.6 Types of Interaction among Community Members

Having considered some general issues regarding communities, let's now turn to the subject of how members of a community might *interact* with each other. Here's a short list of ways:

Competition
Predation and parasitism
Mutualism and commensalism

Before continuing with the exploration of these modes of interaction, let's go over a couple of concepts that will apply to them all.

Two Important Community Concepts: Habitat and Niche

A **habitat** is the physical surroundings in which a species can normally be found. Though two populations of a species may be widely separated, they can generally be found dwelling in similar natural surroundings. A habitat is sometimes described as a species' "address," but a more accurate metaphor might be a species' preferred type of neighborhood.

The term **niche** has been defined in several ways, but it is useful to think of it in terms of a simple metaphor: A niche is an organism's occupation. How and where does the organism make a living? What does it do to obtain resources? How does it deal with competition for these resources? The horseshoe crab has found a niche walking the bottom of shallow coastal waters, feeding on food items that range from algae to small invertebrates. Note that this is about more than what the horseshoe crab eats. It includes specific surroundings (shallow ocean waters) specific behaviors (ocean-floor crawling), and perhaps seasonal or daily feeding times

among other things. If you were to specify all the things that define a horseshoe crab's niche, the odds are that no other organism would exactly fit into it.

Competition among Species in a Community

With these two concepts covered, let's look at the ways organisms interact in communities, starting with a familiar type of interaction—competition.

Despite the fact that niches tend to be specific to given organisms, some species—particularly closely related species—have niches that *overlap* to some degree in a community. A large proportion of both species' diet may be made up of a given organism or both species will occupy similar kinds of spaces on rocks or branches or pond surfaces. What arises in such instances is **interspecific competition**, meaning competition between two or more species.

What may come to mind with interspecific competition is a never-ending series of physical battles between two species, but things seldom work like this. For one thing, competition tends to be indirect—it often is a competition for *resources*, in which the winner generally triumphs not by fighting but by being more efficient at doing something, such as acquiring food. Lots of animals, including birds and many large mammals, are **territorial**—meaning they will attempt to keep other creatures out of a territory they have laid claim to—but this usually is a matter of them trying to repel members of their own species.

No Two Species Can Share the Same Vital Resource for Long

When the niches of two species greatly overlap, it's unlikely that you'd see *long-standing* competition among them, for the simple reason that one species or the other is likely to *win* such a competition in fairly short order. Because the competition is for nothing less than vital resources, the result is that the losing species will be driven to local extinction.

It was a laboratory experiment, performed by the Russian biologist G.F. Gause in the 1930s, that pointed the way on this latter principle. Gause grew two species of *Paramecium* protists, *P. caudatum* and *P. aurelia*, in culture and found that, in time,

P. aurelia was always the sole survivor. True to what was just noted, *P. aurelia* wasn't eating or wounding *P. caudatum*; instead, it grew faster and thus used more of the surrounding resources. Nevertheless, the outcome was always the death of all the *P. caudatum*. This led Gause to formulate what came to be called the **competitive exclusion principle**: When two populations compete for the same limited, vital resource, one will always outcompete the other and thus bring about the latter's local extinction (**see Figure 28.17a**).

Ecologists wouldn't expect to witness the competitive exclusion principle operating much in nature however, because as noted, the *Paramecium* scenario is likely to be played out quickly.

Nevertheless, competitive exclusion has been observed in nature, often when humans have a hand in things. For example, humans introduced a Chinese vine, the kudzu plant (*Pueraria lobata*), to the American South in the 1930s. Growing at a rate of up to 30 centimeters or nearly 1 foot per day, kudzu has now taken over millions of acres in the South, locally eliminating most of the plants in its path (**see Figure 28.18**).

Resource Partitioning Is Common in Natural Environments

Competitive exclusion notwithstanding, there are instances in nature in which two related species will use the same kinds of resources from the same habitat over a long period of time. So, why isn't one of them eliminated? The answer is contained in another experiment conducted by Gause. He took the successful species from the earlier experiment, *P. aurelia*, and placed it in a test tube with a different paramecium, *P. bursaria*. This time, neither species was eliminated (see Figure 28.17b). Instead, they divided up the habitat, *P. aurelia* feeding in the upper part of the test tube and *P. bursaria* flourishing in the lower part. (*P. bursaria* had an advantage in the lower, oxygen-depleted water because it has symbiotic algae that grow with it, producing oxygen.) In a nutshell, this describes a situation that often exists in nature: **Coexistence** through **resource partitioning**, meaning the dividing up of scarce resources among species that have similar requirements. If you look at **Figure 28.19**, you can see how this works among some species that are a little more familiar—several varieties of warbler. The main message here is that species often flourish by specializing, deriving resources by feeding from different locations (or during different times) within the same habitat.

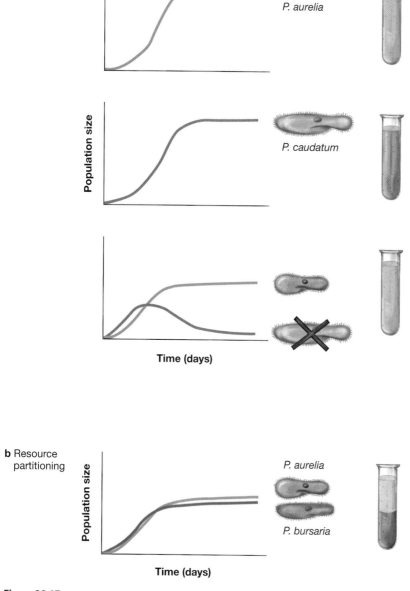

a Competitive exclusion

P. aurelia

P. caudatum

Population size

Time (days)

b Resource partitioning

P. aurelia

P. bursaria

Population size

Time (days)

Figure 28.17
Competition for Resources among Species
Laboratory experiments by G. F. Gause showed that competition for resources between two species can have two possible outcomes.

a When two species compete for the same limited, vital resource, one will always drive the other to local extinction—as the paramecium *P. aurelia* did to the paramecium *P. caudatum*. This is the competitive exclusion principle at work.

b Conversely, when Gause put *P. aurelia* together with another paramecium, *P. bursaria*, the two species divided up the habitat and both survived. This is a demonstration of resource partitioning.

Other Modes of Interaction: Predation and Parasitism

It is one thing for two species to compete for resources; it is another for one species to *be* a resource for another—to be eaten or used by another species. Thus do ecologists distinguish the first mode of interaction in communities, competition, from the second, predation, and a variety of it, parasitism.

Predation can be defined as one free-standing organism feeding on parts or all of a second organism. Note that the prey here can be plants, protists, animals—whatever is preyed upon. Predation is generally thought of in terms of animals killing other animals, but note that its definition includes such things as animals grazing on plants. **Parasitism** is a variety of predation in which the predator feeds on prey, but does not kill the prey immediately—and may not ever kill it. Parasites that come to mind tend to be animals, but there are an estimated 3,000 species of parasitic plants, the most familiar of which are probably the mistletoes that make a living by tapping into the "plumbing" or vascular systems of trees (**see Figure 28.20** on page 630).

The Value of Predation and Parasitism

Predation and parasitism are the features of nature that many nature lovers can't stand. Why this should be so is a fascinating question, though beyond the scope of this book. If one values more complex organisms (such as ourselves), predation and parasitism have had a clear value, in that they have spurred on evolution by culling the less capable members from populations and by stimulating the "arms race" that has resulted in evolutionary adaptations such as vision and flight. Whatever we may think of it, predation is simply a fact of nature, and it is impossible to imagine life without it. But it's also true that to admire the beauty of nature is in significant part to admire the handiwork of predation, since it has done so much to shape the living world.

Apart from these more abstract considerations, the predation practiced by higher-level predators has a down-to-earth effect on communities: It may promote species diversity. You've seen this in action already in the predator sea star *Pisaster* on the Washington coast. *Remove* a top predator such as *Pisaster*, and what you're likely to see is a species that was formerly held in check by the predator "taking over" to a significant extent—crowding out several other species. This was the case with the mussels in the *Pisaster* community and the

Figure 28.18
Kudzu Vines Run Wild
The kudzu plant was introduced in the South in the 1930s and has since spread at a rapid rate, locally eliminating many plants in its path. This is competitive exclusion in action. Here kudzu has overgrown a house in Mississippi.

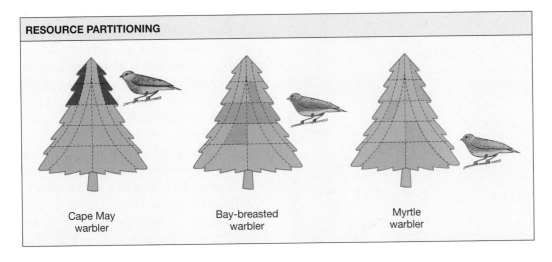

RESOURCE PARTITIONING

Cape May
warbler

Bay-breasted
warbler

Myrtle
warbler

Figure 28.19
Resource Partitioning
Ecologist Robert MacArthur spent long stretches of time over several years in the 1950s observing the feeding patterns of several species of warblers. All of them ate caterpillars, but from substantially different, though overlapping, parts of the tree.

Purring Predators: Housecats and Their Prey

Housecats are famous for bringing home animals they've killed or captured, of course; intrigued by the carnage brought into their own homes in the 1980s, Peter Churcher and John Lawton decided to make a scientific study of the predatory behavior of the domestic cat. To do so, they enlisted the help of 172 households in the small English village of Bedfordshire, where Churcher lived. The two researchers asked the locals to "bag the remains of any animal the cat caught," and turn the evidence over to them once a week. This process went on for a year with a high degree of cooperation from the cat owners.

Some of the results were unsurprising. Young cats hunt more than older cats; small mammals, such as mice, are the favored prey; and cat hunting is not based purely on hunger—all the cats were fed by their owners and yet hunting was widespread.

What was remarkable, however, was the *scale* of the killing. Concentrating on the village's house sparrow population, Churcher and Lawton found that cats were responsible for between a third and half of all sparrow deaths in the village, a figure they believed no other single predator could match. When they looked at all animals killed, and projected the village figures onto the whole of Great Britain, they calculated that cats were responsible for about 70 million deaths a year. Given that birds account for somewhere between 30 and 50 percent of these kills, this means that cats kill at least 20 million birds a

> What was remarkable was the scale *of the killing.*

year in Britain. The "at least" here may be an important qualifier. An American biologist, the researchers note, found that cats bring home only about half of the food they catch.

Figure 1
Predator on the Loose
A domestic cat carrying its prey.

Figure 28.20
Plants Parasitizing Plants
The oval green mass at the top of the tree is mistletoe, a parasitic plant that lives off of food and water taken from a host tree.

result was fewer species. You can read about a very familiar predator, the house cat, in "Purring Predators: Housecats and Their Prey" above.

Predator-Prey Dynamics

It may be obvious that, while predators attack prey, predators are also *dependent* on prey as a food source. An ongoing question in ecological research is how tight this linkage is. An attractive theory is that when the prey population flourishes, the resulting bounty of available food makes the *predator* numbers grow; this, however, means that more predators will hunt, which will bring the prey population's numbers down. When this happens, there may be a steep decline in the predator population, because a large part of its food supply is now gone. The result of this, of course, would be a series of linked population ups and downs, or cycles.

If you look at **Figure 28.21**, you can see a series that has occurred with one linked pair of species, the moose of Michigan's Isle Royale National Park and the wolf that preys on it. At first glance, these figures seem to provide evidence for the scenario just described, as the moose and wolf numbers move generally in opposite directions. In reality, however, things are more complex than this. The huge crash in the wolf population that began in 1981 was caused in part by a canine virus that invaded the island and perhaps by a lack of genetic diversity among the wolves. Suffice it to say that population cycles generally are the result of many factors, only one of which is a given predator-prey relationship.

Parasites: Making a Living from the Living

Let's look now at the special variety of predation called *parasitism*. Some familiar parasites, such as leeches and ticks, have a straightforward strategy: Get on, hold on, and consume material from the prey, known as the **host**. But the relationship between parasites and hosts can be much more sophisticated than this. Take, for example, the barely visible roundworm *Strongyloides stercoralis*, which lives throughout the tropics and has human beings in the southeastern United States as one of its hosts. The worm comes into the human body from the soil in a threadlike larval form, then makes its way into the veins, which it travels through to get to the lungs. From there it moves up the trachea and then down into the digestive system, ending up in the small intestine. There it lays eggs that develop into larvae, that this time move *out*, with feces, back into the Earth to start the whole life cycle over again. *Strongyloides* is not just using humans as an incubator, however; it can worm its way, you might say, directly from the small intestines into the bloodstream, bringing about a continuous cycle of infection whose symptoms range from severe diarrhea to lung problems. (Lest all of this seem too disturbing, take heart; there is a medicine that can clear this pest from human bodies.) So, note the complexity here: *Strongyloides* not only feeds on a host, but uses the host in its reproduction cycle, something that actually is fairly common among parasitic worms and protozoans.

Were you to start examining the *range* of host-parasite dynamics, you would find lots of interactions as elaborate—and chilling—as this. To give but one example, Nancy Beckage and her colleagues have been studying a species of wasp, *Cotesia congregata*, that lays its eggs inside its host, in this case a caterpillar. There's nothing special about this; there are legions of these "parasitoid" species whose offspring *develop* in the host, feeding on it and ultimately killing it (**see Figure 28.22**). The difference here is that the eggs the wasp lays are coated with a virus that cripples the immune system of the caterpillar, thus allowing the eggs to develop free of a caterpillar immune system attack. This virus is produced in the female wasp's ovary, with the genetic instructions for its production coming from the wasp's own DNA.

Figure 28.21
Predator and Prey Populations
The population dynamics of a predator species (the wolf) and its prey (the moose) in Michigan's Isle Royale National Park over several decades. (Redrawn and adapted from an illustration in *Science*, August 27, 1993, p. 1115.)

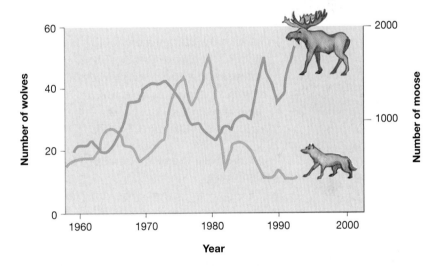

Figure 28.22
Parasitism Can Be Very Specialized
When laying its eggs in a caterpillar, the parasitoid wasp *Cotesia congregata* also injects a special virus that coats the eggs, preventing the caterpillar's immune system from destroying them. The wasp larvae feed on the caterpillar while inside it, eventually bursting forth from the caterpillar, which then dies.

The Effect of Predator-Prey Interactions on Evolution

With the interactions between *Cotesia* and the caterpillar, you can begin to see some *evolutionary* effects of predator-prey interactions. Through evolution, *Cotesia* developed a new weapon that conferred greater reproductive fitness on it. Such an adaptation is part of what was referred to earlier as the arms race in evolution, meaning the continual adaptation of species in response to pressure from other species. Warm-water "frogfish" have evolved a spine on their dorsal fins that is tipped with a piece of flesh that looks for all the world like a small worm floating free in the ocean. When a would-be predator tries to snatch up this "worm," however, it finds *itself* the prey (**see Figure 28.23**). On the prey side, look at **Figure 28.24**, to see the amazing changes in camouflaging coloration that a black skimmer chick can go through to hide itself from predators. Needless to say, predator-prey evolution goes on within all the kingdoms of life. Plants have not only developed spines, thorns, and other protective structures but also a vast array of chemical compounds—more than 15,000 have been characterized so far—that protect them from predators.

Mimicry Is a Theme in Predator-Prey Evolution

The ability to fool an opponent by faking is as valuable in nature as it is in sports, as you've seen with the frogfish. Another way such de-

Figure 28.23
Fooling Predators about Prey
The predatory frogfish uses a modified spine resembling a tasty worm to lure its prey. Here a tasseled frogfish clearly shows its lure while swimming near Edithburg, South Australia.

Figure 28.24
Avoiding Predation through Camouflage
A black skimmer chick blends into it surroundings in Botswana.

ception takes place in nature is through **mimicry**, a phenomenon through which one species has evolved to assume the appearance of another. One form of mimicry involves at least three participants: a model, a mimic, and a dupe. If you look at **Figure 28.25**, you can see a model on the left, the yellowjacket wasp, which obviously can provide a painful lesson to any animal that tries to eat it. On the right, you can see the mimic, the clearwing moth, which is harmless, but *looks* a great deal like a wasp. For the dupe, you could select any predator species that sees a clearwing moth and passes it by, believing it to be a dangerous yellowjacket. This is so-called **Batesian mimicry**: the evolution of one species to resemble a species that has a superior protective capability.

In a second type of mimicry, **Müllerian mimicry**, several species that *have* protection against predators come to resemble each other (**see Figure 28.26**). This creates a visual warning that becomes known to an array of predators. The benefit is that the learning a would-be predator does on one species will keep it away from *all* the look-alike species, meaning that fewer individuals in any of these species will be disturbed or killed.

Beneficial Interactions: Mutualism and Commensalism

The competitive, predatory, and parasitic modes of interaction you've reviewed have an each-organism-against-all quality to them, but other kinds of relationships exist in nature as well. Consider **mutualism**, meaning an interaction between two species that is *beneficial* to both. There's a great example of this in Chapter 20 (page 406), with the linkage between plant roots and the slender, belowground extensions of fungi called *hyphae*. What the fungi get from growing into the plant roots is food that comes from the photosynthesizing plant; what the plants get is minerals, growth hormones, and water, all absorbed by the network of hyphae. Some plants, such as orchids, cannot grow without this association. Perhaps the most famous example of mutualism is between the rhinoceros and oxpecker birds. Sitting on the back of the rhino, the oxpecker removes parasites and pests (ticks and flies), thus getting food and a safe place to perch, while the rhino gets relief from its tiny adversaries (**Figure 28.27**).

There is also **commensalism**, in which one species benefits while the other is neither harmed nor helped. Birds can make their nests in trees, and benefit from this, but generally don't affect the trees in any way. **Table 28.3** summarizes all the kinds of interactions you've gone over in this chapter.

Coevolution: Species Driving Each Other's Evolution

Earlier you read about the effect that predator-prey interactions have had on the evolutionary arms race among species. Yet it's clear that organisms have affected each other's evolution through mutualism as well. Indeed, species that are tightly linked in any way over a long period of time are likely to shape one another's evolution. Perceiving this, scientists Paul Ehrlich and Peter Raven developed the concept of **coevolution**, meaning the interdependent evolution of two or more species. Some of the clearest examples of coevolution involve flowering plants and the animals that pollinate them. Recall from Chapter 21 that many flowers have both fragrances that attract insects and ultraviolet color patterns (invisible to us) that guide the insects to the

Table 28.3
Community Interactions

In This Type of Interaction...	One organism...	While the other...
Competition	Is harmed	Is harmed
Predation and parasitism	Gains	Is harmed
Mutualism	Gains	Gains
Commensalism	Gains	Is unaffected

Figure 28.25
Harmless, but Looking Dangerous
In an example of Batesian mimicry, the clearwing moth, on the right, has no sting but has evolved to look like an insect that does, the yellowjacket wasp, on the left. The yellowjacket is the model and the moth is the mimic.

a

b

Figure 28.26
Müllerian Mimicry
(a) A yellow jacket and **(b)** a cicada-killer wasp are two of the many varieties of stinging insects that have similar patterns of warning coloration. This form of mimicry sends a message to predators that they should keep away from all organisms that have this color pattern—or pay a price.

Figure 28.27
Mutually Beneficial

a Several oxpecker birds sit atop a black rhinoceros, ridding the rhino of ticks and other pests while the rhino provides a safe habitat for the birds. This is a demonstration of mutualism—an interaction between two species that is beneficial to both.

b The orange-spotted shrimp goby, on the right, and the snapping shrimp on the left also exhibit mutualism. Here the shrimp goby stands guard near the snapping shrimp, which digs out the burrow on the sea floor that the two creatures share.

a

b

What Is It?
See p. 635.

proper spot for pollination. These features are nature's own homing signals and landing lights, communicating the messages "Food lies this way" and "Land here" (**see Figure 28.28**). On the other side of the coin, consider that the color vision of honey bees is most sensitive to the colors that exist in the very flowering plants they pollinate. What seems likely is that honeybee eyesight evolved in response to plant coloration, and that plant coloration evolved to be maximally attractive to the pollinating insects. These groups of organisms coevolved, in other words.

Having considered how community members interact with one another, let's now go on to consider how communities change *over time*.

Figure 28.28
Coevolution of Plants and Their Pollinators
This flower has an ultraviolet color pattern that is not normally visible to humans, but that attracts bees. It is likely that these color patterns evolved in the flowers because they aided in attracting the bees, while the bees developed vision that was sensitive to the colors exhibited by these same plants.

a what we see

b what a bee sees

Figure 28.29
Rebound from Disaster
The Lang Ridge near the Mount St. Helens volcano, which exploded in 1980. On the left is the ridge in the year of the eruption; on the right is the same area in 1989. This is a demonstration of ecological succession in action.

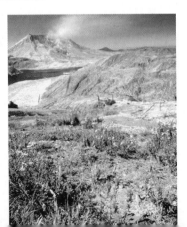

28.7 Succession in Communities

When Washington State's Mount St. Helens volcano erupted in May of 1980, it first collapsed inward, thus sending most of the mountain's north face sliding downhill in the largest avalanche in recorded history. Then came the actual eruption, which sent a huge volume of rock and ash hurtling not straight up, but out at an angle, toward the north. Some stands of forest in the path of this blast were instantly incinerated down to bare rock; another 35,000 hectares (86,000 acres) of trees were snapped in two like so many twigs. Then came the mudslides caused by the vast expanse of snow and ice melted in the explosion; then came the fall of hundreds of millions of tons of ash, some of it landing as far away as Wyoming. After viewing the devastated area around the blast, President Jimmy Carter said that it "makes the surface of the moon look like a golf course."

But today? No one would claim that the Mount St. Helens area has returned to anything like its former state, but look at the pictures in **Figure 28.29** to get an idea of the transition that has occurred at one location around Mount St. Helens.

The exact form of this rejuvenation may surprise us, but don't we intuitively expect something like this? Just from our everyday experience of watching, say, an abandoned urban lot get progressively more weed-filled, don't we expect this would generally be the case? As it turns out, our intuition is right, because any parcel of land or water that has been either abandoned by humans or devastated by physical forces will always be "reclaimed" by nature, at least to some degree.

The question is, how does this proceed? Areas that are rebounding don't just get one type of vegetation and then retain it. Instead, the vegetation changes through time—one type of growth *succeeds* another—with a general movement toward more and larger greenery as time goes by. Ecologists call this phenomenon **succession**, by which is meant a series of replacements of community members at a given location until a stable final state is reached. Within this framework, there are two kinds of succession. The first is **primary succession**, in which the starting state is one of little or no life. This is the kind of succession that took place at the "blast-zone" locations at Mount St. Helens. Then there is **secondary succession**, in which the final state of a habitat has first been disturbed by some outside force, after which the site has been left alone. The classic example of this is land that has first been cleared for farming but later abandoned by the farmer.

The stable community that develops at the end of any process of succession is called a **climax community**. There will be some shifts over time in a climax community, just as there are in any other. A prolonged drought will cause some change in the mix of the community's animal and plant life, for example. But, with some exceptions, what does not happen is a shift to a *fundamentally* different mix of life-forms, barring a major change in climate. Grassland stays grassland, and forest stays forest.

Succession can be fairly predictable within small geographical areas. Two abandoned farm plots that lie close together will have very similar kinds of succession—so much so that local farmers or naturalists can sometimes tell to within a few years how long it's been since a field was abandoned. As you may be able to tell, succession generally is thought of in terms of the various *plant* communities that succeed one another, though modern ecology is attempting to bring animals into the picture as well.

An Example of Primary Succession: Alaska's Glacier Bay

One of the best examples of primary succession comes from Alaska's Glacier Bay, where glaciers have been retreating for some 200 years. Such withdrawal initially leaves behind a rocky terrain that is completely devoid of organic matter—no fallen leaves or branches, no decomposing animals. Indeed, what's left is "till" (not soil) that has a high pH and no available nitrogen—not a promising environment for life to establish itself. But establish itself it does. Glacier Bay's succession particulars—what species succeeded what—are not important to us; what's important is the general process that occurred. Here's what happened in Glacier Bay after the glacier retreated (**see Figure 28.30**).

**Figure 28.30
One Example of Primary Succession: Glacier Bay, Alaska**

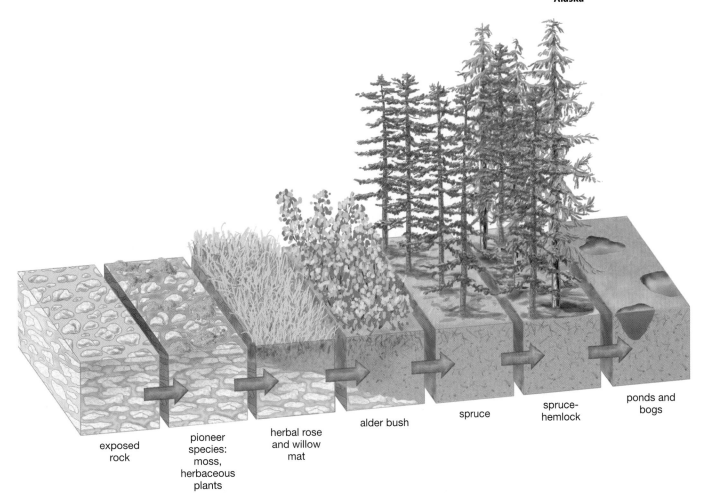

exposed rock → pioneer species: moss, herbaceous plants → herbal rose and willow mat → alder bush → spruce → spruce-hemlock → ponds and bogs

The first "pioneer" species to come were mosses and a non-woody plant that had the ability to grow essentially on rocks. Then came more plants, including an Arctic herbal rose, *Dryas*, that grows in a mat. Willows that hug the ground followed, making a combined *Dryas* and willow mat that began to build up the decayed organic matter that helps produce soil. This community was succeeded by a plant called the *alder bush* that excluded almost everything else—it was a highly successful dominant, in other words. Fifty years after the glacier's retreat, alders were growing in stands that were 10 meters, or about 33 feet tall. With their growth, the terrain got a great deal of nitrogen and, through acidic alder leaves that fell to the ground, a rapid lowering of pH.

The alder sowed the seeds of its own elimination, however. With the lowered soil pH, an extensive population of Sitka spruce trees could develop; when they did, they "shaded out" the smaller alders by blocking their sunlight. Some 120 years after the glacier's retreat a spruce forest existed, one that, within 30 more years, started becoming a combined spruce-hemlock forest.

You might think that this would be the end of things; that the spruce-hemlock trees and the foliage around them would constitute the climax community. Instead, it was this forest that was replaced. Water-absorbing mosses eventually started to grow and, with them, the forest floor became watery, the tree roots couldn't absorb the oxygen they needed, and the trees died. What eventually was left was land that was filled with ponds and watery bogs. Whether *this* is the climax community will become known only with the passage of more time.

Common Elements in Primary Succession

Successions often are presented as ending with a mixture of tall forests and undergrowth, but there obviously are lots of different climax communities, of which bogs and ponds may be one. In the Glacier Bay succession you can, however, see developments that apply to many primary successions:

- An arrival of wind-borne "pioneer" plants in the form of seeds (or spores in other cases).

- A steady increase in biomass, compared to the starting state. This increase eventually stops in all successions; in Glacier Bay it actually was reversed with the invasion of the late-stage mosses.
- A general movement toward longer-lived species.
- A general trend toward an increase in species diversity, though not without some reversals (the alder bushes and the late-stage moss community).
- The **facilitation** of the growth of some later species (the Sitka spruce here) through the actions of earlier species (the shedding of leaves by the alder bush).
- The competitive driving out of some species (the alder bush) by the actions of later species (the blocking of sunlight by the spruce trees).

Ecologists have been studying succession for about 100 years now, but it remains a field filled with basic questions. This is understandable because for one thing, community interactions are very complex; for another, successions are long-lasting events compared to the lives of ecologists. Spirited debate has gone on for years about what factors hold true in all instances of succession, and what models of development best describe succession. (Is facilitation more important than competition?) What remains to be seen is how much agreement will be achieved in the future on these questions.

On to Ecosystems and Biomes

In this chapter, you have looked at populations and communities. In populations, you saw the building blocks of communities. In communities, you looked through a lens of the interactions of living things. But it's obvious that *nonliving* forces and factors—weather, soil chemistry—interact with communities. When living things and nonliving factors are considered together, the resulting whole is called an ecosystem—the subject of the first part of Chapter 29. Very large-scale ecosystems are called biomes; coverage of them forms the concluding portion of Chapter 29 and will bring to a close this book's coverage of ecology.

Chapter Review ▰▰▰▰▰▰▰▰▰▰

Summary

28.1 The Study of Ecology

- Ecology is the study of the interactions living things have with each other and with their environment.

- Ecology is not the same thing as environmentalism. The function of ecology is to describe interactions that affect the living world, not to work on behalf of environmentalism.

- There are five scales of life that concern ecology: physiology, populations, communities, ecosystems, and the biosphere. A population is all the members of a single species that live together in a specified geographical area; a community is all the species living in a single area; an ecosystem is a community and all the nonliving elements that interact with it; the biosphere is the interactive collection of all the Earth's ecosystems.

28.2 Populations: Size and Dynamics

- Ecologists employ several means to estimate the size of populations of living things, among them counting animal droppings or surveying bird populations as they migrate.

- An arithmetical increase occurs when, over a given interval of time, an unvarying number of new units is added to a population. An exponential increase occurs when the number of new units added to a population is proportional to the number of units that exists. Populations of living things are capable of increasing exponentially, because living things are capable of giving rise to more living things.

- The rapid growth that sometimes characterizes living populations is referred to as *exponential growth*, or as the *J-shaped growth form*. Populations that initially grow, but whose growth later levels out, have experienced logistic growth, sometimes referred to as the *S-shaped growth form*.

- The size of living populations is kept in check by environmental resistance, defined as all the forces of the environment that act to limit population growth.

- Exponential growth in living populations can be calculated by subtracting a population's death rate from its birth rate, which yields the population's growth rate. Denoted as *r*, this rate is also known as the population's intrinsic rate of increase; it can be thought of as the population's potential for growth.

- Carrying capacity, denoted as *K*, is the maximum population density of a given species that can be sustained within a defined geographical area over an extended period of time.

28.3 *r*-Selected and *K*-Selected Species

- Different species have different reproductive strategies, meaning characteristics that have the effect of increasing the number of fertile offspring a species member will bear.

- Some species are said to be *K*-selected, or equilibrium, species. These species tend to be large, to experience their environment as relatively stable, and to lavish a good deal of attention on relatively few offspring. The pressures on *K*-selected species tend to be density dependent, meaning that as a population's density goes up, factors that limit the population's growth assert themselves ever more strongly.

- Other species are said to be *r*-selected or opportunist species. These species tend to be small, to experience their environment as relatively unstable, and to give little or no attention to the numerous offspring they produce. The pressures on *r*-selected species tend to be density independent, meaning pressures that are unrelated to the population's density.

- Survivorship curves describe how soon species members tend to die within the species' life span. There are three idealized types of survivorship curves: late-loss, constant loss, and early loss.

28.4 Thinking about Human Populations

- Survivorship curves are created from lifetables, which set forth the probabilities of a member of a species being alive after given intervals of time.

Key Terms

arithmetical increase 616

Batesian mimicry 632

biodiversity 626

biome 615

biosphere 615

carrying capacity 619

climax community 635

coevolution 633

coexistence 628

commensalism 633

community 615

competitive exclusion principle 628

density-dependent 621

density-independent 621

ecological community 625

ecological dominants 625

ecology 614

ecosystem 615

emigration 623

environmental resistance 618

equilibrium species 620

exponential growth 617

exponential increase 616

facilitation 636

habitat 627

host 631

immigration 623

interspecific competition 627

intrinsic rate of increase 619

J-shaped growth form 617

keystone species 625

K-selected 621

life table 622

logistic growth 618

mimicry 632

Müllerian mimicry 632

mutualism 632

niche 627

opportunist species 621

- An important step in calculating the future growth of human populations is to learn what proportion of the population is at or under reproductive age. A population pyramid displays this proportion. Populations whose pyramids are heavily weighted toward younger age groups are likely to experience relatively large growth.

- Immigration can have a large impact on the human population of a given country. Because of immigration, the population of the United States is projected to grow much more rapidly in the coming decades than it would without immigration.

- The populations of less-developed nations are in general growing at a much faster rate than the populations of developed nations.

- The Earth's human population has increased greatly in the last 50 years and now stands at about 6 billion people. There is a debate over whether the Earth has reached its carrying capacity for the human population.

- Some scientists believe that there is no greater single threat to the environment than the continued growth of the human population. Others argue that a more important concern is the use of natural resources per person.

28.5 Communities: Looking at the Interactions of Many Populations

- An ecological community is all the populations that inhabit a given area, though the term can be used to mean a collection of populations in a given area that potentially interact with one another.

- Most communities tend to be dominated by only a few species; the few species that are abundant in a given area are called *ecological dominants.*

- Keystone species are species who may not be numerous in a given area, but whose absence would bring about significant change in the community.

- Some keystone species are top predators in community, meaning species who prey upon other species, but who are not preyed upon themselves. Keystone species need not be top predators, however, nor does every community have a keystone species.

- Biodiversity is variety among living things. It can take three primary forms: a diversity of species in a given area; a distribution of species across the Earth; and genetic diversity within a species.

28.6 Types of Interaction among Community Members

- There are three primary types of interaction among community members: competition, predation and parasitism, and mutualism and commensalism.

- *Habitat* is the physical surroundings in which a species can normally be found.

- *Niche* can be defined metaphorically as an organism's occupation, meaning what the organism does to obtain the resources it needs to live.

- The competitive exclusion principle states that when two populations compete for the same limited, vital resource, one always outcompetes the other and thus brings about the latter's local extinction.

- There are numerous instances in nature in which two related species use the same kinds of resources from the same habitat over an extended period of time. Neither species undergoes local extinction in this interaction, because neither is superior in competing for all resources; this leads to coexistence through resource partitioning.

- *Predation* is one free-standing organism feeding on parts or all of a second organism. *Parasitism* is a variety of predation in which the predator feeds on prey, but does not kill it immediately, if ever.

- Predation has done much to shape the natural world, and the practice of it by higher-level predators promotes species diversity.

- The population dynamics of a predator and its prey can be linked, but predator-prey interaction generally is only one of several factors controlling the population level of either group.

- The prey of parasites is known as the host. Hosts may be used by parasites not only for food, but to facilitate parasite reproduction as well.

- *Mimicry* is a phenomenon by which one species has evolved to assume the appearance of another. Species that have evolved in this way generally suffer less predation as a result. *Batesian mimicry* is the evolution of one species to resemble a species that has a superior protective capability. In *Müllerian mimicry*, several species that have protection against predators come to resemble each other.

- Mutualism is an interaction between two species that is beneficial to both. Commensalism is an interaction in which one species benefits while the other is neither harmed nor helped.

- Coevolution is the interdependent evolution of two or more species. Flowers have evolved colors and fragrances that function to attract bees, for example, while bees have evolved vision capabilities that make them more sensitive to colors possessed by the flowers they pollinate.

28.7 Succession in Communities

- Parcels of land or water that have been abandoned by humans or devastated by physical forces will always be reclaimed by nature to some degree. The process by which this takes place is called *succession*: a series of replacements of community members at a given location until a stable final state is reached.

- Primary succession proceeds from an original state of little or no life. Secondary succession occurs when a final state of habitat is first disturbed by some outside force, after which the site is left alone. The final community in any process of succession is known as the *climax community*.

- A common set of developments occurs in most instances of primary succession, including the facilitation of the growth of some later species through the actions of earlier species, and the competitive driving out of some species by the actions of later species.

Understanding the Basics

Multiple-Choice Questions

1. A(n) _____ consists of members of a single species that live in a specific geographic region, while a _____ consists of the populations of all species that live in a single region.
   a. ecological dominant, community
   b. ecological dominant, biosphere
   c. community, ecosystem
   d. population, community
   e. population, ecosystem

2. A community and all the nonliving elements that interact with it (sunlight, nutrients, soil, etc.) is called
   a. a population group   d. a biome
   b. a biosphere   e. a social system
   c. an ecosystem

3. When population growth increases in proportion to the number of members of a population, _____ growth is said to have occurred
   a. density dependent
   b. density independent
   c. *K*-selected
   d. arithmetical
   e. exponential

4. The use of pesticides in many agricultural communities has led to a great reduction in many "targeted" insects. However, many other "nontarget" insects have declined as well. In particular, ecologists and fruit growers have noticed that certain species of bees that pollinate valuable plants have declined, leading to a decrease in many bee-pollinated species. In communities where great change has occurred due to the decline in this ecosystem function, bees may be
   a. ecological dominants
   b. bottleneck species
   c. keystone species
   d. top predators
   e. parasitoids

5. An interaction between two species in which one species benefits while the other is neither harmed nor helped is called
   a. parasitism
   b. mutualism
   c. mimicry
   d. commensalism
   e. cooperation

6. In the process of nitrogen fixation, plants contribute carbon to bacteria that are capable of fixing atmospheric N2 gas, and the bacteria can supply the plants with a usable form of nitrogen. This is an example of

a. parasitism   **d.** mimicry
b. mutualism   **e.** cooperation
c. fertilization

7. The process by which one species evolves features that result in its taking on the appearance of another species is called
a. mimicry
b. mutualism
c. behaviorism
d. parasitism
e. natural deception

8. The largest scale of life, which can be thought of as the interaction of all of Earth's ecosystems, is called
a. the biome
b. the ecosphere
c. the biosphere
d. the environment zone
e. the ionosphere

Brief Review

1. Name the four levels of group organization of living things in order of increasing size.

2. Compare and contrast *r*-selected and *K*-selected species, and give an example of each. Are humans *r*-selected or *K*-selected?

3. Explain what is meant by environmental resistance and its relationship to population growth.

4. The red-spotted newt *Notophthalmus viridescens* secretes toxins from its skin that make predators avoid it. The red salamander *Pseudotriton ruber* has no such secretions but resembles the red-spotted newt and gains protection from this resemblance. What is this an example of? Who is the model, the mimic, and the dupe?

5. A volcano erupted in Hawaii, and lava covered land that had been farmed for centuries. Eventually, new lichens, and then plants, grew on the lava. Is this an example of primary or secondary succession?

Applying Your Knowledge

1. Few species can continue exponential growth for very long, because "environmental resistance" will assert itself. However, the human population has been growing exponentially in the last few decades. Do you think that human populations are truly without an upper bound, or do you think that carrying capacity (*K*) is just always rising? What evidence do you see that humans might eventually reach an absolute upper limit (other than that of pure physical space)? How does the emergence of disease, or epidemics, factor into this discussion?

2. What is a keystone species? How do you think we can tell if a particular endangered species is a keystone species—such as the spotted owl in the Pacific Northwest, or the California condor? Should a species' role as a keystone species in a community make us increase efforts to protect it? Explain, and give several arguments for (or against) laws that protect individual species.

3. Some scientists argue that no single factor poses a greater threat to the environment than the continued growth of the human population, while others believe the use of resources per person is a more critical factor. What arguments can you think of for or against either proposition?

4. In steep, mountainous terrain, logging of forests is thought to cause an increase in the chance of landslides. When landslides occur, most of the top soil and parent rock fragments slide downhill. Is the regrowth on this newly exposed terrain an example of primary or secondary succession? Would you expect the vegetation to grow faster or slower than the nearby land that was harvested, but that did not experience a landslide?

Group Activity

The "Green Revolution" allowed the human population to increase the productivity of farmland dramatically and thus provide food to sustain high population growth. Using resources in your library and on the Internet, research the scientific basis of this revolution. Based on the information you gathered, do you think that this increase in productivity will continue in the future? Why or why not?

MediaLab
Can Earth Support All of Us? Population Growth Patterns

Nineteenth-century economist Thomas Malthus noted that populations increase geometrically until they encounter a factor that limits continued population growth. We know some populations grow rapidly, others slowly. Since 1800 the number of people on Earth has increased dramatically, with more yet to come.

How can we accurately predict the effect of this continued growth? This *CD Activity* reviews concepts about population growth. The following *Web Investigation* lets you examine a variety of population growth situations and the problems they cause. In *Communicate Your Results*, you can apply these concepts.

This *MediaLab* can be found in Chapter 28 on your CD-ROM and Companion Website (http://www.prenhall.com/krogh).

CD ACTIVITY

Species can differ greatly in the type of population growth they can experience. Equilibrium species, like elephants, increase their numbers slowly; opportunist species, like houseflies, increase rapidly. In this *CD Activity*, you'll investigate the factors controlling population growth between these two types of species.

CHARACTERISTICS OF K-SELECTED AND r-SELECTED SPECIES

K-selected equilibrium species

r-selected opportunist species

population size | time

population size | time

population size:
- limited by carrying capacity (K)
- density dependent
- relatively stable

organisms:
- larger, long lived
- produce fewer offspring
- provide greater care for offspring

population size:
- limited by reproductive rate (r)
- density independent
- relatively unstable

organisms:
- smaller, short-lived
- produce many offspring
- provide no care for offspring

Activity

1. *First, you will investigate humans, an equilibrium or K-selected species, which means our species is limited by the carrying capacity (K) of the environment. We tend to live longer, providing greater care for fewer offspring; thus, our populations increase slowly.*

2. *Then, you will look at the population growth of an opportunist or r-selected species, which is limited by the reproductive rate (r) of the organisms. Creatures of this type tend to be smaller, short-lived, and provide no care for their numerous offspring.*

3. *Finally, you will see a visual demonstration of how these two populations increase geometrically until they encounter a limiting factor (space, food, etc.).*

The following *Web Investigations* present examples of population growth in both human and animal populations, so that you can see how the characteristics of a species' growth influence how slowly or rapidly its population increases.

WEB INVESTIGATION

Investigation 1
Estimated time for completion = 15 minutes

Select the Keyword **GROWTH RATE** on your CD or Website for this *MediaLab* and read about the population growth rate. Notice how the human population has changed over time. Would you say human population is growing arithmetically (straight line) or geometrically (curved line)? What factors are increasing or slowing the rate of increase?

Investigation 2
Estimated time for completion = 10 minutes

Tour an online exhibit about population growth by selecting the Keyword **BIRTH RATE** on your CD or Website for this *MediaLab*. What factors affect the number of children a woman can have in your area of the planet? Sign in and choose another continent, and see how that might affect the number of children a woman might have.

Investigation 3

Estimated time for completion = 20 minutes

"So what—we have more people on the planet. That doesn't concern me!" Perhaps it should. Select the Keyword **OVERPOPULATION** on your CD or Website for this *MediaLab* and read the project description. This project describes some issues that ought to concern all of us, no matter what country we are living in. Consider recent news reports. Are there any signs that population pressures, as detailed at this site, might be factors in recent aggression?

Now that you've seen how populations grow and how increasing populations might be factors in social and ecological problems, do the following exercises to determine the severity of increasing populations.

COMMUNICATE YOUR RESULTS

Exercise 1

Estimated time for completion = 5 minutes

You have seen how populations grow. Plot the following data on a sheet of graph paper (or enter it into a program).

Year	Estimated population (in millions)
8000 BC	5
1750 AD	75
1950	2,500
1975	3,700
1990	5,600

From the *CD Activity*, you will recall that humans are a *K*-selected species. Does the graph you produced look like a *K*-selected species? What developments/events might account for the shape of the human population graph?

Exercise 2

Estimated time for completion = 5 minutes

Since the 1970s, the Chinese government has promoted a policy of one child per family. Apply what you learned in the first two sections of this *MediaLab* to reconstruct the reasons why the Chinese, who traditionally valued children and family, undertook such a harsh policy. Would that policy work where you live? What measures would be needed to make such a policy work?

Exercise 3

Estimated time for completion = 5 minutes

What should be done about overpopulation? Birth rates are declining in most countries, so perhaps there is no reason to worry. Get together with some of your classmates, divide into teams and take "sides" about human population growth, presenting evidence in a limited debate.

29

An Interactive Living World

Ecosystems and the Biosphere

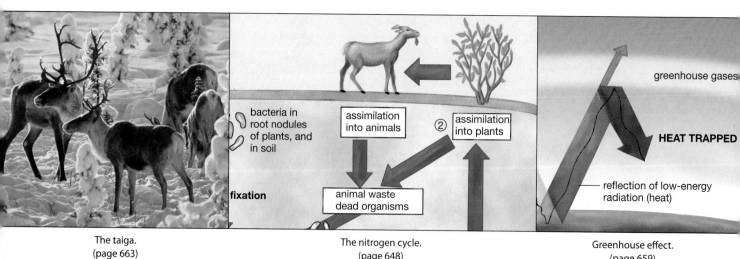

The taiga.
(page 663)

The nitrogen cycle.
(page 648)

Greenhouse effect.
(page 659)

The word "system" is much in use today. We hear about a given country's political system or its economic system. The term was first used in this context by academics who needed a way to refer to complex entities composed of many interacting parts. This way of looking at things is applicable in ecology as well as in economics or politics. In Chapter 28, when you went over ecological communities, you actually were studying something in isolation: the interactions of all the *living* things in a given area. It goes without saying that these living things also interact with many nonliving entities. To get a complete picture of how the living world functions, then, it's necessary to understand how living things interact not only with each other, but with these nonliving elements.

29.1 The Ecosystem Is the Fundamental Unit of Ecology

What do nonliving elements add to life? All communities need an original source of energy, generally meaning the sun; they all need a supply of water and a supply of nutrients, such as nitrogen and phosphorus; and all organisms need to exchange gases with their environment (think of your own breathing). The sun's en-ergy is making a one-way trip through the community, ultimately being transformed into heat; but water and nutrients are being *cycled*—from the Earth into organisms and then back again into the Earth. In this quality, we can begin to perceive the reality of ecological *systems*: of functional units that tightly link both living or **biotic** factors and nonliving or **abiotic** factors to form a whole.

Given this, the fundamental unit of ecology is the **ecosystem**, which can be defined as a self-sustaining community of organisms and the physical environment with which they interact. This chapter begins with a look at small-scale ecosystems and concludes with a look at the very large-scale ecosystems called biomes.

29.2 Abiotic Factors Are a Major Component of Any Ecosystem

In Chapter 28 you learned a good deal about two of the biotic factors of the ecosystem, which is to say individual populations of living things and the communities they make up. This chapter focuses on some abiotic factors of an ecosystem. In overview, these abiotic factors fall into two categories. First are the *conditions* in which an ecosystem exists, such as the

Zones in a lake.
(page 671)

King penguins off Antarctica.
(page 669)

Making weather.
(page 660)

wet rising air
(drops rain as it rises)

falling air (dry)

wet rising air (drops rain as it rises)

average temperature; second are the *resources* that exist in the ecosystem, such as water and nutrients. Resources will be considered first in this chapter and conditions later.

The Cycling of Ecosystem Resources

Those of you who went through Chapter 2, on chemistry, know what a chemical **element** is: a substance that is "pure" in that it cannot be broken down into any other component substances by chemical means. The 92 stable elements include such familiar substances as gold and helium, but only 30 or so of these elements are vital to life and are thus called **nutrients**. Some nutrients, such as iron or iodine, are needed only in small or "trace" quantities by living things, but others are needed in large quantities, among them carbon, oxygen, nitrogen, and phosphorus. Living things also have a great need for water, which is not an element but a molecule composed of two elements, hydrogen and oxygen. Water and nutrients move back and forth between the biotic and abiotic realms, with the term for this movement being a mouthful: **biogeochemical cycling**. Let's look now at the cycling of two elements, carbon and nitrogen, and then at the cycling of water.

Carbon as One Example of Ecosystem Cycling

A certain amount of cosmic debris makes its way to Earth's surface in the form of meteorites, but this material is pretty sparse. What else comes to Earth from space? The sun's precious rays, to be sure, as well as light from distant stars, and some other forms of radiation. But when we're talking about chemical elements, the Earth really is a spaceship: It carries a fixed amount of resources with it. Nothing comes to Earth *from* the outside, and relatively little leaves Earth *for* the outside. For an ecologist, this self-containment simplifies things: What we possess in terms of elements is all we'll ever possess. Thus the question becomes not so much what we have, but where we have it.

In recent years there has been considerable concern about the buildup of the heat-trapping gas carbon dioxide (CO_2) in the Earth's atmosphere, as this increase is now regarded by most scientists as a cause of global warming. Well, if Earth has this fixed quantity of elements—in this case the carbon and oxygen that make up CO_2—how could there be a buildup of carbon dioxide in the atmosphere? The answer is that carbon has been *transferred* from one place on

Earth to another. It has moved from the "fossil fuels" of coal and oil, where it was locked up, into the atmosphere.

Lock-up and transfer; these two concepts are fundamental to biogeochemical cycling. When we look a little deeper into carbon cycling, we find other players involved in lock-up and transfer. Plants need CO_2 to perform photosynthesis, and they take in great quantities of it for that purpose, producing their own food as a result. While the plants are alive, they use part of the carbon they take in to grow—to build up leaves and stems and roots. And as long as these structures exist, they will lock up carbon within them, just as coal and oil do. (Thus, one of the proposed solutions to global warming is to grow a huge number of trees, each of which could be thought of as a kind of piggy bank for carbon.) Eventually the plants die, of course, after which their decomposition by bacteria and fungi releases carbon into the soil and atmosphere, again as CO_2. With this, you can begin to see the whole of the carbon cycle in the natural world (**see Figure 29.1**).

Some plants will, of course, be eaten by *animals*, who need carbon-based molecules for tissues and energy, just as plants do. The difference is that animals get these molecules *from plants*—in the form of seeds and leaves and roots—rather than from the air. These animals will likewise lock up carbon for a time, only to return it to the soil with their death and decomposition. Note, however, the critical difference between animals and plants: Carbon comes into the living world only through plants (and their fellow photosynthesizers algae and some bacteria). An animal might take in some carbon dioxide in breathing, but nothing life-sustaining happens as a result. Conversely, a plant that takes in carbon dioxide can *incorporate* it into a life-sustaining molecule, in this case a sugar that is the initial product of photosynthesis. With this sugar as a food source, plants grow, and this bounty ultimately feeds everything else. Look at this simplified cycle:

$$\text{atmospheric } CO_2 \rightarrow \text{plants} \rightarrow \text{animals} \rightarrow$$
$$\text{decomposers} \rightarrow \text{atmospheric } CO_2$$

What does it track, the path of carbon or the path of food? The answer is both. This is one of the reasons that biologists refer to life on Earth as "carbon-based."

How Much Atmospheric CO_2? In following the carbon trail, it's probably apparent that, just as money is a medium of exchange in our economy, CO_2 is the living world's medium of exchange for carbon; photosynthesizers take *in* carbon as CO_2, and decomposers yield *up* carbon as CO_2. (All organisms also release CO_2 as a by-product of carrying out life's basic processes, as with your own exhalations in breathing.) Given how much **biomass** or living material there is in the world, you might think there would have to be a great deal of CO_2 in the atmosphere, but as a proportion of atmospheric gases, CO_2 is a bit player, making up about 0.035 percent of the atmosphere. But what an important player it is! It's critical in feeding us all, and it may be that relatively small changes in its atmospheric concentration can lead to significant changes in global temperature.

The Nitrogen Cycle

Like carbon, nitrogen is an element; and like carbon, nitrogen cycles between the biotic and abiotic domains. Nitrogen makes up only a small proportion of the tissues of liv-

ing things, but it is a critical proportion because it is required for DNA, RNA, and all proteins. All living things need this element, then; the question is, how do they get it?

The source for all the nitrogen that enters the living world is atmospheric nitrogen, which exists in great abundance. You've seen that carbon dioxide makes up less than 1 percent of the atmosphere, and it turns out that oxygen makes up about 21 percent. But nitrogen makes up 78 percent of the atmosphere. So rich is the atmosphere in this element that a typical garden plot has tons hovering above it. The problem? All this nitrogen is in the form of pairs of nitrogen atoms (N_2) that have a great tendency to stay together, rather than combining with anything else. Thus, in a case of so near and yet so far, plants have no direct access to atmospheric nitrogen. So how do they take it up? In the natural world, the answer essentially is: through the actions of bacteria. It is bacteria that are carrying out the process of

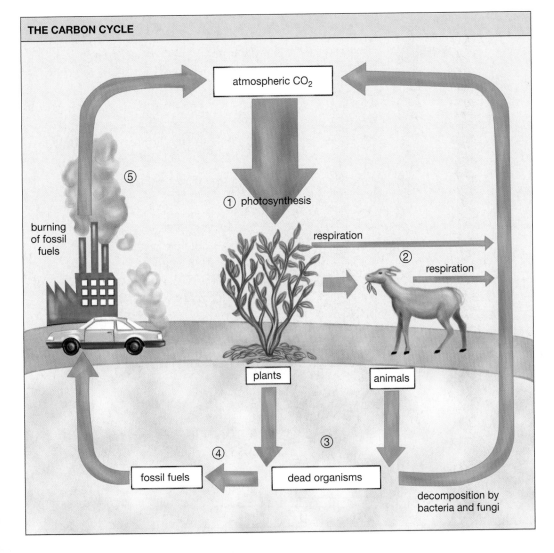

THE CARBON CYCLE

atmospheric CO_2

⑤

burning of fossil fuels

① photosynthesis

respiration

②

respiration

plants

animals

④

③

fossil fuels

dead organisms

decomposition by bacteria and fungi

Figure 29.1
The Carbon Cycle

1. Plants and other photosynthesizing organisms take in atmospheric carbon dioxide (CO_2) and convert or "fix" it into molecules that become part of the plant.

2. The physical functioning or respiration of organisms converts the carbon in their tissues back into CO_2.

3. Plants and animals die and are decomposed by fungi and bacteria. Some CO_2 results, which moves back into the atmosphere.

4. Some of the carbon in the remains of dead organisms becomes locked up in carbon-based compounds such as coal or oil.

5. The burning of these fossil fuels puts this carbon into the atmosphere in the form of CO_2.

nitrogen fixation, meaning the conversion of atmospheric nitrogen into a form that can be taken up and used by living things.

The Process of Bacterial Fixation of Nitrogen

You can see nitrogen fixation and the rest of the nitrogen cycle diagrammed in **Figure 29.2**. The essence of the cycle is that several types of nitrogen-fixing bacteria take in atmospheric nitrogen (N_2) and convert it into ammonia (NH_3). This ammonia then is converted (either in water or by other bacteria) into two types of nitrogen-containing compounds that plants can *assimilate*—can take up and use. And as you've seen, what comes into the plant world will come into the animal world, when animals eat plants. The two usable nitrogen compounds produced in the cycle are the ammonium ion (NH_4^+) and nitrate (NO_3^-). Some of the nitrate that results from this process is used by yet another kind of bacteria, denitrifying bacteria, which can convert nitrate back into atmospheric nitrogen, completing the cycle.

Nitrogen as a Limiting Factor in Food Production

If you were to look at the nitrogen story 100 years ago, there would be little more to it than what you've just gone over. In other words, through most of human history, nitrogen was fixed solely by bacteria (with a small assist from lightning, which can produce a usable form of nitrogen that falls to the Earth in rain). Human beings had a problem with this single route to nitrogen, however, because nitrogen is critical for all plant growth, including *agricultural* plant growth. Indeed, for centuries a lack of nitrogen was a primary limiting factor in getting more food from a given amount of land. Long before anyone knew what nitrogen was, farmers were trying to get more of it to crops in two ways. One was by applying organic fertilizer, such as rotting organic material, to their fields. (Remember how the Pilgrims were taught by the Indians to bury dead fish around their corn plants?) The second was by planting crops that carried their own nitrogen-fixing bacteria within them. This is the case with certain

Figure 29.2
The Nitrogen Cycle
Some bacteria "fix" nitrogen, meaning they convert atmospheric nitrogen (N_2) into an organic form that can be used by other living things.

1. Nitrogen-fixing bacteria convert N_2 into ammonia (NH_3), which converts in water into the ammonium ion (NH_4^+). The latter is a compound that plants can assimilate into tissues. In the diagram, bacteria living symbiotically in plant root nodules have produced NH_4^+, which their plant partners have taken up and used. Meanwhile, free-standing bacteria living in the soil have likewise produced NH_4^+.

2. Other plants take up NH_4^+ that has been produced by soil-dwelling bacteria and assimilate it. Animals eat plants and assimilate the nitrogen from the plants.

3. Animal waste and the tissues of dead animals and plants are decomposed by fungi and by other bacteria, which turn the organic nitrogen back into NH_4^+.

4. Other "nitrifying" bacteria convert NH_4^+ into nitrate (NO_3^-), which likewise can be assimilated by plants.

5. Some nitrate, however, is converted by "denitrifying" bacteria back into atmospheric nitrogen, completing the cycle.

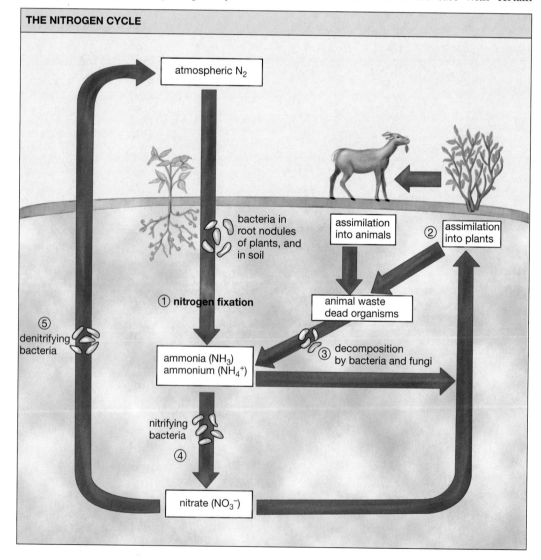

THE NITROGEN CYCLE

atmospheric N_2

bacteria in root nodules of plants, and in soil

assimilation into animals

② assimilation into plants

① **nitrogen fixation**

animal waste dead organisms

⑤ denitrifying bacteria

③ decomposition by bacteria and fungi

ammonia (NH_3) ammonium (NH_4^+)

nitrifying bacteria

④

nitrate (NO_3^-)

legumes, such as soybeans, which carry symbiotic bacteria in their root nodules.

Early in this century, however, a momentous change came about when the German chemist Fritz Haber developed an *industrial* process for turning atmospheric nitrogen into ammonia. In essence, human beings became nitrogen fixers and, by the 1960s, had taken up this activity on a grand scale. In 1990 about 80 million metric tons of biologically active nitrogen were manufactured for use as fertilizer; in addition, human beings planted crops that resulted in the production of another 40 million tons (**see Figure 29.3**). This human-driven production was roughly equal to the amount of active nitrogen that all of nature produced on its own.

It is hard to overstate the importance of the industrial fixation of nitrogen. Put simply, the world's human population could not be fed *without* such fixation. Only with synthetic nitrogen can the Earth's arable land produce enough food to feed the Earth's burgeoning human population.

The massive human intervention in the nitrogen cycle has a downside as well, however. First, remember that human-manufactured nitrogen ends up being used in a very *concentrated* way; it is poured onto the relatively small patches of earth we call farms. A good

deal of this nitrogen does not end up in the crops it was intended for, but instead departs as runoff. Later in this chapter you'll consider a phenomenon called eutrophication, in which lakes can contain too many nutrients (such as nitrogen), sometimes resulting in enormous fish kills. Nitrogen application can also make the soil too acidic, and it contributes to smog and to global warming.

The Cycling of Water

Now let's turn from nutrient to water cycling. Those who went through the material on water in Chapter 2 know how important water is to life. The human body is about 66 percent water by weight, and any living thing that does not have a supply of water is doomed in the long run. As with nitrogen or carbon, all water is either being cycled or is locked up. Carbon may be locked up in coal or trees, but water can be locked up in ice— in glaciers or in polar ice, for example.

The oceans hold more than 95 percent of Earth's water, but thankfully for us, when the oceans' salt water evaporates, it falls back to Earth as fresh water. We can also be grateful that not all ocean water falls back to Earth on the oceans; instead, about 10 percent falls on land (**see Figure 29.4**). This water represents about 40 percent of the precipitation that land

Figure 29.3
Nutrients beyond What Nature Provides
A helicopter applies fertilizer to a sugar beet crop in California.

Figure 29.4
The Hydrologic Cycle
More than 95 percent of Earth's water is stored in the oceans. When water evaporates from the ocean, 90 percent returns to the ocean directly by way of precipitation. The other 10 percent falls on land. There, the water either runs back into the ocean, is stored in such structures as glaciers, or is moved by transpiration and evaporation back into the atmosphere.

THE HYDROLOGIC CYCLE

90%   water vapor   10%

precipitation over ocean   evaporation   precipitation over land   transpiration, evaporation

surface runoff

ocean   groundwater

groundwater runoff

areas get; the other 60 percent comes from a form of cycling reviewed in Chapter 21: *transpiration*, meaning water that is taken up by the roots of land plants, then moved up through their stems and out through their leaves as water vapor. After it evaporates into the atmosphere, it returns to Earth as rain, fog, snow—all the forms of precipitation possible. The driving force behind the hydrologic cycle is the energy of the sun; its heat powers both evaporation and transpiration.

Figure 29.5
Earth's Water
Although there is a tremendous amount of water on our planet, only a small fraction of it is available to us as fresh water.

salt water = 97.5%

oceans

fresh water = 2.5%

ice caps and glaciers 1.97%

ground-water 0.5%

lakes, rivers, and soil 0.03%

atmosphere 0.001%

source of water for confined aquifer

source of water for unconfined aquifer

wells

unconfined aquifer

confined aquifer

impermeable rock layers

Figure 29.6
Aquifers Store Groundwater
When water seeps into the ground, it moves freely through layers of sand and porous rock but moves either extremely slowly or not at all through layers of impermeable rock. As a result, water becomes trapped in different underground layers, from which humans can draw water. Unconfined aquifers receive water directly from a large surface area and therefore tend to be more vulnerable to pollution by chemicals such as fertilizers and pesticides. Confined aquifers are located between layers of impermeable rock. They are replenished more slowly but tend to contain purer water.

With more than 95 percent of Earth's water in the oceans, perhaps as little as 2.5 percent of the Earth's water is fresh water at any given time. Of this fresh water, more than 75 percent is locked up in glaciers and other forms of ice (**see Figure 29.5**). Thus any snapshot we would take of Earth would show that as little as 0.5 percent of its water is available as fresh water, but even here we have to qualify things. About 25 percent of this fresh water is **groundwater**, meaning water that moves down through the soil until it reaches porous rock that is saturated with water. In line with this, about 20 percent of the water used in the United States comes from groundwater. The porous rock that groundwater is stored in is called an **aquifer**, which you can see in **Figure 29.6**. Major aquifers can be enormous, as you can see from the map of the High Plains Aquifer in **Figure 29.7**.

Human Use of Water

Given the sheer number of people on Earth and the importance of water to each of them, it will not surprise you to learn that human beings are laying claim to a great deal of Earth's fresh water; civilization now uses about half the world's accessible supply. But from what you've seen so far, think of what this means. There is no less *water* than there ever was. Like nitrogen or carbon, all the water we have is either being recycled or is locked up. What's changed is that the human population is using an ever-greater proportion of the water that's available. Human beings fight among themselves for water, of course, because it is a scarce commodity in locations throughout the world (**see Figure 29.8**). Close to home, parts of California's Owens Valley that once were farming communities now are desert-like because of the diversion of Owens River water to Los Angeles.

But human beings are also capable of diverting water from species that have no say in this division of resources. California's natural populations of coho salmon have been reduced by some 94 percent since the 1940s, in significant part through diversions of the freshwater streams that the salmon swim up in order to spawn.

If you look at **Figure 29.9**, you can get a quick idea of humanity's impact on the three resources just reviewed—carbon, nitrogen, and water.

Human Beings Are Not Separate from the Earth They Live on

One final thing about water and nutrient cycling. The average person, who has no knowledge of these cycling processes, may regard human beings as a kind of independent line that stretches back in time. Under this view, water and food may pass through us, but they are as separate from us as gasoline is from a car engine.

From your reading of this section, however, it may now be apparent to you that this is not true. Some of the carbon in the bread you ate this morning could have been contained in a tree that decomposed in China 100 years ago. After many cycles, carbon from this tree was taken up by a wheat plant in America whose seeds were ground up, yielding flour that ended in the bread on your shelf. This carbon may be burned by your body to supply the energy it needs, it's true, but it may also become part of your bones or your nerves. Then, when you die, the material you are made of will go *back* to the Earth to be recycled in the ways you've

been reading about, perhaps to be part of a tree or a person in a future generation. We are intertwined with the Earth in some basic, physical ways.

29.3 How Energy Flows through Ecosystems

Having looked at the abiotic elements of nutrients and water, let's now turn to another important component of any ecosystem, energy. Those of you who read through Chapter

a

b

Figure 29.8
Fresh Water Is a Limited Resource

a Women carry drinking water through a polluted slum in Port-au-Prince, Haiti.

b Squatters wash clothes on the shores of the Vaigai River in southern India.

Figure 29.7
Enormous Store of Underground Water
The High Plains Aquifer underlies an expanse of land stretching across eight states. Billions of gallons of water are withdrawn from it each day, mostly for agriculture. (Modified from Gutentag, E. D., Heimes, F. J., Krothe, N. C., Luckey, R. R., and Weeks, J. B., 198. *Geohydrology of the High Plains Aquifer in Part of Colorado, Kansas, Nebraska, New Mexico, Oklahoma, South Dakota, Texas, and Wyoming.* U.S. Geological Survey Professional Paper 1400-B, 63 p.)

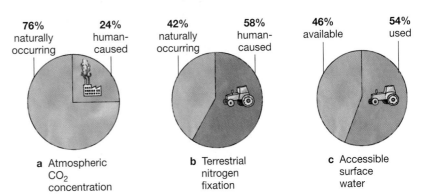

a Atmospheric CO_2 concentration

b Terrestrial nitrogen fixation

c Accessible surface water

Figure 29.9
Human Impact on Resources

a Nearly a quarter of the current atmospheric carbon dioxide concentration is produced by human activity, primarily through the burning of fossil fuels.

b Over half of terrestrial nitrogen fixation comes about because of human activity, including the manufacture of fertilizer for agriculture.

c Over half of the Earth's accessible surface water is now diverted for use by humans, mostly for agriculture. Percentages are approximate.

6, on energy, know that one of energy's in-flexible laws—the first law of thermodynamics—is that energy is never gained or lost, but only transformed. The sun's energy is not used up by green plants; rather, some of it is *converted* by plants into chemical form—initially into the chemical bonds of the sugar that plants make in photosynthesis.

The second law of thermodynamics is that energy spontaneously flows in only one direction: from more ordered to less ordered. The carbohydrate molecules in bread are very ordered things—so many atoms of carbon, oxygen, and hydrogen in a precise spatial relationship to one another. Contrast this with another form of energy, heat, which is the *random* motion of molecules—clearly a disordered form of energy compared to a carbohydrate. Indeed, heat is the last stop on the energy line, because it is the least-ordered form of energy. And, because all energy spontaneously moves toward less order, heat is the ultimate *fate* of all energy. Energy relentlessly ratchets down toward the form of heat because, like a casino that must get a "cut" of each bet laid down, heat gets a cut of each energy transaction. The chemical bonds in gasoline can be broken through combustion and thus be used to power a car engine, but not all the energy released from the combustion drives the engine. Some dissipates as heat, and this happens *every* time energy is used, whether we're talking about a piston firing or a cell dividing.

All of this provides a framework to conceptualize an important part of ecosystems, which is how energy flows through them. Think of the sun's energy, with the energetic rays that leave it as a starting point; now think of the ultimate destiny of all this energy, which is heat, randomly dispersed in the universe. Looked at one way, life on Earth *intervenes* in this flow. Life is an enormous energy collection and storage enterprise, gathering some of the sun's energy

and locking it up for a time in the form of chemical bonds. These bonds can then be broken—think of digesting a muffin—which *releases* this stored energy so that an organism can grow and reproduce. Life does not stop the march of the sun's energy into heat, but it does intervene in it by transforming it into chemical bonds that have order and stability.

Once we see life in these terms, we can begin to think of the *flow* of energy through it. That's what you'll be looking at in the sections that follow.

Producers, Consumers, and Trophic Levels

One way to look at food and energy is in terms of production and consumption: Plants and other photosynthesizers (algae, some bacteria) are an ecosystem's **producers**, while the organisms that eat plants are its **consumers**. But as you know, there are consumers *of* consumers—grass is eaten by zebras, but zebras are eaten by lions. Thus do we get to the concept of feeding *levels* or, as ecologists put it, **trophic levels**. Producers are one level, and then there several levels of consumers. Here's a list of trophic levels through four stages.

First trophic level:	**Producers** (photosynthesizers)
Second trophic level:	**Primary consumers**— plant predators (herbivores/omnivores)
Third trophic level:	**Secondary consumers**— herbivore predators (carnivores/omnivores)
Fourth trophic level:	**Tertiary consumers**— organisms that feed on secondary consumers (carnivores/omnivores)

It is the sun's energy that is locked up in leaves and stems, and it is this energy that is being passed along through these trophic

Figure 29.10
Trophic Levels
Plants, algae, and some bacteria are called *producers* because they use the sun's energy to produce their own food through photosynthesis. Organisms at all other trophic levels ultimately derive their energy from these photosynthesizers.

1st trophic level   2nd trophic level   3rd trophic level   4th trophic level

Producers (photosynthesizers)   **Primary consumers** (plant predators)   **Secondary consumers** (herbivore predators)   **Tertiary consumers** (carnivore predators)

levels. Note that the levels are not just categories that could be put in any order. We must have primary consumers to have secondary consumers; sharks cannot exist on algae (**see Figure 29.10**). You may remember that an animal that eats only plants or algae is a **herbivore**; then there are **carnivores**, who eat only meat, and **omnivores**, who eat both plants and meat.

You have thus far thought in terms of food chains, which is to say trophic *lines* in which a single organism follows another. In reality, of course, nature is much more complex than this; what really exists are complex feeding patterns called **food webs**, one example of which you can see in **Figure 29.11**.

A Special Class of Consumers: Detritivores

To this scheme of trophic levels, ecologists add a special class of consumers, the **detritivores**. These are consumers who feed on detritus, which in normal usage simply means a collection of debris. In ecology, however, detritus is the remains of dead organisms or cast-off material from living organisms (waste, a fallen branch from a living tree). A worm or a dung beetle feeding on such material would fit this description of a detritivore.

Of particular interest, however, is a special kind of detritivore: A **decomposer**, which is an organism that, in feeding on dead or cast-off organic material, breaks it down into its inorganic components, which then can be recycled through the ecosystem. *Inorganic* here can be thought of as building blocks; more technically, it means an element or compound that does not contain carbon bound covalently to hydrogen. A simple rule about decomposers is that they are all fungi or bacteria. A number of detritivores usually are responsible for breaking

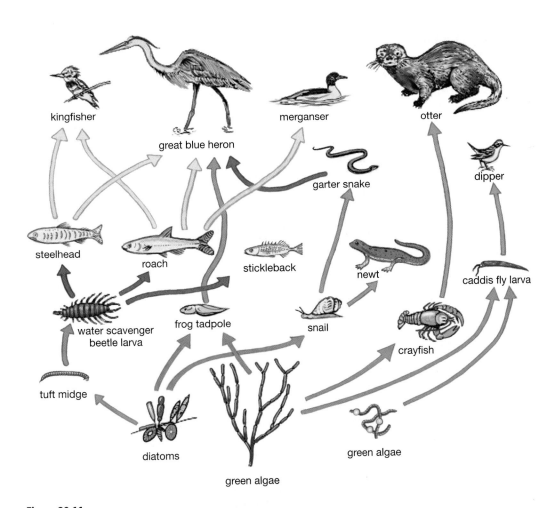

Figure 29.11
A River Food Web
The diagram indicates who eats whom along a portion of the Eel River in Northern California. The arrows have been color-coded by trophic level. Note that there are up to five trophic levels in the web. (Adapted with permission from an original drawing by Mary E. Power, University of California, Berkeley.)

up organic material, first into pieces and then into its inorganic components (**see Figure 29.12**).

Stepping back from this, it's worth noting that if a factory made such complete use of its materials as nature does, the factory would be considered a wonder of the world. Think of it this way: What is thrown away in the feeding and decomposing chain you've been looking at? Nothing, which is all to the good, because if there were any natural garbage of this sort, spaceship Earth would sooner or later be filled up with this useless material. This very thought ought to give us some pause about the compounds human beings currently are putting into the environment that are not **biodegradable**, meaning compounds that can't be decomposed by living organisms. Of particular concern here are plastics, many of which will remain intact for hundreds of years following their production.

Accounting for Energy Flow through the Trophic Levels

All material substances eventually are recycled in ecosystems, but this is not true of energy, which is just passing through, on its way to heat. Two interesting questions arise from this: How *much* of the sun's energy does the living world collect in the first place, and then how much of the collected energy makes it through to each successive trophic level?

This may seem like a routine set of questions, but the first ecologists who thought in these terms brought about a revolution in ecology. Working in the 1940s at Minnesota's Cedar Bog Lake, ecologist Raymond Lindeman was the first to propose the **energy-flow model** of ecosystems. Part of the impetus for

his research was to answer a very down-to-earth question, succinctly expressed by ecologist Paul Colinvaux: Why are big, fierce animals rare? (If you doubt that they are, think of this. It's utterly routine to see plants, insects, or small birds, but it's often the experience of a lifetime to see a bear or a shark or an eagle.) The energy-flow model explained the small numbers of fierce animals, as you'll see, but it then went on to shed light on all kinds of ecological relationships. Why? Because energy is a kind of currency in the natural world; all organisms use it, and it is transferred among them. Moreover, it can be *measured*, usually in units called **kilocalories** (the amount of heat it takes to raise 1 liter of water one degree Celsius). If we know that a field mouse must use 68 percent of the kilocalories it consumes just to stay alive—to move, to keep a constant body temperature—but that a weasel must use 93 percent of its kilocalories for this purpose, that gives us a meaningful basis for comparison between these two animals. Beyond this, weasels *eat* field mice, and with the energy-flow model we have some basis for *tracking* this important commodity as it moves through an ecosystem.

Available Energy Is Greatly Reduced from One Trophic Level to the Next

So, what has energy-flow research told us? One critical insight is that the organisms in each succeeding trophic level bear a comparison with children who inherit far more than they will ever pass along. Recall that the sun is the source of almost all energy on Earth, and that photosynthesizers are the entry point for this energy into the living world. So how much of the sun's energy do plants capture? As you can see in **Figure 29.13**, very lit-

Figure 29.12
How Organic Material Is Broken Down
Different kinds of detritivores consume the organic material in a branch that has fallen from a tree. Most of these animals are consuming pieces of the log, but the detritivore called a *decomposer* (a fungus in the figure) is breaking organic materials into their inorganic components, thus returning nutrients to the soil.

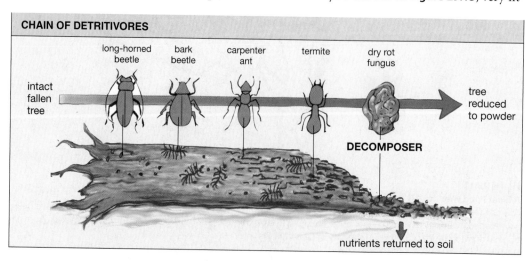

CHAIN OF DETRITIVORES

long-horned beetle   bark beetle   carpenter ant   termite   dry rot fungus

intact fallen tree   tree reduced to powder

DECOMPOSER

nutrients returned to soil

tle; only about 2 percent of the sun's energy that falls around plants is taken in for photosynthesis. This raises an important point, which is noting the difference between the energy that *comes to* an organism and what the organism *does with* that energy. The proportion of energy received that a plant assimilates—that it takes in for photosynthesis—is known as its **gross primary production**. But plants also have energy *costs* and energy *losses*. It takes energy to perform photosynthesis, to move sap from one end of the plant to the other, and so forth. Such costs are generally subsumed under the heading of "cellular respiration," but you could also think of them as the plant's *overhead*; they are the price of doing business. Apart from this, we know that some of the energy in every overhead transaction is lost as heat. After subtracting all these costs, we are left with the energy that a plant uses to build itself up with and to reproduce—to build trunks, leaves, and roots, with much of this tissue being storage material. This material is known as the plant's **net primary production**. Plants turn out to have a net primary production that ranges from 30 to 85 percent. In other words, at the low end of this range, only 30 percent of the energy that a plant assimilates is made into *tissue* that is even potentially available to consumers.

Up a Level, to Primary Consumers

Now, what happens when you jump up one trophic level, to those primary consumers? Well, in the first place, they can't *get to* a lot of the tissue that's been produced—it's in roots that are buried or in leaves that are too high to reach. A good deal of the material is inedible in any event, except by detritivores. Critically, most of the energy these scavengers take in is making an *exit* from the trophic chain; it won't be available to primary consumers, which means it won't be available to secondary or tertiary consumers either. Adding all this up, you can begin to see that of the energy locked up by plants, very little is assimilated by organisms at the next step up—the primary consumers.

This pattern of energy reduction is then repeated in the successive trophic levels. But the reductions tend to get more severe, because animals have higher energy costs than plants: They excrete a lot of the material they consume and the warm-blooded or "endothermic" animals we're most familiar with spend a great deal of energy keeping a constant body temperature—producing heat, which obviously is energy lost to the food chain. So, while a plant's net primary production might be 50 percent, for endothermic animals the figure is about 2 percent, meaning they use up 98 percent of the energy they assimilate just to stay alive, rather than to produce tissue with energy locked up inside it.

All of this leads to what you see in **Figure 29.14**: A reduction in available energy, and, in most cases, a reduction in the number of organisms at each trophic level. The gist of this is that each succeeding trophic level has a great deal less energy available to it than did the level before it. The amount that is passed on varies by community, but there is a rule of thumb in ecology that for each jump up in trophic level, the amount of available energy drops by 90 percent. Take all the energy converted into tissue by, say, a second trophic level herbivore, and of this, only 10 percent ends up being converted into tissue in a *third* trophic level carnivore. Hence the children's inheritance analogy. There is a great variability in conversion rates, however, and 10 percent probably is more like a maximum passed on.

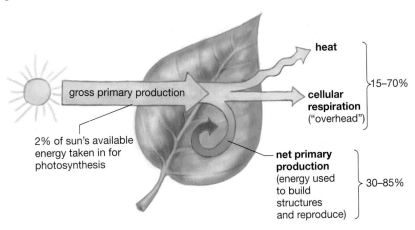

gross primary production

heat

cellular respiration ("overhead")

2% of sun's available energy taken in for photosynthesis

net primary production (energy used to build structures and reproduce)

15–70%

30–85%

Figure 29.13
Absorption and Use of Solar Energy by Plants
Only 2 percent of the solar energy reaching the Earth's surface is taken in by plants for photosynthesis. Of this energy, only a fraction is used to build plant tissues. It is this fraction (the net primary production) that is available to the next trophic level in a food chain. The remaining energy is either dissipated as heat or used by photosynthesizers to power their own physical processes.

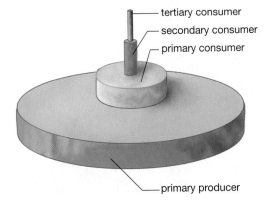

tertiary consumer
secondary consumer
primary consumer
primary producer

Figure 29.14
Energy Pyramid
Only a small fraction (10 percent or less) of energy at each trophic level is available to the next higher trophic level.

The Effects of Energy Loss through Trophic Levels

So, what are the consequences of this energy reduction at trophic levels? First, *this* is the reason that big, fierce animals are rare. If you had a large field that could feed 100 field mice, how many weasels could the field support, if those weasels were existing solely on the mice? Remembering that, at most, only 10 percent of a trophic level's energy is likely to be passed along to the next level, you might guess that 10 weasels could make a living in the field, but even this number would be too high. For 10 predators to make a living from 100 mice, these predators would have to be no *bigger* than the mice, which is certainly not true of weasels. Extending this further, not even a single hawk could survive on the available weasels in the field. Why are big, fierce animals rare? Because of the loss of available energy at each step in the trophic chain. Because weasels are not about to start eating the same things as mice, weasels are locked into a trophic level with a more limited energy supply. When there's only so much energy, there can be only so many weasels.

Beyond this, the energy flow model allows us to see the *limits* to top predators. Why couldn't we have a land carnivore more big and fierce than a grizzly bear? Well, imagine that you found some awesome creature that fit this bill somewhere on Earth, and that you plopped it down in Alaska. What would happen? If it could feed only on grizzly bears, it would starve to death. There simply is not enough energy locked up in grizzly bears for there to be a trophic level above them. This is why, in most food webs, there are so few trophic levels above primary producers; the energy simply runs out. This question of energy loss with each trophic level has relevance

for a human issue, which is eating meat that comes from grain-fed cattle, as you can see in "A Cut for the Middleman," on page 658.

Primary Productivity Varies across the Earth by Region

Primary productivity concerns the production of *biomass* or living material. It will come as no surprise that some areas of the Earth produce more than others. If you look at **Figure 29.15**, you can get an idea of the differences around the globe in connection with plant biomass. The reasons for these differences will be explored in this chapter's section on the large ecosystems called biomes. For now, suffice it to say that much of what you see in Figure 29.15 is intuitive. Productivity is very low in such desert areas as North Africa and very high in such tropical rainforest areas as equatorial Africa. Yet why should one part of the Earth be a rainforest and another part a desert? The answer has to do with Earth's large-scale physical environment.

29.4 Earth's Physical Environment

Earth's Atmosphere

Earth itself exists within an environment, called the **atmosphere**, which is the layer of gases surrounding the Earth. One of the main things to realize about Earth's atmosphere is how it differs from outer space: There's something *to* Earth's atmosphere (a mix of gases) while outer space really is *space*; there's almost nothing in it.

The atmosphere is divided into several layers, but here you need be concerned with only two of them, which you can see in **Figure 29.16**. The lowest layer of the atmosphere, called the **troposphere**, starts at sea level and extends upward about 12 kilometers (or 7.4 miles), and it contains the bulk of the gases in the atmosphere. Nitrogen and oxygen make up 99 percent of the troposphere, but carbon dioxide exists there in small amounts, as do some other gases, such as argon and methane. After a transitional zone, the next layer above the troposphere is the **stratosphere**. Of greatest concern in it is a gas that reaches its greatest density at about 20 to 35 kilometers (13 to 21 miles) above sea level. The oxygen we breathe comes in the form of two oxygen molecules bonded together (O_2). When ultraviolet sunlight strikes O_2 in the stratosphere, however, it can

Figure 29.15
Primary Productivity by Geographic Region
Primary productivity around the globe. This composite satellite image shows how the concentration of plant life varies on both land and sea. On land, the areas of lowest productivity are tan-colored, as with the enormous swath of land that begins with the eastern Sahara Desert in North Africa. More productive land areas are yellow, more productive yet are light green, and the most productive of all are dark green. On the oceans, the gradient runs from dark blue (the least productive), though lighter blue, green, yellow, and orange-red (the most productive). The ocean measurements are of concentrations of phytoplankton—the tiny, photosynthesizing organisms that drift on the water and form the base of the marine food web. Note the high productivity of the oceans in both the far northern and southern latitudes. Antarctica itself is gray because the imaging technique did not work well for it.

put it in a form of three oxygen molecules bonded together (O_3). This is the gas known as **ozone**. Created initially as a *product* of life—since atmospheric oxygen came from photosynthesis—this ozone layer ultimately came to protect life by blocking some 99 percent of the ultraviolet radiation the sun showers on the Earth. It was this blockage that allowed life to come onto land some 460 million years ago, and it is this blockage that protects life now; the UV radiation that pours from the sun can wreak havoc on living tissue, bringing about cancer and immune system problems in people, and damage to vegetation as well.

The Worrisome Issue of Ozone Depletion

It's well known that some human-made chlorine compounds, notably **chlorofluorocarbons (CFCs)**, have the effect of destroying ozone, and they have been doing that for years. CFCs have been used as coolants in refrigerators, as industrial cleaning compounds, and as propellants in spray cans. As a result of their release into the atmosphere, we now have the annual spectacle of a report on how big the seasonal "ozone holes" are over the North and South Poles, and there is concern about a general thinning of the ozone layer. However, there is some good news on this issue. An international agreement, reached in Montreal in 1987 to phase out CFCs, seems to be working. Chlorine levels in the lower atmosphere peaked in 1994 and have been headed downward since then. It will take years to see this reflected in the width of the ozone layer—it will get thinner before it gets thicker—but in the large scheme of things the trend seems good.

The Worrisome Issue of Global Warming

A second environmental issue involving our atmosphere is that of global warming. It's worth noting at the outset that there is still some difference of opinion about whether human-caused global warming is taking place at all. What is not in doubt is that the levels of several heat-trapping gases (notably

carbon dioxide) have been rising in the atmosphere, and that the reason for this is mostly human activities, such as the burning of coal and oil (**see Figure 29.17**). We also have evidence that the Earth has been getting warmer over the very time scales we would expect if human-caused global warming

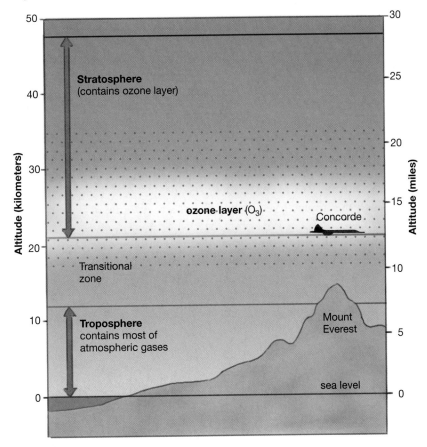

Figure 29.16
Earth's Atmosphere
The two lowest layers of the atmosphere are shown with Mount Everest (the tallest mountain) and the Concorde (the highest-flying passenger airplane) for scale. The troposphere contains most of the atmosphere's gases—mostly nitrogen and oxygen, with some carbon dioxide, methane, and other gases. The stratosphere contains the ozone layer, which blocks 99 percent of the sun's harmful ultraviolet radiation.

Figure 29.17
Global Warming
Atmospheric carbon dioxide concentration is increasing, and global temperature appears to be as well. It is difficult to be certain that the CO_2 increase is driving the increase in temperature because, over long periods of time, there are great natural fluctuations in global temperature.

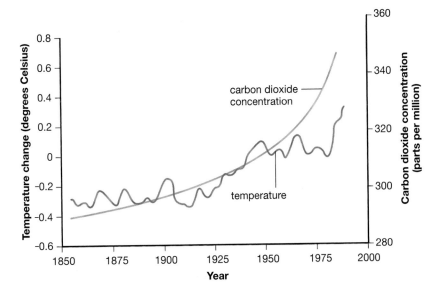

A Cut for the Middleman: Livestock and Food

It may be difficult to think of cows in a field as occupants of one of the trophic levels you've been looking at, but that's just what they are, with some provocative consequences for human beings and food. People can't survive on grass the way cows and sheep can, and it follows that these animals are turning food that is useless to human beings into products that are use*ful* to human beings—meat, wool, and milk. In this sense, then, these animals are a boon to the efficiency of worldwide agriculture in that they allow us to get more human food and other products from the same amount of land.

There is another side to this story, however, which has to do with the *loss* of food that occurs as we move up through the trophic levels. Recall that as a rough rule of thumb, 90 percent of the energy that is locked up in one trophic level is lost to the organisms in the next trophic level. If cattle and other domesticated animals were doing nothing but *foraging*—if they were being fed solely on the grass available in pastures—this would not be a matter of concern, because the grass is not useful to humans as a food source anyway. But in developed countries such as the United States, much of the diet of domesticated animals comes from *grain* that is grown on farms. A large portion of this animal feed could go directly to human beings, but it instead passes through the trophic level of animals, with a predictable result: It takes 9 pounds of feed to produce 1 pound of beef ready for human consumption. What does this mean in practice? A harvest of 2,000 pounds of grain will feed five people for a year. Take roughly this same harvest, feed it to livestock and then feed the *livestock* products to people, and you will feed only *one* person for a year.

This calculation takes on added importance when considering the scale of the grain-fed operation. In the feedlot, while being fattened for slaughter, an average steer consumes up to 2,700 pounds of grain. It may be apparent that such consumption adds up quickly; some 70 percent of the grain produced in the United States is fed to livestock. There are about 70 million cattle in the United States and Canada alone.

But does this mean that human hunger would be reduced if domestic animals were eliminated as a "middleman"? Not necessarily. In a simple calculation of calories-produced versus calories-needed, the world *already*

It takes 9 pounds of feed to produce 1 pound of beef ready for human consumption.

---

produces enough food to feed all its people. The current problem, then, is one of food distribution, not food production. As an often-used phrase has it, "Hunger is real, scarcity is not." The burden that grain-fed animals pose for the future may have to do with the total amount of food produced, in that, as countries get richer, their human populations eat more meat, which means less total food available for humans worldwide.

Figure 1
Fattening at the Feedlot
About 40,000 head of cattle are fed for meat production at this commercial feedlot in Imperial Valley, California.

were taking place: 1997 seems to have been the warmest year of the twentieth century, nine of the ten hottest years since 1961 have occurred since 1987, and, in the Northern Hemisphere, the twentieth century seems to have been the warmest in the last 600 years. The best estimates we have are that global temperature has risen about 0.5° Celsius (0.9° Fahrenheit) since the middle of the nineteenth century. Some researchers believe, however, that the temperature changes we've seen might be explainable as a matter of the random temperature variations that can take place on Earth over very long time scales. It's fair to say, however, that most experts now believe that the Earth is indeed warming because of human activity.

What Is the Greenhouse Effect?

So why should the concentration of gases such as carbon dioxide have anything to do with a warmer planet? Sunlight that is not filtered out by ozone comes to Earth in the form of very energetic, short waves that can easily pass through the atmosphere (**see Figure 29.18**). Once this energy reaches the land and ocean, most of it is quickly transformed into heat that does all the things you've just read about: warms the planet, drives the water cycle, etc. This heat ultimately is radiated back toward space. But heat is not short-wave radiation; it is *long*-wave radiation, and it can be *trapped* by certain compounds, among them carbon dioxide and methane.

There currently is a great deal of research going on regarding what the consequences of global warming might be. This is largely a matter of "modeling" the warming, which is a very complex undertaking. Researchers first have to assume a given amount of warming over a given period, and then do their best to construct a chain of consequences from it: So much ice will melt (and water expand), causing ocean levels to rise by so much; the productivity in some farming areas will increase while in others it will decrease. This modeling has resulted in predictions that range from catastrophic to benign. One thing that can be said for certain is that, if global warming is taking place, humanity is conducting a kind of massive, unplanned experiment with the Earth as the test tube.

Earth's Climate: Why Are Some Areas Wet and Some Dry, Some Hot and Some Cold?

Did you ever look at a globe of the Earth and wonder why it is tilted? Earth exists in space, after all, so what could it be tilted against? When we say that a rod stuck in the ground is tilted, we mean that it is not *perpendicular* to the ground. Here, of course, we're thinking of the ground as a flat surface—a plane. Earth, too, can be viewed as existing on a plane, in this case the plane of its orbit around the sun. (We generally think of this orbit as looking like a large hula hoop with the sun as a yellow ball in the middle.) Now think of the spherical Earth

on this plane as having a rod sticking through it in the form of its north-south axis—the imaginary line that runs from the North Pole straight through to the South Pole. The critical thing is that this rod does not stick straight up and down with respect to the plane of the Earth's orbit; it is *tilted*, at an angle of 23.5°. This tilt dictates a good deal about Earth's climate, and Earth's climate dictates a great deal about life on Earth. Once again, in other words, the sun sets the basic conditions for life.

You can see the effects of Earth's tilt in **Figure 29.19**: In June, our Northern Hemisphere tilts toward the sun, while in January it tilts away from it. Thus the sun's rays strike us more directly in June, and the days are warmer (and longer) than days in January. The angle at which the sun's rays strike a given portion of the Earth is very important. Relative to, say, Brazil, sunlight strikes the Arctic at an indirect or "oblique" angle. Thus, the far north gets less of the solar energy that powers photosynthesis.

Figure 29.18
The Greenhouse Effect
The high-energy rays of the sun can easily penetrate the layer of gases in the troposphere. However, the lower-energy radiation (heat) that reflects from the Earth's surface cannot penetrate the layer of gases as easily. Carbon dioxide and methane thus take on the role of glass panes in a greenhouse: They let solar energy in, but do not let heat out.

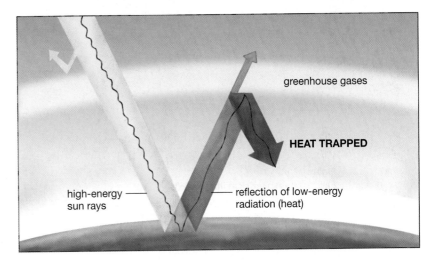

greenhouse gases

HEAT TRAPPED

high-energy sun rays

reflection of low-energy radiation (heat)

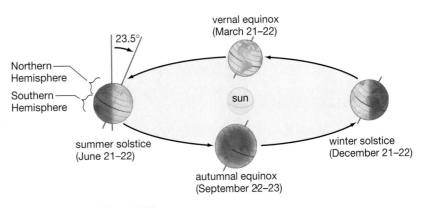

23.5°

Northern Hemisphere
Southern Hemisphere

vernal equinox (March 21–22)

sun

summer solstice (June 21–22)

autumnal equinox (September 22–23)

winter solstice (December 21–22)

Figure 29.19
Earth's Tilt and the Seasons
The Northern Hemisphere gets more sunlight in June and less in January, while the reverse is true for the Southern Hemisphere. This is the reason for seasonal climate variations over large portions of the globe.

The Circulation of the Atmosphere and Its Relation to Rain

The variation in temperature caused by these sunlight variations is the most important factor in the *circulation* of Earth's atmosphere. Near the equator Earth simply gets more warmth and, critically, warm air *rises*. Warm air also can retain more moisture than cold air. You've seen an example of this whenever you've looked at the beads of "sweat" that form on the outside of a cold glass in summer. Moisture-laden warm air comes into contact with the cooler air immediately around the glass; thus cooled, the air cannot hold as much moisture and releases it onto the glass.

This same thing happens with the warm air that rises on both sides of the equator. Because of the heat of the tropics, air is rising in quantity from the tropical oceans; because this air is warm, it is carrying with it a great deal of moisture. The air cools as it rises, however, and then drops much of its moisture on the tropics, which is why they're so wet.

Following this a step further, this volume of air is now cooler, drier, and moving toward the poles in both directions from the equator (**see Figure 29.20**). At about 30° north and south of the equator, it descends, warming as it drops and actually absorbing moisture *from* the land. The land will be dry at these latitudes because this is where the dry, hot air descends.

The air that has descended now flows in two directions from the 30° point. In the Northern Hemisphere this means north toward the North Pole and south toward the equator. Traveling at a fairly low altitude, this air eventually picks up moisture, rises, and deposits its moisture, this time at about 60° north and near the equator, with the same events happening at the same locations in the Southern Hemisphere.

What you get from this rising and falling is what you see in Figure 29.20. A set of interrelated "circulation cells" of moving air, each existing all the way around the globe at its latitude, and each acting like a conveyor belt that is dropping rain on the Earth where it rises but drying the Earth where it descends.

The Impact of Earth's Circulation Cells

The full meaning of this becomes obvious only when you look at a map such as the one

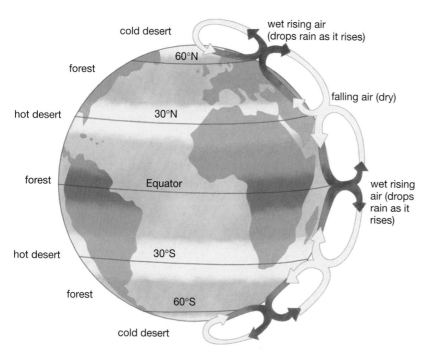

Figure 29.20
Earth's Atmospheric Circulation Cells
Because the Earth receives its most direct sunlight at the equator, this region is warm—and warm air holds moisture and rises. This air moves north and south from the equator, cooling as it rises, and then loses its moisture as rain, with the result that the land and ocean immediately north and south of the equator get lots of rain. As the now-dry air moves farther north and south, it cools and sinks again at about 30° in both hemispheres. Thus, the land at these latitudes tends to be dry. Two other bands of circulation cells exist at higher latitudes, operating under the same principles.

Figure 29.21
Earth's Atmospheric Circulation Cells and Precipitation

The wettest regions of the Earth lie along the equator, while the driest regions lie along the latitudes 30° N and S. Compare these regions to the circulation cells shown in Figure 29.20. The variations of rainfall patterns within cells are explained by factors such as the presence of mountain ranges.

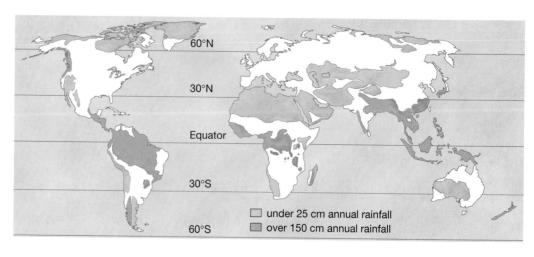

in **Figure 29.21** (or better yet, a globe). Draw your finger across the equatorial latitude on the map and look at how much green there is—Southeast Asia, equatorial Africa, the Amazon. This is no surprise, because we expect the equator to be wet and green. Now, however, do the same thing at the dry 30° north latitude. Look at how much desert there is at this latitude all around the globe—it's where we find the Sahara, the Arabian, and the Sonoran deserts, among others. And more desert territory exists at 30° south, with the Australian desert and Africa's Namib. From this you can see that the climatic fate of Earth's various regions is largely written in our globe's large-scale wind patterns.

Mountain Chains Affect Precipitation Patterns

Latitude doesn't tell the whole story about precipitation on Earth, however. Mountain ranges force air upward and it too cools as it rises, dropping its moisture on the windward side of the range. Then the air descends on the opposite (leeward) side of the range, only this time it is dry air and is thus *picking up* moisture from the ground, rather than depositing it (**see Figure 29.22**). A dramatic example of this "rain shadow" effect can be seen on the Pacific Coast of the United States in southern Oregon, where the western side of the Cascade Mountains is lush with greenery all year round, while the basin below the eastern slopes is a desert.

Beyond this, we know that it's possible to be in a warm latitude, such as that of Ecuador in South America, and still be very cold—if you go to the top of the Ecuadorian Andes. As altitude increases, temperature drops, in other words. The changes in climate that occur as you move from flatland to mountain peak are similar to the changes that occur as you move from the equator to one of the poles.

The Importance of Climate to Life

The factors you have just reviewed are among the most important in shaping the large-scale **climates** or average weather conditions we find on Earth. It's obvious that climate has a

great deal of influence on life, since we never see polar bears in Panama or monkeys in Montana, but this influence is even greater than you might think. Looking over the globe, you see different forms of dominant vegetation in different areas—grassland in America's Great Plains, rainforest in northern Brazil. So strongly does climate affect this pattern that Earth's vegetation regions essentially are defined by Earth's climate regions. Far in Canada's north, there is a fairly distinct boundary at which Canada's vast coniferous forest gives way to the mat-like vegetation called *tundra*. North of this line there is tundra; south of it are the forests. What is this line? In summer it is the far southern reach of a mass of cold air called the *arctic frontal zone* (**see Figure 29.23**).

**Figure 29.22
Rain Shadows**
Mountain ranges force air to rise. This cools the air, causing it to lose its moisture on the windward side of the range. This creates a rain shadow, meaning a lack of precipitation, on the leeward side of the range.

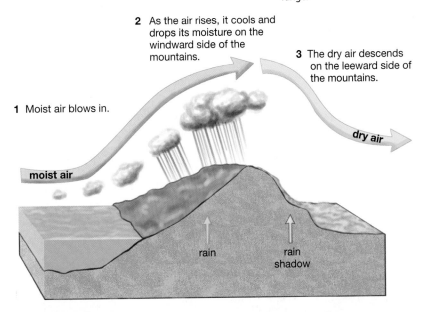

2 As the air rises, it cools and drops its moisture on the windward side of the mountains.

3 The dry air descends on the leeward side of the mountains.

1 Moist air blows in.

dry air

moist air

rain

rain shadow

**Figure 29.23
Line of Transition**
In Alaska's Denali National Park, a region of forested taiga vegetation gives way to a region of grassy tundra vegetation in a fairly abrupt way.

Vegetative formations generally overlap one another to a greater extent than is seen with Canada's tundra and coniferous forest, so that the closer we get to a vegetation boundary the less distinct it looks. Nevertheless, there are discrete regions of vegetation on Earth whose boundaries are largely determined by climate—particularly by temperature and rainfall. Vegetative exceptions *within* these regions tend to be mountains, whose climates can vary greatly from base to peak.

29.5   Earth's Biomes

When these large, land vegetation formations are looked at as *ecosystems*, complete with animals, fungi, nutrient cycling, and all the rest, they go by the name **biomes**. Grasslands, whether found in Russia or in the American Midwest, are ecologically very similar to one another and thus constitute one type of biome. The tropical rain forests found in Africa, South America, and Asia constitute another. The world can be divided into any number of different biome types— this is a human classification system overlaid on nature—but six are recognized as the minimum: tundra, taiga, temperate deciduous forest, temperate grassland, desert, and tropical rainforest. See **Figure 29.24** for an overview of locations of these biome types throughout the globe. Polar ice and moun-

tains often are recognized as separate biomes as well, as is another grassland variation, the savanna, found most famously in equatorial Africa. There is also chaparral, a shrub-dominated vegetation formation that is found in "Mediterranean" climates such as those in Southern California's coastal region.

The fact that two biomes are of the same type does not mean that they are identical. Two areas that are broadly the same in climate can end up with some significant differences in life-forms. Let's look briefly now at each of Earth's major biome types and then at its aquatic ecosystems.

Cold and Lying Low: Tundra

Along with polar ice, tundra is the biome of the far north, stretching in a vast, mostly frozen ring around the northern rim of the world. So inaccessible is tundra that the average person may never have heard of it, yet it occupies about a fourth of Earth's land surface. The word *tundra* comes from a Finnish word that means "treeless plain," and the description is apt; its flat terrain stretches out for mile after mile with little change in the vegetational pattern of low shrubs, mosses, lichens, grasses, and the grass-like sedges (**see Figure 29.25**).

Tundra is a paradoxical biome in that it averages about 25 centimeters (or 10 inches) of rain per year, which almost puts it in the

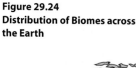

**Figure 29.24
Distribution of Biomes across the Earth**

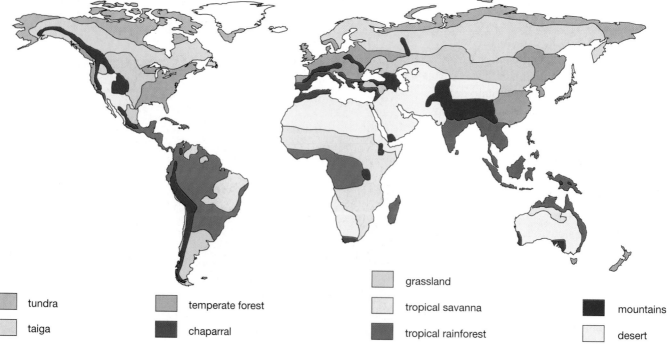

tundra	temperate forest	grassland	
		tropical savanna	mountains
taiga	chaparral	tropical rainforest	desert

desert category, and yet it is marshy in summer, with numerous shallow lakes dotting its surface. How can both things be true? Once the summer rain falls, it essentially has no place to go. No more than a meter below the surface, the ground is permanently frozen down to a depth of as much as 500 meters or a quarter-mile. This is the **permafrost**, or permanent ice, which acts as a boundary beyond which neither water nor roots can go. Because water cannot drain far into the soil, it sits on top of the tundra in summer, which makes for the bogs and small lakes. The low-lying tundra vegetation makes the most of a short growing season—lasting as little as 50 days—by producing a year's worth of food through photosynthesis in only a few weeks.

The tundra has a good deal of animal life, particularly in the summer when migrating animals move north to feed on the vegetation. Large mammals such as caribou and musk oxen live there. Arctic hares and foxes are year-round residents, as are arctic lemmings—herbivores that are key to much of the tundra's animal food web. (Contrary to popular belief, these lemmings do not commit suicide by jumping off cliffs. They merely *migrate* in great numbers at certain times.)

Northern Forests: Taiga

The **taiga** or boreal (northern) forest includes the enormous expanse of coniferous trees mentioned earlier that juts up against the tundra. Conditions are still fairly dry and cold, and the growing season is short in this northern biome, but in all three respects the taiga is less severe than the tundra. Note in Figure 29.24 that, as with tundra, taiga is found only in the Northern Hemisphere. This is so because there is little or no land in the Southern Hemisphere at the extreme latitudes of these biomes. This in turn is related to a more general point, which is that most of Earth's land mass is in the Northern Hemisphere.

The taiga is a model of species uniformity as only a few types of trees—spruce, fir, and pine—are the ecological dominants, but they are present in great numbers (**see Figure 29.26**). The taiga's vegetation supports lots of animals, including large mammals such as moose, bear, and caribou. When thinking of the fur trade, you might have in mind some

freezing, coniferous territory in the far north. The taiga turns out to be it. Sables, minks, beavers, arctic hares, wolverines: This harsh biome has an abundance of these furred creatures.

Hot in Summer, Cold in Winter: Temperate Deciduous Forest

Temperate deciduous forest is a biome familiar throughout much of the United States as it exists roughly from the Great Lakes nearly to the Gulf of Mexico and from the western

Figure 29.25
Tundra
Grizzly bears forage for fall berries in Alaska in the shrub-and-grass biome known as the tundra

Figure 29.26
Taiga
Alaskan caribou in winter in the northern forest biome known as taiga.

Great Lakes to the Atlantic Ocean. The pine forests that cover large parts of the southeastern United States generally are considered to be temperate deciduous forest because the pines are a successional stage *leading to* deciduous forest. (Remember, from Chapter 28, the concept of ecological succession, under which species in a community are replaced by other species over time until a final, stable state is reached.) Meanwhile, the warm Gulf Coast is not temperate deciduous forest per se, but a variant on it called *temperate evergreen forest*. Much of Europe and eastern China fit within the temperate deciduous classification, though large parts of Europe have few forests left, and China has very little.

Temperate deciduous forests are the forests of maple and beech and oak and hickory (**Figure 29.27**). The "deciduous" in the name refers to trees that exhibit a pattern of loss of leaves in the fall and regrowth of them in the spring. Trees do not dominate these forests to the extent they do the taiga, however. Making use of the good soil, a robust "understory" of woody and herbaceous

plants will spring up on the floor of the deciduous forest. Some of this activity occurs early in the growing season, before the trees have "leafed-out," and blocked the sunlight.

Temperate forests grow where there is plenty of water: between 75 and 200 centimeters (or 30 to 78 inches) per year, with the lower amount being three times what the tundra gets.

It's worth noting that while the tundra has no reptiles or amphibians and the taiga has only a few, the temperate deciduous forest has an environment that is wet and warm enough to support a variety of both animal forms. Many species of the large mammals that once populated the deciduous forests in the United States have either been greatly reduced (black bears, wild pigs) or nearly eliminated altogether (wolves), though in the case of the black bear, fairly robust conservation and restoration efforts are underway.

Dry but Sometimes Very Fertile: Grassland

There is an irregular line in western Indiana that forms the dividing line between deciduous forest and prairie. What is "prairie" in the United States is known as "steppes" in Russia, "pampas" in Argentina and Uruguay, and "veldt" in South Africa. These are all names for the biome that ecologists call *temperate grassland* (**Figure 29.28**).

Looked at one way, grasslands are what often literally lie between forests and deserts. Whereas the precipitation range for deciduous forests is from 75 to 200 centimeters per year (30 to 78 inches), for grassland the range is from 25 to 100 centimeters per year (10 to 39 inches). The North American grassland has its own internal division based on the amount of rainfall. It ranges from tall-grass prairie in Illinois to a drier mixed-grass prairie that begins at about the middle of Kansas and Nebraska—this is where the Great Plains begin—to a short-grass prairie that lies in the rain shadow of the Rocky Mountains. Other American grassland areas can be found in California's Central Valley and eastern Washington and Oregon.

A type of *tropical* grassland, the **tropical savanna**, exists in Australia, Africa, and South America. Tropical savanna is characterized by seasonal drought, small changes in the generally warm temperatures

Figure 29.27
Temperate Deciduous Forest
Plenty of water and seasonally warm temperatures lead to abundant plant and animal life in the biome known as temperate deciduous forest. This forest is in the Great Smoky Mountains National Park in Tennessee.

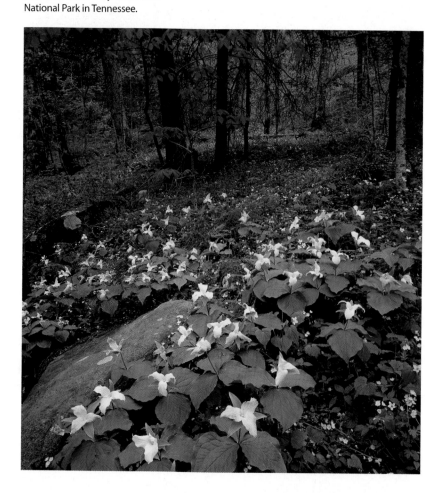

throughout the year, and stands of naturally occurring trees that punctuate the grassland. This is the homeland of many of the great mammals we associate with Africa, such as lions and zebras.

When Europeans first beheld the American grasslands, they could scarcely believe what they saw: A territory that had essentially no trees, but instead a sea of grasses that could grow up to 3 meters or 10 feet tall in the wetter tall-grass prairie. Prairie soil turned out to be some of the most fertile on Earth, however, and that fact eventually spelled the end of natural prairie in the United States and most of the rest of the world. Almost all the tall-grass and mixed-grass prairie in the United States eventually was plowed under for cultivation, while the short-grass prairie is now used for crops and cattle grazing. Prior to the coming of Europeans, 22 million acres of Illinois was prairie; today that figure stands at slightly more than 2,000 acres. From the former prairie, however, the United States feeds itself and some of world outside its borders. Illinois and Kansas and Nebraska are still grasslands, but the grasses that grow there now are not the native big bluestem or needlegrass; they are corn and wheat.

The American grasslands once were home to huge herds of one of the largest of Earth's grazing animals, the American bison (*Bison bison*). Better known as the buffalo, this animal was shot nearly to extinction—hunted is not the right word—but it is now rebounding somewhat in protected areas.

Chaparral: Rainy Winters, Dry Summers

Chaparral is not grassland, but instead is a biome dominated by evergreen shrub vegetation that is dotted with pine and scrub oak trees. Chaparral is found in areas that have a Mediterranean climate: the region that rings the Mediterranean Ocean, western Australia, parts of Chile and South Africa and coastal Southern California. What most of these areas have in common is that they are on the west coast of a land mass. This placement results in mild, wet winters and very dry summers.

The Challenge of Water: Deserts

Deserts are defined in various ways, one of them being: an area in which rainfall is less

than 25 centimeters or 10 inches per year. Note that temperature has nothing to do with this definition. Thus, there can be temperate deserts (the Mojave in Southern California), cold deserts (the Gobi in China) and hot weather deserts, such as you can see in **Figure 29.29**. There are areas in some deserts, such as Death Valley in California, that achieve the dubious distinction of having no life whatsoever in them, but this is rare. Most deserts harbor a variety of life-forms, and what we think of as desert can shade imperceptibly into dry grassland.

Figure 29.28
Temperate Grassland
A white tail deer fawn stands among the grass and flowers in a tall grass prairie in Missouri. Such tall grass is one kind of growth found in the biome known as temperate grassland.

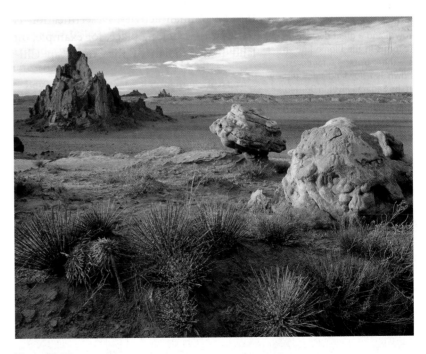

Figure 29.29
Desert
When water gets sparse, life does too. Shown is a desert biome in Monument Valley, Arizona.

zone extends from the point on the shore where the ocean's waves reach at high tide to a point out at sea where an ocean-floor formation called the continental shelf drops off (**see Figure 29.31**). Beyond this point there is the **open sea**. *Within* the coastal zone, there is an area bordered on one side by the ocean's low-tide mark and on the other by its high-tide mark. This is the **intertidal zone**.

If you look next at the ocean strictly in its vertical dimension, all of the water from the ocean's surface to its floor is called the **pelagic zone**, while the ocean floor itself is called the **benthic zone**. *Within* the pelagic zone, from sea level down to a depth of at least 100 meters (about 330 feet), the sun's rays can penetrate strongly enough to drive photosynthesis. This zone of photosynthesis is known as the **photic zone**.

More Productive Near the Coasts, Less Productive on the Open Sea

Each of these zones marks off an area that is meaningful in terms of life. Start first at the intertidal zone. This is a world *in between*, we might say, because the creatures in it live part of the time submerged in the ocean and part of the time exposed to the air. It turns out that the intertidal zone is extremely productive for a reason that you might think would make it *unproductive*: the ocean waves, which bring in nutrients, carry away wastes, and expose more of the surface area of photosynthesizers to sunlight.

Ocean life continues to be abundant out past the intertidal zone and in general re-mains so all the way to the end of the coastal zone. In short, ocean life is most productive in its shallower depths (**see Figure 29.32**). Once past the continental shelf and into the open ocean, productivity can drop off to levels that are less than those of a desert. Why should this be so? Remember that sunlight strong enough to power photosynthesis penetrates ocean waters only to a depth of 100 meters or so. Meanwhile, the average depth of the ocean is about 3,000 meters or 1.8 miles. Thus there is a vast depth of open ocean that simply isn't very productive. In addition, coastal areas benefit from higher levels of nutrients, primarily as a result of nutrient run-off from the land.

One measure of the coastal zone's importance is that it is the location of all the world's great fisheries. At one time, it was thought that these areas were places of such endless bounty that they could never be depleted. The 1990s proved this wrong, however. Modern industrial fishing techniques have left one commercial fish species after another seriously depleted—haddock and bluefin tuna, Atlantic cod and swordfish. Depletion is not the same as endangerment, but it should be of real concern if only from a self-interested point of view: Almost 20 percent of the animal protein that humans consume worldwide comes from fish. For a perspective on environmental issues, see "Good News about the Environment," on page 670.

More Productive Near the Poles, Less Productive Away from Them

Apart from the shallow-to-deep transition, the ocean also has one other gradient with respect to the amount of life it harbors. In the pelagic zone, there is relatively more ocean life toward the poles, with this concentration decreasing with movement toward the equator—the exact opposite of what we find with land life. The oceans that surround Antarctica are filled with life. **Phytoplankton**, meaning floating, microscopic photosynthesizers such as algae and cyanobacteria, are the base producers for it all. These are consumed by the tiny floating animals known as **zooplankton**. Shrimp-like crustaceans called krill feed mostly on the phytoplankton and serve as the principal source of food for the whales and seals we associate with this area.

**Figure 29.31
Ocean Zones**

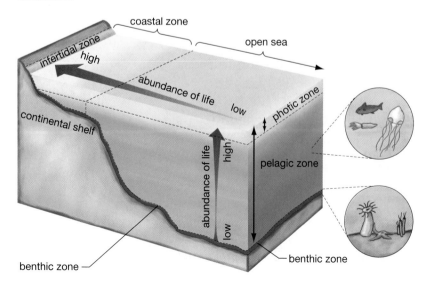

Cities of Productivity: Coral Reefs

Despite the richness of the colder oceans, the tropical oceans do boast marine communities that are as productive as any in the ocean—indeed, as any on Earth. These are the **coral reefs**, which lie in shallow waters at the edge of continents or islands at tropical latitudes. Coral reefs actually are the piled-up remains of many generations of coral animals, with each individual animal known as a polyp. With the help of a type of red algae that grows on them, the coral animals leave behind limestone deposits that make up the structure of the reef. You can get an idea of how many generations of coral are piled up by considering that most reefs are between 5,000 and 8,000 years old and some are millions of years old. What the reef as a whole creates is *habitat*: nooks and crannies and tunnels and surfaces that are home to an amazing array of living things. So rich is this habitat that it covers only 2 percent of the ocean's floor, but is home to perhaps 25 percent of the ocean's species. One coral reef, Australia's Great Barrier Reef, also is one of the largest structures created by any living thing, including humans. It measures some 2,000 kilometers or 1,200 miles in length.

Freshwater Systems

Whereas the oceans cover almost three-quarters of Earth's surface, freshwater ecosystems—inland lakes, rivers, and other and running water—together cover only about 2.1 percent of Earth's surface. Like the ocean, lakes can be divided into biological zones. The most productive zone in a lake is the shallow-water area along its edge, the **littoral zone**, whose outer boundary is defined as the point at which the water is so deep that rooted plants can no longer grow. Then there is the **limnetic zone**, which starts at the surface of the lake and extends down to the point at which sunlight no longer penetrates

a

b

c

**Figure 29.32
Diversity in Ocean Life**

a A man and two children investigate tidepools, a part of the ocean's intertidal zone. The deeper water beyond them can be seen both above and below the surface.

b King penguins swim underwater near a breeding colony in the fertile waters off the coast of Antarctica.

c Fish called sweepers and fairy basslets feed in the currents above a coral plate covered with animals called crinoids. The structure is off the coast of Papua New Guinea.

Good News about the Environment

A sense of gloom about the environment may arise from between the lines of these pages as the list of current or potential environmental problems piles up: global warming, nitrogen buildup, overuse of water resources, tropical rainforest destruction. These are real problems, but their existence should not blind us to the fact that there has been a great deal of good news about the environment in the last 30 years, much of it showing how effective government can be when prodded into action by its citizens.

If there was a time for unrestrained gloom about the environment in the United States, it probably would have been in the late 1960s, when the country had plenty of endangered species but no Endangered Species Act; when levels of smog in Los Angeles were almost twice what they are now; when there was plenty of household trash to be recycled, but almost no recycling going on; when only a third of the bodies of water in the country were safe for fishing and swimming, whereas today two-thirds are.

As a symbol of change, consider Cleveland, Ohio, whose Cuyahoga River was so polluted in 1969 that it *caught fire*, while southwestern Lake Erie, into which the Cuyahoga flows, was essentially a dead zone for fish during the late summer. The lake came to this low state in part because of the nearly 80 metric tons of phosphorus—then regularly added to washing detergents—that were dumped into it each day. That figure has now been reduced by perhaps 85 percent, and Lake Erie is biologically very active and a major freshwater fishery. Meanwhile the Cuyahoga River has been cleaned up enough

that an area near the river's mouth has been transformed into an entertainment district complete with restaurant tables near the water's edge—not bad for a small river running through a big city. This does not mean that the Cuyahoga now looks as it did when Europeans first saw it; toxic sediment from the river, mostly the legacy of the

---

Environmentalism has entered not only our lawbooks but, more important, our consciousness.

---

past, still flows into Lake Erie. And the lake itself is now beset by another environmental calamity in the form of an invasion of a small, voracious mollusk known as the zebra mussel (*Dreissena polymorpha*). But if we look at the *trend*, there is no doubt that things have gotten better in Cleveland, not worse.

This same thing could be said of most of the United States. Simply put, there has been a profound change with respect to the environment over the past 30 years. From pollution controls, to species protection, to personal recycling, environmentalism has entered not only our lawbooks but, more important, our consciousness. And support for the environment has grown not just in America but in most developed countries. The cleanup of Lake Erie required the cooperation of the United States and Canada; the heartening trend in preservation of the

strongly enough to drive photosynthesis. Finally, there is the area beneath the limnetic zone, the **profundal zone**, in which photosynthesis can't be performed (**Figure 29.33**).

Nutrients in Lakes

Lakes can be naturally **eutrophic**, meaning nutrient-rich, or **oligotrophic**, meaning nutrient-poor. A lake that has few nutrients will have relatively few photosynthesizers, and this in turn means few animals. Oligotrophic lakes generally are clear (and deep), while eutrophic lakes often have an abundant algal cover.

Nutrient enrichment can happen naturally over time, or it can take place because of human intervention. The latter variety is

known as **artificial eutrophication**, a process that sometimes is undertaken intentionally just to increase the yield of fish from a lake or pond. There is such a thing as having too *many* human-added nutrients in a lake, however, and this can be big trouble for this ecosystem. An overabundance of such nutrients as phosphorus or nitrogen brings on an overabundance of algae—a so-called **algal bloom**—and this eventually means an overabundance of dead algae falling to the bottom of the lake (**see Figure 29.34**, page 672). When this happens, decomposing bacteria flourish, and they are using *oxygen*—so much of it that the fish in a lake can suffocate. How do these harmful levels of phosphorus or nitrogen get into earth's ozone layer is the product of an international agreement.

None of this is to say that mechanisms are in place such that we need not worry about the environment. Far from it; largely because of the growth in human population and the greater per capita use of natural resources, the environment is threatened on thousands of fronts and

will remain so as far out as we can see into the future. But the record of the last 30 years is not just one of environmental loss; it is one of great environmental accomplishment as well. Perhaps the period's most important lesson is that there is nothing inevitable about environmental degradation. Concerned, hard-working people have made a great deal of difference in this issue.

2000 ← Pounds of hydrocarbons produced by an average new passenger car during 100,000 miles of driving.

50

1965   1998

Figure 1
Moving in the Right Direction
Over the last 30 years, the amount of car emissions in California has decreased 40-fold, as measured in pounds of hydrocarbon emissions produced by an average new passenger car in California during 100,000 miles of driving. One result is that cities like Los Angeles have experienced dramatic decreases in smog levels, despite steep increases in population.

aquatic ecosystems? Fertilizers from lawns or agriculture, construction that disturbs bedrock, sewage, and detergents are some of the sources (though in many states, phosphate-containing detergents have been banned for decades). This is one form of human environmental impact that in some instances is reversible in a straightforward way: Reduce the nutrients that are flowing into the ecosystem.

Figure 29.33
Zones in a Lake
The littoral zone starts at water's edge and extends out to the point at which rooted plants can no longer grow. The limnetic zone starts at the lake's surface and extends down to the point at which sunlight no longer drives photosynthesis. The profundal zone is the area in the lake in which photosynthesis cannot be performed.

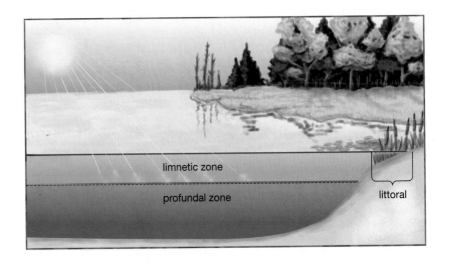

limnetic zone

profundal zone

littoral

Figure 29.34
Too Many Nutrients
A pond with an overabundance of nutrients has experienced an overgrowth of algae—an algal bloom—which may be detrimental to other life-forms in the pond.

Figure 29.35
Where Salt Water and Fresh Water Meet
The Chesapeake Bay is one of the largest and most productive estuaries in the world. In it, the fresh water of 19 principal rivers meets the salt water of the ocean, forming zones of different degrees of saltiness. Each zone supports specific communities of organisms. (Redrawn from A. J. Lippson and R. L. Lippson. *Life in the Chesapeake Bay.* Baltimore: Johns Hopkins University Press, 1984.)

Estuaries and Wetlands: Two Very Productive Bodies of Water

There is one important type of aquatic water system that always straddles the line between a freshwater and saltwater habitat. This is the **estuary**: an area where a stream or river flows into the ocean. Estuaries rank with tropical forests and coral reefs in being the most productive ecosystems on Earth. The cause of this productivity is the constant movement of water—the same force at work in the ocean's intertidal zone. Ocean tides and river flow are constantly stirring up estuary silt that is rich in nutrients. Plants and algae thus get an abundance of these substances, grow in abundance, and pass their bounty up the food chain. In **Figure 29.35** you can see an image of an important estuary in the United States, the Chesapeake Bay.

Another important type of aquatic ecosystem is **wetlands**, which are just what they sound like: lands that are wet for at least part of the year. "Wet" here covers a variety of conditions, from soil that is merely temporarily waterlogged—a prairie "pothole" in Minnesota, for example—to a cover of permanent water that may be several feet in depth (the bayous of Louisiana). Wetlands go by such specific names as swamps, bogs, and marshes and tidal marshes; the overwhelming majority of them are freshwater **inland wetlands**, rather than the ocean-abutting **coastal wetlands**, which can be fresh water or salt water (**see Figure 29.36**). Wetlands are sites of great biological productivity, and they serve as vital habitat for migratory birds. It's been estimated that the amount of wetlands in the United States has been reduced by 55 percent since the arrival of Europeans—120 million of the 215 million acres that were originally wetlands have been drained or paved over. Currently, the state of Florida and the federal government are working on one of the most far-reaching and expensive habitat restoration efforts in history, as they try to bring the largest expanse of wetlands in the nation, the Florida Everglades, to a healthier condition.

Life's Largest Scale: The Biosphere

You've been reviewing life on a very large scale, which is to say the aquatic ecosystems and biomes of the Earth. But there is one larger scale of life yet, which is the collection of *all* the world's ecosystems, the **biosphere**. Much could be said about the biosphere, but here's just one observation about life on this global scale. The Earth is about 40,000 kilometers or a little more than 24,000 miles in circumference, with life existing along any line you could draw all around the globe. If we look at life in its *vertical* dimension, however, it has been found to exist only about 3 kilometers or 1.8 miles below Earth's surface and 8.3 kilometers or about 5.2 miles above sea level (in the upper Himalayas). This means that all of life on Earth exists in a band that is at most about 10 kilometers or 6.2 miles thick, with the vast bulk existing in a more narrow span yet. Actually, this is all the life that we know of in the *universe*, so that if we could imagine ourselves traveling by cosmic elevator, we could potentially come up from the molten center of Earth, pass through the bacteria lodged between underground rocks, and then move through a tissue-thin layer of a green living world, after which we might never see anything like it again, no matter how far out into the universe we traveled. Taking this perspective, the really remarkable thing is not how much life there is, but how little.

On to . . . the Rest of Life

Almost every chapter of this book has ended with a little preview of what is to come in the next chapter. In the current case, there is no such thing as a next chapter because you have reached the end of the book. As this chapter has perhaps made clear, science has a great relevance to issues such as the environment that confront society every day. Apart from such practical considerations, however, this chapter may also have helped elucidate what is so fundamentally interesting and important about science. Nature has rules; people called scientists are trying to figure out what those rules are. How wonderful.

a

b

Figure 29.36
Inland Wetland and Coastal Wetland

a A bird's nest in a prairie wetland in the United States.

b A view from an observation tower across the sawgrass prairie of Florida's Everglades National Park.

Chapter Review

Key Terms

abiotic 645

algal bloom 670

aquifer 650

artificial
    eutrophication 670

atmosphere 656

benthic zone 668

biodegradable 654

biogeochemical
    cycling 646

biomass 647

biome 662

biosphere 673

biotic 645

carnivore 653

CFCs 657

climate 661

coastal wetlands 672

coastal zone 667

consumer 652

coral reef 669

decomposer 653

desertification 666

detritivore 653

ecosystem 645

element 646

energy-flow model 654

estuary 672

eutrophic 670

food web 653

gross primary
    production 655

groundwater 650

herbivore 653

inland wetlands 672

intertidal zone 668

kilocalorie 654

limnetic zone 669

littoral zone 669

net primary
    production 655

nitrogen fixation 648

nutrients 646

oligotrophic 670

omnivore 653

open sea 668

ozone 657

Summary

29.1 The Ecosystem Is the Fundamental Unit of Ecology

- An ecosystem is a self-sustaining community of organisms and the physical environment with which these organisms interact.

- Very large-scale ecosystems are known as *biomes*, which are defined by the types of vegetation within them.

29.2 Abiotic Factors Are a Major Component of Any Ecosystem

- The approximately 30 chemical elements that are vital to life are known as *nutrients*. Along with water, nutrients move back and forth between abiotic (nonliving) and biotic (living) domains on Earth in a process called *biogeochemical cycling*.

- Elements can be locked up in living things, transferred between them, or transferred between them and the abiotic sphere.

- Carbon comes into the living world through the plants, algae, and bacteria that take in atmospheric carbon dioxide to perform photosynthesis. Animals obtain their carbon from these photosynthesizing organisms. Carbon moves back into the atmosphere in the form of carbon dioxide, which plants and animals give off as a product of their respiration, and as a product of their decomposition after death.

- Carbon dioxide makes up a small but critical proportion of the Earth's atmosphere, because it is vital to life and may greatly affect global temperature.

- Prior to this century, nitrogen entered the biotic domain mostly through the action of certain bacteria that have the ability to convert atmospheric nitrogen into forms that can be used by living things. Other bacteria have the ability to convert this organic nitrogen back into atmospheric nitrogen, completing the cycle. Human beings now use industrial processes to convert nitrogen into a biologically usable form, thus greatly expanding agricultural productivity.

- As with carbon or nitrogen, all of Earth's water is either being cycled or is locked up—in such forms as glaciers or polar ice. As little as 0.5 percent of Earth's water is available as fresh water, with about 25 percent of this being groundwater.

29.3 How Energy Flows through Ecosystems

- Plants and other photosynthesizers are an ecosystem's producers, while the organisms that eat plants are its consumers. Every ecosystem has a number of feeding or trophic levels, with producers forming the first trophic level and consumers forming several additional levels.

- A detritivore is a class of consumer who feeds on the remains of dead organisms or cast-off material from living organisms. A decomposer is a special kind of detritivore that breaks down dead or cast-off organic material into its inorganic components, which can then be recycled through an ecosystem.

- The energy-flow model of ecosystems provides ecologists with a powerful analytical tool; it measures energy as it is used by and transferred among different members of an ecosystem.

- Very little of the energy that a given trophic level receives is passed along to the next trophic level. A rule of thumb in ecology is that for each jump up in trophic level, the amount of available energy drops by 90 percent. This explains why there are many plants on Earth relative to the number of large animals.

29.4 Earth's Physical Environment

- The atmosphere is the layer of gases surrounding the Earth. The lowest layer of the atmosphere, the troposphere, contains the bulk of the atmosphere's gases. Nitrogen and oxygen make up 99 percent of the troposphere, with carbon dioxide and small amounts of such gases as argon and methane making up the rest.

- The gas called *ozone* screens out 99 percent of the sun's potentially harmful ultraviolet radiation. Human-made compounds such as chlorofluorocarbons (CFCs), used in various consumer and industrial products, can destroy ozone. International action taken on this issue in the 1980s seems to be reducing the amount of ozone-destroying compounds being put into the atmosphere.

- There is a debate about whether human-caused global warming is taking place, but most scientists now believe this is the case. The probable cause of this warming is the greenhouse effect, in which higher atmospheric concentrations of gases such as carbon dioxide—released through fossil-fuel burning—are trapping additional quantities of the heat that comes to the Earth from the sun.

- Earth is tilted at an angle of 23.5° relative to the plane of its orbit around the sun, a fact that dictates much about climate on Earth, which in turn dictates much about life on Earth.

- Sunlight strikes the equatorial region of the Earth more directly than the polar regions. The differential warming that results produces a set of enormous interrelated circulation cells of moving air, each existing all the way around the globe at its latitude. Each of these cells drops rain on the Earth where it rises, but dries the Earth where it descends. This is why some regions of the Earth get so much more rainfall than others.

- More rain will be deposited on the windward side of a mountain range than the opposite leeward side. This rain-shadow effect can cause opposite sides of a mountain range to have dramatically different vegetation patterns.

- A climate is an average weather condition in a given area. Large vegetative formations essentially are defined by climate regions.

29.5 Earth's Biomes

- A biome is an ecosystem defined by a large-scale vegetation formation.

- Six types of biomes are recognized at a minimum: tundra, taiga, temperate deciduous forest, temperate grassland, desert, and tropical rainforest. Polar ice and mountains often are recognized as separate biomes, as are the tropical grasslands called tropical savannas, and the dry, shrub-dominated formations called chaparral.

- Tundra is the biome of the far north, frozen much of the year, but with a seasonal vegetation formation of low shrubs, mosses, lichens, grasses, and the grass-like sedges.

- Taiga is another biome of the north; it includes the enormous expanse of coniferous trees that lies south of the tundra at northern latitudes. The taiga exhibits a great deal of species uniformity because only a few types of trees—spruce, fir, and pine—are the ecological dominants. The region supports large populations of fur-bearing animals.

- Temperate deciduous forests grow in regions of greater warmth and rainfall than is the case with tundra or taiga. These forests, existing over much of the eastern United States, are composed of an abundance of trees such as maple and oak, complemented by a robust understory of woody and herbaceous plants.

- Temperate grassland goes by several names around the world, including prairie and steppes. This biome is characterized by less rainfall than that of temperate forest and by grasses as the dominant vegetation formation. Such regions can be very fertile agricultural land.

- Chaparral is a biome dominated by evergreen shrub vegetation. It is found in a few, relatively small regions of the world that have a Mediterranean climate. These regions have mild, rainy winters and very dry summers.

- Deserts are characterized by rainfall that is low and sporadic. Deserts may be hot, cold, or temperate, but all desert life is shaped by the need to collect and conserve water. Desertification refers to the process by which an area may be transformed into a desert, sometimes through human activity.

- The tropical rainforest biome is characterized by warm, stable temperature, abundant moisture, great biological productivity, and great species diversity. Rainforest productivity is concentrated above the forest floor, often in the canopy high aboveground. Found in Earth's equatorial region, rainforests are being greatly reduced through cutting and burning.

29.6 Life in the Water: Aquatic Ecosystems

- Ocean or marine ecosystems are most biologically productive near the coasts, with the deep open oceans having a productivity that can be less than that of deserts. Coastal areas benefit from wave actions that bring in nutrients, carry away wastes, and expose more of the surface area of photosynthesizers to sunlight.

- Modern fishing techniques have seriously depleted many once-abundant commercial fish species.

- There is relatively more pelagic ocean life toward the poles with this concentration decreasing with distance from the equator. The food webs in the oceans surrounding Antarctica are based on photosynthesizing phytoplankton, meaning floating microscopic organisms such as algae and cyanobacteria.

- Coral reefs are warm-water marine structures composed of the piled-up remains of generations of coral animals. Coral reefs provide a habitat that results in a rich species diversity.

- Freshwater ecosystems, which include inland lakes, rivers, and other running water, cover only about 2.1 percent of Earth's surface.

- Freshwater lakes are most productive near their shores and near their surface. Lakes can be naturally eutrophic, meaning nutrient-rich, or oligotrophic, meaning nutrient-poor. The human activity known as *artificial eutrophication* can sometimes introduce too many nutrients into a lake, which can cause some species to flourish (algae, bacteria) while harming other species (some fish).

- Estuaries are areas where streams or rivers flow into the ocean. They are characterized by high biological productivity because of the constant movement of water within them.

- Wetlands, also known as swamps or marshes, are lands that are wet for at least part of the year. Wetland soil may merely be waterlogged for part of the year or under a permanent, relatively deep cover of water. Wetlands are very productive and are important habitats for migratory birds. The amount of wetlands in the United States is estimated to have been reduced by 55 percent since the arrival of the Europeans.

Understanding the Basics

Multiple-Choice Questions

1. In nature, it is _____ that are primarily responsible for converting atmospheric nitrogen into a form that living things can use
   a. plants   d. animals
   b. algae   e. fungi
   c. bacteria

2. In recent years there has been substantial concern about global warming, which is caused to a great extent by the build-up of the heat-trapping gas
   a. oxygen
   b. nitrogen dioxide
   c. sulfur dioxide
   d. carbon dioxide
   e. neon

3. You walk in on the middle of a nature program on television and see aerial shots of huge expanses of one or two types of coniferous trees. The type of biome you are looking at probably is
   a. chaparral
   b. taiga
   c. tundra
   d. savanna
   e. temperate forest

4. There is a simple food chain in an eastern Oregon desert grassland: mice are consumed by several species of snakes, which themselves are consumed by hawks. The hawk is a _____ and occupies the _____ trophic level in this ecosystem.

a. tertiary consumer, fourth
b. secondary consumer, fourth
c. tertiary consumer, third
d. secondary consumer, third
e. primary consumer, third

5. A dung beetle (can you guess what it eats?) is an example of a
a. detritivore
b. omnivore
c. carnivore
d. autotroph
e. homeotherm

6. The so-called greenhouse effect is caused by
a. incoming long-wave radiation that is trapped as short-wave radiation by the atmosphere
b. short-wavelength light that is re-radiated as long-wave radiation and is trapped by certain gases
c. heat that is stored as short-wave radiation by rough surfaces of the earth and then re-radiated as long-wavelength light
d. cold air masses that mix with the warm gases and produce warm atmospheric gases
e. reactions in the stratosphere between these greenhouses gases and ozone, producing long-wave heat

7. Donna is measuring rainfall over a large elevation gradient in the Cascade Mountains of Oregon. As she walks up the west side, she notices that rainfall increases and that when she reaches the top and walks down the east side, it is much drier. This is because
a. dry air from the equator flows east at this latitude, dropping moisture on the west side
b. wet air from the tropical equator flows east at this latitude, dropping moisture on the west, and then becomes dry by the time it reaches the eastern side

c. air is flowing from the west and drops moisture as it rises and cools, then absorbs moisture from the land as it descends and warms
d. warm air from 30° latitudes picks up moisture as it travels to 60° north, then drops this moisture as it cools
e. the rain shadow effect means that cool, moist air drops rainfall as it warms as elevation rises and the air travels closer to the sun

8. Microscopic floating photosynthesizers such as algae that are the base producers in the oceans are known as
a. pelagic producers
b. secondary producers
c. littoral consumers
d. zooplankton
e. phytoplankton

9. A eutrophic lake will have
a. high nutrient levels, large phytoplankton populations, and low oxygen levels at depth
b. high levels of nutrients, low phytoplankton levels, high oxygen levels in surface waters
c. low nutrient levels, large phytoplankton populations, and low oxygen levels at depth
d. low nutrient levels, low phytoplankton populations, and high oxygen levels at depth
e. low nutrient levels, large zooplankton populations, and low phytoplankton levels

Brief Review

1. In a large reserve in Idaho, Forest Service Wildlife scientists measured wolf populations as well as elk populations, and found that the wolf population was much lower than that of the elk population. If wolves feed on elk, why aren't the population sizes similar?

2. Explain why most deserts across the globe occur at about 30° north and south of the equator.

3. What relationship does climate have to the large-scale vegetation patterns that are seen on Earth?

4. All of a sudden a great many dead fish are found floating on the surface of a pond. What might be the cause of such a fish-kill?

5. Name several ways in which the tundra and taiga biomes differ.

Applying Your Knowledge

1. You have learned that ozone is essential to protect life against ultraviolet radiation, and that CFCs deplete ozone, with the result that CFCs have been banned in order to protect ozone. But in Los Angeles and other large cities, there are strict air emission laws to stop excess production of ozone. What is going on here?

2. One solution that has been proposed to the problem of global warming is to "grow trees." What is the basis of this argument? Knowing that there are different possible fates for different trees, explain if growing trees is a reasonable solution to global warming.

3. Why have some ecologists argued that organic fertilizer ("green manures"—plant debris, compost, or animal wastes) are better for the soil than Haber-process fertilizer?

4. Slash-and-burn agriculture is common in tropical regions: Farmers cut rainforest trees, burn them, plant crops for 1–2 rotations, and then move on. Why is this so common: In other words, why are the trees burned and not sold, and why must the farmers then move on?

Group Activity

Desertification is usually defined as the transformation of an area into a desert, perhaps due to human activity. Recent research suggests that desertification is the result of complex interactions between physical, chemical, biological, socio-economic and political climate of a region, and that people not only cause desertification, they are also its victims. In small groups, research different aspects of global desertification; United Nations reports on desertification are available on the Internet. Have one group research climate variability and desertification; one group should explore the putative link between overgrazing and desertification in the southwestern U.S.; one group can research land-use, population growth, and other socio-economic aspects of desertification.

Appendix

Metric–English System Conversions

Length

1 inch (in.) = 2.54 centimeters (cm)
1 centimeter (cm) = 0.3937 inch (in.)
1 foot (ft) = 0.3048 meter (m)
1 meter (m) = 3.2808 feet (ft) = 1.0936 yard (yd)
1 mile (mi) = 1.6904 kilometer
1 kilometer (km) = 0.6214 mile (mi)

Area

1 square inch (in.^{2}) = 6.45 square centimeters (cm^2)
1 square centimeter (cm^2) = 0.155 square inch (in.^{2})
1 square foot (ft^2) = 0.0929 square meter (m^2)
1 square meter (m^2) = 10.7639 square feet (ft^2) = 1.1960 square yards (yd^2)
1 square mile (mi^2) = 2.5900 square kilometers (km^2)
1 acre (a) = 0.4047 hectare (ha)
1 hectare (ha) = 2.4710 acres (a) = 10,000 square meters (m^2)

Volume

1 cubic inch (in.^{3}) = 16.39 cubic centimeters (cm^3 or cc)
1 cubic centimeter (cm^3 or cc) = 0.06 cubic inch (in.^{3})
1 cubic foot (ft^3) = 0.028 cubic meter (m^3)
1 cubic meter (m^3) = 35.30 cubic feet (ft^3) = 1.3079 cubic yards (yd^3)
1 fluid ounce (oz) = 29.6 milliliters (mL) = 0.03 liter (L)
1 milliliter (mL) = 0.03 fluid ounce (oz) = 1/4 teaspoon (approximate)
1 pint (pt) = 473 milliliters (mL) = 0.47 liter (L)
1 quart (qt) = 946 milliliters (mL) = 0.9463 liter (L)
1 gallon (gal) = 3.79 liters (L)
1 liter (L) = 1.0567 quarts (qt) = 0.26 gallon (gal)

Mass

1 ounce (oz) = 28.3496 grams (g)
1 gram (g) = 0.03527 ounce (oz)
1 pound (lb) = 0.4536 kilogram (kg)
1 kilogram = 2.2046 pounds (lb)
1 ton (tn), U.S. = 0.91 metric ton (t or tonne)
1 metric ton = 1.10 tons (tn), U.S.

Metric Prefixes

Prefix	Abbreviation	Meaning
giga-	G	$10^9 = 1{,}000{,}000{,}000$
mega-	M	$10^6 = 1{,}000{,}000$
kilo-	k	$10^3 = 1{,}000$
hecto-	h	$10^2 = 100$
deka-	da	$10^1 = 10$
		$10^0 = 1$
deci-	d	$10^{-1} = 0.1$
centi-	c	$10^{-2} = 0.01$
milli-	m	$10^{-3} = 0.001$
micro-	μ	$10^{-6} = 0.000001$

Answers

CHAPTER 1
Multiple Choice
1. c, 2. b, 3. b, 4. d, 5. e.

Brief Review
1. Science is a body of knowledge and a formal process for acquiring that knowledge. It is thus a collection of unified insights about the physical world, and also a way of learning. Like belief systems such as religion, science is an attempt to understand and explain the physical reality of the world around us and our experience of it. Science differs from religious belief, however, in that it deals exclusively with physical phenomena and does not claim to describe a non-material or metaphysical world. Indeed, the existence of such worlds is not a question taken up by science. Further, science does not accept descriptions of physical reality that are based on faith, nor does it recognize any entity as an unquestionable authority. Scientific information is not immutable; it is open to challenge by all, and is regularly challenged by its own practitioners, scientists.
2. An experiment is said to be controlled when the experimenter controls the factors in it that will change and those that will be kept constant. Suppose a scientist wanted to test the effect of the insecticide DDT on the reproductive development of lab rats. In this case, the experiment might consist of feeding DDT-laced food to one group of rats, while another similar group—called the control group—would receive the same food without DDT. It is important that the two groups of rats be as closely matched as possible in regard to all other factors. For instance, both groups must consist of the same strain of lab mice; both must be the same gender or mix of genders; they must be given the same amount of food and water, and housed in identical environments. Suppose differences were observed in, for example, the size of the offspring born to one group of rats as opposed to the other. Were the starting conditions not controlled in the experiment, there would be no way to tell whether the DDT or some other factor (such as different housing) was responsible for these differences.
3. Pasteur investigated a widely held notion of the time that living things can arise spontaneously from basic chemicals. He hypothesized that instances where life appeared to arise in medium believed to lack it could be readily explained as the result of contamination from unseen airborne microscopic organisms. He sterilized meat broth in a flask with a long, S-shaped neck, by heating both the broth and the flask. Although the mouth of the long tube remained open, any incoming dust particles laden with microscopic organisms would be trapped in the bent neck and could not gain access to the meat broth. Under these conditions, the broth remained free of any visible signs of life. In other tests, the S-shaped neck was broken off or the flask tilted so that broth made contact with the neck of the flask; in these instance, the broth came "alive" with teeming bacteria and fungi. Pasteur inferred from these experiments that the appearance of life forms depended on whether the broth was contaminated by airborne microscopic organisms or not. In instances in which the air-borne organisms had no access to the broth, however, it remained sterile. There was no evidence, in other words, of life coming about through spontaneous generation.
4. All living things, as we know them on this planet, acquire, store, and use energy through metabolism; they respond to stimuli in their environment; they maintain a relatively constant internal environment (homeostasis); they reproduce, using DNA as a means for transmitting hereditary information; they exhibit cellular organization; and they show evidence of having evolved, or changed over time.
5. Life can be analyzed at the minutest level of organization—atoms and molecules—or at successively larger and more complex levels. Atoms and molecules are components of organelles, which in turn are constituents of cells, the basic functional unit of life. In multicellular organisms, cells are organized into distinct tissue types. Collections of various tissues compose the organ systems of plants and animals. The total assemblage of organ systems constitutes an individual organism. All the individuals of the same species living in a defined area are referred to as a population of that organism. Populations of different types of organisms living in a defined area compose a community. All the communities of the Earth, together with their physical environment, make up the biosphere.

CHAPTER 2
Multiple Choice
1. b, 2. b, 3. c, 4. a, 5. d, 6. b, 7. e, 8. a.

Brief Review
1. The forms are called isotopes. Isotopes of an element differ from one another in accordance with the number of neutrons they have. A regular carbon atom has 6 neutrons in its nucleus while carbon-14 atom has eight neutrons.
2.

Ionic Bonds   Polar Covalent   Nonpolar

Covalent ———————————— No charge
(complete +
or – charge)

3. Free radicals exist when one of the atoms in a covalently bonded molecule has an unpaired electron. One free radical can produce many more free radicals in a chain reaction. Free radicals can do damage by irritating artery walls or by causing mutations in DNA.
4. An atom's electrons move through volumes of space outside the atom's nucleus. An entire atom is about 100,000 times larger than its nucleus. Thus, most of an atom is the space through which electrons move.
5. When the valence (outer) shell is filled, the atom is at a lower, more stable energy state.

CHAPTER 3
Multiple Choice
1. e, 2. a, 3. e, 4. c, 5. e, 6. b, 7. b, 8. d, 9. a.

Brief Review
1. Water absorbs tremendous amounts of heat from the sun, and releases this heat slowly, with the onset of night's colder temperature. The perspiration human beings throw off carries with it a great deal of heat, which has been absorbed by the water.
2. LDL is associated with deposition of cholesterol in the arteries, so you should be concerned about high levels. HDL is associated with carrying cholesterol to the liver, away from the arteries where it does damage. A high HDL count relative to LDL count is in general a good sign.
3. Table 3.2 lists several protein functions, including enzymes, transport proteins, structural proteins, contractile proteins, etc. Proteins can do many different kinds of jobs because they take on so many different shapes. Each shape allows the protein to do a different job.
4. The water molecules cling together so tightly that when each molecule evaporates from a plant leaf, it draws up a chain of water molecules that extends all the way down to the root.
5. Strong hydrogen bonds help explain why water is so extremely cohesive, has such a strong surface tension, and has such a high specific heat.

CHAPTER 4
Multiple Choice
1. e, 2. d, 3. a, 4. b, 5. c, 6. d, 7. b, 8. e.

Brief Review
1. The cell would have difficulty obtaining materials from the external environment because its surface-to-area ratio would be too low.
2. Eukaryotes have a nucleus, bound within a thin membrane, that contains almost all their DNA. The DNA of prokaryotes is localized (in a "nucleoid" region), but it is not bound within a membrane. Eukaryotes are usually much larger than prokaryotes. Eukaryotes are also quite often multicelled organisms, while prokarotes are single-celled. Most eukaryotes need oxygen to exist, whereas many prokaryotes can survive without oxygen.
3. The DNA code is copied into an RNA molecule, which leaves the nucleus through the pores in the nuclear membrane. The RNA is then "read" and translated into a molecule similar to insulin in a ribosome on the rough ER. This molecule is dumped into the cisternae of the ER, where it folds and may be modified. This molecule is then encased in a membrane for shipment to the Golgi complex. The Golgi complex further modifies the molecule and packages it for export out of the cell in a vesicle.
4. Cilia and flagella are cell extensions formed by microtubules that are either a means of propulsion or, in the case of some cilia, a means of moving material around a cell. Cilia exist as a profusion of hair-like growths extending from cells that move back and forth very rapidly, while only a few flagella sprout from a given cell—indeed, there will often be just a single tail-like flagellum. In the human body, cilia are found in the nasal passages and in the lungs. The only animal cell that has a flagellum is a sperm cell.
5. Each cell would be isolated from each other cell by two thick cell walls. Thus plants could not actually be an organism if there were no plasmodesmata. Each cell would have to be pretty much independent of the rest.
6. A gap junction consists of clusters of protein structures that shoot through the plasma membrane of a cell from one side to the other. When these structures line up in adjacent cells, they form a channel for passage of small molecules and electrical signals.

CHAPTER 5
Multiple Choice
1. c, 2. a, 3. b, 4. e, 5. d, 6. b, 7. b, 8. c, 9. a, 10. c.

Brief Review
1. There are receptor proteins on the surface of the taste bud cell that recognize sugar molecules. When this recognition occurs, the cell changes in a way that allows a message to travel to the brain and convey that what you ate was sweet. Because salt receptors are not on that cell, it responds to sugar but not salt.
2. Because steroids are lipids, they pass easily across the phospholipid bilayer by dissolving in it. Thus they can affect every single cell of the body.
3. Because the phospholipid "heads" are polar or hydrophilic, they are attracted to the polar water molecules on the outer edges (outside and inside) of the membrane. Because the "tails" are very hy-

drophobic, they tend to be pushed away from the water to the inside of the membrane.

4. Receptor proteins are not the same for every cell. Thus a cell can be "tailored" to respond to a specific hormone or group of hormones by the specific receptor proteins on its surface.

5. You would not expect that communication and recognition, and possibly transport, would be as important inside the cell as on the outside. Thus the glycocalyx would be unnecessary, and the proteins in the membranes would serve different functions.

6. When you shake the tray gently, the marbles move until they are fairly evenly spaced across the tray. In diffusion, molecules or other small particles such as ions hit against each other until they eventually bump each other to places that are roughly equidistant. This is purely a random process. Because temperature is a measure of the speed of molecules and is the result of the energy of the molecules, shaking the tray is roughly like changing the temperature of molecules—the harder you shake, the faster the rate of diffusion.

7. Your stomach cells carry out active transport, which requires energy. In this process, the cells may also need a pumping protein in the membrane.

CHAPTER 6
Multiple Choice
1. a, 2. c, 3. d, 4. d, 5. e, 6. e, 7. a, 8. c.

Brief Review

1. Plants abide by the second law of thermodynamics in that, in building themselves up, they are contributing to the total amount of entropy in the universe; in building a leaf from carbon dioxide and water, heat will be released. A plant leaf represents only a local increase in order.

2. This is a coupled reaction in that the breakdown of ATP to ADP is an exergonic or downhill reaction that powers an endergonic or uphill reaction—the movement of sodium ions against their concentration gradient to the outside of the cell.

3. The energy stored and released by ATP comes from food. ATP is an energy "currency" molecule in that it facilitates an organism's expenditure of energy on various processes. Just as the value inherent in labor is put into a medium of exchange—money—that can be easily expended, so the energy in food needs to be put into the form of ATP to be easily used by living things.

4. Enzymes are involved with every aspect of cellular life. They are necessary for breaking things down, but they are also used to make larger molecules out of smaller ones.

5. Allosteric regulation of an enzyme occurs when the product of a reaction facilitated by an enzyme binds to the enzyme itself and changes its shape, thus making the enzyme unable to bind to its substrate.

CHAPTER 7
Multiple Choice
1. b, 2. d, 3. e, 4. a, 5. d, 6. b, 7. d, 8. a.

Brief Review

1. To get a trampoline to bounce you into the air, you put energy into it by jumping first. The trampoline stops bouncing when you quit jumping. The electrons from glucose are losing energy (just like you do when you jump). As they lose energy, some of it is transferred to ADP to make ATP (just as the energy from your jump is transferred to the trampoline, and from there to your body, to shoot you back into the air).

2. Muscles require a lot of energy. Thus they need mitochondria to carry out the Krebs cycle and electron transport, which extract the most energy from glucose.

3. We cannot make ATP without a source of energy. Foods supply the energy so that we can make ATP.

4. Water is produced in respiration at the very end of the process, when the electrons from the ETC are passed to oxygen, which then combines with hydrogen to make water.

5. Like NAD^+, FAD serves to carry electrons (and the energy stored in them) from the Krebs cycle to the ETC. Therefore, if your supply of FAD is inadequate, the Krebs cycle and electron transport cease, and you will be able to get only small amounts of energy out of food.

CHAPTER 8
Multiple Choice
1. e, 2. a, 3. c, 4. d, 5. e, 6. c, 7. b.

Brief Review

1. Yes, plants are as dependent on cellular respiration as we are. Plants make food—carbohydrates—in photosynthesis. Just as we have to break down the food we eat to get the energy out and into ATP, so plants have to break down the food they make to get the energy out and into ATP.

2. The sunlight that makes it to the Earth's surface is composed of a spectrum energetic rays (measured by their "wavelengths") that range from very short, ultraviolet rays, through visible light rays, to the longer and less energetic infrared rays. The photosynthesis we are concerned with is driven by part of the visible light spectrum—mainly by blue and red light of certain wavelengths.

3. In a photosystem, several hundred pigment (chlorophyll *a* and accessory pigment) molecules serve as antennae to capture light energy. The molecules pass their excited electrons, still in the excited state, to two chlorophyll *a* molecules, called the reaction center. These pass the excited electrons to a receptor molecule, called the primary electron acceptor. All of these parts are necessary: If there were no antennae molecules, the light would have to strike the reaction center directly each time, slowing photosynthesis significantly. The reaction center is essential because its molecules pass the electrons on. Without the electron acceptor, energy would not be made available to make sugars.

4. There would soon be no life, at least as we know it. Because most organisms on Earth depend on the energy harnessed in photosynthesis, soon all animals and humans would die off. Algae would die along with plants. Fungi and many bacteria would eventually die when all the other dead organisms had been decomposed.

5. In a bright, moderately hot environment, the C_4 plant would grow most rapidly. But the C_3 plant would grow fastest in cloudy or shady conditions, in which the C_4 plant would be spending too much energy moving compounds around in the plant leaf. The CAM plant, which always photosynthesizes slowly, would grow most rapidly in the desert. In other words, each plant grows most rapidly in the area to which it is best adapted.

6. 1. Rubisco adds CO_2 to RuBP to form two molecules of 3PGAL. 2. Energy is added to convert to G3P. 3. G3P is used to supply energy and/or matter to the plant for all the processes that plants carry out. 4. Most of the G3P is used to make more RuBP so that the cycle can continue. The first two steps add matter (CO_2) to the energy and hydrogens brought over from the light-dependent reactions. The third step allows the plant to use this energy. The fourth step allows the process to continue. Energy is like money to the cell; plants and animals are very unlikely to waste it. Occasionally, cells waste energy because there is no simple way to avoid it, but typically organisms guard their use of energy carefully.

CHAPTER 9
Multiple Choice
1. a, 2. d, 3. d, 4. c, 5. b, 6. a, 7. d, 8. b, 9. b, 10. e, 11. c.

Brief Review

1. The cell has been arrested in metaphase. There are eight chromatids present in this stage, and each daughter cell will have four chromosomes.

2. Mitosis serves to equally divide a cell's genetic material, while cytokinesis divides the cell's cytoplasm.

3. binary fission

4. Interphase. Interphase is the longest phase of the cell cycle. During this time, the cell may be in a period of "rest," or cell growth without division, called G_1. If the cell is an actively dividing cell, chromosome duplication will occur during the S phase, followed by additional cell preparations for mitosis in the G_2 phase.

   Prophase. Duplicated chromosomes condense and become visible, centrosome duplicates, centrioles migrate to poles, nuclear membrane breakdown, and the mitotic spindle is formed.

   Metaphase. Chromosome pairs align along the center of the cell, forming the metaphase plate, and attach to the microtubules extending from the cellular poles.

   Anaphase. Separation of sister chromatids paired along the metaphase plate, and their migration to the separate poles of the cell.

   Telophase. Telophase is the final stage of mitosis. It involves the unwinding of the condensed chromosomes at their respective poles, and the formation of new nuclear membranes. At this point, division of genetic material is complete. The cell is ready for cytokinesis.

5. |-A-T-|
   |-A-T-|
   |-C-G-|
   |-T-A-|
   |-A-T-|
   |-G-C-|

6. Plant cells have cell walls, animal cells do not. To compensate, membrane lined vesicles migrate to the center of the dividing cell, near the metaphase plate. These vesicles form the flat cell plate. As the flat cell plate grows toward the original cell wall, the membrane of the flat cell plate fuses with the membrane of the original cell. This action completes cell division, resulting in two new daughter cells and a new cell wall.

7. CDK stands for cyclin-dependent kinases. The CDK complex acts to control cell division.

8. Prokaryotes only have one circular chromosome, while eukaryotes have several linear chromosomes.

9. (a) four different nitrogen-containing bases (adenine, thymine, guanine, and cytosine); (b) in the nucleus; (c) "messenger"; (d) mRNA is used as a cellular "messenger," copying the DNA information in the nucleus and carrying it to the cytoplasm, where the message is used by ribosomes to assemble amino acids into proteins. (e) amino acids; (f) in the cytoplasm.

CHAPTER 10
Multiple Choice
1. b, 2. a, 3. c, 4. a, 5. b, 6. a, 7. e, 8. d, 9. a, 10. e.

Brief Review

1. Chromosome reduction takes place during meiosis I.

2. No. Asexual reproduction also occurs in single celled prokaryotes, as well as some eukaryotes. Regeneration occurs in higher animals.

3. The cells in males that go through meiosis, giving rise to immature sperm, are called primary spermatocytes and are themselves the product of cells called spermatogonia. These spermatogonia can give rise not only to primary spermatocytes but to more spermatogonia as well.

4. Meiotic cytokinesis differs from mitotic cytokinesis in the selective shunting of cytoplasm. In mitosis, cytokinesis divides the cytoplasmic contents equally, while in meiotic cytokinesis, cytoplasm is preferentially shunted to the cell destined to be the ovum, and away from the three polar bodies.

5. Somatic cells are diploid cells that undergo mitosis. They include every cell in the body except the reproductive cells. Gametes are haploid reproductive cells, resulting from the process of meiosis: eggs and sperm in the female and male, respectively.

6. (a) metaphase I, meiosis I; (b) anaphase II, meiosis II; (c) anaphase, mitosis; (d) metaphase II, meiosis II; (e) metaphase, mitosis

7. (a) n=3, metaphase of meiosis II

(b) 2n=8, prophase meiosis I

8. Anaphase of meiosis II is most like anaphase of mitosis, because both involve the separation of sister chromatids—in anaphase of meiosis I, homologous chromosomes separate, but not sister chromatids.

9. Sperm contribute only DNA, packaged in chromosomes, to the succeeding generation. In addition, they must make a journey to an egg in order to fertilize it, thus making small size an advantage. By contrast, eggs need to have stored within them a rich mixture of nutrients and other materials to support the newly formed offspring in its early stages.

10. (a) gamete: a reproductive cell (sperm and eggs); (b) crossing over (or recombination): a process during meiosis in which homologous chromosomes exchange reciprocal parts with one another. (c) asexual reproduction: production of a new organism without fertilization. (d) haploid: a cell or organism containing half the original genetic material. Each cell contains only one chromosome of each original "pair." Examples: gametes, gametophytes.

CHAPTER 11
Multiple Choice
1. a, 2. a, 3. e, 4. c, 5. d, 6. b, 7. b, 8. c, 9. d, 10. e.

Brief Review
1. Phenotype is the representation of visible traits in an organism (that is, yellow vs. green peas), while genotype is the genetic composition responsible for the phenotype of a trait (that is, YY or Yy = yellow peas, yy = green peas).

2. Dominant is used to refer to a trait that is equally expressed in either the homozygous or heterozygous genotype (i.e., YY or Yy = yellow peas). Recessive is the term used to describe a trait that will be expressed only in the homozygous form. Recessive traits are masked in the heterozygous genotype (i.e. yy = green peas, Yy (green peas).

3. Codominance refers to the condition where a heterozygote will produce an equal amount of product from each allele it carries. This is exhibited in blood types, where an AB individual will produce both A and B cell surface molecules. Incomplete dominance is displayed when the heterozygote displays a phenotype that is an intermediate of the dominant and recessive phenotypes. An example is the generation of pink F_1 flowers resulting from a parental cross of one red-flowering and one white-flowering plant.

4. True.

5. The fragile-x syndrome demonstrates the genetic principle of pleiotropy, in which one gene has more than one effect. This differs from what Mendel observed in that, in his pea plants, one gene had one effect. There was a gene for flower color, for example, that came in alleles that either brought about purple or white flowers, but that did not then go on to affect other characters in the plant. By contrast, the gene connected to fragile-x syndrome has multiple effects—on intelligence and various physical characteristics.

6. In Mendelian genetics, two different genotypes may represent a dominant phenotype: YY and Yy.

7. Fifty percent of the progeny will be yellow peas.

8. Mendel's "element of inheritance" is now identified as the gene.

9. Mendel's first law (law of segregation): Members of a gene pair will segregate from each other during gamete formation such that each gamete receives one and only one of the original two genes from the gene pair. Mendel's second law (law of independent assortment): During gamete formation, gene pairs assort independently of one another.

10. (a) $Aa \times aa$; (b) $Zz \times Zz$; (c) $Bbcc \times bbCc$; (d) $EeFf \times EeFf$

CHAPTER 12
Multiple Choice
1. c, 2. e, 3. a, 4. d, 5. e, 6. d, 7. c, 8. e, 9. d, 10. a.

Brief Review
1. (a) autosomal dominant; (b) chromosomal deletion; (c) aneuploidy; (d) aneuploidy; (e) X-linked; (f) autosomal recessive; (g) aneuploidy; (h) X-linked; (i) X-linked

2. Male, because sex-linked diseases are often associated with the X chromosome. Because females have two copies of the X chromosome, if one X chromosome carries a mutation, the "normal" copy masks the effects of the mutated gene. However, males carry only one copy of the X chromosome, so if they have a mutated gene on their X, they don't have a normal copy to mask its effects.

3. The genes for red and green color vision are very closely aligned on the X chromosome, while the gene for blue vision is on chromosome 3. It would be possible to have dysfunctional genes for red and green but a functional gene for blue.

4. Deletions—a portion of the chromosome is completely lost.

   Inversions—a portion of the chromosome is broken off and rejoins the original chromosome, but in a reversed order.

   Translocations—occur when two chromosomes that are not homologous exchange pieces, leaving both with improper gene sequences.

   Duplications—a portion of a chromosome is duplicated, most often by gaining a chromosome piece during abnormal crossing over.

5. (a) non-disjunction; (b) polyploidy

6. Because people who have Klinefelter syndrome have a Y chromosome, which confers the male sex, but also have two copies of the X chromosome, rather than the one copy that males normally possess. Females normally have two copies of the X chromosome. Thus it makes sense that those with Klinefelter syndrome would exhibit some female characteristics.

7. A heterozygous genotype for hemoglobin would be advantageous in areas of mosquito infestation while neither homozygous genotype would be advantageous in such areas. Those who are homozygous for hemoglobin A would not have the increased resistance to malaria; those who are homozygous for hemoglobin S would inherit sickle-cell anemia. Those who are heterozygous would have malaria resistance and yet would not inherit sickle-cell anemia.

Genetics Problems
1. (a-I) 50 percent; aII. 0 percent; (b-I) $Xx \times XY$; (b-II) $xY \times XX$; (c-I) 50 percent; (c-II) 100 percent

2. (a) $bbEe \times bbEe$; (b) black, 0 percent; brown, 75 percent; golden, 25 percent; (c) no black puppies, 9 brown, 3 golden

3. (a)

   (b) Dominant. We know that both Barry's mother and father are affected by the condition. If it were a recessive condition, both parents would have to be homozygous for the trait, meaning all their offspring would have to inherit the condition as well. Yet they have produced one offspring (Barry) who is not affected. (c) genotypes: Barry's mom A; Barry's sister Aa or AA; Barry aa; Jan: aa. (d) There is no chance that Barry and Jan will have a child with the syndrome.

4. Mother, AO. (Her blood type can't be AA; if she were, the child would have to have at least one A allele.) Father, BO. (Same reasoning as above.) Child, OO. (A person who is type O cannot have inherited either the A or B allele.)

5. (a) Debbie is hh, June is probably Hh. (It is likely that Debbie's and June's mother is hh, because this is a rare disease, but one can't be sure. Thus, June could be HH.) (b) Debbie 0 percent, June 100 percent; (c) Debbie 0 percent, June 50 percent.

6. (a) 25 percent; (b) 50 percent; (c) 25 percent (males only); (d) 100 percent if either parent is homozygous dominant, 75 percent if both parents are heterozygous.

7. (a) Person I: A or a; Person II: a; Person III: aB and ab; Person IV: AB, Ab, aB, ab; (b) Person I with person II: $Aa \times aa$ will produce 50 percent Aa and 50 percent aa. Person III with person IV: $AaBb \times aaBb$ will produce about 13 percent $AaBB$, 25 percent $AaBb$, 13 percent $Aabb$, 13 percent $aaBB$, 25 percent $aaBb$, and 13 percent $aabb$.

8. (a) autosomal recessive, aa; (b) X-linked recessive, xY.

9. The parental generation cross is $TT \times tt$. The F_1 cross is $Tt \times Tt$. Remembering Mendel's 3:1 ratio, the outcome would be 150 tall plants (TT, Tt, Tt) and 50 short plants (tt). There would be no intermediate height plants, because there was no incomplete dominance in Mendel's crosses.

CHAPTER 13
Multiple Choice
1. b, 2. e, 3. d, 4. e, 5. c, 6. a, 7. b, 8. e, 9. d, 10. b.

Brief Review
1. "Something old, something new" indicates that the cell replicates its DNA in such a way that each of the two helixes formed contains one "old" handrail, and its associated bases and one "new" complementary set of bases and a handrail.

2. A point mutation may arise through an error in DNA replication (as described in this chapter); some of the chromosomal mutations described in Chapter 12 arose through an error in the process of crossing over while others arose from the faulty "disjoining" of chromosomes or chromatids during meiosis.

3. False—DNA polymerases have an editing function, moving backward to check for and correct mistakes.

4. False—a small proportion of mutations act to improve or change protein function, therefore contributing to the evolutionary process.

5. Even though there are only four bases, the order they occur in as they line up to form the DNA can vary extensively, thereby allowing for the production of a huge number of proteins.

6. Two examples given:
   Cancer point mutations—uncontrolled replication;
   Sickle-cell anemia—one A → T substitution leads to the disease.

CHAPTER 14
Multiple Choice
1. d, 2. d, 3. b, 4. c, 5. a, 6. e, 7. e, 8. d, 9. a, 10. b.

Brief Review
1. The replication of DNA and the production of proteins from DNA.

2. (a) mRNA (messenger RNA), which takes DNA information to ribosomes; (b) rRNA (ribosomal RNA), which helps make up the ribosomes that act the site of protein synthesis; and (c) tRNA (transfer RNA), which brings and amino acids to the ribosomes and binds with mRNA codons there.

3. Transcription is the process by which the information encoded in DNA sequences is put into the form of mobile mRNA sequences. Translation is the process by which the information encoded in mRNA sequences is put into the form of amino acid sequences; this results in the production of a protein.

4. Viruses require the use of host genetic machinery in order to function and replicate. If we used a different genetic code than viruses, we would be immune to virus invasion. As it is, we are susceptible because of the common code.

5. Introns are the sequences of mRNA that do not code for amino acid sequences. Existing in eukaryotes, they are edited out of the original mRNA sequence in the nucleus prior to the transport of mRNA out of the nucleus. Exons are sequences of mRNA that code for amino acids. They are the sequences that remain after mRNA is edited in the nucleus.

6. (a) Inducer—the substance that initiates the activity of genes not constantly expressed. (b) Operon—the complex genetic system that guides the activity of genes not constantly expressed. (c) Promoter—the sequence of DNA that binds the RNA polymerase upstream from the gene to be transcribed. (d) Operator—the region of DNA between the promoter and the gene to be transcribed. When the operator is bound by a repressor, the gene will not be transcribed; when it is not bound, the gene will be transcribed. (e) Repressor—the molecule that binds to the operator region of DNA, thus preventing transcription. Capable also of binding with the inducer, in which case it is released from the operator, which allows transcription to go forward.

7. Three, one.

8. Because the genetic code—this DNA triplet coding for that amino acid—is nearly universal in all living things, we infer that it must have been operating in an ancient organism that is ancestral to all the organisms alive today. Because DNA encodes information, and because we share in the genetic code, we share in an informational linkage that stretches back billions of years.

CHAPTER 15
Multiple Choice
1. c, 2. e, 3. b, 4. a, 5. d, 6. d, 7. c, 8. e, 9. e, 10. a.

Brief Review
1. It is useful because some human beings cannot turn out sufficient quantities of their own supply of needed human proteins; this is so with some diabetics and the protein insulin, for example. Meanwhile, the protein may available only in limited quantity—or with limited purity—from other sources. Thus it can be useful for human beings to

manufacture supplies of given proteins through the processes of biotechnology.

2. The DNA fragments differ in their length. Those fragments poured into a lane that make it farthest through the gel are the shortest fragments, while those that go least far through the gel are the longest fragments.

3. PCR is valuable because it allows scientists to quickly make a large quantity of a given segment of DNA from a very small starting sample. The text notes that this is useful in producing enough crime-scene DNA for testing, or for producing quantities of DNA from small amounts extracted from ancient artifacts.

4. Some limitations of the Human Genome Project are that (a) it provides the means of getting more complete knowledge, but not all the answers about human genetics; and (b) it may identify genes, but it does not identify their function or their effect on other genes.

5. Human gene transfer, or gene therapy, is the process of inserting working genes into people who suffer from a mutation, or absence of a functional gene. Procedure: Introduce gene into vector, vector introduces gene into human cells within the affected individual, human begins producing protein from the introduced gene.

6. Sticky ends are useful in restriction enzymes because they allow for the linking of specific sequences of DNA with one another.
   Fragment sequences:
   ATCG   GATCCTCCG
   TAGCCTAGGAG   GC

7. Dolly looked like the sheep who donated the udder cell, rather than the sheep who donated the egg, because the udder cell retained the DNA in its nucleus while the DNA in the egg was removed. Thus, all the instructions for Dolly's development came ultimately from the DNA in the udder cell.

CHAPTER 16
Multiple Choice
1. b, e; 2. a, b, d; 3. a, c, d, e; 4. Lyell—Geological forces observable today caused changes in Earth. Wallace—Natural selection is differential survival or reproduction. Lamarck—Inheritance of acquired characters. Cuvier—Catastrophic extinction and new creations explain fossil record. 5. Homologous: a, c, d; Don't belong together: b, e. 6. d; 7. a, b, d, e; 8. a, c, d; 9. a, c, d.

Brief Review
1. The term homology describes structures that are similar due to common ancestry. Homology provides evidence for evolution because it shows descent with modification.

2. Figure 16.13 shows that there are fewer nucleotide changes between humans and pigs than between humans and yeast.

3. (a) Placement of fossils agrees with dates of fossils from radiometric dating. (b) More nucleotide differences can be found between organisms that shared a common ancestor long ago (fossil record) and those that are more recently diverged.

4. Darwin brought back many birds that he thought were the blackbirds, wrens, and warblers with which he was familiar in Britain, but he found they were in fact all species of finches. Darwin perceived this as an example of descent with modification. Without the *Beagle* voyage, Darwin would not have been exposed to such extravagant examples of branching evolution, nor would he have made various other biological observations that spurred his thinking about evolution by natural selection.

CHAPTER 17
Multiple Choice
1. c; 2. gene pool—the set of all alleles in a population; allele—variant form of a particular gene;

allele frequency—the proportion of a given form of a gene in a population; genotype—the genetic makeup of an organism; gene flow—exchange of genes between populations; 3. a, c; 4. a, c, e; 5. d; 6. b, d; 7. b; 8. mutation—changes in the genetic material; gene flow—movement of alleles between populations by migration; natural selection—differential survival and reproduction; genetic drift—chance alterations of allele frequencies in a population; founder effect—genetic drift due to a few colonizing genotypes; 9. d; 10. c; 11. c.

Brief Review
1. An individual may be selected for survival—and hence greater reproduction—because of the qualities that individual has, such as coloration or height. However, this individual's characteristics do not change because it possesses these qualities. An individual does not evolve, in other words. An individual can pass on more of its genes relative to other individuals in a population, however, causing the population's allele frequencies to change, meaning the population has evolved.

2. A gene pool is the sum total of alleles that exist in a population. The inheritable characteristics of living things are based on the alleles they possess. Thus, for a population to change in a way that is passed on from one generation to the next—for the population to evolve—the mixture of alleles in that population must change. Thus, the evolution of a population is based at root on changes in that population's gene pool.

3. The five causes of microevolution are: 1. Mutation: changes the form of the genetic material that result in new alleles. 2. Gene flow: brings alleles from other populations into the target population. 3. Non-random mating: mating in which a given member of a population is not equally likely to mate with any other given member. Can change the genetic contribution individuals make to a succeeding generation. 4. Genetic drift: The chance alteration of allele frequencies in a small population. May occur if a new population is founded by only a few individuals (founder effect), or if the size of a population is decreased suddenly (the bottleneck effect). 5. Natural selection: differential survival and hence reproduction of individuals based on their particular phenotypes. If the phenotypic variation is inherited, the population will evolve to be like the individuals that survive longest and reproduce most.

4. An individual in a population has to have greater reproductive success relative to other members of the population in order to contribute more to the gene pool of the succeeding generation.

5. Mutation and genetic drift alter allele frequencies in random directions, and hence have nothing necessarily to do with adaptation. Gene flow doesn't necessarily bring in alleles that will be suited to the new environment, and non-random mating concerns mate choice, rather than to an organism's fit with its environment. Through natural selection, however, phenotypes of individuals that survive and reproduce best in a given environment spread through the population.

6. Evolutionary fitness—the relative success in contributing genes to the next generation—depends crucially on the environment. It is the fit between environment and organism that determines which organisms will have the greatest success in survival and reproduction. Thus, the evolutionary fitness of an individual could be expected to vary greatly from one environment to another. In the Grant's study of the Galapagos finches, a deeper bill was favored after the drought, but this situation later reversed.

CHAPTER 18
Multiple Choice
1. b; 2. b, c, d; 3. b, c, d; 4. a, c, d; 5. temporal isola-

tion—mating occurs at different times of day or year; ecological isolation—populations live in different environments; geographical isolation—populations are separated by physical barriers; gametic isolation—egg and sperm do not recognize one another; hybrid sterility—offspring are unable to reproduce; **6.** b, c, e; **7.** species, genus, family, class, phylum; **8.** b, d, e.

Brief Review

1. Convergent evolution is the evolution of a very similar character in two unrelated organisms, caused when the organisms experience similar types of natural selection in their environments. For example, many unrelated species of desert plants have succulent leaves that store water. However, these similar characters are not homologous (they are analogous), and so they cannot be used to determine relationships in a phylogenetic study.

2. In the gradualistic view, evolutionary changes in form are gradual through time, with no sudden changes and no long periods of constancy. In the punctuated equilibria view, organisms remain the same for long periods, and then evolutionary change occurs in a rapid burst, often associated with speciation.

3. Anagenesis is non-branching evolution, in which a new species may evolve that presumably would be reproductively isolated from its ancestral species. Cladogenesis is branching evolution, in which a single species may evolve into two or more species, the parent species continuing, while others branch off from it.

4. Allopatric speciation occurs when populations are separated by a physical barrier to gene flow. This allows independent evolution in the two groups, which eventually leads to the evolution of one or more intrinsic reproductive isolating mechanisms. In sympatric speciation, intrinsic isolating mechanisms evolve in the absence of geographical separation. Both forms of speciation require the evolution of one or more intrinsic reproductive isolating mechanisms.

5. In the biological species concept, two organisms are in the same species if they actually or potentially interbreed successfully in nature, while being reproductively isolated from other groups.

6. Without the evolution of intrinsic isolating mechanisms, two populations would simply interbreed, remaining one species, if the physical barrier to interbreeding were removed. Thus, they can't be called separate species unless they possess intrinsic reproductive isolating mechanisms.

CHAPTER 19
Multiple Choice
1. c, g, d, h, a, e, b, i, f; **2.** a, c, d ; **3.** b, e; **4.** b, c, e; **5.** a, b, d; **6.** False; **7.** b, c, d; **8.** a, b, d; **9.** b, e, g; **10.** a.

Brief Review

1. Eukarya have membrane-bound nuclei and cell organelles, and may be multicellular, whereas bacteria have no nuclei or organelles and are essentially single-celled.

2. The amniotic egg evolved in the Reptilia. It is an egg in which the embryo lives in a watery environment (albumin, the "white") while being enclosed in a protective leathery or hard shell. Special membranes facilitate gas exchange while protecting the embryo from drying out. This was a huge breakthrough for terrestrial animals, because it allowed them to lay eggs outside an aquatic environment without risking drying.

3. Primates have forward-looking eyes and an opposable thumb. This permits the evolution of sharp vision to the front, and the ability to pick up and manipulate objects, paving the way for the use of tools.

4. Humans and chimps.

5. No.

6. Life may have begun as early as about 3.9 billion years ago. Land was not colonized until about 460

million years ago. Between those times, all life existed the oceans, the land having no life on it.

7. Mammals have hair, which protects the skin. They have internal development of young in the uterus, and they have mammary glands that enable them to feed their offspring with milk.

CHAPTER 20
Multiple Choice
1. b, c; **2.** b, c; **3.** a, c, d, e; **4.** hyphae—filamentous strands of porous material; mycorrhizae—symbiotic infection of plant roots; cell walls—found only in plants; archaea—often occupy extreme environments; protists—ancestors of animals, plants, and fungi; **5.** c; **6.** False; **7.** a, b, e; **8.** a, b; **9.** False; **10.** False.

Brief Review

1. An extremophile is a prokaryotic organism (bacteria or archaea) that lives in an extremely harsh environment, such as a hydrothermal vent. Extremophiles live under conditions that are much too harsh for eukaryotes.

2. Viruses stimulate the immune system to produce specific antibodies and T cells to recognize and eliminate a specific virus. Even after the infection is cleared, some of these highly specific proteins and cells remain. If the same virus invades again, it is wiped out before it has a chance to become established. Vaccines made from killed virus or from weakened virus can provoke the immune response and thus offer protection from the virus.

3. Viruses cannot replicate themselves without being in a host cell. The ability to self-replicate is generally considered to be a fundamental, defining feature of life.

4. Phytoplankton do much of the photosynthesis on Earth, and they are at the base of many aquatic food chains.

5. Some fungi interact with plant roots, assisting them in their water uptake. Fungi obtain food from the plant in this process. Other specialized fungi are able to form a close association with algal or bacterial cells in forming a lichen. In a lichen, the fungus obtains food from photosynthesis, and the algae or bacteria obtain protection from drying out.

6. Plants make their own food through photosynthesis (they are photoautotrophs). Fungi and animals do not make their own food. Fungi digest food outside their bodies (heterotrophic, external digestion), and animals digest food inside their bodies (heterotrophic, internal digestion).

7. Raw plants have cell walls made out of cellulose, and each cell has a large, water-filled vacuole. Chewing breaks the cell walls, releases the water, and makes the plant crunch.

CHAPTER 21
Multiple Choice
1. b, **2.** c, **3.** e, **4.** c, **5.** b, **6.** c, **7.** a, **8.** b, **9.** b, **10.** e.

Brief Review

1. Distinguishing features of the main plant groups:

Plant Group	Feature
bryophytes	dependent on water for fertilization (have swimming sperm)
seedless vascular plants	dependent on water for fertilization; have vascular tissues; have true roots, stems and leaves
gymnosperms	not dependent on water for fertilization (produce pollen); have vascular tissue; have true roots, stems and leaves; produce seeds
angiosperms	all the features of gymnosperms; also produce flowers and fruits

2. Roots serve to anchor the plant to the substratum (usually soil) and they are the main absorptive organs of the plant. They absorb water and minerals from their surroundings and make

it available to the above ground parts of the plant. In some species roots also store substantial reserves of food. Tap roots and fibrous roots are the two main types of roots produced by flowering plants.

3. Parts of a flower and main functions of each:

Flower Part	Function
pedicel	Flower stalk; lifts flower up to make it noticeable to pollinators.
sepals	Leaf-like structures that protect flower buds before they open; commonly, also augment food supplies because they are usually green and photosynthetic.
petals	Generally most conspicuous part of the flower; often brightly colored; sometimes fragrant; attract pollinators.
stamens	Male reproductive structures; consist of a stalk called a filament, which holds aloft the pollen-producing sacs called anthers.
pollen	is the male gametophyte, and the vehicle that conducts male sperm to the female reproductive structures.
pistil	Female reproductive structure; consists of the ovary, style, and stigma; the ovary produces and protects the egg-containing female gametophyte, and after fertilization turns into the fruit; the stigma is the receptive surface that captures pollen and promotes germination of the right kind of pollen; the slender style raises the stigma to make it accessible to pollinators, and also serves as a "racetrack" that screens for pollen vigor.

4. Perennial plants live over many growing seasons. They can be extremely long-lived. Some species such as the redwoods, bristlecone pines, Sitka spruces, and Douglas firs can live for a few thousand years. Even commonplace orchard trees can grow to a hundred or more years. People are among the longest-lived animals, and can achieve a life span of 100 years or slightly more. However, few other animals live to a hundred.

    While people and other animals grow over their whole body (general surface growth of all organ systems), plants confine their growth to distinct zones called meristems. Almost all of the cell division in an adult plant is confined to the meristem. Meristems are perpetually maintained as regions specialized for cell division; they do not turn into (differentiate into) other, more specialized types of cells. Plants, therefore, grow at the ends of the roots and shoots, not all over. Because meristems can be active throughout the life span of a plant, plants continue to grow all through their lives. Animals, in contrast, typically cease growth on reaching maturity.

5. Plants enter into cooperative arrangements with members of nearly all the kingdoms of life. The roots of some plants—those of legumes such as beans and peas, for example—harbor nitrogen-fixing bacteria; the plant supplies carbohydrate to the bacteria, which in return manufactures hard-to-obtain nitrogen containing compounds that the plant can use. In nature, the roots of most plants are colonized by certain types of fungi (mycorrhizal fungi) that allows the plant to absorb nutrients such as phosphates with greater ease, and receive carbohydrates fixed by the plant in exchange.

6. The gametophyte is the multicellular haploid phase of a plant; that is, each cell in the gametophyte has just one set of chromosomes. The gametophyte produces gametes (egg and sperm). The sporophyte is the multicellular diploid phase

of the plant; the sporophyte produces the spore that gives rise to the gametophyte.

CHAPTER 22
Multiple Choice
1. a, 2. b, 3. c, 4. c, 5. b, 6. a, 7. e, 8. b, 9. c and d, 10. c.

Brief Review
1. Differences between monocots and dicots:

	Monocots	Dicots
number of cotyledons (food-storing parts of the embryo)	one	two
arrangement of flower parts	in sets of three	in sets of four or five
leaf vein pattern	parallel veins	net-like veins
arrangement of vascular bundles	scattered	in a discrete ring
type of root system	fibrous root	taproot

2. The three main cell types in a plant and the functions performed by each:

Cell Type	Function
parenchyma	living cells; perform a wide variety of metabolic functions such as photosynthesis, food storage, wound repair
collenchyma	living cells with irregularly thickened walls; perform some metabolic functions such as photosynthesis, but also impart mechanical strength to tissues experiencing active growth
sclerenchyma	usually dead at maturity; main role is structural support for the plant body, and water conduction in some cases

3. Meristems are tissues specialized for cell division alone. After embryonic development, meristems are the source of almost all new cells that contribute to the growth of the plant body. Meristematic cells do not acquire new, more specialized cell functions. Meristems are perpetually maintained. Apical meristems at the tips of shoots and roots contribute to the length or height of a plant. Lateral meristems are of two main kinds: the vascular cambium gives rise to secondary xylem and phloem, and cork cambium creates a protective layer of cells known as bark; both types contribute to an increase in the girth of the plant.

4. Xylem-conducting elements—tracheids and vessel elements—are dead at maturity. They have thick, lignified walls. Tracheids have many small perforations, vessels one or a few relatively large perforations. Xylem-conducting elements provide mechanical support, and conduct water and dissolved minerals to all parts of the plant.

Phloem-conducting elements—sieve tubes—are living, but lack nuclei. They have relatively thin walls. Sieve tubes are composed of several sieve tube members (individual cells) connected through cytoplasmic channels. Each sieve tube member is closely associated with, and totally dependent on, living cells called companion cells. Sieve tubes conduct carbohydrates, mainly sucrose, from sites of production (sources) to locales where they will be utilized or stored (sinks) within the plant. Some other compounds, such as amino acids and plant hormones, are also transported in phloem sap.

5. Growth rings are created by abrupt changes in the size and structure of xylem elements laid down earlier in the growing season versus later in the growing season. The first-formed wood, when the growing season resumes, is called early or spring wood. Early wood is composed of xylem elements with a wide diameter. It is more porous and less dense, and appears lighter because there's proportionately more air space relative to thick cell walls. Late wood is formed later in the season, when soil water becomes depleted and trees may experience water shortage. The tracheids and vessels produced then have a narrower bore and proportionately thicker walls. Late wood therefore appears denser and darker.

Each band of early and late wood constitutes an annual ring, and represents the wood deposited in a single growing season. In temperate regions, that means a single year's worth of secondary growth. Because the dark, late wood of the current season lies next to the following season's lighter, early wood, the annual rings produced in successive seasons are easy to see. Thus the total number of annual rings in a tree trunk is equivalent to the number of seasons, or years, that tree has been growing.

6. Double fertilization refers to the two fusion events that take place when pollen tubes deliver the two sperm cells to the embryo sac (haploid female gametophyte, in the ovule). One of the two sperm cells fuses with the egg to give rise to the diploid zygote and future embryo. The second sperm cell fuses with a set of two haploid nuclei that are suspended within a large central cell; this fusion event gives rise to a triploid structure, the endosperm.

CHAPTER 23
Multiple Choice
1. c, 2. a, 3. b, 4. b, 5. e, 6. b, 7. e, 8. c, 9. c, 10. d, 11. d.

Brief Review
1. The sequence of the levels of organization of the human body from the simplest level to the most complex is molecular, cellular, tissue, organ, organ system, organism (the individual).
2. blood and lymph
3. Keratin is a durable and water-resistant protein. It fills the outer epidermal cells, making a waterproof skin. (It also makes up the nails of the hand and foot; and also forms the outer layer of hairs.)
4. Three of the functions of the skeletal system are support, storage of minerals and lipids, and blood cell production.
5. Three primary functions of the muscular system are to produce skeletal movement, to maintain body posture and body position, and to maintain body temperature.
6. The endocrine system is made up of ductless glands whose secretions (hormones) are carried throughout the body, where they alter the activities of organ systems.
7. Cells are lost through abrasion (scraping), and the new cells that are formed through cell division push up on the layers above them.
8. Concentric layers of bone around a central canal indicate a Haversian system (osteon) that makes up compact bone. Because the ends (epiphyses) of long bones are mostly spongy bone, this sample most likely came from the shaft (diaphysis).
9. Skeletal muscle appears striated, or striped, under a microscope because it is composed of thin (actin) and thick (myosin) protein filaments arranged in such a way that they produce a banded appearance in the muscle.

CHAPTER 24
Multiple Choice
1. c, 2. d, 3. a, 4. e, 5. c, 6. a, 7. d, 8. e, 9. e, 10. b.

Brief Review
1. The three functional groups are sensory neurons, motor neurons, and interneurons. The sensory neurons carry information about the internal and external environments to the central nervous system (CNS). Motor neurons carry information from the CNS to other tissues, organs, or organ systems that can change their activities. Interneurons analyze the sensory information and coordinate motor responses.
2. Dual innervation of a bodily organ by both the sympathetic and parasympathetic divisions of the autonomic nervous system produces opposite effects on organ function. This helps maintain relatively rapid control of the organ in question, stimulating or relaxing it.
3. The cerebrospinal fluid provides cushioning and support for the brain. It also transports nutrients, chemical messenger molecules, and waste substances.
4. In communication by the nervous system, the source and destination are quite specific and the effects are short-lived. In endocrine communication, the effects are slow to appear, and they often persist for days. A single hormone can alter the metabolic activities of multiple tissues and organs simultaneously.
5. The nonspecific defenses are present at birth and do not discriminate between one threat and another. Specific defenses provide protection against threats on an individual basis.
6. Even though the medulla oblongata is small, it contains many vital reflex centers including those that control breathing and regulate the heart and blood pressure. Damage to the medulla oblongata can result in cessation of breathing or changes in heart rate and blood pressure that are incompatible with life.
7. In negative feedback, elements that bring about a response have their activity decreased by that response. In positive feedback, elements that bring about a response have their activity increased by that response.
8. The pituitary gland releases six hormones from its anterior portion and two from the posterior portion. Many of the pituitary's hormones stimulate, or "turn on," other endocrine glands. As a result, the pituitary is called the "master" endocrine gland.
9. Helper T cells promote B-cell division, the maturation of plasma cells, and the production of antibody by the plasma cells. Without the helper T cells, the antibody-mediated immune response would take much longer to occur and would not be as efficient.
10. A decrease in the number of monocyte-forming cells in the bone marrow would result in a decreased number of macrophages in the body, because all the different macrophages are derived from monocytes.

CHAPTER 25
Multiple Choice
1. b, 2. c, 3. a, 4. b, 5. c, 6. c, 7. e, 8. d, 9. c, 10. b.

Brief Review
1. Transportation: of dissolved gases, nutrients, hormones, and metabolic wastes. Regulation: pH and electrolyte composition of interstitial fluids throughout the body, stabilization of body temperature, and restriction of fluid losses through damaged vessels. Protection: defense against toxins and pathogens.
2. Capillaries are only one cell thick. Small gaps between adjacent endothelial cells permit the diffusion of water and small solutes into the surrounding interstitial fluid but prevent the loss of blood cells and plasma proteins. The walls of arteries and veins are several cell layers thick and are not specialized for diffusion.
3. External respiration is the diffusion of gases between the blood in the lung capillaries and air in the alveoli of the lungs.
4. The four layers of the digestive tract, beginning with the innermost layer, are the mucosa, submucosa, muscularis externa, and serosa.
5. The urinary system performs vital excretory functions and eliminates the organic waste products generated by cells throughout the body, while conserving needed materials.
6. Drinking this much fluid would likely cause your body to reduce its production of antidiuretic hormone, thus allowing the excess fluid to be retained by the kidney tubules and flow into the bladder.

7. Most of the carbon dioxide released by body tissues enters red blood cells where it is converted into bicarbonate ions that then diffuse into the blood plasma. At the lungs the bicarbonate ions are converted back into carbon dioxide within the red blood cells and released into the alveoli of the lungs.

8. The small intestine has several adaptations that increase surface area to increase its capacity to absorb food. Its internal wall consists of folds that are covered by fingerlike projections of tissue called villi. The cells that cover the villi, in turn, have smaller fingerlike projections called microvilli. In addition, the small intestine has a rich blood and lymphatic supply to transport the absorbed nutrients.

9. Blood cells and large protein molecules are too large to pass out of the circulatory system and into the filtrate in the kidneys.

CHAPTER 26

Multiple Choice

1. d, 2. b, 3. d, 4. c, 5. e, 6. d, 7. a, 8. c, 9. c, 10. b, 11. e.

Brief Review

1. After fertilization, the cleavage stage of development begins. This results in the formation of a hollow ball of cells. The embryo does not change in size during cleavage. Next, these cells undergo rearrangement and movement, in the process called gastrulation. Now the embryo has a body cavity and three layers of cells: the endoderm, ectoderm and mesoderm.

2. After gastrulation in the chick embryo, a small group of cells is induced to form the wing of the chick. This primitive wing, the limb bud, is transformed into the wing. A portion of the developing wing, called the zone of polarizing activity (ZPA), sends out a morphogen that directs the proper order of digit production in the chick's wing.

3. There are two types of mesenchyme cells in the sea urchin embryo. Primary mesenchyme cells detach from the vegetal pole of the blastula and begin to migrate to the animal pole. This begins the process of invagination and eventual production of the primitive gut. The secondary mesenchyme cells send out filopodia that help to pull the archenteron up to the animal pole of the embryo. These cells then migrate to other positions in the embryo to become mesodermal cells.

4. Cell adhesion molecules underlie all the processes of development that involve cell movement and reorganization. Gastrulation and neural crest cell migration are just two examples of processes that rely on this class of proteins.

5. A concentration gradient refers to the distribution of a molecule within an embryo in a pattern of high concentration to low. The morphogen bicoid is distributed in a concentration gradient with high concentrations in the anterior of the embryo and progressively lower concentrations in the posterior regions. High concentrations of bicoid result the production of head structures where as absence of bicoid allows the production of posterior structures.

6. Cells become determined through all the developmental processes discussed in this chapter. The production of the sheep Dolly provides evidence that the DNA in determined cells can be reprogrammed to produce another organism.

7. Cell death is a normal process of the development of most organisms. In humans, cell death "trims" down the tissue between each finger so that the fingers are no longer webbed when the baby is born.

CHAPTER 27

Multiple Choice

1. c, 2. e, 3. a, 4. c, 5. d, 6. c, 7. b, 8. a, 9. e, 10. c.

Brief Review

1. Gonadotropin-releasing hormone, which is released by the hypothalamus, stimulates the release of follicle stimulating hormone(FSH) and lutenizing hormone (LH). Both of these hormones are released by the anterior pituitary gland. FSH promotes the development of the follicles and LH stimulates the production of estrogen. As estrogen levels increase, FSH levels at first begin to decrease, but as estrogen levels begin to peak due to the development of the tertiary follicle, both FSH and LH levels spike. This results in ovulation. The corpus luteum then begins to synthesize progesterone, which acts on the hypothalamus to reduce FSH and LH to original levels, resulting in degeneration of the endometrium and menstruation.

2. Sperm development is dependent on temperature. If things get too hot, then sperm production is impaired. Similarly, cold temperatures also prevent sperm development. The structure of the testes acts to keep sperm production at the correct temperature. The sperm are protected from the high body temperatures because of the location of the testis outside the body. In addition, a muscle that surrounds the testes contracts when temperatures drop, so that the testes are pulled in closer to the warm body.

3. Scientists hypothesize that menstruation serves to remove harmful pathogens, such as bacteria, that may have entered along with sperm during intercourse. Also, the cost of maintaining an embryonic environment may exceed the monthly buildup of uterine tissue.

4. The tertiary follicle and the corpus luteum are variations of the same structure. Once ovulation has occurred and the tertiary follicle releases the oocyte, it then becomes the corpus luteum. Both structures release hormones.

5. The fast block to polyspermy occurs when the first sperm has fertilized the egg. Membrane channels in the egg open, allowing positively charged sodium ions to enter the egg. This changes the electrical potential of the egg, which prevents additional sperm from fertilizing the egg. This is necessary to prevent the introduction of additional sperm, with their additional sets of genetic information.

6. Prior to implantation, the fertilized egg divides to form a morula. This tightly packed ball of cells continues to divide forming a hollow ball of cells called the blastocyst. The blastocyst will then implant in the uterine wall.

7. Any baby born before 37 weeks is premature. Premature births may have many complications due to the incomplete development of organs such as the lungs.

8. Birth control pills work using a combination of synthetic estrogen and progestin. Normally, low estrogen levels trigger the beginning of follicle development. The pill creates high estrogen levels that suppress the release of FSH from the hypothalamus. The progestin in the pill suppresses the release of LH in order to prevent ovulation of any follicle that might develop in the presence of high estrogen levels.

CHAPTER 28

Multiple Choice

1. d , 2. c , 3. e, 4. c , 5. d , 6. b, 7. a, 8. c.

Brief Review

1. Population, community, ecosystem, biosphere.
2. K-selected species spend most of their lives in populations that are close to carrying capacity (K), and thus they put energy into longevity, competition, survival, and parental care of offspring. In contrast, r-selected species (species whose population is limited primarily by reproductive rate) put little energy into longevity and survival, but rather produce many offspring, with little energy or parental care invested in each individual offspring. Humans and elephants are K-selected: there is a great deal of parental care invested in each offspring. Flies or dandelions are r-selected species.

3. Environmental resistance is the term used to describe all the forces of the environment that act to limit population growth. Organisms will run out of food or have their sunlight blocked; temperatures will plunge or rain will cease; greater numbers of predators will discover the population; the wastes produced by the organisms will begin to be toxic to the population.

4. This is an example of Batesian mimicry, in which one species (the red salamander) has evolved to acquire the appearance of a species with superior protective capabilities (the red newt). The red newt is the model, the red salamander is the mimic and the dupe is any predator species that would avoid the red salamander, believing it to be a red newt.

5. Because this begins from a state of little or no life, this is an example of primary succession.

CHAPTER 29

Multiple Choice

1. c, 2. d, 3. b, 4. a, 5. a, 6. b, 7. c, 8. e, 9. a.

Brief Review

1. According to the energy flow model of ecosystems (and the second law of thermodynamics), the energy contained in each trophic level is much less than the energy contained in the trophic level below it, because much of the energy in each trophic level is used for metabolic processes (running, staying warm) and much energy from these processes is lost from the system as heat. In addition, not all of the prey are eaten by the predators.

2. Warm air (1) rises and (2) can carry more moisture than cold air. Near the equator, the Earth gets warm due to the direct rays coming from the sun. As this warm air rises, it cools and drops moisture. This cooler air moves toward the poles in both directions, and descends at about 300, warming as it goes, thus absorbing moisture from the land, which is dry as a result.

3. Climate essentially defines the large-scale vegetation patterns that are seen on Earth. Variations within large-scale climate regions tend to be mountains, whose climates can vary greatly from base to peak.

4. It is likely that eutophication has occurred in the pond. If so, large algal blooms can result from the excess nutrients. As this dead organic matter falls to the lower depths, decomposers consume the organic matter, but deplete the water of oxygen, which kills fish.

5. Permafrost makes the tundra wet during summer months, even with low rainfall, and it is too cold for trees to grow. Taiga has a less severe climate and a longer growing season, and no permafrost; coniferous trees grow slowly there and support a larger animal community.

Glossary

1n Haploid; having one chromosome of each type per cell instead of a pair of chromosomes of each type.

2n Diploid; having two chromosomes of each type per cell.

abiotic Pertaining to nonliving things.

abscission The detachment of plant parts, such as leaves, petals, or fruits, from the main plant.

accessory pigment A type of pigment found in chloroplasts that, along with chlorophyll *a*, absorb some wavelengths of sunlight in the first step of photosynthesis.

acid Any substance that yields hydrogen ions when in solution. An acid has a number lower than 7 on the pH scale.

acrosome A structure, located on the front end of the head of a vertebrate sperm cell, that contains enzymes that help the sperm penetrate the accessory cells surrounding the oocyte.

action potential A transient change in a cell's membrane potential. In neurons, this results in transmission of a nerve signal.

activation energy The energy required to initiate a chemical reaction. Enzymes lower the activation energy of a reaction, thereby hugely speeding up the rate of the reaction.

active site The area of an enzyme, usually a groove or pocket, that binds the substrate and changes its shape to affect its reactivity.

actively acquired immunity Immunity resulting from natural exposure to an antigen in the environment or from deliberate exposure to an antigen.

active transport Transport of materials across the plasma membrane in which energy must be expended. Through active transport, solutes can be moved against their concentration gradient. The sodium-potassium pump is an example of active transport.

adaptation A modification in the structure or behavior of organisms over generations that makes them better suited to their environment.

adaptive radiation The rapid emergence of many species from a single species that has been introduced to a new environment. The different species specialize to fill available niches in the new environment.

adipose tissue Also known as fat, this type of loose connective tissue insulates the body and stores energy in addition to providing cushioning.

adrenocorticotropic hormone (ACTH) A hormone, secreted by the anterior pituitary, that stimulates release of steroid hormones from the outer cells of the adrenal gland.

afferent division The division of the peripheral nervous system that carries sensory information towards the central nervous system, having gathered information about the body or environment.

AIDS Acquired immune deficiency syndrome, a disease caused by the human immunodeficiency virus that attacks the body's own immune system and leaves it unable to fight off even mild infections.

alcoholic fermentation The process by which yeasts, undergoing glycolysis in an oxygenless environment, produce alcohol as a waste product.

algal bloom An overabundance of algae in a lake as a result of an excess of nutrients. The many dead algae that fall to the bottom allow decomposing bacteria to flourish, using up so much oxygen that fish can suffocate.

alimentary canal Another term for the digestive tract, the muscular tube that runs through the body from mouth to anus and through which food passes.

alkaline Basic, as in solutions. Alkaline (basic) solutions have numbers above 7 on the pH scale.

allele One of the alternative forms of a single gene. In pea plants, a single gene codes for seed color, and it comes in two alleles: one codes for yellow seeds, the other for green seeds.

allergen A foreign substance that initiates an allergic reaction in the body. These substances are usually derived from living things, including pollen, dust mites, foods, and fur.

allergy A body's reaction to a foreign substance in which histamines and heparin are released. When this attempt to deal with invaders is excessive, we call it an allergy.

allopatric speciation The type of speciation that occurs when a population is divided by geographical barriers and the resulting populations become separate species. This type of speciation is thought to be the most common in nature.

allosteric Denoting a type of chemical reaction in which a substance is changed to a different shape, rendering it temporarily inactive. Allosteric reactions are important in negative feedback when the product of a reaction changes the shape of the activating enzyme, thus temporarily turning off the production of more product.

alpha helix A common secondary structure of proteins that resembles a corkscrew or spiral staircase. Hair, nails, and horns are proteins made up almost entirely of alpha helices.

alternation of generations Plant life cycle in which a haploid phase alternates with a diploid phase. A haploid gametophyte, through mitosis, produces gametes; fertilization of gametes forms a diploid zygote that grows into a diploid sporophyte which, through meiosis, produces haploid spores; the spores grow into a haploid gametophyte.

alveolus One of the many tiny air sacs that form the terminal branches in the lungs. Gases are exchanged with the capillaries through the walls of the alveoli.

amniocentesis A medical procedure in which amniotic fluid is withdrawn from a mother's abdomen with a needle and the fetal epidermal cells that are in the fluid are tested for genetic abnormalities.

amniotic egg An egg with a hard outer casing and an inner series of membranes and fluids that form a padding around the growing embryo. Reptiles and birds have amniotic eggs, and the evolution of this type of egg freed reptiles from the constraint of developing in water.

amniotic fluid A protective and nutritive fluid that surrounds the fetus of mammals, including filling the lungs.

anagenesis A form of evolution in which a species changes over time, eventually having changed enough that it would be reproductively isolated from its ancestral species. In this nonbranching evolution, the original species changed into the new species and itself disappeared.

analogy A structure found in different organisms that shares a function and a superficial appearance, but is not due to inheritance from a common ancestor. Analogies must be distinguished from homologies to get a true picture of evolutionary relationships.

ancestral character A character that existed in the common ancestor of a group of organisms. Cladistics distinguishes ancestral from derived characters and uses these characters to determine evolutionary relationships.

aneuploidy A condition in which either more or fewer chromosomes occur than normal for that organism, so the individual has the wrong number of chromosomes in its diploid set. A different genetic disorder results depending on which chromosome is missing or extra.

angiosperm A flowering plant. The angiosperms are the dominant and the most diverse group of plants on Earth today.

animal pole The end of a zygote with relatively less yolk and lying closer to the cell's nucleus. The location of the egg's poles define the orientation in which the embryo develops.

annual A type of plant that goes through its entire life cycle, from germination of the seed, through growth, flowering, and death, in one year or less. Many crop plants, such as tomatoes and grains, are annuals.

annual ring The ring formed in wood, showing the abrupt change between the secondary xylem of late summer, when cells get less water and thus are smaller and have thicker walls, and the secondary xylem of early spring, when the cells get more water and are much larger.

anterior pituitary A region of the pituitary that secretes six different hormones involved in reproduction and growth. The anterior pituitary is under the control of the hypothalamus, which secretes releasing factors into the capillary network that links the two.

anther The part of a flower that produces pollen grains. The anther is on top of a filament, and together they are called the stamen.

antibiotic Medicine used to kill disease-causing agents. When we speak of antibiotics, we usually mean medicines directed against disease-causing bacteria.

antibody A protein of the immune system, found on the surface of a B cell or circulating free in the plasma, that is formed in response to a particular antigen and reacts with it specifically, deactivating it.

antibody-mediated immunity Immune system response that uses production of lots of antibodies to fight infection. This type of immunity provides a kind of immune system memory that protects against reinfection by the same bacterium or virus.

anticodon The end of the transfer RNA molecule that can form a base pair with a particular codon on the mRNA transcript.

antidiuretic hormone (ADH) Hormone, released by the posterior pituitary, that acts to decrease the amount of water lost by the kidney, helping to conserve water.

antigen Any foreign substance that elicits a response by the immune system. Certain proteins on the surface of an invading bacterial cell, for instance, act as antigens that trigger an immune response.

antigen-presenting cell (APC) A macrophage that has ingested an infected cell and displays fragments of proteins from the invader on its surface. The APCs bind with helper T cells and activate them, stimulating secretion of interleukin-2, which has multiple functions in mounting further attacks against the invader.

aorta The major artery in the circulatory system. The aorta has branches that send blood to all of the body's other tissues except the lungs.

apical dominance The inhibition of lateral buds by the apical meristem (growing tip) of a plant.

apical meristem The group of plant cells at the tips of the roots and shoots that are able to keep dividing, and from which all tissues ultimately develop.

appendicular skeleton The division of the skeletal system consisting of the bones of the paired appendages, including the pelvic and pectoral girdles to which they are attached.

aqueous solution A solution in which water is the solvent. Because living things contain so much water, nearly all solutions important in biological systems are aqueous solutions.

aquifer The porous underground rock in which groundwater is stored.

Archaea The domain (major group) that contains only microscopic, single-celled organisms superficially similar to bacteria but genetically quite different. Many of the Archaea live in extreme environments such as boiling-hot vents on the ocean floor.

arithmetical increase An increase in numbers by an addition of a set number in each time period. The number added does not depend on the number already in existence, and natural populations do not show this type of increase.

arteriole The smallest type of artery, which branches into the capillary beds, where gases and nutrients are delivered to tissues.

artery A blood vessel that carries blood away from the heart.

artificial eutrophication Nutrient enrichment of a lake through human intervention. This can be intentional, or it can occur accidentally when, for example, fertilizers from lawns or agriculture are washed into the lake.

asexual reproduction Reproduction that occurs without the union of two reproductive cells (sexual reproduction). Offspring of asexual reproduction are genetically identical to their parents.

assortative mating Mating in which males and females that share a particular characteristic, such as short height, tend to mate with one another. Such mating tends to bring similar alleles together but does not directly alter allele frequencies in the population.

atmosphere The layer of gases that surround the Earth. The atmosphere is nonliving, but its presence enables life to exist on Earth.

atom A unit of matter, once thought to be the smallest particle in nature but now known to be made up of many smaller particles including electrons, and protons and neutrons which are in turn made up of quarks.

atomic number The number of protons in the nucleus of an atom. This number defines the element, so gold is gold because it has an atomic number of 79. The elements are ordered on the periodic table according to atomic number.

atomic weight The combined weight (mass) of protons and neutrons in an atom's nucleus.

ATP synthase The enzyme that enables the movement of hydrogen ions back across the mitochondrial inner membrane to power the production of ATP in the final process of the electron transport chain. The hydrogen ions flow through the enzyme, causing part of the enzyme to spin, thus driving the synthesis of ATP from ADP and phosphate.

autonomic nervous system A subdivision of the efferent division of the peripheral nervous system that provides involuntary regulation of the smooth muscles, cardiac muscles, and glands.

autosomal dominant The condition in which a single faulty allele on an autosome can cause damage, even when the individual's second allele for that gene codes for normal products.

autosomal recessive disorder A disorder in which the gene or genes lie on one or more autosomes, rather than on the sex chromosomes, and two recessive, nonfunctional alleles are present. Sickle-cell anemia is one example.

autotroph Any organism that manufactures its own food. Plants are the most important autotrophs on Earth.

axial skeleton The part of the skeletal system that forms the central column, including the skull, vertebral column, and rib cage.

axon A single, large extension of the cell body of a neuron that carries signals away from the cell body toward other cells. The axons of some neurons in humans may be several feet long.

bacillus A rod-shaped bacterium.

Bacteria The domain (major group) that contains single-celled, microscopic organisms that superficially resemble archae but are genetically quite different. These are the bacteria that are familiar to us.

bacteriophage A type of virus that is used in biotechnology to put foreign DNA into bacterial cells, therefore serving as a cloning vector.

ball-and-stick model A picture showing the three-dimensional structure of a molecule, with the atoms drawn as balls and the bonds between them drawn as sticks. This type of representation clearly shows the relationship of the atoms to each other.

bark Everything outside the vascular cambium in woody plants, consisting of cork cambium, cork, phelloderm, and secondary phloem.

base Any substance that accepts hydrogen ions in solution. A base has a number higher than 7 on the pH scale.

base pair Linked pair of nitrogenous bases (adenine, guanine, cytosine, or thymine) forming the "steps" in the DNA double helix. The bases link according to rules: A links with T, G with C. The order of bases defines the alleles and determines what proteins are built.

basement membrane A dense network of protein fibers and filaments to which epithelial cells are attached. This membrane provides a barrier that keeps large molecules from passing through.

Batesian mimicry A type of mimicry in which one species evolves to resemble a species that has superior protection against predators.

benthic zone The ocean floor, where bottom-dwellers live.

beta pleated sheet A common secondary structure of proteins that resembles the folds of an accordion. Silk is a protein made up almost entirely of beta pleated sheets lying stacked like pancakes on top of each other.

biennial A type of plant that goes through its life cycle in about two years, flowering in the second year.

bilateral symmetry The type of symmetry in which an organism's right and left sides are very similar, though the top and bottom and front and back may differ. Animals generally are bilaterally symmetrical.

bile A substance produced by the liver that aids in the digestion of fats. Bile can enter the small intestine through ducts directly from the liver, or from a storage organ, the gallbladder.

binary fission The form of cell division carried out by prokaryotic cells in which the chromosome replicates and the cell pinches between the attachment points of the two chromosomes to form two new cells.

binding site An area of a protein, located at the surface of the plasma membrane, with a particular conformation that recognizes only one molecule or a closely related group of molecules. Binding sites enable specificity of cell communication.

binomial nomenclature The system of naming species that uses two names (genus and species) for each species. This system helps identify groupings among living things.

biodegradable Capable of being broken down by living organisms.

biodiversity The diversity of living things. This includes species diversity, diversity of distributions, and genetic diversity.

biogeochemical cycling The movement of water and nutrients back and forth between biotic and abiotic systems.

biological species concept A definition of species that relies on the breeding behavior of populations in nature. Accordingly, groups of populations that actually or potentially interbreed in nature and are reproductively isolated from other such groups constitute a species.

biology The study of living things.

biomass The weight of all the living organisms in a given area.

biome An ecosystem dominated by a large vegetation formation, whose boundaries are largely determined by climate. The same biome type can occur on two different continents and have different species, but the two regions will bear striking similarities.

biopharmaceutical A drug that has been produced through recombinant DNA technology.

biosphere All of life on Earth, plus the physical environment that the living things interact with. The entire biosphere interacts, so destruction of the rainforest in Brazil affects the climate in the rest of the world.

biotechnology The use of living organisms to create products or facilitate processes.

biotic Pertaining to living things.

birth control pill A pill that prevents pregnancy by disrupting the normal hormonal cycle in a way that suppresses development of the follicle and also prevents ovulation. Without ovulation of a mature oocyte, there is no egg in a position to be fertilized.

blade In plants, the major, broad part of a leaf.

blastocoel The fluid-filled cavity in a blastula.

blastula A ball of cells with an interior, fluid-filled cavity (the blastocoel), that is produced during animal development by cell rearrangement at the end of the cleavage phase.

blood-brain barrier The covering over tiny blood vessels in the brain, made of neuroglia cells, that prevents some substances from passing from the general circulation into the brain. This barrier helps protect the brain from potentially damaging substances.

B-lymphocyte (B cell) A type of lymphocyte that is central to antibody-mediated immunity. B cells bearing a specific antibody reproduce rapidly, then both produce additional antibody to fight the infection and also retain some cells that remember the antigen so a reinfection can be prevented.

bone A type of supportive connective tissue that contains mineral deposits, primarily calcium, which, along with flexible collagen fibers, make the tissue both strong and resistant to shattering.

bottleneck effect The chance alteration of allele frequencies during a sharp reduction in a population's size. Disease or natural catastrophe can greatly reduce the population size, and chance can have a profound effect on the genetic makeup of small populations.

Bowman's capsule A hollow structure in the vertebrate kidney that receives the filtrate from the blood and passes it into the proximal tubule of the nephron. This structure is the point where wastes are transferred from the blood into the kidney.

bryophytes The most primitive land plants. Bryophytes have no vascular structure, so they cannot grow very tall. Mosses are the most familiar bryophytes.

bud An undeveloped plant shoot, composed mostly of meristematic tissue.

buffering system A solution that helps to keep pH at optimal levels. These solutions are generally weak acids or bases that can work to neutralize sudden infusions of acid or base by accepting or donating hydrogen ions. Buffering systems help living beings function by keeping pH at whatever level is best in different parts of an organism.

bulbourethral gland A gland that, in human males, contributes fluids to the semen as it proceeds down the urethra in the process of ejaculation.

bundle-sheath cells Special cells, wrapped around leaf veins, that receive CO_2 from a four-carbon molecule in the C_4 pathway. In C_4 plants, the CO_2 concentration is high and the Calvin cycle takes place in these cells.

Calvin cycle The set of steps in photosynthesis through which carbon dioxide, with energy from ATP and electrons from NADPH, becomes an energetic carbohydrate. This cycle is the stage of photosynthesis where building occurs.

CAM metabolism The type of photosynthesis used by succulents and some other plants from hot climates. CO_2 is fixed at night, and stomata are opened during the day to get energy to make ATP and complete photosynthesis. This process allows plants in hot environments to open their stomata for a shorter time during the day than in other types of photosynthesis, thus losing less water.

capillary The smallest type of blood vessel, connecting the arteries and veins out in the body's tissues. Gases, nutrients, and wastes are exchanged between the blood and the tissues through the thin walls of capillaries.

capillary bed An interconnected network of capillaries that lies in the body's tissues. The network brings the capillaries close to each part of the tissues and provides lots of surface area for more efficient diffusion.

carbohydrase The type of pancreatic enzyme, secreted through ducts, that breaks down sugars and starches.

carbohydrate A class of biological molecule that contains carbon, oxygen, and hydrogen, usually with twice as much hydrogen as oxygen. Carbohydrates are composed of monosaccharide (simple sugar) building blocks. Starch, cellulose, and glycogen, and are carbohydrates.

cardiac muscle The type of striated muscle tissue that forms the muscles of the heart.

cardiovascular system The organ system that transports nutrients, dissolved gases, and hormones to tissues throughout the body, and carries wastes to sites of filtration. This system consists of the heart, blood vessels, and blood, and the bone marrow where blood cells are formed.

carpel The female reproductive structure of a flower, consisting of an ovary, a style, and a stigma.

carrier An individual who does not express the symptoms of a disorder, but who carries an allele and therefore can pass it on to offspring.

carrying capacity The maximum population of a species that can be sustained in a given geographical area over time. In ecology, this is often denoted as *K*.

cartilage A type of supportive connective tissue that provides support and flexibility, as in the outer portions of the nose and ear, between bones of the spine, and between limb bones.

catalyst A substance that retains its original chemical composition while bringing about a change in a substrate. Enzymes are catalysts in chemical reactions, and one enzyme can carry out hundreds or thousands of chemical transformations.

cecum An expanded chamber at the beginning of the large intestine, adjacent to the small intestine. In some animals, especially herbivores, the cecum is a fermentation chamber for gut microorganisms that digest cellulose.

cell A highly organized structure that consists of a cell membrane with cytoplasm (cell fluid) and DNA inside. A cell can be an entire organism on its own or can be a building block for tissues and other structures in a multi-celled organism like a human. Cells are the structural units of all organisms.

cell adhesion The sticking together of cells. This adhesion is used during animal development to attach layers of cells, thus enabling specific movements and shaping of the embryo.

cell cycle The repeating pattern of growth, genetic duplication, and division that many types of cells undergo. The cell cycle has two main phases: interphase, in which cells grow, duplicate chromosomes, and work, and mitotic phase, in which cells divide.

cell junction A linkage between animal cells. A gap junction is a type of cell junction that facilitates cell communication.

cell-mediated immunity An immune system response that recognizes and kills the body's own cells that have been invaded by foreign organisms.

cell movement In development, the organized movement of cells to form specific layers and structures.

cell plate A new piece of cell wall that forms by fusion of vesicles near the metaphase plate of a plant cell. Plants form a cell plate to accomplish cytokinesis because the cell wall of plants is too rigid to be constricted by a contractile ring.

cell wall A structure composed mostly of cellulose and lignin that surrounds the plasma membrane in plants. The cell wall provides structure and helps regulate water intake.

cellular respiration The process of extracting energy from a food molecule such as glucose to form ATP, the most useful form of energy in a cell. Three stages are included: glycolysis, Krebs cycle, and the electron transport chain.

cellulose A complex carbohydrate that is the largest single component of plant cell walls and is also found in the cell walls of many other organisms. Cellulose is dense and rigid and provides structure for much of the natural world. Mammals cannot digest cellulose, so it serves as insoluble dietary fiber that helps move food through the digestive tract.

central cell Cell with two nuclei in the embryo sac (mature female gametophyte) in a flower. Upon fertilization, this cell will develop into the endosperm, the food for the embryo.

central nervous system (CNS) The part of the nervous system that integrates and coordinates sensory data and sends out commands. This division consists of the brain and the spinal cord.

central vacuole An organelle in a plant cell, containing mostly water but also other substances, that may occupy up to 90 percent of the volume of the cell. The vacuole stores nutrients, retains or degrades waste, and is involved in metabolism.

centrosome A microtubule organizing center found in the cytoplasm just outside the nucleus of cells. The centrosomes organize and orient the mitotic spindle for separating duplicated chromosomes.

cerebellum Region of the vertebrate brain that adjusts ongoing movements on the basis of sensory information and memory of past movements. The cerebellum helps us both maintain balance and remember how to ride a bicycle.

cerebral cortex The highly convoluted layer of gray matter at the surface of the brain of vertebrates. This layer consists largely of cell bodies of sensory and motor neurons.

cerebral hemisphere Half of the cerebrum of the vertebrate brain. This is the site of many functions, including conscious thought, sensations, and memory.

cerebrospinal fluid The fluid that circulates in the spine and brain, supplying nutrients, hormones, and white blood cells, and cushioning the nervous tissue from jarring injury.

cerebrum The top portion of the vertebrate brain, divided into two cerebral hemispheres. Here, conscious thought, memory, sensations, and complex motor patterns originate.

cervix The lower part of the uterus, a narrow neck which opens into the vagina.

character An attribute of an organism, such as stem length, seed color, or seed shape in Mendel's pea plants. Mendel studied the traits of these characters (green or yellow seeds, for example) in offspring of crosses he controlled.

chemical bonding General term for a bond created when electrons of two atoms interact and rearrange into a new from that allows the atoms to become attached to each other. Ionic, covalent, and hydrogen bonds are all chemical bonds. Chemical bonding is what enables atoms to stick together and form visible substances and structures.

chemoautotrophy Nutritional mode in which an organism gets its carbon from carbon dioxide, but gets the energy to make food from this carbon by oxidizing inorganic materials. Some bacteria and some archaea are chemoautotrophs.

chemoheterotrophy Nutritional mode in which an organism gets its carbon and its energy from organic materials, referred to as food. Animals are chemoheterotrophs.

CFCs A class of human-made chlorine compounds that destroys the atmospheric ozone that protects life on land from damaging ultraviolet radiation.

chitin A complex carbohydrate that gives shape and strength to the external skeleton of arthropods, including insects, spiders, and crustaceans.

chlorofluorocarbons A class of human-made chlorine compounds that destroys the atmospheric ozone that protects life on land from damaging ultraviolet radiation.

chlorophyll *a* The primary pigment of chloroplasts, found embedded in its membranes. Together with the accessory pigments, chlorophyll *a* absorbs some wavelengths of sunlight in the first step of photosynthesis.

chloroplast A type of plastid (an organelle) in a plant cell that contains chlorophyll and is the site of photosynthesis. Chloroplasts are power plants in plant cells.

chorionic villus sampling A medical procedure to test for genetic abnormalities in a fetus. Cells derived from the embryo are suctioned from the villi (extensions of the chorionic membrane surrounding the placenta) and examined. This alternative to amniocentesis can be done somewhat earlier and results are faster, but it is riskier for the fetus.

chromatid One of the two identical strands of chromatin (DNA plus associated proteins) that make up a chromosome in its duplicated state.

chromatin Substance consisting of DNA plus the protein material around which DNA is wrapped. Chromatin is the material in chromosomes.

chromosome Package made up of chromatin, which is DNA and associated proteins. All the genes of organisms are in the DNA of chromosomes.

chyme The soupy mixture of food and gastric juices that is passed from the stomach to the duodenum (upper portion of the small intestine) during digestion.

cilium A hair-like extension of a cell, made of microtubules. Many cilia will occur on the surface of a given cell, and they move rapidly back and forth to propel the cell or to move material around the cell.

cisterna A flattened sac of the rough endoplasmic reticulum (ER). The many cisternae increase the surface area of the ER so that there is more space for the ribosomes to synthesize proteins.

cisternal space Space in the middle of the cisternae (flattened sacs) of the rough endoplasmic reticulum. Amino acid chains, newly synthesized on the ribosomes, enter the cisternal space and then fold into their protein conformations.

citric acid cycle Another name for the Krebs cycle, named for the first product of the cycle, citric acid.

cladistics The branch of systematics that uses shared derived characters to determine the order of branching events in speciation and therefore which species are most closely related. Cladistics is concerned only with evolutionary relationships, not classification.

cladogenesis A form of evolution that occurs by species diverging into two species, with the original species and the new species both persisting. This form is referred to as branching evolution because the new species has branched off from the original species.

cladogram An evolutionary tree constructed using the cladistic system.

class A taxonomic grouping including several related orders. This category is below phylum and above order. Humans are in the class Mammalia.

cleavage The cellular division of a zygote during animal development, resulting in a ball of smaller cells which then differentiate.

cleavage furrow The groove visible on the surface of an animal cell at the site of the contractile ring during cytokinesis.

climate The average weather conditions, including temperature, precipitation, and wind, in a particular region.

climax community The stable community that develops at the end of any process of ecological succession.

clone An exact genetic copy. Also, used as a verb, to make one of these copies. A single gene or a whole, complex organism can be cloned.

cloning vector A vehicle that enables biotechnologists to transfer DNA between organisms. Usually, bacteria or bacteriophages are used.

closed circulation Type of circulatory system, found in all vertebrates and in some invertebrates, in which blood travels only within vessels.

clotting factor A group of substances, found in blood platelets, that are important in the process of blood-clotting.

coastal wetlands Lands adjacent to the ocean that are wet for at least part of the year. These can be fresh or salt water wetlands.

coastal zone The region lying between the point on shore where the ocean's waves reach at high tide to the point off shore where the continental shelf drops off.

coccus A spherical bacterium.

codominant The condition in which two alleles in a heterozygous organism are both expressed, so the phenotype has characteristics of both of the phenotypes of individuals homozygous for either allele. Blood type (A, B, AB, O) is an example of codominance.

codon An mRNA triplet that codes for a single amino acid or a start or a stop command in the translation stage of protein synthesis.

coelom An internal body cavity that is not open to the external environment.

coenzyme A type of accessory molecule that participates at the active site of an enzyme and allows it to function. Many vitamins are important coenzymes.

coevolution The interdependent evolution of two or more species. Coevolution can benefit both species, as in flowering plants and their animal pollinators, or it can be an arms race between species, as in a plant and its predators.

coexistence The condition in which two species can live in the same habitat, dividing up resources in a way that allows both to survive.

collecting duct A tube in the nephron that receives urine from several nephron tubules. Some additional reabsorption, especially of water, takes place in the collecting duct, resulting in more concentrated urine.

collenchyma cell The type of plant cell that provides support to allow stretching and elongating the growing parts of the plant.

colon The longest region of the large intestine, between the cecum and the rectum.

colonial multicellularity A condition in which aggregated cells are all capable of the same functions; cells are not specialized. This is the only type of multicellularity that occurs in prokaryotes.

commensalism A close association between organisms of two species in which one organism benefits while the other is neither helped nor harmed.

commitment In development, a cell that is unable to alter its developmental fate has undergone commitment.

common descent with modification The modification of species in successive generations, with such change sometimes resulting in the formation of new, separate species. This principle was one of Darwin's major insights.

community All living things in a defined area. The different individuals and types of individuals will interact with and depend on each other.

companion cell Cells that are closely associated with sieve elements in phloem. The companion cells provide housekeeping needs of the sieve elements, which have lost their nuclei to provide room for faster conduction of phloem sap.

competitive exclusion principle When two populations compete for the same limited, vital resource, one will always outcompete the other and thus bring about the local extinction of the latter species.

complement A group of proteins in the blood that supplements the action of antibodies by enhancing phagocytosis, destroying the membranes of invading cells, and promoting inflammation.

complex carbohydrate A carbohydrate built of many monosaccharides. Also known as polysaccharides, these compounds include starch, glycogen, cellulose, and chitin.

concentration gradient A gradient within a given medium defined by the difference between the highest and lowest concentration of a solute. The solute will have a natural tendency to move from the areas of higher concentration to lower, thus diffusing.

consumer Any organism that eats other organisms rather than producing its own food.

continental drift The lateral movement of continental plates over the globe, allowing continents to divide and rejoin in different patterns. This process can separate populations of organisms, providing the geographic barriers that can result in speciation.

contractile ring Two sets of protein filaments that work together to constrict the equator of an animal cell in cytokinesis, thus dividing the cell.

control A treatment in an experiment in which no parts are varied, so that other treatments can be compared to this standard and the effect of the variables can be determined.

convergent evolution Evolution that occurs when nature shapes two separate evolutionary lines in analogous ways, so they appear more similar than would be expected from their evolutionary relationship.

coral reef The piled up limestone deposits that are the remains of many generations of coral animals. These reefs occur in tropical oceans and provide habitat for a rich diversity of organisms.

cork Cells, dead in their mature state, that form the outermost covering of woody plants. These cells are infused with a waxy substance that protects the plant from drying out and from invaders.

cork cambium Secondary meristematic tissue in woody plants that forms the outer living covering of woody plants. This cambium produces the cork cells that, in their dead, mature state, protect the outside of the trunk and branches.

coronary artery An artery that delivers oxygenated blood to the muscles of the heart. Blockage of coronary arteries causes heart attack.

corpus luteum The structure that develops in the mammalian ovary from the ruptured follicle after ovulation. The corpus luteum secretes hormones that help prepare the reproductive tract for pregnancy. If pregnancy does not occur, this body degenerates.

cortex In dicotyledonous plants, the area of ground tissue in the stem that is outside the ring of vascular bundles.

cotyledon An embryonic leaf. A major division in plants is between the monocots, which have one embryonic leaf, and the dicots, which have two embryonic leaves.

coupled reaction The situation in which an exergonic and an endergonic reaction are brought together, with the exergonic reaction releasing energy that is used to power the endergonic reaction.

covalent bond A type of chemical bond in which two atoms are linked because they share electrons.

Cretaceous Extinction A mass extinction event that occurred at the boundary between the Cretaceous and Tertiary periods. This event, which included an asteroid impact, resulted in the end of the dinosaurs along with many other organisms.

cross-bridge Link formed when a head from a thick myosin filament binds with an adjacent thin actin filament in skeletal muscle tissue. The myosin head pivots at its base, sliding the thin filament towards the center of the sarcomere and thus contracting the muscle.

cross-pollinate Pollinate one plant with pollen of another plant. Mendel used this technique in conducting his experiments to uncover rules of heredity.

crossing over The exchange of pieces of chromosomes that occurs when homologous chromosome pairs intertwine during prophase I. This process, also called recombination, accounts for some of the diversity in individual gametes.

cuticle In plants, the waxy outer covering of leaves and shoots. The cuticle forms a protective layer over the dermal tissue, keeping water in and infecting bacteria and fungi out.

cytokinesis The stage of cell division in which a cell's cytoplasm divides.

cytoplasm The region inside the plasma membrane and outside the nucleus. Usually, this region is filled with the jelly-like cytosol containing the cell's extra-nuclear organelles.

cytoskeleton The set of protein strands that function as internal scaffolding for a cell. These protein strands give the cell shape, anchor its structures, act as monorails for particles moving within the cell, and allow whole cells to move. Microfilaments, intermediate filaments, and microtubules are all parts of the cytoskeleton.

cytosol The protein-rich, jelly-like fluid in which a cell's organelles that are outside the nucleus are immersed.

deciduous Refers to plants that show a coordinated, seasonal loss of leaves. This strategy allows plants to conserve water during a time they could perform little photosynthesis anyway.

decomposer A type of detritivore that, in feeding on dead or cast-off organic material, breaks it down into its inorganic components, nutrients that can then be taken up by plants. Decomposers are all fungi or bacteria.

defecation The excretion of solid wastes through the rectal opening (anus).

deletion A chromosomal condition in which a piece of a chromosome has been lost. This occurs when a chromosomal fragment that breaks off and does not rejoin any chromosome.

denatured Pertaining to a protein that has lost its precise conformation, and therefore also its function.

dendrite An extension of the cell body of a nerve cell that generally receives information that it transmits to the neuron. Typically, a neuron has many dendrites.

density-dependent Having an effect that is proportional to the size of a population. These are biological factors, such as food supply and disease, that act to limit the population size.

density-independent Having an effect that is not related to the size of a population. These are physical forces, such as temperature and rain, that act to limit the population size.

deoxyribonucleic acid (DNA) A molecule, composed of a huge number of nucleotides, that provides the information for the construction of proteins and RNA. DNA can replicate itself, and it carries genetic information between cells and between generations.

derived character A character that is found in the descendents of a common ancestor, but was not found in that ancestor. Cladistics distinguishes ancestral from derived characters and uses these characters to determine evolutionary relationships.

dermal tissue The epidermis, or outer layer of cells of a plant. In addition to covering the plant, dermal tissue forms trichomes such as root hairs.

dermis The thick layer of the skin, composed mostly of connective tissue, that underlies, nourishes, and supports the epidermis.

desertification The transformation of an area into a desert, sometimes through human activity.

determined Referring to a cell during development whose fate is predictable.

detritivore An organism that feeds on the remains of dead organisms or the cast-off material from living organisms.

dicotyledon A type of plant that has two embryonic leaves within the seed. More than three-quarters of all flowering plants are dicotyledons.

diffusion The movement of molecules or ions from areas of their higher concentration to areas of their lower concentration. Over time, the random movement of molecules will result in the even distribution of the material.

digestive system The organ system that transports food into the body, secretes digestive enzymes that help break down food to allow it to be absorbed by the body, and excretes waste products. This system consists of esophagus, stomach, and large and small intestines, plus the accessory glands that produce the enzymes along the way.

digestive tract The muscular tube that runs through the body from mouth to anus, through which food passes.

dihybrid cross An experimental cross in which the plants used differ in two of their characters.

dikaryotic Having two haploid nuclei in one cell. A dikaryotic phase occurs in the life cycle of fungi because there is a delay between cytoplasm fusion and fusion of the nuclei, so all cells of a growing mycelium are dikaryotic.

diploid Having two chromosomes of each type per cell. Animal cells, except gametes, are diploid.

directional selection The type of selection that moves a character toward one of its extremes. This process could lead to taller and taller redwood trees.

disruptive selection The type of selection that moves a character toward both of its extremes, operating against individuals with the average for that character. This type of selection seems to be less common in nature than either stabilizing or directional selection.

distal tubule In the nephron tubule of the kidney, the convoluted portion that lies between the loop of Henle and the collecting duct. Some reabsorption of salt and water occurs here.

DNA ligase An enzyme that acts during DNA replication to strengthen the bonds between the newly matched base pairs.

DNA polymerase An enzyme that is active in DNA replication, separating strands of DNA, bringing bases to the parental strands, and correcting errors by removing and replacing incorrect base pairs.

domain A major taxonomic grouping of organisms. There are only three domains: Archaea, Bacteria, and Eukarya.

dominant disorder A disorder in which a single faulty allele can cause damage, even when the individual's second allele for that gene is normal.

dormancy A state in which growth is suspended and there is a prolonged low level of metabolic activity. Dormancy allows organisms to conserve energy during times of unfavorable environmental conditions.

dorsal root ganglion The collection of cell bodies of the sensory neurons that send their axons into the spinal cord through the dorsal root, carrying sensory information into the central nervous system.

double bond Type of covalent bond formed when two pairs of electrons are shared between atoms.

double fertilization In plants, the fusion of sperm with the egg and with the central cell at the same time, one resulting in the zygote, the other in endosperm to provide food for the embryo. Double fertilization occurs almost exclusively in flowering plants.

Down syndrome A disorder in humans in which affected individuals usually have three copies of chromosome 21. Individuals with this syndrome have short stature, shortened life span, and low IQ.

duodenum The portion of the small intestine in vertebrates that is closest to the stomach.

ecological community All living things in a defined area. The different individuals and types of individuals will interact with and depend on each other.

ecological dominant A species that is abundant and obvious in a given community. In any community, a few species, usually plants, will dominate in numbers.

ecology The study of the interaction of organisms with each other and with their environment.

ecosystem A self-sustaining community of organisms and the nonliving elements with which they interact.

ectopic pregnancy An abnormal pregnancy in which the blastocyst attaches inside the uterine tube or to the cervix, rather than in the normal position in the dorsal uterine wall. The embryo usually does not survive in this type of pregnancy.

ectothermic Having the primary source of body heat outside of the body. For example, lizards are ectotherms that often bask in the sun to warm up.

effector A muscle or gland that responds to instructions received from a motor neuron, thus effecting change in the body.

efferent division The division of the peripheral nervous system that carries motor commands from the central nervous system (CNS), towards the effectors (muscles and glands).

electrical charge A quantity of unbalanced electricity. For a molecule, the electrical charge is written as one or more times that of an electron. Electrical charge enables some types of chemical bonds to occur.

electron A basic constituent of an atom that has negative electrical charge. Electrons are distributed in an atom at a distance from the nucleus. Electrons interact to form chemical bonds between atoms, and the electronic structure determines the chemical properties of an atom.

electron transport chain (ETC) The third stage of cellular respiration, occurring within the inner membrane of the mitochondria, in which the majority of the ATP are formed. In this process, electrons are transported along a chain of molecules, providing energy to pump hydrogen ions against their concentration gradient. When these hydrogen ions return down their concentration gradient, the energy released drives the synthesis of up to 32 ATP per molecule of glucose.

electronegativity The measure of the strength of attraction a molecule has for electrons. A molecule with higher electronegativity will tend to pull electrons away from molecules with lower electronegativity.

element A substance that cannot be reduced to any simpler set of components through chemical processes. An element is defined by the number of protons in its nucleus.

embryo An early developmental stage of an organism, before it has emerged from the egg, seed, or body of its mother.

embryo sac The mature female gametophyte plant, consisting of seven cells produced by the megaspore. These cells include the egg and the central cell, which contains two nuclei.

embryology The study of how animals develop, from fertilization to birth. Comparing embryology in different organisms can provide insights into their evolutionary relationships.

emigration The movement of individuals out of a population. The individuals that leave remove their alleles from the gene pool of that population.

emulsification The breaking of a large droplet into many small droplets, greatly increasing the surface area on which digestive enzymes can act.

endergonic Denoting a chemical reaction in which the products contain more energy than the reactants, so energy is stored in the reaction. Energy must be put into such a reaction to make it go.

endocrine gland A gland that releases its materials directly into surrounding tissues or into the bloodstream, without using ducts. Many hormones are produced by endocrine glands.

endocrine system The organ system that sends signals throughout the body using hormones, which generally travel through the blood. This system consists of the endocrine glands and other organs that produce hormones.

endocytosis The process by which the plasma membrane folds inward and pinches off, bringing relatively large materials into the cell enclosed inside a vesicle.

endomembrane system The network of membrane spheres within a cell that can bud, move through the cytosol, and fuse with other membranes.

endometrium The tissue lining the interior of the uterus in mammals, which thickens in response to progesterone secretion during ovulation and is shed during menstruation. If pregnancy occurs, this tissue houses the embryo.

endosperm The tissue, serving as food storage, that surrounds an angiosperm embryo in the seed. The rice and wheat grains that we eat consist mostly of endosperm.

endothelium The innermost layer of blood and lymph vessels, made up of a type of flat epithelial cells. Capillary walls consist of one layer of endothelial cells, and exchange of gases, nutrients, and wastes occurs through these cells and the junctions between them.

endothermic Having the primary source of body heat come from inside the body, from the organism's own metabolism. For example, mammals are endotherms that generate their own body heat internally.

energy The capacity to bring about movement against an opposing force.

energy flow model A way of looking at an ecosystem that tracks and quantifies energy as it is passed through components of that ecosystem.

environmental resistance All the forces in the environment that act to level off the rate of growth and limit population size. Limited food or sunshine, low temperature or rainfall, or predators are some components of environmental resistance.

enzyme A chemically active type of protein that speeds up, or in practical terms enables, chemical reactions in living things.

epicotyl All tissue of an embryonic or seedling plant above the cotyledons. The epicotyl gives rise to the first true leaves.

epidermis The outermost layer of skin in animals, or the outermost cell layer in plants.

epididymis The mass of tubules near the testis in which sperm complete their development and are stored.

epoch A division of a geologic period, which is in turn a division of an era. The Pleistocene is an epoch.

equilibrium species A species whose population size stays relatively stable and at or near its carrying capacity. Also known as a *K*-selected species.

era A large division of geologic time. The Precambrian, Paleozoic, Mesozoic, and Cenozoic are eras. An era includes several periods.

erythrocyte A blood cell, also known as a red blood cell, that transports oxygen to and carries carbon dioxide from every part of the body.

esophagus The muscular tube that begins at the pharynx and ends at the stomach, forming a portion of the digestive tract.

estrogen A class of hormones, produced by cells of the ovary, that supports egg development, growth of uterine lining, and development of female sex characteristics.

estuary An area where a river flows into the ocean, bringing freshwater and saltwater habitat together. Estuaries are among the most productive ecosystems on Earth.

Eukarya The domain (major group) that includes all organisms whose cells have a nucleus. This group includes plants, animals, fungi, and protists.

eukaryotic cell A cell containing a nucleus that is bound within a thin membrane and contains almost all of the cell's DNA. Eukaryotic cells have internal specializations, and this allows different cells to specialize in a multicellular organism. Most familiar living things—your pets, your plants, you—are eukaryotes.

eutrophic Pertaining to a lake that has many nutrients, and therefore sustains many photosynthesizers. Eutrophic lakes often have abundant algal cover, cutting down visibility in the water.

evolution The modification of populations of living things over time, sometimes resulting in the development of new species. Evolution is of central importance in biology because every living thing has been shaped by evolution.

exergonic Denoting a chemical reaction in which the reactants contain more energy than the products, so energy is released in the reaction. Exergonic reactions provide the energy for other, endergonic processes.

exocrine gland A gland that secretes its materials through ducts (tubes). For example, sweat glands conduct perspiration through ducts to the skin.

exocytosis The process in which a transport vesicle fuses with the plasma membrane of a cell and the contents of the vesicle are ejected outside the cell. Exocytosis can be used to expel waste products in single-celled organisms or to export protein products in multi-celled organisms.

exon A portion of an mRNA transcript (or its complementary DNA strand) that codes for amino acids. Before editing of eukaryotic mRNA, there are many introns (noncoding sequences) interspersed within the exons.

exponential growth Growth in numbers that occurs from exponential increase, in which the number produced is proportional to the number already in existence. This type of growth in the size of a population can produce huge numbers of individuals over time.

exponential increase An increase in numbers that is proportional to the number already in existence. This type of increase occurs in populations of living things, and it carries the potential for enormous growth of populations.

extremophile Type of organism that flourishes in extreme conditions such as high heat, high pressure, high salt, or extreme pH. Many archaea and some bacteria are extremophiles, and the enzymes that allow them to thrive in their environments can be used by biotechnologists to make products useful to humans.

extrinsic isolating mechanism A barrier to interbreeding of populations that is not an inherent characteristic of the organisms. Geographic separation of populations is an extrinsic isolating mechanism.

facilitated diffusion Diffusion that requires transport protein channels through the plasma membrane to allow polar molecules larger than water to pass through the membrane down their concentration gradients.

facilitation In primary succession, the actions or qualities of earlier species that in some way assist growth of later species.

Fallopian tube The tube that transports the ovulated egg from the ovary to the uterus in humans. Fertilization occurs inside this tube.

family A taxonomic grouping of related genera. This category lies below order and above genus. Humans are in the family Hominidae.

fast block to polyspermy The mechanism that quickly prevents additional sperm from entering an egg after it is fertilized. Membrane channels in the egg open and sodium ions rush in, changing the electrical potential of the egg membrane and keeping sperm out.

fatty acid A long chain of carbon and hydrogen atoms punctuated at the end with an acid carboxyl (COOH) group. Fatty acids are found in glycerides (common lipids) attached to the alcohol "head."

fertilization The fusion of two gametes to form a zygote. In humans (and most other organisms), the haploid sperm and egg fuse, and the resulting zygote is diploid.

fetus A vertebrate embryo that has completed the earliest, major features of development. A developing human from the end of the first trimester until birth is referred to as a fetus.

fever A sustained, elevated body temperature which speeds up the activities of the immune system and may therefore more quickly rid the body of an infection. In humans, fever is defined as a continued body temperature greater than 37.2°C (99°F).

fiber In skeletal muscle, a single elongated muscle cell, containing hundreds of long, thin myofibrils that run the length of the cell.

fibrinogen A type of protein in blood plasma that, under certain conditions, forms into long strands of fibrin, the framework molecule of a blood clot.

fibrous root system A plant root system that consists of many roots all about the same size.

filament The part of a stamen (male reproductive part of a flower) that is shaped like and functions as a stalk, and has an anther at the top.

filtrate The fluid that is passed from the blood in the glomerulus into Bowman's capsule of the nephron in the first stage of urine formation in the kidney.

filtration The process of passing fluid from the blood into the kidney at Bowman's capsule. Materials such as red blood cells, proteins, and lipids are too large to pass out of the capillaries of the glomerulus, and these remain in the blood.

fimbria A finger-like projection of the funnel-shaped entrance to the uterine tube in mammals. The cilia on these fimbriae sweep the oocyte into the tube.

first filial generation The offspring of the parental generation in an experimental cross.

first law of thermodynamics Energy cannot be created or destroyed, but only transformed from one form to another. The process of transformation of energy is never totally efficient, and some energy is always converted into heat.

fission The type of asexual reproduction in which the cell or whole body divides into two or more parts that each become whole cells or bodies.

fitness The success of an organism, relative to other members of its population, in passing on its genes to offspring. Fitness is a relative concept only; some organisms are better than others at passing on their genes in a given environment at a given point in time.

fixation The process of a gas being incorporated into an organic molecule.

flagellum A long, whip-like cell extension, made of microtubules, that helps the cell move. Only one or a few flagella will occur on any given cell. The tail of a sperm cell is a flagellum that enables it to swim to an unfertilized egg.

fluid-mosaic model The widely accepted view of the cell membrane as a mosaic of proteins moving within the fluid that is the phospholipid bilayer.

follicle In the vertebrate ovary, the complex of the oocyte (developing egg) and the cells and fluids that surround and nourish the oocyte.

follicle stimulating hormone (FSH) A hormone, secreted by the anterior pituitary, that promotes egg development and stimulates secretion of estrogens in women, and supports sperm production and testosterone secretion in men.

food web A complex set of interactions among organisms by which energy and materials are passed among them. A food web includes organisms at all trophic levels.

forensic DNA typing Popularly known as genetic fingerprinting, this technique compares the number of copies of short, repeated stretches of DNA between DNA samples. These DNA patterns are highly variable among individuals, so a sample can be matched to a suspect.

founder effect The effect on the genetic makeup of a small subpopulation migrating to a new area and starting a new population. The genetic makeup of the few founder individuals defines the gene pool for the new population.

fraternal twins Two siblings that develop at the same time, but from separate eggs and sperm. Multiple ovulations, fertilizations, and implantations must take place for fraternal twins to develop.

free radical A molecule with an unpaired electron, usually existing for only a very brief time. Although free radicals are natural products of biological processes, these molecules can be destructive in living tissues. One theory of aging states that the damage from free radicals builds up over time in structures including DNA.

free ribosome A type of ribosome that is located in the cytosol of a cell, not attached to the endoplasmic reticulum. Generally, proteins that will reside in the cell's plasma membrane and proteins that will be exported out of the cell are synthesized on the free ribosomes.

fruit The mature ovary of any flowering plant. Many fruits protect the underlying seeds, and many attract animals that will eat the fruit and disperse the seeds.

gallbladder A muscular sac, located beneath the right lobe of the liver in humans, that stores and concentrates bile, a digestive substance produced by the liver.

gamete A haploid reproductive cell, either egg or sperm.

gametophyte generation The generation in a plant's life cycle that produces gametes (sperm or egg). This generation is microscopic in angiosperms.

ganglion Any collection of nerve cell bodies in the peripheral nervous system.

gap junction A type of cell junction that facilitates cell communication, made up of clusters of protein structures extending from one side to the other of the cell's plasma membrane. When these structures line up in adjacent cells, the channel formed allows small molecules and electrical signals to pass through.

gastrointestinal (GI) tract Generally refers to the lower portion of the digestive tract (the stomach and intestines).

gastrulation The stage of animal development in which the zygote's cells move in a programmed fashion to form three layers, each of which will give rise to particular organs or tissues.

gel electrophoresis A technique used to separate DNA fragments by size, using electrical charge to run the fragments through a gel. Smaller fragments move faster, and the different lengths of fragments arrange in bands according to size.

gene A sequence of DNA nucleotides that encodes a molecule of RNA. The gene may contain instructions for making an individual protein, or it may have a regulatory function.

gene flow The movement of genes from one population to another. Gene flow can introduce new alleles to a population, and it also can prevent two populations from becoming genetically separate from each other.

gene pool All alleles that exist in a population. The gene pool is the raw material that evolution works with.

gene therapy The process of inserting working genes directly into cells of a person who lacks working copies of that gene. The concept is still largely theoretical because it has proven difficult to carry out successfully. It is also known as human gene transfer.

genetic code The inventory that specifies which nucleotide triplets code for which particular amino acids or for start or stop commands in protein synthesis. With few exceptions, this genetic code is universal in all living things.

genetic drift The chance alteration of allele frequencies in populations. When populations are small, chance can have a profound effect on their genetic makeup.

genetics The study of genes and heredity.

genome The complete haploid set of an organism's chromosomes.

genotype The genetic makeup of an organism, including all the genes that lie along its chromosomes.

genus A taxonomic grouping of related species. This category is below family and above species. Humans are in the genus *Homo*.

germ-line cell A cell that becomes an egg or a sperm cell. Mutations in germ-line cells are heritable because the affected gamete may become part of an offspring.

glomerulus The ball of capillaries that sits inside the Bowman's capsule of a vertebrate kidney. Filtrate passes from the blood in the glomerulus into Bowman's capsule in the first stage of urine formation.

glycocalyx An outer coat of a cell consisting of carbohydrate and sugar chains protruding from the proteins and phospholipids of the membrane, plus some peripheral proteins.

Golgi complex The organelle in the cytosol that receives proteins via transport vesicles from the rough ER, processes them by adding or trimming side groups, and sorts and sends them to the correct destinations. The Golgi complex thus acts as a processing and distribution center for protein products of the cell.

glyceride The most common type of lipid, consisting of a head made of an alcohol (usually glycerol) attached to fatty acid subunits.

glycerol A particular kind of alcohol, frequently found in the "head" of glyceride.

glycogen A complex carbohydrate that serves as the major form of carbohydrate storage for animals. Glycogen is stored in muscle cells and in the liver.

glycolysis The first stage of cellular respiration, occurring in the cytosol. Each molecule of glucose is converted to two molecules of pyruvic acid that go on to the Krebs cycle. Two NADH that go directly to the electron transport chain and two ATP are also produced. Glycolysis is the sole process for extracting energy from food molecules in organisms such as bacteria. In more complex organisms, it is a necessary precursor to the other two stages of cellular respiration.

glycoprotein A molecule that combines protein and carbohydrate. Glycoproteins play important roles as cell receptors and some types of hormones, among other functions.

gonadotropin-releasing hormone A hormone released in tiny amounts from the brain's hypothalamus that stimulates the anterior pituitary to release two other hormones (follicle-stimulating hormone and luteinizing hormone) important in reproduction.

granum A stack of flattened vesicles (thylakoids) inside a chloroplast. The light-dependent reactions of photosynthesis occur here.

gravitropism The bending of a plant's roots or shoots in response to gravity. This tendency helps a plant orient properly with its roots in the ground.

gray matter The tissue within the central nervous system that appears gray and consists largely of cell bodies, plus interneurons and neuroglia.

gross primary production The proportion of the sun's energy received that a plant assimilates. This is a rough measure of efficiency of a plant, before subtracting the plant's energy costs and losses.

groundwater Water that is found underground in porous rock saturated with water.

growth hormone In humans, this hormone stimulates overall body growth through cell growth. Also known as somatotropin.

gymnosperm A type of plant that is vascular and has seeds, but lacks flowers. These were the first seed plants, liberated from water because sperm move to the egg inside pollen grains, which can be carried by the wind.

habitat The surroundings in which individuals of a species are normally found.

hair follicle A tiny complex organ, formed from epidermis, from which hair grows.

hair papilla A peg-like structure of connective tissue, containing capillaries and nerves, at the base of a hair follicle.

half-life The time needed for one-half of a given amount of a radioactive element to decay to another isotope or element. This time differs among elements but is uniform for each element. Half-life can be used, along with measured proportions of elements, to radiometrically date a rock layer.

haploid Having only one chromosome of each type per cell instead of a pair of chromosomes of each type. Cells that have undergone meiosis are haploid, and some organisms are always haploid.

heart attack An episode of heart failure in which, due to blockage of a coronary artery, the blood supply is cut off to a group of cardiac muscles. The interruption in blood supply can cause the muscles to stop beating or to beat in a wild, irregular rhythm.

heartwood An area of dead, darker colored, inner wood in a tree trunk, where xylem tubes no longer conduct sap and are filled with oils, lignin, and microbe-fighting substances for strength and resistance to bacteria and fungi.

HDL High density lipoprotein. A lipoprotein, carried in the blood, that has a relatively high ratio of protein to lipid. HDLs carry cholesterol away from outlying tissues to the liver, and a high proportion of HDL is a predictor of keeping a healthy heart.

helper T cell A type of T cell, active in antibody-mediated immunity, that binds to specific antigen-presenting cells and secretes a protein (interleukin-2) that has multiple functions in mounting the immune system's attack on the invader.

hemoglobin The iron-containing protein in red blood cells that binds to oxygen. Hemoglobin gives our blood its red color.

herbaceous plant A plant that never develops wood (secondary xylem), and therefore has relatively thin and weak shoots. Herbaceous plants do not undergo secondary growth (thickening).

herbivore An animal that eats only plants.

heterotroph A type of organism that cannot manufacture its own food, so it must get food from elsewhere. Animals are heterotrophs.

heterozygous Possessing two different alleles of a gene for a given character. A heterozygous individual could pass on either allele to its offspring, so its offspring may differ in that character.

HIV Human immunodeficiency virus, the causative agent of the disease AIDS. This virus attacks the body's immune system, eventually leaving it unable to fight off even minor infections.

homeostasis A process that maintains a relatively stable internal environment of an organism, even when the outside environment varies. An example of homeostasis is that humans maintain a body temperature of about 98.6°F, even when the outside temperature fluctuates.

homologous In anatomy, having the same structure owing to inheritance from a common ancestor. Forelimb structures in whales, bats, cats, and gorillas are homologous.

homologous chromosomes A matched pair (usually) of chromosomes that are the same in shape and function. One chromosome of each pair comes from each parent. The sequence of genes is the same on both chromosomes of the pair, but the allele (such as blue or brown eye color) may be different.

homology A structure that is shared in different organisms owing to inheritance from a common ancestor. Homologies are used to help decipher evolutionary relationships.

homozygous Having two identical alleles of a gene for a given character. A homozygous individual only possesses one allele, and therefore can pass only that allele to its offspring.

homozygous dominant Having two identical alleles of a gene, with the allele being dominant, so that this allele is the one expressed in the heterozygous condition.

homozygous recessive Having two identical alleles of a gene, with that allele being recessive, or the one not expressed in the heterozygous condition.

hormone A chemical messenger made in one part of an organism that influences the functioning of another part of the organism. Both plants and animals have hormones.

host The prey in a parasitic relationship.

human gene transfer Also known as gene therapy, this is the process of inserting working genes directly into cells of a person who lacks working copies of that gene.

Human Genome Project A massive, multinational effort to completely sequence the genome of humans, identifying genes and the protein products of the genes.

hydrocarbon A compound made of hydrogen and carbon. Hydrocarbons are nonpolar covalent molecules and therefore cannot be dissolved in water.

hydrochloric acid The acid, secreted in the vertebrate stomach, that lowers the pH of the gastric juice, kills microorganisms, and breaks down cell walls and connective tissues in food.

hydrogen bond A relatively weak bond formed when a hydrogen atom, already covalently bonded to another atom, is attracted to an electronegative atom. The hydrogen can be linked to another atom in the same molecule, in which case the hydrogen bond helps hold a molecule in a certain shape, or it can be linked to an atom in another molecule, thus weakly holding the two molecules together.

hydrolysis The splitting of bonds in a molecule by reaction with water.

hydrophilic Property of a compound indicating that it will interact with water. Table salt (NaCl), which dissolves readily in water, is hydrophilic.

hydrophobic Property of a compound indicating that it will not interact with water. Water excludes these compounds by bonding with other water molecules, not with the compound. Oil is hydrophobic and will never dissolve in water.

hydroxide ion The OH⁻ ion. Compounds that yield hydroxide ions are strongly basic, so they can be used to counteract acids and shift solutions towards neutral or basic on the pH scale.

hypha A slender tube structure of a fungus that is one of many making up the mycelium. The hyphae are the bulk of the fungus, and they dissolve the food.

hypothalamus Small section of the vertebrate brain, lying below the thalamus and above the pituitary. The hypothalamus is linked by a narrow stalk to the pituitary, making it the main link between the nervous and endocrine systems. Cell bodies involved in emotions, autonomic functions, and hormone production are also located here.

hypothesis A tentative, testable explanation of an observed phenomenon. By proposing and testing new hypotheses, scientists advance their understanding of the world.

identical twins Two siblings that develop from the same zygote, so they are genetically identical. The splitting of one zygote into two can happen at either the zygote stage or the blastocyst stage.

immigration The movement of individuals into a population. Immigration can add genetic diversity to the population.

immunity Specific resistance, or long-lasting protection that results from reaction of the specific defenses of the immune system to foreign chemical compounds or pathogens.

incompletely dominant The condition in which neither of two alleles in a heterozygous organism is completely dominant to the other, so the phenotype expressed is different from what the phenotype would be if the individual were homozygous for either allele.

independent assortment The random alignment of homologous chromosomes at the metaphase plate during metaphase I of meiosis. This random alignment assures that different gametes will have different alleles, accounting for some of the diversity among offspring.

inducer Substance that prompts the activity of an inducible gene. This system allows genes to be turned on only when they are needed to perform their function.

induction The capacity of some embryonic cells to direct the development of other embryonic cells. For example, the notochord induces the tissue above it to form the neural tube. Induction helps organize development.

inflammation A localized tissue response to injury. Inflammation interferes with the spread of pathogens and initiates a wide range of defenses to overcome the pathogens and repair damaged tissues.

inland wetlands Lands away from the ocean that are wet at least part of the year. The overwhelming majority of wetlands are inland and freshwater.

inner cell mass In mammalian development, the group of cells in the embryo that will develop into the baby, rather than into the placenta.

inorganic Pertaining to nonliving things, or compounds made up of nonliving things. Also, pertaining to the chemistry of compounds that do not contain carbon. Inorganic fertilizer is made of pure elements within binding materials, made by chemical processes.

integral protein A protein of the plasma membrane that is attached to the hydrophobic interior. Such proteins may extend out one or both sides of the membrane and may be involved in cell recognition, cell communication, and transport of materials through the membrane.

integumentary system The organ system that protects the body from the external environment and assists in regulation of body temperature. This system consists of the skin and associated structures, such as glands, hair, and nails.

integuments The outer layers of tissue of an ovule, surrounding the embryo sac in a flower. These tissues develop into the seed coat that protects the embryonic plant.

intercalary meristem Meristematic tissue found at the base of each growing node in grasses. These multiple growing regions permit fast growth, and also allow grasses to keep growing even when their top is removed by grazing or mowing.

interferon A small protein that causes normal, uninfected cells to produce compounds that interfere with viral replication. Interferon is released by activated lymphocytes and macrophages, and by tissues infected with viruses.

intermediate electron carrier A molecule in the electron transport chain of photosynthesis that serves to accept and then in turn donate electrons down the energy hill.

intermediate filament The protein component of the cytoskeleton that is intermediate in size but is the most permanent. The positions of the nucleus and organelles are stabilized by intermediate filaments.

interneuron A type of neuron, located entirely within the brain and spinal cord, that connects other neurons. These neurons are responsible for the analysis of sensory inputs and the coordination of motor commands.

interphase The stage of the cell cycle in which a cell simultaneously carries out its work, grows, and duplicates its chromosomes. Cells spend most of their time in interphase.

interspecific competition Competitive interaction between individuals of two different species. The competition tends to be indirect, and the winner triumphs by being more efficient at acquiring something, such as food.

interstitial fluid Fluid that has been pushed out of the smallest blood vessels into the surrounding body tissues. This fluid will later enter the lymphatic system.

intertidal zone The region within the coastal zone of the ocean that extends from the ocean's low-tide mark to its high-tide mark.

intraspecific competition Competitive interaction between individuals of the same species.

intrinsic isolating mechanism A barrier to interbreeding of populations that is an inherent characteristic of the organisms. Examples include temporal, behavioral, and gametic isolation.

intrinsic rate of increase The rate at which a population would grow if there were no external limits on its growth. In ecology, this is often denoted as r.

intron A series of nucleotides that lies within a gene sequence but does not code for amino acids. Before editing of eukaryotic mRNA, there are many introns interspersed within the exons (coding sequences).

inversion Flipping of a chromosome fragment before it rejoins the chromosome it came from, so the order of the fragment's chemical sequence is reversed from normal. Inversion can affect the phenotype.

ion An atom that has positive or negative charge because it has fewer or more electrons than protons. Ions will be attracted to other ions with the opposite charge and can form ionic compounds.

ionic bonding A type of bond in which two atoms are attracted to each other by their opposite charge.

ionic compound A substance in which two or more atoms are held together through ionic bonding. Table salt (NaCl) is a familiar example.

isomer A form of a molecule that has the same chemical formula but a different spatial arrangement than another isomer. Because of their different shapes, different isomers can have different properties.

isotope A form of an element containing a specified number of neutrons. Different isotopes of an element have different numbers of neutrons but the same number of protons. Some isotopes may be more stable than others, a property that is used in medicine when radioactive isotopes are used to diagnose illness.

J-shaped growth Exponential growth. When plotted on a graph, this type of increase in numbers can be so extreme that it looks like the letter J.

junk DNA DNA sequences that have no known function and can have huge numbers of copies within the genome. Junk DNA can lie within gene sequences (forming introns) or between gene sequences. Some "junk DNA" may actually form binding sites for regulatory proteins or have other, still undetermined functions.

karyotype A pictorial arrangement of a full set of an organism's chromosomes. A karyotype can help locate chromosomal abnormalities such as an extra piece of a chromosome. Such abnormalities can have profound consequences for the organism.

keratin A tough, fibrous, water-resistant protein formed by some epidermal tissues. In skin, it keeps water in and invaders out. Also found in claws, feathers, hair, and hooves.

keystone species A species whose absence would bring about a significant change in the community.

kidney In vertebrates, the organ that regulates water balance in the blood and excretion of metabolic wastes as urine.

kinetic energy The energy of motion. A rolling rock has kinetic energy.

killer T cell A type of T cell, active in antibody-mediated immunity, that attaches to infected cells in the body and punctures their outer membranes, causing them to rupture.

kilocalorie The amount of heat it takes to raise one liter of water one degree Celsius.

kingdom A taxonomic grouping of related phyla of organisms. Kingdom is a major category, with only a domain including more species. Humans are in the kingdom Animalia.

knot In a tree trunk, a branch that died out when the tree was younger and then was surrounded by secondary xylem.

Krebs cycle The second stage of cellular respiration, occurring in the inner compartment of mitochondria. This cycle of reactions yields 6 NADH and 2 FADH$_2$ to enter the electron transport chain, plus two ATP per molecule of glucose. Krebs cycle is the major source of electrons for the electron transport chain.

K-selected species Also known as an equilibrium species, this is a species whose population size tends to be at or near its carrying capacity.

labor The regular contractions of the uterine muscles that sweep over the fetus, creating pressure that opens the cervix and expels the baby and the placenta. Hormones released by the fetus cause labor to begin.

lactate fermentation The process in animal cells by which pyruvic acid, the product of glycolysis, accepts electrons from NADH to form lactic acid. The lactic acid is a waste product for animals.

lactic acid A waste product formed in animal cells when pyruvic acid, the product of glycolysis, accepts electrons from NADH. Lactic acid buildup causes muscle burn when a person exercises faster than aerobic respiration can keep up.

lateral bud An undeveloped plant shoot, located between the stem and a leaf, that may give rise to a branch, a flower, or even take over the role of shoot apex if the apical meristem becomes damaged. All buds are composed mostly of meristematic tissue.

law of conservation of mass A fundamental rule of nature stating that matter is neither created nor de-stroyed. Therefore, the mass of the products in a chemical reaction will equal the mass of the reactants.

Law of Segregation Mendel's insight that organisms have two genetic elements that separate when gametes are formed. The physical basis for this law is the separation of homologous chromosomes during meiosis, and it produces much of the variation in traits among offspring.

Law of Independent Assortment Mendel's insight that during gamete formation, gene pairs assort independently of one another. The physical basis for this law is the independent assortment of chromosomes at the metaphase plate during meiosis, resulting in traits coded in genes on different chromosomes being inherited independently.

LDL Low-density lipoprotein, a lipoprotein, carried in the blood, that has a relatively low ratio of protein to lipid. LDLs carry cholesterol to the outlying tissues, and a high proportion of LDL is a risk factor for heart attack.

leukocyte Another term for white blood cell. This class of blood cells is critical to immune function.

life cycle The repeating series of steps that occur in the reproduction of an organism.

life science All science dealing with living things, including biology, medicine, botany, agriculture, and fisheries.

life table A table showing how likely it is for an average species member to survive a given unit of time. These tables are used by insurance companies to set insurance rates.

light-dependent reaction The first stage of photosynthesis, which requires light. In this stage, energy from the sun is absorbed and transmitted through an electron transfer chain to yield the ATP used in the second stage of photosynthesis. Without this stage, there would be no energy for the synthesis of carbohydrate in the second stage.

light-independent reaction The second stage of photosynthesis, not dependent on light, in which the high-energy sugar G3P is formed. This product of photosynthesis can be used for energy or plant growth.

lignin A noncarbohydrate molecule that is a major component of cell walls of plants. Lignin provides stiffening by binding and supporting cellulose fibers.

limnetic zone The vertical zone in a lake extending from the surface to a depth where sunlight no longer penetrates strongly enough to drive photosynthesis.

lipase The type of pancreatic enzyme, secreted through ducts, that breaks down lipids.

lipid A type of biological molecule, made up of carbon, oxygen, and hydrogen, that has much more hydrogen relative to oxygen than do the carbohydrates. Pure lipids are insoluble in water. Oils, fats, cholesterol, and some hormones are lipids.

lipoprotein A molecule that combines lipid and protein. Lipoproteins transport fat molecules through the bloodstream to all parts of the body. Low-density lipoproteins (those containing relatively less protein compared to lipid) carry cholesterol out to tissues in the body including arteries, and thus contribute to heart disease.

littoral zone The shallow-water zone in a lake that extends from the edge to the point where water is too deep for rooted plants to grow.

logistic growth A type of growth in numbers that starts like exponential growth, but eventually the rate of growth slows, then stops. This type of growth occurs in natural populations, and the population stabilizes at a certain level, called the carrying capacity.

luteinizing hormone (LH) A hormone, secreted by the anterior pituitary, that, in women, induces ovulation and stimulates the ovary to secrete estrogens and progestins to prepare the body for possible pregnancy. In men, it stimulates the testes to produce androgens such as testosterone.

lymph The fluid carried by the lymphatic vessels. This fluid will be checked by cells of the immune system for signs of injury or infection before it is returned to the blood.

lymphatic Vessel of the lymphatic system, through which lymph flows.

lymphatic system The organ system that collects intracellular fluid, transports it as lymph through vessels, checks the fluid for infection, can destroy invaders, and delivers the fluid to blood vessels. This system consists of the vessels that transport lymph and the lymphoid organs that produce or support the specialized immune cells.

lyse Burst a cell. When a cell invaded by a virus lyses, it releases the many viral clones that have been produced in the cell.

lysogenic cycle A viral life cycle in which a virus invades a cell and integrates its DNA into the host's DNA. The viral DNA is then replicated each time the host cell divides, but the virus does not lyse the cell.

lysosome Membrane-bound organelle in an animal cell that breaks down, exports, or stores old materials and processes new materials. Lysosomes thus clean up a cell and help recycle its parts.

lytic cycle A viral life cycle in which a virus invades a cell, induces the cell to make copies of the virus, lyses (bursts) the cell, and the new viruses then invade more cells.

macroevolution The large-scale patterns produced by microevolution. Looking at these patterns, such as the evolution of mammals from reptiles, helps us understand the history of life on Earth.

macrophage A large type of phagocyte (a type of white blood cell) that can be found in most tissues of the body, where they engulf infected cells and cell fragments.

mass A measure of the resistance of an object to being moved, takes into account density and volume of the object. For most biological purposes, mass is equivalent to weight.

mast cell A special type of connective tissue cell that releases histamines and heparin into the interstitial fluid, thus initiating the process of inflammation.

mechanical work The work done in moving an object through a space. Muscle contraction is mechanical work.

medulla oblongata The section of the vertebrate brain that is connected to the spinal cord. This structure contains major centers involved in autonomic functions, including heart rate, and respiration, and also relays sensory information to the brain stem and thalamus.

megaspore A haploid cell that results from meiosis inside the ovary of a flower. The megaspore then gives rise to several types of cells, including the egg.

megaspore mother cell A single diploid cell in the ovary of a flower that undergoes meiosis, producing megaspores, one of which will develop into the embryo sac containing the egg.

meiosis A process in which a single diploid cell divides to produce haploid reproductive cells. Meiosis produces the gametes for sexual reproduction.

memory B cell The type of B cell produced in antibody-mediated immunity that remains in the body, ready to produce more plasma cells and quickly mount a defense in the event of another invasion by the same invader.

menopause The cessation of the monthly ovarian cycle that occurs when women reach about 50 years of age. The scarcity of follicles remaining in the ovary is the main factor that brings about menopause.

menstruation The release of blood and the specialized uterine lining, the endometrium, occurring about once every 28 days in humans, except when the oocyte is fertilized.

messenger RNA (mRNA) A type of RNA, synthesized in the cell nucleus, onto which DNA's instructions for protein synthesis are copied (transcribed). The mRNA then moves out of the nucleus to ribosomes where the instructions are translated and polypeptide chains are formed. Thus, mRNA carries DNA's instructions for protein synthesis.

metabolic pathway A sequential set of enzymatically controlled reactions in which the product of one reaction serves as the substrate for the next. By such a pathway, complex processes with many steps are carried out.

metabolism The sum of all chemical reactions carried out by a cell or larger organism.

metaphase plate The cellular equator, along which chromosomes align before they separate to move to the cellular poles.

microevolution The change of allele frequencies in populations over time. This change is the essential process of evolution.

microfilament The smallest-sized component of the cytoskeleton, composed of the protein actin. Actin is a support structure of cells, helps a cell grow rapidly at one end to move or capture prey, and, along with the protein myosin, enables muscles to contract.

micrograph A picture taken with the aid of a microscope. Micrographs enable us to see cells and their structures.

micrometer (μm) A millionth of a meter. Most small bacterial cells are 1 to 10 μm in length, and most plant and animal cells are 10 to 100 μm in length.

microphage A small type of phagocyte (a type of white blood cell) that generally circulates in the blood but leaves the bloodstream to enter injured or infected tissues. Neutrophils and eosinophils are the two types of microphages.

micropyle A tiny opening in the integuments that cover the ovule in the female gametophyte of a flower. The pollen tube will grow through the micropyle to fertilize the egg.

microspore A haploid cell that results from meiosis in the anthers of flowers. Microspores develop into pollen grains inside the anthers.

microspore mother cell A diploid cell type, found within the anthers of flowers, that undergoes meiosis to produce haploid microspores that develop into pollen grains.

microtubule The largest-sized component of the cytoskeleton. Microtubules determine the shape of the cell, act as monorails along which transport vesicles travel, form the mitotic spindle, and form cilia and flagella for propulsion.

midbrain The part of the vertebrate brain that processes information about sight and hearing, generates involuntary motor responses to maintain posture, and has centers that help maintain consciousness.

migration The movement of individuals from one population to the territory of another. Both animals and plants (as seeds and pollen) can migrate, and this process facilitates gene flow.

mimicry A phenomenon in which one species has evolved to resemble another species. The mimic species benefits from this similarity because predators avoid it, based on its appearance.

mitochondrion A membrane-bound organelle that converts energy to a form easily used by cells. Mitochondria are the power plants of cells.

mitosis The stage of cell division in which a cell's chromosomes divide.

mitotic phase The stage of the cell cycle in which the duplicated chromosomes separate and the whole cell splits in two.

mitotic spindle The microtubules active in cell division, including those that align and move the chromosomes to evenly divide them in cell division.

mobile Able to move from one location to another, as are most animals.

modern synthesis The unified evolutionary theory that resulted from the convergence of several lines of biological research between 1937 and 1950. Advances in taxonomy, genetics, and paleontology all contributed to this deeper understanding of how evolution and natural selection work.

molecular biology The study of life at the level of its individual molecules, their constituent parts, and how they interact.

molecular formula A notation that specifies the elements in a molecule, with the number of each shown as a subscript. This notation shows us the makeup of the molecule, but not its structure.

molecule A structure with a defined number of atoms in a certain spatial arrangement. The atoms and their arrangement determine how the molecule will interact with other molecules.

monocotyledon A type of plant that has one embryonic leaf within the seed. Although comprising only one-quarter of all flowering plants, most important food plants are monocotyledons.

monohybrid cross An experimental cross in which the plants used differ in one of their characters.

monomer Small molecules that are building blocks for polymers.

monosaccharide The building block (monomer) of carbohydrates. Also known as simple sugars, glucose and fructose are familiar examples.

monounsaturated fatty acid A fatty acid with one double bond, so a hydrogen atom could be added. At room temperature, these fats are generally soft solids.

morphogen A diffusible substance whose concentration in an area of the embryo affects development in that region.

morphology The study of physical forms that organisms take. Comparing these forms can provide insights into the evolutionary relationships of organisms.

morula A tightly packed ball of cells that results from repeated cleavage of the zygote, before gastrulation begins during animal development.

motor neuron A neuron that carries instructions from the central nervous system to an effector (muscle or gland) in the body.

Müllerian mimicry A type of mimicry in which several species that have protection against predators come to look alike. All of the species benefit, because once a predator learns what one distasteful individual looks like, it will avoid all the similar-looking species.

multiple alleles Three or more alternative forms of a gene occurring in a population. One individual will have only two of these alleles, but multiple alleles account for the great range of variation in traits such as skin color in humans.

muscular system The organ system that produces movement and maintains posture of the body. This system consists of all the skeletal muscles of the body that are under voluntary control.

mutation A permanent alteration in a cell's DNA base sequences. Mutations arise as mistakes and are usually neutral and sometimes harmful, but they provide important new genetic sequences that serve as the raw material for natural selection.

mutualism A close association between organisms of two species that benefits both.

mycelium A web of fungal hyphae that makes up the major part of a fungus.

mycorrhiza An association of plant root and fungal hyphae that benefits both. The fungal hyphae absorb minerals, growth hormones, and water that are then available to the plant, and the fungus gets carbohydrates for food from the plant. The association is so important that some plants cannot live without their fungal partners.

myelin The white, fatty covering over axons of vertebrates that enables faster transmission of nerve impulses. Myelin is made of neuroglia cells called Schwann cells, which wrap their cell membranes around the axons.

myofibril The contractile structure of a skeletal muscle cell, containing thick and thin filaments arranged in sarcomeres.

nanometer (nm) A billionth of a meter. A nanometer is 1000 times smaller than a micrometer.

natural killer (NK) cell Type of lymphocyte (a type of white blood cell) that kills abnormal cells by secreting proteins that destroy their cell membranes.

natural selection A difference in survival or reproduction among individuals based on how well their traits suit them for their environment. This concept was one of Darwin's major insights.

negative feedback A system of control in which the product of a process turns off the process, thus preventing more of itself from being formed. Negative feedback is usually temporary and prevents overproduction of a product.

nephron The functional unit of the mammalian kidney, consisting of a nephron tubule, its associated blood vessels, and the interstitial fluid in which both are immersed.

nephron tubule A convoluted tubule in the kidney of birds and mammals. Filtrate passes through the tubule, where reabsorption and secretion occur during the formation of urine.

nerve A bundle of axons in the peripheral nervous system that transmits information to or from the central nervous system.

nervous system The organ system that monitors an animal's internal and external environment, integrates the sensory information received, and coordinates the animal's responses. This system consists of all the body's neurons, plus the supporting neuroglia cells.

net primary production The amount of the energy received by a plant that is made into plant tissues. This value reflects efficiency and the plant's energy costs and losses.

neural crest Embryonic cells, unique to vertebrates, that break away from the periphery of the neural tube as it is folding over during development. These cells migrate to different locations where they develop into different organs and tissues.

neural tube An enclosed, hollow tube that forms from ectodermal cells in the developing animal embryo. Eventually, this structure gives rise to the brain and spinal cord.

neuroglia Associated helper cells of neurons, providing physical support, nourishment, insulation, and defense from infection.

neuron A nerve cell, typically possessing a cell body with a nucleus, a long extension called an axon that transmits information from the neuron, and many dendrites that transmit information to the neuron.

neurotransmitter A substance that is released into a synaptic cleft by a presynaptic neuron and binds to receptors on the postsynaptic neuron or other effector. The neurotransmitter causes the action potential to continue in the postsynaptic cell.

neutron A basic constituent of an atom with no electrical charge, found in the nucleus of the atom. The number of neutrons in the atom defines the isotope.

niche A characterization of an organism's way of making a living that includes its habitat, food, and behavior.

nicotinamide adenine dinucleotide (NAD) The most important intermediate electron carrier in cellular respiration, it picks up a solo electron and a hydrogen atom to form NADH, then donates them to another substance that has greater potential to accept electrons than it does, thus returning to NAD^+ form.

nitrogen fixation The conversion of atmospheric nitrogen into a form that can be taken up by living things. Bacteria fix nitrogen, which is essential to life.

nondisjunction The failure of homologous chromosomes or sister chromatids to separate during meiosis, resulting in unequal numbers of chromosomes in the daughter cells. Nondisjunction results in aneuploidy.

nonpolar covalent bond A type of covalent bond in which electrons are shared equally between atoms, resulting in no positive or negative side of the molecule.

nonrandom mating A type of mating in which a given member of a population is not equally likely to mate with any other given member. Assortative mating and sexual selection both result in nonrandom mating.

notochord A rod-shaped support organ that forms in the developing vertebrate embryo. The notochord influences the tissue above it to form the neural tube, which later gives rise to the brain and spinal cord.

nuclear envelope The concentric, double membrane that forms the outer layer of a cell's nucleus. The nuclear envelope contains nuclear pores that enable substances to pass in and out of the nucleus.

nuclear pore A channel through the nuclear envelope surrounding the cell's nucleus. Materials pass in and out of the nucleus through the thousands of pores in the nuclear envelope.

nucleolus A structure within the nucleus where ribosomes are synthesized. The unfinished ribosomes are passed out of the nucleus and finished in the cytoplasm. Ribosomes are continually wearing out and new ones must be synthesized by the nucleolus.

nucleotide The building block of DNA, this molecule consists of a phosphate group, a sugar, and a nitrogen-containing base. Nucleotides link to form chains, and two chains link together in a ladder and twist to form the double helix of DNA.

nucleus of atom The central core of an atom, containing its protons and neutrons. Nearly all of the mass of the atom resides in its nucleus.

nucleus of cell The membrane-bound organelle in a eukaryotic cell that contains nearly all of its DNA. The nucleus is considered the control center of the cell, and it is the site of DNA synthesis.

nutrient A chemical element that is used by living things to sustain life.

oil Fat in liquid form. Polyunsaturated fatty acids naturally occur as oils at room temperature.

oligotrophic Pertaining to a lake that has few nutrients, and therefore few photosynthesizers and few animals. Oligotrophic lakes are generally clear and deep.

omnivore An animal that eats both plants and animals.

oncogene A cancer-causing gene that acts by prompting excessive cell division. Generally, both oncogenes and faulty-tumor suppressing genes must be active for cancers to start.

one-gene, one-enzyme hypothesis The insight by Beadle and Tatum that each gene calls up production of one enzyme, which undertakes a particular metabolic task.

oocyte The precursor cell of eggs in the vertebrate ovary. Oocytes develop from oogonia before the female is born. In mammals, it is the oocyte stage that is ovulated and fertilized.

oogonium The diploid cell that is the starting female cell in gamete production. Oogonia become primary oocytes while the female is still an embryo, and only a few of these will ever mature in the ovary.

open circulation A type of circulatory system, found in many invertebrates, in which arteries carry blood into open spaces called sinuses instead of into capillary beds. The blood returns to the heart or hearts through openings in the veins.

open sea The region of the ocean that begins where the continental shelf drops off and extends out to sea. This region includes all of the ocean except the coastal zone.

operon A group of genes that are transcribed together onto a single mRNA transcript. The operon governs a specific function, such as lactose metabolism, and contains sites that enable regulation of gene transcription.

opportunist species Species whose population size tends to fluctuate widely in reaction to a highly variable environment. These species have a high intrinsic rate of increase and are also known as r-selected species.

order A taxonomic grouping of related families. This category is below class and above family. Humans are in the order Primates.

organ A highly organized functioning unit within an organism, performing one or more functions. An organ can contain several types of tissues. Kidneys, heart, lungs, and liver are all familiar examples of organs in humans.

organ system A group of interrelated organs and tissues that serve a particular set of functions in the

body. For example, the digestive system consists of mouth, stomach, and intestines, and functions in digestion of food and elimination of waste.

organelle A highly organized, membrane-bound structure inside a cell that performs one or more functions for the cell, like a tiny organ. Organelles are essential to the function of cells. A cell's nucleus and its mitochondria are examples of organelles.

organic Pertaining to organisms, or compounds made up of organisms, as organic fertilizer. Also, pertaining to the chemistry of compounds that contain carbon.

organic chemistry the chemistry of compounds that contain carbon. This branch of chemistry is directly relevant to biology because life on Earth is based on carbon compounds.

organism A living being. An organism can be one-celled, like a bacterium, or can contain many cells, like a dog or a rose bush.

organismal biology The study of whole organisms.

osmosis The net movement of water across a semipermeable membrane from an area of lower solute concentration to an area of higher solute concentration.

osteoblast An immature bone cell that secretes organic material that becomes bone matrix, thus producing new bone.

osteon The functional unit of compact bone, consisting of osteoblasts and the concentric layers of bone produced by them, forming a cylinder of bone with a hole in the center, paralleling the long axis of the bone.

osteoporosis The decrease in bone mass that occurs when bone is broken down faster than new bone is produced. Osteoporosis is especially common in postmenopausal women, and estrogen replacement therapy can help prevent fractures.

ovarian follicle The complex of the oocyte (developing egg) and the cells and fluids that surround and nourish the oocyte in the ovary of vertebrates.

ovary In flowers, the area, located at the base of the carpel, where fertilization of the egg and early development of the embryo occurs. In animals, the female reproductive organ, where eggs develop.

ovulation The release of an oocyte from the ovary in animals.

oxidized An oxidized substance is one that has lost one or more electrons.

oxytocin A hormone, secreted by the posterior pituitary, that stimulates smooth muscle contractions in the uterus and special cells of the mammary glands in women and the walls of the prostate gland in men.

pancreatic duct The tube through which the digestive enzymes and buffers secreted by the pancreas are carried to the duodenum (upper portion of the small intestine).

parasitism A type of predation in which the predator gets nutrients from the prey but does not kill the prey immediately and may never kill it. A parasite can be a plant or an animal.

parasympathetic division The division of the autonomic nervous system that generally has relaxing effects on the body.

parenchyma cell The most abundant type of cell in plants, these cells have thin cell walls, are usually alive at maturity, and serve numerous functions within the plant, including giving rise to the other plant cell types.

parental generation The generation that begins an experimental cross between organisms. Such a cross is made to study genetics and heredity of traits.

parthenogenesis A type of reproduction in which offspring develop from an egg that has not been fertilized by sperm.

passive transport Transport of materials across the plasma membrane in which no energy is expended. Simple and facilitated diffusion are examples of passive transport.

pathogen A disease-causing organism.

pedigree A family tree, used to help trace the genetic basis for and characteristics of a medical condition.

pelagic zone The vertical zone of the ocean including all of the water from the surface to the floor.

perennial A type of plant that lives for many years, such as trees, woody shrubs, and many grasses.

peripheral nervous system (PNS) The part of the nervous system that includes all of the neural tissue outside the central nervous system. The PNS brings information to and carries it from the CNS, and provides voluntary control of the skeletal muscles, plus involuntary control of the smooth muscles, cardiac muscles, and glands.

period A division of geologic time that is a part of an era and in turn includes several epochs. The Cambrian and the Jurassic are periods.

peripheral protein A protein of the plasma membrane that lies on the inside or outside of the membrane but is not attached to the hydrophobic interior. Usually they are attached to integral proteins at the surface of the membrane. Functions may include structural support, cell recognition, and cell communication.

peristalsis A wave of muscle contractions that pushes materials through a tubular organ, such as the digestive tract.

permafrost The permanently frozen ground that begins about a meter below the ground in the tundra biome of the far north. Neither roots nor water can penetrate this layer.

Permian Extinction The greatest mass extinction event in Earth's history. This event, in which up to 96 percent of all species on Earth were wiped out, occurred about 245 million years ago.

petal The colorful, leaflike structure of a flower; petals attract pollinators.

petiole The stem of a leaf, attaching it to a branch or trunk.

pH scale The scale, ranging from 0 to 14, that quantifies the concentration of hydrogen atoms in solution. The lower the pH number, the more acidic the solution; the higher the pH, the more basic the solution. The scale is logarithmic, so a pH difference of 1 means a 10-fold difference in acidity.

phagocyte A type of white blood cell that can ingest other cells, parts of cells, or other particles. These cells remove both pathogens and parts of the body's own cells that have become cellular debris.

phagocytosis Also known as cell eating, in this form of endocytosis, a cell sends out extensions of its plasma membrane (pseudopodia) that engulf large particles of food and fuse their ends together, thus bringing the food inside the cell encased in a vesicle.

pharynx The region at the rear of the mouth that forms the passageway between the mouth and both the rest of the digestive tract and the rest of the respiratory system.

phenotype A physical function, bodily characteristic, or action of an organism that is the observable outcome of its genotype. The phenotype interacts with the environment, so natural selection acts on the phenotype, affecting the likelihood that the underlying genotype will be passed on.

phloem The tissue through which the food produced in photosynthesis, along with some hormones and other compounds, is conducted in vascular plants.

phloem sap The material, consisting mostly of sucrose and water, with some hormones, amino acids, and other compounds, that is conducted by phloem.

phosphate group A phosphorus atom surrounded by four oxygen atoms. This hydrophilic group is found attached to the glycerol head in a phospholipid.

phospholipid A type of lipid consisting of glycerol and a phosphate group forming the "head" with two fatty acid "tails" attached to the head. The dual hydrophilic and hydrophobic nature of these lipids defines the structure and nature of cell membranes.

phospholipid bilayer The basic structure of a cell's plasma membrane, in which two layers of phospho-lipids align with the hydrophobic "tails" pointing inward, toward each other, and the hydrophilic "heads" pointing outward to the watery environment both inside and outside the cell. Only very small hydrophilic molecules can pass through without help, but hydrophobic molecules pass through fairly freely.

photic zone The vertical layer of the ocean into which the sun's rays can penetrate and in which photosynthesis can therefore take place.

photoautotrophy The nutritional mode in which an organism gets its carbon from carbon dioxide and its energy to make food from the sun's rays. Plants are photoautotrophs.

photoheterotrophy The nutritional mode in which an organism gets its carbon from organic material and its energy to make food from the sun's rays. A few bacteria and archaea are photoheterotrophs.

photoperiodism The ability of a plant to respond to intervals of day and night, allowing plants to make seasonal adaptations. Most plants in the temperate zone exhibit photoperiodism.

photorespiration The process in which rubisco binds oxygen instead of carbon dioxide, thus preventing carbohydrate from being produced and blocking photosynthesis. The C_4 pathway allows plants in warm climates to avoid this blockage, but at an energetic cost.

photosystem Located in the thylakoid membranes in chloroplasts, the combination of the antennae pigments, which absorb energy from the sun, and the reaction center, which receives this energy and turns it into chemical energy.

phototropism The bending of a plant's shoots in response to light. Generally, this tendency helps a plant grow towards the sun to get the most available sunlight.

phylogeny A hypothesis about the evolutionary relationships of organisms.

phylum A taxonomic grouping of related classes. The category is below kingdom and above class. A phylum is a major group of organisms that share the same body plan. Humans are in the phylum Chordata.

physical science The natural sciences not concerned with life. Because organisms live in the physical world, understanding the physical sciences is essential to understanding biology.

physiology The study of the physical functioning of animals and plants, largely concerning the mechanisms of functioning of tissues and organs.

phytoplankton Microscopic photosynthesizing organisms that drift in the upper layers of oceans or bodies of fresh water. Phytoplankton grow in abundance and are a major food resource in water-based food chains.

pinocytosis A form of endocytosis in which the substances brought into the cell are mostly water plus some solutes.

pith The area of ground tissue in the stem of dicotyledonous plants that is inside the ring of vascular bundles. The pith cells are specialized for storage.

placenta The complex network of maternal and embryonic blood vessels and membranes in mammalian pregnancy that allow nutrients and oxygen to diffuse to the embryo from the mother, while carbon dioxide and waste diffuse from the embryo to the mother.

plankton Microscopic organisms, including both phytoplankton and zooplankton, that live in the upper layers of bodies of water. They provide the basis for food chains in the water.

plasma The watery portion of the blood in which the formed elements are suspended.

plasma cell A type of B cell produced in antibody-mediated immunity that exists in the lymph nodes and produces lots of antibody specific to the current infection. This free antibody moves through the bloodstream and directly fights the invader.

plasma membrane The outer boundary of a cell, made up of a phospholipid bilayer.

plasmid A ring of DNA that lies outside the chromosome in bacteria. Plasmids can be incorporated into the chromosomal DNA, and they provide both a natural way that bacteria incorporate new genes from other sources and a useful tool in genetic engineering.

plasmodesma One of a series of tiny channels in a plant's cell wall that enables communication between cells by making the cytoplasm between cells continuous.

platelet A small fragment that has pinched off from larger, complete blood cell. Platelets contain enzymes and clotting factors.

pleiotropy The capacity of one gene to have many effects. This relationship between genes and characters is common because many processes in living things are interrelated.

point mutation Mutation at a single location in the genome, rather than a whole chromosome aberration.

polar body A very small haploid cell produced during meiosis in a female. During meiosis, the cells divide unevenly to produce one richly endowed egg cell and three very small polar bodies. The polar bodies will eventually disintegrate into their constituent parts.

polar covalent bond A type of covalent bond in which electrons are shared unequally between atoms, so one end of the molecule will have a slightly greater negative charge and the other end a slightly greater positive charge. These ends can then be attracted to oppositely charged atoms or molecules, forming other chemical bonds.

polarity A difference in electrical charge at one end of the molecule from the other. These ends can then interact with other molecules.

pollination The transfer of pollen, by wind or carried by an animal, from a stamen to a stigma.

polygenic Having multiple genes affecting a given character, such as height in humans.

polymer Large compound made up of many repeating molecular units called monomers.

polymerase chain reaction A technique for generating many copies of a DNA sequence. Essentially, the ingredients needed for DNA replication are put together with the DNA segment, and replication is repeated, with the amount of DNA doubling each time.

polypeptide A chain of ten or more amino acids. When the chain is folded into its three-dimensional form, it is a protein.

polyploidy The condition of having more than two sets of chromosomes. Polyploidy has severe consequences for humans, but many plants and some other organisms can survive and thrive as polyploids.

polysaccharide A carbohydrate built of many monosaccharides. Also known as complex carbohydrates, these compounds include starch, glycogen, cellulose, and chitin.

polyunsaturated fatty acid A type of fatty acid in which there are two or more double bonds, so hydrogen atoms could be added. At room temperature, these fats are liquid, known as oils.

pons A part of the vertebrate brain that contains relay centers to connect several other brain structures. The pons also contains cells involved in involuntary control of breathing.

population The members of a species living in a geographic region. Individuals in a population will interact with each other, and the population is the smallest unit that can evolve.

positive feedback A system of control in which the product of a process further stimulates the process. This unstable process must be turned off by an outside force or process.

posterior pituitary A region of the pituitary that stores and releases oxytocin and antidiuretic hormone, which are produced in the hypothalamus and transported to the posterior pituitary within nerve fibers.

postsynaptic neuron The neuron that receives the action potential's signal via neurotransmitter released by the presynaptic neuron at a synapse. In effect, this is the downstream neuron.

potential energy Stored energy. A rock perched at the top of a hill has potential energy.

predation The act of one free-standing organism feeding on parts or all of another organism. Predation includes animals eating other animals and animals eating plants.

presynaptic neuron The neuron that transmits the action potential to a synapse and releases neurotransmitter into the synaptic cleft to communicate with the postsynaptic cell.

pressure-flow model The hypothesis that the force behind the movement of phloem sap comes from water flowing into the phloem by osmosis, following solutes that have been pumped into the phloem.

primary consumer Any organism that eats producers (organisms that make their own food). A primary consumer occupies the second trophic level.

primary follicle In humans, a stage of development undergone by some follicles in which the follicular cells surrounding the primary oocyte have become larger. A few primary follicles will go on to develop into secondary follicles during each menstrual cycle.

primary growth The type of growth in plants that occurs at the tips of their roots and shoots and mainly increases their length.

primary oocyte A diploid cell produced in the female embryo that may later mature in the ovary. After the female reaches puberty, an average of one oocyte per month is selected to complete maturation in the ovary.

primary spermatocyte A diploid cell in a male that will undergo meiosis to produce haploid sperm cells.

primary structure The sequence of amino acids in a protein. This sequence dictates the final shape of the protein because electrochemical bonding and repulsion forces act on the structure to create the folded-up protein.

primary succession The type of succession in which the starting state is little or no life, and life must colonize a new physical environment. The parts of Mount St. Helens that were incinerated down to bare rock became sites of primary succession.

producer Any organism that manufactures its own food. Plants are the most important producers on Earth. By converting the sun's energy into biomass, producers package the energy that is then passed along in food webs. Producers occupy the first trophic level.

product A substance formed in a chemical reaction. The products are written on the right side of a chemical equation.

profundal zone The vertical zone in a lake that occurs beneath the limnetic zone and in which not enough light penetrates for photosynthesis to occur.

progestin A class of hormones, produced by the ovaries, that prepare the body for possible pregnancy.

programmed cell death During development, the organized death of cells through their own metabolic processes, creating spaces that define shapes. Programmed death of the cells between the fingers gives rise to the human hand.

prokaryote An organism whose cells lack nuclei. Bacteria and Archaea are prokaryotes.

prokaryotic cell A cell that has DNA in a localized region (the nucleoid) that is not bound by a membrane. Prokaryotes are microscopic forms of life, and all are either bacteria or archaea.

prolactin A hormone, secreted by the anterior pituitary, that stimulates the development of the mammary glands and their production of milk in female mammals.

promoter The segment of a DNA operon to which RNA polymerase binds to initiate transcription. The promoter's sole function is regulatory, to permit or block transcription.

prostate gland A gland that, in human males, surrounds the urethra near the urinary bladder and contributes fluids to the semen. Enlargement of this gland constricts the urethra, reducing urine flow.

protease The type of pancreatic enzyme, secreted through ducts, that break down proteins.

protein A polypeptide folded into a specific three-dimensional form. Proteins make up about half of the dry weight of most cells. Speeding up chemical reactions, signaling within the body, allowing muscles to contract, and providing support for structures are just some examples of the many critical roles played by proteins.

protein conformation The three-dimensional shape of a protein, determined by its order of amino acids and their interactions with one another in folding and twisting to link to each other. This shape is critical in allowing a protein to interact properly with other molecules to perform its function.

proton A basic constituent of an atom, found in the nucleus of the atom and having positive electrical charge. The number of protons in the atom defines the element. The neutrons and protons together make up the bulk of the mass of the atom.

proximal tubule In the nephron tubule of the kidney, the convoluted portion that is closest to Bowman's capsule. Some reabsorption of salt and water occurs here.

pseudopodium A "false foot," or cell extension formed when a portion of the cell is thrust forward, anchored, and then cytoplasm streams into the extension. Amoebas move by using pseudopodia.

pulmonary circulation The system that circulates blood between the heart and the lungs. This system brings oxygen into and takes carbon dioxide away from the body.

pulmonary ventilation The physical movement of air into and out of the lungs through the respiratory passages, also known as breathing.

pyloric sphincter The circular muscle that regulates the flow of chyme (food mixed with gastric juices) between the stomach and small intestine.

quaternary structure The way in which two or more polypeptide chains come together to form a protein.

radial symmetry The type of symmetry in which body parts are distributed evenly about a central axis. Starfish are radially symmetrical.

radiometric dating A technique used to date rocks by comparing the proportion of a radioactive element to its nonradioactive decay product, calculating time using the uniform rate at which the decay occurs. With this technique, fossil layers can be dated to determine when the animals and plants in those layers lived.

reabsorption In the kidney, the process of drawing water and nutrients back from the fluid in the nephron tubules into the blood supply, conserving these valuable materials.

reactant A substance (atoms, molecules, or compounds) that goes into a chemical reaction. The reactants interact to form the product(s). The reactants are written on the left side of a chemical equation.

reaction center A molecular complex in a chloroplast that transforms solar energy into chemical energy. This center contains a pair of special chlorophyll *a* molecules that receive energy from the pigments of the antennae system, which trapped the energy from the sun. The reaction center turns the sun's energy into a form useful for the plant.

reaction wood Extra wood formed in reaction to pressure from one side, to resist being pushed over and continue to grow up towards the sun.

receptor-mediated endocytosis (RME) A form of endocytosis in which receptors bind to a nutrient, then move laterally within the plasma membrane to join other proteins of their type that have bound to the same kind of nutrients. These bound receptors congregate in a pit that deepens and pinches off, creating a vesicle with the nutrients inside.

receptor protein A protein with a binding site at the surface of the plasma membrane. The binding site attaches only a specific molecule or group of molecules, causing a change in the cell's activity. Receptors enable specificity of communication within the body.

recessive condition A condition that occurs when only recessive, nonfunctional alleles are present. Red-green color blindness is an example of a recessive condition, because only one set of functional alleles need be present for normal color vision.

recombinant DNA Two or more segments of DNA that have been combined by humans into a sequence that does not exist in nature.

recombination The exchange of pieces of chromosomes that occurs when homologous chromosome pairs intertwine during prophase I. This process, also called crossing over, accounts for some of the diversity in individual gametes.

recognition sequence A series of DNA bases that is the site recognized by a particular restriction enzyme as the place for it to cut the DNA strand.

rectum The terminal portion of the large intestine, and therefore the end of the digestive tract. In defecation, feces pass from the colon through the rectum and out the anus.

red blood cell (RBC) The blood cells, also known as erythrocytes, that transport oxygen to and carry carbon dioxide from every part of the body.

redox reaction A combination of a reduction and an oxidation reaction, in which the electrons lost in the oxidation are gained in the reduction reaction. The substance being oxidized is going from a more ordered to a less ordered state, so it is releasing energy.

reduction The state of having gained one or more electrons in a chemical reaction. Such a substance has reduced its positive charge by gaining negative charge from electrons.

reflex An automatic nervous system response, triggered by a specific stimulus, that helps the organism avoid danger or preserve a stable physical state. The knee jerk response is a well-known reflex.

regeneration The process by which an animal can reproduce itself from a part of itself. An arm attached to part of the central disk from a sea star can regenerate into a whole sea star. The term can also be used to refer to the repair of damaged tissue.

renal cortex The outer region in the vertebrate kidney, where the glomeruli, Bowman's capsules, proximal and distal tubules, and the beginning of the collecting ducts lie.

renal medulla The inner region of the vertebrate kidney, where the loops of Henle lie. Most of the reabsorption of water and nutrients from the nephron tubules back into the blood vessels takes place here.

renal pelvis The chamber inside the kidney of birds and mammals that receives urine from the collecting ducts and channels it into the ureter.

reproductive cloning The cloning of whole, complex living things so that a genetically identical organism is produced.

reproductive isolating mechanism Any factor that, in nature, prevents interbreeding between individuals of the same or closely related species. These factors keep species separate.

reproductive system The organ system that develops gametes and delivers them to a location where they can fuse with another gamete (fertilization) to produce a new individual. This system consists of reproductive organs and ducts, different for males and females.

resource partitioning Dividing up scarce resources among species that have similar requirements. Such partitioning allows species to coexist in the same habitat.

respiratory system The organ system that brings oxygen into the body and expels carbon dioxide from the body. In humans, this system includes the lungs and passageways that carry air to the lungs.

restriction enzyme A type of enzyme, occurring naturally in bacteria, that recognizes a specific series of DNA bases (restriction sequence) and cut the DNA strand at that site. Restriction enzymes are used in biotechnology to cut DNA in specific places.

reversible reaction A chemical reaction that can go in both directions, both forward and backward as written on paper.

ribonucleic acid (RNA) A class of nucleic acids that, depending on the specific type, carries the instruction for synthesis or helps in the assembly of proteins. RNA also is the genetic material of many viruses.

ribosomal RNA (rRNA) The type of RNA that, along with some proteins, forms the ribosomes.

ribosome A structure, either free in the cytosol or embedded in the rough endoplasmic reticulum, where proteins are synthesized in a cell.

ribozyme Molecule composed of RNA that can both encode information and act as an enzyme. Molecules similar to these would have been necessary early in the evolution of life.

RNA polymerase The enzyme which unwinds the DNA double helix and puts together a chain of RNA nucleotides that are complementary to the DNA strands in the transcription stage of protein synthesis.

root cap A collection of cells at the very tip of a root that protects the adjacent meristematic cells and also secretes a lubricant to help the root move through its growing medium.

root hair A threadlike extension of the main roots, these are a type of trichome. The many root hairs greatly increase the surface area of roots that can absorb water and nutrients.

rough endoplasmic reticulum A cell organelle that is a folded continuation of the cell membrane studded with ribosomes. The ribosomes of the rough ER are the site of synthesis of proteins that will reside in the cell's plasma membrane and also of proteins that will be transported out of the cell.

r-selected species Species whose population size tends to fluctuate widely in reaction to a highly variable environment. These species have a high intrinsic rate of increase and are also known as opportunist species.

rubisco The large enzyme that brings CO_2 and the sugar RuBP together in the first step of the Calvin cycle of photosynthesis. In this step, plants take carbon dioxide from the air and begin to turn it into sugars.

rule of addition In probability theory, the principle that, when an outcome can occur in two or more different ways, the probability of that outcome is the sum of the respective probabilities.

rule of multiplication In probability theory, the principle that the probability of any two events happening is the product of their respective probabilities.

sapwood The lighter colored, outer wood in a tree trunk where xylem tubes conduct xylem sap.

sarcomere The functional unit of a striated muscle, which contracts when thin filaments slide past thick filaments. The sarcomeres shorten, thus contracting the whole muscle.

saturated fatty acid The type of fatty acid in which there are no double bonds, so no more hydrogen atoms can be added. At room temperature, these fats will be solid. Saturated fatty acids have been linked with heart disease and are found primarily in animal products and tropical oils.

science Both a process of learning about nature by observation and experiment, and a collection of knowledge and insights about nature.

scientific method The testing of hypotheses through observation and experiment. Our understanding of science advances through use of the scientific method.

sclerenchyma cell A type of plant cell that is dead at maturity, has thick cell walls containing cellulose, and helps the plant return to its original shape after it has been deformed. Sclerenchyma cells provide strong support for the plant.

sebaceous gland A type of gland in the skin that produces a waxy, oily secretion (sebum) that lubricates the hair shaft and inhibits bacterial growth in the surrounding area.

sebum The waxy, oily secretion, produced by sebaceous glands in the skin, that lubricates the hair shaft and inhibits bacterial growth in the surrounding area.

second law of thermodynamics Energy transfer always results in a greater amount of disorder (entropy) in the universe.

secondary consumer Any organism that eats a primary consumer.

secondary follicle In humans, a stage of development undergone by a few primary follicles during each menstrual cycle in which the oocyte becomes larger and more layers of follicular cells develop around the oocyte. Only one secondary follicle during each cycle will continue developing to the stage where the oocyte is ovulated.

secondary growth The type of lateral growth in plants that thickens the plant. Secondary growth occurs in woody plants, such as trees, and thickens their trunks and branches.

secondary phloem Phloem tissue produced to the outside of the vascular cambium in woody plants as part of secondary growth, or thickening.

secondary spermatocyte A haploid cell formed after a primary spermatocyte undergoes meiosis I. After meiosis II, this cell will become a spermatid and then a mature sperm cell.

secondary structure The small-scale twisting or bending of a protein. The alpha helix and beta pleated sheet are two examples of secondary structure.

secondary succession The type of succession that occurs after the final state of a habitat has been disturbed by an outside force and the disturbance stops. A farmer's field that has been abandoned is a site of secondary succession.

secondary xylem Xylem tissue produced to the inside of the vascular cambium in woody plants. This tissue, also called wood, is responsible for much of the thickening of woody plants.

secretion In the kidney, the process of moving hydrogen and potassium ions, as well as other compounds, from the capillaries into the distal tubules of the nephron.

secretory protein A protein that is exported out of a cell. These proteins are synthesized on the ribosomes of the rough ER. Protein hormones are one type of secretory protein.

seed A plant reproductive structure that includes a plant embryo, its food supply, and a tough protective casing. Seeds protect vulnerable offspring, allowing seed plants to live in dry habitats and survive harsh conditions.

seed coat The tough outer layer of a seed, derived from the integuments of the ovule, which protect the embryo from mechanical damage and water loss.

seedless vascular plant A plant that has a vascular system, enabling it to grow tall, but does not produce seeds, so its offspring are vulnerable. These plants are tied to water because their sperm must move through water to fertilize the egg. Ferns are a familiar example.

selfish DNA DNA whose purpose seems to be only to replicate itself. These sequences, however, may have taken on regulatory and stabilizing roles.

semen The mixture of sperm and glandular materials that is ejaculated from the human male through the urethra.

seminal vesicle A gland that, in human males, contributes fluids to the semen as the semen moves down the urethra in the process of ejaculation.

seminiferous tubule A convoluted tubule inside the testes where sperm development begins. Eventually, immature sperm are released into the interior cavity and travel to the epididymis where sperm maturation is completed.

sensory neuron A neuron that senses conditions both within and outside the body and conveys that information to neurons in the central nervous system. The body gathers its information through these neurons.

sepal The leaflike structure that, with other sepals, protects the flower before it opens.

serum The fluid left in blood after clotting.

sessile Fixed in location, as are some organisms including mushrooms and sponges.

sex chromosome The X or Y chromosome. The two sex chromosomes in a cell determine the individual's sex. In mammals, two X chromosomes determine a female, and an X and a Y chromosome determine a male. Sex chromosomes are the only nonhomologous pair of chromosomes in humans.

sexual selection A difference in mating rate among individuals of the same sex in a population. Unequal distribution of mating can result from females choosing certain male characteristics or from some males outcompeting others for opportunities to mate. The result is that some individuals will have more offspring and therefore pass on more of their genes.

shoot apical meristem The growing tip of plant shoots that produce the cells responsible for vertical growth of the plant.

sieve element A type of cell in phloem that conducts nutrients. These cells are stacked on top of each other and have ends studded with small holes, forming tubes for nutrient flow. The cells are alive at maturity but lack nuclei, making room for rapid conduction of nutrients.

sieve tube A tube, formed by a stack of sieve elements, that conducts nutrients in phloem.

simple diffusion Diffusion through the plasma membrane that does not require special protein channels. Water, oxygen, carbon dioxide, and steroid hormones can all cross the membrane to areas of lower concentration by simple diffusion.

simple sugar The building block of carbohydrates. Also known as a monosaccharide, glucose and fructose are familiar examples.

single bond A type of covalent bond in which a single pair of electrons is shared between two atoms.

sink The unloading site of a flow system, where the material ends up.

skeletal muscle tissue The muscle tissue responsible for movement of the body. This tissue is made of long cells, each typically containing hundreds of nuclei.

skeletal system The organ system that forms an internal supporting framework for the body and protects delicate tissues and organs. This system consists of all the bones and cartilages in the body, and the connective tissues and ligaments that connect the bones at the joints.

skin An organ consisting of two layers, epidermis and dermis, and covering the outside of an animal. The skin protects the body and receives signals from the environment.

smooth endoplasmic reticulum The part of the endoplasmic reticulum membrane, farther out from the cell nucleus, that has no ribosomes. Depending on the cell type, different lipids are synthesized here and potentially harmful substances may be detoxified.

smooth muscle tissue Made of nonstriated muscle tissue, this tissue is responsible for contraction of blood vessels, air passages, and hollow organs such as the uterus, and is not under voluntary control.

solute The ingredient being dissolved in a solvent to form a solution. For example, sugar is the solute in the sugar-water nectar you put in your hummingbird feeder.

solution A homogeneous mixture of two or more substances in the same phase (gas, liquid, or solid). Frequently, solutions consist of a solute dissolved in water, and these are called aqueous solutions.

solvent The substance in which a solute is dissolved to form a solution. In an aqueous solution, the solvent is water.

somatic cell Any cell that is not and will not become an egg or sperm cell. A mutation in a somatic cell is not heritable because somatic cells are not passed on to offspring.

somatic nervous system The subdivision of the efferent division of the peripheral nervous system that provides voluntary control over the skeletal muscles.

somite One of the blocks of mesodermal tissue in the vertebrate embryo that lie on both sides of the notochord and give rise to muscles and the vertebrae that enclose the spinal cord.

source The loading site of a flow system, where the material comes from.

space-filling model A picture of a molecule showing the three-dimensional structure, particularly showing how big atoms are in relation to each other and the overall shape of the whole molecule.

speciation The formation of a new species from an ancestral form.

species A group of organisms whose individuals may interbreed with one another in nature, but who are reproductively isolated from other such groups.

specific heat The amount of energy needed to raise the temperature of one gram of a substance by one degree Celsius. Water serves as an insulator because it has a high specific heat compared to other substances, and the oceans therefore help stabilize the Earth's temperature.

sperm cell In a plant, the cell in a pollen grain which may eventually fertilize an egg.

spermatid An immature sperm cell. This haploid cell will become a mature sperm cell after developing a tail-like flagellum and ridding itself of nearly all cellular organelles.

spermatocyte An early stage in the development of a sperm cell. Spermatocytes develop from spermatogonia; they develop into spermatids, and eventually into mature sperm.

spermatogonium Diploid cell that is the starting cell in sperm production in males. Some spermatogonia will produce more spermatogonia, while others will continue to develop into sperm.

spinal nerve Paired nerves that connect the spinal cord to different parts of the body, including the head, neck, diaphragm, arms, chest, abdominal muscles, legs, bladder, bowels, sexual organs, and feet. In humans, there are 31 pairs of spinal nerves.

spirochete A type of spiral-shaped bacterium.

spongy bone Type of bone that is porous and less dense than compact bone. Spongy bone fills the expanded ends of long bones.

spore A single, asexual reproductive cell that, without fusing with anything, has the ability to grow into an adult plant or fungus.

sporophyte generation The generation in a plant's life cycle that produces spores. This generation is the dominant, visible generation in flowering plants.

S-shaped growth Logistic growth. When plotted on a graph, this type of increase in numbers looks like the letter S. Growth begins exponentially, but eventually the rate of growth slows, then stops, with the population stabilizing at a certain level.

stabilizing selection The type of selection in which intermediate forms of a given character are favored over either extreme. This process tends to maintain the average for the character.

stamen The male reproductive structure in a flower, consisting of a filament (stalk) with an anther, bearing pollen, at the top.

starch A complex carbohydrate that serves as the major form of carbohydrate storage in plants. Starches are important sources of energy found in food plants such as potatoes, rice, carrots, and corn.

stasis A situation in which there is no change. In geographical time, this refers to long periods in the fossil record when there seems to be no or only tiny change in species.

steroid A type of lipid that has a linked set of four carbon rings to which various side chains can be attached. Cholesterol, testosterone, and estrogen are steroids.

steroid hormone A type of hormone built from cholesterol molecules. These hormones have important roles in reproductive and behavioral functions in the body. Testosterone and estrogen are examples.

stigma The tip end of the carpel of a flower, where pollen grains are deposited before fertilization occurs.

stimulus Any signal that influences the activity of an organism or a part of the organism.

stoma One of many microscopic pores in the epidermal cells on the underside of a leaf. Carbon dioxide enters and water leaves through these openings. Stomata can be closed to conserve a plant's water.

stratosphere The layer of the Earth's atmosphere that lies above the troposphere, at about 20 to 35 kilometers (13 to 21 miles) above sea level. The ozone layer lies in this level.

stroma The liquid material of chloroplasts.

structural formula A two-dimensional representation of a molecule that shows how the atoms are arranged and the types of bonds between them. Knowing the spatial arrangement helps to understand how molecules will interact with each other.

style The slender tube structure in the carpel of a flower, connecting the stigma and the ovary.

substrate The substance that is worked on by an enzyme.

succession In ecology, a series of replacements of community members at a given location until a stable final state is reached.

surface tension The strength of the surface of water where it meets air, given by the number and direction of hydrogen bonds. Water at the surface is pulled down and sidewards by these bonds, packing the molecules closely together so that the surface has cohesion. Small animals can walk on water because water has surface tension.

sympathetic division The division of the autonomic nervous system that generally has stimulatory effects on the body.

sympatric speciation The type of speciation that occurs when a population is not geographically separated. Polyploidy is one form of sympatric speciation, and another is temporal isolation in breeding seasons.

synapse A site where a neuron communicates with another cell. A synapse consists of the presynaptic neuron, a space (synaptic cleft), and a postsynaptic cell, which can be a neuron or another type of effector.

suppressor T cell The type of T cell that inhibits the production of cytotoxic (killer) T cells once the immediate infection has passed.

synaptic cleft The gap between a presynaptic neuron and the postsynaptic cell. Neurotransmitter is released into this cleft by the presynaptic neuron.

synaptic terminal The structure at the end of an axon or branch of the axon that forms a synapse with another neuron or effector. Neurotransmitter is released from the synaptic terminal and binds to receptors on the postsynaptic cell, effecting communication between the cells.

synthetic work The work done in building up complex molecules from their simpler components.

systematics The field of biology that deals with the diversity and relatedness of organisms. Systematists study the evolutionary history of groups of organisms.

systemic circulation The system that circulates blood between the heart and the body, exclusive of the lungs. This system allows exchange of gases, nutrients, and wastes with the tissues.

taproot system The type of plant root system that consists of a large central root and a number of smaller lateral roots.

target cell A cell, bearing specific receptor molecules, that will be affected by a specific hormone.

taxon A group of organisms in a category of the taxonomic system. In order of increasing inclusiveness, these categories are: species, genus, family, order, class, phylum, kingdom, and domain.

taxonomic system A set of groupings used to classify every species of living thing on earth.

tendon The connective tissue that attaches a skeletal muscle to a bone.

tendril A thin, modified leaflet that wraps around another object to help a plant climb upwards towards light.

terminal bud Another term for the shoot apex, or the main apical meristem of a plant, used especially when it is dormant.

territorial Pertaining to organisms that occupy and defend an area, attempting to exclude others, especially members of their own species.

tertiary consumer Any organism that eats secondary consumers.

tertiary follicle In humans, a stage of development undergone by one follicle during each menstrual cycle. This follicle nurtures the oocyte until ovulation from the ovary.

tertiary structure The large-scale twists and turns in a protein conformation.

testis The organ in the male reproductive system where sperm begin development and testosterone is produced.

tetrad The grouping formed by two homologous chromosomes pairing up, as they do in prophase I of meiosis. The four chromatids in this grouping give it the name tetrad. Recombination, which contributes to diversity, occurs while the chromosomes are in this conformation.

thalamus A part of the vertebrate brain stem that contains relay and processing centers for incoming sensory information.

theory A general set of principles, supported by evidence, that explains some aspect of nature. A theory is formulated to explain how things work. Although some details of a theory may be in dispute, a theory is well supported and widely accepted.

theory of punctuated equilibrium The idea that evolution acts in rapid bursts after long periods of stasis. This pattern would explain the lack of transitional forms in the fossil record.

thermodynamics The study of energy.

thigmotropism The bending of a plant in response to touch. This tendency allows tendrils wrap around other objects, thus helping a plant climb upward towards light.

thylakoid compartment The interior water space of thylakoids.

thylakoid A flattened, membrane-bound sac in the interior of a chloroplast. The light-dependent reactions of photosynthesis occur in the thylakoids.

thyroid stimulating hormone (TSH) A hormone, secreted by the anterior pituitary, that acts on the thyroid gland to trigger the release of thyroid hormones.

tissue An organized assemblage of similar cells that serves a common function. Nervous, epithelial, and muscle tissue are some familiar examples.

T-lymphocyte (T cell) A class of lymphocytes that plays a central role in cell-mediated immunity. T cells come in several varieties that play specific roles in recognizing and killing the body's infected cells.

top predator A species that preys on other species but is not preyed upon by any species.

totipotent Ability of a cell to develop into any kind of cell in the organism. Early embryonic cells are totipotent; their developmental fate is not yet determined.

tracheid A slender, tapered cell of the xylem that has perforations (bordered pits) that match up with adjacent cells to allow water to move between them. Tracheids are dead and empty of cellular contents at maturity, allowing conduction of xylem sap.

trait A variation in a character, such as green or yellow seed color in Mendel's pea plants. These character traits were studied in offspring of known parental crosses to uncover rules of heredity.

transcript A length of mRNA that results from complementary base pairing with a strand of DNA. The transcript, with instructions encoded in its base pair sequence, then moves to the cytoplasm and undergoes translation to complete protein synthesis.

transcription The process in which DNA's information is copied onto mRNA. This process is the first stage in protein synthesis and occurs in the cell nucleus.

transfer RNA (tRNA) The type of RNA that brings amino acids to ribosomes to form polypeptides. One end of a particular tRNA bonds with a specific amino acid, the other end (anticodon) recognizes the codon in mRNA. tRNA acts as a bridge between mRNA and amino acids.

transformation A cell's incorporation of genetic material from outside its boundary. Some bacteria readily undergo this process, and others can be induced to for uses in biotechnology.

translation The process in which a polypeptide chain is produced by a ribosome from mRNA's instructions, encoded in the sequence of base pairs. This process is the second stage in protein synthesis and occurs in the cytoplasm.

translocation The swapping of fragments by non-homologous chromosomes, resulting in gene sequences that are out of order on both chromosomes. Translocation can have phenotypic effects.

transpiration The process by which plants lose water when water vapor leaves the plant through open stomata. More than 90 percent of the water that enters a plant evaporates into the atmosphere via transpiration, providing the suction that moves water through xylem.

transport protein In a cell membrane, transport proteins form hydrophilic channels through the hydrophobic interior of the plasma membrane, allowing hydrophilic materials to pass down their concentration gradient through the membrane. In blood, transport proteins carry compounds that would otherwise be filtered out or would not dissolve in the blood.

transport vesicle A sphere of membrane, carrying proteins and other molecules, that moves through the cytosol of a cell. Transport vesicles can fuse with other membranes and deliver their contents within or out of the cell.

transport work The work done in transporting molecules against their concentration gradient. The sodium-potassium pump does transport work.

triglyceride Form of glyceride in which three fatty acids link up with a single molecule of glycerol. Most substances we call fats are triglycerides because about 90 percent of the weight of lipids in foods consists of triglycerides.

trimester A period lasting about three months during a human pregnancy. There are three trimesters during the 9-month pregnancy.

triple bond A covalent bond in which three pairs of electrons are shared between atoms.

triplet code The code used by DNA to instruct protein synthesis. Each three DNA base pairs specify three mRNA base pairs, but each mRNA triplet specifies a single amino acid or a start or stop command.

trophic level The position of a species in a food web or chain, based on where it gets its energy. A plant is a producer and occupies the first trophic level, while a cow eats grass, is a primary consumer, and occupies the second trophic level.

trophoblast The cells at the periphery of the developing mammalian embryo that establish physical links with the mother's uterine wall and eventually develop into the fetal portion of the placenta.

tropical savanna A type of tropical grassland that is characterized by seasonal drought, small seasonal changes in the generally warm temperatures, and stands of trees that punctuate the grassland. Tropical savanna exists in Africa, South America, and Australia.

troposphere The lowest layer of the Earth's atmosphere, extending from sea level to about 12 kilometers (7.4 miles) above sea level. This layer contains most of the gases in the atmosphere.

tube cell A cell in a pollen grain that, after the pollen grain has landed on a stigma, must germinate and form a pollen tube that grows down through the style so that the sperm cells may pass through and fertilization of the egg can occur.

tumor-suppressing gene Genes that inhibit growth. Faulty tumor-suppressing genes contribute to cancers by failing to turn off cell division. Generally, both these and oncogenes (initiating the excessive cell division) must be active for cancers to start.

umbilical cord The tissue that links the fetus with the placenta in mammals and contains the fetal blood vessels that carry blood between the fetus and the placenta.

urea An organic waste product produced in the liver of vertebrates and excreted in the urine.

ureter One of a pair of tubes that carries urine from the kidneys to the urinary bladder in reptiles, birds, and mammals.

urethra The tube which carries urine from the urinary bladder to the outside of the body of mammals. The urethra in males also carries semen in the process of ejaculation.

urinary bladder The muscular organ in vertebrates that stores urine prior to excretion.

urinary system The organ system that eliminates waste products from the blood through formation of urine. In mammals, this system consists of the kidneys where the urine is formed, and ureters, urinary bladder, and urethra, which transport the urine, eventually, outside the body.

urine The liquid waste filtered from the blood by the kidneys. In mammals, urine is temporarily stored in the bladder prior to excretion through the urethra.

uterine tube The tube that transports an ovulated egg from the ovary to the uterus in mammals. Fertilization occurs inside this tube. In humans, it is called the Fallopian tube.

uvula The fleshy structure suspended from the rear of the soft palate in humans that, with the base of the tongue, marks the boundary between the oral cavity and the pharynx.

variable An element of an experiment that is changed (varied). By changing one thing at a time, we can determine the effect of that variable on the result.

vascular bundle Collections of xylem and phloem tubes that run in parallel in the stems of plants, forming a transport system.

vascular cambium The thin layer of tissue in woody plants between the primary xylem and phloem cylinders that generates the secondary xylem and phloem tissues. This meristematic tissue gives rise to secondary growth in woody plants.

vas deferens In humans, the tube that carries the sperm from the epididymis on top of the testis to the urethra for ejaculation.

vasoconstriction The decrease in diameter caused by contraction of the smooth muscles in the wall of a blood vessel. Together with vasodilation, this process helps regulate blood pressure.

vasodilation The increase in diameter caused by relaxation of the smooth muscles in the wall of a blood vessel. Together with vasoconstriction, this process helps regulate blood pressure.

vegetal pole The end of the zygote with relatively more yolk and lying farther from the cell's nucleus. The location of the egg's poles define the orientation in which the embryo develops.

vegetative reproduction The form of asexual reproduction in which a cutting from a plant, when properly planted, can grow into an independent plant that is an exact genetic replica of the original plant and has the ability to take part in sexual reproduction.

vein A blood vessel that carries blood back to the heart.

venule A small diameter vein, which receives blood from the capillaries and sends it on to larger diameter veins, on its way back to the heart.

vertebrate An animal with a skull surrounding its brain. Most vertebrates also have a vertebral column made up of individual bones (vertebrae) that protect the spinal cord.

vessel element A type of cell in xylem that stacks on top of each other, with walls between adjacent cells broken down, forming tubes for rapid conduction of water. These cells are dead and empty of cellular contents at maturity, making room for conduction of water.

villus In vertebrates, one of the many fingerlike projections of the lining of the small intestine that greatly increases the surface area for absorption.

viroid A virus-like entity that consists of only a small strand of infectious RNA, lacking even a protein coat.

viscosity The measure of a fluid's resistance to flow.

vitamin An organic compound required in a very small quantity for critical chemical reactions in the body.

wetlands Lands that are wet for at least part of the year. Wetlands are sites of great biological productivity, and they provide vital foraging grounds for migrating birds.

white blood cell (WBC) A class of blood cells that are critical to immune function. Types of white blood cells include T cells, B cells, neutrophils, eosinophils, and macrophages, each playing a specific role in immune responses.

white matter The tissue within the central nervous system that has a white appearance from the myelin sheaths surrounding the axons that run through the area.

wood Secondary xylem. This material in woody plants is responsible for most of their thickening.

woody plant A plant that has wood, or secondary xylem.

xylem The tissue through which water and dissolved minerals flow in vascular plants.

xylem sap The water and dissolved minerals that are conducted through xylem in plants.

zero population growth State of a population in which births exactly equal deaths in a given period, so there is no change in numbers.

zone of cell division In plants, the area just behind the apical meristem where more cells are produced. Cells then move on to the zones of elongation and differentiation.

zone of differentiation In plants, the area just behind the zone of elongation, near the apical meristem, where primary tissue types finish taking shape.

zone of elongation In plants, the area between the zones of cell division and differentiation, near the apical meristem, where most primary plant growth occurs through the elongation of cells.

zooplankton Tiny animals and other heterotrophs that drift or move weakly in the upper layers of bodies of water.

zygote A fertilized egg. A zygote is one diploid cell that will develop into a whole organism.

Photo and Web Credits

Index

MINIMUM SYSTEM REQUIREMENTS:

Windows

Pentium processor
Windows 95/98/NT 4.0x
16 MB RAM
4X CD-ROM drive
256-Color monitor with 640x480 pixel resolution
Internet browser 4.0 or above with Flash Plug-in
QuickTime 4.0
(both included on CD-ROM)

Macintosh

PowerPC
System 7.0 or above
16 MB RAM
4X CD-ROM drive
256-Color monitor with 640x480 pixel resolution
Internet browser 4.0 or above with Flash Plug-In
QuickTime 4.0
(both included on CD-ROM)